Third Edition

Essentials of Anatomy & Physiology

Kenneth S. Saladin
Georgia College

Robin K. McFarland
Cabrillo College

Digital Author
Christina A. Gan
Highline College

ESSENTIALS OF ANATOMY & PHYSIOLOGY, THIRD EDITION

Published by McGraw Hill LLC, 1325 Avenue of the Americas, New York, NY 10121. Copyright ©2022 by McGraw Hill LLC. All rights reserved. Printed in the United States of America. Previous editions ©2018 and 2014. No part of this publication may be reproduced or distributed in any form or by any means, or stored in a database or retrieval system, without the prior written consent of McGraw Hill LLC, including, but not limited to, in any network or other electronic storage or transmission, or broadcast for distance learning.

Some ancillaries, including electronic and print components, may not be available to customers outside the United States.

This book is printed on acid-free paper.

1 2 3 4 5 6 7 8 9 LWI 26 25 24 23 22 21

ISBN 978-1-260-26640-5 (bound edition)
MHID 1-260-26640-0 (bound edition)
ISBN 978-1-264-34986-9 (loose-leaf edition)
MHID 1-264-34986-6 (loose-leaf edition)

Senior Portfolio Manager: *Matthew Garcia*
Product Developer: *Melisa Seegmiller*
Marketing Manager: *Valerie Kramer*
Senior Content Project Manager: *Vicki Krug*
Lead Content Project Manager: *Brent dela Cruz*
Senior Buyer: *Laura Fuller*
Lead Designer: *David W. Hash*
Senior Content Licensing Specialist: *Lori Hancock*
Cover Image: *Buena Vista Images/Getty Images*
Compositor: *MPS Limited*

All credits appearing on page or at the end of the book are considered to be an extension of the copyright page.

Library of Congress Cataloging-in-Publication Data

Names: Saladin, Kenneth S., author. | McFarland, Robin, author. |
  Gan, Christina A., digital author.
Title: Essentials of anatomy & physiology / Kenneth S. Saladin, Georgia
  College, Robin K. McFarland, Cabrillo College; digital author,
  Christina A. Gan, Highline Community College.
Other titles: Essentials of anatomy and physiology
Description: Third edition. | New York : McGraw Hill LLC, [2022] |
  Includes bibliographical references and index.
Identifiers: LCCN 2020053032 (print) | LCCN 2020053033 (ebook) |
  ISBN 9781260266405 (hardcover) | ISBN 9781264349869 (spiral bound) |
  ISBN 9781264349852 (epub)
Subjects: LCSH: Human physiology—Textbooks. | Human anatomy—Textbooks.
Classification: LCC QP34.5 .S232 2022 (print) | LCC QP34.5 (ebook) |
  DDC 612—dc23
LC record available at https://lccn.loc.gov/2020053032
LC ebook record available at https://lccn.loc.gov/2020053033

The Internet addresses listed in the text were accurate at the time of publication. The inclusion of a website does not indicate an endorsement by the authors or McGraw Hill LLC, and McGraw Hill LLC does not guarantee the accuracy of the information presented at these sites.

mheducation.com/highered

Brief Contents

About the Authors iv
Table of Contents v
Taking Anatomy & Physiology to New Heights x
Making Anatomy & Physiology Intriguing and Inspiring xii
Acknowledgments xxi

PART ONE
Organization of the Body

1. The Study of Anatomy and Physiology 1
2. Life, Matter, and Energy 30
3. Cells 60
4. Histology—The Tissue Level of Organization 93

PART TWO
Support and Movement

5. The Integumentary System 125
6. The Skeletal System 143
7. The Muscular System 197

PART THREE
Internal Coordination

8. The Nervous System I: Nerve Cells, the Spinal Cord, and Reflexes 248
9. The Nervous System II: The Brain, Cranial Nerves, and Autonomic Nervous System 280
10. The Sense Organs 319
11. The Endocrine System 356

PART FOUR
Circulation and Defense

12. The Circulatory System I: Blood 385
13. The Circulatory System II: The Heart and Blood Vessels 410
14. The Lymphatic System and Immunity 462

PART FIVE
Intake and Output

15. The Respiratory System 495
16. The Urinary System 522
17. The Digestive System 555
18. Nutrition and Metabolism 593

PART SIX
Human Life Cycle

19. The Reproductive System 625
20. Human Development and Aging 669

Appendix A: Answer Keys A-1
Appendix B: Health Science Careers A-8
Appendix C: Symbols, Weights, and Measures A-10
Appendix D: Biomedical Word Roots, Prefixes, and Suffixes A-12
Appendix E: Periodic Table of the Elements A-16
Glossary G-1
Index I-1

About the Authors

Tim Vacula

KENNETH S. SALADIN is Distinguished Professor of Biology, Emeritus, at Georgia College in Milledgeville, Georgia, where he taught from 1977 to 2017. Ken's courses included human anatomy and physiology, introductory medical physiology, histology, animal behavior, parasitology, and natural history of the Galápagos Islands. Ken is a member of the Human Anatomy and Physiology Society, American Physiological Society, American Association for Anatomy, Society for Integrative and Comparative Biology, and Textbook and Academic Authors Association. He is the author of the best-selling textbooks *Anatomy & Physiology: The Unity of Form and Function* and *Human Anatomy*. Ken's textbooks support several scholarships—as well as an endowed chair in biomedical science; an endowed annual lecture in science, medicine, and society; and ecosystem conservation in the Galápagos Islands.

©Robin McFarland

ROBIN MCFARLAND has taught anatomy and physiology at Cabrillo College in Aptos, California, since 1998. She earned a Ph.D. in physical (biological) anthropology from the University of Washington, where she studied the relationship between body fat and reproduction in primates. Robin subsequently conducted research on comparative ape anatomy with colleagues at the University of California, Santa Cruz. Robin is a member of the Human Anatomy and Physiology Society, and a member of the American Association of Physical Anthropology. She is passionate about teaching anatomy and physiology. She and her husband Jeff have two children: Reid and Madeleine. Robin enjoys hiking and climbing mountains.

Yuen Lui Studios/Chris Gan

CHRISTINA A. GAN, digital author for Connect, has been teaching anatomy and physiology, genetics, microbiology, and general biology at Highline College near Seattle, Washington, since 2004. Before that, she taught at Rogue Community College in Medford, Oregon for 6 years. She earned her M.A. in biology from Humboldt State University where she studied population genetics, and is a member of the Human Anatomy and Physiology Society. When she is not in the classroom or developing digital media, she is climbing, mountaineering, skiing, kayaking, sailing, cycling, and mountain biking throughout the Pacific Northwest.

Dedicated to the memory of my most important teacher and mentor, Donald R. Sly (1931–2019).   K.S.S.
This book is dedicated to my students, who inspire and delight me.   R.K.M.

The authors would enjoy hearing from colleagues and students alike who use this book and may wish to offer suggestions for our next edition, or encouragement to continue doing certain things the way we have. Such feedback is invaluable for improving a textbook, and the authors will endeavor to answer all correspondence.

Kenneth S. Saladin, Ph.D.
Milledgeville, Georgia
ksaladin2@windstream.net

Robin K. McFarland, Ph.D.
Cabrillo College, Aptos, California
romcfarl@cabrillo.edu

Table of Contents

About the Authors iv
Table of Contents v
Taking Anatomy & Physiology to New Heights x
Making Anatomy & Physiology Intriguing and Inspiring xii
Acknowledgments xxi

PART ONE
Organization of the Body

Eye of Science/Science Source

Chapter 1
The Study of Anatomy and Physiology 1

1.1 Anatomy—The Structural Basis of Human Function 2
1.2 Physiology—Dynamic Processes in the Living Body 8
   Perspectives on Health 9
1.3 The Human Body Plan 13
1.4 The Language of Medicine 21
   Career Spotlight: Radiologic Technologist 26
   Study Guide 27

Chapter 2
Life, Matter, and Energy 30

2.1 Atoms, Ions, and Molecules 31
2.2 Water, Acids, and Bases 37
2.3 Organic Compounds 40
   Perspectives on Health 51
2.4 Energy and Chemical Reactions 53
   Career Spotlight: Medical Technologist 56
   Study Guide 57

Chapter 3
Cells 60

3.1 The General Structure of Cells 61
3.2 The Cell Surface 63
3.3 The Cell Interior 75
3.4 The Life Cycle of Cells 85
   Perspectives on Health 89
   Career Spotlight: Cytotechnologist 90
   Study Guide 91

Chapter 4
Histology—The Tissue Level of Organization 93

4.1 The Study of Tissues 94
4.2 Epithelial Tissue 97
4.3 Connective Tissue 104
4.4 Nervous and Muscular Tissue—The Excitable Tissues 112
   Perspectives on Health 115
4.5 Glands and Membranes 116
4.6 Tissue Growth, Repair, and Death 119
   Career Spotlight: Histotechnician 121
   Study Guide 122

PART TWO
Support and Movement

©Eye of Science/Science Source

Chapter 5
The Integumentary System 125

- 5.1 The Skin and Subcutaneous Tissue 126
  Perspectives on Health 132
- 5.2 Accessory Organs 134
  Aging of the Integumentary System 139
  Career Spotlight: Dermatology Nurse 139
  Connective Issues 140
  Study Guide 141

Chapter 6
The Skeletal System 143

- 6.1 Skeletal Structure and Function 144
- 6.2 Bone Development and Metabolism 150
  Perspectives on Health 154
- 6.3 The Axial Skeleton 155
- 6.4 The Appendicular Skeleton 171
- 6.5 Joints 180
  Aging of the Skeletal System 192
  Career Spotlight: Orthopedic Nurse 192
  Connective Issues 193
  Study Guide 194

Chapter 7
The Muscular System 197

- 7.1 Muscular Tissue and Cells 198
- 7.2 Physiology of Skeletal Muscle 206
- 7.3 Cardiac and Smooth Muscle 215
- 7.4 Anatomy of the Muscular System 218
  Perspectives on Health 220
  Aging of the Muscular System 243
  Career Spotlight: Massage Therapist 243
  Connective Issues 244
  Study Guide 245

PART THREE
Internal Coordination

Sciepro/Science Photo Library/Getty Images RF

Chapter 8
The Nervous System I: Nerve Cells, The Spinal Cord, and Reflexes 248

- 8.1 Cells and Tissues of the Nervous System 249
- 8.2 The Physiology of Neurons 258
  Perspectives on Health 264
- 8.3 The Spinal Cord, Spinal Nerves, and Reflexes 265
  Career Spotlight: Occupational Therapist 276
  Study Guide 277

Chapter 9

The Nervous System II: The Brain, Cranial Nerves, and Autonomic Nervous System 280

9.1 Overview of the Brain 281
9.2 Principal Divisions of the Brain 288
9.3 Multiregional Brain Functions 295
9.4 The Cranial Nerves 302
9.5 The Autonomic Nervous System 304
Perspectives on Health 312
Aging of the Nervous System 313
Career Spotlight: Electroneurodiagnostic Technologist 314
Connective Issues 315
Study Guide 316

Chapter 10

The Sense Organs 319

10.1 Receptors and Sensations 320
10.2 The General Senses 323
10.3 The Chemical Senses—Taste and Smell 326
10.4 The Ear—Equilibrium and Hearing 330
10.5 The Eye and Vision 339
Perspectives on Health 350
Aging of the Sense Organs 351
Career Spotlight: Optician 352
Study Guide 353

Chapter 11

The Endocrine System 356

11.1 Overview of the Endocrine System 357
11.2 Endocrine Physiology 360
11.3 The Hypothalamus and Pituitary Gland 363
11.4 Other Endocrine Glands and Tissues 369
Perspectives on Health 378
11.5 Stress Physiology 379
Aging of the Endocrine System 380
Career Spotlight: Diabetes Educator 380
Connective Issues 381
Study Guide 382

PART FOUR
Circulation and Defense

Dr. Yorgos Nikas/Science Source

Chapter 12

The Circulatory System I: Blood 385

12.1 Introduction 386
12.2 Erythrocytes 390
12.3 Leukocytes 396
Perspectives on Health 399
12.4 Platelets and Control of Bleeding 401
Career Spotlight: Phlebotomist 405
Connective Issues 406
Study Guide 407

Chapter 13

The Circulatory System II: The Heart and Blood Vessels 410

13.1 Overview of the Cardiovascular System 411
13.2 Gross Anatomy of the Heart 414
Perspectives on Health 420
13.3 Physiology of the Heart 421
13.4 General Anatomy of Blood Vessels 430
13.5 Physiology of Circulation 435
13.6 Circulatory Routes and Blood Vessels 441
Aging of the Circulatory System 457
Career Spotlight: Electrocardiographic Technician 457
Connective Issues 458
Study Guide 459

Chapter 14
The Lymphatic System and Immunity 462

14.1 The Lymphatic System 463
14.2 Innate Immunity 472
14.3 Features of Adaptive Immunity 477
14.4 Cellular and Humoral Immunity 483
Perspectives on Health 489
Aging of the Lymphatic and Immune Systems 490
Career Spotlight: Public Health Nurse 490
Connective Issues 491
Study Guide 492

PART FIVE
Intake and Output

©Susumu Nishinaga/Science Source

Chapter 15
The Respiratory System 495

15.1 Functions and Anatomy of the Respiratory System 496
15.2 Pulmonary Ventilation 506
Perspectives on Health 511
15.3 Gas Exchange and Transport 512
Aging of the Respiratory System 517
Career Spotlight: Respiratory Therapist 517
Connective Issues 518
Study Guide 519

Chapter 16
The Urinary System 522

16.1 Functions of the Urinary System 523
16.2 Anatomy of the Kidney 525
16.3 Glomerular Filtration 531
16.4 Tubular Reabsorption and Secretion 536
16.5 Water Conservation 539
Perspectives on Health 541
16.6 Urine Storage and Elimination 541
16.7 Fluid, Electrolyte, and Acid–Base Balance 546
Aging of the Urinary System 550
Career Spotlight: Dialysis Technician 550
Connective Issues 551
Study Guide 552

Chapter 17
The Digestive System 555

17.1 Overview of the Digestive System 556
17.2 The Mouth Through Esophagus 560
17.3 The Stomach 564
Perspectives on Health 571
17.4 The Liver, Gallbladder, and Pancreas 571
17.5 The Small Intestine 576
17.6 Chemical Digestion and Absorption 580
17.7 The Large Intestine 584
Aging of the Digestive System 588
Career Spotlight: Dental Hygienist 588
Connective Issues 589
Study Guide 590

Chapter 18
Nutrition and Metabolism 593

18.1 Nutrition 594
Perspectives on Health 602
18.2 Carbohydrate Metabolism 603
18.3 Lipid and Protein Metabolism 607
18.4 Metabolic States and Metabolic Rate 611
18.5 Appetite Regulation 614
18.6 Body Heat and Thermoregulation 617
Career Spotlight: Dietitian 621
Study Guide 622

PART SIX
Human Life Cycle

©David M. Phillips/Science Source

Chapter 19
The Reproductive System 625

19.1 Essentials of Sexual Reproduction 626
19.2 The Male Reproductive System 627
19.3 The Female Reproductive System 635
19.4 The Production and Union of Sex Cells 643
Perspectives on Health 657
19.5 Pregnancy, Childbirth, and Lactation 658
Aging of the Reproductive System 663
Career Spotlight: Midwife 664
Connective Issues 665
Study Guide 666

Chapter 20
Human Development and Aging 669

20.1 Fertilization and Preembryonic Development 670
20.2 The Embryonic and Fetal Stages 677
Perspectives on Health 687
20.3 The Neonate 688
20.4 Aging, Senescence, and Death 693
Career Spotlight: Genetic Counselor 697
Study Guide 698

Appendix A: Answer Keys A-1
Appendix B: Health Science Careers A-8
Appendix C: Symbols, Weights, and Measures A-10
Appendix D: Biomedical Word Roots, Prefixes, and Suffixes A-12
Appendix E: Periodic Table of the Elements A-16
Glossary G-1
Index I-1

Taking Anatomy & Physiology to New Heights

Audience

Essentials of Anatomy & Physiology, third edition, is intended for students in associate degree, certification, and career-training programs; students in high-school advanced placement classes; students who are seeking a general education science class; and those who may not have set foot in a college classroom for many years. The prose and vocabulary in *Essentials of Anatomy & Physiology* are appropriate to serve this broad spectrum of readers.

Keeping in mind that many students are interested in exploring medical professions, a "Career Spotlight" feature has been included in every chapter, and references to further career information are found in appendix B.

Theme

The goal of this book is to help students succeed. Like climbing a mountain, learning essentials of anatomy and physiology takes place one step at a time, and, similar to the feeling of exhilaration when one has arrived at a mountain peak, there is a powerful sense of achievement upon mastering the knowledge of anatomy and physiology. The mountain-climbing theme is echoed in visual and pedagogical features of the book. A base camp on the first page of each chapter lists key information students need to understand in preparation for navigating the chapter. Throughout the chapter, there are milestones that mark progress, such as the Before You Go On checkpoints with an image of a climber steadily moving upward. The study guide at the end of each chapter features a victorious climber on the summit, a celebration that mirrors the student's mastery of the chapter information.

Before You Go On

Answer these questions from memory to test your understanding of the previous section. Reread the material if you can't answer the questions.

1. Which term refers to all the cell contents between the plasma membrane and nucleus: cytosol, cytoplasm, tissue fluid, or extracellular fluid?
2. About how big would a cell have to be for you to see it without a microscope? Are any cells actually this big? If so, which ones?
3. Explain why cells cannot grow to an indefinitely large size.

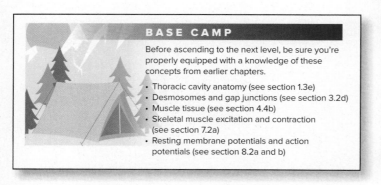

BASE CAMP

Before ascending to the next level, be sure you're properly equipped with a knowledge of these concepts from earlier chapters.

- Thoracic cavity anatomy (see section 1.3e)
- Desmosomes and gap junctions (see section 3.2d)
- Muscle tissue (see section 4.4b)
- Skeletal muscle excitation and contraction (see section 7.2a)
- Resting membrane potentials and action potentials (see section 8.2a and b)

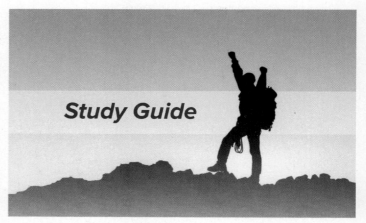

Study Guide

What's New in the Third Edition?

The new edition of *Essentials of Anatomy & Physiology* by Saladin and McFarland has been significantly updated. A hallmark of previous editions, according to both students and reviewers, is the exceptionally clear writing. In this new edition, we continue our commitment to students, with approachable language and relatable examples and analogies. We present current, solid scientific information. We have included numerous updates based on recent, peer-reviewed journal articles, as well as updated clinical examples and disease statistics. We have expanded discussions of health and disease to help students apply concepts of anatomy and physiology to their daily lives.

Updated Science and Enhanced Content

Some examples of updated or new scientific information, as well as enhanced anatomical and physiological concepts, follow:

- Chapter 1—new information about use of PET scans to diagnose Alzheimer disease
- Chapter 3—updated discussion of genomic medicine; new summary table of features and functions of organelles
- Chapter 4—expanded discussion of epithelial functions; new research on brown and white fat and endocrine functions of adipose tissue; new Clinical Application and photo of diabetic foot ulcers; new information on regenerative medicine and stem cells
- Chapter 5—updates on pathology and immunotherapy of melanoma
- Chapter 6—updates on the endocrine function of bones and their widespread effects on the body; updates on functions of osteocytes; new Clinical Application on rickets
- Chapter 7—new information on causes of muscle fatigue; streamlined chapter by omitting minor muscles; expanded discussion of health benefits of exercise
- Chapter 8—expanded discussion of organization of the nervous system and added a figure; expanded discussion of functions of the nervous system; added table of functions of neuroglial cells; expanded explanation of myelin
- Chapter 9—new information on meningitis and traumatic brain injury; updates on the basal nuclei
- Chapter 10—added evolutionary perspectives on taste preference and present-day obesity and comparative, evolutionary perspectives on the sense of smell; new research on smell and Alzheimer disease and depression
- Chapter 11—new summary tables for pituitary gland hormones and hormones from other sources; updated discussion of oxytocin; added new information about leptin
- Chapter 12—new summary table of ABO blood group; updated discussion of research on gene therapy and sickle-cell disease
- Chapter 13—updates on the interaction between heart and brain; streamlined discussion of coronary circulation; updated blood pressure guidelines
- Chapter 14—expanded discussion of microbiome; new research about neutrophil extracellular traps
- Chapter 15—expanded discussion of surfactant and premature infants; new information about effects of e-cigarettes (vaping) and smoking marijuana on respiratory health; updated statistics regarding tobacco cigarette smoking
- Chapter 17—updates on enteric nervous system; updates on rotavirus vaccine and childhood mortality
- Chapter 18—updates on sugar consumption in the United States and obesity
- Chapter 19—new information about prostate cancer; new information on the role the fetus plays in initiating labor
- Chapter 20—added new information about the zica virus

Making *Anatomy & Physiology* Intriguing and Inspiring

Essentials of Anatomy & Physiology crafts the facts of A&P into art and prose in a way that makes the book exciting and rewarding to read.

Captivating Art and Photography

A&P is a highly visual subject; beautiful illustrations pique the curiosity and desire to learn. *Essentials of Anatomy & Physiology*'s illustrations set a new standard in the A&P Essentials market, where many students regard themselves as visual learners.

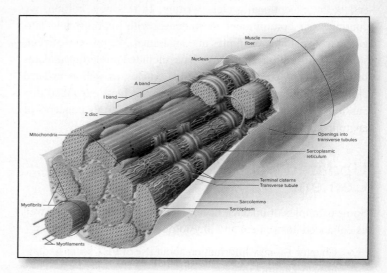

Cognitive Skill Building

Essentials of Anatomy & Physiology asks questions that not only test memory, but also exercise and expand the student's thinking skills at multiple levels of Bloom's Taxonomy of Learning Outcomes. Within Connect™ there is also the opportunity to assess student understanding of the Learning Outcomes by leveraging question filters that allow the curation of custom assignments and efficient reporting for administrative assessment purposes.

Testing Your Comprehension

1. Most osteocytes of an osteon are far removed from blood vessels, but are still able to respond to hormones in the blood. Explain how it is possible for hormones to reach and stimulate these cells.

2. How does the regulation of blood calcium concentration exemplify negative feedback and homeostasis?

3. Name the action that would occur at each of the following joints in the indicated situation. (For example, the shoulder in picking up a suitcase. Answer: elevation.) (a) The arm when you raise it to rest your hand on the back of a sofa on which you're sitting. (b) Your neck when you look up at a plane in the sky. (c) Your tibia when you turn the toes of one foot to touch the heel of the other foot. (d) Your humerus when you reach up to scratch the back of your head. (e) A bowler's backswing. (f) A basketball player's foot as she makes a jump shot. (g) Your shoulder when you pull back on the oars of a rowboat. (h) Your elbow when lifting a barbell. (i) A soccer player's knee when kicking the ball. (j) Your index finger when dialing an old rotary telephone. (k) Your thumb when you pick up a tiny bead between your thumb and index finger.

Expected Learning Outcomes

This book provides a ready-made course outline of course objectives and means of assessment with its "Expected Learning Outcomes" presented at the start of each chapter section.

Assess Your Learning Outcomes

The parallel "Assess Your Learning Outcomes" at the end of each chapter provides a comprehensive overview of key points in the chapter. Study guide questions probe understanding of concepts and highlight for students what they need to review.

Before You Go On/ Apply What You Know

Intermediate aids such as "Before You Go On" and "Apply What You Know" provide an easy means for meeting the requirements of an outcome-driven curriculum and also work to encourage active learning over passive reading.

3.4 The Life Cycle of Cells

Expected Learning Outcomes

When you have completed this section, you should be able to

a. describe the stages of a cell's life cycle and list the events that define each stage; and
b. name the stages of mitosis and describe what occurs in each.

Before You Go On

Answer these questions from memory to test your understanding of the previous section. Reread the material if you can't answer the questions.

1. *Which term refers to all the cell contents between the plasma membrane and nucleus: cytosol, cytoplasm, tissue fluid, or extracellular fluid?*
2. *About how big would a cell have to be for you to see it without a microscope? Are any cells actually this big? If so, which ones?*
3. *Explain why cells cannot grow to an indefinitely large size.*

Apply What You Know
Physical exercise obviously increases cardiac output. Do you think it achieves this through heart rate, contraction strength, or both? Explain.

Making *Anatomy & Physiology* Intriguing and Inspiring

We have seen how a nerve signal is initiated; now we examine how it travels to its final destination. The action potential is a voltage spike over a limited area of plasma membrane. However, it triggers another action potential in the membrane immediately ahead of it, and that action potential triggers another, and so forth. Thus, we get a chain reaction of one action potential after another along the length of a nerve fiber. This chain reaction constitutes the **nerve signal.** An illuminating analogy to this is standing up a long row of dominoes and pushing the first one over. When that domino falls, it pushes over the second, and so forth—and the chain reaction produces a wave of energy traveling to the end of the line. No one domino moves to the other end of the line; a falling domino is a local event. Similarly, an action potential is a local event, but it triggers the next one and, like the row of falling dominoes, we get a wave of energy traveling from one end of the axon to the other. That traveling wave is the nerve signal (fig. 8.11). Action potentials do not travel; nerve signals do.

Stimulating Prose

Far more than "just the facts," *Essentials of Anatomy & Physiology*'s narrative style weaves relevant details into an engaging story of human form and function. Vivid analogies that captivate the imagination make complex concepts easier to understand.

Figure Legend Questions

Thought questions in many figure legends encourage students to think analytically about the art, not merely view it. These questions are also great for in-class discussion.

Figure 5.5 Structure of a Hair and Its Follicle. (a) Anatomy of the follicle and associated structures. (b) Light micrograph of the base of a hair follicle.
APR
b: Ed Reschke/Getty Images

❓ *In light of your knowledge of hair, discuss the validity of an advertising claim that a shampoo will "nourish your hair." Where and how does a hair get its sole nourishment?*

Building Vocabulary

The plethora of medical terms in A&P is one of a student's most daunting challenges. Chapter 1 teaches core principles of how to break words down into familiar roots, prefixes, and suffixes, making medical terminology less intimidating while teaching the importance of precision in spelling *(ilium/ileum, malleus/malleolus)*.

- An end-of-book "Glossary" provides clear definitions of the most important or frequently used terms, and "Appendix D: Biomedical Word Roots, Prefixes, and Suffixes" defines nearly 400 Greek and Latin roots, which make up about 90% of today's medical terms.
- *Footnoted word origins* show how new terms are composed of familiar word roots.
- *Pronunciation guides* that appear throughout chapters make it easier to pronounce key terms, and make these words more likely to be remembered and understood.

1.4a Analyzing Medical Terms

There is a simple trick to becoming more comfortable with the technical language of medicine. Those who, at first, find scientific terms confusing and difficult to pronounce, spell, and remember often feel more confident once they realize the logic of how such terms are composed. A term such as *hyponatremia* is less forbidding once we recognize it is composed of three common word elements: *hypo-* (below normal), *natr-* (sodium), and *-emia* (blood condition). Thus, hyponatremia is a deficiency of sodium in the blood. Those three word elements appear over and over in many other medical terms: *hypothermia, natriuretic, anemia,* and so on. Once you learn the meanings of *hypo-, natri-,* at least partially understand hundreds

cholecystokinin (CCK) (CO-leh-SIS-toe-KY-nin) A polypeptide employed as a hormone and neurotransmitter, secreted by some brain neurons and cells of the small intestine. In the digestive system, stimulates contraction of the gallbladder, release of bile, and secretion of pancreatic enzymes.

[5]*oss* = bone; *icle* = little
[6]*malleus* = hammer, mallet
[7]*incus* = anvil
[8]*stapes* = stirrup
[9]Bartholomeo Eustachio (1520–74), Italian anatomist

Study Guide

The "Study Guide" at the end of each chapter provides an overview of key points, as well as a variety of self-testing question formats to effectively reinforce the material. A student who masters these study guides should do well on an exam.

Study Guide

Assess Your Learning Outcomes

To test your knowledge, discuss the following topics with a study partner or in writing, ideally from memory.

3.1 The General Structure of Cells
1. Fundamental components of a cell
2. Intracellular and extracellular fluids
3. The typical size range of human cells and what factors limit cell size

3.2 The Cell Surface
1. Molecular components and organization of the plasma membrane
2. Varieties and functions of the plasma membrane proteins
3. The composition, location, and functions of a cell's glycocalyx
4. Structural and functional distinctions between microvilli, cilia, flagella, and pseudopods
5. Structural distinctions and respective advantages of three types of cell junctions—tight junctions, desmosomes, and gap junctions
6. The eight modes of transport through a plasma membrane and how they differ with respect to the use of carrier proteins, direction of movement of the transported substances, and demand for ATP

3.3 The Cell Interior
1. Components and functions of the cytoskeleton
2. Types of cell inclusions and how inclusions differ from organelles
3. What organelles have in common and how they differ, as a class, from other cellular components
4. Structure of the nucleus, particularly of its nuclear envelope, chromatin, and nucleoli
5. Two forms of endoplasmic reticulum, their spatial relationship, their structural similarities and differences, and their functional differences
6. The composition, appearance, locations, and function of ribosomes
7. Structure of the Golgi complex and its role in the synthesis, packaging, and secretion of cell products
8. [Similarities and differences between] lysosomes and peroxisomes
9. [Evolutionary] origin of mitochondria, [mitochondrial] DNA
10. [Structure] of centrioles
11. The processes of genetic transcription and translation, including the roles of mRNA, rRNA, and tRNA
12. How the amino acid sequence of a protein is represented by the codons of mRNA
13. How proteins are processed and secreted after their assembly on a ribosome

3.4 The Life Cycle of Cells
1. Four phases of the cell cycle and the main events in each phase
2. How DNA is replicated in preparation for mitosis
3. Functions of mitosis
4. Four stages of mitosis; changes in chromosome structure and distribution that occur in each stage; and the role of centrioles and the mitotic spindle
5. The mechanism and result of cytokinesis

Testing Your Recall

1. The clear, structureless gel in a cell is its
   a. nucleoplasm.
   b. endoplasm.
   c. cytoplasm.
   d. neoplasm.
   e. cytosol.

2. New nuclei form and a cell begins to pinch in two during
   a. prophase.
   b. metaphase.
   c. interphase.
   d. telophase.
   e. anaphase.

3. The amount of _____ in a plasma membrane affects its fluidity.
   a. phospholipid
   b. cholesterol
   c. glycolipid
   d. glycoprotein
   e. integral protein

4. Cells specialized for absorption of matter from the extracellular fluid are likely to show an abundance of
   a. lysosomes.
   b. microvilli.
   c. mitochondria.
   d. secretory vesicles.
   e. ribosomes.

6. The word root *phago-* means
   a. eating.
   b. drinking.
   c. emitting fluid.
   d. intracellular.
   e. extracellular.

7. The amount of DNA in a cell doubles during
   a. prophase.
   b. metaphase.
   c. anaphase.
   d. the S phase.
   e. the G_2 phase.

8. Fusion of a secretory vesicle with the plasma membrane and release of the vesicle's contents is
   a. exocytosis.
   b. receptor-mediated endocytosis.
   c. active transport.
   d. pinocytosis.
   e. phagocytosis.

9. Most cellular membranes are made by
   a. the nucleus.
   b. the cytoskeleton.
   c. enzymes in the peroxisomes.
   d. the endoplasmic reticulum.
   e. replication of existing membranes.

10. Which of the following is/are not involved in protein synthesis?
    a. ribosomes
    b. centrioles
    c. mRNA
    d. rough endoplasmic reticulum
    e. codons

11. Most human cells are 10 to 15 _____ wide.

12. When a hormone cannot enter a cell, it binds to a _____ at the cell surface.

13. _____ are channels in the plasma membrane that open or close in response to various stimuli.

14. Cells are somewhat protected from mechanical trauma by a carbohydrate surface coat called the _____.

15. The separation of chromatids from each other marks the _____ stage of mitosis.

16. The majority of molecules that compose the plasma membrane are _____.

17. Two human organelles that are surrounded by a double membrane are the _____ and _____.

What's Wrong with These Statements?

Briefly explain why each of the following statements is false, or reword it to make it true.

1. A cell specialized for absorption would be expected to have a high density of cilia on its surface.
2. DNA replication occurs during mitosis.
3. A cell can release its secretory products by exocytosis, phagocytosis, or pinocytosis.
4. In the plasma membrane, the phosphate heads of the phospholipid molecules cluster together in the middle of the membrane and the fatty acid tails are pointed toward the ICF and ECF.
5. Cells of the digestive glands store enzymes in their lysosomes and release them into the digestive tract when needed to digest food.
6. As a carrier-mediated transport process, facilitated diffusion requires ATP.
7. Osmosis is a type of active transport involving water.
8. White blood cells can move about in the tissues by means of either cilia or pseudopods.
9. Desmosomes enable solutes to pass from cell to cell.
10. Ribosomes and the Golgi complex play similar roles in the synthesis of proteins.

Answers in Appendix A

Testing Your Comprehension

1. Breast milk contains both sugar (lactose) and proteins (albumin and casein). Identify which organelles of the mammary gland cells are involved in synthesizing and secreting these components, and describe the structural pathway from synthesis to release from the cell.
2. A person with lactose intolerance cannot digest lactose, so instead of being absorbed by the small intestine, this sugar passes undigested into the large intestine. Here, it causes diarrhea among other signs. Which of the membrane transport processes do you think is most directly involved in the diarrhea? On that basis, explain why the diarrhea occurs.
3. Consider a cardiac muscle cell, an enzyme-producing pancreatic cell, a phagocytic white blood cell, and a hormone-secreting cell of the ovary. Which of these would you expect to show the greatest number of lysosomes? Mitochondria? Rough endoplasmic reticulum? Smooth endoplasmic reticulum? Explain each answer.

Multiple Question Types

- "Testing Your Recall" questions check for simple memory of terms and facts.
- The false assertions in "What's Wrong with These Statements?" require students to analyze the validity of ideas and to explain or rephrase each false statement.
- "Testing Your Comprehension" questions necessitate insight and application to clinical and other scenarios.

Making *Anatomy & Physiology* Intriguing and Inspiring

BASE CAMP

Before ascending to the next level, be sure you're properly equipped with a knowledge of these concepts from earlier chapters.

- Thoracic cavity anatomy (see section 1.3e)
- Desmosomes and gap junctions (see section 3.2d)
- Muscle tissue (see section 4.4b)
- Skeletal muscle excitation and contraction (see section 7.2a)
- Resting membrane potentials and action potentials (see section 8.2a and b)

Tying It All Together
Base Camp

- "Base Camp" lists key concepts from earlier chapters that a student should know before embarking on the new one, and effectively ties all chapters together into an integrated whole.

Connective Issues

- No organ system functions in isolation. The "Connective Issues" tool shows how every organ system affects all other body systems, and generates a more holistic understanding of human function.

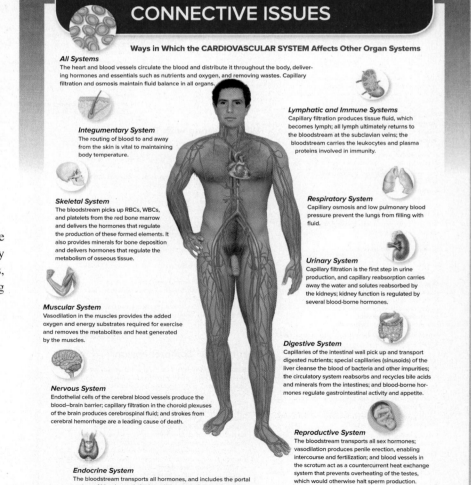

CONNECTIVE ISSUES

Ways in Which the CARDIOVASCULAR SYSTEM Affects Other Organ Systems

All Systems
The heart and blood vessels circulate the blood and distribute it throughout the body, delivering hormones and essentials such as nutrients and oxygen, and removing wastes. Capillary filtration and osmosis maintain fluid balance in all organs.

Integumentary System
The routing of blood to and away from the skin is vital to maintaining body temperature.

Skeletal System
The bloodstream picks up RBCs, WBCs, and platelets from the red bone marrow and delivers the hormones that regulate the production of these formed elements. It also provides minerals for bone deposition and delivers hormones that regulate the metabolism of osseous tissue.

Muscular System
Vasodilation in the muscles provides the added oxygen and energy substrates required for exercise and removes the metabolites and heat generated by the muscles.

Nervous System
Endothelial cells of the cerebral blood vessels produce the blood–brain barrier; capillary filtration in the choroid plexuses of the brain produces cerebrospinal fluid; and strokes from cerebral hemorrhage are a leading cause of death.

Endocrine System
The bloodstream transports all hormones, and includes the portal system of blood vessels for hypothalamo–pituitary communication.

Lymphatic and Immune Systems
Capillary filtration produces tissue fluid, which becomes lymph; all lymph ultimately returns to the bloodstream at the subclavian veins; the bloodstream carries the leukocytes and plasma proteins involved in immunity.

Respiratory System
Capillary osmosis and low pulmonary blood pressure prevent the lungs from filling with fluid.

Urinary System
Capillary filtration is the first step in urine production, and capillary reabsorption carries away the water and solutes reabsorbed by the kidneys; kidney function is regulated by several blood-borne hormones.

Digestive System
Capillaries of the intestinal wall pick up and transport digested nutrients; special capillaries (sinusoids) of the liver cleanse the blood of bacteria and other impurities; the circulatory system reabsorbs and recycles bile acids and minerals from the intestines; and blood-borne hormones regulate gastrointestinal activity and appetite.

Reproductive System
The bloodstream transports all sex hormones; vasodilation produces penile erection, enabling intercourse and fertilization; and blood vessels in the scrotum act as a countercurrent heat exchange system that prevents overheating of the testes, which would otherwise halt sperm production.

Career Spotlight

- "Career Spotlight" features provide a relevant career example in every chapter with basic information on educational requirements, and expand student awareness of opportunities in allied health professions. "Appendix B" refers students to online sources of further information about 20 career fields and a list of 83 more health-care career ideas.

Clinical Application

- "Clinical Application" essays apply basic science to interesting issues of health and disease.

Perspectives on Health

- "Perspectives on Health" essays make basic science relevant to the student's interest in health and disease.

Aging of Body Systems

- "Aging of [Body Systems]" is a section within systems chapters that describes how each organ system changes over time, especially in old age. This discussion expands anatomical and physiological understanding beyond the prime of life, and is highly relevant to patient treatment, since older patients constitute most of the health-care market.

Instructors: Student Success Starts with You

Tools to enhance your unique voice

Want to build your own course? No problem. Prefer to use our turnkey, prebuilt course? Easy. Want to make changes throughout the semester? Sure. And you'll save time with Connect's auto-grading too.

65%
Less Time Grading

Laptop: McGraw Hill; Woman/dog: George Doyle/Getty Images

Study made personal

Incorporate adaptive study resources like SmartBook® 2.0 into your course and help your students be better prepared in less time. Learn more about the powerful personalized learning experience available in SmartBook 2.0 at **www.mheducation.com/highered/connect/smartbook**

Affordable solutions, added value

Make technology work for you with LMS integration for single sign-on access, mobile access to the digital textbook, and reports to quickly show you how each of your students is doing. And with our Inclusive Access program you can provide all these tools at a discount to your students. Ask your McGraw Hill representative for more information.

Padlock: Jobalou/Getty Images

Solutions for your challenges

A product isn't a solution. Real solutions are affordable, reliable, and come with training and ongoing support when you need it and how you want it. Visit **www.supportateverystep.com** for videos and resources both you and your students can use throughout the semester.

Checkmark: Jobalou/Getty Images

Students: Get Learning that Fits You

Effective tools for efficient studying

Connect is designed to make you more productive with simple, flexible, intuitive tools that maximize your study time and meet your individual learning needs. Get learning that works for you with Connect.

Study anytime, anywhere

Download the free ReadAnywhere app and access your online eBook or SmartBook 2.0 assignments when it's convenient, even if you're offline. And since the app automatically syncs with your eBook and SmartBook 2.0 assignments in Connect, all of your work is available every time you open it. Find out more at **www.mheducation.com/readanywhere**

> *"I really liked this app—it made it easy to study when you don't have your textbook in front of you."*
>
> - Jordan Cunningham, Eastern Washington University

Calendar: owattaphotos/Getty Images

Everything you need in one place

Your Connect course has everything you need—whether reading on your digital eBook or completing assignments for class, Connect makes it easy to get your work done.

Learning for everyone

McGraw Hill works directly with Accessibility Services Departments and faculty to meet the learning needs of all students. Please contact your Accessibility Services Office and ask them to email accessibility@mheducation.com, or visit **www.mheducation.com/about/accessibility** for more information.

Top: Jenner Images/Getty Images, Left: Hero Images/Getty Images, Right: Hero Images/Getty Images

McGraw-Hill empowers students to learn and succeed in the Anatomy and Physiology course.

50% of the country's students are unable to pass the A&P course*

SMARTBOOK®

SmartBook provides personalized learning to individual student needs, continually adapting to pinpoint knowledge gaps and focus learning on concepts requiring additional study. The result? Students are highly engaged in the content and better prepared for lecture.

LEARNSMART PREP®

LearnSmart Prep helps students thrive in college-level A&P by helping solidify knowledge in the key areas of cell biology, chemistry, study skills, and math. The result? Students are better prepared for the A&P course.

PhILS

Ph.I.L.S. 4.0 (Physiology Interactive Lab Simulations) software is the perfect way to reinforce key physiology concepts with powerful lab experiments. The result? Students gain critical thinking skills and are better prepared for lab.

Concept Overview Interactives are ground-breaking interactive animations that encourage students to explore key physiological processes and difficult concepts. The result? Students are engaged and able to apply what they've learned while tackling difficult A&P concepts.

Practice ATLAS

Practice Atlas for A&P is an interactive tool that pairs images of common anatomical models with stunning cadaver photography, allowing students to practice naming structures on both models and human bodies, anytime, anywhere. The result? Students are better prepared, engaged, and move beyond basic memorization.

*Statistic courtesy of The New England Journal of Higher Education

Virtual Labs

Connect Virtual Labs helps connect the dots between lab and lecture, boosts student confidence and knowledge, and improves student success rates. The result? Students are engaged, prepared, and utilize critical thinking skills.

Anatomy & Physiology Revealed® 4.0

Anatomy & Physiology Revealed® (APR) 4.0 is an interactive cadaver dissection tool to enhance lectures and labs, which students can use anytime, anywhere. The result? Students are prepared for lab, engaged in the material, and utilize critical thinking.

Acknowledgments

I gratefully acknowledge the team at McGraw-Hill who have provided excellent ideas and unfailing encouragement throughout this project. I am immensely grateful to my coauthor Ken Saladin for a rewarding collaboration and firm friendship. I appreciate my colleagues at Cabrillo College who inspire me every day with their dedication to student success. Finally, I wish to thank my husband Jeff and my children Reid and Madeleine.

Robin McFarland

My heartfelt appreciation goes to our team at McGraw-Hill who have provided such friendship, collegiality, and support over my 27-year history in textbooks; to Robin for adding this new dimension and stimulating collaboration to my writing career; to my colleagues at Georgia College for an atmosphere that supports and rewards such work; and to Diane for her steadfast love and encouragement.

Ken Saladin

Our grateful thanks are extended to these reviewers, who read early drafts of these chapters and provided instructive comments to help shape the content within these pages.

Darlene R. Dulin
Ivy Tech Community College, Valparaiso Campus

Amy Goode
Illinois Central College

Kristin A. Jacobson Flex
Illinois Central College

Elizabeth A. Mays
Illinois Central College

Tanya Z. Quiroz
Central Carolina Technical College

Barbara Reitz
Pensacola State College

Susan Rohde
Triton College

Rachael L. Smith
Greenville Technical College

Sherry S. Stewart
Navarro College

Heather Stottman
Navarro College

Remote Proctoring & Browser-Locking Capabilities

 +

New remote proctoring and browser-locking capabilities, hosted by Proctorio within Connect, provide control of the assessment environment by enabling security options and verifying the identity of the student.

Seamlessly integrated within Connect, these services allow instructors to control students' assessment experience by restricting browser activity, recording students' activity, and verifying students are doing their own work.

Instant and detailed reporting gives instructors an at-a-glance view of potential academic integrity concerns, thereby avoiding personal bias and supporting evidence-based claims.

Design Element Credits: (Camping): monkographic/Shutterstock; (Abseiling): Image Source; (Stethoscope): MedicalRF.com; (Clouds): Pakhnyushchy/shutterstock.com; (Doctors): Purestock/SuperStock; (Mountain climber): McGraw-Hill Education.

PART 1 Organization of the Body

The Study of Anatomy and Physiology

Chapter **1**

Chapter Outline

1.1 Anatomy—The Structural Basis of Human Function
- 1.1a The Study of Anatomy
- 1.1b Examination of the Body
- 1.1c Techniques of Medical Imaging
- 1.1d Anatomical Variation

1.2 Physiology—Dynamic Processes in the Living Body
- 1.2a The Physiological Sciences
- 1.2b Essential Life Functions
- 1.2c Homeostasis and Feedback
- 1.2d Physiological Variation

1.3 The Human Body Plan
- 1.3a Levels of Human Structure
- 1.3b Anatomical Position
- 1.3c Anatomical Planes
- 1.3d Major Body Regions
- 1.3e Body Cavities and Membranes
- 1.3f Organ Systems

1.4 The Language of Medicine
- 1.4a Analyzing Medical Terms
- 1.4b Singular and Plural Forms
- 1.4c Directional Terminology

Clinical Applications/Perspectives on Health
- Clinical Application 1.1: Men in the Oven
- Clinical Application 1.2: Peritonitis
- Perspectives on Health

End of Chapter
- Career Spotlight: Radiologic Technologist
- Study Guide

A full-body image made by magnetic resonance imaging (MRI). MRI is one of several ways of viewing the interior of the body without surgery.
Simon Fraser/Science Source

Module 1 Body Orientation

No branch of science hits as close to home as the science of our own bodies. We're grateful for the dependability of our hearts, we're awed by the capabilities of joints and muscles displayed by athletes, and we are curious about the mysteries of mind and emotion. We want to know how our body works, and when it malfunctions, we want to know what's happening and what we can do about it. In recent decades, scientists have revealed a wealth of information about our bodies, but fascination with the science of the body is nothing new. Ancient texts and art are testaments to humanity's timeless quest to know and heal the body and mind.

This book introduces the essentials of human structure and function. It will give you a deeper understanding of the healthy body, as well as accurate, up-to-date insights into disease processes. The disciplines of anatomy and physiology are fundamental to health-care professionals, as well as to those who study human performance, fitness, and nutrition. Beyond that, however, the study of anatomy and physiology provides a deeply satisfying sense of self-understanding.

In this chapter, we introduce the disciplines of anatomy and physiology. We discuss criteria that define life and consider a core concept called *homeostasis,* a vital process necessary for maintaining life. We look at the body's general structural plan and levels of organization. Finally, because one of the greatest challenges to beginning students is to master vocabulary associated with anatomy and physiology, we end the chapter with tools to help you effectively learn and use the language of the body.

1.1 Anatomy—The Structural Basis of Human Function

Expected Learning Outcomes

When you have completed this section, you should be able to

a. define *anatomy* and *physiology;*
b. describe some of the subfields of human anatomy;
c. explain the importance of dissection;
d. describe some methods of examining a living patient;
e. discuss the principles and applications of some medical imaging methods; and
f. discuss the significance of variations in human anatomy.

Anatomy is the study of the structure of the body, with an emphasis on how it relates to function. *Physiology* is the study of dynamic processes in the living body. The two disciplines are very much intertwined, and both are necessary to understand the totality of the body.

1.1a The Study of Anatomy

There are many approaches to the study of human anatomy, both in research for the purposes of discovery and understanding, and in clinical settings for diagnosis and treatment. **Gross anatomy** is structure visible to the naked eye, either by surface observation or dissection. Ultimately, though, body functions result from individual cells. To see those, we usually take tissue samples, thinly slice and stain them, and observe them under the microscope. This approach is called **histology.**[1] **Histopathology** is the microscopic examination of tissues for signs of disease.

Surface anatomy is the external structure of the body, and is especially important in conducting a physical examination of a patient. **Systemic anatomy** is the study of one organ system at a time; this is the approach taken by introductory textbooks such as this one. **Regional anatomy** is the study of multiple organ systems at the same time in a given region of the body, such as the head or chest. Medical schools and anatomical atlases typically teach anatomy from this perspective, because it is more logical to dissect all structures of the head and neck, the chest, or a limb, than to try to dissect the entire digestive system, then the cardiovascular system, and so forth. Dissecting one system almost inevitably destroys organs of other systems that stand in the way.

> **Apply What You Know**
>
> Do you think that a surgeon thinks more in terms of systemic anatomy or regional anatomy? Explain your answer.

You can study human anatomy from an atlas; yet, as fascinating and valuable as anatomy atlases are, they teach almost nothing but the locations, appearances, and names of structures. This book is much different; it deals with what biologists call **functional morphology**[2]—not simply describing structures but also analyzing how they function.

Functional morphology draws heavily on **comparative anatomy,** the study of more than one species. Such comparisons reveal similarities and differences, highlight evolutionary trends, and clarify structure–function relationships. Often, human structure makes sense only when we compare it to the structure of other animals. The human pelvis, for example, has a unique bowl-shaped configuration that can be best understood by comparison with animals such as a chimpanzee, whose pelvis is adapted to walking on four legs rather than two.

1.1b Examination of the Body

There are many ways to examine the body, the simplest of which is **inspection**—simply looking at the surface as physicians do during a physical examination. A deeper understanding depends on **dissection**[3]—carefully cutting and separating tissues to reveal relationships between structures. The word *anatomy*[4] literally means "cutting apart." Historically, the study of anatomy relied on dissections of dead human bodies, or **cadavers,**[5] to accurately map the human body. Cadaver dissection remains an essential part of the training of many health-science students.

---

[1]*histo* = tissue; *logy* = study of
[2]*morpho* = form; *logy* = study of
[3]*dis* = apart; *sect* = cut
[4]*ana* = apart; *tom* = cut
[5]*cadere* = to fall or die

Dissection, of course, is not the method of choice when examining a living patient! Some additional methods of clinical examination include the following:

- **Palpation**[6] is feeling structures with the fingertips, such as palpating a swollen lymph node or taking a pulse.
- **Auscultation**[7] (AWS-cul-TAY-shun) is listening to the natural sounds made by the body, such as heart and lung sounds.
- **Percussion** is tapping on the body and listening to the sound for signs of abnormalities such as pockets of fluid or air.
- **Medical imaging** includes methods of viewing the inside of the body without surgery. Anatomy learned in this way is called **radiologic anatomy,** and those who use radiologic methods for clinical purposes include **radiologists** and **radiologic technologists** (see Career Spotlight at the end of the chapter).

1.1c Techniques of Medical Imaging

It was once common to diagnose disorders through *exploratory surgery*—opening the body and taking a look inside to see what was wrong. Most exploratory surgery has been replaced by imaging techniques that allow physicians to see inside the body without cutting, posing much less risk to the patient. Medical imaging methods are called *noninvasive* if they involve no penetration of the skin or body orifices. *Invasive* techniques may entail inserting ultrasound probes into the esophagus, vagina, or rectum to get closer to the organ to be imaged, or injecting substances into the bloodstream or body passages to enhance image clarity.

Anatomy students today must be acquainted with the basic methods of imaging and their advantages and limitations. Many images in this book have been produced by the following techniques. Most of these methods produce black and white images; those in the book are colorized to enhance detail or for esthetic appeal.

Radiography (fig. 1.1a, b) is the process of photographing internal structures with X-rays, a form of high-energy radiation. The term *X-ray* also applies to a photograph *(radiograph)* made by this method. X-rays are absorbed by dense structures

---

[6]*palp* = touch, feel
[7]*auscult* = listen

Figure 1.1 Radiologic Images of the Head. (a) An X-ray (radiograph) of the head. (b) A colorized cerebral angiogram, made by injecting a substance opaque to X-rays into the circulation and then taking an X-ray of the head to visualize the blood vessels. (c) A CT scan of the head at the level of the eyes. (d) An MRI scan of the head at the level of the eyes. The optic nerves appear in red and the muscles that move the eyes in green. (e) A PET scan of the brain of an unmedicated schizophrenic patient. Red areas indicate regions of high metabolic rate. In this patient, the visual center of the brain (at bottom of photo) was especially active.
a: ©U.H.B. Trust/Tony Stone Images/Getty Images; **b:** Zephyr/Science Source; **c:** Miriam Maslo/Science Source; **d:** UHB Trust/Getty Images **e:** ISM/Sovereign/Medical Images

? *Why is a PET scan considered invasive, whereas MRI is noninvasive?*

(a) X-ray (radiograph)

(b) Cerebral angiogram

such as bone, teeth, and tumors, which produce a lighter image than soft tissues. Radiography is commonly used in dentistry; mammography; diagnosis of fractures; and examination of the digestive, respiratory, and urinary tracts. Some disadvantages of radiography are that images of overlapping organs can be confusing; slight differences in tissue density are not detected well; and, although the risk of harm is very low, X-rays can potentially cause mutations and cancer.

Computed tomography[8] (the **CT scan**) (fig. 1.1c) is a more sophisticated application of X-rays. The patient is moved through a ring-shaped machine that emits low-intensity X-rays on one side and receives them with a detector on the opposite side. A computer analyzes signals from the detector and produces an image of a "slice" of the body about as thin as a coin. CT scanning has the advantage of imaging thin sections of the body, so there is little organ overlap and the image is much sharper than a conventional X-ray. CT scanning is useful for identifying tumors, aneurysms, cerebral hemorrhages, kidney stones, and other abnormalities.

Magnetic resonance imaging (MRI) (fig. 1.1d) is better than CT for visualizing soft tissues. The patient lies in either a tube or an open-sided scanner with a powerful electromagnet. Hydrogen atoms in the patient's tissues alternately align themselves with this magnetic field and with a radio-frequency field turned on and off by the technologist. These changes in hydrogen alignment generate signals that are analyzed by computer to produce an anatomical image. MRI can "see" clearly through the skull and spine to produce images of the nervous tissue within, and it is better than CT for distinguishing between soft tissues such as the white and gray matter of the brain. It has some disadvantages, however, such as the claustrophobic feeling some patients experience in the scanner, and long exposure times that prevent sharp images being made of the constantly moving stomach and intestines. *Functional MRI (fMRI)* is a form of MRI that visualizes moment-to-moment changes in tissue function. fMRI scans of the brain, for example, show shifting patterns of activity as the brain applies itself to a specific task. This method has been very useful in clarifying which parts of the brain are involved in emotions, thought, language, sensation, and movement.

---

[8]*tomo* = section, cut, slice; *graphy* = recording process

(c) Computed tomographic (CT) scan

(d) Magnetic resonance image (MRI)

(e) Positron emission tomographic (PET) scan

Positron emission tomography (the **PET scan**) (fig. 1.1e) is used to assess the metabolic state of a tissue and to distinguish which areas are most active. It uses an injection of radioactively labeled glucose to highlight which tissues are most actively consuming energy at the moment of the scan. In cardiology, for example, PET scans can show the extent of tissue death from a heart attack. Because damaged tissue consumes little or no glucose, it appears dark. PET scans are widely used to diagnose cancer and evaluate tumor status. It is now possible to diagnose Alzheimer disease using PET scans. Until recently, a definitive diagnosis required analysis of brain tissue after death. Diagnosing living people with PET scans makes earlier diagnosis possible, potentially leading to more effective management of the disease. The PET scan is an example of **nuclear medicine**—the use of radioisotopes to treat disease or to form diagnostic images of the body.

Sonography[9] (fig. 1.2) uses a handheld device placed firmly against the skin; it emits high-frequency ultrasound and receives signals reflected back from internal organs. Sonography avoids the harmful effects of X-rays, and the equipment is relatively inexpensive and portable. It also is very useful for imaging motion, such as operation of the heart valves, ejection of blood from the heart, and fetal movements. It is the method of choice in obstetrics, where the image *(sonogram)* can be used to locate the placenta and evaluate fetal age, position, and development. *Echocardiography* is the sonographic examination of the beating heart. The primary disadvantages of sonography are that it does not produce a very sharp image and it cannot penetrate bone.

1.1d Anatomical Variation

A quick look around any classroom is enough to show that no two humans look exactly alike; on close inspection, even identical twins exhibit differences. Anatomy atlases and textbooks can easily give you the impression that everyone's internal anatomy is the same, but this simply is not true. Someone who thinks that all human bodies are the same internally would be a very confused medical student or an incompetent surgeon. Books such as this one teach only the most common structural patterns—the anatomy seen in approximately 70% or more of people.

Some people completely lack certain organs. For example, most of us have a *palmaris longus muscle* in the forearm, but not everyone does. Most of us have one spleen, but some people have two. Most have two kidneys, but some have only one. **Figure 1.3** shows some common variations in human anatomy, and Perspectives on Health (in section 1.2) describes a particularly dramatic variation.

---

[9]*sono* = sound; *graphy* = recording process

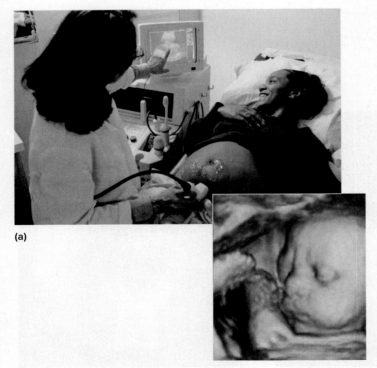

Figure 1.2 Sonography. (a) Producing a sonogram. (b) Three-dimensional sonogram of a fetus at 32 weeks of gestation.
a: Kevin Brofsky/Getty Images; b: ©Ken Saladin

❓ Why is this procedure safer than radiography for fetal assessment?

Figure 1.3 Variations in Human Anatomy. Not all humans have the usual "textbook structure."

Before You Go On

Answer these questions from memory to test your understanding of the previous section. Reread the material if you can't answer the questions.

1. What is the difference between gross anatomy and histology?
2. In a routine physical examination, a physician might inspect you by palpation and auscultation. What is the difference between these procedures?
3. What are the advantages of CT over sonography? Conversely, what are the advantages of sonography over CT?

1.2 Physiology—Dynamic Processes in the Living Body

Expected Learning Outcomes

When you have completed this section, you should be able to

a. identify some subdisciplines of physiology;
b. describe the characteristics that define an organism as alive;
c. define *homeostasis,* explain its significance, and discuss how it is maintained by negative feedback;
d. discuss positive feedback and its effects on the body; and
e. discuss the significance of variation in human physiology.

Physiology[10] is the study of the body's life processes. The term comes from Aristotle, who believed in both supernatural and natural causes of human disease. He called the supernatural causes *theologi* and natural causes *physiologi.* For centuries, physicians were called "doctors of physick."

1.2a The Physiological Sciences

Physiology uses the methods of experimental science to determine how the body functions. It has many subdisciplines such as *neurophysiology* (physiology of the nervous system), and *pathophysiology* (mechanisms of disease). Partly because of limitations on experimentation with humans, much of what we know about how the body functions is from studies of **comparative physiology** that explore how different species have solved problems of life such as water balance, respiration, and reproduction. Comparative physiology is also the basis for the development of most new medications and procedures. For example, new drugs are tested for safety in laboratory mammals such as rats before proceeding to trials with human subjects.

1.2b Essential Life Functions

Whereas anatomy views the body as a set of interconnected structures, physiology views it as a set of interconnected processes. Collectively, we call these processes *life*. But what exactly is life? Why do we consider a growing child to be alive, but not a growing crystal? Is abortion the taking of a human life? If so, what about a contraceptive foam that kills only sperm? As a patient is dying, at what point does it become ethical to disconnect life-support equipment and remove organs for donation? (See Perspectives on Health that follows.) If these organs are alive, as they must be to be useful to someone else, then why isn't the donor considered alive? Such questions have no easy answers, but they demand a concept of what life is—a concept that may differ with one's biological, medical, religious, or legal perspective.

From a biological viewpoint, life is not a single property. It is a collection of qualities that help to distinguish living from nonliving things:

- **Organization.** Living things exhibit a far higher level of organization than the nonliving world around them. They expend a great deal of energy to maintain order, and disease and death result from a breakdown in this order.

---

[10]*physio* = nature; *logy* = study of

PERSPECTIVES ON HEALTH

Situs Inversus and Other Unusual Anatomy

Two particularly striking examples of anatomical variation are situs (SITE-us) perversus and situs inversus. In *situs perversus,* an organ occupies an atypical locality; for example, a kidney might be located low in the pelvic cavity instead of high in the abdominal cavity (see fig. 1.3), or a parathyroid gland might be found in the root of the tongue instead of on the posterior surface of the thyroid gland.

In most people, the heart tilts toward the left, the spleen and sigmoid colon are on the left, and the gallbladder and appendix are on the right. But in *situs inversus,* occurring in about 1 out of 8,000 people, the organs of the thoracic and abdominal cavities are reversed between right and left. Selective right–left reversal of the heart is called *dextrocardia.* Conditions such as dextrocardia can cause serious medical problems. Complete situs inversus, however, usually causes no functional problems because all of the viscera, though reversed, maintain their normal relationships to each other.

Defining the End of Life

Earlier in this chapter, we saw that *life* is a difficult property to define. That being the case, so is defining the end of life—and yet we're often forced to make decisions on that issue. How do we decide when to "let go" of a terminally ill loved one, perhaps to disconnect life-support equipment?

There is no easily defined instant of biological death. Some organs function for an hour or longer after the heart stops beating. During this time, even if a person is declared legally dead, living organs may be removed for transplantation. For legal purposes, death was once defined as the loss of a spontaneous heartbeat and respiration. Now that cardiopulmonary functions can be restarted and artificially maintained for years, this criterion is less useful. Clinical death is now widely defined in terms of *brain death*—a lack of any detectable electrical activity in the brain, including the brainstem, accompanied by coma, lack of unassisted respiration, and lack of brainstem reflexes (such as pupillary, blinking, or coughing reflexes). A judgment of death is generally accepted only upon finding a complete lack of brain activity for a period ranging from 2 to 24 hours, depending on state laws. The permanent lack of cerebral activity is called a *persistent vegetative state.* Controversy has lingered, however, over the question of whether death of the entire brain (including the brainstem) should be required as a criterion of clinical death, or whether death may be declared upon lack of activity in only the cerebrum (the upper level of the brain that houses consciousness, sensation, and thought).

Medical educators, ethicists, philosophers, and theologians struggle continually with the difficulty of defining life and the moment of its cessation. The demand for organs for transplant pressures physicians to make delicate decisions as to when the life of the whole person is irretrievable, yet individual organs are still in sufficiently healthy condition to be useful to a recipient. Theologians, on the other hand, may wish for moral certainty that death has overtaken the whole person, and may see the "culture of organ donation" as incompatible with religious values.

- **Cells.** Living matter is always compartmentalized into one or more cells.
- **Metabolism.**[11] Living things take in molecules from the environment and chemically change them into molecules that form their own structures, control their physiology, or provide energy. Metabolism is the sum of all this internal chemical change. There is a constant turnover of molecules in the body. Although you sense a continuity of personality and experience from

---

[11]*metabol* = change; *ism* = process

your childhood to the present, nearly every molecule of your body has been replaced within the past year.
- **Development.** Development is any change in form or function over the lifetime of the organism. It involves two major processes: (1) **differentiation,** the transformation of cells with no specialized function to ones that are committed to a particular task; and (2) **growth,** an increase in size. Some nonliving things grow, but not in the way your body does. If you let a sugar solution evaporate, crystals will grow from it, but not through a change in the composition of the sugar. Sugar molecules are simply added from the solution to the crystal surface. The growth of the body, by contrast, occurs through metabolic change; for the most part, the body is not composed of the molecules one eats, but of molecules made by chemically altering the food.
- **Excitability.** The ability to sense and react to *stimuli* (changes in the environment) is called *excitability* or *irritability*. It occurs at all levels, from the cell to the entire body, and it characterizes all living things from bacteria to humans. Excitability is especially obvious in animals because of nerve and muscle cells that exhibit high sensitivity to stimuli, rapid transmission of information, and quick reactions.
- **Homeostasis.**[12] Although the environment around an organism changes, the organism maintains relatively stable internal conditions—for example, a stable temperature, blood pressure, and body weight. This internal stability, called *homeostasis,* is discussed in greater depth in section 1.2c.
- **Reproduction.** Living organisms produce copies of themselves, thus passing their genes on to new, younger "containers"—their offspring.
- **Evolution.** All living species exhibit genetic change from generation to generation, and therefore evolve. This occurs because new variations are inevitably introduced by *mutations* (changes in the genes), and environmental conditions favor some variations over others, thus perpetuating some genes and eliminating others. Evolution simply means genetic change in the population over time. Unlike the other characteristics of life, evolution is a characteristic seen only in the population as a whole. No single individual evolves over the course of its life. Evolution, however, holds the explanation for why human structure and function are as they are. *Evolutionary medicine* is a science that interprets human disease and dysfunction in the context of the biological history of the species.

1.2c Homeostasis and Feedback

The quality of life most frequently addressed in this book is homeostasis (ho-me-oh-STAY-sis)—the ability to maintain internal stability. Homeostatic mechanisms stabilize such variables as body temperature, blood pressure, body weight, electrolyte balance, and pH.

Homeostasis has been one of the most enlightening concepts in physiology. The term was introduced by American physiologist Walter Cannon in his book *The Wisdom of the Body* (1932), but the concept that the body maintains internal

---

[12]*homeo* = the same; *stasis* = stability

stability was around long before that. Physiology centers around mechanisms that maintain this stability, and the loss of homeostatic control usually leads to illness or death. Pathophysiology is essentially the study of unstable conditions that result when our homeostatic controls fail.

Negative Feedback and Stability

The fundamental mechanism that maintains homeostasis is **negative feedback**—a process in which the body senses a change and activates mechanisms that negate (reverse) it. Negative feedback does not produce absolute constancy in the body, but maintains physiological values within a narrow range of a certain **set point**—an average value such as 37°C (98.6°F) for body temperature. Conditions fluctuate slightly around the set point. Thus, negative feedback is said to maintain a *dynamic equilibrium*—not a total lack of change, but a state of ever-changing balance within limits. Variables regulated by negative feedback mechanisms include blood pressure, blood glucose (sugar), and many others. By maintaining physiological equilibrium, negative feedback is the key mechanism for maintaining health.

Let's consider blood pressure regulation as an example of negative feedback. When you first rise from bed in the morning, gravity causes some of your blood to drain away from your head and upper torso, resulting in falling blood pressure in this region (fig. 1.4). The resulting imbalance in homeostasis is detected by sensory nerve endings called *baroreceptors* in large arteries near the heart. The baroreceptors transmit nerve signals to the brain, where we have a cardiac center that regulates the heart rate. The cardiac center responds by sending nerve signals to the heart, which speed it up. The faster heart rate raises the blood pressure and restores normal homeostasis. In elderly people, this feedback loop is sometimes insufficiently responsive, and they might feel dizzy as they rise from a reclining position. The drop in blood pressure may result in decreased blood to the brain and this sometimes causes fainting.

This correction of blood pressure illustrates three common, although not universal, components of a feedback loop: a receptor, an integrating center, and an effector. The **receptor** is a structure that senses a change in the body, such as the baroreceptors that monitor blood pressure. The **integrating (control) center,** such as the cardiac center of the brain, processes this information, relates it to other available information (for example, comparing what the blood pressure is with what it should be), and "makes a decision" about what the appropriate response should be. The **effector** is the cell or organ that carries out the final corrective action. In the blood pressure example, the effector is the heart. The **response,** such as the restoration of normal blood pressure, is then sensed by the receptor, and the feedback loop is complete.

Figure 1.4 Negative Feedback in Response to Drop in Blood Pressure.

Clinical Application 1.1

MEN IN THE OVEN

English physician Charles Blagden (1748–1820) staged a rather theatrical demonstration of homeostasis long before Cannon coined the word. In 1775, Blagden spent 45 minutes in a chamber heated to 127°C (260°F)—along with a steak, a dog, and some research associates. Being dead and unable to maintain homeostasis, the steak was cooked. But being alive and capable of evaporative cooling, the dog panted, the men sweated, and all of them survived. History does not record whether the men ate the steak in celebration or shared it with the dog.

Positive Feedback and Rapid Change

Positive feedback is a self-amplifying cycle in which a physiological change leads to even greater change in the same direction, rather than producing the self-corrective effects of negative feedback. Positive feedback is sometimes a normal way of producing rapid change. During childbirth, for example, the head of the fetus pushes against a woman's cervix (the neck of the uterus) and stimulates its nerve endings (**fig. 1.5**). Nerve signals travel to the brain, which stimulates the pituitary gland to secrete the hormone *oxytocin*. Oxytocin travels in the blood and stimulates the uterus to contract. This pushes the fetus downward, stimulating the cervix still more and causing the positive feedback loop to be repeated. Labor contractions therefore become more and more intense until the fetus is expelled. Other cases of beneficial positive feedback occur in blood clotting and the generation of nerve signals.

More often, however, positive feedback is a harmful or even life-threatening process. This is because its self-amplifying nature can quickly change the internal state of the body to something far from its homeostatic set point. Consider a high fever, for example. A fever triggered by infection is beneficial up to a point, but if the body temperature rises much above 40°C (104°F), it may create a dangerous positive feedback loop: The high temperature raises the metabolic rate, which makes the body produce heat faster than it can get rid of it. Thus, temperature rises still further, increasing the metabolic rate and heat production still more. This "vicious circle" becomes fatal at approximately 45°C (113°F). Positive feedback loops often create dangerously out-of-control situations that require emergency medical treatment.

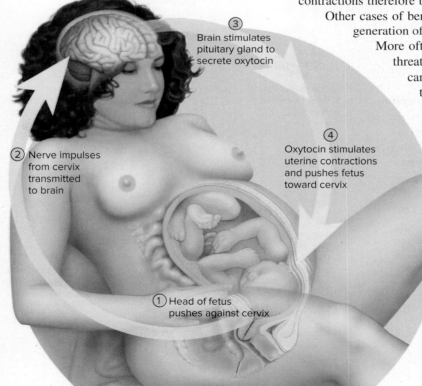

Figure 1.5 Positive Feedback in Childbirth. Repetition of this cycle of events has a self-amplifying effect, intensifying labor contractions until the infant is born. This is one case in which positive feedback has a beneficial outcome.

❓ *Could childbirth as a whole be considered a negative feedback event? Discuss.*

Apply What You Know

In a heart attack, the death of cardiac muscle reduces the heart's pumping effectiveness. Thus, blood flow throughout the body slows down. This leads to widespread blood clotting. Blood clots block the flow of blood to the cardiac muscle, so the muscle is less nourished and still more of it dies. Is this positive feedback, negative feedback, or neither? Explain.

1.2d Physiological Variation

Earlier we considered the clinical importance of variations in human anatomy, but physiology is even more variable. Physiological variables differ with sex, age, weight, diet, degree of physical activity, and environment, among other things. If an introductory textbook states a typical human heart rate, blood pressure, red blood cell count, or body temperature, it is generally assumed that such values are for a healthy young adult unless otherwise stated. Health-care professionals take sex and age into account when they perform medical assessments and determine drug dosages, for example. Failure to do so can lead to medical mistakes such as overmedication of the elderly or medicating women on the basis of research that was done on men.

Before You Go On

Answer these questions from memory to test your understanding of the previous section. Reread the material if you can't answer the questions.

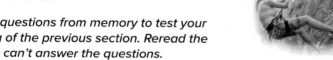

 4. What is the difference between growth and development?
 5. Explain why positive feedback is likely to cause a loss of homeostasis, and why negative feedback can restore homeostasis.
 6. Why is it better to define homeostasis as a dynamic equilibrium than to define it as a state of internal constancy?

1.3 The Human Body Plan

Expected Learning Outcomes

When you have completed this section, you should be able to

  a. list the levels of human complexity in order from the whole organism down to atoms;
  b. define or demonstrate the *anatomical position* and explain its importance in descriptive anatomy;
  c. define the three major anatomical planes of the body;
  d. identify the major anatomical regions of the body;
  e. describe the body's cavities and the membranes that line them; and
  f. name the 11 organ systems, their principal organs, and their functions.

This section gives a broad overview of the structural organization of the human body, providing a vocabulary and a context for the study of specific regions and organ systems in the chapters to follow. We also preview the major organs and functions of the body's 11 organ systems.

1.3a Levels of Human Structure

The human body is organized into levels ranging from the whole organism to atoms and molecules. Consider for a moment this analogy: The English language, like the human body, is very complex, yet an infinite variety of ideas can be conveyed with a limited number of words. All words in the English language are, in turn, composed of various combinations of just 26 letters. Between an essay and the alphabet are successively simpler levels of organization: paragraphs, sentences, words, and syllables. Humans have an analogous hierarchy of complexity (fig. 1.6), as follows:

The organism is composed of organ systems,
  organ systems are composed of organs,
    organs are composed of tissues,
      tissues are composed of cells,
        cells are composed (in part) of organelles,
          organelles are composed of molecules, and
            molecules are composed of atoms.

The **organism** is a single, complete individual.

An **organ system** is a group of organs that carry out a basic function such as circulation, respiration, or digestion. The human body has 11 organ systems, defined and illustrated later in this chapter. Usually, the organs of a system are physically interconnected, such as the esophagus, stomach, small intestine, and large intestine that form the digestive system.

An **organ** is a structure composed of two or more tissue types that work together to carry out a particular function. Organs have definite anatomical boundaries and are visibly distinguishable from adjacent structures. They include not only what people traditionally think of as the "internal organs," such as the heart and kidneys, but less obvious examples such as the skin, muscles, and bones. Most familiar organs are visible to the naked eye. However, there are organs within organs—the large organs contain smaller organs, some of which are microscopic. The skin, for example, is the body's largest organ. Included within it are thousands of smaller organs: hair follicles, nails, sweat glands, nerves, and blood vessels.

A **tissue** is a group of similar cells and cell products that forms a discrete region of an organ and performs a specific function. The body is composed of only four primary classes of tissue—epithelial, connective, nervous, and muscular tissues. *Histology*, the study of tissues, is the subject of chapter 4.

Cells are the smallest units of an organism that carry out all the basic functions of life; nothing simpler than a cell is considered alive. A cell is a microscopic compartment enclosed in a film called the *plasma membrane*. It usually contains one nucleus and a variety of other organelles. *Cytology*, the study of cells, is the subject of chapter 3.

Organelles[13] are microscopic structures that carry out a cell's individual functions. Examples include nuclei, mitochondria, centrioles, and lysosomes.

Figure 1.6 The Structural Hierarchy of the Human Body.

---

[13]*elle* = little

Organelles and other cellular components are composed of **molecules.** A molecule is a particle composed of at least two **atoms.** Atoms and molecules are described in chapter 2.

1.3b Anatomical Position

In describing the human body, anatomists assume that it is in **anatomical position** (fig. 1.7)—that of a person standing upright with the feet flat on the floor, arms at the sides, and the palms and face directed toward the observer. Without such a frame of reference, to say that a structure such as the sternum or thyroid gland is "above the heart" would be vague, because it would depend on whether the subject was standing, lying face down *(prone),* or lying face up *(supine).* From the perspective of anatomical position, however, we can describe the thyroid gland as *superior* to the heart, and the sternum as *anterior (ventral)* to it. These descriptions remain valid regardless of the subject's position.

Unless stated otherwise, assume that all anatomical descriptions refer to anatomical position. Bear in mind that if a subject is facing you in anatomical position, the subject's left will be on your right and vice versa. In most anatomical illustrations, for example, the appendix appears on the left side of the page, though it is located on the right side of the abdomen.

1.3c Anatomical Planes

Many views of the body are based on real or imaginary "slices" called sections or planes. *Section* implies an actual cut or slice to reveal internal anatomy, whereas *plane* implies an imaginary flat surface passing through the body. The three major anatomical planes are *sagittal, frontal,* and *transverse* (fig. 1.7).

A **sagittal**[14] **plane** (SADJ-ih-tul) extends vertically and divides the body or an organ into right and left portions. The **median (midsagittal) plane** passes through the midline of the body and divides it into *equal* right and left halves. Other planes parallel to this but off-center are called *parasagittal*[15] *planes* and divide the body into unequal right and left portions. The head and pelvic organs are commonly illustrated on the median plane (fig. 1.8a).

A **frontal (coronal**[16]**) plane** also extends vertically, but it is perpendicular to the sagittal plane and divides the body into anterior (front) and posterior (back) portions. A frontal section of the head, for example, would divide it into one portion bearing the face and another bearing the back of the head. Contents of the thoracic and abdominal cavities are commonly shown in frontal section (fig. 1.8b).

A **transverse (horizontal) plane** passes across the body or an organ perpendicular to its long axis; therefore, it divides the body or organ into superior (upper) and inferior (lower) portions. Many CT and MRI scans are transverse sections (fig. 1.8c; see also fig. 1.1c–e).

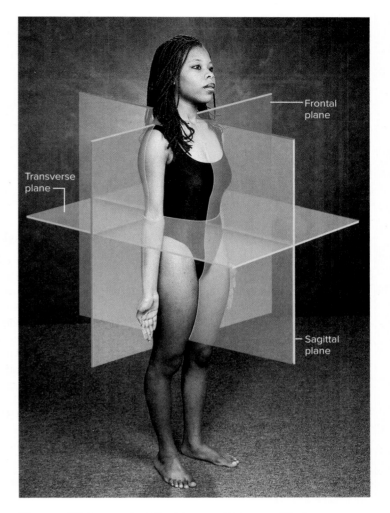

Figure 1.7 Anatomical Position and Planes of Reference.
Ken Saladin/Joe DeGrandis, photographer

---

[14]*sagitta* = arrow
[15]*para* = next to
[16]*corona* = crown; *al* = like

Figure 1.8 Sections of the Body in the Three Primary Anatomical Planes. (a) Sagittal section of the pelvic region. (b) Frontal section of the thoracic region. (c) Transverse section of the head at the level of the eyes.

(a) Sagittal section (b) Frontal section (c) Transverse section

1.3d Major Body Regions

Knowledge of the external anatomy and landmarks of the body is important in performing a physical examination, reporting patient complaints, and many other clinical procedures. The body is divided into two major regions called the *axial* and *appendicular regions,* each with many smaller regions.

The **axial region** consists of the **head, neck (cervical[17] region)**, and **trunk.** The trunk is further divided into the **thoracic region** above the diaphragm and the **abdominal** and **pelvic regions** below it.

One way of referring to the locations of abdominopelvic structures is to divide the region into quadrants. Two perpendicular lines intersecting at the umbilicus (navel) divide the area into a **right upper quadrant (RUQ), right lower quadrant (RLQ), left upper quadrant (LUQ),** and **left lower quadrant (LLQ)** (**fig. 1.9a, b**). The quadrant scheme is often used to describe the site of an abdominal pain or abnormality.

[17]*cervic* = neck

Figure 1.9 Quadrants of the Abdominopelvic Region. (a) The four quadrants of the abdominopelvic region. (b) Internal anatomy correlated with the four quadrants.

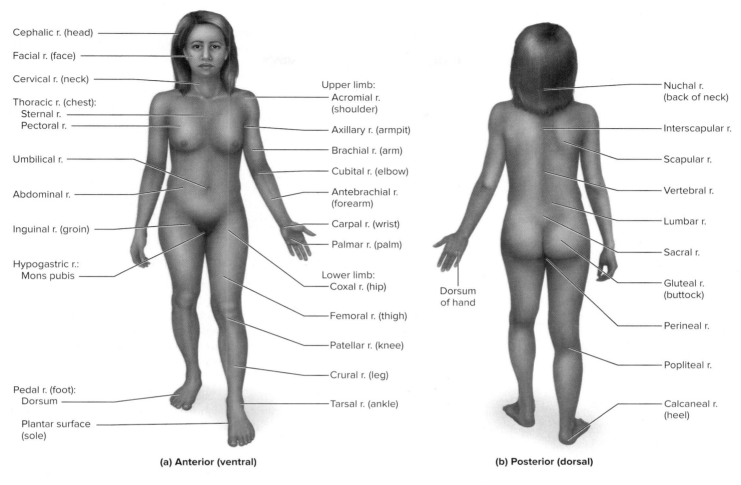

Figure 1.10 External Body Regions (r. = region). APR

The **appendicular region** (AP-en-DIC-you-lur) of the body consists of the upper and lower limbs (also called *appendages* or *extremities*). The **upper limb** includes the **arm (brachial region)** (BRAY-kee-ul), **forearm (antebrachial[18] region)** (AN-teh-BRAY-kee-ul), **wrist (carpal region), hand,** and **fingers (digits).** The lower limb includes the **thigh (femoral region), leg (crural region), ankle (tarsal region), foot,** and **toes (digits).** In strict anatomical terms, *arm* refers only to that part of the upper limb between the shoulder and elbow. *Leg* refers only to that part of the lower limb between the knee and ankle. Figure 1.10 identifies several smaller body regions.

A **segment** of a limb is a region between one joint and the next. The arm, for example, is the segment between the shoulder and elbow joints, and the forearm is the segment between the elbow and wrist joints. When you flex your fingers, you can easily see that your thumb has two segments (proximal and distal—terms defined in table 1.3), whereas the other four digits have three segments (proximal, middle, and distal). The segment concept is especially useful in describing the locations of bones and muscles and the movements of the joints.

---

[18]*ante* = fore, before; *brachi* = arm

Table 1.1 Body Cavities and Membranes

Name of Cavity	Principal Viscera	Serous Membranes
Cranial Cavity	Brain	Meninges
Vertebral Canal	Spinal cord	Meninges
Thoracic Cavity		
Pleural cavities (2)	Lungs	Pleurae
Pericardial cavity	Heart	Pericardium
Abdominopelvic Cavity		
Abdominal cavity	Digestive organs, spleen, kidneys	Peritoneum
Pelvic cavity	Bladder, rectum, reproductive organs	Peritoneum

1.3e Body Cavities and Membranes

The axial region of the body has cavities containing the "internal organs" or **viscera** (VISS-er-uh) (singular, *viscus*[19]) (**table 1.1**). These cavities are lined by thin **serous membranes,** which secrete a lubricating film of moisture similar to blood serum (hence the name *serous*). These membranes are distinct from the mucous membranes that line the digestive, respiratory, urinary, and reproductive tracts—passages open to the exterior. Serous and mucous membranes are further described in chapter 4.

The **cranial cavity** (CRAY-nee-ul) is enclosed by the skull and contains the brain. The **vertebral canal,** continuous with the cranial cavity, is a space about as wide as your finger that passes down the vertebral column (spine) (**fig. 1.11a**). Both of these cavities are lined by three membranes called **meninges** (meh-NIN-jeez). Among other functions, they protect the delicate nervous tissue from the hard protective bone that encloses it. The meninges are discussed in more detail in chapter 9.

A muscular sheet, the **diaphragm,** separates the **thoracic cavity** above it from the **abdominopelvic cavity** below. The thoracic cavity is additionally divided into right, left, and median portions by a partition called the **mediastinum**[20] (ME-dee-ah-STY-num) (**fig. 1.11b**). The mediastinum is the region between the lungs extending from the base of the neck to the diaphragm. It contains the esophagus, trachea, and the heart and major blood vessels attached to it. The heart is enveloped by a two-layered serous membrane called the **pericardium.**[21] The anatomy of the pericardium and its relationship with the heart are further described in chapter 13. The right and left sides of the thoracic cavity contain the lungs, which are each enfolded in another two-layered serous membrane, the **pleura**[22] (PLOOR-uh) (plural, *pleurae*). The anatomy of the pleurae and their relationship with the lungs are detailed in chapter 15.

The abdominopelvic cavity can be subdivided into the **abdominal cavity** and **pelvic cavity,** although the two form one continuous space; they are not separated by a wall the way the abdominal and thoracic cavities are separated by the diaphragm. The dividing line between them is the margin of the pelvic inlet (see fig. 6.30a in section 6.4). The abdominal cavity contains most of the digestive organs, as well as the spleen, kidneys, and ureters. The pelvic cavity is markedly narrower and its lower end tilts posteriorly (fig. 1.11a). It contains the lowermost portion of the large intestine, the urinary bladder and urethra, and the reproductive organs.

(a) Left lateral view

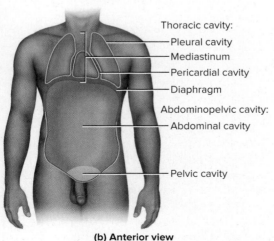

(b) Anterior view

Figure 1.11 The Major Body Cavities.

---

[19]*viscus* = body organ
[20]*mediastinum* = in the middle
[21]*peri* = around; *cardi* = heart
[22]*pleur* = rib, side

The abdominopelvic cavity is lined by a two-layered serous membrane called the **peritoneum**[23] (PERR-ih-toe-NEE-um). The outer layer, lining the abdominal wall, is called the *parietal peritoneum.* Along the posterior midline of the abdominal wall, it turns inward and becomes another layer, the *visceral peritoneum,* suspending certain abdominal viscera from the body wall and covering their outer surfaces (fig. 1.12). The space between the parietal and visceral peritoneum is called the **peritoneal cavity.** The visceral peritoneum is also called a **mesentery**[24] (MESS-en-tare-ee) at points where

---

[23]*peri* = around; *tone* = stretched
[24]*mes* = in the middle; *enter* = intestine

Figure 1.12 The Peritoneal Cavity and Retroperitoneal Organs. (a) Cross section at the upper abdominal level shown in the figure on the right. The gray space is the peritoneal cavity (with most organs omitted), enclosed in peritoneum (the thin blue line). Note that the peritoneum forms the outer layer (serosa) of some viscera. Viscera toward the top of the figure (kidneys, inferior vena cava, aorta) are in the abdominal cavity but not in the peritoneal cavity; they are retroperitoneal. The intestine and liver are intraperitoneal. (b) Left lateral view showing abdominal viscera suspended by mesenteries from the body wall. APR

Clinical Application 1.2

PERITONITIS

Peritonitis is inflammation of the peritoneum. It is a critical, life-threatening condition necessitating prompt treatment. The most serious cause of peritonitis is a perforation in the digestive tract, such as an abdominal bullet wound or a ruptured appendix. Digestive juices cause immediate chemical inflammation of the peritoneum, followed by microbial inflammation as intestinal bacteria invade the body cavity. Anything that perforates the abdominal wall can also lead to peritonitis, such as abdominal trauma or surgery. So, too, can free blood in the abdominal cavity, as from a ruptured aneurysm (a weak point in a blood vessel) or ectopic pregnancy (implantation of an embryo anywhere other than the uterus); blood itself is a chemical irritant to the peritoneum. Peritonitis tends to shift fluid from the circulation into the abdominal cavity. Death can follow within a few days from severe electrolyte imbalance, respiratory distress, kidney failure, and a widespread blood clotting called disseminated intravascular coagulation.

Figure 1.13 Mesentery. This is a translucent serous membrane associated with the small intestine and other abdominal organs. Mesenteries contain blood vessels, lymphatic vessels, and nerves supplying the viscera.
MedicImage/Universal Images Group/Getty Images

it forms a membranous curtain suspending and anchoring the viscera (fig. 1.13), and a **serosa** (seer-OH-sa) at points where it enfolds and covers the outer surfaces of organs such as the stomach and small intestine. These relationships of the peritoneum, serosa, and peritoneal cavity with the abdominal digestive organs are detailed in chapter 17.

Some organs of the abdominal cavity lie against the posterior body wall and are covered by peritoneum only on the side facing the peritoneal cavity. They are said to have a **retroperitoneal**[25] position. Examples include the kidneys, aorta, and inferior vena cava, as shown in figure 1.12a and the duodenum and pancreas, as shown in figure 1.12b. Organs that are encircled by peritoneum and suspended from the posterior body wall by mesenteries, such as the loops of small intestine shown in those figures, are **intraperitoneal.**[26]

1.3f Organ Systems

The human body has 11 organ systems (fig. 1.14) and an immune system, which is better described as a population of cells that inhabit multiple organs rather than as an organ system. These systems are classified in the following list by their principal functions, but this is an unavoidably flawed classification. Some organs belong to two or more systems—for example, the male urethra is part of both the urinary and reproductive systems; the pancreas is part of the digestive and endocrine systems; and the mammary glands can be considered part of the integumentary and female reproductive systems.

---

[25]*retro* = behind
[26]*intra* = within

The human organ systems

Systems of protection, support, and movement
  Integumentary system
  Skeletal system
  Muscular system
Systems of internal communication and integration
  Nervous system
  Endocrine system
Systems of fluid transport
  Circulatory system
  Lymphatic system
Systems of intake and output
  Respiratory system
  Urinary system
  Digestive system
Systems of reproduction
  Male reproductive system
  Female reproductive system

Before You Go On

Answer these questions from memory to test your understanding of the previous section. Reread the material if you can't answer the questions.

7. Rearrange the following alphabetical list in order from the largest, most complex components of the body to the smallest, simplest ones: cells, molecules, organelles, organs, organ systems, tissues.
8. Examine figures 1.1 and 1.12 and identify whether each figure shows the body on a frontal, sagittal, or transverse plane.
9. Identify each of the following regions as belonging to the upper limb, lower limb, or axial region of the body: cubital, pectoral, gluteal, antebrachial, calcaneal, patellar, nuchal, thoracic, and left upper quadrant.
10. Name the body cavity in which each of the following viscera is found: spinal cord, liver, lung, spleen, heart, pancreas, gallbladder, and kidney.

1.4 The Language of Medicine

Expected Learning Outcomes

When you have completed this section, you should be able to

a. explain why precision is important in the use of medical terms;
b. demonstrate how to break medical terms into their roots, prefixes, and suffixes;
c. identify the relationships between singular and plural forms of a medical term; and
d. define directional terms for the locations of anatomical structures relative to each other.

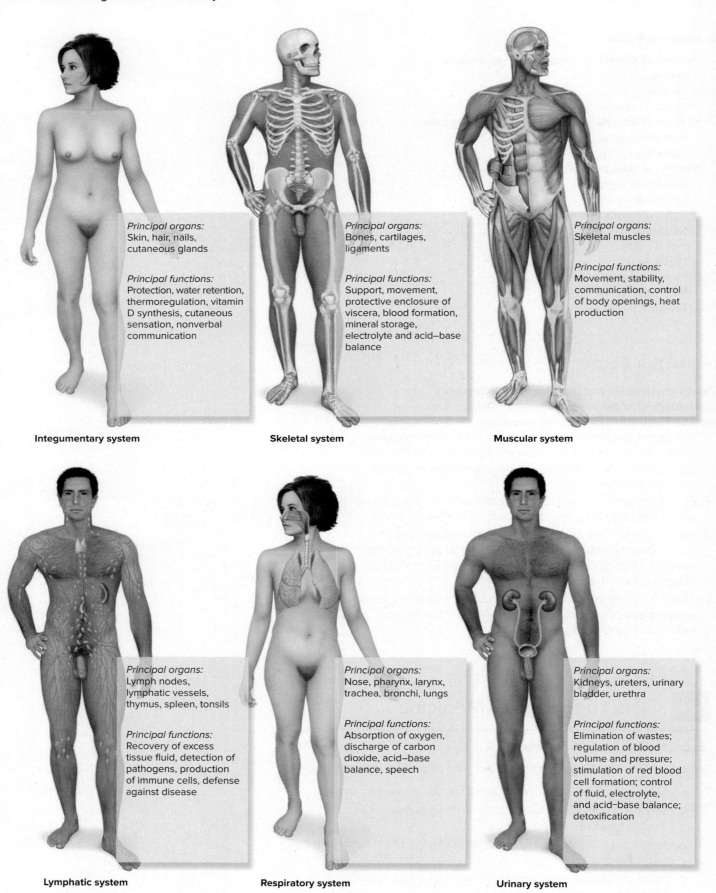

Figure 1.14 The Human Organ Systems.

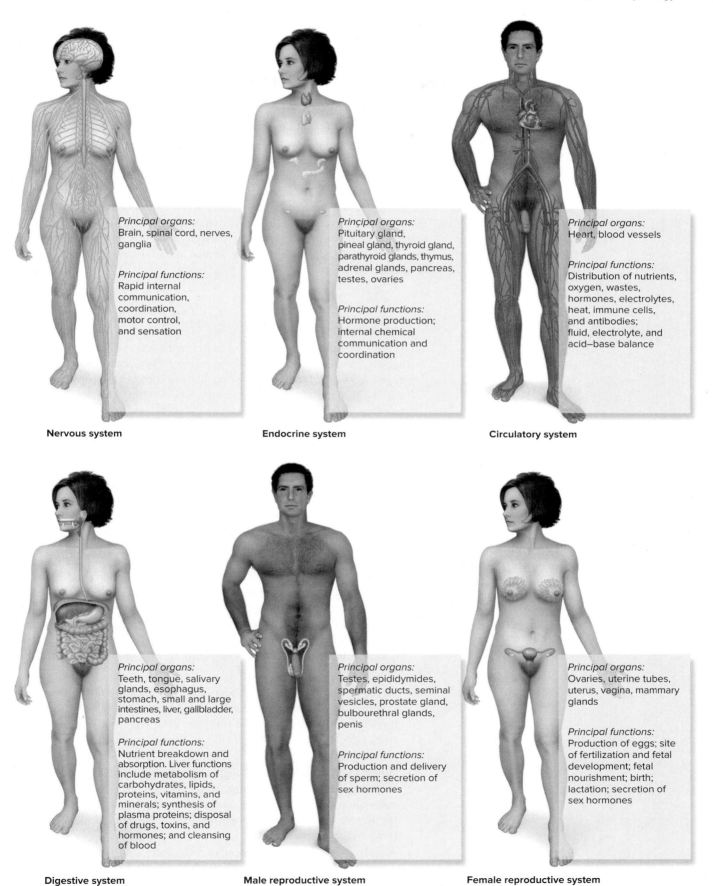

Figure 1.14 The Human Organ Systems (*continued*).

One of the greatest challenges faced by students of anatomy and physiology is the vocabulary. In this book, you will encounter such Latin terms as *corpus callosum* (a brain structure) and *extensor carpi radialis longus* (a forearm muscle). You might wonder why structures aren't called by more common names that everyone can understand, and how you will ever remember such formidable terms. This section will give you some answers to these questions and some useful tips on mastering anatomical terminology.

It is of greatest importance to use medical terms precisely. It may seem trivial if you misspell *trapezius* as *trapezium,* but in doing so, you would be changing the name of a back muscle to the name of a wrist bone. A "little" error such as misspelling *malleus* as *malleolus* changes the name of a middle-ear bone to the name of a protuberance of the ankle. A difference of only one letter distinguishes *gustation* (the sense of taste) from *gestation* (pregnancy). The health professions demand precision and accuracy. People's well-being may depend on it, and indeed many patients die simply because of written and oral miscommunication among hospital staff.

> ### Apply What You Know
> A student means to write about part of the small intestine called the ileum, but misspells it *ilium*. In complaining about the points lost, the student says, "I was only one letter off!" The instructor says, "But you changed the entire meaning of the word." With the help of a dictionary, explain the instructor's reasoning.

1.4a Analyzing Medical Terms

There is a simple trick to becoming more comfortable with the technical language of medicine. Those who, at first, find scientific terms confusing and difficult to pronounce, spell, and remember often feel more confident once they realize the logic of how such terms are composed. A term such as *hyponatremia* is less forbidding once we recognize it is composed of three common word elements: *hypo-* (below normal), *natr-* (sodium), and *-emia* (blood condition). Thus, hyponatremia is a deficiency of sodium in the blood. Those three word elements appear over and over in many other medical terms: *hypothermia, natriuretic, anemia,* and so on. Once you learn the meanings of *hypo-, natri-,* and *-emia,* you already have the tools to at least partially understand hundreds of other biomedical terms.

Scientific terms are typically composed of one or more of the following elements:

- At least one *root (stem)* that bears the core meaning of the word. In *cardiology,* for example, the root is *cardi-* (heart). Many words have two or more roots. In *adipocyte,* the roots are *adip-* (fat) and *-cyte* (cell). Word roots are often linked through an *o* or other vowel to make the word more pronounceable.
- A *prefix* may be present at the beginning of a word to modify its core meaning. For example, *gastric* (pertaining to the stomach or to the belly of a muscle) takes on a variety of new meanings when prefixes are added to it: *epigastric* (above the stomach), *hypogastric* (below the stomach), and *endogastric* (within the stomach).
- A *suffix* may be added to the end of a word to modify its core meaning. For example, *microscope, microscopy, microscopic,* and *microscopist* have different meanings because of their suffixes alone.

Consider another word, *gastroenterology*, a branch of medicine dealing with the stomach and small intestine. It breaks down into

gastro/entero/logy:
gastro = "stomach,"
entero = "small intestine," and
logy = "the study of."

"Dissecting" words in this way and paying attention to the word-origin footnotes throughout this book can help make you more comfortable with the language of anatomy. Breaking a word down and knowing the meaning of its elements make it far easier to pronounce it, spell it, and remember its definition. In appendix D, you will find a lexicon of the word roots, prefixes, and suffixes most frequently used in this book.

1.4b Singular and Plural Forms

A point of confusion for many beginning students is how to recognize the plural forms of medical terms. Few people would fail to recognize that *ovaries* is the plural of *ovary*, but the connection is harder to make in other cases: For example, the plural of *cortex* is *cortices* (COR-ti-sees), and the plural of *corpus* is *corpora*. Table 1.2 will help you make the connection between common singular and plural noun terminals.

1.4c Directional Terminology

In "navigating" the human body and describing the locations of structures, anatomists use a set of standard **directional terms** (table 1.3). You will need to be very familiar with these in order to understand anatomical descriptions later in this book. The terms assume that the body is in anatomical position.

Intermediate directions are often indicated by combinations of these terms. For example, a structure that is *superomedial* to another is above and medial to it; the bridge of the nose is superomedial to the flare of the nostrils.

Because of the bipedal (two-legged), upright stance of humans, some directional terms have different meanings for humans than they do for other animals. *Anterior*, for example, denotes the region of the body that leads the way in normal

Table 1.2 Singular and Plural Forms of Some Noun Terminals

Singular Ending	Plural Ending	Examples
-a	-ae	pleura, pleurae
-ax	-aces	thorax, thoraces
-en	-ina	lumen, lumina
-ex	-ices	cortex, cortices
-is	-es	testis, testes
-is	-ides	epididymis, epididymides
-ix	-ices	appendix, appendices
-ma	-mata	carcinoma, carcinomata
-on	-a	ganglion, ganglia
-um	-a	septum, septa
-us	-era	viscus, viscera
-us	-i	villus, villi
-us	-ora	corpus, corpora
-x	-ges	phalanx, phalanges
-y	-ies	ovary, ovaries
-yx	-yces	calyx, calyces

Table 1.3 Directional Terms in Human Anatomy APR

Term	Meaning	Examples of Usage
Ventral	Toward the front* or belly	The aorta is *ventral* to the vertebral column.
Dorsal	Toward the back or spine	The vertebral column is *dorsal* to the aorta.
Anterior	Toward the ventral side*	The sternum is *anterior* to the heart.
Posterior	Toward the dorsal side*	The esophagus is *posterior* to the trachea.
Superior	Above	The heart is *superior* to the diaphragm.
Inferior	Below	The liver is *inferior* to the diaphragm.
Medial	Toward the midsagittal plane	The heart is *medial* to the lungs.
Lateral	Away from the midsagittal plane	The eyes are *lateral* to the nose.
Proximal	Closer to the point of attachment or origin	The elbow is *proximal* to the wrist.
Distal	Farther from the point of attachment or origin	The fingernails are at the *distal* ends of the fingers.
Superficial	Closer to the body surface	The skin is *superficial* to the muscles.
Deep	Farther from the body surface	The bones are *deep* to the muscles.

*In humans only; definition differs for other animals.

CAREER SPOTLIGHT

Radiologic Technologist

A radiologic technologist is a person who produces medical images for the purposes of diagnosing and treating illnesses and injuries, or who administers radiation therapy. (The profession is not to be confused with *radiologic technician*—one who sets up, maintains, and repairs radiologic equipment.) Radiologic technologists work closely with radiologists—physicians who interpret the images, make diagnoses, and prescribe courses of treatment. Programs in radiologic technology range from a 2-year associate degree to bachelor's and master's degrees. The training of a radiologic technologist is a mixture of medical and physical sciences, including anatomy, physiology, genetics, pathology, chemistry, general and nuclear physics, medical terminology, patient examination and positioning, radiologic instrumentation, and medical imaging sciences. Writing, speech, and figure drawing also provide useful experience; one must also have good analytical thinking skills, attention to detail, compassion, and patience. Entry into the profession requires passing a board examination; earning certification, although optional, is a further step that enhances one's career mobility. Many employers prefer to hire only certified radiologic technologists. Career-long continuing education courses are necessary to keep one's certification current. Specialties in radiologic technology include radiography (X-ray technology), mammography, sonography, fluoroscopy, CT, MRI, PET, nuclear medicine, medical dosimetry, and bone densitometry. Some radiologic technologists go into clinical administration or radiologic education, or work for equipment manufacturers. See appendix B for additional career information.

ERproductions Ltd/Blend Images

locomotion. For a four-legged animal such as a cat, this is the head; for a human, however, it is the front of the chest and abdomen. Thus, *anterior* has the same meaning as *ventral* for a human but not for a cat. *Posterior* denotes the region of the body that comes last in normal locomotion—the tail end of a cat but the dorsal side of a human. These differences must be kept in mind when dissecting other animals for comparison to human anatomy.

Before You Go On

Answer these questions from memory to test your understanding of the previous section. Reread the material if you can't answer the questions.

11. Following the example given for gastroenterology, break each of the following words down into their roots, prefixes, and suffixes: electrocardiography, brachiocephalic, hyperkalemia, substernal, periodontal.
12. Write the plural form of each of the following terms: stoma, lacuna, nucleus, epithelium, diagnosis. *Write the singular form of each of the following:* larynges, carpi, ampullae, matrices, ova.
13. Examine figure 1.10 and use the proper directional term from table 1.3 to describe (a) the location of the axillary region relative to the cubital region; (b) the location of the sacral region relative to the lumbar region; (c) the location of the sternal region relative to the scapular region; and (d) the location of the umbilical region relative to the lumbar region.

Study Guide

Assess Your Learning Outcomes

To test your knowledge, discuss the following topics with a study partner or in writing, ideally from memory.

1.1 Anatomy—The Structural Basis of Human Function

1. The distinction between gross anatomy and histology, and the relationship of histology to histopathology
2. How surface anatomy, systemic anatomy, regional anatomy, and functional morphology differ in their perspectives
3. Why the study of other animal species is important for understanding human functional morphology
4. Distinctions between dissection, palpation, auscultation, percussion, and medical imaging as methods of studying human structure
5. How each of the following methods of medical imaging is performed, and why one might be chosen over the others for specific purposes: radiography, computed tomography (CT), magnetic resonance imaging (MRI), positron emission tomography (PET), and sonography
6. Examples of individual anatomical variation and why such variations are important in the practice of medicine

1.2 Physiology—Dynamic Processes in the Living Body

1. The meaning of *physiology,* and the importance of studying the physiology of other species for understanding humans
2. The properties that define something as alive and the difficulty of defining *life* or the moment of death
3. The meaning of *homeostasis,* the role of negative feedback loops in maintaining homeostasis, and the contrast between the set point and a dynamic equilibrium
4. Examples of negative feedback and homeostasis
5. The fundamental components seen in many negative feedback loops
6. How positive feedback differs from negative feedback, and examples of beneficial and harmful effects of positive feedback
7. Why it is clinically important to be aware of individual physiological variation

1.3 The Human Body Plan

1. The eight levels of structural complexity from organism to atoms, and the definitions of such levels as *organ, tissue, cell, organelle,* and others
2. The definition of *anatomical position,* and why anatomical position is such an important frame of reference
3. Definition of the three principal anatomical planes of the human body
4. Components of the axial and appendicular regions of the human body
5. Landmarks that define the four quadrants of the abdomen; names of the quadrants; and some organs that lie within each quadrant
6. Terminology of the surface regions of human anatomy
7. Meaning of the *segments* of an appendage, and examples
8. Differences between the serous and mucous membranes of the body and examples of locations
9. The principal body cavities; names of the membranes that line them; and the most important viscera found in each
10. The meanings of *parietal* and *visceral peritoneum*
11. The location, appearance, and function of the mesenteries, and how the mesenteries are related to the peritoneum and the serosa of the viscera
12. The distinction between intraperitoneal and retroperitoneal organs, and some examples of each
13. The 11 organ systems of the human body, functions of each system, and principal organs of each

1.4 The Language of Medicine

1. Examples of why precision is important in anatomical terminology, and why spelling errors can be more significant than they might seem
2. How to break down biomedical terms into roots, prefixes, and suffixes, and why it is helpful to make a habit of seeing terminology in this way
3. Relationships between the singular and plural forms of the same biomedical nouns
4. The distinctions between *dorsal* and *ventral; anterior* and *posterior; superior* and *inferior; medial* and *lateral; proximal* and *distal;* and *superficial* and *deep*
5. Why *anterior* and *posterior, dorsal* and *ventral* have different meanings for a human than for a cat

Testing Your Recall

1. Structure that can be observed with the naked eye is called
   a. gross anatomy.
   b. histology.
   c. ultrastructure.
   d. comparative anatomy.
   e. cytology.

2. The method of medical imaging that exposes a person to radio waves is
   a. a PET scan.
   b. an MRI scan.
   c. radiology.
   d. sonography.
   e. a CT scan.

3. The tarsal region is _____ to the popliteal region.
   a. dorsal
   b. distal
   c. superior
   d. proximal
   e. appendicular

4. Which of the following regions is *not* part of the upper limb?
   a. palmar
   b. antebrachial
   c. cubital
   d. carpal
   e. popliteal

5. Which of the following is *not* an organ system?
   a. muscular system
   b. endocrine system
   c. immune system
   d. lymphatic system
   e. integumentary system

6. The simplest structures considered to be alive are
   a. cells.
   b. organisms.
   c. the tissues.
   d. the organelles.
   e. organs.

7. Why is the stomach considered to be an organ?
   a. It is part of the digestive system.
   b. It has a specific, complex physiological function.
   c. It is composed of two or more tissues.
   d. It consists of a mass of similar cells that form a specific secretion.
   e. It is physically interconnected with other organs.

8. The lining of the abdominal cavity is
   a. a mucous membrane.
   b. the peritoneum.
   c. the meninges.
   d. the pleura.
   e. the serosa.

9. The word root *patho-* means
   a. doctor.
   b. medicine.
   c. organ.
   d. health.
   e. disease.

10. The prefix *epi-* means
    a. next to.
    b. below.
    c. above.
    d. within.
    e. behind.

11. When a doctor presses on the abdomen to feel the size and texture of the liver, he or she is using a technique of physical examination called _____.

12. A method of medical imaging that uses X-rays and a computer to generate an image of a thin slice of the body is called _____.

13. Most physiological mechanisms serve the purpose of _____, maintaining a stable internal environment in the body.

14. A/an _____ is the simplest body structure to be composed of two or more types of tissue.

15. The carpal region is more commonly known as the _____, and the tarsal region is more commonly known as the _____.

16. In standard directional terms, the sternal region is _____ to the pectoral region.

17. The layer of peritoneum facing the body wall is called the _____ layer, and the layer on the surface of an internal organ is called the _____ layer.

18. Homeostasis is maintained by a cycle of events called a _____, in which the body senses a change and activates mechanisms to minimize or reverse it.

19. The directional terms of human anatomy assume that a person is in _____, which means standing upright with the feet together and the palms, face, and eyes forward.

20. The elbow is said to be _____ to the wrist because it is closer to the upper limb's point of origin (the shoulder).

Answers in Appendix A

What's Wrong with These Statements?

Briefly explain why each of the following statements is false, or reword it to make it true.

1. It is possible to see both eyes in a median section of the head.
2. The diaphragm is ventral to the lungs.
3. A PET scan is a noninvasive method of medical imaging.
4. The pleural and pericardial cavities are lined by mucous membranes.
5. Abnormal skin color or dryness is one piece of information that could be obtained by auscultation.
6. A CT scan is a simpler and safer way than sonography to monitor fetal development.
7. Histopathology is a subdiscipline of gross anatomy.
8. Negative feedback is more often harmful than beneficial to the body.
9. There are more cells than organelles in the human body.
10. The pericardial sac is an internal lining of the heart.

Answers in Appendix A

Testing Your Comprehension

1. Identify which anatomical plane—sagittal, frontal, or transverse—is the only one that could *not* show (a) both the cerebrum and tongue, (b) both eyes, (c) both the hypogastric and gluteal regions, (d) both the sternum and vertebral column, and (e) both the heart and uterus.
2. Name one structure or anatomical feature that could be found in each of the following locations relative to the ribs: medial, lateral, superior, inferior, deep, superficial, posterior, and anterior. Try not to use the same example twice.
3. For each of the following nonbiological processes, state whether you think it is analogous to physiological positive feedback, negative feedback, or neither of these, and justify each answer: (a) a flushed toilet tank refilling to its original resting level; (b) a magnifying glass focusing the sun's rays and catching a piece of paper on fire; (c) a house fire in which heat from the flames ignites adjacent flammable material until, if unchecked, the whole house is consumed; (d) the increasingly loud howl of a loudspeaker as a band is setting up the sound stage for a concert and gets a microphone too close to the speaker.

Chapter 2

Life, Matter, and Energy

Urea crystals seen with a polarizing microscope. Urea, a product of protein metabolism, is the principal organic waste product of urine.
Eye of Science/Science Source

Chapter Outline

2.1 Atoms, Ions, and Molecules
- 2.1a Chemical Elements
- 2.1b Atoms
- 2.1c Isotopes and Radioactivity
- 2.1d Ions, Electrolytes, and Free Radicals
- 2.1e Molecules and Chemical Bonds

2.2 Water, Acids, and Bases
- 2.2a Water
- 2.2b Acids, Bases, and pH

2.3 Organic Compounds
- 2.3a Carbohydrates
- 2.3b Lipids
- 2.3c Proteins
- 2.3d Nucleic Acids (DNA and RNA)
- 2.3e Adenosine Triphosphate (ATP)

2.4 Energy and Chemical Reactions
- 2.4a Forms of Energy
- 2.4b Chemical Reactions
- 2.4c Metabolism, Oxidation, and Reduction

Clinical Applications/Perspectives on Health
- Clinical Application 2.1: pH and Drug Action
- Clinical Application 2.2: *Trans* Fats and Cardiovascular Health
- Perspectives on Health

End of Chapter
- Career Spotlight: Medical Technologist
- Study Guide

BASE CAMP

Before ascending to the next level, be sure you're properly equipped with a knowledge of these concepts from chapter 1.

- The concept of metabolism (see "Essential Life Functions" in section 1.2b)
- "Levels of Human Structure" in section 1.3a

Anatomy & Physiology Revealed 4.0
Module 2
Cells & Chemistry

Why is too much sodium or cholesterol harmful? Why does an iron deficiency cause anemia and a calcium deficiency weaken the bones? Why do some pregnant women suffer convulsions after several days of vomiting? How can radiation cause cancer as well as cure it?

None of these questions can be answered, nor would the rest of this book be intelligible, without understanding the chemistry of life. A little knowledge of chemistry can help you choose a healthy diet, use medications more wisely, evaluate health fads and avoid fraudulent claims, and understand medical treatments and procedures. And so we begin our study of the human body with basic chemistry, the simplest level of the body's structural organization.

2.1 Atoms, Ions, and Molecules

Expected Learning Outcomes

When you have completed this section, you should be able to

a. recognize elements of the human body from their chemical symbols;
b. distinguish between chemical elements and compounds;
c. state the functions of minerals in the body;
d. explain the basis for radioactivity and the uses and hazards of ionizing radiation;
e. distinguish between ions, electrolytes, and free radicals; and
f. define the types of chemical bonds.

2.1a Chemical Elements

A chemical **element** is the simplest form of matter to have unique chemical properties. Water, for example, has unique properties, but it can be broken down into two elements, hydrogen and oxygen, each with unique properties of their own. If we break them down any further, however, we find that hydrogen and oxygen are made of protons, neutrons, and electrons—none of which are unique. A proton of gold is identical to a proton of oxygen. Therefore, the elements hydrogen and oxygen are the simplest unique components of water.

There are 91 naturally occurring elements on earth, 24 of which play normal roles in humans. Table 2.1 groups these 24 according to their abundance in the body. Six of them account for 98.5% of the body's weight: oxygen, carbon, hydrogen, nitrogen, calcium, and phosphorus. The next 0.8% consists of another six elements: sulfur, potassium, sodium, chlorine, magnesium, and iron. The remaining 12 total only 0.7% of body weight; thus, these are known as **trace elements.** Despite their minute quantities, trace elements play vital roles in physiology. Iodine, for example, is an essential component of thyroid hormone. Other elements without natural physiological roles can contaminate the body

Table 2.1 Elements of the Human Body

Name	Symbol		Percentage of Body Weight
Major Elements (Total 98.5%)			
Oxygen	O		65.0
Carbon	C		18.0
Hydrogen	H		10.0
Nitrogen	N		3.0
Calcium	Ca		1.5
Phosphorus	P		1.0
Lesser Elements (Total 0.8%)			
Sulfur	S		0.25
Potassium	K		0.20
Sodium	Na		0.15
Chlorine	Cl		0.15
Magnesium	Mg		0.05
Iron	Fe		0.006
Trace Elements (Total 0.7%)			
Chromium	Cr	Molybdenum	Mo
Cobalt	Co	Selenium	Se
Copper	Cu	Silicon	Si
Fluorine	F	Tin	Sn
Iodine	I	Vanadium	V
Manganese	Mn	Zinc	Zn

and severely disrupt its functions, as in heavy metal poisoning with lead or mercury.

The elements are represented by one- or two-letter symbols, usually based on their English names: C for carbon, Mg for magnesium, Cl for chlorine, and so forth. A few symbols are based on Latin names, such as K for potassium *(kalium),* Na for sodium *(natrium),* and Fe for iron *(ferrum).* The periodic table of the elements (appendix E) summarizes information on all the natural chemical elements and their relative abundance in the human body.

Several elements are classified as **minerals**—substances extracted from the soil by plants and passed up the food chain to humans and other organisms. Minerals constitute about 4% of the human body by weight. Nearly three-quarters of this is Ca and P; the rest is mainly Cl, Mg, K, Na, and S. Minerals contribute significantly to body structure. The bones and teeth consist partly of crystals of Ca, P, Mg, F, and sulfate ions. In addition, sulfur is a component of many proteins; phosphorus is a major component of DNA and cell membranes; and iron is a component of hemoglobin.

2.1b Atoms

Each chemical element is composed of a unique type of **atom.** At the center of an atom is the **nucleus,** composed of positively charged **protons (p^+)** and uncharged **neutrons (n^0).** The nucleus is orbited by **electrons (e^-),** tiny particles with a negative charge that swarm about the nucleus in concentric regions called *energy levels (electron shells)* (**fig. 2.1**). Each shell holds a limited

Figure 2.1 Two Representative Elements.

number of electrons. Those in the outermost shell, **valence electrons,** determine the formation of chemical bonds. Electrons have a very low mass, even when compared to protons. A person who weighs 64 kg (140 lb) contains less than 24 g (1 oz) of electrons. And yet, these tiny particles determine the chemical properties of atoms and thereby govern chemical reactions and the formation of molecules. Despite their small mass, electrons play a big role in cell structure and function.

2.1c Isotopes and Radioactivity

Not every atom of an element is identical; all elements have two or more varieties called **isotopes,**[1] which differ from each other only in number of neutrons. Most hydrogen atoms, for example, have a nucleus composed of only one proton; this isotope is symbolized ^{1}H. Hydrogen has two other isotopes: *deuterium* (^{2}H) with one proton and one neutron, and *tritium* (^{3}H) with one proton and two neutrons. Over 99% of carbon atoms have a nucleus of six protons and six neutrons, and are called carbon-12 (^{12}C), but a small percentage of carbon atoms are ^{13}C, with seven neutrons, and ^{14}C, with eight.

Many isotopes are unstable and *decay* (break down) to more stable isotopes by giving off radiation. This process of decay is called **radioactivity,** and unstable isotopes are therefore called **radioisotopes.** Every element has at least one radioisotope.

Radioactivity is one form of **ionizing radiation** (ultraviolet radiation and X-rays are others). Ionizing radiation can be very damaging to tissues and is capable of causing cancer, birth defects, or in high doses, immediate death. In controlled, targeted doses, however, it is useful for such purposes as radiography, PET scans, and cancer radiation therapy.

2.1d Ions, Electrolytes, and Free Radicals

Ions are charged particles with unequal numbers of protons and electrons. Ions form because elements with one to three valence electrons tend to give them up, and those with four to seven electrons tend to gain more. If an atom of the first kind is exposed to an atom of the second, electrons may transfer from one to the other and turn both of them into ions. This process is called *ionization.* The particle that gains electrons acquires a surplus negative charge and is called an **anion** (AN-eye-on). The one that loses electrons is left with an excess positive charge (from a surplus of protons) and is called a **cation** (CAT-eye-on).

Consider, for example, what happens when sodium and chlorine meet (**fig. 2.2**). Sodium has a total of 11 electrons: 2 in its first (inner) shell, 8 in the second, and 1 in the third (valence) shell. If it gives up the electron in the third shell, its second shell becomes the valence shell and has a stable configuration of 8 electrons. Chlorine has 7 electrons in its valence shell (17 in all). If it can gain one more electron, it can fill the third shell with 8 electrons and become stable. Sodium and chlorine seem "made for each other"—one needs to lose an electron and the other needs to gain one. This is just what they do. When an electron transfers from sodium to chlorine, sodium is left with 11 protons

[1]*iso* = same; *top* = place (same position in the periodic table)

Figure 2.2 Ionization. A sodium atom donates an electron to a chlorine atom. This electron transfer converts the atoms to a positive sodium ion (Na$^+$) and a negative chloride ion (Cl$^-$). Attraction of these two oppositely charged ions to each other then constitutes an ionic bond.

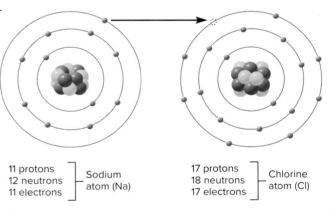

11 protons
12 neutrons — Sodium atom (Na)
11 electrons

17 protons
18 neutrons — Chlorine atom (Cl)
17 electrons

① Transfer of an electron from a sodium atom to a chlorine atom

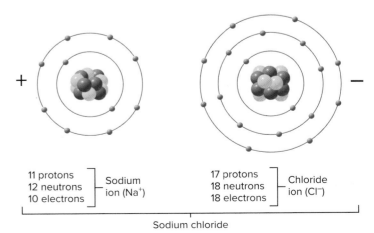

11 protons
12 neutrons — Sodium ion (Na$^+$)
10 electrons

17 protons
18 neutrons — Chloride ion (Cl$^-$)
18 electrons

Sodium chloride

② The charged sodium ion (Na$^+$) and chloride ion (Cl$^-$) that result but only 10 electrons; thus, it has a positive charge, and we symbolize the sodium ion Na$^+$. Chlorine is changed to the chloride ion with a surplus negative charge, symbolized Cl$^-$. Ions are not always single atoms that have become charged; some are groups of atoms such as phosphate (PO_4^{3-}) and bicarbonate (HCO_3^-) ions, or even molecules as large as a protein.

Ions with opposite charges are strongly attracted to each other. They often unite in *ionic bonds,* described shortly, and they tend to follow each other through the body. Thus, when Na$^+$ is excreted in the urine, Cl$^-$ tends to follow it. The attraction of cations and anions to each other is important in maintaining the excitability of muscle and nerve cells, as we shall see in later chapters.

Electrolytes are salts that ionize in water and form solutions capable of conducting electricity (table 2.2). Electrolytes are important for their chemical reactivity (as when calcium phosphate becomes incorporated into bone), osmotic effects (influence on water content and distribution in the body), and electrical effects (which are essential to nerve and muscle function). Electrolyte balance is one of the most important considerations in patient care. Electrolyte imbalances, discussed in chapter 16, have effects ranging from muscle cramps and brittle bones to coma and cardiac arrest.

Free radicals are chemical particles with an odd number of electrons. For example, oxygen normally exists as a stable molecule composed of two oxygen atoms, O_2, but if an additional electron is added, it becomes a free radical called the *superoxide anion,* symbolized $O_2^-\cdot$. The dot on the right symbolizes the odd electron.

Free radicals are produced continuously by normal metabolic reactions in cells (such as the ATP-producing reactions in mitochondria), and by external agents such as ionizing radiation, cigarette smoke, and some chemicals (for example, the food preservatives called nitrites). Free radicals are unstable and short-lived. They combine quickly with molecules such as fats, proteins, and DNA, converting them into free radicals and triggering chain reactions that destroy still more molecules. The swath of molecular destruction caused by free radicals is implicated in numerous diseases, including myocardial infarction (the death of heart tissue), spinal cord death in spinal injuries, Alzheimer disease, and cancer. One theory of aging is that it results in part from lifelong cellular damage by free radicals.

Because free radicals are so common and destructive, the body has multiple mechanisms for counteracting them. **Antioxidants** neutralize free radicals and limit their destruction. Some antioxidants are produced naturally by our own bodies as a defense against the inevitable production of free radicals. Others are found in fruits and vegetables, including selenium, vitamin E (α-tocopherol), vitamin C (ascorbic acid), and carotenoids (such as β-carotene). Dietary deficiencies of antioxidants are associated with increased incidence of heart attacks, sterility, muscular dystrophy, and other disorders.

Table 2.2 Major Electrolytes and the Ions Released by Their Dissociation APR

Electrolyte		Cations	Anions
Calcium chloride ($CaCl_2$)	→	Ca^{2+}	2 Cl$^-$
Disodium phosphate (Na_2HPO_4)	→	2 Na$^+$	HPO_4^{2-}
Magnesium chloride ($MgCl_2$)	→	Mg^{2+}	2 Cl$^-$
Potassium chloride (KCl)	→	K$^+$	Cl$^-$
Sodium bicarbonate ($NaHCO_3$)	→	Na$^+$	HCO_3^-
Sodium chloride (NaCl)	→	Na$^+$	Cl$^-$

2.1e Molecules and Chemical Bonds

Molecules form when two or more atoms are united by a chemical bond. The atoms may be identical, as in nitrogen (N_2), or different, as in glucose ($C_6H_{12}O_6$). Molecules composed of two or more *different* elements are called **compounds**. For example, oxygen (O_2) and carbon dioxide (CO_2) are both molecules because both consist of at least two atoms, but only CO_2 is a compound.

Molecules are represented by *molecular formulae* that list each element only once and show how many atoms of each are present. Molecules with identical formulae but different arrangements of their atoms are called **isomers**[2] of each other. For example, both ethanol (grain alcohol) and ethyl ether (once used as an anesthetic) have the molecular formula C_2H_6O, but they are certainly not interchangeable! To show the difference between them, we use *structural formulae* that show the location of each atom, or *condensed structural formulae* that show all the elements on one line but in groups reflecting the unique structural formula of each isomer (fig. 2.3).

A molecule is held together, and molecules are attracted to each other, by forces called **chemical bonds**. The three bonds of greatest physiological interest are *ionic bonds, covalent bonds,* and *hydrogen bonds.*

An **ionic bond** is the attraction of a cation to an anion. Sodium and chloride ions (Na^+ and Cl^-), for example, are attracted to each other and form the compound sodium chloride (NaCl), common table salt (see fig. 2.2). Ionic bonds are weak and easily dissociate (break up) in the presence of something more attractive, such as water. The ionic bonds of NaCl break down easily as salt dissolves in water, because both Na^+ and Cl^- are more attracted to water molecules than they are to each other.

Figure 2.3 Two Isomers, Ethanol and Ethyl Ether. The molecular formulae are identical, but the structures and chemical properties are different.

> **Apply What You Know**
> Do you think ionic bonds are common in the human body? Why or why not?

Covalent bonds form by the sharing of electrons. For example, two hydrogen atoms share valence electrons to form a hydrogen molecule, H_2 (fig. 2.4a). The two electrons, one donated by each atom, swarm around both nuclei in a

---
[2] *iso* = same; *mers* = parts

Figure 2.4 Covalent Bonding. (a) Two hydrogen atoms share a single pair of electrons to form a hydrogen molecule. (b) A carbon dioxide molecule, in which a carbon atom shares two pairs of electrons with each oxygen atom, forming double covalent bonds.

Figure 2.5 Nonpolar and Polar Covalent Bonds. (a) A nonpolar covalent bond between two carbon atoms, formed by electrons that spend an equal amount of time around each nucleus, as represented by the symmetric blue cloud. (b) A polar covalent bond, in which electrons orbit one nucleus significantly more than the other, as represented by the asymmetric cloud. This results in a slight negative charge (δ−) in the region where the electrons spend most of their time, and a slight positive charge (δ+) at the other pole.

dumbbell-shaped cloud. A *single covalent bond,* the sharing of a single pair of electrons, is symbolized by a single line between atomic symbols, for example H—H. A *double covalent bond,* the sharing of two pairs of electrons, is symbolized by two lines—for example, in carbon dioxide, O=C=O, where each oxygen shares two electrons with the central carbon (fig. 2.4b).

When shared electrons spend approximately equal time around each nucleus, they form a *nonpolar covalent bond* (fig. 2.5a). Carbon atoms bond to each other with nonpolar covalent bonds. If shared electrons spend significantly more time orbiting one nucleus than they do the other, they lend their negative charge to the region where they spend the most time, and they form a *polar covalent bond* (fig. 2.5b). For example, in water, H—O—H, the electrons are more attracted to the oxygen nucleus and orbit it more than they do the hydrogen. This makes the oxygen slightly negative (symbolized δ−) and the hydrogen slightly positive (δ+).

A **hydrogen bond** is a weak attraction between a slightly positive hydrogen atom ($H^{\delta+}$) in one molecule and a slightly negative oxygen ($O^{\delta-}$) or nitrogen ($N^{\delta-}$) in another. Water molecules are weakly attracted to each other by hydrogen bonds (fig. 2.6). Hydrogen bonds also form between different regions of the same molecule, especially in very large molecules such as proteins and DNA that can bend back on themselves. Hydrogen bonding causes such molecules to fold

Figure 2.6 Hydrogen Bonding of Water. The polar covalent bonds of water molecules enable each oxygen to form a hydrogen bond with a hydrogen of a neighboring molecule. Thus, the water molecules are weakly attracted to each other.

? *Why would a liquid without hydrogen bonding (such as liquid nitrogen) have a lower boiling point than water?*

or coil into precise three-dimensional shapes. Hydrogen bonds are represented by dotted or broken lines between atoms: —C=O⋯H—N—.

Table 2.3 summarizes the types of chemical bonds.

Table 2.3 Types of Chemical Bonds APR

Bond Type	Definition and Characteristics
Ionic Bond	Relatively weak attraction between an anion and a cation. Easily disrupted in water, as when salts dissolve.
Covalent Bond	Sharing of one or more pairs of electrons between nuclei.
Single covalent	Sharing of one electron pair.
Double covalent	Sharing of two electron pairs. Often occurs between carbon atoms, between carbon and oxygen, and between carbon and nitrogen.
Nonpolar covalent	Covalent bond in which electrons are equally attracted to both nuclei. May be single or double. Strongest type of chemical bond.
Polar covalent	Covalent bond in which electrons are more attracted to one nucleus than to the other, resulting in slightly positive and negative regions in one molecule. May be single or double.
Hydrogen Bond	Weak attraction between polarized molecules or between polarized regions of the same molecule. Important in the three-dimensional folding and coiling of large molecules. Weakest of all bonds; easily disrupted by temperature and pH changes.

Before You Go On

Answer these questions from memory to test your understanding of the previous section. Reread the material if you can't answer the questions.

1. Consider ammonia (NH_3), iron (Fe), and hydrogen gas (H_2). Which of these is/are atoms? Which of them is/are molecules? Which of them is/are compounds? Explain.
2. Where do free radicals come from? What harm do they do? What protections do we have against free radicals?
3. How does an ionic bond differ from a covalent bond?
4. What is a hydrogen bond? Why do hydrogen bonds depend on the existence of polar covalent bonds? What role do hydrogen bonds play in the three-dimensional structure of large molecules?

2.2 Water, Acids, and Bases

Expected Learning Outcomes

When you have completed this section, you should be able to

a. describe the biologically important properties of water;
b. define *acid* and *base;* and
c. interpret the pH scale.

2.2a Water

We all know that we need water to live, but why? Our bodies are 50% to 75% water, depending on age, sex, body fat, and other factors—but again, why? Why not some other liquid, or less water? Its structure, simple as it is, has profound biological effects. Water is a polar molecule, as discussed previously. The slight negative charge on the oxygen and a slight positive charge on each hydrogen explain why. Because of this polarity, water molecules are attracted to each other by hydrogen bonds. This gives water a set of properties that account for its ability to support life:

- A solvent is a substance in which other materials dissolve. Water is called the **universal solvent** because more substances (**solutes**) dissolve in water than in any other liquid. Therefore, it is the ideal medium for the chemical reactions that must occur in the body, and it is ideally suited for transporting substances from place to place, as in the blood. Substances that dissolve in water, such as sugars and salts, are said to be **hydrophilic**[3] (HY-dro-FILL-ic); the relatively few that do not, such as fats, are **hydrophobic**[4] (HY-dro-FOE-bic).
- Water has the quality of **adhesion,** the tendency to cling to other substances. In the body, it adheres to membranes such as those that surround the lungs and heart, where it acts as a lubricant and reduces friction during breathing and the heartbeat.
- Water also exhibits **cohesion,** an attraction of its molecules to each other. Its hydrogen bonds make water molecules cling together like little magnets. This is why water forms puddles and hangs from a faucet in drops, and it accounts for the next property of water.
- The cohesion of water gives it relatively great **thermal stability,** or resistance to temperature changes. To increase in temperature, the molecules of a substance must move around more rapidly. The cohesion of water molecules inhibits their movement, so water can absorb a given amount of heat without changing temperature (molecular motion) as much. Water thus helps to stabilize the body temperature and serves as a very effective coolant. When water evaporates, it carries a large amount of heat with it—an effect that is very apparent when you are sweaty and stand in front of a fan. Indeed, the base unit of heat is defined in reference to water: one **calorie**[5] **(cal)** is the amount of heat that raises the temperature of 1 g of water 1°C. One thousand calories is a **kilocalorie (kcal),** also known as 1 dietary Calorie (with a capital C).
- The **chemical reactivity** of water is its ability to participate in chemical reactions. Not only does water ionize many other chemicals such as acids and salts, but water itself ionizes into hydrogen and hydroxide ions: H^+ and OH^-. These ions can be incorporated into other molecules, or released from them, in the course of chemical reactions such as *hydrolysis* and *dehydration synthesis,* described later in this chapter.

Apply What You Know

The lungs have millions of tiny air sacs with moist walls where oxygen and carbon dioxide are exchanged. The air sacs produce a soaplike substance called surfactant that breaks the cohesion between water molecules. Why do you think this is necessary?

---

[3]*hydro* = water; *philic* = loving, attracted to
[4]*hydro* = water; *phobic* = fearing, avoiding
[5]*calor* = heat

Body fluids are complex mixtures of water and other chemicals. Much of our physiology is aimed at maintaining the proper amount of dissolved matter in the body fluids because deviations can have profound effects, with consequences as severe as seizures, cardiac arrest, and death. It is often important to know the **concentration** of a solution—how much solute is present in a given volume of water. Solute concentrations are commonly measured in body fluids such as blood and urine, as well as in intravenous fluids and other liquids given to patients. There are several ways of expressing concentration, suitable for different purposes. Appendix C defines and explains the most commonly used measures of concentration: weight per volume, percentage, molarity, osmolarity, milliequivalents per liter, and pH.

2.2b Acids, Bases, and pH

Most people have some sense of what acids and bases are. Advertisements speak of excess stomach acid and pH-balanced shampoo. We know that drain cleaner (a strong base) and battery acid can cause serious chemical burns. But what exactly do *acid* and *base* mean, and how can they be quantified?

An **acid** is any *proton donor,* a molecule that releases a proton (H^+) in water. A **base** is a proton acceptor. Because hydroxide ions (OH^-) accept H^+, many bases are substances that release hydroxide ions—sodium hydroxide (NaOH), for example. A base doesn't have to be a hydroxide donor, however. Ammonia (NH_3) is also a base. It doesn't release hydroxide ions, but it readily accepts hydrogen ions to become the ammonium ion (NH_4^+).

Acidity is expressed in terms of **pH,** a measure derived from the concentration of H^+ (see appendix C). The pH scale (**fig. 2.7**) was invented in 1909 by Danish biochemist and brewer Sören Sörensen to measure the acidity of beer. The scale extends from 0.0 to 14.0. A solution with a pH of 7.0 is **neutral;** solutions with pH below 7 are **acidic;** and solutions with pH above 7 are **basic (alkaline).** The lower the pH value, the more hydrogen ions a solution has and the more acidic it is. Each step of the scale represents a 10-fold change in H^+ concentration. For example, a solution of pH 5.0 has 10 times the H^+ concentration as one with a pH of 6.0; a pH of 4.0 represents 100 times the H^+ concentration at 6.0.

Slight disturbances of pH can seriously disrupt physiological functions and alter drug actions (see Clinical Application 2.1), so it is important that the body carefully controls its pH. Blood, for example, normally has a pH ranging from

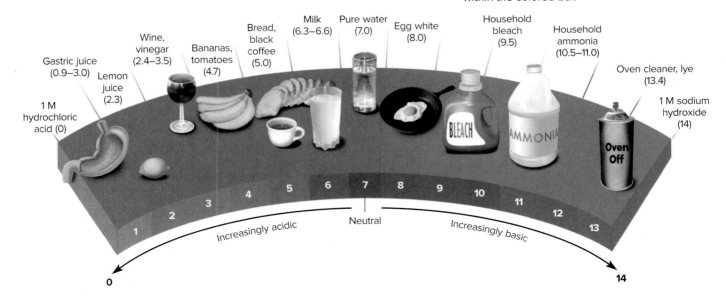

Figure 2.7 The pH Scale. The pH is shown within the colored bar.

>
> ### Clinical Application 2.1
> #### pH AND DRUG ACTION
> The pH of our body fluids has a direct bearing on how we react to drugs. Depending on pH, drugs such as aspirin, phenobarbital, and penicillin can exist in charged (ionized) or uncharged forms. Whether a drug is charged or not can determine whether it will pass through cell membranes. When aspirin is in the acidic environment of the stomach, for example, it is uncharged and passes easily through the stomach lining into the bloodstream. Here it encounters a basic pH, whereupon it ionizes. In this state, it is unable to pass back through the membrane, so it accumulates in the blood. This effect, called *ion trapping* or *pH partitioning,* can be controlled to help clear poisons from the body. The pH of the urine, for example, can be clinically manipulated so that poisons become trapped there and are more rapidly excreted from the body.

7.35 to 7.45. Deviations from this range cause tremors, fainting, paralysis, or even death. Chemical solutions that resist changes in pH are called **buffers.** Buffers and pH regulation are considered in detail in chapter 16.

Before You Go On

Answer these questions from memory to test your understanding of the previous section. Reread the material if you can't answer the questions.

5. What is meant by the terms hydrophilic and hydrophobic?
6. What physical property of water makes it resistant to temperature changes? What specific type of chemical bond is responsible for that physical property?
7. Why is it important to understand various measures of concentration of body fluids?
8. Define acid and base, and state what ranges of pH values are considered acidic and basic.

2.3 Organic Compounds

Expected Learning Outcomes

When you have completed this section, you should be able to

a. distinguish between organic and inorganic compounds, and define *biochemistry;*
b. describe the structural properties that distinguish carbohydrates, lipids, proteins, and nucleic acids from each other;
c. describe or define the subclasses of each of those categories of biomolecules;
d. discuss the roles that each of these categories of molecules play in the body;
e. explain how enzymes function; and
f. describe the structure, production, and function of ATP.

Compounds of carbon are called **organic compounds,** distinguishing them from the **inorganic compounds,** such as water, that we have considered up to this point. The study of organic compounds is called **organic chemistry. Biochemistry** is the subdiscipline of organic chemistry that relates organic compounds to the processes of life. Biochemists classify the larger organic molecules of life into four main categories: *carbohydrates, lipids, proteins,* and *nucleic acids.*

Carbon is a uniquely suitable element to serve as the basis for the broad variety of biological molecules. Carbon atoms can bond with each other and form long chains, branched molecules, and rings. They also bind readily to hydrogen, oxygen, nitrogen, sulfur, and other elements. Because carbon can form long chains, some biological molecules attain enormous sizes and are called *macromolecules.* Most macromolecules are **polymers**[6]—molecules made of a repetitive series of identical or similar subunits called **monomers.** Starch, for example, is a polymer of about 3,000 glucose monomers.

Living cells use a process called **dehydration synthesis** to join monomers together to form polymers (**fig. 2.8a**). They remove a hydrogen (—H) from one monomer and a hydroxyl group (—OH) from another, producing water as a by-product. The two monomers become joined by a covalent bond. This is repeated for each monomer added to the chain.

The opposite of dehydration synthesis is **hydrolysis**[7] (**fig. 2.8b**). In hydrolysis, a water molecule ionizes into OH$^-$ and H$^+$. The cell breaks a covalent bond linking one monomer to another and adds the H$^+$ to one monomer and the OH$^-$ to the other one. All chemical digestion consists of hydrolysis reactions.

Figure 2.8 Synthesis and Hydrolysis Reactions. (a) In dehydration synthesis, a hydrogen atom is removed from one monomer, and a hydroxyl group is removed from another. These combine to form water as a by-product. The monomers become joined by a covalent bond. Repetition of this process can produce long molecular chains such as polymers. (b) In hydrolysis, a covalent bond between two monomers is broken. Water donates a hydrogen atom to one monomer and a hydroxyl group to the other. Polymers such as starch are digested in this manner.

2.3a Carbohydrates

Carbohydrates[8] include sugars, starch, and glycogen. They all consist of carbon and a 2:1 ratio of hydrogen to oxygen; their very name, *carbo + hydrate,* loosely means "carbon plus water." For example, glucose (blood sugar) has the formula $C_6H_{12}O_6$—essentially six carbons and six waters. Carbohydrates are hydrophilic—they readily absorb or mix with water. The names of individual carbohydrates are often built on the word root *sacchar-* or the suffix *-ose,* both of which mean "sugar" or "sweet."

Monosaccharides

The simplest carbohydrates are called **monosaccharides**[9] (MON-oh-SAC-uh-rides), or simple sugars. The three of primary importance are **glucose, fructose,** and **galactose,** all with the molecular formula $C_6H_{12}O_6$; they are isomers of each other. We obtain these sugars mainly by digesting more complex carbohydrates, but many processed foods contain large amounts of high-fructose corn syrup. Glucose (**fig. 2.9a**) is commonly called *blood sugar.*

---

[6]*poly* = many; *mer* = part
[7]*hydro* = water; *lysis* = splitting apart
[8]*carbo* = carbon; *hydr* = water
[9]*mono* = one; *sacchar* = sugar

Figure 2.9 Carbohydrate Structures. (a) Glucose (blood sugar), our most abundant monosaccharide. (b) Sucrose (table or cane sugar), a disaccharide, composed of glucose and fructose (the tan sugar). (c) Glycogen, a polysaccharide, composed of long branched chains of glucose. Each angle in the rings represents a carbon atom except the one where oxygen is shown. This is a conventional way of representing carbon in the structural formulae of organic compounds.

Disaccharides

Disaccharides are sugars composed of two monosaccharides bonded to each other. The most important ones are **sucrose** (made of glucose + fructose), **lactose** (glucose + galactose), and **maltose** (glucose + glucose). Sucrose (fig. 2.9b) is common table sugar, lactose is milk sugar, and maltose is mainly a product of starch digestion, but is present in a few foods such as malt beverages and germinating grains.

Polysaccharides

Polysaccharides (POL-ee-SAC-uh-rides) are polymers of glucose. The most important of these are glycogen, starch, and cellulose. **Glycogen**[10] is an energy-storage polysaccharide made by the liver, muscles, uterus, and a few other organs. It is a long branched polymer (fig. 2.9c). The liver stores glycogen after a meal, when the blood glucose level is high, and then breaks it down between meals to maintain blood glucose levels when there is no food intake. Muscle stores glycogen for its own energy needs, and the uterus uses it in pregnancy to nourish the embryo. **Starch** is the corresponding energy-storage polysaccharide of plants. It is the only significant digestible polysaccharide in the human diet. **Cellulose** is a structural polysaccharide of plants, and is the most abundant organic compound on earth. It is the principal component of wood, cotton, and paper. Cellulose is indigestible to us, yet it is important as dietary fiber (*bulk* or *roughage*). It swells with water in the digestive tract and promotes movement of other materials through the intestines.

Carbohydrates are, above all, a source of energy that can be quickly mobilized, but they have other functions as well (table 2.4). They are often covalently bonded to lipids and proteins to form components of mucus, tissue gel, cell membranes, the tough matrix of cartilage, the lubricating fluid in joints, and the gelatinous filler in the eyeball.

---

[10]*glyco* = sugar; *gen* = producing

Table 2.4 Carbohydrate Functions

Type	Function
Monosaccharides	
Glucose	Blood sugar—energy source for most cells
Galactose	Converted to glucose and metabolized
Fructose	Converted to glucose and metabolized
Disaccharides	
Sucrose	Cane sugar—digested to glucose and fructose
Lactose	Milk sugar—digested to glucose and galactose; important in infant nutrition
Maltose	Malt sugar—product of starch digestion, further digested to glucose
Polysaccharides	
Cellulose	Structural polysaccharide of plants; dietary fiber
Starch	Energy storage in plant cells; dietary source of energy for humans
Glycogen	Energy storage in animal cells (liver, muscle, uterus)

2.3b Lipids

Lipids include triglycerides (fats and oils), phospholipids, steroids (such as cholesterol, estrogen, and testosterone), and others (table 2.5). They are hydrophobic (it is well known that oil and water do not mix, for example). Like carbohydrates, lipids are usually composed only of carbon, hydrogen, and oxygen, but they have a higher ratio of hydrogen to oxygen. A fat called *tristearin* (tri-STEE-uh-rin), for example, has the molecular formula $C_{57}H_{110}O_6$—more than 18 hydrogens per oxygen. Lipids have more calories per gram than carbohydrates. Beyond these criteria, it is difficult to generalize about lipids; they are much more variable in structure than the other macromolecules we are considering.

Triglycerides and Fatty Acids

Triglycerides (try-GLISS-ur-ides) are commonly called **fats** if they are solid at room temperature and **oils** if they are liquid. A triglyceride is built by dehydration synthesis reactions linking a three-carbon molecule called **glycerol** with three long zigzag carbon chains called **fatty acids** (fig. 2.10). Fatty acids and the fats made from them are classified as *saturated* or *unsaturated*. A **saturated fatty acid** has as much hydrogen as it can carry, such as the palmitic and stearic acids in the figure. Carbon atoms are limited to forming four covalent bonds, and in saturated fatty acids, no more hydrogen could be added without exceeding that four-bond limit;

Table 2.5 Some Lipid Functions

Type	Function
Triglycerides	Energy storage; thermal insulation; filling space; binding organs together; cushioning organs
Fatty acids	Precursor of triglycerides; source of energy
Phospholipids	Major component of cell membranes
Cholesterol	Component of cell membranes; precursor of other steroids
Steroid hormones	Chemical messengers between cells
Bile acids	Steroids that aid in fat digestion and nutrient absorption

that is, the molecule is "saturated" with hydrogen. In **unsaturated fatty acids**, such as the linoleic acid shown, some carbon atoms are joined by double covalent (C=C) bonds. Each of these could potentially share one pair of electrons with another hydrogen atom instead of the adjacent carbon, so hydrogen could be added to this molecule. **Polyunsaturated fatty acids** are those with many C=C bonds. A *saturated fat* is a triglyceride made with all saturated fatty acids; an *unsaturated* or *polyunsaturated fat* has at least one unsaturated fatty acid. Saturated fats and *trans* fats are major contributors to cardiovascular disease (see Clinical Application 2.2).

The primary function of triglycerides is energy storage, but when concentrated in *adipose (fat) tissue*, they also provide thermal insulation and act as a shock-absorbing cushion for vital organs.

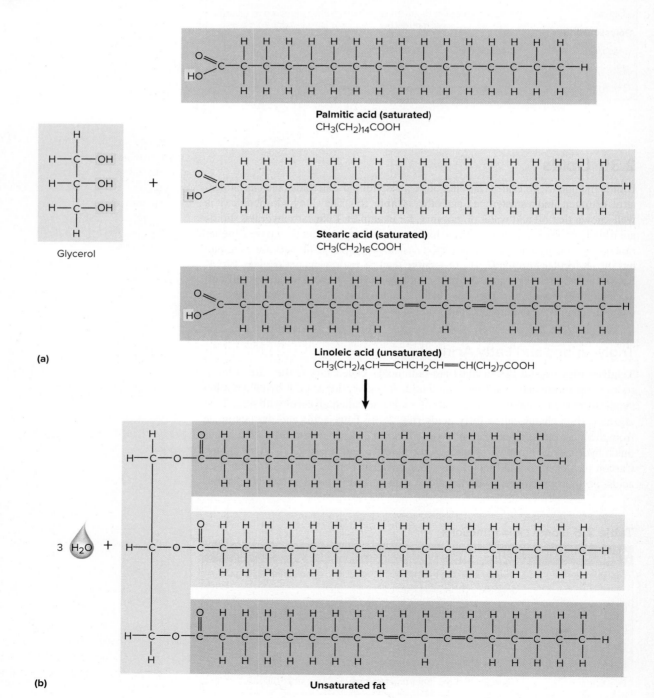

Figure 2.10 Fatty Acid and Triglyceride (Fat) Structure. (a) Glycerol and three fatty acids, the building blocks of a triglyceride. (b) The triglyceride formed by dehydration synthesis, with a glycerol "backbone" and three fatty acid "tails." This triglyceride is shown with two saturated fatty acids and one unsaturated fatty acid.

Figure 2.12 Phospholipid Structure. (a) A space-filling model that gives some idea of the actual shape of the molecule. (b) A simplified representation of a phospholipid used in diagrams of cell membranes.

Phospholipids

Phospholipids are similar to triglycerides except that in place of one fatty acid, they have a phosphate group that, in turn, is linked to another organic group (**fig. 2.12**). The glycerol and phosphate-containing group constitute a hydrophilic "head," and the two fatty acids constitute hydrophobic zigzag "tails" of the molecule. Together, the head and two tails of a phospholipid give it a shape like a clothespin. A phospholipid is said to be **amphipathic**[11] (AM-fih-PATH-ic)—part of it is attracted to water and part of it is repelled. The most important function of phospholipids is to serve as the structural foundation of cell membranes. (See section 3.2, where the significance of this amphipathic property will be more fully explained.)

Steroids

Steroids are lipids with 17 of their carbon atoms arranged in four rings (**fig. 2.13**). **Cholesterol** is the "parent" steroid from which the others are synthesized.

---

[11]*amphi* = both; *pathic* = feeling

Figure 2.13 Cholesterol. All steroids have this basic four-ringed structure, with variations in the side groups and locations of double bonds within the rings.

Clinical Application 2.2

TRANS FATS AND CARDIOVASCULAR HEALTH

Trans fats have been a significant part of the American diet for more than a century. Yet, in 2015, the U.S. Food and Drug Administration (FDA) ordered food manufacturers to discontinue their use. So what are *trans* fats, why were they so popular for so long, and why are they now almost entirely banished?

A *trans* fat is a triglyceride containing one or more *trans*-fatty acids. Ordinarily, when a fatty acid has a C=C double bond, the bond creates a kink in the chain (fig. 2.11a). These molecules are called *cis*-fatty acids. The kinks prevent triglyceride molecules from packing too closely together, making them liquid (oily) at room temperature. Small amounts of *trans* fats occur naturally in meat and dairy products, but in the early 1900s, the food industry developed a way to modify these into *trans*-fatty acids. These have straight fatty acid chains (fig. 2.11b) that allow the triglycerides to pack closely together, making them solid at room temperature. Crisco vegetable shortening was the first commercial product composed of *trans* fats. *Trans* fats are much easier to work with in making pie crusts, biscuits, and other baked goods. They were used abundantly in snack foods, baked goods, fast foods such as French fries, and fried chicken.

The disadvantage of *trans* fats is that they resist digestion, remain in circulation longer, and have more tendency to deposit in the arteries than saturated fats and *cis* fats do. Therefore, they raise the risk of coronary heart disease (CHD). From 1980 to 1994, medical scientists tracked a cohort of 80,082 female nurses (the Nurses' Health Study II) and, among other things, correlated their incidence of CHD with their self-reported diets. They concluded that for every 2% increase in calories from *trans* fats in place of carbohydrate calories, the women had a 93% higher risk of CHD. As the dangers of *trans* fats became widely known, their consumption in the United States greatly declined, and they are now banned from processed and restaurant foods.

Figure 2.11 *Cis*- and *Trans*-Fatty Acids. Each example is unsaturated at the C=C double bond in red. (a) In a *cis*-fatty acid, the bonds angle in the same direction. (b) In a *trans*-fatty acid, the bonds on opposite sides of the C=C bond angle in opposite directions, like bicycle pedals (arrows). The straight chains of *trans*-fatty acids allow fat molecules to pack together more tightly, thus remaining solid at room temperature.

Others include cortisol, progesterone, estrogen, testosterone, and bile acids (table 2.5). These differ from each other in the location of C=C bonds within the rings and in the functional groups attached to the rings.

The average adult contains over 200 g (half a pound) of cholesterol, of which about 85% is synthesized in one's own body and 15% comes from animal products in the diet. (Plants contain insignificant traces of cholesterol.) Cholesterol has a bad reputation as a factor in cardiovascular disease, and it is true that hereditary and dietary factors can elevate blood cholesterol to dangerously high levels. Nevertheless, cholesterol is a natural product of the body and we must remember that if we did not have it, we also would not have certain stress hormones, sex hormones, or other vital products derived from it.

2.3c Proteins

The word *protein* is derived from the Greek word *proteios*, "of first importance." Proteins are the most versatile molecules in the body, and many discussions in this book will draw on your understanding of protein structure and behavior. Table 2.6 summarizes several of their very diverse functions.

Table 2.6 Protein Functions

Function	Examples
Structure	*Keratin* is a tough protein that strengthens nails, hair, and the skin surface. *Collagen* is a structural foundation of the bones, cartilage, teeth, and deeper layer (dermis) of the skin.
Communication	Some hormones and other cell-to-cell signals are proteins, as are the receptors to which the signal molecules bind in the receiving cell.
Membrane transport	Some proteins form channels in cell membranes that govern what can pass through the membrane and when. Others act as carriers that briefly bind to solute particles and move them to the other side of the membrane. Among other roles, membrane transport proteins turn nerve and muscle activity on and off.
Catalysis	Most metabolic pathways of the body are controlled by enzymes, which are globular proteins that function as biological catalysts.
Recognition and protection	Cell surface proteins function in immune recognition; antibodies attack and neutralize toxins and organisms that invade the body; clotting proteins seal broken blood vessels and reduce blood loss.
Movement	Proteins are the basis for all movement, ranging from transport of molecules and organelles from place to place within a cell to the contractions of muscles and beating of cilia. Some proteins are called *molecular motors* for their ability to repeatedly change shape and produce movement.
Cell adhesion	Proteins bind cells to each other and to the extracellular material, prevent tissues from falling apart, enable the egg and sperm to join during fertilization, and enable immune cells to bind to enemy cells such as microbes and cancer cells.

A **protein** is a polymer of amino acids. An **amino acid** is a small organic molecule with a central carbon, an **amino group** (—NH_2), and a **carboxyl group** (—COOH) (**fig. 2.14**). The 20 amino acids used to make proteins differ only in a third functional group, called the R group, attached to the central carbon (like the R_1 and R_2 in the figure). R groups can range from a single hydrogen atom to chains and rings of carbon and other atoms.

Amino acids can be linked together by dehydration synthesis reactions to form *peptides* (a dipeptide for a chain of two amino acids, tripeptide for a chain of three, and so forth) (**fig. 2.14b**). The link formed between the C═O of one

Figure 2.14 Amino Acids and Peptide Bonding. (a) Two representative amino acids. Note that they differ only in the R group, shaded in pink. (b) The joining of two amino acids by a peptide bond, forming a dipeptide. Side groups R_1 and R_2 could be the groups indicated in pink in part (a), among other possibilities.

amino acid and —NH of the next is called a **peptide bond.** Chains larger than 10 or 15 amino acids are called *polypeptides,* and a polypeptide of 50 amino acids or more is considered to be a protein.

You could think of a protein as loosely analogous to a bead necklace with 20 different-colored beads corresponding to the 20 amino acids from which proteins may be made. Proteins differ from one another in the order of amino acids—their *amino acid sequence.* This sequence is specified by the *genetic code* in one's DNA.

Some proteins can function only when a non–amino acid component is bound to them. Hemoglobin, for example, requires an iron-containing ring called a *heme* group (fig. 2.15c), without which it cannot transport oxygen; this is one reason you must have iron in your diet. Many proteins must have zinc, magnesium, or other ions bound to them in order to function.

Protein Shapes

As a protein is assembled, interactions between the amino acids and water, and between one amino acid and another, cause the chain to coil and fold into complex three-dimensional shapes. One of the most common shapes is a springlike *alpha helix* (fig. 2.15a), which is stabilized by hydrogen bonds between the amino acids of the chain. Many proteins are bent or folded into globular or fibrous shapes. Globular proteins resemble a wadded ball of yarn, a shape that is well-suited for proteins that move freely in the body fluids, such as antibodies and hemoglobin (fig. 2.15b). Ropelike *fibrous* proteins provide structural strength to tissues, such as the *collagen* that makes up tendons, ligaments, the dermis of the skin, and part of the bone tissue.

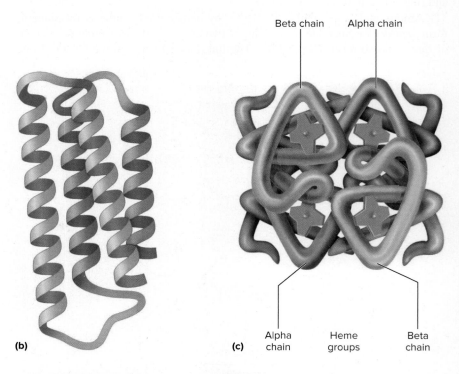

Figure 2.15 Protein Shapes. (a) Collagen, a fibrous protein, forming much of the substance of bones, tendons, ligaments, and dermis of the skin. This is composed of three alpha helices intertwined into a long ropy triple helix. (b) A protein composed of four regions of alpha helix connected by short nonhelical regions. (c) Hemoglobin, the red blood pigment. This is a compact globular protein composed of two identical alpha chains and two identical beta chains, all four of which are alpha helices. Each chain possesses a nonprotein oxygen-carrying *heme* group.

The three-dimensional shape *(conformation)* of a protein is critical to its function. An important property of proteins is their ability to change shape. Conformation changes can be triggered by electrical changes on a cell membrane or the binding of a hormone to a protein. Subtle, reversible changes in conformation are important to processes such as enzyme function, muscle contraction, nerve signaling, and the opening and closing of pores in cell membranes. **Denaturation** is a more drastic conformational change in response to conditions such as extreme heat or pH. It is seen, for example, when you cook an egg and the egg white (albumen) turns from clear and runny to opaque and stiff. Denaturation makes a protein unable to perform its normal function. It is sometimes reversible, but usually it permanently destroys protein function.

Enzymes

Enzymes are proteins that function as biological catalysts; a catalyst is any substance that enables a chemical reaction to go faster or at lower temperature but that is not consumed by that reaction. As catalysts, enzymes permit biochemical reactions to occur rapidly at normal body temperatures.

Enzymes were initially given somewhat arbitrary names such as *pepsin* and *trypsin*. The modern system of naming enzymes, however, is more uniform and informative. It identifies the substance the enzyme acts upon, called its **substrate**; it sometimes refers to the enzyme's action; and it adds the suffix *-ase*. For example, *amylase* digests starch (*amyl-* = starch) and *carbonic anhydrase* removes water (*anhydr-*) from carbonic acid.

An enzyme has surface pockets called **active sites** that bind specific substrates. A substrate fits its active site somewhat like a key fitting a lock. Just as no other key will fit a given lock, no other substrates will fit a given enzyme. An enzyme that acts on glucose, for example, will not act on the similar sugar fructose. Once a substrate is bound, the enzyme chemically changes it to one or more reaction products, releases the products, and is then available to act on another molecule of the same substrate (fig. 2.16). Because enzymes are not consumed by the reactions they catalyze, one enzyme molecule can convert millions of substrate molecules—and at astonishing speeds. A single molecule of carbonic anhydrase, for example, breaks carbonic acid (H_2CO_3) down to H_2O and CO_2 at a rate of 36 million molecules per minute.

Figure 2.16 The Three Steps of an Enzymatic Reaction. One or more substrate molecules bind to the enzyme's active sites. The substrates and enzyme form a temporary enzyme–substrate complex and the substrates react chemically with each other. The enzyme releases the reaction products and is available to catalyze the same reaction again.

Figure 2.17 Nucleotides and Nitrogenous Bases. (a) The structure of a nucleotide found in DNA and RNA. (b) The five nitrogenous bases found in DNA and RNA nucleotides. DNA has the bases A, T, C, and G. In RNA, U takes the place of T.

Factors that change the shape of an enzyme—notably temperature and pH—can alter or destroy the ability of the enzyme to bind its substrate. They disrupt the weak forces that hold the enzyme in its proper conformation, essentially changing the shape of the "lock" so that the "key" no longer fits. Enzymes vary in optimum pH according to where in the body they normally function. For example, a digestive enzyme that works in the mouth functions best at a neutral pH of 7, whereas one that works in the more acidic environment of the stomach functions best at a pH around 2. Our internal body temperature is nearly the same everywhere, however, and all human enzymes function best at a temperature of about 37° to 40°C.

2.3d Nucleic Acids (DNA and RNA)

Nucleic acids are organic polymers that serve as the cell's genetic machinery—**deoxyribonucleic acid (DNA),** which resides in the nucleus (hence the term *nucleic*) and contains the individual's genes, and a few forms of **ribonucleic acid (RNA),** which carry out the "orders" given by the DNA. Both of these are composed of monomers called nucleotides.

A **nucleotide** consists of a monosaccharide, a phosphate group, and a **nitrogenous base,** which is a single or double ring of carbon and nitrogen atoms (fig. 2.17a). There are five nitrogenous bases. Three of them have a single carbon–nitrogen ring and are named **cytosine (C), thymine (T),** and **uracil (U).** The other two have double rings and are named **adenine (A)** and **guanine (G)** (fig. 2.17b).

DNA consists of two nucleotide chains twined together in a *double helix,* shaped like a twisted ladder (fig. 2.18). It is the body's largest polymer, a chain of typically 100 million to 1 billion nucleotides. The "backbone" of each helix, like the vertical rails of a ladder, consists of phosphate groups alternating with a sugar called *deoxyribose.* The rungs of the ladder are pairs of nitrogenous bases. Wherever one chain has an A, the one across from it has a T; where one has C, the other has G. Determination of the base sequence of one strand by the sequence on the other is called the **law of complementary base pairing.** A–T and C–G are called the **base pairs** of DNA. Note that a single-ringed base is always paired with a double-ringed one, giving the DNA molecule a uniform diameter throughout.

DNA directs the synthesis of proteins by means of its smaller cousins, the ribonucleic acids. There are three types of RNA involved in carrying out the cell's genetic instructions and making protein. All of them differ from DNA in that they are much smaller; they are composed of just one nucleotide chain rather than two chains twined around each other; and they employ the sugar **ribose** instead of deoxyribose and contain uracil in place of thymine. DNA normally stays in the nucleus, giving orders from there, whereas RNA functions mainly in the cytoplasm, carrying out the instructions encoded in the DNA. The actions of DNA and RNA in protein synthesis and cell division will be described in the next chapter.

Figure 2.18 DNA Structure. (a) A space-filling molecular model of DNA giving some impression of its actual geometry. (b) The "twisted ladder" structure of DNA. The two sugar–phosphate backbones (blue) twine around each other to form the double helix (lower half of figure), and the nitrogenous bases (colored bars) face each other in the center. (c) A small segment of DNA showing backbones of alternating deoxyribose (green) and phosphate (yellow), and the complementary pairing of the nitrogenous bases. **APR**

PERSPECTIVES ON HEALTH

Anabolic Steroids

The Effect of Heavy Use of Anabolic Steroids.
Master1305/Shutterstock

Anabolic steroids are compounds related to testosterone that are used to promote tissue growth and repair. In limited doses, anabolic steroids have legitimate medical uses such as treatment of anemia, breast cancer, osteoporosis, and some muscle diseases. Some amateur and professional athletes, however, use them to stimulate muscle growth and to accelerate the repair of tissues damaged in training or competition.

The doses used by athletes are 10 to 1,000 times higher than those prescribed for medical purposes. Although such doses bulk up the muscles, they have hidden, devastating effects on the body. They raise cholesterol levels and thus promote coronary artery disease, heart and kidney failure, and stroke. Deteriorating blood circulation can lead to gangrene and has necessitated many limb amputations among steroid abusers. Liver disease, immune suppression, behavioral disorders, atrophy of the testes, impotence, and overgrowth of the male breasts *(gynecomastia)* are other common effects. In women, anabolic steroids tend to cause atrophy of the breasts and uterus; menstrual irregularity; and overgrowth of facial hair, body hair, and the clitoris.

Because of these severe effects, anabolic steroid use for performance enhancement is condemned by the American Medical Association, American College of Sports Medicine, International Olympic Committee, National Football League, and National Collegiate Athletic Association.

Collagen Diseases

Collagen is the most abundant protein in the body. Defects in collagen synthesis therefore cause a wide range of disorders collectively called *collagen diseases* or *collagenopathies*. Humans have about 28 types of collagen, which are subject to more than 1,000 mutations that can affect its chemical structure (amino acid sequence) or the quantity synthesized. Such defects can have severe effects on the many tissues and organs that depend on collagen for strength—tendons, ligaments, bones, blood vessels, eyes, skin, and viscera.

Ehlers–Danlos[12] *syndrome (EDS)*, for example, has six main types that vary in effect from mild to life-threatening. Its most common signs are loose ligaments and hyperextensible joints that look like one is "double-jointed" to an extreme degree, and highly elastic skin that can stretch several centimeters but is very fragile. The skin bruises easily and even minor trauma can cause wide, gaping wounds. It is very difficult for surgeons to successfully suture the skin of such patients, because the sutures tear out so easily. Depending on the mutational type, people with EDS can also show spinal deformation, frequent hip dislocations, fragile eyeballs, extreme muscle weakness, and fragile blood vessels that might spontaneously rupture. Virtuoso violinist Niccolò Paganini (1782–1840) wrote compositions that many other accomplished violinists find nearly impossible to play because they require such extraordinarily wide finger placements. It is thought that Paganini was able to play such pieces because of EDS, exceptionally long fingers, and the associated hypermobility of his finger and wrist joints.

Another collagen disease is *osteogenesis imperfecta (OI)*, or *brittle bone disease*. The colloquial name comes from the fact that collagen-deficient bones fracture very easily. Infants with the most severe forms of OI often die before, during, or shortly after birth, and are born with severely deformed limbs and multiple fractures from birth trauma. Various forms of OI can be characterized by fragile and misplaced teeth, spinal deformity, short stature, barrel chest, and hearing loss. The white capsule of the eyeball often is so thin that internal blood vessels show through and the "whites of the eyes" have a blue-gray to purple tint. Many persons with OI, however, can enjoy a normal life span and high professional achievement. Among them, British actor, playwright, and novelist Nabil Shaban (1953–) has made a fascinating documentary, "The Strangest Viking" (see YouTube), about speculation that the dwarfish but fearsome Viking known as Ivar the Boneless had OI.

Both EDS and OI appear in infancy and early childhood and are sufficiently rare that they are often unrecognized or misdiagnosed. There is an added tragedy stemming from the frequent bruising of EDS and bone fractures of OI: Parents have been falsely accused of child abuse and been taken before child protective agencies and family courts before being exonerated by a correct medical diagnosis.

---

[12] Edvard Ehlers (1863–1937), Danish dermatologist; Henri-Alexandre Danlos (1844–1912), French dermatologist

2.3e Adenosine Triphosphate (ATP)

Adenosine triphosphate (ATP) is a nucleotide built from the nitrogenous base *adenine,* the sugar ribose, and three phosphate groups (fig. 2.19). It functions as the body's most important energy-transfer molecule. It briefly stores energy gained from reactions such as glucose oxidation, and releases it within seconds for physiological work such as synthesizing proteins, muscle contraction, and pumping ions through cell membranes.

The second and third phosphate groups of ATP are attached to the rest of the molecule by high-energy covalent bonds traditionally indicated by a wavy line (P~O) in the structural formula. Because phosphate groups are negatively charged, they repel each other. It requires a high-energy bond to overcome that repulsive force and hold them together—especially to add the third phosphate group to a chain that already has two negatively charged phosphates. Most energy transfers to and from ATP involve adding or removing that third phosphate.

When ATP breaks down, it loses one phosphate group as *inorganic phosphate (Pi)*. The remainder of the molecule becomes adenosine diphosphate (ADP). Most of the energy given off is lost as heat, a major contributor to body temperature. A small but life-sustaining fraction of the energy, however, is used to do many kinds of work in the body.

$$ATP + H_2O \longrightarrow ADP + P_i + Energy \nearrow Heat \searrow Work$$

ATP is a short-lived molecule, usually consumed within 60 seconds of its formation. The entire amount of ATP in the body would support life for less than 1 minute if it were not continually replenished. At a moderate rate of physical activity, a full day's supply of ATP would weigh twice as much as you do. Even if you never got out of bed, you would need about 45 kg (99 lb) of ATP to stay alive for a day. The reason cyanide is so lethal is that it halts ATP synthesis.

Figure 2.19 Adenosine Triphosphate (ATP). ATP transfers energy from one chemical reaction to another. The transferable energy is carried in the last two P~O covalent bonds, indicated by wavy lines.

Before You Go On

Answer these questions from memory to test your understanding of the previous section. Reread the material if you can't answer the questions.

9. Which reaction—dehydration synthesis or hydrolysis—converts a polymer to its monomers? Which one converts monomers to a polymer? Explain your answer.
10. What is the chemical name of blood sugar? What carbohydrate is polymerized to form starch and glycogen?
11. What is the main chemical similarity between carbohydrates and lipids? What are the main differences between them?
12. Do proteins ever change their shape when the body is functioning normally? Explain your answer.
13. Use the lock-and-key analogy to explain why excessively acidic body fluids (acidosis) could destroy enzyme function.
14. How does ATP change structure in the process of releasing energy?

2.4 Energy and Chemical Reactions

Expected Learning Outcomes

When you have completed this section, you should be able to

a. understand how chemical reactions are symbolized by chemical equations;
b. list and define the fundamental types of chemical reactions;
c. define *metabolism* and its two subdivisions; and
d. define *oxidation* and *reduction* and relate these to changes in the energy content of a molecule.

The essence of life is chemical reactions and energy transfers—in short, **metabolism.** We conclude this chapter by examining the basic types of chemical reactions and how they relate to metabolism and energy transfer.

2.4a Forms of Energy

Energy is the capacity to do work—to move matter or change its structure. Biological work is as diverse as digesting starch, moving chemicals through a cell membrane, building a protein, circulating blood, or running a marathon. Energy has two principle forms—potential and kinetic.

Potential energy is stored energy, energy that is not doing work at the moment but that has the potential to be released and do work. The energy stored in chemical bonds, such as the calories in a candy bar, is potential energy. So is energy that exists

Figure 2.20 Decomposition and Synthesis Reactions. (a) In a decomposition reaction, large molecules are broken down into simpler ones. (b) In a synthesis reaction, smaller molecules are joined to form larger ones.

by virtue of the position of matter—for example, the energy stored in a flashlight battery or in water backed up behind a dam, which has the potential to flow through a turbine and generate electricity.

Kinetic energy is the energy of motion or change—for example, the electrical current produced when you turn on a flashlight, the energy of sodium ions flowing through the membrane of a nerve cell, the energy of muscle contraction, and the heat generated by exercising muscles.

Energy is routinely converted from one form to another, often through chemical reactions. In muscle contraction, for example, the chemical (potential) energy of ATP is released by chemical reactions that produce muscular motion and heat. Some energy is lost as heat in every chemical reaction, and then is no longer available to do useful work.

2.4b Chemical Reactions

A **chemical reaction** is a process in which a covalent or ionic bond is formed or broken. The course of a chemical reaction is symbolized by a **chemical equation,** which typically shows the **reactants** on the left, the **products** on the right, and an arrow pointing from the reactants to the products. For example, consider the following example. If you open a bottle of wine and let it stand for several days, it turns sour. Wine "turns to vinegar" because oxygen gets into the bottle and reacts with ethanol to produce acetic acid and water. Acetic acid gives the tart flavor to vinegar and spoiled wine. The equation for this reaction is

$$CH_3CH_2OH + O_2 \longrightarrow CH_3COOH + H_2O$$
$$\text{Ethanol} \quad \text{Oxygen} \quad \text{Acetic acid} \quad \text{Water}$$

Ethanol and oxygen are the reactants, and acetic acid and water are the products. Not all reactions are shown with the arrow pointing from left to right. In complex biochemical equations, reaction chains are often written vertically or even in circles.

Chemical reactions can be classified as decomposition or synthesis reactions. In **decomposition reactions,** a large molecule breaks down into two or more smaller ones; symbolically, AB → A + B. When you eat a potato, for example, digestive enzymes decompose its starch into thousands of glucose molecules (**fig. 2.20a**), and most cells further decompose glucose to water and carbon dioxide. One molecule of starch ultimately yields about 36,000 molecules of H_2O and CO_2.

Synthesis reactions are the opposite—two or more small molecules combine to form a larger one; symbolically, A + B → AB (**fig. 2.20b**). When the body synthesizes proteins, for example, it typically combines 50 or more amino acids into one protein molecule (fig. 2.20b).

Reversible reactions can go in either direction under different circumstances and are represented with paired or double-headed arrows. The direction in which a reversible reaction goes is determined by the relative abundance of substances on each side of the equation. If there is a surplus of CO_2, this reaction proceeds left to right and produces bicarbonate and hydrogen ions. If bicarbonate and hydrogen ions are present in excess, the reaction proceeds right to left and generates CO_2 and H_2O.

$$CO_2 + H_2O \rightleftharpoons H_2CO_3 \rightleftharpoons HCO_3^- + H^+$$
$$\text{Carbon dioxide} \quad \text{Water} \quad \text{Carbonic acid} \quad \text{Bicarbonate ion} \quad \text{Hydrogen ion}$$

This *carbonic acid reaction* appears in this book more often than any other, especially where we discuss respiratory, urinary, and digestive physiology.

2.4c Metabolism, Oxidation, and Reduction

All the chemical reactions in the body are collectively called *metabolism*. Metabolism has two divisions—catabolism and anabolism. **Catabolism**[13] (ca-TAB-oh-lizm) consists of energy-releasing decomposition reactions. Such reactions break covalent bonds, produce smaller molecules from larger ones, and release energy that can be used for other physiological work. Energy-releasing reactions are called *exergonic*[14] reactions. If you break down energy-storage molecules (fats and carbohydrates) to run a race, you get hot. The heat signifies that exergonic reactions are occurring. The earlier-described decomposition of ATP to $ADP + P_i$ is an example of catabolism.

Anabolism[15] (ah-NAB-oh-lizm) consists of energy-storing synthesis reactions, such as the production of protein or fat. Reactions that require an energy input, such as these, are called *endergonic*[16] reactions. Anabolism is driven by the energy that catabolism releases, so endergonic and exergonic processes, anabolism and catabolism, are inseparably linked.

Oxidation is any chemical reaction in which a molecule gives up electrons and releases energy. A molecule is *oxidized* by this process, and whatever molecule takes the electrons from it is an **oxidizing agent (electron acceptor)**. The term *oxidation* stems from the fact that the electron acceptor is often oxygen. The rusting of iron, for example, is a slow oxidation process in which oxygen is added to iron to form iron oxide (Fe_2O_3). Many oxidation reactions, however, do not involve oxygen at all. For example, when yeast ferments glucose to alcohol, no oxygen is required; indeed, the alcohol *contains less oxygen* than the sugar originally did, but it is *more oxidized* and contains less energy than the sugar:

$$C_6H_{12}O_6 \longrightarrow 2\ CH_3CH_2OH + 2\ CO_2$$
$$\text{Glucose} \qquad \text{Ethanol} \qquad \text{Carbon dioxide}$$

Reduction is a chemical reaction in which a molecule gains electrons and energy. When a molecule accepts electrons, it is said to be *reduced;* a molecule that donates electrons to another is therefore called a **reducing agent (electron donor)**. The oxidation of one molecule is always accompanied by the reduction of another.

Table 2.7 summarizes the variety of energy-transfer chemical reactions in the body.

---

[13]*cata* = down, to break down
[14]*ex, exo* = out; *erg* = work
[15]*ana* = up, to build up
[16]*end* = in; *erg* = work

Table 2.7 Energy-Transfer Reactions in the Human Body

Exergonic Reactions	Reactions in which there is a net release of energy. The products have less total energy than the reactants did.
Oxidation	An exergonic reaction in which electrons are removed from a reactant. The product is then said to be oxidized.
Decomposition	A reaction such as digestion or ATP hydrolysis, in which larger molecules are broken down into smaller ones.
Catabolism	The sum of all decomposition reactions in the body.
Endergonic Reactions	Reactions in which there is a net input of energy. The products have more total energy than the reactants did.
Reduction	An endergonic reaction in which electrons are donated to a reactant. The product is then said to be reduced.
Synthesis	A reaction such as protein and glycogen synthesis, in which two or more smaller molecules are combined into a larger one.
Anabolism	The sum of all synthesis reactions in the body.

CAREER SPOTLIGHT

Medical Technologist

Will & Deni McIntyre/Science Source

A medical technologist (MT) is one who performs diagnostic tests on body fluids and tissue specimens to provide patient information to physicians and other clinicians. MTs work in hospital laboratories, doctors' offices, independent diagnostic laboratories, and the biotechnology industry. They work with such specimens as blood, urine, cerebrospinal fluid, pericardial fluid, and bone marrow. Using microscopic examination, analytical instruments, and other methods, they perform blood counts, liver function tests, urinalysis, comprehensive metabolic panels, blood and tissue typing, coagulation panels, lipid profiles, bacteriological cultures, and other assays. Most MTs are generalists, but in large institutions they may specialize in such areas as immunology, hematology, or cytogenetics.

Education usually consists of a 4-year undergraduate program that leads to a bachelor's degree in medical technology. Typically, it is a "3 + 1" program with 3 years of classroom instruction and 1 year of clinical rotation; the latter consists of hands-on training in various laboratory departments. Such programs are taught in universities and hospitals. There are multiple national certifying organizations (the American Society for Clinical Pathology and others), but states have their own certification requirements and some states require a state license in addition to the national one. Graduate education is available for career advancement.

For further information on medical technology programs and careers, see appendix B.

Apply What You Know

When sodium chloride is formed, an electron transfers from a sodium atom to a chlorine atom (see fig. 2.2). Therefore, which element—sodium or chlorine—is oxidized? Which one is reduced? Explain your answers.

Before You Go On

Answer these questions from memory to test your understanding of the previous section. Reread the material if you can't answer the questions.

15. Other than the examples already given in this chapter, give an example of potential energy and how it might be transformed to kinetic energy.
16. Distinguish between a decomposition reaction and a synthesis reaction, and state one role played by each type in the human body.
17. Define metabolism, catabolism, *and* anabolism.
18. What does oxidation *mean?* What does reduction *mean?* Which of them is endergonic and which is exergonic?
19. Suppose you eat a hamburger and your body digests the starch of the bun to glucose and digests the protein and fat to amino acids and fatty acids. Identify which of the following terms pertain to these digestive processes—catabolism, reduction, oxidation, anabolism, endergonic, exergonic.

Study Guide

Assess Your Learning Outcomes

To test your knowledge, discuss the following topics with a study partner or in writing, ideally from memory.

2.1 Atoms, Ions, and Molecules

1. The definition of a chemical *element,* and the most common elements in the human body
2. The significance of minerals in human structure and function
3. How protons, neutrons, and electrons are arranged in the structure of an atom, and the special significance of valence electrons
4. How chemical isotopes differ from each other, the relationship of radioactivity to certain isotopes, and the medical uses and risks of ionizing radiation
5. The differences between an atom and an ion, and between a cation and an anion
6. The definition of *electrolytes,* their physiological roles, and the most significant electrolytes in human physiology
7. The definition of *free radicals,* their relevance to human physiology, and antioxidants as protection from free radical damage
8. Distinctions between atoms, molecules, and compounds
9. How isomers resemble and differ from each other
10. Three types of chemical bonds and the differences between them
11. Why polar covalent bonds are necessary for the existence of hydrogen bonds, and the role of hydrogen bonds in molecular structure

2.2 Water, Acids, and Bases

1. The properties of water that make it so important to life, and how these arise from the polarity of its covalent bonds
2. The concepts of hydrophilic and hydrophobic substances
3. How *calorie* is defined in relation to water
4. The meanings of *acid* and *base*
5. The pH scale and how acidity and alkalinity are related to their respective numerical ranges on the scale
6. Why it is important that pH be maintained within a relatively narrow range in the body

2.3 Organic Compounds

1. The definitions of *molecules, organic molecules,* and *biochemistry*
2. The meanings of *monomer* and *polymer,* how they relate to each other, and the roles of dehydration synthesis and hydrolysis in converting one to the other
3. The defining characteristics and functions of carbohydrates
4. The most relevant monosaccharides, disaccharides, and polysaccharides, and their dietary sources and functions
5. The defining characteristics and functions of lipids
6. The distinction between saturated and unsaturated fatty acids
7. The structure of triglycerides and phospholipids; the similarity and difference between the two; and their respective functions
8. The structure and functions of steroids
9. The characteristics that all amino acids have in common, and how amino acids differ from each other
10. The relationships between amino acids, peptides, polypeptides, and proteins
11. The various three-dimensional shapes that proteins assume after their amino acid sequence is assembled
12. How the shape of a protein is related to its function and to denaturation
13. What enzymes are and their universal role in cellular function
14. The lock-and-key analogy for enzyme–substrate interaction
15. The reason enzymes are so sensitive to variations in temperature and pH
16. The characteristics that nucleotides have in common and how nucleotides differ from each other
17. The double-helix structure and complementary base pairing of DNA
18. How RNA differs structurally and functionally from DNA
19. Three major components of adenosine triphosphate (ATP) and what role ATP plays in cellular metabolism

2.4 Energy and Chemical Reactions

1. The definition of *energy* and the distinction between potential and kinetic energy
2. How chemical reactions are represented; the meanings of *reactant* and *product*
3. Reversible reactions and chemical equilibrium
4. The difference between catabolism and anabolism, and an example of each
5. The essence of oxidation and reduction, how they relate to catabolism and anabolism, and how they affect the energy content of a molecule

Testing Your Recall

1. A substance that _____ is considered to be a chemical compound.
   a. contains at least two different elements
   b. contains at least two atoms
   c. has covalent bonds
   d. has any type of chemical bond
   e. has a stable valence shell

2. An ionic bond is formed when
   a. two anions meet.
   b. two cations meet.
   c. an anion interacts with a cation.
   d. electrons are unequally shared between nuclei.
   e. electrons transfer completely from one atom to another.

3. The ionization of a sodium atom to produce Na^+ is an example of
   a. oxidation.
   b. reduction.
   c. catabolism.
   d. anabolism.
   e. decomposition.

4. The weakest chemical bonds, easily disrupted by temperature and pH changes, are
   a. polar covalent bonds.
   b. nonpolar covalent bonds.
   c. hydrogen bonds.
   d. ionic bonds.
   e. disulfide bonds.

5. A substance capable of dissolving freely in water is
   a. hydrophilic.
   b. hydrophobic.
   c. hydrolyzed.
   d. hydrated.
   e. amphipathic.

6. A carboxyl group is symbolized
   a. —OH.
   b. —NH_2.
   c. —CH_3.
   d. —CH_2OH.
   e. —COOH.

7. The only polysaccharide synthesized in the human body is
   a. cellulose.
   b. glycogen.
   c. cholesterol.
   d. starch.
   e. triglyceride.

8. Which is the most abundant type of protein in the body?
   a. keratin
   b. antibodies
   c. collagen
   d. pepsin
   e. hemoglobin

9. Which of the following functions is more characteristic of carbohydrates than of proteins?
   a. contraction
   b. energy storage
   c. catalyzing reactions
   d. immune defense
   e. intercellular communication

10. The feature that most distinguishes a lipid from a carbohydrate is that a lipid has
    a. more phosphate.
    b. more sulfur.
    c. a lower ratio of carbon to oxygen.
    d. a lower ratio of oxygen to hydrogen.
    e. a greater molecular weight.

11. When an atom gives up an electron and acquires a positive charge, it becomes a/an _____.

12. Dietary antioxidants are important because they neutralize _____.

13. Any substance that increases the rate of a reaction without being consumed by it is a/an _____. In the human body, _____ serve this function.

14. All the synthesis reactions in the body form a division of metabolism called _____.

15. A chemical reaction that produces water as a by-product is called _____.

16. The suffix _____ denotes a sugar, whereas the suffix _____ denotes an enzyme.

17. The amphipathic lipids of cell membranes are called _____.

18. _____ is a nucleotide highly important in transferring energy from one chemical reaction pathway to another.

19. Metabolic reactions that break down large molecules into smaller ones and release energy are called _____.

20. A substance acted upon and changed by an enzyme is called the enzyme's _____.

Answers in Appendix A

What's Wrong with These Statements?

Briefly explain why each of the following statements is false, or reword it to make it true.

1. The monomers of a polysaccharide are called amino acids.
2. Water molecules cling together because they are nonpolar.
3. ATP provides better long-term storage of energy than triglycerides do.
4. Enzymes are often consumed by the reactions they catalyze.
5. Peptide bonds link glucose molecules into long polymers such as starch and glycogen.
6. A saturated fat is defined as a fat to which no more carbon can be added.
7. Catabolism is an energy-storing oxidation process that combines small, low-energy molecules into large, high-energy molecules.
8. Cellulose is an important source of dietary calories.
9. Two isomers have identical chemical properties because their chemical behavior depends on the number and types of atoms present, not on how the atoms are arranged.
10. A solution of pH 8 has 10 times the hydrogen ion concentration of a solution with pH 7.

Answers in Appendix A

Testing Your Comprehension

1. How would the important life-sustaining properties of water change if it had nonpolar covalent bonds instead of polar covalent? Explain.
2. In one form of radioactive decay, a neutron breaks down into a proton and electron and emits a gamma ray. Is this an endergonic or exergonic reaction, or neither? Is it an anabolic or catabolic reaction, or neither? Explain both answers.
3. Some metabolic conditions such as diabetes mellitus cause disturbances in the acid–base balance of the body, which give the body fluids an abnormally low pH. Explain how this could affect the ability of enzymes to control biochemical reactions in the body.

Chapter 3: Cells

Dividing cancer cells from an adenocarcinoma. Adenocarcinoma is a tumor arising from glands in the mucous membrane of an organ such as the lung.
Eye of Science/Science Source

Chapter Outline

3.1 The General Structure of Cells
- 3.1a Cell Components
- 3.1b Cell Sizes

3.2 The Cell Surface
- 3.2a The Plasma Membrane
- 3.2b The Glycocalyx
- 3.2c Cell Surface Extensions
- 3.2d Cell Junctions
- 3.2e Transport Through Cellular Membranes

3.3 The Cell Interior
- 3.3a The Cytoskeleton
- 3.3b Inclusions
- 3.3c Organelles
- 3.3d Protein Synthesis

3.4 The Life Cycle of Cells
- 3.4a The Cell Cycle
- 3.4b Mitosis

Clinical Applications/Perspectives on Health
- Clinical Application 3.1: When Desmosomes Fail
- Clinical Application 3.2: Calcium Channel Blockers
- Perspectives on Health

End of Chapter
- Career Spotlight: Cytotechnologist
- Study Guide

BASE CAMP

Before ascending to the next level, be sure you're properly equipped with a knowledge of these concepts from earlier chapters.

- "Levels of Human Structure" in section 1.3a
- "Phosphoilipids" in section 2.3b
- Protein structure and function ("Proteins" in section 2.3c)
- "Nucleic Acids (DNA and RNA)" in section 2.3d
- "Adenosine Triphosphate (ATP)" in section 2.3e

Anatomy & Physiology Revealed 4.0

Module 2
Cells & Chemistry

All organisms are composed of cells. Most organisms are single cells, like bacteria. Complex organisms like us are composed of trillions of cells that form the varied structures in our body and enable functions as diverse as movement and digestion. In recent decades, technological advances and new methods of study have dramatically increased our understanding of cellular structure and processes, which in turn has led to significant advances in our understanding and treatment of countless diseases. Our study of cell structure and function in this chapter lays the foundation for understanding the rest of this book.

3.1 The General Structure of Cells

Expected Learning Outcomes

When you have completed this section, you should be able to

a. outline and define the major structural components of a cell;
b. name the fluids inside and outside a cell; and
c. state the size range of human cells and explain why cell size is limited.

Cytology,[1] the study of cellular structure and function, got its start in the seventeenth century when inventors Robert Hooke (1635–1703) and Antony van Leeuwenhoek (1632–1723) crafted microscopes adequate for seeing individual cells. Cytology made little further progress, however, until improved optics and tissue-staining techniques were developed in the nineteenth century. Even then, the material between the nucleus and cell surface was thought to be little more than a gelatinous mixture of chemicals and vaguely defined particles. When electron microscopes were developed in the mid-twentieth century, their vastly superior magnification and resolution revealed the inner landscape of cells: a crowded maze of passages, compartments, and fibers. **Figure 3.1** depicts a very generalized cell. In reality, there are about 200 kinds of cells in the human body, with a variety of shapes, sizes, and functions.

3.1a Cell Components

The major components of a cell are as follows:

Plasma membrane
Cytoplasm
    Cytoskeleton
    Organelles
    Inclusions
    Cytosol

The **plasma membrane (cell membrane)** forms a cell's surface boundary. The cell's largest organelle is typically the nucleus. The material between the plasma membrane and the nucleus is the **cytoplasm.**[2] It contains the *cytoskeleton,* a supportive framework of protein filaments and tubules; an abundance of *organelles,* diverse structures that perform various metabolic tasks for the cell; and *inclusions,* which are accumulated cell products such as lipids and pigments or internalized

---

[1]*cyto* = cell; *logy* = study of
[2]*cyto* = cell; *plasm* = formed, molded

Figure 3.1 Structure of a Generalized Cell. The cytoplasm is usually more crowded with organelles than is shown here. The organelles are not all drawn to the same scale. Components of the cytoskeleton are shown in greater detail in figure 3.12. **APR**

foreign matter such as dust and bacteria. The cytoskeleton, organelles, and inclusions are embedded in a clear gel called the **cytosol.**

The cytosol is also called **intracellular fluid (ICF).** All body fluids not contained in the cells are collectively called the **extracellular fluid (ECF).** The ECF located between the cells is also called **tissue fluid.** Extracellular fluids also include blood plasma, lymph, and cerebrospinal fluid.

3.1b Cell Sizes

The most useful unit of measurement for designating cell sizes is the **micrometer (μm)**—one-millionth (10^{-6}) of a meter, one-thousandth (10^{-3}) of a millimeter. The smallest objects most people can see with the naked eye are about 100 μm, which is about one-quarter the size of the period at the end of this sentence. A few human cells fall within this range, such as egg cells and some fat cells, but most human cells are about 10 to 15 μm wide. The longest human cells are nerve cells (sometimes over a meter long) and muscle cells (up to 30 cm long), but these are usually too slender to be seen with the naked eye. With a good light microscope, one can see objects as small as 0.5 μm. Objects smaller than that generally require an electron microscope.

For a variety of reasons, cells cannot attain unlimited size. If a cell grew excessively large, it would rupture like an overfilled water balloon. Also, molecules could not diffuse from place to place fast enough to support its metabolism. The time required for diffusion is proportional to the square of distance, so if cell diameter doubled, the travel time for molecules within the cell would increase fourfold. For example, if it took 10 seconds for a molecule to diffuse from the surface to the center of a cell with a 10 μm radius, then we increased this cell to a radius of 1 mm, it would take 278 hours to reach the center—far too slow to support the cell's life activities. Having organs composed of many small cells instead of fewer large ones has another advantage: The death of one or a few cells is of less consequence to the structure and function of the whole organ.

Before You Go On

Answer these questions from memory to test your understanding of the previous section. Reread the material if you can't answer the questions.

1. Which term refers to all the cell contents between the plasma membrane and nucleus: *cytosol, cytoplasm, tissue fluid,* or *extracellular fluid*?
2. About how big would a cell have to be for you to see it without a microscope? Are any cells actually this big? If so, which ones?
3. Explain why cells cannot grow to an indefinitely large size.

3.2 The Cell Surface

Expected Learning Outcomes

When you have completed this section, you should be able to

a. identify the components of the plasma membrane and state their functions;

b. describe the composition, appearance, and functions of the glycocalyx that coats each cell;

c. describe the structure and functions of microvilli, cilia, flagella, and pseudopods;

d. name the different types of junctions that connect cells to each other, and describe their functions; and

e. explain the processes for moving material through cellular membranes.

A great deal of physiology takes place at the cell surface—for example, the binding of signal molecules such as hormones, the stimulation of cellular activity, the attachment of cells to each other, and the transport of materials into and out of cells. This, then, is where we begin our study of cellular structure and function. Like an explorer discovering a new island, we will examine the interior of the cell only after we have investigated its boundary.

3.2a The Plasma Membrane

The plasma membrane defines the boundary of a cell and governs its interactions with other cells. It controls the passage of materials into and out of the cell and maintains differences in chemical composition between the ECF and ICF. Several organelles are enclosed in membranes that are structurally similar to the plasma membrane, but the term *plasma membrane* refers exclusively to the cell surface.

Membrane Lipids

The plasma membrane is an oily, two-layered lipid film with proteins embedded in it (fig. 3.2). About 75% of the lipid molecules are **phospholipids.** In chapter 2, we saw that phospholipids are *amphipathic*—they have a hydrophilic phosphate head and two hydrophobic fatty acid tails. The heads face the water on both the inside and outside of the cell, thus forming a sandwichlike *phospholipid bilayer*. The tails form the middle of the sandwich, as far away from the surrounding water as possible. The phospholipids are not stationary but highly fluid—drifting laterally from place to place, spinning on their axes, and flexing their tails.

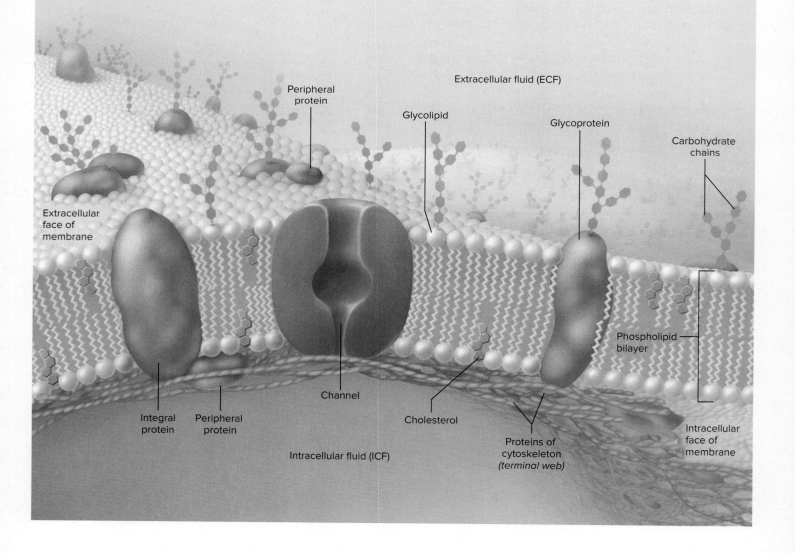

Figure 3.2 The Plasma Membrane. Review the relationship between the yellow phospholipid symbols here and the phospholipid structure in figure 2.12. APR

? Why do all the phospholipid heads face the ECF and ICF and not the middle of the membrane?

About 20% of the lipid molecules are **cholesterol,** which affects membrane fluidity. If there is too much cholesterol, it inhibits the action of enzymes and other proteins in the membrane; too little, and plasma membranes become excessively fragile. This is one of several reasons why cholesterol, in spite of its bad reputation in health science, is indispensable to human survival.

The remaining 5% of the lipids are **glycolipids**—phospholipids with short carbohydrate chains bound to the extracellular surface.

Membrane Proteins

The types of proteins associated with the plasma membrane vary considerably from cell to cell, in contrast to the lipid portion, which has the same basic composition regardless of cell type. Proteins give membranes specific abilities and contribute greatly to the functional differences between cell types. Some proteins adhere only to the inner surface of the plasma membrane, whereas others penetrate all the way through. Most of the latter are **glycoproteins,** which, like glycolipids, have carbohydrate chains attached to them. **Channel proteins** have pores that selectively allow certain solutes to enter or leave the cell; some of these *channel proteins* act as *gates* that open or close to allow material to pass through only at specific times (**fig. 3.3**). Gates are important in the timing of the heart's pacemaker, conduction of nerve signals, and contraction of muscles. Other proteins actively bind to chemicals in the ECF or ICF, such as glucose, amino acids, and inorganic ions, and transport them through the membrane. Some membrane proteins act as *receptors* for hormones and other chemical messengers from other cells. Still others are enzymes that carry out chemical reactions at the cell surface, adhesion molecules that bind cells together in a tissue, and cell-identity markers that enable the immune system to distinguish our own cells from invaders that should be attacked. The functions of membrane proteins are highly diverse and are among the most interesting aspects of cell physiology.

Figure 3.3 Some Functions of Membrane Proteins.

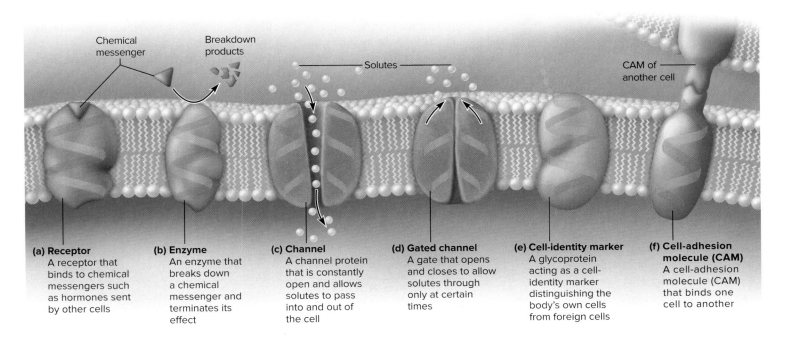

(a) **Receptor**
A receptor that binds to chemical messengers such as hormones sent by other cells

(b) **Enzyme**
An enzyme that breaks down a chemical messenger and terminates its effect

(c) **Channel**
A channel protein that is constantly open and allows solutes to pass into and out of the cell

(d) **Gated channel**
A gate that opens and closes to allow solutes through only at certain times

(e) **Cell-identity marker**
A glycoprotein acting as a cell-identity marker distinguishing the body's own cells from foreign cells

(f) **Cell-adhesion molecule (CAM)**
A cell-adhesion molecule (CAM) that binds one cell to another

3.2b The Glycocalyx

All cells are covered with a fuzzy carbohydrate coat called the **glycocalyx**[3] (GLY-co-CAY-licks) (fig. 3.4). It consists of short chains of sugars belonging to the glycolipids and glycoproteins. Essentially, every one of our cells is "sugar-coated." The glycocalyx has multiple functions. It includes the cell-adhesion molecules that bind cells together and prevent tissues from falling apart. It also cushions the plasma membrane and protects it from physical and chemical injury, somewhat like the styrofoam "peanuts" in a shipping carton. It is chemically unique in everyone except identical twins, and acts like an identification tag that enables the body to distinguish its own healthy cells from diseased cells, invading organisms, and transplanted tissues. Human blood types and transfusion compatibility are determined by the glycocalyx.

3.2c Cell Surface Extensions

Most cells have surface extensions of one or more types called *microvilli, cilia, flagella,* and *pseudopods.* These aid in absorption, movement, sensory processes, and phagocytosis.

Microvilli

Microvilli[4] (MY-cro-VIL-eye; singular, *microvillus*) are extensions of the plasma membrane, about 1 to 2 μm long, that serve primarily to increase its surface area (fig. 3.4). They are best developed in cells specialized for absorption, such as the epithelial cells of the small intestine and kidney tubules. Microvilli give a cell up to 40 times more absorptive surface area. On cells of the taste buds and inner ear, they are well developed but less numerous, and serve sensory roles instead of absorptive functions.

Cilia

Cilia[5] (SIL-ee-uh; singular, *cilium*) are hairlike processes about 7 to 10 μm long. They are supported by a core of microtubules that, in cross section, looks a bit like a Ferris wheel. Each pair of microtubules is equipped with little motor proteins that produce the beating motion of the cilium. Figure 3.5 shows this arrangement and contrasts the size of cilia with that of microvilli.

Cells of the respiratory tract and uterine (fallopian) tubes typically have about 50 to 200 cilia each (fig. 3.5a). They beat in waves that sweep across the surface of an epithelium. In the respiratory tract, they move mucus from the lungs and trachea up to the throat, where it is swallowed. In the uterine tubes, they move an egg or embryo toward the uterus, like people in a stadium passing a beach ball overhead from hand to hand.

Nearly every cell of the body also has a nonmotile, usually solitary *primary cilium,* which often serves as the cell's "antenna" for monitoring nearby conditions. In the inner ear, they play roles in the senses of motion and balance; in the retina of the eye, they are highly elaborate and form the light-absorbing part of sensory cells; and in kidney tubules, they monitor fluid flow. Cilia on the sensory cells of the nose bind odor molecules.

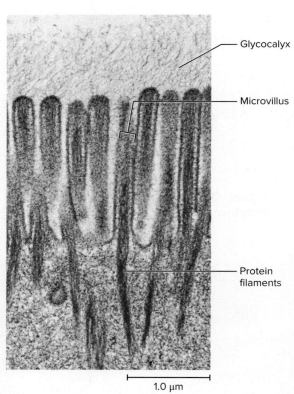

Figure 3.4 The Glycocalyx and Brush Border. The microvilli are anchored by protein filaments, which occupy the core of each microvillus and project into the cytoplasm. The glycocalyx is composed of short carbohydrate chains bound to the membrane phospholipids and proteins.

Don W. Fawcett/Science Source

❓ *Name an organ of the body from which this cell might have come. Explain your answer.*

---

[3]*glyco* = sugar; *calyx* = cup, vessel
[4]*micro* = small; *villi* = hairs
[5]*cilia* = hairs

Figure 3.5 The Structure of Cilia. (a) Epithelium of the uterine (fallopian) tube (SEM). The short, mucus-secreting cells between the ciliated cells show bumpy microvilli on their surfaces. (b) Cross section of a few cilia and microvilli (TEM). Note the relative sizes of cilia and microvilli.
a: SPL/Science Source b: Don Fawcett/Science Source

Flagella

A **flagellum**[6] (fla-JEL-um) resembles a long solitary cilium. The only functional flagellum in humans is the whiplike tail of a sperm cell, measuring about 50 μm long (see fig. 19.13). It has a core of microtubules like a cilium, but it is stiffened by a sheath of filaments in a space between those and the plasma membrane. Sperm use the tail to crawl like a snake up the mucous membrane of the uterus and uterine tubes in their quest to find an egg.

---

[6]*flagellum* = whip

> **Apply What You Know**
> Kartagener syndrome is a hereditary disease in which cilia and flagella lack the motor protein and therefore cannot move. How do you think Kartagener syndrome will affect a man's ability to father a child? How might it affect his respiratory health? Would it affect a woman's fertility? Explain your answers.

Pseudopods

Pseudopods[7] are cytoplasm-filled extensions of the cell ranging from filamentous to blunt fingerlike processes (fig. 3.6). They change shape continually; some form anew as the cell surface bubbles outward and cytoplasm flows into it, usually at the leading end of a moving cell, while others at the "tail" end are retracted into the cell. Some white blood cells use pseudopods to crawl about in the tissues, and when they encounter a bacterium or other foreign particle, they reach out with pseudopods to surround and engulf it. Tissue cells called *macrophages* snare bacteria with thin filamentous pseudopods and "reel them in" to be digested by the cell. Like little janitors, macrophages and white blood cells keep our tissues cleaned up and defend us against bacteria and other invaders.

3.2d Cell Junctions

In complex multicellular organisms, no cell is independent. Most cells are linked to one another and to the extracellular environment by **cellular junctions** composed of proteins. These attachments enable cells to grow and divide normally, resist stress, and communicate with each other. The three principal types of connections between cells are tight junctions, desmosomes, and gap junctions (fig. 3.7).

A **tight junction** completely encircles an epithelial cell near its upper end and joins it securely to adjacent cells, somewhat like the plastic harness on a six-pack of soda cans. The junctions are formed by the fusion of the plasma membranes of neighboring cells, and seal off the intercellular space. Substances are restricted from passing *between* the epithelial cells, but rather must pass *through* them, ensuring that the epithelial cells chemically process materials that travel across. In the stomach and intestines, for example, tight junctions prevent digestive juices from leaking between epithelial cells and digesting the underlying tissue, and they help prevent bacteria from invading the tissues.

A **desmosome**[8] (DEZ-mo-some) is a patch of interlocking proteins that holds cells together at a specific point somewhat like a snap on a pair of jeans. Desmosomes keep cells from pulling apart and thus enable a tissue to resist mechanical stress. Desmosomes are common in such tissues as the epidermis and cardiac muscle. Without them, your skin would peel off when you rubbed it and a heartbeat would pull your cardiac muscle cells apart.

Figure 3.6 Pseudopods. (a) *Amoeba,* a freshwater organism that crawls and captures food by means of pseudopods. (b) A neutrophil (white blood cell) that uses pseudopods for locomotion and capturing bacteria. (c) A macrophage extending slender pseudopods to snare and "reel in" bacteria.

---

[7]*pseudo* = false; *pod* = foot
[8]*desmo* = band, bond, ligament; *som* = body

Apply What You Know

Why wouldn't desmosomes be suitable as the sole type of intercellular junctions between epithelial cells of the stomach?

A **gap junction** is formed by a ring of six proteins arranged somewhat like the segments of an orange, surrounding a water-filled channel. Ions, glucose, amino acids, and other small solutes can diffuse through the channel directly from the cytoplasm of one cell into the next. In the human embryo, nutrients pass from cell to cell through gap junctions until the circulatory system forms and takes over the role of nutrient distribution. In cardiac muscle, gap junctions allow electrical excitation to pass directly from cell to cell so that the cells contract in near unison.

3.2e Transport Through Cellular Membranes

The human body is compartmentalized by various cellular membranes—the *epithelia* (sheets of cells) that line the blood vessels, digestive and respiratory tracts, and other surfaces; the *plasma membranes* that form the outer boundaries of all cells; and *intracellular membranes* of the various organelles detailed later. These membranes cannot be impenetrable, like brick walls, but must allow chemicals to pass through. Otherwise, nutrients and oxygen could never leave the bloodstream to reach the tissues, cells could not take up needed substances or get rid of wastes, and organelles could not exchange materials with the rest of the cytoplasm. These membranes are all **selectively permeable,** allowing some substances to pass through while excluding others. Here we examine nine methods by which materials move through the body's various membranes.

Figure 3.7 Cell Junctions.

❓ *Which type of junction allows materials to pass directly from cell to cell?*

Clinical Application 3.1

WHEN DESMOSOMES FAIL

We often get our best insights into the importance of a structure from the dysfunctions that occur when it breaks down. Desmosomes are destroyed in a disease called *pemphigus vulgaris*[9] (PEM-fih-gus vul-GAIR-iss), in which the immune system launches a misguided attack on the desmosome proteins, especially in the skin. The resulting breakdown of desmosomes between the cells leads to widespread blistering of the skin and oral mucosa, loss of tissue fluid, and sometimes death. The condition can be controlled with drugs that suppress the immune system, but such drugs compromise the body's ability to fight off infections.

---

[9]*pemphigus* = blistering; *vulgaris* = common

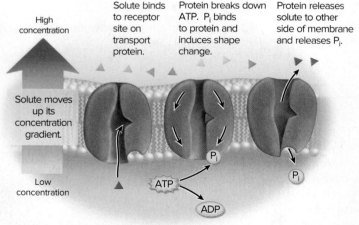

Figure 3.8 Some Modes of Membrane Transport. APR

? *Which of these processes can also occur in synthetic membranes such as cellophane or dialysis tubing?*

Filtration

Filtration is a process in which a physical pressure drives fluid through a selectively permeable barrier that allows water and small particles to pass through but prevents the passage of large particles. An everyday example is the process of brewing coffee, where water pressure (weight of the water) drives water and dissolved matter through the paper filter while holding the coffee grounds back. The most important example in the human body is blood pressure driving water and small solutes through gaps in the walls of the capillaries, while holding blood cells and large molecules such as proteins in the bloodstream. Filtration is the mechanism by which salts, nutrients, and other blood solutes are delivered to the cells of surrounding tissues, and how the kidneys selectively filter wastes from the blood.

Simple Diffusion

Simple diffusion (fig. 3.8a) is the net movement of particles from a place of high concentration to a place of low concentration—in other words, *down a concentration gradient*. This is how oxygen and steroid hormones enter cells and carbon dioxide leaves them, for example. The cell doesn't expend energy to achieve this; all molecules are in spontaneous random motion, and this alone provides the energy for their diffusion through space. Molecules can diffuse through both living membranes (the plasma membrane) and nonliving ones (such as dialysis tubing and cellophane) if the membrane has large enough gaps or pores. Nonpolar solutes such as oxygen and carbon dioxide, and hydrophobic substances such as steroids, diffuse through the lipid regions of the plasma membrane; hydrophilic solutes such as salts, however, can diffuse only through the water-filled protein channels of the membrane.

Osmosis

Osmosis[10] (oz-MO-sis) is the net movement of water through a selectively permeable membrane from the side where there is a relatively low concentration of solutes (the "more watery" side) to the side where there is a higher solute concentration (the "less watery" side) (fig. 3.9). The reason it occurs is that when water molecules move to the high-solute side, they tend to loosely cling to the solute molecules. Becoming less mobile, they are not as able to break free and cross back to the low-solute side. In essence, solute particles on one side of the membrane draw water away from the other side, and water accumulates on the side with the most solute. Thus, if the fluids on two sides of a cell membrane differ in the concentration of dissolved matter (and these solutes cannot penetrate the membrane), water tends to move by osmosis from the more dilute to the less dilute side. Osmosis plays a key role in many aspects of homeostasis. For example, blood capillaries absorb fluid from the tissues by osmosis, thereby removing metabolic wastes from the tissues and preventing them from swelling.

---

[10]*osm* = push, thrust; *osis* = condition, process

Osmotic balance is crucial to the survival and function of living cells. If a cell is in balance with the surrounding fluid, it gains and loses water at equal rates and maintains a stable volume and internal pressure. If there is an osmotic imbalance, however, the cell may shrivel and die from loss of water or swell with excess water and potentially burst. Either of these can be fatal to the person, as seen in the many cases of people who have died of dehydration or, conversely, because they lost water and electrolytes through urine and sweat and tried to replace their losses by drinking large volumes of plain water.

The ability of a solution to affect intracellular pressure and volume is called **tonicity.** It is determined by the concentration of solutes, inside the cell and out, that cannot pass through the plasma membrane *(nonpermeating solutes)* but drive the cell's gain or loss of water. If the intracellular and extracellular fluids (ICF and ECF) have the same tonicity, they are said to be **isotonic**[11] with each other. In an isotonic solution, cells gain and lose water at equal rates and cell volume remains constant (fig. 3.10b). The saline sometimes given to patients in intravenous fluid therapy, 0.9% sodium chloride (NaCl), is isotonic saline. But if the ECF has a higher concentration of nonpermeating solutes than the ICF, such as 2% NaCl, it is **hypertonic**[12] to the cells and can cause cellular damage and death by drawing water out of them (fig. 3.10c). However, hypertonic intravenous fluid is sometimes given deliberately to patients to reduce swelling of the brain (cerebral edema) by drawing excess water out of the cells. If the ECF has a lower concentration of nonpermeating solutes than the ICF, it is **hypotonic**[13] to the cells and causes cellular swelling and potentially ruptures them (fig. 3.10a).

Apply What You Know

In dehydration, the relative solute concentration of the extracellular fluid (ECF) increases. How does this affect the volume (sizes) of cells? Explain.

[11]*iso* = equal; *ton* = tension
[12]*hyper* = more; *ton* = tension
[13]*hypo* = less; *ton* = tension

Figure 3.9 Osmosis. The dashed line represents a selectively permeable membrane dividing the chamber in half. The large particles on side B represent any solute, such as protein, too large to pass through the membrane. The small particles are water molecules. Water passes predominantly from side A to side B and aggregates around the solute particles. This inhibits the water from moving back to side A. Fluid level rises in side B and falls in side A.

(a) Hypotonic (b) Isotonic (c) Hypertonic

Figure 3.10 Effects of Tonicity on Red Blood Cells (RBCs). (a) An RBC in a hypotonic medium, such as distilled water, absorbs more water from the medium than it loses to it and therefore swells and may rupture. (b) An RBC in an isotonic medium gains and loses water at equal rates and therefore maintains its normal shape and volume. (c) An RBC in a hypertonic medium loses more water than it gains and therefore shrivels. Cells in situations (a) and (c) are likely to die.
a–c: David M. Phillips/Science Source

Clinical Application 3.2
CALCIUM CHANNEL BLOCKERS

Membrane channels may seem only an abstract concept until we see how they relate to disease and drug design. For example, drugs called *calcium channel blockers* are often used to treat high blood pressure (hypertension). How do they work? The walls of the arteries contain smooth muscle that constricts to narrow the vessels and raise blood pressure, or relaxes to let them widen and reduce blood pressure. Excessive, widespread vasoconstriction (vessel narrowing) can cause hypertension, so one approach to the treatment of hypertension is to inhibit vasoconstriction. In order to constrict, smooth muscle cells open calcium channels in the plasma membrane. The inflow of calcium activates the proteins of muscle contraction. Calcium channel blockers act, as their name says, by preventing calcium channels from opening and thereby preventing constriction.

Facilitated Diffusion

The next two processes, facilitated diffusion and active transport, are called *carrier-mediated transport* because they employ carrier proteins in the plasma membrane. **Facilitated**[14] **diffusion** (see fig. 3.8b) is the movement of a solute through a cellular membrane, down its concentration gradient, with the aid of a carrier. It does not involve any ATP (energy) expenditure by the cell. The carrier binds a particle on the side of a membrane where the solute is more concentrated, then releases it on the side where it is less concentrated. The carrier transports solutes that otherwise couldn't pass through the membrane or would pass less efficiently. One use of facilitated diffusion is to remove sugar (glucose) from the blood and transport it into cells. The process is very rapid, with one protein carrier transporting as many as 1,000 glucose molecules per second across the plasma membrane.

Active Transport

Active transport (see fig. 3.8c) is a process that employs a carrier protein and uses energy from ATP to move a solute through the membrane *up its concentration gradient*—that is, from the side where it is less concentrated to the side where it is already more concentrated. Active transport requires ATP because moving particles up a gradient requires an energy input, like pedaling a bicycle uphill. If a cell stops producing ATP, owing to cell death or poisoning, active transport ceases immediately.

An especially important active-transport process is the **sodium–potassium (Na^+–K^+) pump.** Sodium is normally much more concentrated in the ECF than in the ICF, and potassium is more so in the ICF. Yet cells continually pump more Na^+ out and more K^+ into the cell. The Na^+–K^+ pump binds three sodium ions from the ICF and ejects them from the cell, then binds two potassium ions from the ECF and releases these into the cell. It repeats the process over and over, using one ATP molecule for each cycle. The Na^+–K^+ pump plays roles in controlling cell volume (water follows Na^+ by osmosis); generating body heat (ATP consumption releases heat); maintaining the electrical excitability of your nerves, muscles, and heart; and providing energy for other transport pumps to draw upon in moving such solutes as glucose through the plasma membrane. About half of the calories that you "burn" every day are used just to operate your Na^+–K^+ pumps.

Vesicular Transport

All of the processes discussed up to this point move molecules or ions individually through the plasma membrane. In **vesicular transport,** however, cells move larger

---
[14]*facil* = easy

particles or droplets of fluid through the membrane in bubblelike *vesicles*. Like active transport, all forms of vesicular transport require ATP. **Figure 3.11** illustrates the principal modes of vesicular transport.

All vesicular processes that bring matter into a cell are called **endocytosis**[15] (EN-doe-sy-TOE-sis). There are three forms of endocytosis: *phagocytosis, pinocytosis,* and *receptor-mediated endocytosis*.

---

[15]*endo* = into; *cyt* = cell; *osis* = process

(a) Phagocytosis

Figure 3.11 Modes of Vesicular Transport. (a) Phagocytosis. A white blood cell engulfing bacteria with its pseudopods. (b) Pinocytosis. A cell taking in droplets of extracellular fluid. (c) Receptor-mediated endocytosis. The Ys in the plasma membrane represent membrane receptors. The receptors bind a solute in the extracellular fluid, then cluster together. The membrane caves in at that point until a vesicle pinches off into the cytoplasm bearing the receptors and bound solute. (d) Exocytosis. A cell releasing a secretion or waste product. **APR**

(b) Pinocytosis

(c) Receptor-mediated endocytosis

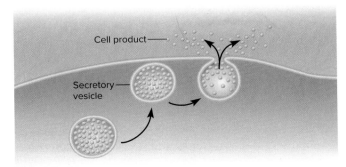

(d) Exocytosis

In **phagocytosis**[16] (FAG-oh-sy-TOE-sis), or "cell eating," a cell reaches out with its pseudopods and surrounds a particle such as a bacterium or a bit of cell debris. It engulfs the particle and enzymatically degrades it (fig. 3.11a). Phagocytosis is carried out especially by white blood cells and *macrophages.* Some macrophages consume as much as 25% of their own volume in material per hour, thus living up to their name[17] and playing a vital role in cleaning up the tissues.

Pinocytosis[18] (PIN-oh-sy-TOE-sis), or "cell drinking," occurs in all human cells. In this process, dimples form in the plasma membrane and progressively cave in until they pinch off as vesicles containing droplets of extracellular fluid (ECF) (fig. 3.11b). One use of pinocytosis is seen in kidney tubule cells; they use this method to reclaim the small amount of protein that filters out of the blood, thus preventing protein from being lost in the urine.

Receptor-mediated endocytosis (fig. 3.11c) is more selective. It enables a cell to take in specific molecules from the ECF with a minimum of unnecessary fluid. Chemical messengers in the ECF bind to specific receptor proteins on the plasma membrane. The receptors then cluster together and the membrane sinks in at this point, creating a pit. The pit soon pinches off to form a vesicle in the cytoplasm. One use of receptor-mediated endocytosis is the absorption of insulin from the blood. Hepatitis, polio, and AIDS viruses "trick" cells into admitting them by receptor-mediated endocytosis.

Exocytosis[19] (fig. 3.11d) is the process of discharging material from a cell by vesicular transport. It is used, for example, by digestive glands to secrete enzymes, by breast cells to secrete milk, and by sperm cells to release enzymes for penetrating an egg. It resembles endocytosis in reverse. A secretory vesicle in the cell migrates to the surface and fuses with the plasma membrane. A pore opens up that releases the product from the cell, and the empty vesicle usually becomes part of the plasma membrane. In addition to releasing cell products, exocytosis is the cell's way of replacing the bits of membrane removed by endocytosis.

Before You Go On

Answer these questions from memory to test your understanding of the previous section. Reread the material if you can't answer the questions.

4. What property of phospholipid molecules causes them to organize themselves into a bilayer?
5. Which of these is important in determining blood transfusion compatibility: cell-adhesion molecules, membrane carriers, membrane cholesterol, the glycocalyx, or microvilli? Explain.
6. Compare and contrast microvilli and cilia in terms of their structure, function, and general location.
7. Which type of cell junction best serves to keep food from scraping away the lining of your oral cavity? Which type enables a cardiac muscle cell to electrically stimulate the neighboring cell? Which type best serves to keep your digestive enzymes from eroding the tissues beneath your intestinal lining?
8. Which membrane transport processes get all the necessary energy from the spontaneous movement of molecules? Which ones require ATP as a source of energy? Which processes are carrier-mediated? Which ones are not?

---

[16]*phago* = eating; *cyt* = cell; *osis* = process
[17]*macro* = big; *phage* = eater
[18]*pino* = drinking; *cyt* = cell; *osis* = process
[19]*exo* = out of; *cyt* = cell; *osis* = process

3.3 The Cell Interior

Expected Learning Outcomes

When you have completed this section, you should be able to

a. describe the cytoskeleton, its three components, and its functions;
b. give some examples of cell inclusions and explain how inclusions differ from organelles;
c. list the main organelles of a cell and explain their functions; and
d. describe how cells synthesize, process, package, and secrete proteins.

We now probe more deeply into the cell to study the structures in the cytoplasm. These are classified into three groups—cytoskeleton, inclusions, and organelles—all embedded in the clear, gelatinous cytosol. If we think of a cell as being like an office building, the cytoskeleton would be the steel beams and girders that hold it up and define its shape and size; the plasma membrane and its channels would be the building's exterior walls and doors; and organelles would be the interior rooms that divide the building into compartments with different functions.

3.3a The Cytoskeleton

The **cytoskeleton** is a network of protein filaments and tubules. It often forms a very dense supportive web in the cytoplasm (fig. 3.12). It structurally supports a cell, determines its shape, organizes its contents, and—going beyond our office building analogy—transports substances within the cell and contributes to movements of the cell as a whole. It is connected to proteins of the plasma membrane and they in turn are connected to proteins outside the cell, so there is a strong structural continuity from the cytoplasm to the extracellular material.

The cytoskeleton is composed of *microfilaments, intermediate filaments,* and *microtubules.* **Microfilaments** are about 6 nanometers (nm) thick and are made of the protein *actin*. They form a dense fibrous mesh called the **terminal web** on the internal side of the plasma membrane. The oily plasma membrane is spread out over the terminal web like butter on a slice of bread. It is thought that the membrane would break up into little droplets without this support. Actin is also the supportive protein in the cores of the microvilli (see fig. 3.4) and plays a role in muscle contraction and other cell movements.

Intermediate filaments (8–10 nm in diameter) are thicker and stiffer than microfilaments. They contribute to the strength of the desmosomes and include the tough protein *keratin* that fills the cells of the epidermis and gives strength to the skin.

Microtubules (25 nm in diameter) are hollow cylinders of protein. They hold organelles in place, form bundles that maintain cell shape and rigidity, and act somewhat like monorails to guide organelles and molecules to specific destinations in a cell. They form the cores of cilia and flagella, as well as the *centrioles* and *mitotic spindle* that guide chromosome movement during cell division (described later). Microtubules are not permanent structures, but may assemble and disassemble rapidly as vesicles and organelles move about the cell or as the cell changes shape.

If you think of intermediate filaments as being like stiff uncooked spaghetti, you could, by comparison, think of microfilaments as being like fine angel-hair pasta and microtubules as being like tubular penne pasta.

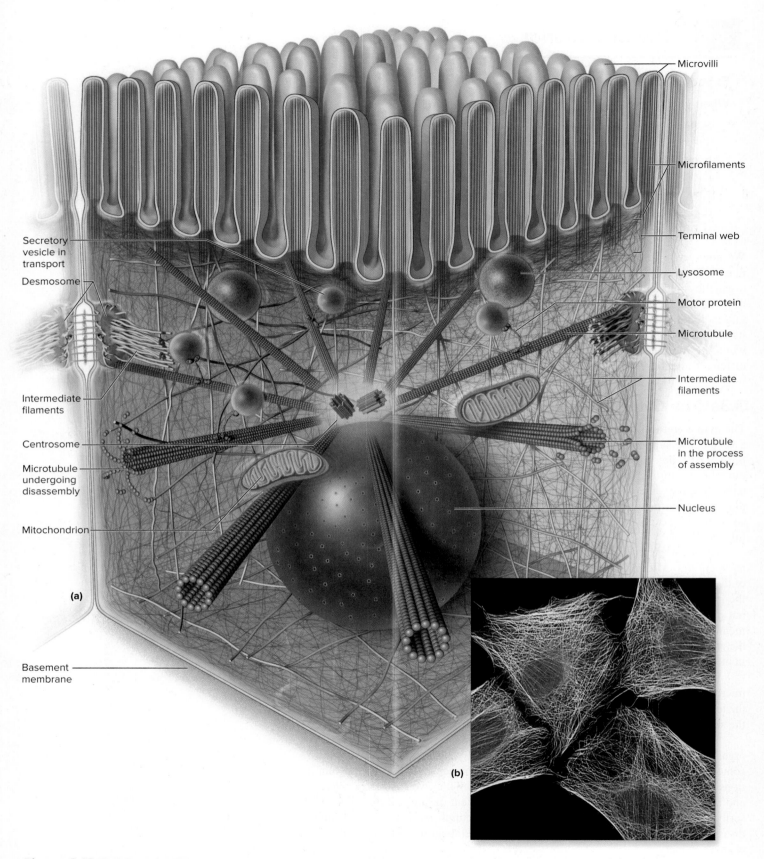

Figure 3.12 The Cytoskeleton. (a) Artist's conception of the cytoskeleton. Most organelles are omitted in order to emphasize the cytoskeletal filaments and microtubules. (b) Cells with their cytoskeleton selectively labeled with a fluorescent dye. These show that the cytoskeleton is actually far denser than shown in part (a).
b: Dr. Torsten Wittmann/Science Source

3.3b Inclusions

Inclusions are of two kinds: stored cellular products such as pigments, fat globules, and glycogen granules; and foreign bodies such as viruses, bacteria, and dust particles or other debris phagocytized by a cell. Inclusions are never enclosed in a membrane, and unlike the organelles and cytoskeleton, they are not essential to cell survival.

3.3c Organelles

Organelles (literally "little organs") are to the cell what organs are to the body—metabolically active structures that play individual roles in the survival of the whole (see fig. 3.1). A cell may have 10 billion protein molecules, some of which are potent enzymes with the potential to destroy the cell if they're not contained and isolated from other cellular components. You can imagine the enormous problem of keeping track of all this material, directing molecules to the correct destinations, and maintaining order against the incessant tendency toward disorder. Cells maintain order partly by compartmentalizing their contents in organelles. **Figure 3.13** shows the most important ones.

The Nucleus

The **nucleus** (fig. 3.13a) is the largest organelle and usually the only one visible with the light microscope. Most cells have only one nucleus, but there are exceptions. Mature red blood cells have none, whereas some cells have dozens of nuclei, such as skeletal muscle cells and certain bone-dissolving cells.

The nucleus is surrounded by a **nuclear envelope** consisting of two parallel membranes. The envelope is perforated with **nuclear pores** formed by a ring-shaped complex of proteins. These proteins regulate molecular traffic into and out of the nucleus and bind the two membranes together.

The nucleus contains the **chromosomes**[20]—threadlike bodies of DNA and protein—and is therefore the genetic control center of cellular activity. In nondividing cells, they are in the form of very fine filaments broadly dispersed throughout the nucleus, visible only with the electron microscope. Collectively, this material is called **chromatin** (CRO-muh-tin). The nucleus also usually exhibits one or more dense masses called **nucleoli** (singular, *nucleolus*), where subunits of the *ribosomes* (described shortly) are made before they are transported out to the cytoplasm.

Endoplasmic Reticulum

The term *endoplasmic reticulum (ER)* literally means "little network within the cytoplasm." The ER is a system of interconnected channels called **cisternae**[21] (sis-TUR-nee) enclosed by a membrane (fig. 3.13e). In areas called **rough endoplasmic reticulum,** the cisternae are flat, parallel, and covered with ribosomes, which give it its rough or granular appearance. The rough ER is continuous with the outer membrane of the nuclear envelope, and adjacent cisternae are connected by transverse bridges. In areas called **smooth endoplasmic reticulum,** the membrane lacks ribosomes, the cisternae are more tubular in shape, and they branch more extensively. The cisternae of the smooth ER are continuous with those of the rough ER, so the two are functionally different parts of the same network.

---

[20]*chromo* = color; *some* = body
[21]*cistern* = reservoir

Figure 3.13 Major Organelles. APR

❓ *Which of these six organelles directly participate in protein synthesis and secretion?*

The endoplasmic reticulum synthesizes steroids and other lipids, detoxifies alcohol and other drugs, and manufactures the cell's internal and surface membranes. The rough ER produces the phospholipids and proteins of the plasma membrane. It also synthesizes proteins that are destined to be secreted from the cell or packaged in organelles called *lysosomes*. Rough ER is most abundant in cells that synthesize large amounts of protein, such as the cells that produce digestive enzymes and antibodies.

Most cells have only scanty smooth ER, but it is relatively abundant in cells that engage extensively in detoxification, such as liver and kidney cells. Long-term abuse of alcohol, barbiturates, and other drugs leads to tolerance partly because the smooth ER proliferates and detoxifies the drugs more quickly. Smooth ER is also abundant in cells that synthesize steroid hormones, for example in the testes and ovaries. Skeletal and cardiac muscle contain extensive networks of smooth ER, which store calcium ions when the muscle is at rest. Upon stimulation, the smooth ER releases calcium to trigger muscle contraction.

Ribosomes

Ribosomes are small granules of protein and **ribonucleic acid (RNA);** they are produced in the nucleus, but most of them are then exported into the cytoplasm and function there. Ribosomes "read" coded genetic messages from the nucleus and assemble amino acids into proteins specified by the code. Many ribosomes are attached to the surface of the nuclear envelope and the rough ER, and make proteins that will either be packaged in lysosomes or, as in digestive enzymes, secreted from the cell. Others lie free in the cytoplasm and make enzymes and other proteins for internal use by the cell. Some occur also within the nucleus and mitochondria, making proteins for internal use by those organelles.

Golgi Complex

The **Golgi**[22] **complex** (GOAL-jee) is a small cluster of cisternae that synthesize carbohydrates and put the finishing touches on protein and glycoprotein synthesis (fig. 3.13d). The complex resembles a stack of pita bread. Typically, it consists of about six cisternae, slightly separated from each other, each of them a flattened, slightly curved sac with swollen edges.

The edges of the Golgi cisternae pinch off membranous sacs called **Golgi vesicles,** like the waxy globules that break away from the main mass in a lava lamp. These are filled with newly synthesized protein. Some of them become lysosomes, the organelles described shortly. Some migrate to the plasma membrane and fuse with it, contributing fresh protein and phospholipid to the membrane. Still others become **secretory vesicles** that store a cell product, such as breast milk, mucus, or digestive enzymes, for later release by exocytosis.

Lysosomes

A **lysosome**[23] (LY-so-some) (fig. 3.13c) is a package of enzymes enclosed in a membrane. These enzymes break down proteins, nucleic acids, carbohydrates, phospholipids, and other substances. When white blood cells called *neutrophils* phagocytize bacteria, a lysosome releases enzymes to digest and destroy the microbes. Lysosomes also digest worn-out mitochondria and other organelles

---

[22]Camillo Golgi (1843–1926), Italian histologist
[23]*lyso* = loosen, dissolve; *some* = body

and contribute enzymes that carry out a sort of "cell suicide" called *apoptosis* (AP-oh-TOE-sis) or *programmed cell death* (see "Tissue Shrinkage and Death" in section 4.6).

Peroxisomes

Peroxisomes (not illustrated) resemble lysosomes, but contain different enzymes and are not produced by the Golgi complex. They are especially abundant in liver and kidney cells. Their primary function is to break down fatty acids into two-carbon molecules that can be used as an energy source for ATP synthesis. They also neutralize free radicals, detoxify alcohol and other drugs, and kill bacteria. They are named for the hydrogen peroxide (H_2O_2) they produce in the course of these reactions.

Mitochondria

Mitochondria[24] (MY-toe-CON-dree-uh) (singular, *mitochondrion*) are organelles specialized for ATP synthesis and are therefore often called the "powerhouses" of the cell. They have a variety of shapes: spheroid, rod-shaped, bean-shaped, or threadlike (fig. 3.13f), and are surprisingly mobile, squirming about like organisms in their own right. Like the nucleus, a mitochondrion is surrounded by a double membrane. The inner membrane usually has folds called **cristae**[25] (CRIS-tee), which project like shelves across the organelle. Cristae bear the enzymes that produce most of the ATP. The space between cristae, called the **mitochondrial matrix,** contains enzymes, ribosomes, and small DNA molecules called *mitochondrial DNA (mtDNA)*.

Mitochondria evolved from free-living bacteria that were internalized by another primitive cell, became permanent residents, and eventually became indispensable to the life of the host cell. Unlike other organelles, mitochondria can reproduce themselves and have their own DNA, separate from nuclear DNA. Mitochondrial DNA is a circular molecule that resembles bacterial DNA more than it does the linear DNA of the cell nucleus. We hear a lot these days about mtDNA in the news and even in television crime dramas, because it is often used to link a suspect to crime-scene evidence. Mutations in mtDNA are responsible for some muscle, heart, and eye diseases.

Centrioles

A **centriole** (SEN-tree-ole) is a short cylindrical assembly of microtubules (fig. 3.13b). Near the nucleus, most cells have a small, clear patch of cytoplasm called the **centrosome**[26] containing a pair of centrioles (see fig. 3.1). These centrioles play a role in cell division described later. For a complete list of the features and functions of the various organelle types, see table 3.1.

3.3d Protein Synthesis

One of the most significant activities of a cell is to produce proteins, and we are now in a position to understand how several of the organelles interact to do this. The proteins of a cell can be for internal structural use, such as proteins of the cytoskeleton and cellular membranes, or they may be enzymes that control the cell's internal metabolism. In many cases, they are secreted from the cell to serve elsewhere, as in the case of digestive enzymes, antibodies, and some hormones such as insulin and growth hormone.

---

[24]*mito* = thread; *chondr* = grain
[25]*crista* = crest
[26]*centro* = central; *some* = body

Table 3.1 Summary of Organelles

Organelle	Structural Features	Function
Nucleus (fig. 3.13a)	Largest organelle in most cells, surrounded by double membrane with nuclear pores	Genetic control center of cell; directs protein synthesis; shelters DNA
Smooth ER (fig. 3.13e)	Network of membranous tubules with smooth surface (no ribosomes)	Lipid synthesis, detoxification, calcium storage
Rough ER (fig. 3.13e)	Sheets of parallel membranes with ribosomes on outer surface	Protein synthesis and manufacture of cellular membranes
Ribosomes (fig. 3.13e)	Small dark granules on surface of rough ER or free in cytosol	Interpret the genetic code and synthesize polypeptides
Golgi complex (fig. 3.13d)	Several closely packed, membranous cisterns with thick edges, often with many Golgi vesicles nearby	Receives and modifies newly synthesized polypeptides, synthesizes carbohydrates, adds carbohydrates to some proteins; packages cell products into Golgi vesicles
Lysosomes (fig. 3.13c)	Round to irregular sacs near Golgi comlex, often with dark featureless interior but sometimes with protein crystals	Contain enzymes for intracellular digestion and programmed cell death
Peroxisomes (not illustrated)	Similar to lysosomes; often lighter in color	Contain enzymes for detoxification of free radicals, alcohol, and other drugs; break down fatty acids for use in ATP synthesis
Mitochondria (fig. 3.13f)	Rod-shaped, bean-shaped, or round structures enclosed in double membrane with shelflike infoldings called cristae	ATP synthesis
Centrioles (fig. 3.13b)	Short cylindrical bodies composed of microtubules	Form mitotic spindles during cell division

To understand protein synthesis, it may help to start with the big picture. DNA in the nucleus contains codes (genes) for how to construct the needed proteins. Synthesis begins when a small copy of a gene called *messenger RNA (mRNA)* is made. This migrates into the cytoplasm, where ribosomes read the code and assemble amino acids in the right order to make that protein. The step from DNA to mRNA is called *transcription,* and the step from mRNA to protein is called *translation.* The details of these two steps follow.

Transcription

Protein synthesis begins with **transcription** in the cell's nucleus (fig. 3.14, panel ①). At the site of a particular gene, an enzyme unzips the double helix of DNA and exposes the gene's nitrogenous bases. Another enzyme reads these bases (panel ②) and creates a parallel molecule of **messenger RNA (mRNA),** which is more or less a mirror image of the gene. Where the enzyme finds a thymine (T) on the DNA, it

① An enzyme (not shown) uncoils the DNA double helix at the site of the relevant gene, exposing bases of the coding strand.

② **Transcription.** A second enzyme reads bases of the coding strand, draws RNA nucleotides from a free nucleotide pool, and assembles these in the order specified by the gene to make mRNA.

③ **Translation.** A two-part ribosome binds to the mRNA and moves along it, reading its codons. tRNA binds free amino acids and delivers them to the ribosome. The ribosome assembles the amino acids in the order specified by the genetic code in the codons. After delivering its amino acid, the free tRNA leaves the ribosome to repeat the process.

④ The ribosome spools the new protein into the cisterna of the rough ER, where it may be further processed and packaged in transport vesicles. Transport vesicles take it to the Golgi complex for further processing.

Figure 3.14 Protein Synthesis. APR

adds the complementary base adenine (A) to the mRNA. Where it finds a guanine (G) on the DNA, it adds cytosine (C) to the mRNA. Where it finds adenine, however, it adds uracil (U) to the mRNA; remember from "Nucleic Acids (DNA and RNA)" in section 2.3 that RNA does not contain any thymine, but substitutes uracil for this base. Therefore, if the enzyme had read a DNA base sequence of TACC-GTCCA, it would produce a complementary mRNA sequence of AUGGCAGGU. A typical mRNA is several hundred bases long and some are up to 10,000.

The mRNA is "edited" in the nucleus to remove some noncoding segments. The coding segments are spliced together to make a mature, coding mRNA molecule, which is then usually exported from the nucleus, through a nuclear pore, into the cytoplasm. (Some remains in the nucleus and is translated there.)

Translation

The next step of protein synthesis is **translation** (fig. 3.14, panel ③). This is where ribosomes enter the story. They are composed of **ribosomal RNA (rRNA)** and enzymes, forming a complex that can read the mRNA and translate its code into the amino acid sequence of a protein.

The genetic code is in the form of **codons**—three-base segments of mRNA. There are 64 of these, most of which stand for a particular amino acid to be inserted into a protein. For example, the sequence AUG represents methionine, GCA represents alanine, and GGU represents glycine.

Each time the ribosome reads a particular codon, it binds a smaller RNA molecule called **transfer RNA (tRNA)** with a complementary base series called an *anticodon*. For example, if the ribosome reads GGU on the mRNA, it binds a tRNA with anticodon CCA. Each tRNA brings a particular amino acid with it, which the enzymes in the ribosome add to the growing protein chain. The ribosome then reads the next codon, adds the next amino acid, and so on. When it reaches a special base sequence called a *stop codon,* like the period at the end of a sentence, the protein is finished. The ribosome releases the new protein and detaches from the mRNA.

Protein Processing and Secretion

Several organelles are involved in producing, packaging, and secreting proteins (fig. 3.14, panel ④). If a protein is destined to be packaged in a lysosome or secreted from the cell, the ribosome–mRNA complex "docks" on the surface of the rough endoplasmic reticulum and the protein reels into the ER cisterna as it is assembled. In the rough ER, enzymes cut certain amino acid segments from the protein, splice other segments together, and perform other alterations.

The altered protein is then shuffled into **transport vesicles**—bubblelike organelles that bud off the ER and carry the protein to the nearest cisterna of the Golgi complex (fig. 3.15). The Golgi complex sorts these proteins and passes them along from one cisterna to the next. It may further cut and splice them or add other components such as carbohydrate chains. The Golgi complex finally buds off Golgi vesicles, which can remain in the cell as lysosomes or release products from the cell by exocytosis, as described earlier.

Before You Go On

Answer these questions from memory to test your understanding of the previous section. Reread the material if you can't answer the questions.

9. Which cytoskeletal component gives support to the plasma membrane? Which component can be compared to a monorail on which intracellular materials can travel from place to place in the cytoplasm? Which type lends toughness to your skin surface?

10. Which two organelles are surrounded by a double membrane? Which of these has shelflike infoldings of the inner membrane? What is the purpose of that organelle?
11. A red blood cell (RBC) escapes from the circulation into the tissues, where a macrophage finds it and phagocytizes it. What is that RBC to the macrophage at this point—an organelle, an inclusion, or neither? What is the basis for your decision?
12. Name three organelles involved in protein synthesis and describe their respective roles. Explain the difference between transcription and translation and state where in the cell each of these occurs.

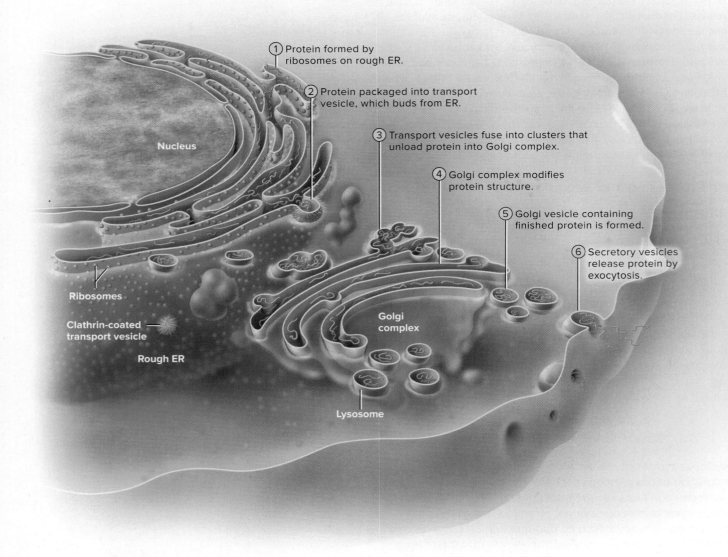

Figure 3.15 Organelles of Protein Synthesis and Secretion. The steps in protein synthesis and secretion are numbered 1 through 6.

3.4 The Life Cycle of Cells

Expected Learning Outcomes

When you have completed this section, you should be able to

a. describe the stages of a cell's life cycle and list the events that define each stage; and
b. name the stages of mitosis and describe what occurs in each.

3.4a The Cell Cycle

A basic principle of cell biology is that all cells arise from existing cells, both in an evolutionary sense, reaching across billions of years to the earliest cell, and in our lifetimes, when we grow from a single-celled fertilized egg to a fetus, infant, and adult. Most cells have a finite life span, during which they must accurately duplicate their DNA and separate the copies into two daughter cells. The life cycle of a cell extends from one division to the next. This **cell cycle** is divided into four main phases: G_1, S, G_2, and M (fig. 3.16).

The **first gap (G_1) phase** is an interval between cell division (the "birth" of two new cells from a parent cell) and DNA replication. During this time, a new cell synthesizes proteins, grows, and carries out its preordained tasks for the body, such as secreting enzymes if it's a digestive gland cell. Cells in G_1 also begin to replicate their centrioles in preparation for the next division, and they accumulate the materials needed in the next phase to replicate their DNA.

The **synthesis (S) phase** is a period in which a cell replicates its DNA, doubling it in preparation for the upcoming cell division. Remember the *law of complementary base pairing* in DNA (see "Nucleic Acids [DNA and RNA]" in section 2.3). If you knew that one strand of the double helix had a base sequence ATCGCA, you could predict from this law that the one across from it must read TAGCGT. This predictability also enables a cell to produce duplicate copies of its DNA. In the S phase, each DNA molecule in the nucleus "unzips" at several places along its length, separating the double helix into two strands (fig. 3.17, panel ①). At each of these sites, an enzyme called *DNA polymerase* (green in panel ②) reads the base sequence on one strand and assembles nucleotides in the right order to make a complementary strand. At each replication site, two of these enzymes work simultaneously, moving along their DNA strands in opposite directions. The end result of this process going on in numerous places at once on the 46 chromosomes is that, by the end of the S phase, the cell has two complete sets of identical DNA molecules. Each DNA molecule is composed of one strand from the old DNA and one that has been newly made by the polymerase (panel ③). Each chromosome, at that point, has two identical DNA molecules, which are available to be divided up between daughter cells at the next cell division.

The **second gap (G_2) phase** is a relatively brief interval between DNA replication and cell division. In G_2, a cell finishes replicating its centrioles and synthesizes enzymes that control cell division.

The **mitotic (M) phase** is the period in which a cell undergoes mitosis—it replicates its nucleus and divides its DNA into two identical sets (one per nucleus). At the end of the M phase, the cell undergoes

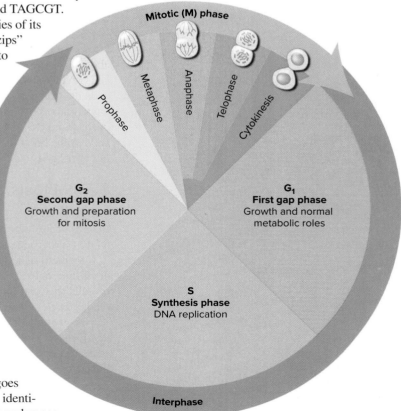

Figure 3.16 The Cell Cycle.

Figure 3.17 DNA Replication. ① Enzyme unzips DNA. ② DNA polymerase assembles nucleotides to make a complementary strand. ③ The process results in two identical DNA molecules. **APR**

cytokinesis[27] (SY-toe-kih-NEE-sis)—it pinches in half to form two genetically identical daughter cells. The details of this phase are considered in the next section. Phases G_1, S, and G_2 are collectively called **interphase**—the time between cell divisions.

The length of the cell cycle varies greatly from one cell type to another. Cultured connective tissue cells called *fibroblasts* divide about once a day and spend about 18 to 24 hours in interphase and 1 to 2 hours in mitosis. Stomach and skin cells divide rapidly, bone and cartilage cells slowly, and skeletal muscle and nerve cells not at all. Some cells leave the cell cycle and stand by without dividing for days, years, or the rest of one's life. The balance between cells that are actively cycling and those on standby is an important factor in determining the number of cells in the body. An inability to stop cycling and go on standby is characteristic of cancer cells (see Perspectives on Health at the end of this chapter).

3.4b Mitosis

Cells divide by two mechanisms called mitosis and meiosis. Meiosis is restricted to one purpose, the production of eggs and sperm, and is therefore treated in chapter 19 on reproduction. **Mitosis** serves all the other functions of cell division: the development of a fertilized egg (one cell) into an individual composed of some 40 trillion cells; continued growth of all the organs after birth; the replacement of cells that die; and the repair of damaged tissues. Four phases of mitosis are recognizable—*prophase, metaphase, anaphase,* and *telophase* (fig. 3.18).

In **prophase,**[28] the chromosomes coil into short, dense rods that are easier to distribute to daughter cells than the long, delicate chromatin of interphase. At this stage, a chromosome consists of two genetically identical bodies called **chromatids,** joined together at a pinched spot called the **centromere** (fig. 3.19). There are

---

[27]*cyto* = cell; *kinesis* = action, motion
[28]*pro* = first

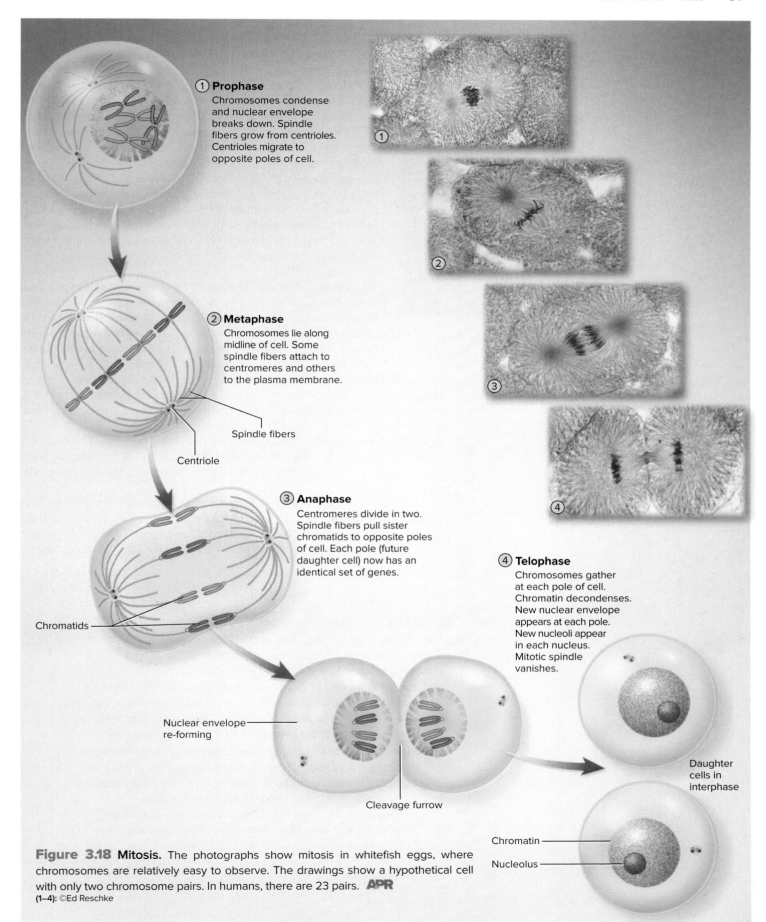

Figure 3.18 Mitosis. The photographs show mitosis in whitefish eggs, where chromosomes are relatively easy to observe. The drawings show a hypothetical cell with only two chromosome pairs. In humans, there are 23 pairs. **APR**
(1–4): ©Ed Reschke

46 chromosomes, two chromatids per chromosome, and one molecule of DNA in each chromatid—or 92 DNA molecules in all. The nuclear envelope disintegrates during prophase and releases the chromosomes into the cytosol. The centrioles begin to sprout elongated microtubules called **spindle fibers,** which push the centrioles apart as they grow. Eventually, one pair of centrioles comes to lie at each pole of the cell. Microtubules grow toward the chromosomes and attach to the centromeres. The spindle fibers then tug the chromosomes back and forth until they line up along the midline of the cell.

In **metaphase,**[29] the chromosomes are aligned on the cell equator, oscillating slightly and awaiting a signal that stimulates the next step: anaphase. The spindle fibers now form a lemon-shaped array called the **mitotic spindle,** with long microtubules reaching out from each centriole to the chromosomes, and shorter microtubules anchoring the assembly to the inside of the plasma membrane at each end of the cell.

Anaphase[30] begins with activation of an enzyme that splits each centromere in two, separating the sister chromatids from each other. Each chromatid is now regarded as a separate, single-stranded *daughter chromosome*. One daughter chromosome migrates to each pole of the cell, with its centromere leading the way and the arms trailing behind. Because the two chromatids of a chromosome are genetically identical, and because each daughter cell receives one chromatid from each chromosome, the daughter cells of mitosis are genetically identical. Almost every cell of your body is genetically identical (with a few exceptions such as eggs, sperm, and certain immune cells).

In **telophase,**[31] the chromatids cluster on each side of the cell. The rough ER produces a new nuclear envelope around each cluster, and the chromatids begin to uncoil and return to the thinly dispersed chromatin form. The mitotic spindle breaks up and vanishes.

Telophase is the end of mitosis proper (nuclear division), but overlaps with cytokinesis, division of the cytoplasm. Cytokinesis starts with a crease called the *cleavage furrow* around the equator of the cell, and the cell eventually pinches in two. Interphase has now begun for these new cells; the life cycle begins anew.

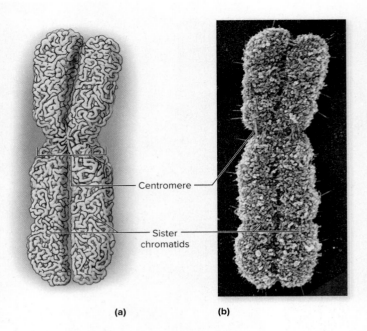

Figure 3.19 Chromosomes. (a) Drawing of a metaphase chromosome. (b) Scanning electron micrograph.

b: Biophoto Associates/Science Source

❓ What are the chromatids composed of besides DNA?

Before You Go On

Answer these questions from memory to test your understanding of the previous section. Reread the material if you can't answer the questions.

13. Identify the stage of the cell cycle where each of the following occurs: (a) DNA replication, (b) synthesis of proteins, including enzymes for export, (c) conclusion of centriole replication, (d) separation of the two sets of DNA into separate nuclei, and (e) splitting of the cytoplasm into two separate cells.

14. Identify the stage of mitosis in which each of the following occurs: (a) double-stranded chromosomes align on the cell equator awaiting a signal to split in two, (b) chromosomes coil and thicken and their chromatids become visible, (c) the centromeres split and the chromatids separate, and (d) new nuclear envelopes form around each chromosome set.

---

[29]*meta* = next in a series
[30]*ana* = apart
[31]*telo* = end, final

PERSPECTIVES ON HEALTH

Cancer

Of all things that can go wrong with cellular function, cancer is one of the most dreaded. It's typically the first thing that comes to mind when one is found to have a tumor, or *neoplasm*.[32] The next question to arise in a worried patient's mind is often: Is it benign or malignant? If a tumor is *benign*[33] (be-NINE), this means the cellular mass is surrounded by a fibrous capsule, the cells can't freely break loose and spread to other sites in the body, and the tumor is relatively slow-growing. Even some benign tumors can be deadly, but most can be surgically removed, producing full recovery.

A *malignant*[34] tumor is fast-growing, lacks a capsule, and has cells capable of breaking free and spreading to other organs *(metastasizing*[35]*)*. The word *cancer* refers only to malignant tumors. Cancer was named by Hippocrates, who compared the distended veins in some breast tumors to the outstretched legs of a crab.[36]

Most cancer is caused by mutations (changes in DNA or chromosome structure), which can be induced by chemicals, viruses, or radiation, or simply occur through errors in DNA replication during the cell cycle. Agents that cause mutation are called *mutagens,*[37] and those that induce cancer are also called *carcinogens.*[38] Cancer stems from mutations mainly in two gene families, the oncogenes and tumor-suppressor genes. *Oncogenes*[39] are mutated genes that promote the secretion of excessive growth factors (chemicals that stimulate cell division) or excessive sensitivity of target cells to growth factors. *Tumor-suppressor (TS) genes* inhibit the development of cancer by opposing oncogenes, promoting DNA repair, and other means. Some cancers result from the inability of mutated TS genes to perform this protective function. Oncogenes are like an accelerator to the cell cycle, whereas TS genes are like a brake. With either a "stuck accelerator" or "faulty brakes," cells divide at a rate exceeding cell death, thus amassing to form a tumor.

Untreated cancer is almost always fatal. Tumors destroy healthy tissue; they can grow to block major blood vessels or respiratory airways; they can damage blood vessels and cause hemorrhaging; they can compress and kill brain tissue; and they tend to drain the body of nutrients and energy as they hungrily consume a disproportionate share of the body's oxygen and nutrients. Treatments for cancer include surgery, chemotherapy, and radiation therapy.

---

[32]*neo* = new; *plasm* = growth, formation
[33]*benign* = mild, gentle
[34]*mal* = bad, evil
[35]*meta* = beyond; *stas* = being stationary
[36]*cancer* = crab
[37]*muta* = change; *gen* = to produce
[38]*carcino* = cancer; *gen* = to produce
[39]*onco* = tumor

Genomic Medicine

Genomic medicine is the application of our knowledge of the genome—one individual's complete set of genes—to the prediction, diagnosis, and treatment of disease. It is relevant to disorders as diverse as cancer, Alzheimer disease, schizophrenia, obesity, and even a person's susceptibility to nonhereditary diseases such as AIDS and tuberculosis.

The technology of gene sequencing has improved to such an extent that a person's entire genome can now be sequenced for no more than $1,000. Why would we want to? Because knowing one's genome could dramatically change clinical care. It will allow clinicians to assess a person's risk of disease, to detect diseases earlier, and to precisely tailor effective clinical interventions. Drugs that are safe for most people can have serious side effects in others, owing to genetic variations in drug metabolism. Genomics is influencing medical approaches to conditions as diverse as cancer, Parkinson disease, depression, and diabetes. Genomics is also playing a new role in drug design; *pharmacogenomics* is the science of examining the effects of genetic variation on drug action.

Knowing the sites of disease-producing mutations expands the potential for *gene-substitution therapy.* This is a procedure in which cells are removed from a patient with a genetic disorder, supplied with a normal gene in place of the defective one, and reintroduced to the body. The hope is that these genetically modified cells will proliferate and provide the patient with a gene product that he or she was lacking—perhaps insulin for a patient with diabetes or a healthy hemoglobin gene for a patient suffering from sickle cell anemia. Early attempts at gene therapy were marred by some tragic setbacks but, although scientists still face great technical difficulties, recent successes show promise for the future of gene therapy.

Genomics has generated high hopes that the twenty-first century may be an era of *personalized medicine,* with treatments tailored to the genetic constitution of the individual. Yet genomic medicine is introducing new problems in medical ethics and law. Should your genome be a private matter between you and your physician? Or should an insurance company be entitled to know your genome before issuing health or life insurance to you, so it can know your risk of contracting a catastrophic illness, adjust the cost of your coverage, or even deny coverage? Should a prospective employer have the right to know your genome before offering employment? These are areas in which biology, politics, and law converge to shape public policy.

CAREER SPOTLIGHT

Cytotechnologist

Rich Legg/Getty Images

A cytotechnologist is a person who examines tissue specimens for signs of cancer and other diseases. Cytotechnologists work under the supervision of pathologists in hospitals, clinics, commercial diagnostic laboratories, universities, and public health agencies. Using automated tissue preparation, microscopy, computer-assisted screening, and other techniques, a cytotechnologist scans a slide; marks areas of cellular abnormality such as changes in cell size, shape, color, or nuclear volume; and refers abnormal specimens to the pathologist for diagnosis. The tissue specimens studied by cytotechnologists include Pap smears and biopsies of breast, lymph node, liver, thyroid, lung, and many other tissues. To become a cytotechnologist, one must first earn a baccalaureate degree with coursework in biology, chemistry, and mathematics or statistics; attend a 1- to 2-year program in cytotechnology training at a university or hospital; and pass a board certification exam. The advanced training includes tissue-preparation techniques, microscopy, cytochemistry, anatomy and physiology, histology, immunology, and other subjects. Sophisticated molecular diagnostic methods also are a fast-growing area of cytotechnology education and practice. See appendix B for further career information.

Study Guide

Assess Your Learning Outcomes

To test your knowledge, discuss the following topics with a study partner or in writing, ideally from memory.

3.1 The General Structure of Cells

1. Fundamental components of a cell
2. Intracellular and extracellular fluids
3. The typical size range of human cells and what factors limit cell size

3.2 The Cell Surface

1. Molecular components and organization of the plasma membrane
2. Varieties and functions of the plasma membrane proteins
3. The composition, location, and functions of a cell's glycocalyx
4. Structural and functional distinctions between microvilli, cilia, flagella, and pseudopods
5. Structural distinctions and respective advantages of three types of cell junctions—tight junctions, desmosomes, and gap junctions
6. The eight modes of transport through a plasma membrane and how they differ with respect to the use of carrier proteins, direction of movement of the transported substances, and demand for ATP

3.3 The Cell Interior

1. Components and functions of the cytoskeleton
2. Types of cell inclusions and how inclusions differ from organelles
3. What organelles have in common and how they differ, as a class, from other cellular components
4. Structure of the nucleus, particularly of its nuclear envelope, chromatin, and nucleoli
5. Two forms of endoplasmic reticulum, their spatial relationship, their structural similarities and differences, and their functional differences
6. The composition, appearance, locations, and function of ribosomes
7. Structure of the Golgi complex and its role in the synthesis, packaging, and secretion of cell products
8. Similarities and differences between lysosomes and peroxisomes in structure, contents, and functions
9. Structure, function, and evolutionary origin of mitochondria, and the significance of mitochondrial DNA
10. Structure, locations, and functions of centrioles
11. The processes of genetic transcription and translation, including the roles of mRNA, rRNA, and tRNA
12. How the amino acid sequence of a protein is represented by the codons of mRNA
13. How proteins are processed and secreted after their assembly on a ribosome

3.4 The Life Cycle of Cells

1. Four phases of the cell cycle and the main events in each phase
2. How DNA is replicated in preparation for mitosis
3. Functions of mitosis
4. Four stages of mitosis; changes in chromosome structure and distribution that occur in each stage; and the role of centrioles and the mitotic spindle
5. The mechanism and result of cytokinesis

Testing Your Recall

1. The clear, structureless gel in a cell is its
   a. nucleoplasm.
   b. endoplasm.
   c. cytoplasm.
   d. neoplasm.
   e. cytosol.

2. New nuclei form and a cell begins to pinch in two during
   a. prophase.
   b. metaphase.
   c. interphase.
   d. telophase.
   e. anaphase.

3. The amount of _____ in a plasma membrane affects its fluidity.
   a. phospholipid
   b. cholesterol
   c. glycolipid
   d. glycoprotein
   e. integral protein

4. Cells specialized for absorption of matter from the extracellular fluid are likely to show an abundance of
   a. lysosomes.
   b. microvilli.
   c. mitochondria.
   d. secretory vesicles.
   e. ribosomes.

5. A _____ serves as a mechanical linkage between adjacent cells but does not obstruct the movement of materials through the space between cells.
   a. glycocalyx
   b. phospholipid bilayer
   c. tight junction
   d. gap junction
   e. desmosome

6. The word root *phago-* means
   a. eating.
   b. drinking.
   c. emitting fluid.
   d. intracellular.
   e. extracellular.

7. The amount of DNA in a cell doubles during
   a. prophase.
   b. metaphase.
   c. anaphase.
   d. the S phase.
   e. the G_2 phase.

8. Fusion of a secretory vesicle with the plasma membrane and release of the vesicle's contents is
   a. exocytosis.
   b. receptor-mediated endocytosis.
   c. active transport.
   d. pinocytosis.
   e. phagocytosis.

9. Most cellular membranes are made by
   a. the nucleus.
   b. the cytoskeleton.
   c. enzymes in the peroxisomes.
   d. the endoplasmic reticulum.
   e. replication of existing membranes.

10. Which of the following is/are not involved in protein synthesis?
    a. ribosomes
    b. centrioles
    c. mRNA
    d. rough endoplasmic reticulum
    e. codons

11. Most human cells are 10 to 15 _____ wide.

12. When a hormone cannot enter a cell, it binds to a _____ at the cell surface.

13. _____ are channels in the plasma membrane that open or close in response to various stimuli.

14. Cells are somewhat protected from mechanical trauma by a carbohydrate surface coat called the _____.

15. The separation of chromatids from each other marks the _____ stage of mitosis.

16. The majority of molecules that compose the plasma membrane are _____.

17. Two human organelles that are surrounded by a double membrane are the _____ and _____.

18. Liver cells can detoxify alcohol with two organelles, the _____ and _____.

19. A macrophage would use the process of _____ to engulf a dying tissue cell.

20. To crawl about in the body's tissues or engulf foreign matter, a macrophage would use structures called its _____.

Answers in Appendix A

What's Wrong with These Statements?

Briefly explain why each of the following statements is false, or reword it to make it true.

1. A cell specialized for absorption would be expected to have a high density of cilia on its surface.
2. DNA replication occurs during mitosis.
3. A cell can release its secretory products by exocytosis, phagocytosis, or pinocytosis.
4. In the plasma membrane, the phosphate heads of the phospholipid molecules cluster together in the middle of the membrane and the fatty acid tails are pointed toward the ICF and ECF.
5. Cells of the digestive glands store enzymes in their lysosomes and release them into the digestive tract when needed to digest food.
6. As a carrier-mediated transport process, facilitated diffusion requires ATP.
7. Osmosis is a type of active transport involving water.
8. White blood cells can move about in the tissues by means of either cilia or pseudopods.
9. Desmosomes enable solutes to pass from cell to cell.
10. Ribosomes and the Golgi complex play similar roles in the synthesis of proteins.

Answers in Appendix A

Testing Your Comprehension

1. Breast milk contains both sugar (lactose) and proteins (albumin and casein). Identify which organelles of the mammary gland cells are involved in synthesizing and secreting these components, and describe the structural pathway from synthesis to release from the cell.
2. A person with lactose intolerance cannot digest lactose, so instead of being absorbed by the small intestine, this sugar passes undigested into the large intestine. Here, it causes diarrhea among other signs. Which of the membrane transport processes do you think is most directly involved in the diarrhea? On that basis, explain why the diarrhea occurs.
3. Consider a cardiac muscle cell, an enzyme-producing pancreatic cell, a phagocytic white blood cell, and a hormone-secreting cell of the ovary. Which of these would you expect to show the greatest number of lysosomes? Mitochondria? Rough endoplasmic reticulum? Smooth endoplasmic reticulum? Explain each answer.

Chapter 4

Histology—The Tissue Level of Organization

Chapter Outline

4.1 The Study of Tissues
- 4.1a The Primary Tissue Classes
- 4.1b Interpreting Tissue Sections
- 4.1c Cell Shapes

4.2 Epithelial Tissue
- 4.2a Simple Epithelia
- 4.2b Stratified Epithelia

4.3 Connective Tissue
- 4.3a Overview
- 4.3b Fibrous Connective Tissue
- 4.3c Adipose Tissue
- 4.3d Supportive Connective Tissue
- 4.3e Blood

4.4 Nervous and Muscular Tissue—The Excitable Tissues
- 4.4a Nervous Tissue
- 4.4b Muscular Tissue

4.5 Glands and Membranes
- 4.5a Glands
- 4.5b Membranes

4.6 Tissue Growth, Repair, and Death
- 4.6a Tissue Growth and Repair
- 4.6b Tissue Shrinkage and Death

Clinical Applications/Perspectives on Health
- Clinical Application 4.1: Marfan Syndrome—A Connective Tissue Disease
- Clinical Application 4.2: Diabetic Foot Ulcers
- Perspectives on Health

End of Chapter
- Career Spotlight: Histotechnician
- Study Guide

Cilia on a cell of the uterine (fallopian) tube. The synchronized beating of such cilia serves to move an egg or embryo toward the uterus.
Steve Gschmeissner/Science Source

BASE CAMP

Before ascending to the next level, be sure you're properly equipped with a knowledge of these concepts from earlier chapters.

- "Levels of Human Structure" in section 1.3a
- "Body Cavities and Membranes" in section 1.3e
- Cilia and microvilli (see "Cell Surface Extensions" in section 3.2c)
- Secretory vesicles and exocytosis (see "Vesicular Transport" in section 3.2e)

Anatomy & Physiology Revealed 4.0
Module 3 Tissues

With its **50 trillion cells** and thousands of organs, the human body may seem to be a structure of forbidding complexity. Fortunately for our health, longevity, and self-understanding, the biologists of past generations were not discouraged by this, but discovered patterns that make it more understandable. For example, these trillions of cells belong to only 200 types or so, and are organized into tissues that fall into just four primary categories—epithelial, connective, nervous, and muscular.

An organ is a structure with discrete boundaries that is composed of two or more tissue types. Organs derive their function not from their cells alone but from how the cells are organized into tissues. This chapter describes the structural and functional characteristics of the major human tissues, and is a bridge between the previous chapter on cells and later chapters that describe the organ systems.

4.1 The Study of Tissues

Expected Learning Outcomes

When you have completed this section, you should be able to

a. name the four primary classes of adult tissues;
b. visualize the three-dimensional shape of a structure from a two-dimensional tissue section; and
c. interpret descriptive terms for various shapes of cells.

Table 4.1 The Four Primary Tissue Classes

Type	Characteristics	Representative Locations
Epithelial	Tissue composed of layers of closely spaced cells; covers organ surfaces, forms glands, and serves for protection, secretion, and absorption	Epidermis, lining of digestive tract, liver, and other glands
Connective	Tissue with usually more matrix than cell volume; often specialized to support, bind, and protect organs	Tendons, ligaments, cartilage, fat, bone, blood
Nervous	Tissue containing excitable cells specialized for rapid transmission of information to other cells	Brain, spinal cord, nerves
Muscular	Tissue composed of elongated, excitable cells specialized for contraction and movement	Skeletal muscles; heart; walls of uterus, bladder, intestines, and other internal organs

Epithelial, Nervous, Muscular: Ed Reschke/Photolibrary/Getty Images
Connective: Dennis Strete/McGraw-Hill Education

4.1a The Primary Tissue Classes

Histology[1] is the study of tissues and how they are arranged into organs. A **tissue** is a group of similar cells and cell products that arise from the same region of an embryo and work together to perform a specific structural or physiological role in an organ. The four *primary tissues*—epithelial, connective, nervous, and muscular (table 4.1)—differ from each other in the types and functions of their cells, the characteristics of the **matrix** (extracellular material) that surrounds the cells, and the relative amount of space occupied by cells versus matrix.

The matrix is nonliving matter secreted by the tissue cells. It is composed of fibrous proteins and, usually, a clear gel variously known as **ground substance** or **extracellular fluid (ECF)**. In summary, a tissue is composed of cells and matrix, and the matrix is composed of protein fibers and ground substance.

---
[1]*histo* = tissue; *logy* = study of

4.1b Interpreting Tissue Sections

Students of anatomy and physiology often study **histological sections**—very thin tissue slices, generally one or two cells thick, that are mounted on microscope slides and dyed with various **stains.** These stains lend contrast to such structures as cell nuclei, cytoplasm, and extracellular fibers, but one must remember that these colors are entirely artificial; they do not represent the natural colors of the organs.

A tissue section shows a three-dimensional object as a two-dimensional image. You must keep this in mind and try to translate the microscopic image into a mental image of the whole structure, like trying to imagine what a whole loaf of bread looks like if you had seen only a few slices. Like the boiled egg and elbow macaroni in **figure 4.1,** an object may look quite different when it is cut at various levels, or *planes of section.* A coiled tube, such as a gland of the uterus (**fig. 4.1c**), is often broken up into multiple portions because it meanders in and out of the plane of section. With experience, however, one recognizes that the separated pieces are parts of a single tube winding its way to the organ surface.

Figure 4.1 Three-Dimensional Interpretation of Two-Dimensional Images. (a) A boiled egg. Note that grazing sections (upper left and right) would miss the yolk, just as a tissue section may miss the nucleus of a cell and create an illusion that the cell did not have one. (b) Elbow macaroni, which resembles many curved ducts and tubules. A section far from the bend would give the impression of two separate tubules; a section near the bend would show two interconnected lumina (interior spaces); and a section still farther down could miss the lumen completely. (c) A coiled gland in three dimensions and as it would look in a vertical tissue section.

Apply What You Know
How would a cylindrical blood vessel look in two dimensions if cut lengthwise (a longitudinal section); if cut perpendicular to this (a cross section); and if cut on a slant between these two (an oblique section)?

4.1c Cell Shapes

Descriptions of organ and tissue structure often refer to the shapes of the cells by the following terms (**fig. 4.2**):

- **Squamous**[2] (SKWAY-mus)—a thin, flat, scaly shape. Squamous cells line the esophagus and form the surface layer (epidermis) of the skin.

---

[2]*squam* = scale; *ous* = characterized by

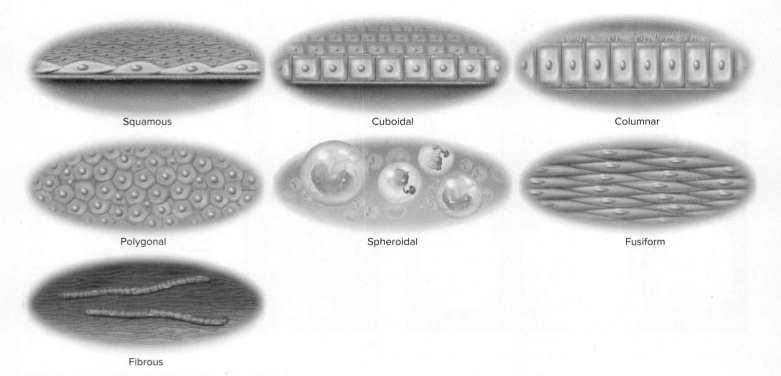

Figure 4.2 Terminology of Cell Shapes.

- **Cuboidal**—squarish and about equal in height and width. Good examples are found in the kidney tubules and liver.
- **Columnar**—distinctly taller than wide, such as the inner lining cells of the stomach and intestines.
- **Polygonal**[3]—having irregular, angular shapes with four, five, or more sides. Squamous, cuboidal, and columnar cells often look polygonal when viewed from above rather than from the side.
- **Spheroidal** to **ovoid**—round to oval, as in egg cells and white blood cells.
- **Fusiform**[4] (FEW-zih-form)—elongated, with a thick middle and tapered ends, as in smooth muscle cells.
- **Fibrous**—long, slender, and threadlike, as in skeletal muscle cells.

Before You Go On

Answer these questions from memory to test your understanding of the previous section. Reread the material if you can't answer the questions.

1. Classify each of the following into one of the four primary tissue classes: the epidermis, fat, the spinal cord, most heart tissue, bones, tendons, blood, and the inner lining of the stomach.
2. What are tissues composed of in addition to cells?
3. What is the term for a thin, stained slice of tissue mounted on a microscope slide?

---

[3]*poly* = many; *gon* = angles
[4]*fusi* = spindle; *form* = shape

4.2 Epithelial Tissue

Expected Learning Outcomes

When you have completed this section, you should be able to

a. describe the properties that distinguish epithelium from other tissue classes;
b. list and classify eight types of epithelium, distinguish them from each other, and state where each type can be found in the body;
c. discuss how the structure of each type of epithelium relates to its function; and
d. recognize epithelial types from specimens or photographs.

Epithelial tissue, or **epithelium,**[5] consists of a flat sheet of closely spaced cells, one or more cells thick, like bricks in a wall. The upper surface is usually exposed to the environment or to an internal space in the body. Epithelium covers the body surface, lines body cavities, forms the external and internal linings of many organs, and makes up most gland tissue. Epithelia contain no blood vessels, but they almost always lie on a layer of loose connective tissue and depend on its blood vessels for nourishment and waste removal. The functions of epithelial tissue include

- **Protection.** Epithelia protect deeper tissues from invasion and injury. The epidermis of the skin, for example, is a barrier to infection and water loss from the body.
- **Secretion.** Epithelia produce mucus, sweat, enzymes, hormones, and most of the body's other secretions; glands are composed largely of epithelial tissue.
- **Excretion.** Epithelia remove wastes from the tissues. CO_2 is moved across the pulmonary epithelium to be exhaled, for example.
- **Absorption.** Epithelia absorb material; nutrients, for example, are absorbed through the epithelium of the small intestine.
- **Filtration.** All substances leaving the blood are selectively filtered through the epithelium that lines the blood vessels; all urinary waste is filtered through epithelia of the kidneys.
- **Sensation.** Epithelia have nerve endings that sense stimulation such as touch on the skin or irritation of the stomach.

Epithelia contain no blood vessels, but they usually lie on a layer of blood-rich connective tissue that supplies nourishment and removes waste. Between an epithelium and the underlying connective tissue is a layer called the **basement membrane,** composed mainly of protein. It anchors the epithelium to the connective tissue, regulates the exchange of materials between the epithelium and the underlying tissues, and binds growth factors from below that regulate epithelial development and maintenance.

The surface of a cell that attaches to the basement membrane is its **basal surface;** the opposite, upper surface, that faces the body surface or the **lumen** (internal cavity) of an organ is the **apical surface;** and the sides of the cell are the **lateral surfaces.** You could compare these to the floor, roof, and walls of a house, respectively. These surfaces often differ in their functions (such as absorption, secretion, attachment, or intercellular communication) and therefore differ in the components of their plasma membranes. Epithelial cells often have cilia, microvilli, or both on their apical surface.

---

[5]*epi* = upon; *theli* = nipple, female

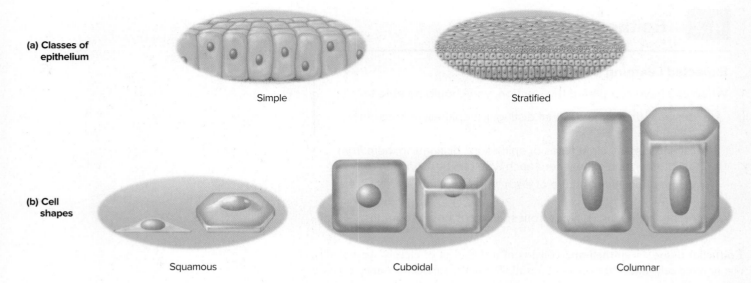

Figure 4.3 Epithelial Types and Cell Shapes. (a) Simple epithelia are composed of one layer of cells whereas stratified epithelia have two or more cell layers. (b) Cell shapes vary in different types of epithelia. **APR**

Epithelia are classified into two broad categories—*simple* and *stratified*—with four types in each category. In a simple epithelium, every cell touches the basement membrane, whereas in a stratified epithelium, some cells rest on top of other cells (like a multistory apartment building) and do not extend to the basement membrane (fig. 4.3).

4.2a Simple Epithelia

A simple epithelium has only one layer of cells. Three types of simple epithelia are named for the shapes of their cells: **simple squamous** (thin scaly cells), **simple cuboidal** (squarish or rounded cells), and **simple columnar** (tall narrow cells). In the fourth type, **pseudostratified columnar,** not all cells reach the free surface; the shorter cells are covered over by the taller ones, but all of them reach the basement membrane—much like trees in a forest that reach different heights but are all anchored in the soil below. Simple columnar and pseudostratified epithelia often produce protective mucous coatings. The mucus is secreted by wineglass-shaped **goblet cells.** Table 4.2 (figs. 4.4 to 4.7) illustrates and summarizes the structural and functional differences between the four types of simple epithelium. The magnifications given in these tables, such as ×400, are the magnifications at which the tissue would show this level of detail if you were at the microscope; they are not the magnifications attained by enlarging the images to the size presented on the page.

4.2b Stratified Epithelia

Stratified epithelia range from 2 cell layers to 20 or more. Three of the stratified epithelia are named for the shapes of their surface cells: **stratified squamous, stratified cuboidal,** and **stratified columnar epithelia.** The deeper cells may be of a different shape than the surface cells. The fourth type, **transitional epithelium,** was named when it was thought to represent a transitional stage between stratified squamous and stratified columnar epithelium. This is now known to be untrue, but the name has persisted. Stratified columnar epithelium is rare (occurring in short transitional zones where one epithelium type grades into another) and we will not consider it any further. The other three types are illustrated and summarized in table 4.3.

Table 4.2 Simple Epithelia

Figure 4.4 Simple Squamous Epithelium in the Air Sacs of Lungs (×400). APR
a: Al Telser/McGraw-Hill Education

Microscopic appearance: Single layer of thin cells, shaped like fried eggs with a bulge where the nucleus is located; nucleus somewhat flattened in the plane of the cell, like an egg yolk; cytoplasm may be so thin it is hard to see in tissue sections; in surface view, cells have angular contours and nuclei appear round

Representative locations: Air sacs (alveoli) of lungs; glomerular capsules of kidneys; some kidney tubules; inner lining (endothelium) of heart and blood vessels; serous membranes of stomach, intestines, and some other viscera; surface layer (mesothelium) of pleurae, pericardium, peritoneum, and mesenteries

Functions: Allows rapid diffusion or transport of substances through membranes; secretes lubricating serous fluid

Figure 4.5 Simple Cuboidal Epithelium in the Kidney Tubules (×400). APR
a: Dennis Strete/McGraw-Hill Education

Microscopic appearance: Single layer of squarish or rounded cells; in glands, cells often pyramidal and arranged like segments of an orange around a central space; spherical, centrally placed nuclei; often with a brush border of microvilli in some kidney tubules; ciliated in bronchioles of lung

Representative locations: Liver, thyroid, mammary, salivary, and other glands; many gland ducts; most kidney tubules; bronchioles

Functions: Absorption and secretion; production of protective mucous coat; movement of respiratory mucus

(continued on next page)

Table 4.2 Simple Epithelia *(continued)*

Simple Columnar Epithelium

Figure 4.6 Simple Columnar Epithelium on the Internal Surface (Mucosa) of the Small Intestine (×400). APR
a: Ed Reschke/Photolibrary/Getty Images

Microscopic appearance: Single layer of tall, narrow cells; oval or sausage-shaped nuclei, vertically oriented, usually in basal half of cell; apical portion of cell often shows secretory vesicles visible with TEM; often shows a brush border of microvilli; ciliated in some organs; may possess goblet cells

Representative locations: Inner lining of stomach, intestines, gallbladder, uterus, and uterine tubes; some kidney tubules

Functions: Absorption and secretion; movement of egg and embryo in uterine tube; secretion of mucus

Pseudostratified Columnar Epithelium

Figure 4.7 Ciliated Pseudostratified Columnar Epithelium in the Mucosa of the Trachea (×400). APR
a: Dennis Strete/McGraw-Hill Education

Microscopic appearance: Looks multilayered; some cells do not reach free surface but all cells reach basement membrane; nuclei at several levels in deeper half of epithelium; often with goblet cells; often ciliated

Representative locations: Respiratory tract from nasal cavity to bronchi; portions of male reproductive tract

Functions: Secretes and propels respiratory mucus; absorbs excess fluid from parts of the male reproductive tract

The most widespread epithelium in the body is stratified squamous epithelium, which deserves further discussion. Its deepest cells are cuboidal to columnar and undergo continual mitosis. Their daughter cells push toward the surface and become flatter (more *squamous,* or scaly) as they migrate farther upward, until they finally die and flake off (fig. 4.12). Their separation from the surface is called **exfoliation;**[6] the study of exfoliated cells is called *exfoliate cytology*—for example, in a Pap smear (see fig. 19.8).

There are two kinds of stratified squamous epithelia—keratinized and nonkeratinized. A **keratinized**[7] epithelium, found in the epidermis, is covered with a layer of compact, dead squamous cells. These cells are packed with the durable protein *keratin* and coated with a water repellent. The skin surface is therefore relatively dry, it retards water loss from the body, and it resists penetration by disease organisms. The tongue, oral mucosa, esophagus, vagina, and a few other internal surfaces are covered with the **nonkeratinized** type, which lacks the surface layer of dead cells. This type provides a surface that is, again, abrasion-resistant, but also moist and slippery. These characteristics are well suited to resist stress produced by the chewing and swallowing of food and by sexual intercourse and childbirth.

Transitional epithelium (also called **urothelium**) is found only in the urinary tract, where it has a unique role. Urine is acidic and hypertonic, and would tend to draw water out of the cells by osmosis and kill them if there were nothing to protect them. Specialized, bulging *umbrella cells* at the surface protect the deeper cells from these effects. Not surprisingly, transitional epithelium is thickest in the place where urine lingers the longest—the urinary bladder (fig. 4.11).

Table 4.3 (figs. 4.8 to 4.11) illustrates and summarizes the characteristics of the most common types of stratified epithelium.

Before You Go On

Answer these questions from memory to test your understanding of the previous section. Reread the material if you can't answer the questions.

4. Distinguish between simple and stratified epithelia, and explain why pseudostratified columnar epithelium belongs in the former category even though it may appear to be stratified.
5. Explain how to distinguish a stratified squamous epithelium from a transitional epithelium.
6. What function do keratinized and nonkeratinized stratified squamous epithelia have in common? What is the structural difference between these two? How is this structural difference related to a functional difference between them?
7. How do the epithelia of the esophagus and stomach differ? How does this relate to their respective functions?

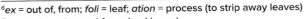

[6]*ex* = out of, from; *foli* = leaf; *ation* = process (to strip away leaves)
[7]*kerat* = horn (named for animal horns)

Table 4.3 Stratified Epithelia

Figure 4.8 Keratinized Stratified Squamous Epithelium on the Sole of the Foot (×400).
a: © Ed Reschke

Microscopic appearance: Multiple cell layers with cells becoming increasingly flat and scaly toward surface; surface covered with a layer of compact dead cells without nuclei; basal cells may be cuboidal to columnar

Representative locations: Epidermis; palms and soles are especially heavily keratinized

Functions: Resists abrasion; retards water loss through skin; resists penetration by pathogenic organisms

Figure 4.9 Nonkeratinized Stratified Squamous Epithelium in the Mucosa of the Vagina (×400).
a: Ed Reschke/Getty Images

Microscopic appearance: Same as keratinized epithelium but without the surface layer of dead cells

Representative locations: Tongue, oral mucosa, esophagus, anal canal, vagina

Functions: Resists abrasion and penetration by pathogenic organisms while providing a moist slippery surface

CHAPTER 4 Histology—The Tissue Level of Organization   103

Table 4.3 Stratified Epithelia *(continued)*

Stratified Cuboidal Epithelium

Figure 4.10 Stratified Cuboidal Epithelium in the Duct of a Sweat Gland (×400). APR
a: © L.V. Bergman/Project Masters

Microscopic appearance: Two or more layers of cells; surface cells squarish or rounded

Representative locations: Sweat gland ducts; egg-producing vesicles (follicles) of ovaries; sperm-producing ducts (seminiferous tubules) of testes

Functions: Contributes to sweat secretion; secretes ovarian hormones; produces sperm

Transitional Epithelium

Figure 4.11 Transitional Epithelium in the Kidney (×400). APR
a: © Johnny R. Howze

Microscopic appearance: Somewhat resembles stratified squamous epithelium, but surface cells are rounded, not flattened, and often bulge above surface; typically five or six cells thick when relaxed, two or three cells thick when stretched; cells may be flatter and thinner when epithelium is stretched (as in a distended bladder); some cells have two nuclei

Representative locations: Limited to urinary tract—part of kidney, ureter, bladder, part of urethra

Functions: Stretches to allow filling of urinary tract; protects deeper cells and tissues from damage by acidic and hypertonic urine

Figure 4.12 Exfoliation of Squamous Cells from the Mucosa of the Vagina.
David M. Phillips/Science Source

What are some other places in the body where you would expect to find similar exfoliation?

4.3 Connective Tissue

Expected Learning Outcomes

When you have completed this section, you should be able to

a. describe the properties that most connective tissues have in common;
b. discuss the types of cells found in fibrous connective tissue;
c. explain what the matrix of a connective tissue is and describe its components;
d. list 10 types of connective tissue, describe their cellular components and matrix, and explain what distinguishes them from each other; and
e. visually recognize each type from specimens or photographs.

4.3a Overview

Connective tissues are the most abundant, widely distributed, and variable of the four primary tissues. Despite their diversity, usually the volume of the extracellular matrix is greater than the volume occupied by cells. Usually the cells are not in direct contact with each other, but are separated by expanses of extracellular material.

The functions of connective tissue include the following:

- **Binding of organs.** Tendons bind muscle to bone, ligaments bind one bone to another, and fat holds the kidneys and eyes in place.
- **Support.** Bones support the body, and cartilage supports the ears, nose, trachea, and bronchi.
- **Physical protection.** Bones protect delicate organs such as the brain, lungs, and heart; fat and fibrous capsules around the kidneys and eyes cushion these organs.
- **Immune protection.** Connective tissue cells attack foreign invaders.
- **Movement.** Bones provide the lever system for body movement, cartilages are involved in movement of the vocal cords, and cartilages on bone surfaces ease joint movements.
- **Storage.** Fat is the body's major energy reserve; bone is a reservoir of calcium and phosphorus that can be drawn upon when needed.
- **Heat production.** Brown fat generates heat in infants and children.
- **Transport.** Blood transports gases, nutrients, wastes, hormones, and blood cells.

Connective tissues fall into four broad categories: fibrous connective tissue, adipose tissue, supportive connective tissues (cartilage and bone), and liquid connective tissue (blood). Each of these will be considered in turn.

4.3b Fibrous Connective Tissue

Fibrous connective tissue is the most diverse type of connective tissue. Despite their varied structures and functions, fibrous connective tissues are grouped together because they all have conspicuous protein fibers. Let's consider the components of fibrous connective tissue:

Fibers

- **Collagenous (col-LADJ-eh-nus) fibers.** These fibers, made of collagen, are tough, flexible, and resist stretching. Collagen is the most abundant protein in the body, comprising about 25% of the total. It is the base for animal products such as gelatin, leather, and glue. Tendons, ligaments, and the deep layer of the skin (dermis) are made mainly of collagen, and collagen fibers also are woven into the matrix of cartilage and bone. In fresh tissue, collagenous fibers have a glistening white appearance, as seen in tendons (fig. 4.13).

- **Reticular**[9] **fibers.** These thin, collagen fibers form a spongelike framework for such organs as the spleen and lymph nodes.
- **Elastic fibers.** Thinner than collagenous fibers, these are composed of a stretchy protein called **elastin.** The coiled structure of elastin allows it to stretch and recoil like a rubber band. Elastic fibers account for the ability of the skin, lungs, and arteries to spring back after they are stretched.

Cells

The cells of fibrous tissues include the following:

- **Fibroblasts.**[10] The dominant cells of fibrous connective tissue are large cells that produce the protein fibers and ground substance that form the matrix. They often taper at the ends and show slender, wispy branches.
- **Macrophages.**[11] These are large phagocytic cells that wander through connective tissues where they engulf and destroy bacteria, other foreign particles, and dead or dying cells of our own body.
- **Leukocytes,**[12] **or white blood cells (WBCs).** WBCs travel briefly in the bloodstream, then spend most of their time in the fibrous connective tissues, crawling about and providing defense against bacteria, toxins, and other foreign agents. The two most common types are *neutrophils* and *lymphocytes*.
- **Fat cells, or adipocytes.** Fat cells appear either in isolation or in small clusters in fibrous connective tissue. When they dominate an area, the tissue is called adipose tissue.

Ground Substance

The cells and fibers of connective tissue are embedded in **ground substance**. This material usually has a gelatinous to rubbery consistency that results primarily from large protein–carbohydrate complexes and the water they absorb and retain. However, the ground substance in bone is hard and stony due to calcium phosphate and other minerals. The ground substance of a fibrous connective tissue absorbs compressive forces and, like the Styrofoam packing in a shipping carton, protects the more delicate cells from mechanical injury.

Fibrous connective tissue is divided into two broad categories according to the relative abundance of fiber: *loose* and *dense*. In **loose connective tissue,** the fibers are widely spaced, running in apparently random directions; the cells tend to be widely separated; and there is an abundance of ground substance, which looks like empty space in tissue specimens. These characteristics are especially conspicuous in *areolar*[13] *tissue* (AIR-ee-OH-lur) (fig. 4.14). In **dense connective tissue,** most of the space is occupied by closely packed fibers. Tables 4.4 and 4.5 (figs. 4.14 to 4.17) summarize the types of loose and dense fibrous connective tissues.

Figure 4.13 Tendons of the Hand. The white glistening appearance results from the collagen of which tendons are composed. The braceletlike band across the wrist is also composed of collagen.
Rebecca Gray/McGraw-Hill Education

---

[8]*colla* = glue; *gen* = producing
[9]*ret* = network; *icul* = little
[10]*fibro* = fiber; *blast* = producing
[11]*macro* = big; *phage* = eater
[12]*leuko* = white; *cyte* = cell
[13]*areola* = little space

Table 4.4 Loose Connective Tissues

Figure 4.14 Areolar Tissue in a Spread of the Mesentery (×400). APR
a: McGraw-Hill Education/Dennis Strete, photographer

Microscopic appearance: Loose, random-looking arrangement of predominantly collagenous and elastic fibers; scattered cells of various types; abundant ground substance; numerous blood vessels

Representative locations: Underlying nearly all epithelia; surrounding blood vessels, esophagus, and trachea; fascia between muscles; mesenteries; visceral layers of pericardium and pleura

Functions: Loosely binds epithelia to deeper tissues; allows passage of nerves and blood vessels through other tissues; provides an arena for immune defense; and its blood vessels provide nutrients and waste removal for overlying epithelia

Figure 4.15 Reticular Tissue of the Spleen (×400). APR
a: Al Telser/McGraw-Hill Education

Microscopic appearance: Loose network of reticular fibers, reticular cells, and fibroblasts, infiltrated with numerous lymphocytes and other blood cells

Representative locations: Lymph nodes, spleen, thymus, bone marrow

Functions: Supportive stroma (framework) for lymphatic organs

Table 4.5 Dense Connective Tissues

Figure 4.16 Dense Regular Connective Tissue of a Tendon (×400).
a: Dennis Strete/McGraw-Hill

Microscopic appearance: Usually has the form of densely packed, parallel, often wavy collagen fibers; slender fibroblast nuclei compressed between collagen bundles; few cells other than fibroblasts; scanty ground substance; few blood vessels; has the form of parallel wavy sheets of elastic tissue in arteries

Representative locations: Tendons, ligaments, vocal cords, arteries

Functions: Ligaments tightly bind bones together, stabilize joints, determine their range of motion, and resist stress; tendons attach muscle to bone and transfer muscular tension to bones; parallel arrangement of fibers is an adaptation to forces acting in a single consistent direction, as when a tendon pulls a bone; stretches and recoils in arteries to accommodate surges in blood pressure and relieve pressure on smaller downstream vessels

Figure 4.17 Dense Irregular Connective Tissue in the Dermis of the Skin (×400).
a: Dennis Strete/McGraw-Hill Education

Microscopic appearance: Densely packed collagen fibers running in seemingly random directions; scanty open space (ground substance); few visible cells; long fibers in the tissue appear as short, chopped-up pieces in thin histological sections, as in photo

Representative locations: Deeper portion of dermis of skin; capsules around viscera such as liver, kidney, spleen; fibrous sheaths around cartilages, bones, and nerves

Functions: Durable, hard to tear; variable orientation of fibers withstands stresses applied in unpredictable directions

Clinical Application 4.1

MARFAN SYNDROME—A CONNECTIVE TISSUE DISEASE

Serious anatomical and functional abnormalities can result from hereditary errors in the structure of connective tissue proteins. *Marfan*[14] *syndrome,* for example, results from the mutation of a gene on chromosome 15 that codes for a glycoprotein called *fibrillin,* the structural scaffold for elastic fibers. Clinical signs of Marfan syndrome include unusually tall stature, long limbs and spidery fingers, abnormal spinal curvature, and a protruding "pigeon breast." Some other signs include hyperextensible joints, hernias of the groin, and visual problems resulting from abnormally long eyeballs and deformed lenses. More seriously, victims exhibit a weakening of the heart valves and arterial walls. The aorta, where blood pressure is the highest, is sometimes enormously dilated close to the heart, and may suddenly rupture. Marfan syndrome is present in about 1 out of 20,000 live births, and most victims die by their mid-30s. Abraham Lincoln's tall, gangly physique and spindly fingers led some authorities to believe that he had Marfan syndrome; however, the evidence is inconclusive. A number of star athletes have died at a young age of Marfan syndrome, including Olympic volleyball champion Flo Hyman, who died of a ruptured aorta during a game in Japan in 1986, at the age of 31.

4.3c Adipose Tissue

Adipose tissue, or **fat,** is connective tissue in which adipocytes are the dominant cell type (table 4.6, fig. 4.18). Adipocytes are rounded cells occupied mainly by a globule of triglyceride (also called *fat,* in the chemical sense), surrounded by a thin rim of cytoplasm. The nucleus is pushed up against the inside of the plasma membrane. Adipocytes are tightly packed and the narrow spaces between them are occupied by areolar tissue, reticular tissue, and blood capillaries.

There are two types of adipose tissue, **white fat** and **brown fat.** White fat is the most abundant, and has either a white or yellow appearance in the body. It serves primarily as an energy reservoir. Adipose tissue was once thought to be relatively inert and inactive but we now know that it is a dynamic tissue where new triglycerides are constantly synthesized or released into circulation, depending on how much energy

---

[14]Antoine Bernard-Jean Marfan (1858–1942), French physician

Table 4.6 Adipose Tissue

Figure 4.18 Adipose Tissue of the Breast (×100).
a: Dennis Strete/McGraw-Hill

a person consumes or uses. In addition to energy storage, white fat provides thermal insulation and anchors and cushions organs such as the eyeballs and kidneys. Adipocytes also secrete hormones that play roles in appetite regulation and metabolism.

Brown fat is a heat-generating tissue that is especially evident in fetuses, infants, and children. It accounts for up to 6% of an infant's weight. It gets its color from an unusual abundance of mitochondria. Recently, it was discovered that human adults have varying amounts of brown fat and there is a great deal of interest in its role in regulating adiposity.

Amounts of adipose tissue vary in individuals. On average, women have more adipose tissue than men, a difference that can be attributed to the increased energy demands of pregnancy and, especially, nursing an infant. Physically active people have less fat than those who are sedentary.

Obesity is an overabundance of adipose tissue. Excess calories result in enlargement of adipose tissue stores primarily due to growth of existing adipocytes, not by adipocyte multiplication. Adipocytes are usually 70 to 120 μm in diameter, but may be five times as large in obese individuals. The number of obese individuals has risen dramatically in recent decades, along with associated diseases such as diabetes. Adipose tissue is the focus of intense research because of the link between obesity and disease.

4.3d Supportive Connective Tissue

Cartilage and bone are strong, durable connective tissues that provide physical support for various organs and the body as a whole.

Cartilage

Cartilage is a supportive connective tissue with a rubbery matrix. You can feel the texture of cartilage by folding your external ear or palpating the tip of your nose or the thyroid cartilage ("Adam's apple") of the larynx at the front of your throat. It is also easily seen at the grocery store—it is the milky-white gristle at the ends of chicken or beef bones, for example. Cartilage shapes and supports the nose and ears, partially encloses the trachea, larynx, and thoracic cavity, and is found at the joint surfaces of bones, where it reduces friction and eases movement. The cells of cartilage are called **chondrocytes**[15] (CON-dro-sites). They secrete the matrix and surround themselves with it until they become trapped in little cavities called **lacunae**[16] (la-CUE-nee). Cartilage lacks blood capillaries, so its nutrition and waste removal depend on diffusion through the stiff matrix. Because this is a slow process, chondrocytes have low rates of metabolism and cell division, and injured cartilage heals slowly. The matrix contains collagen fibers that range in thickness from invisibly fine to conspicuously coarse. Differences in the fibers provide a basis for classifying cartilage into three types: *hyaline*[17] *cartilage* (HY-uh-lin), *elastic cartilage,* and *fibrocartilage* (**table 4.7, figs. 4.19** to **4.21**). The first two types are usually covered with a sheath of dense irregular connective tissue called the **perichondrium**[18] (PERR-ih-CON-dree-um).

Bone

Bone, or **osseous tissue,** is a hard, calcified connective tissue that composes the skeleton. The term *bone* has two meanings—an entire organ such as the femur, or just the osseous tissue. There are two forms of osseous tissue. **Compact (dense) bone** forms the hard surface of bones and appears to the naked eye as a strong, solid material with no visible spaces. (**table 4.8, fig. 4.22**). This tissue is arranged in microscopic cylinders that surround **central canals,** which run longitudinally through

---

[15]*chondro* = cartilage, gristle; *cyte* = cell
[16]*lacuna* = lake, cavity
[17]*hyal* = glass
[18]*peri* = around; *chondri* = cartilage, gristle

Table 4.7 Cartilage

Figure 4.19 Hyaline Cartilage of the Fetal Skeleton (×400). APR
a: Ed Reschke/Getty Images

Microscopic appearance: Clear, glassy matrix, often stained light blue or pink in tissue sections; fine, dispersed collagen fibers, not usually visible; chondrocytes and lacunae often grouped in cell nests; usually covered by perichondrium

Representative locations: Forms a thin articular cartilage over the ends of bones at movable joints; costal cartilages attach ends of the ribs to the breastbone; forms supportive rings and plates around trachea and bronchi; forms a boxlike enclosure around the larynx; forms most of the fetal skeleton

Functions: Eases joint movements; holds airway open during respiration; moves vocal cords during speech; a precursor of bone in the fetal skeleton and growth zones of the long bones of children

Figure 4.20 Elastic Cartilage of the External Ear (×1,000). APR
a: Ed Reschke/Getty Images

Microscopic appearance: Elastic fibers form weblike mesh amid lacunae; always covered by perichondrium

Representative locations: External ear; epiglottis

Functions: Provides flexible, elastic support; gives shape to the ear

Figure 4.21 Fibrocartilage of an Intervertebral Disc (×400). APR
a: © Dr. Alvin Telser

Microscopic appearance: Coarse, parallel bundles of collagen fibers similar to those of tendon; rows of chondrocytes in lacunae between collagen bundles; never has a perichondrium

Representative locations: Pubic symphysis (anterior joint between two halves of pelvic girdle); intervertebral discs that separate bones of vertebral column; menisci, or pads of shock-absorbing cartilage, in knee joint; at points where tendons insert on bones near articular hyaline cartilage

Functions: Resists compression and absorbs shock in some joints; often a transitional tissue between dense connective tissue and hyaline cartilage (for example, at some tendon–bone junctions)

Table 4.8 Compact Bone

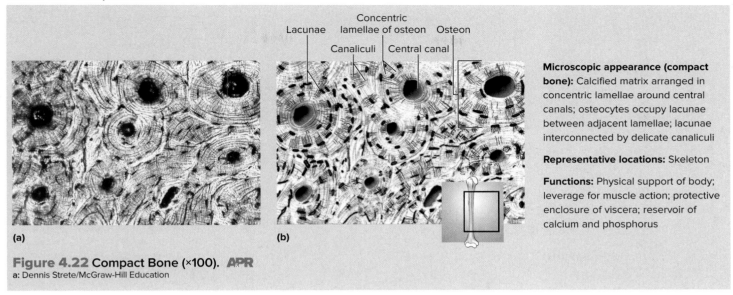

Figure 4.22 Compact Bone (×100).
a: Dennis Strete/McGraw-Hill Education

Microscopic appearance (compact bone): Calcified matrix arranged in concentric lamellae around central canals; osteocytes occupy lacunae between adjacent lamellae; lacunae interconnected by delicate canaliculi

Representative locations: Skeleton

Functions: Physical support of body; leverage for muscle action; protective enclosure of viscera; reservoir of calcium and phosphorus

---

the shafts of long bones such as the femur. Blood vessels and nerves travel through the central canals. The bone matrix is deposited in concentric layers called **lamellae** around each central canal; in cross sections, these look like the layers of an onion slice. Bone cells, called **osteocytes**,[19] occupy **lacunae** between these layers. The structure of compact bone will be examined in further detail in chapter 6.

A more porous type of osseous tissue called **spongy bone** fills the heads of the long bones and forms the middle layer of flat bones such as the sternum and cranial (skull) bones. Although it is calcified and hard, its delicate slivers and plates give it a spongy appearance. Bone marrow occupies the interior spaces of the spongy bone and the hollow shafts of the long bones. The bone as a whole is covered with a tough fibrous **periosteum** (PERR-ee-OSS-tee-um) similar to the perichondrium of cartilage.

4.3e Blood

Blood (table 4.9, fig. 4.23) is a liquid connective tissue that travels through tubular blood vessels. Its primary function is to transport cells and dissolved matter such as oxygen and nutrients from place to place. Blood consists of a ground substance called **plasma,** and cells and cell fragments collectively called **formed elements.** The formed elements are of three kinds: (1) **Red blood cells (RBCs),** or **erythrocytes**[20] (eh-RITH-ro-sites), transport oxygen. (2) **White blood cells (WBCs),** or **leukocytes**, combat infections and other foreign or unwanted agents. (3) **Platelets** function in blood clotting and blood vessel repair and maintenance. All of these are described in more detail in chapter 12.

> *Apply What You Know*
> Connective tissues usually consist of more extracellular material than cell volume. Which connective tissues are exceptions to this generalization?

---

[19]osteo = bone; *cyte* = cell
[20]erythro = red; *cyte* = cell

Table 4.9 Blood

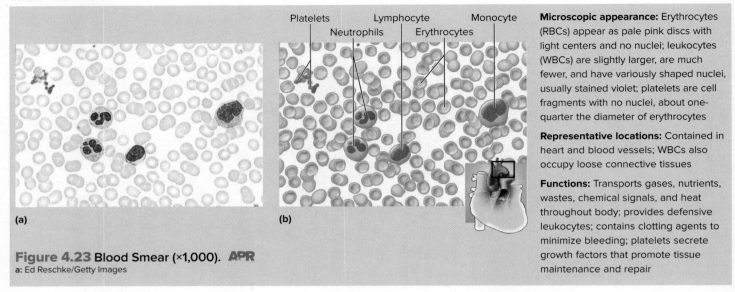

Figure 4.23 Blood Smear (×1,000). APR
a: Ed Reschke/Getty Images

Microscopic appearance: Erythrocytes (RBCs) appear as pale pink discs with light centers and no nuclei; leukocytes (WBCs) are slightly larger, are much fewer, and have variously shaped nuclei, usually stained violet; platelets are cell fragments with no nuclei, about one-quarter the diameter of erythrocytes

Representative locations: Contained in heart and blood vessels; WBCs also occupy loose connective tissues

Functions: Transports gases, nutrients, wastes, chemical signals, and heat throughout body; provides defensive leukocytes; contains clotting agents to minimize bleeding; platelets secrete growth factors that promote tissue maintenance and repair

Before You Go On

Answer these questions from memory to test your understanding of the previous section. Reread the material if you can't answer the questions.

8. What features do most or all connective tissues have in common to set this class apart from others?
9. List the cell and fiber types found in fibrous connective tissues and state their functional differences.
10. How does the matrix of bone differ from that of areolar tissue? Explain how the differences relate to the tissue functions.
11. What is areolar tissue? How can it be distinguished from any other kind of connective tissue?
12. Describe some similarities, differences, and functional relationships between hyaline cartilage and bone.
13. What are the three basic classes of formed elements in blood, and what are their respective functions?

4.4 Nervous and Muscular Tissue— The Excitable Tissues

Expected Learning Outcomes

When you have completed this section, you should be able to

a. explain what distinguishes excitable tissues from other tissues;
b. name the cell types that compose nervous tissue;
c. identify the major parts of a nerve cell;
d. visually recognize nervous tissue from specimens or photographs;
e. name the three kinds of muscular tissue and describe the differences between them; and
f. identify types of muscular tissue from specimens or photographs.

Excitability is a characteristic of all living cells, but it is developed to its highest degree in nervous and muscular tissue, which are therefore described as excitable tissues. The basis for their excitation is an electrical charge on the plasma membrane. Nervous and muscular tissues exhibit quick electrical responses when stimulated. In nerve cells, excitation results in the passage of signals to other cells, whereas in muscle cells, excitation results in contraction, or shortening of the cells.

4.4a Nervous Tissue

Nervous tissue (table 4.10, fig. 4.24) consists of **neurons** (NOOR-ons), or nerve cells, and a much greater number of supporting cells called **glial cells** (GLEE-ul), which protect and assist the neurons. Neurons are specialized to detect stimuli, respond quickly, and transmit information rapidly to other cells. A typical neuron has a prominent **cell body** that houses the nucleus; several short filamentous processes called **dendrites** that receive signals and conduct messages to the cell body; and one long process, the **axon** or **nerve fiber,** that sends outgoing signals to other cells. The structure and functions of neurons and nervous tissue are described in chapters 8 and 9.

4.4b Muscular Tissue

Muscular tissue consists of elongated cells that are specialized to contract and exert a force on other tissues and organs. Not only do movements of the body as a whole depend on muscle, but so do such processes as digestion, waste elimination, breathing, speech, facial expression, and blood circulation. There are three types of muscle—*skeletal, cardiac,* and *smooth* (table 4.11, figs. 4.25 to 4.27). Skeletal and cardiac muscle exhibit fine, transverse, light and dark bands called **striations**[21] (stry-AY-shuns), resulting from the internal arrangement of the proteins involved in their contraction; thus, they are called **striated muscle.** The contractile proteins are less uniformly arranged in smooth muscle, which therefore lacks striations. Skeletal muscle is usually under voluntary control, and is thus called **voluntary muscle.** Cardiac and smooth muscle are called **involuntary muscle** because we usually cannot consciously control their contractions.

---

[21]*striat* = striped

Table 4.10 Nervous Tissue APR

Microscopic appearance: Most nervous tissue sections or smears show a few large neurons, usually with rounded or stellate cell bodies and fibrous processes (axon and dendrites) extending from the somas; neurons are surrounded by a greater number of much smaller glial cells.

Representative locations: Brain, spinal cord, nerves, ganglia

Function: Internal communication

Figure 4.24 A Neuron and Glial Cells of the Spinal Cord (×400). APR
a: Ed Reschke/Photolibrary/Getty Images

Most smooth muscle is found in the walls of hollow organs such as the stomach, intestines, uterus, and urinary bladder, where it is also called **visceral muscle.** The cells of skeletal muscle are commonly called **muscle fibers** because of their extraordinarily long, slender shape, whereas the shorter cells of cardiac and smooth muscle are often called **myocytes.**[22] Skeletal muscle fibers are further discussed in chapter 7 and cardiac myocytes in chapter 13.

Table 4.11 Muscular Tissue

Figure 4.25 Skeletal Muscle (×400).
a: Ed Reschke/Photolibrary/Getty Images

Microscopic appearance: Long, threadlike, unbranched muscle fibers, with relatively parallel appearance in longitudinal tissue sections; striations; each cell with multiple nuclei lying near the plasma membrane

Representative locations: Skeletal muscles, mostly attached to bones but also in the tongue, esophagus, and voluntary *sphincters*[23] (circular muscles) of the eyelids, urethra, and anus

Functions: Body movements, facial expression, posture, breathing, speech, swallowing, control of urination and defecation, and assistance in childbirth; often under voluntary control

Figure 4.26 Cardiac Muscle (×400).
a: Ed Reschke/Getty Images

Microscopic appearance: Short branched cells with a less parallel appearance in tissue sections; striations; thicker dark bands called *intercalated*[24] *discs,* containing mechanical and electrical junctions, where cells meet end to end; usually one nucleus per cell, centrally located and often surrounded by a light zone of glycogen

Representative locations: Heart only

Functions: Pumping of blood; under involuntary control

Figure 4.27 Smooth Muscle of the Intestinal Wall (×1,000).
a: Dennis Strete/McGraw-Hill

Microscopic appearance: Short fusiform myocytes overlapping each other; nonstriated; one nucleus per cell, centrally located

Representative locations: Usually found as sheets of tissue in walls of viscera; also in iris and associated with hair follicles; involuntary sphincters of urethra and anus

Functions: Swallowing; contractions of stomach and intestines; retention and expulsion of feces and urine; labor contractions; control of blood pressure and flow; control of respiratory airflow; control of pupillary diameter; erection of hairs; under involuntary control

---

[22]*myo* = muscle; *cyte* = cell
[23]*sphinc* = squeeze, bind tightly
[24]*inter* = between; *calated* = inserted

PERSPECTIVES ON HEALTH

Regenerative Medicine

The relatively new field of regenerative medicine (also called tissue engineering) seeks to create functional tissues and organs to replace those that are damaged through injury or disease, defective from birth, or worn out by age. In the laboratory, tissue engineering has produced liver, bone, cartilage, cardiac, intestinal and breast tissue, and even a beating rodent heart. To date, regenerative medicine has played a relatively small role in the treatment of patients, although some examples include using engineered skin for covering burns and diabetic foot ulcers, repair of damaged knee cartilages, and transplantation of engineered urinary bladders, small blood vessels, and even an entire trachea. Bioengineered heart valves and arteries for coronary bypass surgery are on the horizon. Tissue engineering holds great promise for not just treatment, but cures, for heart disease and neurodegenerative disorders, as well as other diseases. There is hope that regenerative medicine will someday ease the problem of a shortage of donor organs for those who need life-saving organ transplantations.

One approach to tissue engineering is to start with an organ from a donor body and strip it of all its living cells, leaving only its connective tissue framework, or scaffold. This cell-free scaffold preserves the shape and mechanical properties of the organ, and eliminates the risk of immune rejection by the patient who will receive it. The scaffold is seeded with cells from the patient's body that then repopulate it. The growing organ is maintained in an incubator called a bioreactor, which provides nutrients, oxygen, and growth factors. In another innovative approach to tissue engineering, scientists have built tiny organs-on-a-chip that integrate human cells onto a membrane that mimics the extracellular environment. This technique has produced model organs ranging from hearts, to lungs, to intestines and brains. Organs-on-a-chip technology accelerates the often lengthy path of drug development and testing, and offers the possibility of testing drugs and therapies on a patient's own cells to personalize treatment. In addition, it is an alternative to animal testing, reducing the need for animals used in research.

Stem-Cell Research

One of the most controversial scientific and religious issues in the last several years revolves around histology: stem-cell research. Stem cells are immature cells with the ability to differentiate into one or more types of mature, specialized tissues. Cell biologists, clinicians, and hopeful patients look to stem cells as a possible treatment for diseases arising from the loss of functional tissue. Unfortunately, though, stem-cell research has yet to produce many safe and proven therapies.

Embryonic stem cells compose human embryos up to about 150 cells. They are called pluripotent because they can develop into any type of fetal or adult cell. They are obtained from surplus embryos of about 16 to 32 cells produced by in vitro fertilization (IVF) clinics. Adult stem cells exist in most of the body's tissues, where they multiply and replace older cells that are lost to damage or normal cellular turn over. Conceivably, with the right biochemical coaxing, adult stem cells may be induced to differentiate into various kinds of mature, differentiated cells to replace cardiac muscle damaged by heart attack, restore an injured spinal cord, cure neurodegenerative diseases such as parkinsonism, or cure diabetes by replacing lost insulin-secreting cells.

Embryonic stem cells are less limited than adult stem cells in developmental potential, and therefore have generated great hope that they can be made to grow into new tissues of almost any kind, but their use has generated a significant amount of political, religious, and ethical debate. Some would argue that because the excess embryos of IVF clinics are destined to be destroyed, it would seem sensible to use them to save lives and restore health. Others argue, however, that potential medical benefits cannot justify the destruction of a human embryo. Scientific advances, however, are beginning to defuse the controversy to some degree. Cell biologists have developed methods to make adult stem cells reverse their developmental paths, going back to a pluripotent state that would enable them to go down new roads. In principle, an adult stem cell that was originally destined to become smooth muscle, for example, might be made to revert to a pluripotent state, then be chemically guided down a path leading to nervous tissue for Alzheimer patients or persons paralyzed by spinal cord injury. Such modified adult cells, now called induced pluripotent stem (iPS) cells, have begun to show great promise for both medical benefit and reduced controversy.

Before You Go On

Answer these questions from memory to test your understanding of the previous section. Reread the material if you can't answer the questions.

14. What do nervous and muscular tissue have in common? What is the primary function of each?
15. What two principal cell types compose nervous tissue, and how can they be distinguished from each other?
16. Name the three kinds of muscular tissue, describe how to distinguish them from each other in microscopic appearance, and state a location and function for each one.

4.5 Glands and Membranes

Expected Learning Outcomes

When you have completed this section, you should be able to

a. describe or define various kinds of glands;
b. compare modes of glandular secretion; and
c. describe the structure of mucous and serous membranes.

4.5a Glands

A **gland** is a cell or organ that releases substances for use elsewhere in the body or for elimination from the body. Its product is called a **secretion** if it is useful to the body (such as a digestive enzyme or hormone) and an **excretion** if it is a waste product (such as urine or sweat).

Exocrine and Endocrine Glands

Glands are classified as exocrine or endocrine. **Exocrine**[25] **glands** (EC-so-crin) release a secretion onto a surface by way of a **duct,** an epithelial tube that carries the secretion either to the surface of the body, such as sweat glands, or, more often, into the cavity (lumen) of another organ such as the stomach. Typically, the secretion is produced by a group of cells that form a microscopic secretory sac called an **acinus** (ASS-ih-nus) (**fig. 4.28a, b**) that leads into a duct. These ducts converge until one or a few main ducts exit the gland. Enclosing the gland, there is usually a fibrous **capsule** that sets it off from surrounding tissues. Extensions of the capsule often subdivide the gland into lobes and smaller compartments. Examples of exocrine glands include sweat, mammary, and salivary glands, and the kidneys and liver.

Endocrine[26] **glands** (EN-doe-crin) have no ducts but, instead, release their products into the bloodstream (**fig. 4.28c**). Their secretions, **hormones,** function as chemical messengers to stimulate cells elsewhere in the body. Some examples of endocrine glands are the pituitary, thyroid, and adrenal glands. Some glands, such as the pancreas, ovaries, and liver, have both exocrine and endocrine functions. Endocrine

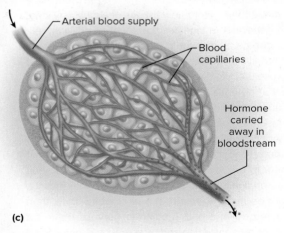

Figure 4.28 Exocrine and Endocrine Glands. (a) General structure of an exocrine gland. The duct branches repeatedly until its finest branches terminate in sacs of secretory cells called acini. (b) Detail of an exocrine acinus and beginning of a duct. (c) General structure of an endocrine gland. The secretions of an endocrine gland (hormones) enter blood capillaries in the gland and are carried away in the bloodstream.

---

[25]*exo* = out; *crin* = to separate, secrete
[26]*endo* = in, into; *crin* = to separate, secrete

Figure 4.29 Modes of Exocrine Secretion. (a) Eccrine secretion in a cell of the mammary gland, secreting milk sugar (lactose) and proteins by exocytosis. (b) Apocrine secretion of fat by a cell of the mammary gland. Fat droplets coalesce in the cytosol and then bud off from the cell surface with a thin coating of cytoplasm and plasma membrane. (c) Holocrine secretion by a sebaceous (oil) gland of the scalp. In this method, entire gland cells break down and become the secretion (sebum).

❓ *Which of these glands would require the highest rate of mitosis in its secretory cells? Why?*

glands are discussed in further detail in chapter 11; the rest of this discussion concerns exocrine glands.

Modes of Secretion

Exocrine glands are classified by the nature of their secretions. **Serous glands** (SEER-us) produce thin, watery fluids such as perspiration, milk, tears, and digestive juices. **Mucous glands,** found in the digestive and respiratory tracts among other places, secrete a glycoprotein called *mucin* (MEW-sin). After it is secreted, mucin absorbs water, swells, and forms sticky *mucus.* (Note that *mucus,* the secretion, is spelled differently from *mucous,* the adjective form of the word.) **Mixed glands,** such as the two pairs of salivary glands in the floor of the mouth, contain both serous and mucous cells and produce a mixture of secretions.

Exocrine glands are further classified according to how they release their secretions. Most of them are **eccrine glands** (EC-rin), which release their products by means of exocytosis (fig. 4.29a). The product is packaged by the Golgi apparatus into vesicles that travel to the cell surface and release it to the exterior. These include the tear glands, pancreas, and most other glands. **Apocrine**[27] **glands** produce their products in the cytosol, not in the Golgi complex, and the products are released by budding from the apical cell surface with a thin coating of cytoplasm and plasma membrane (fig. 4.29b). Mammary glands secrete the sugar and protein of milk by the eccrine method, but secrete the milk fat by the apocrine method. Sweat glands of the axillary (armpit) region were once thought to use the apocrine method as well. They are commonly called *apocrine sweat glands* even though it is now known that they secrete by the ordinary eccrine method.

Holocrine[28] **glands,** in which the gland cells break down entirely and *become* the secretion, are relatively rare (fig. 4.29c). The principal example of these is the oil-producing *sebaceous* (seh-BAY-shus) *glands* of the skin (see chapter 5). Others are found in the margins of the eyelids.

---

[27]*apo* = from, off, away; *crin* = to separate, secrete
[28]*holo* = whole, entire; *crin* = to separate, secrete

4.5b Membranes

Mucous membranes (fig. 4.30a) line tracts of the body that open to the exterior environment: the digestive, respiratory, urinary, and reproductive tracts. A mucous membrane, also called a **mucosa,** consists of an epithelium (usually nonkeratinized stratified squamous, simple columnar, or pseudostratified columnar) overlying a layer of areolar tissue and often a thin layer of smooth muscle. Mucous membranes have absorptive, secretory, and protective functions. They are often covered with mucus secreted by goblet cells, multicellular mucous glands, or both. The mucus traps bacteria and foreign particles, which keeps them from invading the tissues and aids in their removal from the body. This is an important defense for body passages open to the germ- and debris-laden air. The epithelium of a mucous membrane may also include absorptive, ciliated, and other types of cells.

Serous membranes (SEER-us) (fig. 4.30b; see also fig. 4.4) line the thoracic and abdominal cavities and cover the external surfaces of organs such as the stomach and intestines. They include the pleura around each lung, the pericardium around the heart, and the peritoneum of the abdominal cavity. They are composed of a simple squamous epithelium resting on a thin layer of areolar connective tissue. Serous membranes produce watery **serous fluid,** which arises from the blood and derives its name from the fact that it is similar to blood serum in composition. A serous membrane can also be called a **serosa.**

Figure 4.30 Mucous and Serous Membranes.
(a) A mucous membrane (mucosa) such as the inner lining of the trachea. (b) A serous membrane (serosa) such as the external lining of the small intestine.

❓ *Where are some other places one could find each type of membrane?*

Before You Go On

Answer these questions from memory to test your understanding of the previous section. Reread the material if you can't answer the questions.

17. Distinguish between endocrine and exocrine glands, and give an example of each.
18. Contrast the eccrine and holocrine methods of secretion, and name a gland product produced by each method.
19. Distinguish between serous and mucous secretions.
20. Distinguish between serous and mucous membranes, and state where each may be found.

4.6 Tissue Growth, Repair, and Death

Expected Learning Outcomes

When you have completed this section, you should be able to

a. name and describe the modes of tissue growth;
b. distinguish between the ways the body repairs damaged tissues; and
c. name and describe some modes and causes of tissue shrinkage and death.

4.6a Tissue Growth and Repair

Tissues grow either because their cells increase in number or size. Most embryonic and childhood growth occurs through **hyperplasia**[29] (HY-pur-PLAY-zhuh), tissue growth through cell multiplication. Skeletal muscles and adipose tissue, however, grow through **hypertrophy**[30] (hy-PUR-truh-fee), the enlargement of preexisting cells. Even a very muscular or obese adult has about the same number of skeletal muscle fibers or adipocytes as they did in late adolescence, but the cells are much larger. **Neoplasia**[31] (NEE-oh-PLAY-zhuh) is the development of a tumor (neoplasm), whether benign or malignant, composed of abnormal, nonfunctional tissue.

---

[29]*hyper* = above, beyond; *plas* = growth
[30]*hyper* = above, beyond; *trophy* = nourishment
[31]*neo* = new; *plas* = growth

Figure 4.31 Foot Ulcer with Gangrene Caused by Diabetes Mellitus.
William Archibald/CDC

Clinical Application 4.2

DIABETIC FOOT ULCERS

One of the most feared complications of diabetes is the development of ulcers in the feet, wounds that may develop gangrene (fig. 4.31). The underlying causes of diabetic foot ulcers are nerve and arterial damage associated with the disease. Lack of pain sensation means that diabetic patients are often unaware of injuries to the feet, and poor circulation results in slow healing and rapid spread of infection. The spread of infection and tissue death associated with foot ulcers may require amputation of toes or feet, although, with appropriate therapy, foot ulcers heal in many patients, and amputation is averted.

Damaged tissues can be repaired in two ways: *regeneration* or *fibrosis*. **Regeneration** is the replacement of dead or damaged cells by the same type of cells as before, thus restoring normal function to the organ. Most skin injuries (cuts, scrapes, and minor burns) heal by regeneration. The liver also regenerates remarkably well. **Fibrosis** is the replacement of damaged tissue with scar tissue, composed mainly of collagen produced by fibroblasts. Scar tissue helps to hold an organ together, but it does not restore normal function. Examples include the healing of severe cuts and burns, the healing of muscle injuries, scars left in the cardiac muscle by earlier heart attacks, and scarring of the lungs in tuberculosis.

4.6b Tissue Shrinkage and Death

The shrinkage of a tissue through a loss in cell size or number is called **atrophy**[32] (AT-ro-fee). It can result from lack of use of an organ, as when muscles shrink from lack of exercise, or from simple aging.

Necrosis[33] (neh-CRO-sis) is the pathological death of tissue due to trauma, toxins, infection, and so forth. One form of necrosis is **infarction,** the sudden death of tissue that occurs when its blood supply is cut off—for example, the *myocardial infarction (MI)* of a heart attack and the *cerebral infarction* of a stroke. Another form of necrosis is **gangrene,** a more slowly developing necrosis usually resulting from an insufficient blood supply and often involving infection of the tissue (see Clinical Application 4.2). A **bed sore (decubitus ulcer)** is a form of gangrene that occurs when immobilized persons, such as those confined to a hospital bed or wheelchair, are unable to move and continual pressure on the skin cuts off blood flow to an area. They are especially common in the hips, sacral region, and ankles, where bone lies close to the surface and tissues between the skin and bone are easily compressed.

Apoptosis[34] (AP-oh-TOE-sis), or **programmed cell death,** is the normal "prearranged" death of cells that have completed their function and best serve the body by dying and getting out of the way. Enzymes released by the cell's own lysosomes degrade its DNA and proteins. The dying cell bubbles and shrinks and is quickly phagocytized and disposed of by a macrophage. Among other functions, apoptosis disposes of unneeded tissue in embryonic development. For example, the foot and

---

[32]*a* = without; *troph* = nourishment
[33]*necr* = death; *osis* = process
[34]*apo* = away; *ptosis* = falling

hand of a fetus start as paddlelike structures. Separation of the fingers and toes occurs by apoptosis of cells that form the initial webs between them. Apoptosis also shrinks the uterus after a woman gives birth. The uterus typically weighs about 900 g at the end of full-term pregnancy, and shrinks by apoptosis to about 60 g within 5 or 6 weeks after birth.

Before You Go On

Answer these questions from memory to test your understanding of the previous section. Reread the material if you can't answer the questions.

21. Tissues can grow through an increase in cell size or cell number. What are the respective terms for these two kinds of growth?
22. Explain the difference between infarction and gangrene.
23. Distinguish between atrophy, necrosis, and apoptosis, and describe a circumstance under which each of these forms of tissue loss may occur.
24. Distinguish between regeneration and fibrosis. Which process restores normal cellular function? What good is the other process if it does not restore function?

CAREER SPOTLIGHT

Histotechnician

Histotechnicians prepare tissues by sectioning, staining, and other methods to produce specimens like those in the photographs in this chapter. Most histotechnicians work in hospitals, clinics, and diagnostic laboratories, preparing specimens to be examined by a pathologist or cytotechnologist for evidence of disease. A career in this field requires a knowledge of biology and chemistry. Good manual dexterity and hand–eye coordination are needed for work with delicate and tiny specimens and sophisticated laboratory equipment. A histotechnician often must work quickly (but carefully), because a surgical team may be waiting in an operating room for biopsy results that depend on the skills of the histotechnician and pathologist team. Histotechnicians are also highly valued team members in industrial research, veterinary medicine, and forensics laboratories.

To become a histotechnician requires at least an associate degree and training in a hospital or histotechnician program. With a bachelor's degree and further training, one can become a histotechnologist, with responsibility for more advanced techniques and more potential for administrative authority. See appendix B for further information.

Leca/Science Source

Study Guide

Assess Your Learning Outcomes

To test your knowledge, discuss the following topics with a study partner or in writing, ideally from memory.

4.1 The Study of Tissues

1. The scope of histology
2. The definition of *tissue* and the four primary types of tissues
3. Three basic components of a tissue
4. How tissues are prepared for microscopic study, and how the sectioning of tissues relates to one's three-dimensional interpretation of their microscopic appearance
5. Terms that describe common cell shapes

4.2 Epithelial Tissue

1. Characteristics of epithelium as a class
2. Where epithelial tissue is found
3. Functions of epithelial tissue
4. What separates an epithelium from adjacent tissues
5. Terminology of the three surfaces of epithelial cells
6. The distinction between simple and stratified epithelia
7. Four kinds of simple epithelium, defining characteristics of each, where each type can be found, and functional advantages of the various types
8. Four kinds of stratified epithelium, defining characteristics of each, where each type can be found, and functional advantages of the various types
9. Structural and functional differences between keratinized and nonkeratinized stratified squamous epithelium, and where each can be found

4.3 Connective Tissue

1. Characteristics of connective tissue as a class
2. Functions of connective tissues
3. Three principal components of fibrous connective tissue
4. Protein fibers and cell types commonly seen in fibrous connective tissues, and their functions
5. Variations in the ground substance of fibrous connective tissue

6. Types of loose connective tissue, the differences between them, and examples of their locations
7. Types of dense connective tissue, the differences between them, and examples of their locations
8. Structure and functions of adipose tissue, and the functional distinction between the two types of adipose tissue
9. Structure of cartilage, its three types, and their differences in composition and location
10. Structure of compact bone and its relationship with spongy bone, bone marrow, and periosteum
11. Components and functions of the blood

4.4 Nervous and Muscular Tissue— The Excitable Tissues

1. Why nervous and muscular tissues are considered to be "excitable" tissues
2. Two types of cells that compose nervous tissue, and the difference between them
3. The basic structure of a neuron
4. Three types of muscular tissue, where they are found, and how they differ in structure and function

4.5 Glands and Membranes

1. The difference between a secretion and an excretion
2. Differences between endocrine and exocrine glands; examples of each
3. Differences between serous, mucous, and mixed glands; examples of each
4. Differences between eccrine, holocrine, and apocrine modes of secretion; examples of each
5. Differences between mucous and serous membranes; examples of each

4.6 Tissue Growth, Repair, and Death

1. Differences between hyperplasia, hypertrophy, and neoplasia, and which of them is pathological
2. Two modes of tissue repair and how they differ
3. Tissue atrophy and its causes
4. Types of tissue necrosis, and how necrosis differs from apoptosis
5. Some normal developmental functions of apoptosis

Testing Your Recall

1. Transitional epithelium (urothelium) is found in
   a. the urinary system.
   b. the respiratory system.
   c. the digestive system.
   d. the reproductive system.
   e. all of the above.

2. The external surface of the stomach is covered by
   a. a mucous membrane.
   b. a serous membrane.
   c. the pleura.
   d. smooth muscle.
   e. a basement membrane.

3. Which of these is the most widespread type of epithelium?
   a. simple cuboidal
   b. simple columnar
   c. pseudostratified columnar
   d. stratified squamous
   e. stratified cuboidal

4. A seminiferous tubule of the testis is lined with _____ epithelium.
   a. simple cuboidal
   b. pseudostratified columnar
   c. stratified squamous
   d. transitional
   e. stratified cuboidal

5. Which of these cells is specialized to engulf and destroy foreign matter?
   a. an erythrocyte
   b. a fibroblast
   c. a macrophage
   d. an adipocyte
   e. a myocyte

6. Tendons are composed of
   a. areolar tissue.
   b. skeletal muscle.
   c. reticular tissue.
   d. dense irregular connective tissue.
   e. dense regular connective tissue.

7. The collagen of areolar tissue is produced by
   a. macrophages.
   b. fibroblasts.
   c. lymphocytes.
   d. leukocytes.
   e. chondrocytes.

8. Energy storage is a function of
   a. adipose tissue.
   b. areolar tissue.
   c. muscular tissue.
   d. stratified squamous epithelium.
   e. fibrocartilage.

9. The shape of the external ear is due to
   a. skeletal muscle.
   b. elastic cartilage.
   c. fibrocartilage.
   d. articular cartilage.
   e. hyaline cartilage.

10. Any gland that releases its product by exocytosis is called
    a. a mucous gland.
    b. a holocrine gland.
    c. an eccrine gland.
    d. a serous gland.
    e. an apocrine gland.

11. Any form of pathological tissue death is called _____.

12. The peritoneum has a _____ type of epithelium.

13. Osteocytes and chondrocytes occupy little cavities called _____.

14. Bones are covered with a fibrous sheath called the _____.

15. Tendons and ligaments are made mainly of the protein _____.

16. Most of the cells in nervous tissue are called _____.

17. An epithelium rests on a layer called the _____ between its deepest cells and the underlying connective tissue.

18. Fibers and ground substance make up the _____ of a connective tissue.

19. A _____ gland is one in which the secretion forms by complete breakdown of the gland cells.

20. Any epithelium in which every cell touches the basement membrane is called a _____ epithelium.

Answers in Appendix A

What's Wrong with These Statements?

Briefly explain why each of the following statements is false, or reword it to make it true.

1. The esophagus is protected from abrasion by a keratinized stratified squamous epithelium.

2. Mucus-secreting goblet cells can be found in both simple and stratified epithelia.

3. Skeletal muscle is defined by the fact that it is all attached to bones.

4. Fibers are part of the ground substance of a connective tissue.

5. In all connective tissues, the matrix occupies more space than the cells do.

6. Adipocytes are limited to adipose tissue.

7. A single exocrine gland can be either serous or mucous but not both.

8. Weight lifting creates new muscle tissue by the process of neoplasia (*neo* means "new").

9. Nerve cells are the only electrically excitable cells in the body.

10. The tissue degeneration that occurs in gangrene is an example of programmed cell death, or apoptosis.

Answers in Appendix A

Testing Your Comprehension

1. A woman in labor is often told to push. In doing so, is she consciously contracting her uterus to expel the baby? Justify your answer based on the muscular composition of the uterus and what you know about that muscle type.

2. One major tenet of biology is the cell theory, which states in part that all bodily structure and function are based on cells. The structural properties of bone, cartilage, and tendons, however, are due more to their extracellular material than to their cells. Is this an exception to the cell theory? Why or why not?

3. The epithelium of the respiratory tract is mostly of the pseudostratified columnar ciliated type, but in the alveoli—the tiny air sacs where oxygen and carbon dioxide are exchanged between the blood and inhaled air—the epithelium is simple squamous. Explain the functional significance of this histological difference. That is, why don't the alveoli have the same kind of epithelium as the rest of the respiratory tract?

PART 2 Support and Movement

The Integumentary System

Chapter 5

A human hair with a few loose squamous epidermal cells clinging to it. The overlapping scales of the hair surface help prevent hairs from being pulled easily from the scalp.

©Eye of Science/Science Source

Chapter Outline

5.1 The Skin and Subcutaneous Tissue
- 5.1a Functions
- 5.1b Structure

5.2 Accessory Organs
- 5.2a Hair
- 5.2b Nails
- 5.2c Cutaneous Glands

Clinical Applications/Perspectives on Health
- Clinical Application 5.1: Transdermal Absorption
- Clinical Application 5.2: The Epidermis and Vitamin D
- Perspectives on Health

End of Chapter
- Aging of the Integumentary System
- Career Spotlight: Dermatology Nurse
- Connective Issues
- Study Guide

BASE CAMP

Before ascending to the next level, be sure you're properly equipped with a knowledge of these concepts from earlier chapters.

- Desmosomes (see "Cell Junctions" in section 3.2d)
- Stratified squamous epithelium (see table 4.3)
- Collagen (see "Fibrous Connective Tissue" in section 4.3b)
- Areolar tissue (see table 4.4)
- Dense irregular connective tissue (see table 4.5)
- Eccrine, apocrine, and holocrine glands (see "Gland Types" in section 4.5a)

 Module 4 Integumentary

The integumentary[1] **system** consists of the skin *(integument)*, hair, nails, and cutaneous glands. We pay more attention to this organ system than to any other. It is, after all, the most visible one. It is a canvas for cultural and personal expression; make-up, tattoos, scarification, and piercings have been practiced in ancient as well as modern times, as is evident in representations of human figures that date from prehistory. Clinically, physical examination of skin, hair, and nails provides information about a person's state of health, and can give clues about disorders as diverse as liver cancer, anemia, lung disease, nutritional deficiencies, hormonal imbalances, and heart failure. The skin is not only the most visible, but also the most vulnerable, of all of our organs, exposed to the sun's ultraviolet radiation, trauma, infection, and injurious chemicals. It comes as no surprise then, that attention to the integumentary system is an important part of a total plan of patient care.

5.1 The Skin and Subcutaneous Tissue

Expected Learning Outcomes

When you have completed this section, you should be able to

a. describe the functions of the skin;
b. name the cell types of the epidermis and identify their functions;
c. name and describe the layers of the epidermis and dermis;
d. explain what accounts for various normal and pathological skin colors; and
e. name the various physical markings of the skin.

The skin is the body's largest and heaviest organ. It is composed of two layers: a superficial epithelium called the *epidermis* and a deeper connective tissue layer called the *dermis* (**fig. 5.1**). Between the skin and muscles is a looser connective tissue layer called the *hypodermis*. This isn't part of the skin, but is customarily studied in conjunction with it. **Dermatology**[2] is the branch of medicine specializing in the care and treatment of the skin.

5.1a Functions

The skin is much more than a container for the body. Its functions are more diverse than commonly supposed.

- **Resistance to trauma and infection.** The skin bears the brunt of most physical injuries to the body, but it resists and recovers from trauma better than other organs do. The epidermal cells are packed with a tough protein called

---

[1] *integument* = covering
[2] *dermat* = skin; *logy* = study of

Figure 5.1 Structure of the Skin and Accessory Organs. The upper left corner of the epidermis is peeled up to show the dermal–epidermal boundary.
APR

keratin and linked by strong desmosomes. Few infectious organisms can penetrate the intact skin. Bacteria and fungi live on the surface in moderate numbers, but their populations are kept in check by the relative dryness of the skin and by a protective acidic film.

- **Water retention.** The skin is an important barrier to water. It prevents the absorption of too much water when you bathe or swim, but even more importantly, it prevents the body from losing too much water.
- **Vitamin D synthesis.** The skin carries out the first step in synthesizing vitamin D, which is needed for bone development and maintenance (see Clinical Application 5.2).
- **Sensation.** The skin is the body's largest sense organ. It is equipped with a variety of nerve endings that react to heat, cold, touch, texture, pressure, vibration, and injury. The skin senses and nerve endings are more fully described in section 10.2.
- **Thermoregulation.** The skin plays a key role in stabilizing body temperature by regulating heat exchange with the environment. It does this by increasing or reducing blood flow close to the body surface and by sweating in response to overheating, as detailed in section 18.6.
- **Nonverbal communication.** The skin is an important means of communication. We possess numerous small facial muscles that act on the dermis to produce subtle and varied expressions.

5.1b Structure

The skin (fig. 5.2) covers the entire body surface—up to 2 m^2 in adults, and accounts for about 15% of the body weight. Most of the skin is 1 to 2 mm thick—comparable to the cover of a hardcover book—but it ranges from 0.5 mm on the

Figure 5.2 Layers and Cell Types of the Epidermis. Compare this to the photograph in figure 4.8. APR

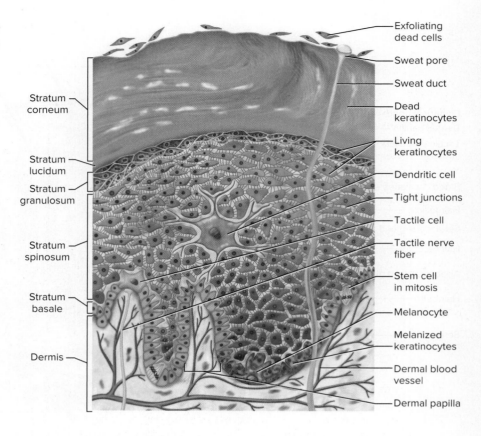

eyelids to 6 mm between the shoulder blades. The difference is due mainly to variation in the thickness of the dermis.

The Epidermis

The **epidermis**[3] is a keratinized stratified squamous epithelium as many as 30 or 40 cells thick in places, composed of the following types of cells (fig. 5.2):

- **Keratinocytes** (keh-RAT-ih-no-sites) are the great majority of epidermal cells. They begin as rounded to cuboidal cells in the deepest layer and, toward the surface, become progressively flatter and packed with the tough protein keratin.
- **Dendritic cells,** scattered among the keratinocytes, probe the epidermis for foreign matter such as toxins and infections and alert the immune system when there is a need to defend the body against invaders.
- **Melanocytes** of the deepest layer of the epidermis produce the dark pigment *melanin,* which protects the deeper tissues from the harmful effects of the ultraviolet rays in sunlight.
- **Tactile cells,** also in the basal layer of the epidermis, play a role in the sense of touch.
- **Stem cells** of the basal layer continually divide and produce new keratinocytes to replace those that die and flake off the skin surface.

The epidermis lacks blood vessels, but the deepest cells receive nourishment from blood vessels in the underlying dermis. As cells move toward the skin surface and away from the blood supply, they die.

---

[3] *epi* = above, upon; *derm* = skin

Clinical Application 5.1

TRANSDERMAL ABSORPTION

The ability of the skin to absorb chemicals makes it possible to administer several medicines as ointments or lotions, or by means of adhesive patches that release the medicine steadily through a membrane. For example, inflammation can be treated with hydrocortisone ointment, nitroglycerine patches are used to relieve heart pain, nicotine patches are used to help overcome tobacco addiction, and other medicated patches are used to control high blood pressure and motion sickness.

Unfortunately, the skin can also be a route for absorption of poisons. These include toxins from poison ivy and other plants; metals such as mercury, arsenic, and lead; and solvents such as acetone (nail polish remover), paint thinner, and pesticides. Some of these can cause brain damage, liver failure, or kidney failure, which is good reason for using protective gloves when handling such substances.

The epidermis is composed of four to five layers. From the deepest layer to the surface, these are as follows:

1. **Stratum basale** (bah-SAIL-ee). This is a deep, single layer of stem cells, keratinocytes, melanocytes, and tactile cells.
2. **Stratum spinosum.** This level consists of many layers of large keratinocytes interspersed with dendritic cells. As keratinocytes are pushed upward by dividing cells below, they cease dividing and synthesize keratin. Keratin accumulation causes the cells to flatten and assume a squamous shape.
3. **Stratum granulosum.** This consists of three to five layers of flat keratinocytes with coarse, dark-staining granules. Keratin filaments in these cells clump together into thick bundles that toughen the epidermis, while the cells become coated with lipid that produces a barrier to water. This is one reason the skin inhibits water loss from the body. The cells die here as the protein barrier cuts them off from nutrients from below.
4. **Stratum lucidum.** This is present only in the thick epidermis of the palms and soles. It is a thin, clear (lucid) layer of dead cells with no nuclei or other visible internal structure.
5. **Stratum corneum.** This consists of up to 40 layers of dead, keratin-packed keratinocytes that give the skin much of its toughness. It is thickest in the palms and soles (*palmar* and *plantar* regions), enabling this skin to withstand the mechanical stresses to which those areas are subjected. The dead cells *exfoliate* (flake off) from the surface as *dander* at a rate compensated by the production of new cells in the stratum basale. A keratinocyte usually lives 30 to 40 days from its "birth" in the stratum basale to its exfoliation from the stratum corneum.

The Dermis

The **dermis** ranges from 0.2 mm thick in the eyelids to 4 mm thick in the palms and soles and 5 or 6 mm thick between the shoulders. It consists mainly of collagen but also contains elastic and reticular fibers and the usual cells of fibrous connective tissue (see section 4.3). It is well supplied with blood vessels, sweat glands, sebaceous glands, and nerve endings. The hair and nails are rooted in the dermis. In the face, skeletal muscles attach to dermal collagen fibers and produce expressions such as a smile or a wrinkle of the forehead. By moving the lips, they also aid in speech.

The boundary between the epidermis and dermis is histologically conspicuous and usually wavy. The upward waves are fingerlike extensions of dermis called **dermal papillae.**[4] They interlock with the epidermis like corrugated cardboard, an arrangement that resists slippage of the epidermis across the dermis and gives the skin more resistance to stress. The whorls and loops of fingerprints are produced by the projecting dermal papillae. If you look closely at the back of your hand, you will see delicate furrows that divide the skin into tiny rectangular to rhomboidal areas. The dermal papillae produce the raised areas between the furrows.

These papillae contain an abundance of blood capillaries, the sole source of nourishment to the bloodless epidermis. If these blood vessels are damaged by such causes as friction from tight shoes, a severe sunburn, or a scald, serous fluid can seep out of the bloodstream and accumulate as a blister, separating the dermis from the epidermis along this boundary.

> **Apply What You Know**
>
> Would you expect to find more prominent dermal papillae in the palms or the forehead? Explain your answer.

The dermis is divided into two layers with only a vague boundary between them. The **papillary**[5] **layer** (PAP-ih-lerr-ee) is a thin zone of areolar tissue directly beneath the epidermis, filling and underlying the dermal papillae. The loose tissue of this layer allows leukocytes to move around easily and attack bacteria or other invaders introduced through breaks in the skin. The deeper **reticular**[6] **layer** is about four-fifths of the dermis. It consists of dense irregular connective tissue (see fig. 4.17), composed mainly of thick bundles of collagen but also containing elastic fibers, fibroblasts, and small clusters of adipocytes. This is a very tough layer; leather is composed of the reticular layer of animal hides. Sweat glands, nail roots, and hair follicles are embedded in this layer.

The Hypodermis

The dermis blends into the underlying **hypodermis**[7] **(subcutaneous tissue).** The hypodermis generally consists of areolar connective tissue and adipose tissue **(subcutaneous fat).** The hypodermis binds the skin to the muscles or other underlying tissues. Subcutaneous fat pads the body, serves as an energy reservoir, and provides thermal insulation. As much as 50% of one's body fat is in the hypodermis. It averages about 8% thicker in women than in men and differs in distribution between the sexes. Drugs are sometimes injected into the hypodermis because it is highly vascular and the medication is quickly absorbed.

Skin Color

The most significant factor in skin color is the brown to black **melanin.**[8] Melanocytes synthesize melanin but transfer it to the keratinocytes, where it accumulates. The dark cells seen in tissue sections are melanized keratinocytes, not melanocytes (fig. 5.3). People of different skin colors have essentially the same number of melanocytes, but in dark skin, the melanocytes produce greater quantities of melanin, the melanin granules in the keratinocytes are more spread out than tightly clumped, and the melanin breaks down relatively slowly. Thus, melanized cells may be seen throughout the stratum basale and stratum spinosum. In light skin, the melanin is

---

[4] *papilla* = little nipple
[5] *ary* = pertaining to, characterized by
[6] *reticul* = little network
[7] *hypo* = below; *derm* = skin
[8] *melano* = black

Figure 5.3 Variations in Skin Pigmentation. (a) The stratum basale shows heavy deposits of melanin in dark skin. (b) Light skin shows little to no visible melanin.

a: (top): ©Tom & Dee Ann McCarthy/Corbis (bottom): Dennis Strete/McGraw-Hill Education b: (top): Creatas/PunchStock (bottom): Dennis Strete/McGraw-Hill Education

❓ Which of the five types of epidermal cells are the melanized cells in part (a)?

less abundant, the granules are more tightly clumped in the keratinocytes so they impart less color to the cells, and it breaks down more rapidly. Melanin is a barrier to ultraviolet (UV) radiation, which otherwise has the potential to cause DNA damage (mutations) and skin cancer. UV rays can also destroy folate, a B vitamin necessary for normal cell division, fertility, and fetal development. A suntan is a protective response to exposure to UV rays, which stimulate melanin synthesis. The melanin accumulates on the "sunny side" of the nucleus of the keratinocytes, where it hangs like a protective umbrella over the DNA. A suntan fades as melanin is degraded in older keratinocytes and as the keratinocytes migrate to the surface and exfoliate. Local concentrations of melanin produce freckles and moles.

Apply What You Know

Skin cancer is relatively rare in people with dark skin. Other than possible differences in behavior, such as intentional suntanning, what could be the reason for this?

Other factors in skin color are hemoglobin and carotene. **Hemoglobin,** the red pigment of blood, imparts reddish to pink hues to the skin as blood vessels show through the white of the dermal collagen. **Carotene**[9] is a yellow pigment acquired from egg yolks and yellow and orange vegetables. It can become concentrated to various degrees in the stratum corneum and subcutaneous fat. It is often most conspicuous in skin with the thickest stratum corneum, such as on the heel and in "corns" or calluses of the feet.

The skin may also exhibit abnormal colors of diagnostic value. **Cyanosis**[10] is blueness of the skin resulting from oxygen deficiency in the blood, as in cases of drowning, emphysema and some other lung diseases, and reduced dermal blood flow in cold weather. **Erythema**[11] (ERR-ih-THEE-muh) is abnormal redness of the skin, as in sunburn, exercise, hot weather, and embarrassment. A **hematoma**[12] is a bruise, a mass of clotted blood showing through the skin. Its causes range from

---

[9] *carot* = carrot
[10] *cyan* = blue; *osis* = condition
[11] *eryth* = red; *em* = blood
[12] *hemat* = blood; *oma* = mass

PERSPECTIVES ON HEALTH

The skin is the most exposed and vulnerable of all of our organs and is subject to a variety of diseases and injuries. Here, we focus on two particularly common and serious disorders: skin cancer and burns.

Skin Cancer

Skin cancer is the most common kind of cancer, affecting about one in five people in the United States. Most cases are due to exposure to UV radiation from the sun, which causes mutations in the DNA of epidermal cells. It is most common in light-skinned people and the elderly, who have had decades of exposure to UV radiation. There are three forms of skin cancer, differing in the cells in which they originate. *Basal cell carcinoma*[13] begins in the stratum basale and eventually invades the dermis. It is the most common but least dangerous form of skin cancer, because it seldom metastasizes. *Squamous cell carcinoma* arises in the stratum spinosum. The chance of recovery is good with early detection and surgical removal, but if neglected, it tends to metastasize to lymph nodes and can be lethal. Basal and squamous cell carcinomas produce similar lesions with raised edges and a concave center (fig. 5.4a and b). *Melanoma* (fig. 5.4c) originates in melanocytes. It is the rarest form of skin cancer (5% of cases) but also the most deadly, because it metastasizes quickly. A useful guideline for recognizing melanoma is the American Cancer Society's *ABCDE rule: A* for asymmetry (one side of the lesion looks different from the other), *B* for border (the contour is not uniform but wavy or scalloped), *C* for color (often a mixture of brown, black, tan, and sometimes red and blue), *D* for diameter (greater than 6 mm, about the diameter of a standard pencil eraser), and *E* for evolving (the mole is changing in size, shape, and color). Depending on the type of cancer and whether it has metastasized, skin cancers are treated by surgical removal, destruction by heat or cold (cryosurgery), radiation, or chemotherapy. Melanoma, particularly, used to have a grim prognosis if it had spread, but newer treatments, including immunotherapies that help the body's immune system attack cancer cells, have improved the outlook for patients.

Burns

Burns can be caused by fires, kitchen spills, excessively hot bath water, electrical shock, strong acids and bases, sunlight, and other forms of radiation. *First-degree burns,* such as most sunburns, damage only the epidermis. They are marked by redness, slight edema (swelling), and pain, but soon heal and seldom leave scars. *Second-degree burns* involve the entire epidermis and part of the dermis. They may be red, tan, or white, and are blistered and very painful. They may take up to several months to heal, and sometimes leave scars. Some sunburns and many scalds are second degree. First- and second-degree burns are also called *partial-thickness burns* because they don't penetrate all the way through the skin. *Third-degree (full-thickness) burns,* a leading cause of accidental death, destroy the epidermis and all of the dermis. Because nothing is left of the dermis in the burned area, the skin can regenerate only from the unburned edges. These burns often require skin grafts and typically result in scarring and sometimes severe disfigurement. The most urgent concerns in treating burn patients are fluid replacement and infection control, because a severe burn destroys the barrier to fluid loss and invading microbes.

(a) Basal cell carcinoma

(b) Squamous cell carcinoma

(c) Melanoma

Figure 5.4 Skin Cancer. (a) Basal cell carcinoma. (b) Squamous cell carcinoma. (c) Melanoma, showing the typical border irregularity, asymmetry, and discoloration.
a: jax10289/Shutterstock b: DR P. Marazzi/Science Photo Library/Alamy c: ©National Cancer Institute

❓ Which of these cancers is the most common? Which is the most dangerous?

---

[13] *carcin* = cancer; *oma* = tumor

Clinical Application 5.2

THE EPIDERMIS AND VITAMIN D

Earlier in this section, we considered some harmful effects of excessive UV radiation on the body. Moderate UV absorption by the skin, however, is necessary for the synthesis of vitamin D, a hormone needed for absorption of dietary calcium. Vitamin D has long been known to be important for skeletal health. Recent studies have shown that it also plays roles in cancer and diabetes prevention, immunity, and cardiovascular health. Epidermal keratinocytes carry out the first step in its synthesis, using the energy of UV radiation to convert a molecule derived from cholesterol to a precursor of vitamin D. The liver and kidneys then convert this to *calcitriol,* the most active form of vitamin D. Inadequate exposure to sunlight can lead to bone deformities such as childhood *rickets* and adult *osteomalacia* ("soft bones"). People of tropical or high-mountain ancestry tend to have dark skin, an adaptation for protection from excess UV. In contrast, people of more northern and southern ancestry tend to have lighter skin that reflects an ancestral adaptation to ensure adequate UV absorption and vitamin D synthesis.

trauma to platelet deficiencies, anticoagulant drugs, and hemophilia. **Pallor** is a temporary pale or ashen color that occurs when there is so little cutaneous blood flow that skin color is dominated by the white dermal collagen. It can result from emotional stress, circulatory shock, cold temperatures, or severe anemia. **Albinism**[14] is a hereditary lack of pigmentation in the skin, hair, and eyes owing to an inability to synthesize melanin. It is characterized by milky white hair and skin and blue-gray eyes. **Jaundice**[15] is yellowing of the skin and whites of the eyes. It occurs in various liver diseases, such as cancer, hepatitis, and cirrhosis, and in some blood diseases with a rapid rate of erythrocyte breakdown.

Before You Go On

Answer these questions from memory to test your understanding of the previous section. Reread the material if you can't answer the questions.

1. How does the skin protect against infectious organisms?
2. Describe two ways in which the skin helps to regulate body temperature.
3. Name the five kinds of epidermal cells and describe their functions.
4. List the five layers of epidermis from deep to superficial. What are the distinctive features of each?
5. What are the two layers of the dermis? What type of tissue composes each layer?
6. Name the pigments responsible for normal skin colors, and explain how certain conditions can produce pathological discoloration of the skin.

---

[14] *alb* = white; *ism* = state, condition
[15] *jaun* = yellow

5.2 Accessory Organs

Expected Learning Outcomes

When you have completed this section, you should be able to

a. describe the histology of a hair and its follicle;
b. explain how a hair grows;
c. discuss the purposes served by various kinds of hair;
d. describe the structure and function of nails;
e. name two types of sweat glands, and describe the structure and function of each; and
f. describe the location, structure, and function of other glands of the skin.

The hair, nails, and cutaneous glands are **accessory organs (appendages)** of the skin. All of them originate in the epidermis of the embryo, but penetrate deeply into the dermis or even the hypodermis of mature skin.

5.2a Hair

A **hair** is a slender filament of keratinized cells that grows from a **hair follicle.** The hair follicle is a diagonal tube of epithelium and connective tissue that penetrates from the skin surface deeply into the dermis or even the hypodermis. Hair and nails are made of a stronger, harder keratin than the keratin of the epidermis. Hair occurs almost everywhere except the lips, nipples, parts of the genitals, palms and soles, and lateral surfaces of the fingers and toes. The density of hairs in a given area does not differ much from one person to another or even between the sexes. Differences in apparent hairiness are due mainly to differences in hair texture, pigmentation, and thickness of individual hairs.

Structure of the Hair and Follicle

The portion of a hair visible above the skin surface is called the **shaft.** The portion below the surface, within the follicle, is the **root.** At the lower end of the root is a swelling called the **bulb** (fig. 5.5a, b). The only living cells of a hair are in or near the bulb. A bit of vascular connective tissue called the **dermal papilla** grows into the bulb and provides the hair with its sole source of nutrition. Just above this is the **hair matrix,** the exclusive site of cellular mitosis and hair growth.

The surface layer of a hair, called the **cuticle,** is composed of thin scaly cells that overlap each other like roof shingles (see this chapter's opening photo). Each scaly cell locks into another cell of the follicle wall that points in the opposite direction, an arrangement that resists pulling the hair out of its follicle when you brush your hair, for example. Beneath the cuticle is the **cortex,** which makes up most of the bulk of the hair. It is composed of several layers of elongated keratinized cells. The thickest hairs, such as the eyebrows and lashes, have a central core called the **medulla,** composed of loosely arranged cells and air spaces. The medulla is lacking from thin scalp and body hairs. In cross section, straight hair is relatively round, whereas tightly curled hair is relatively flat.

The hair follicle has a **root sheath** composed of a layer of epithelium (an extension of the epidermis) alongside the hair and a layer of condensed dermal connective tissue around the epithelial layer. Nerve fibers called **hair receptors** entwine each follicle and respond to hair movements, as when an ant crawls across your arm.

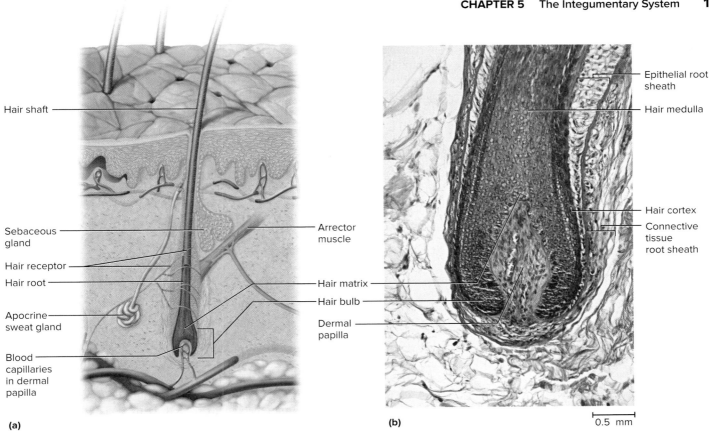

Figure 5.5 Structure of a Hair and Its Follicle.
(a) Anatomy of the follicle and associated structures. (b) Light micrograph of the base of a hair follicle.

b: Ed Reschke/Getty Images

? *In light of your knowledge of hair, discuss the validity of an advertising claim that a shampoo will "nourish your hair." Where and how does a hair get its sole nourishment?*

Each hair has an **arrector muscle,** also known as an *arrector pili*.[16] It contracts in response to cold, fear, and touch, making the hair stand on end and causing "goose bumps." In other mammals, raising the hairs traps an insulating layer of warm air next to the skin or makes the animal appear larger and less vulnerable to a potential enemy. In relatively hairless humans, the effect is not obvious.

Hair color derives from different types and proportions of melanin. Black and brown hair are colored by a form of melanin called *eumelanin.* Another, sulfur-rich melanin called *pheomelanin* causes various shades of blond in moderate amounts, and orange to red hair if more abundant. Gray and white hair result from a combination of air in the medulla and scanty pigment in the cortex.

Hair Growth

Hair grows by mitosis of cells in the hair matrix. Hair and follicle cells together are pushed toward the skin surface as the cells below them multiply. The hair cells become progressively keratinized and die as they are pushed upward, away from the blood supply in the dermal papilla. A typical scalp hair in a young adult grows for 6 to 8 years at a rate of about 1 mm per 3 days (10–18 cm/year). It then stops growing and the follicle atrophies over a period of 2 or 3 weeks. Finally, the hair enters a resting phase 1 to 3 months long, during which it may fall out on its own or be pushed out by a new hair growing beneath it in the same follicle. We typically lose 50 to 100 scalp hairs per day.

Functions of Hair

In comparison to other mammals, the relative hairlessness of humans is so unusual that it raises the question, Why do we have any hair at all? Except on the scalp, it

---

[16] *pilo* = hair

is too sparse to have the heat-retaining function that it serves in other mammals. Scalp hair does, however, help to retain heat in an area where there is an abundant flow of warm blood just under the cranium, but little to no insulating fat. It also protects the scalp from sunburn and skin cancer. Beard, pubic, and axillary hair begin growing at puberty and visually advertise sexual maturity. They also absorb aromatic secretions of specialized scent glands in these areas, discussed shortly. Hair on the trunk and limbs is largely vestigial (an evolutionary remnant of hair that was more functional in our prehuman ancestors), but still serves a sensory role: Stimulation of the hair receptors alerts one to parasites such as ticks and fleas crawling on the skin. The eyebrows accentuate facial expressions produced by muscles that move the skin between and above the eyes. They may also help to reduce the glare of sunlight and keep sweat and debris out of the eyes. Finally, stout protective *guard hairs (vibrissae)* block foreign particles from the nostrils, auditory canals, and eyes. The eyelashes and blink reflex shield the eyes from debris, and the eyelashes reduce drying of the eye by disrupting air currents near the eye surface.

5.2b Nails

Fingernails and toenails are clear, hard derivatives of the stratum corneum. They are composed of thin dead, scaly cells, densely packed together and filled with parallel fibers of hard keratin. Most mammals have claws, whereas humans and other primates have flat nails that are strong keratinized "tools" that can be used for digging, grooming, picking apart food, texting on phones, and other manipulations. The flat nails also support fleshy, sensitive fingertips that let us detect tiny objects and nuanced textures.

The anatomical features of a nail are shown in **figure 5.6**. The **nail plate** is the visible portion covering the fingertip; the **nail bed** is the epidermis underlying most of the plate; the **nail fold** is a thickening of the skin around the margins of the plate; and the **free edge** is the part of the plate that extends beyond the nail bed at the tip of the finger. The **nail matrix** is a growth zone concealed beneath the skin at the proximal edge of the nail. The nail grows by mitosis in the nail matrix, adding new cells to the plate. Fingernails grow at a rate of about 1 mm per week and toenails somewhat more slowly.

The appearance of the fingernails can provide clues for medical diagnoses of conditions such as emphysema and heart disease, among other diseases. Dietary deficiencies may be reflected in nail appearance. For example, iron deficiency may cause nails to become flat or concave (spoonlike) rather than convex.

5.2c Cutaneous Glands

The skin has five types of glands: *eccrine sweat glands, apocrine sweat glands, sebaceous glands, ceruminous glands,* and *mammary glands* (**fig. 5.7**). The two types of sweat glands are also collectively called *sudoriferous glands* (SOO-dor-IF-er-us). Mammary glands are discussed in chapter 19.

Sweat Glands

Most sweat glands are a type called **eccrine (merocrine) sweat glands.** These are widespread in the skin and produce watery perspiration that cools the body. Each begins with a twisted coil in the dermis or hypodermis and an undulating duct that ascends to a sweat pore on the skin surface. Sweat production begins deep in the gland when fluid

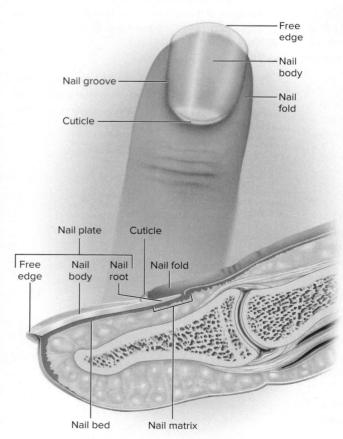

Figure 5.6 Anatomy of a Fingernail. APR

(a) Apocrine gland

Figure 5.7 Cutaneous Glands. (a) Apocrine sweat glands have a large lumen and a duct that conveys their aromatic secretion into a hair follicle. (b) Eccrine sweat glands have a relatively narrow lumen and a duct that opens by way of a pore on the skin surface. (c) Sebaceous glands have cells that break down in entirety to form an oily secretion that is released into the hair follicle.

a–c: Dennis Strete/McGraw-Hill Education

(b) Eccrine gland

(c) Sebaceous gland

from the blood is filtered into the lumen of the gland. Sweat is about 99% water; it contains potassium, sodium, chloride, ammonia, urea, and lactic acid, and it has a pH ranging from 4 to 6. Each day, these glands secrete about 500 mL of *insensible perspiration,* which doesn't produce noticeable wetness of the skin. Under conditions of exercise or heat, however, one may lose as much as a liter of perspiration per hour. Sweating with visible wetness of the skin is called **diaphoresis**[17] (DY-uh-for-EE-sis). Sweating, however, is not the only way the skin loses water. A significant amount is lost by water simply diffusing between the keratinocytes and evaporating from the skin surface—a process called *cutaneous transpiration.*

Apocrine[18] **sweat glands** occur in the groin, anal region, axilla, areola, and male beard area, although they are sparse or absent from some Asian ethnic groups. Their ducts lead into hair follicles rather than directly to the skin surface. The *apocrine* name comes from an earlier misunderstanding about their method of secretion; it is now known that they produce sweat in the same way that eccrine glands do—by exocytosis. The secretory part of an apocrine gland, however, stores up secretion and has a much larger lumen than that of an eccrine gland, so these glands have continued to be referred to as apocrine glands to distinguish them from the eccrine type. Apocrine sweat contains more fatty acids than eccrine sweat, making it thicker and more cloudy.

---

[17] *dia* = through; *phoresis* = carrying
[18] *apo* = away from; *crin* = to separate

Apocrine sweat glands are scent glands that respond especially to stress and sexual stimulation. They are not active until puberty. In women, they enlarge and shrink in phase with the menstrual cycle. These facts, as well as experimental evidence, suggest that apocrine glands secrete *pheromones*—chemicals that influence the physiology or behavior of other members of the species. They apparently correspond to scent glands that develop in other mammals as they approach sexual maturity. Many mammals have apocrine glands associated with specialized tufts of hair. The pubic and axillary hair in humans probably serve to retain the secretion and prevent it from evaporating too quickly, so it can accumulate and exercise a stronger pheromonal effect.

Sebaceous Glands

Sebaceous[19] **glands** (see-BAY-shus) produce an oily secretion called **sebum** (SEE-bum). They are flask-shaped, with short ducts that usually open into a hair follicle. These are holocrine glands whose secretion is formed by the breakdown of entire gland cells (see fig. 4.29c). Sebum keeps the skin and hair from becoming dry, brittle, and cracked. The sheen of well-brushed hair is due to sebum distributed by the hairbrush.

Ceruminous Glands

Ceruminous glands (seh-ROO-mih-nus) are found only in the auditory (ear) canal, where their secretion combines with sebum and dead epidermal cells to form earwax, or **cerumen.**[20] They are simple, coiled, tubular glands with ducts leading to the skin surface. Cerumen keeps the eardrum pliable, waterproofs the auditory canal, and has a bactericidal effect.

No organ system operates in isolation from the others; indeed, every system affects every other system in ways great and small. Beginning with this chapter, every chapter that introduces an organ system for the first time has a Connective Issues feature to underscore the functional connections between that system and the others. These features should be studied for deeper appreciation of the human body as an integrated whole.

Before You Go On

Answer these questions from memory to test your understanding of the previous section. Reread the material if you can't answer the questions.

7. Name and define the three regions of a hair from its base to its tip, and the three layers of a thick hair seen in cross section.
8. State the functions of the hair papilla, hair receptors, and arrector muscle.
9. State a reasonable theory for the functions of hair of the eyebrows, eyelashes, scalp, nostrils, and axilla.
10. Define or describe the nail plate, nail fold, and nail matrix.
11. How do eccrine and apocrine sweat glands differ in structure and function?
12. What other type of gland is associated with hair follicles? How does its mode of secretion differ from that of sweat glands?

---

[19] *seb* = fat, tallow; *aceous* = possessing
[20] *cer* = wax

Aging of the Integumentary System

Aging is especially noticeable in the integumentary system, and nearly all people have concerns or complaints about it as they age. Hair turns grayer and thinner. The skin and hair become drier because of atrophy of the sebaceous glands. The loss of collagen and elastin from the dermis makes the skin thinner, translucent, and looser. Aged skin also has fewer blood vessels than younger skin, and those that remain are more fragile. Consequently, aged skin bruises more easily; the skin can become reddened as broken vessels leak into the connective tissue; and poorer circulation results in slower healing of skin injuries. The density of cutaneous nerve endings declines by two-thirds from the age of 20 to the age of 80. The skin is consequently less sensitive to touch, pressure, and injurious stimuli. Dendritic cells of the skin also decline in number, leaving the skin more susceptible to recurring infections.

The atrophy of cutaneous blood vessels, sweat glands, and subcutaneous fat makes it more difficult to regulate body temperature in old age. Older people are more vulnerable to hypothermia in cold weather and heatstroke in hot weather. Heat waves and cold spells take an especially heavy toll among elderly poor people, who suffer from a combination of reduced homeostasis and inadequate housing.

In addition to all these normal aspects of growing older, the skin also suffers *photoaging*—degenerative changes from a lifetime of exposure to UV radiation. This accounts for more than 90% of the integumentary changes that people find medically troubling or cosmetically disagreeable: age spots, wrinkling, yellowing and mottling of the skin, and skin cancer. All of these affect the face, hands, and arms more than areas of the body that receive less exposure to the sun.

CAREER SPOTLIGHT

Dermatology Nurse

A dermatology nurse is an Advanced Practice nurse, or Clinical Nurse Specialist, with a master's or doctoral degree. Dermatology nurses specialize in skin and wound care; assist dermatologists in the medical, surgical, and cosmetic treatment of the skin; and assist in treating conditions as diverse as psoriasis, burns, skin cancer, skin grafts, and gunshot wounds. Entry into the field of dermatology nursing requires a bachelor's degree and licensure as a registered nurse (RN). With 2 years of experience in dermatology nursing as an RN, one may sit for a board exam to become a *dermatology nurse certified (DNC)* or a *dermatology certified nurse practitioner (DCNP)*. See appendix B for further details on careers in dermatology nursing.

Andrey_Popov/Shutterstock

CONNECTIVE ISSUES

Ways in Which the INTEGUMENTARY SYSTEM Affects Other Organ Systems

All Systems
The integumentary system serves all other systems by providing a physical barrier to environmental hazards and minimizing water loss from the body.

Skeletal System
The role of the skin in vitamin D synthesis promotes calcium absorption needed for bone growth and maintenance.

Muscular System
Vitamin D synthesis promotes absorption of calcium needed for muscle contraction; skin dissipates heat generated by muscles.

Nervous System
Sensory impulses from the skin are transmitted to the nervous system.

Endocrine System
Vitamin D acts as a hormone.

Circulatory System
Dermal vasoconstriction diverts blood from the skin to other organs; the skin prevents loss of fluid from the cardiovascular system.

Lymphatic and Immune Systems
Dendritic cells detect foreign substances and alert the immune system.

Respiratory System
Nasal hairs filter particles that might otherwise be inhaled.

Urinary System
The sweat glands complement the urinary system by excreting salts and some nitrogenous wastes.

Digestive System
Vitamin D promotes calcium absorption by the small intestine.

Reproductive System
Cutaneous receptors respond to sexual stimuli; mammary glands produce milk to nourish infants; apocrine glands produce scents with subtle sexual functions.

Study Guide

Assess Your Learning Outcomes

To test your knowledge, discuss the following topics with a study partner or in writing, ideally from memory.

5.1 The Skin and Subcutaneous Tissue

1. The organs that constitute the integumentary system
2. Six functions of the skin
3. The two layers of skin proper and the layer of tissue immediately beneath the skin
4. The relative thickness and one other structural specialization of the epidermis in the palms and soles
5. Five types of cells in the epidermis, their locations, and the function of each
6. Five layers of the epidermis, characteristics of each one, and which one is missing from thin skin
7. The life history of a keratinocyte from its birth at the base of the epidermis to its death and exfoliation from the epidermal surface; the relationship between these events, the toughness of the epidermis, and the epidermal role as a water barrier
8. Composition and layers of the dermis; dermal papillae and their significance
9. The difference between the hypodermis and dermis; functions of the hypodermis; and an alternative name for the hypodermis in regions where it is mostly adipose tissue
10. The pigments that account for the normal variation in skin colors; names of six pathological changes in skin color and what accounts for each

5.2 Accessory Organs

1. Structures that are considered to be the *accessory organs* of the integumentary system
2. Definitions of *hair* and *hair follicle*
3. How the keratin of hair differs from that of the epidermis
4. The distribution of body hair and relatively hairless regions of the skin
5. Three regions of a hair from base to tip; the sites of hair nutrition, cell division, and growth
6. Three zones of a hair seen in a cross section, and how cross-sectional shape relates to the texture of hair
7. The chemical basis for brown, black, blond, red, gray, and white hair
8. Structure of a hair follicle and its associated nerve ending and muscle
9. The process and duration of hair growth
10. Functions of scalp, beard, pubic, axillary, eyebrow, trunk, eyelash, and nasal hairs
11. Structure of a nail and nail bed
12. Some functions of the nails and the advantage of the flat nails of humans over the claws of other mammals
13. Similarities and differences in structure, function, and locations of eccrine and apocrine sweat glands
14. How sweat is produced, what it contains, and how diaphoresis differs from insensible perspiration
15. How sebaceous glands differ from sweat glands in location, structure, mode of secretion, and function
16. The location and function of ceruminous glands and composition of cerumen

Testing Your Recall

1. Cells of the _____ are keratinized and dead.
    a. papillary layer
    b. stratum spinosum
    c. stratum basale
    d. stratum corneum
    e. stratum granulosum

2. The outermost layer of a hair is called
    a. the medulla.
    b. the cuticle.
    c. the root sheath.
    d. the papilla.
    e. the cortex.

3. Which of the following skin conditions or appearances would most likely result from liver failure?
    a. pallor
    b. erythema
    c. cyanosis
    d. jaundice
    e. melanization

4. Which of the following has/have the thickest epidermis?
    a. the palms
    b. the region between the shoulder blades
    c. the elbows and knees
    d. the chest
    e. the cheeks

5. The integumentary system includes all of the following except
   a. the epidermis.
   b. the nails.
   c. the sweat glands.
   d. the dermis.
   e. the hypodermis.

6. The thickest *living* layer of the epidermis is
   a. the stratum basale.
   b. the stratum spinosum.
   c. the stratum granulosum.
   d. the stratum lucidum.
   e. the stratum corneum.

7. Which of the following is a scent gland?
   a. a holocrine gland
   b. a sebaceous gland
   c. an apocrine gland
   d. a ceruminous gland
   e. an eccrine gland

8. _____ are skin cells with a sensory role.
   a. Tactile cells
   b. Dendritic cells
   c. Stem cells
   d. Melanocytes
   e. Keratinocytes

9. The function of epidermal dendritic cells is
   a. immune defense.
   b. protection from UV radiation.
   c. water retention.
   d. the sense of touch.
   e. mitotic replacement of dead keratinocytes.

10. Red hair gets its color primarily from
    a. eumelanin.
    b. keratin.
    c. hemoglobin.
    d. collagen.
    e. pheomelanin.

11. _____ is sweating without noticeable wetness of the skin.

12. A muscle that causes a hair to stand erect is called a/an _____.

13. Two common word roots that refer to the skin in medical terminology are _____ and _____.

14. Blueness of the skin due to low oxygen concentration in the blood is called _____.

15. Upward projections of the dermis along the dermal–epidermal boundary are called _____.

16. Cerumen is more commonly known as _____.

17. The holocrine glands that secrete into a hair follicle are called _____.

18. The most abundant protein of the epidermis is _____, while the most abundant protein of the dermis is _____.

19. The epidermis, liver, and kidneys work together to synthesize _____, which promotes absorption of dietary calcium.

20. Epidermal melanin is produced by melanocytes but accumulates in the _____.

Answers in Appendix A

What's Wrong with These Statements?

Briefly explain why each of the following statements is false, or reword it to make it true.

1. Secretion of the oil of the skin and hair is called *exfoliation*.

2. The term *integumentary system* means only the skin, but *integument* refers also to the hair, nails, and cutaneous glands.

3. The dermis is composed mainly of keratin.

4. Vitamin D is synthesized by certain cutaneous glands.

5. Cells of the stratum granulosum multiply rapidly to produce the thick stratum corneum of the skin surface.

6. Hair growth results from the mitosis of cells throughout the root, but as these cells emerge from the follicle and are exposed to air, they die and become the hair shaft.

7. The three layers of the skin are the epidermis, dermis, and hypodermis.

8. People of African descent have a much higher density of epidermal melanocytes than do people of northern European descent.

9. Pallor indicates a genetic lack of melanin.

10. Apocrine scent glands serve mainly to help cool the body.

Answers in Appendix A

Testing Your Comprehension

1. Many organs of the body contain numerous smaller organs, perhaps even thousands. Describe an example of this in the integumentary system.
2. Explain how the two histological layers of the dermis reflect two different functions of the dermis.
3. Why is it important for the epidermis to be effective, but not too effective, in screening out UV radiation?

The Skeletal System

Chapter 6

Chapter Outline

6.1 Skeletal Structure and Function
    6.1a Functions of the Skeletal System
    6.1b Osseous (Bone) Tissue
    6.1c Gross Anatomy of Bones

6.2 Bone Development and Metabolism
    6.2a Ossification
    6.2b Growth and Remodeling
    6.2c Mineral Homeostasis

6.3 The Axial Skeleton
    6.3a The Skull
    6.3b The Vertebral Column
    6.3c The Thoracic Cage

6.4 The Appendicular Skeleton
    6.4a The Pectoral Girdle and Upper Limb
    6.4b The Pelvic Girdle and Lower Limb

6.5 Joints
    6.5a Classification
    6.5b Movement
    6.5c Anatomy of Selected Synovial Joints

Clinical Applications/Perspectives on Health

- Clinical Application 6.1: Rickets
- Clinical Application 6.2: Knee Injuries and Arthroscopic Surgery
- Perspectives on Health

End of Chapter

- Aging of the Skeletal System
- Career Spotlight: Orthopedic Nurse
- Connective Issues
- Study Guide

Colorized X-ray of a human knee prosthesis. The durable bonding of such artificial joints with natural bone has been a major achievement of biomedical engineering.
Dr. P. Marazzi/Science Source

BASE CAMP

Before ascending to the next level, be sure you're properly equipped with a knowledge of these concepts from earlier chapters.

- Terminology of body planes and anatomical directions (see "Anatomical Position" and "Anatomical Planes" in sections 1.3b and c)
- Histology of hyaline cartilage (see fig. 4.19 in table 4.7)

Module 5
Skeletal System

In art and history, nothing has symbolized death more than a skull or skeleton.[1] Bones and teeth are the most durable remains of a once-living body and the most vivid reminder of the impermanence of life. The clean dry bones presented for laboratory study suggest that the skeleton is nonliving—an inert scaffold for the body, like the steel beams of a building. Seeing it in such a sanitized form makes it easy to forget that the living skeleton is made of dynamic tissues, full of cells—that it continually remodels itself and interacts physiologically with all of the other organ systems of the body. The skeleton is permeated with nerves and blood vessels, which attests to its sensitivity and metabolic activity. Indeed, if the skeleton were nonliving, a fractured bone would never heal.

This chapter concerns a family of biomedical sciences related to the skeleton—**osteology,**[2] the study of bones; **arthrology,**[3] the study of joints; and **kinesiology,**[4] the study of musculoskeletal movement.

6.1 Skeletal Structure and Function

Expected Learning Outcomes

When you have completed this section, you should be able to

a. name the organs and tissues that compose the skeletal system;
b. state several functions of the skeletal system;
c. describe the gross anatomy of bones;
d. distinguish the two types of bone marrow; and
e. identify features of the microscopic anatomy of bone, including the two types of bone tissue, four types of bone cells, and components of the noncellular matrix.

The **skeletal system** is composed of bones, cartilage, and ligaments joined tightly to form a strong, flexible framework for the body. Cartilage is the precursor of most bones in embryonic and childhood development, and it covers many joint surfaces in the mature skeleton. **Ligaments** hold bones together at the joints. The skeletal system also includes **bone marrow,** the soft bloody or fatty material enclosed in the bones.

6.1a Functions of the Skeletal System

The principal functions of the skeletal system are as follows:

- **Support.** Bones of the limbs and vertebral column support the body; the mandible and maxilla support the teeth; and some viscera are supported by nearby bones.
- **Protection.** Bones enclose and protect the brain, spinal cord, lungs, heart, and pelvic viscera.
- **Movement.** Movements of the limbs, and other movements such as breathing, are produced by the action of muscles on bones.
- **Blood formation.** Red bone marrow is the major producer of blood cells.

---

[1]*skelet* = dried up
[2]*osteo* = bone; *logy* = study of
[3]*arthro* = joint; *logy* = study of
[4]*kinesio* = movement; *logy* = study of

- **Storage.** The skeleton is the body's main reservoir of calcium and phosphorus, and releases them according to the body's physiological needs. The fatty bone marrow present in many bones also serves as one of the body's reserves of stored fuel.

Recently, the skeleton has been found to have broader functions than those listed above. It turns out that the skeleton affects blood sugar regulation, fertility, and brain function in mammals. These new findings underscore the dynamic nature of bone and its relevance to many aspects of health and disease.

6.1b Osseous (Bone) Tissue

Bone, or **osseous**[5] **tissue**, is a connective tissue with a hard, calcified matrix. It exists in two forms: *compact (dense) bone,* which consists of a solid matrix, and *spongy bone,* which is a porous lattice honeycombed with spaces. Spongy bone is always found in the interior of a bone, covered with compact bone of the surface (fig. 6.1). The term *bone* can denote just the osseous tissue, or it can mean an organ such as the femur that is composed of many tissues. Besides osseous tissue, a bone contains blood, nervous tissue, cartilage, adipose tissue, and fibrous connective tissue.

Bone Cells

Like any other connective tissue, bone consists of cells, protein fibers, and ground substance. There are four kinds of bone cells (fig. 6.2):

1. **Osteogenic**[6] **cells** occur on the bone surface, beneath the fibrous connective tissue membrane that covers a bone. They are stem cells that

---

[5]*os, osse, oste* = bone
[6]*osteo* = bone; *genic* = producing

Figure 6.1 Two Types of Osseous Tissue. In this frontal section of the femur, we can see the dense white shell of compact bone on the surface, enclosing more porous spongy bone.
B Christopher/Alamy Stock Photo

? *In this dried specimen, the spongy bone appears full of air spaces. What would occupy that space in a living bone?*

Figure 6.2 Bone Cells and Their Development. (a) Osteogenic cells give rise to osteoblasts, which deposit matrix around themselves and transform into osteocytes. (b) Osteoclasts form by the fusion of bone marrow stem cells.

? *Why do osteoblasts have more rough endoplasmic reticulum than osteogenic cells do?*

(a) Osteocyte development

(b) Osteoclast development

give rise to osteoblasts. They are the only bone cells capable of dividing and making more bone cells.
2. **Osteoblasts**[7] are bone-forming cells that synthesize the organic matter of the bone and promote its mineralization. They have abundant mitochondria and rough endoplasmic reticulum to support these functions. Osteoblasts lie in a single layer on the bone surface, somewhat resembling a cuboidal epithelium.
3. **Osteocytes** make up over 90% of all bone cells. These mature cells are former osteoblasts that have become trapped in the matrix they deposited. They reside in cavities called **lacunae**[8] (la-CUE-nee), which are connected to each other by slender channels called **canaliculi**[9] (CAN-uh-LIC-you-lye). Each osteocyte has delicate cytoplasmic processes that reach into the canaliculi to meet those of neighboring osteocytes. Osteocytes pass nutrients and chemical signals to each other through gap junctions at the tips of their cytoplasmic processes, and pass metabolic wastes to the nearest blood vessel for disposal. Osteocytes have multiple functions. They orchestrate bone remodeling by directing both bone resorption and deposition, and play a key role in the maintenance of bone density. In addition, they sense mechanical stress, such as that produced by weight-bearing exercise, and secrete signals that adjust bone shape and density to adapt. This is why regular weight-bearing exercise helps to preserve bone integrity and strength.
4. **Osteoclasts**[10] are bone-dissolving cells that develop from a separate line of bone marrow stem cells. They are exceptionally large (up to 150 μm in diameter), have multiple nuclei, and lie on the bone surface like osteoblasts. On the side facing the bone, an osteoclast has an unusual comblike row of infoldings of the plasma membrane, called the *ruffled border*. Along this surface, the cell secretes hydrochloric acid and enzymes to dissolve osseous tissue. Osteoclasts often lie in pits that they have eroded into the bone surface.

Bone Matrix

The **matrix** of osseous tissue is the stony matter that surrounds the osteocytes and lacunae. About one-third of it by weight is organic matter, mainly collagen and other proteins. Two-thirds of it is inorganic, mainly calcium phosphate with lesser amounts of other minerals.

The collagen and minerals form a composite material that gives bones a combination of flexibility and strength. The minerals enable the bones to support weight without sagging. When bones are deficient in minerals, they are soft and bend easily. This is seen to some degree in the mineral-deficient, easily deformed bones of childhood *rickets* and the similar adult disease, *osteomalacia* ("soft bones"). Collagen, on the other hand, gives bones a degree of flexibility so the bone can bend slightly without snapping. When you run, for example, each time your body weight comes down on one leg, the long bones in that leg bend slightly. Without collagen, the bones would be so brittle they would simply crack or shatter in such everyday actions as lifting weights and running (see "Collagen Diseases" in Perspectives on Health, section 2.3d).

Spongy and Compact Bone

Spongy (cancellous) bone (fig. 6.3a, b) consists of a porous lattice of slender rods and plates called **trabeculae**[11] (tra-BEC-you-lee). Although calcified and

---
[7] *osteo* = bone; *blast* = form, produce
[8] *lac* = lake, hollow place; *una* = little
[9] *canal* = canal, channel; *icul* = little
[10] *osteo* = bone; *clast* = destroy, break down
[11] *trabe* = plate; *cul* = little

hard, it is permeated by tiny spaces that give it a spongelike appearance. These spaces are filled with bone marrow and small blood vessels. Spongy bone is well designed to impart strength to a bone without adding too much weight.

Compact (dense) bone (fig. 6.3a, c) forms the hard outer shell of the bone. It prevents the bone marrow from seeping out and provides solid attachment surfaces for muscles, tendons, and ligaments. Most of it is organized in elongated cylinders called **osteons**, which are similar to tree trunks. In cross sections, they look like tree rings, with concentric layers called

Figure 6.3 Histology of Osseous Tissue. (a) The three-dimensional structure of compact bone. (b) Microscopic appearance of decalcified spongy bone. (c) Microscopic appearance of a cross section of an osteon of dried compact bone. APR

b: Biophoto Associates/Science Source; c: Custom Medical Stock Photo/Newscom

❓ *Which type of bone, spongy or compact, has more surface area exposed to osteoclast action? Where do you think the most bone tissue would be lost if the osteoclasts were overactive?*

lamellae arranged around a **central canal,** which contains small blood vessels and nerves. The osteocytes occupy lacunae between the lamellae.

The skeleton is richly supplied with blood. Blood vessels enter through small holes that pierce the bone surface and lead to central canals. The wall of a central canal is riddled with innumerable pores through which the innermost osteocytes receive nutrients from the bloodstream. They pass them to neighboring osteocytes through the network of canaliculi. Wastes travel in the other direction, toward the central canal, to be removed by the bloodstream.

6.1c Gross Anatomy of Bones

The most important bones for movement are the *long bones* of the limbs, which are platforms for skeletal muscle attachment. **Figure 6.4** shows the general

Figure 6.4 General Anatomy of a Long Bone.
The femur is shown as an example.

anatomy of a long bone, in this case, the femur. The elongated midsection is called the **shaft** or **diaphysis**[12] (dy-AF-ih-sis), and each expanded end is called a **head** or **epiphysis**[13] (eh-PIF-ih-sis). The shaft provides leverage whereas the heads are enlarged to strengthen a joint and provide added surface area for the attachment of tendons and ligaments. The ends where two bones meet are covered with a thin layer of hyaline cartilage called the **articular cartilage,** which eases joint movements. Each head and nearby region of the shaft is filled with spongy bone. **Bone marrow** occupies the **marrow (medullary) cavity** (MED-you-lerr-ee) of the shaft and the spaces in the spongy bone of the head. Adult bones often exhibit **epiphyseal lines,** a remnant of the childhood growth zone.

Flat bones (fig. 6.5) are shieldlike plates that protect delicate organs, such as the sternum anterior to the heart and the cranial bones that enclose the brain. Some flat bones are platforms for muscle attachment, as in the case of the scapula (shoulder blade) and hip bone. In flat bones, two layers of compact bone enclose a middle layer of spongy bone like a sandwich.

Some bones fit the description of neither long nor flat bones and are sometimes called *short* and *irregular* bones (such as wrist bones and vertebrae, respectively).

Externally, a bone is covered with a fibrous sheath called the **periosteum**[14] (see fig. 6.3a). Collagen fibers of the periosteum penetrate into the bone itself and are also continuous with the tendons that bind muscle to bone. The periosteum provides strong attachment and continuity from muscle to tendon to bone. There is no periosteum over the articular cartilage. The internal surface of a bone is lined with **endosteum,**[15] a thin layer of reticular connective tissue separating the bone from the bone marrow.

There are two types of bone marrow—red and yellow. **Red bone marrow,** which serves to produce blood cells and platelets, fills nearly every bone of a child's skeleton, but has a more limited distribution in adults (fig. 6.6). It is an organ in itself, consisting of a soft but very structured network of delicate blood vessels surrounded by reticular tissue and islands of blood-forming cells. As we mature, much of the red bone marrow is gradually replaced by fatty **yellow bone marrow,** like the fat seen at the center of a ham bone. Yellow marrow dominates the long limb bones of the adult.

Figure 6.5 General Anatomy of a Flat Bone. The example here is the parietal bone from the top of the skull. APR

a: Christine Eckel/McGraw-Hill Education b: Science Photo Library/Alamy Stock Photo

> **Apply What You Know**
>
> Predict at least two functional problems that would result if a mutation caused someone's bones to fail to develop the outer shell of compact bone.

---

[12]*dia* = across; *physis* = growth; originally named for a ridge on the shaft of the tibia
[13]*epi* = upon, above; *physis* = growth
[14]*peri* = around; *oste* = bone
[15]*endo* = within; *oste* = bone

Figure 6.6 Distribution of Red and Yellow Bone Marrow. In an adult, red bone marrow occupies the marrow cavities of the axial skeleton and proximal heads of the humerus and femur. Yellow bone marrow occurs in the long bones of the limbs.

Before You Go On

Answer these questions from memory to test your understanding of the previous section. Reread the material if you can't answer the questions.

1. State three components of the skeleton other than bones and three functions of the skeleton other than support and movement.
2. Name the four types of cells in bone tissue and state their functions.
3. Compare the roles of the collagen and minerals in bone tissue.
4. Describe the differences between compact and spongy bone, and their spatial relationship to each other.
5. Sketch or describe one osteon, including its central canal, lamellae, lacunae, and canaliculi.
6. Sketch or describe a long bone, including its shaft and heads, and indicate where its articular cartilages, periosteum, and endosteum are located.
7. State the differences in location, composition, and function of red and yellow bone marrow.

6.2 Bone Development and Metabolism

Expected Learning Outcomes

When you have completed this section, you should be able to

a. describe the two processes of bone formation from embryonic tissues;
b. describe how bones are remodeled throughout life; and
c. identify hormones and other factors that govern bone deposition, bone resorption, and calcium balance.

6.2a Ossification

The formation of bone is called **ossification** (OSS-ih-fih-CAY-shun). There are two methods of ossification—*intramembranous* and *endochondral*.

Intramembranous Ossification

Intramembranous[16] **ossification** (IN-tra-MEM-bruh-nus) produces the flat bones of the skull. It begins with a sheet of soft embryonic connective tissue—the "membrane" for which the process is named. Embryonic cells deposit collagen in this membrane, then mineral salts are deposited on the collagen to harden the tissue. This creates the thin calcified plates, or trabeculae, typical of spongy bone. In the meantime, the inner and outer surfaces of cranial bones are calcified, producing the sandwichlike arrangement of compact bone on the surface and spongy bone in the middle. When a baby is born, there are still gaps *(fontanels)* between the cranial bones composed of soft fibrous membrane that has not yet calcified.

---

[16]*intra* = within; *membran* = membrane

Endochondral Ossification

Endochondral[17] ossification (EN-doe-CON-drul) produces most of the bones of the body. This process begins with a hyaline cartilage "model" in the approximate shape of the bone to come. It begins around the sixth week of fetal development and continues into a person's 20s. **Figure 6.7** follows the steps of endochondral ossification in the *metacarpal bone* from the palm of the fetal hand, chosen for its anatomical simplicity compared to other long bones such as the femur. Correlate the following steps with the figure as you read the story.

① The future bone begins as hyaline cartilage covered with perichondrium.

② In a *primary ossification center* near the middle of this cartilage, chondrocytes begin to enlarge and die, while the thin walls between them calcify. Meanwhile, osteoblasts of the perichondrium deposit a thin layer of bone around the cartilage model, forming a collar that reinforces the shaft of the future bone. The fibrous sheath around this collar is then considered to be periosteum.

③ Blood vessels pierce the periosteum at the primary ossification center. Stem cells arrive with this blood, differentiate into osteoclasts, and digest calcified tissue, hollowing out the shaft and creating the *primary marrow cavity*. Other stem cells arrive and become osteoblasts, which deposit layers of bone, thickening the shaft. About this time, a *secondary ossification center* develops at one end of the bone, resembling the primary ossification center in the center.

④ By the time of a child's birth, the secondary ossification center has hollowed out a secondary marrow cavity at that end, and another ossification center has begun to appear at the other end of the bone.

⑤ Throughout childhood and adolescence, the primary and secondary marrow cavities are separated by the **epiphyseal plate** (EP-ih-FIZ-ee-ul). It consists of a middle layer of hyaline cartilage bordered on each side by an area where cartilage transitions to bone. The plate functions as a growth zone enabling the individual to grow in limb length and height.

[17]*endo* = within; *chondr* = cartilage

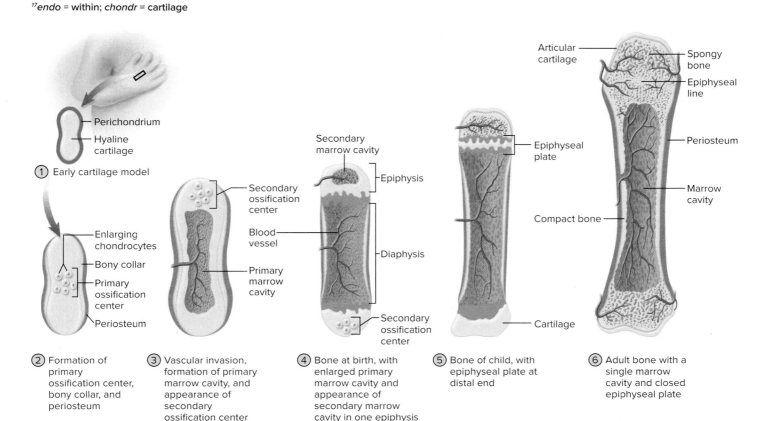

Figure 6.7 Endochondral Ossification. The metacarpal bones of the hand exhibit a relatively uncomplicated case of endochondral ossification. Many other bones have two or more epiphyseal plates.

(6) By the late teens to early twenties, the cartilage in the epiphyseal plate is depleted and the primary and secondary marrow cavities unite into a single medullary cavity. The bones can grow no longer, and one attains his or her maximum adult height.

6.2b Growth and Remodeling

Bones continue to grow and remodel themselves throughout life, changing size and shape to accommodate the changing forces applied to the skeleton. For example, in children the limbs become longer; the curvature of the cranium changes to accommodate a growing brain; and many bones develop bumps, spines, and ridges as a child begins to walk and the muscles exert tension on the bones. Tension on the skeleton stimulates an increase in bone mass. On average, therefore, the bones of people who exercise regularly or who have jobs that involve heavy manual labor have greater mass than the bones of sedentary people. In tennis players, the bones of the racket arm and the clavicle on that side are often more robust than those of the other arm.

Throughout childhood, limb bones elongate as the epiphyseal plates thicken (by the multiplication and enlargement of chondrocytes) and ossify. In children, the epiphyseal plate appears on X-rays as a transparent line across the end of a bone, creating the appearance of a gap separating the epiphysis from the diaphysis. In the late teens to late twenties—at different ages in different bones—the epiphyseal plates close as the gaps fill in with bone. Fusion of the plates prevents further growth and the person has attained full stature.

Throughout life, bones also thicken and change shape by the addition of new tissue to the bone surface. Mature bone is limited to this type of growth. As in the earlier-described intramembranous ossification, surface osteoblasts deposit osteoid tissue, calcify it, and become trapped in it as osteocytes. This tissue is laid down in layers parallel to the surface, not in cylindrical osteons like the tissue deeper in the bone. At the same time that deposition occurs at the outer surface of a bone, osteoclasts dissolve bone on the inner surface and thus enlarge the medullary cavity. **APR**

6.2c Mineral Homeostasis

Even after a bone is fully formed, it remains a metabolically active organ. Not only is it involved in its own maintenance, growth, and remodeling, but it also exerts a profound influence on the rest of the body by exchanging minerals with the extracellular fluid. The skeleton is the body's primary reservoir of calcium and phosphate. These minerals are used for much more than bone structure, and can be withdrawn from bone when needed elsewhere. Phosphate is a component of DNA, RNA, ATP, phospholipids, and other compounds. Calcium is involved in muscle contraction, blood clotting, exocytosis, nervous communication, and cellular responses to hormones, among other processes.

Mineral **deposition** is a crystallization process in which osteoblasts extract calcium, phosphate, and other ions from the blood and deposit them in the osseous tissue. **Resorption** is a process in which the osteoclasts dissolve bone, releasing minerals into the blood and making them available for other uses. There is a critical balance between bone deposition and resorption. If one process outpaces the other, or if both of them occur too rapidly, various bone deformities and other developmental abnormalities occur, some of them quite grotesque and crippling.

The concentration of calcium in the blood and tissue fluid must be regulated within narrow limits. A calcium deficiency, or **hypocalcemia**[18] (HY-po-cal-SEE-me-uh), causes dysfunctions ranging from muscle tremors to tetanus—inability of the muscle to relax. One can suffocate from tetanus in the larynx *(laryngospasm)*. A calcium excess, or **hypercalcemia**,[19] depresses nervous, muscular, and cardiac function.

Calcium homeostasis is regulated mainly by two hormones: calcitriol and parathyroid hormone, both of which raise blood calcium levels. **Calcitriol** is the most active form of vitamin D. It is a hormone produced by a series of steps that begin when

---

[18]*hypo* = below normal; *calc* = calcium; *emia* = blood condition
[19]*hyper* = above normal; *calc* = calcium; *emia* = blood condition

Clinical Application 6.1

RICKETS

Rickets is a childhood disease in which the bones are soft and deformed as a result of vitamin D and calcium deficiency. It is characterized by fragile bones, bone pain, and frequent fractures, as well as deformity of the chest, pelvis, and spine, and bowed legs (fig. 6.8). It is usually preventable by exposure to sunlight—even 15 minutes per day is usually enough for adequate vitamin D synthesis. Vitamin D can also be obtained from fortified milk and dairy products, eggs, oily fish, and some mushrooms.

Rickets is one of the most common childhood diseases of developing countries, where it is often due to dietary calcium deficiency rather than a lack of sunlight exposure. Rickets was common in London during the early 1800s, afflicting 80% to 90% of children because they were kept indoors due to the heavy smog and also were often exploited for long hours of indoor factory work. The United States mandated the addition of Vitamin D to milk in 1930, which greatly reduced the incidence of rickets.

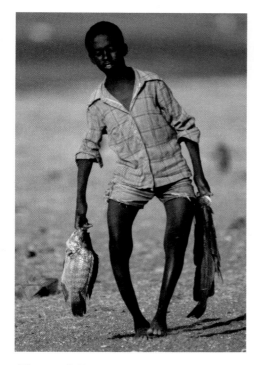

Figure 6.8 Rickets in a Boy from Eastern Kenya.

Jeff Rotman / Alamy Stock Photo

ultraviolet radiation from the sun produces a previtamin D molecule that is then further transformed in the liver and kidney. It promotes calcium absorption in the small intestine, reduces urinary loss of calcium, and stimulates osteoclasts to release calcium from the bones to the blood. The main function of calcitriol is to raise the blood concentration of calcium. Even though it promotes bone resorption, it is also necessary for bone deposition because without its effects on the kidneys and small intestine, blood calcium levels are too low for normal deposition. Low levels of calcitriol results in **rickets** in children, a condition characterized by soft bones (see Clinical Application 6.1).

Parathyroid hormone (PTH) is secreted by the parathyroid glands, which adhere to the posterior surface of the thyroid. The parathyroids respond to a drop in blood calcium level by secreting PTH. PTH then stimulates bone resorption by osteoclasts, promotes calcium reabsorption by the kidneys, and promotes calcitriol synthesis. All of these effects raise blood calcium levels.

Many other hormones, growth factors, and vitamins affect osseous tissue in complex ways that are still not well understood. *Calcitonin* from the thyroid gland plays a role in bone metabolism in children and possibly pregnant women. Growth hormone, estrogen, and testosterone promote ossification and bone growth, especially in adolescence. Bone deposition is also promoted by thyroid hormone, insulin, and at least 12 growth factors that are produced in the bone itself and stimulate nearby bone cells. Not only vitamin D but also vitamins A and C are needed for bone deposition and repair.

Before You Go On

Answer these questions from memory to test your understanding of the previous section. Reread the material if you can't answer the questions.

8. Concisely describe the process that accounts for the sandwichlike, three-layered structure of a cranial bone.

9. Explain how a cartilage model is replaced by osseous tissue to form a long bone such as the humerus. Explain how this process accounts for the medullary cavity, articular cartilage, and epiphyseal plate of a bone.

10. Explain why fully developed bones must still be remodeled, and how such remodeling is achieved.

11. Describe the effects of calcitriol, parathyroid hormone, and other hormones and vitamins on bone and calcium metabolism.

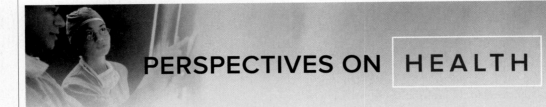

PERSPECTIVES ON HEALTH

Among the most familiar disorders of the skeletal system are fractures, osteoporosis, and arthritis. Other conditions are briefly described in Clinical Applications 6.1 and 6.2.

Fractures

Many of us have broken a bone at one time or another. Perhaps surprisingly, the most commonly broken bone is the clavicle, because it is so close to the surface. Half of clavicle fractures are due to athletic injuries, with the most common being due to bicycle accidents. When a fracture results from an unusual stress on a bone, such as a fall or auto accident, it is called a *stress fracture*. When a bone has been weakened by some other condition, such as bone cancer or osteoporosis, and it fractures under a stress that a healthy bone would withstand, it is called a *pathological fracture*. Any fracture that breaks through the skin is called an *open fracture;* if it does not break the skin, it is a *closed fracture*.

When a bone fractures, it bleeds and the blood clot *(fracture hematoma)* soon becomes infiltrated with fibroblasts, macrophages, osteoclasts, and osteogenic cells, which clean up the tissue fragments and begin the repair process. Deposition of collagen and fibrocartilage turns the hematoma into a *soft callus*. In 4 to 6 weeks, osteoblasts deposit calcified tissue, a *hard callus*, around the soft callus to splint the broken bone pieces together. It is especially important during this time that the bone be immobilized by traction or casting to prevent reinjury. The hard callus persists for 3 to 4 months. During this time, osteoclasts dissolve and remove remaining fragments of broken bone while osteoblasts build spongy bone to bridge the gap between the broken ends. This spongy bone then fills in by intramembranous ossification to solidify the repair.

Clinically, most fractures are set by *closed reduction,* which requires no surgery but uses a cast to immobilize the healing bones. More serious fractures may require *open reduction,* in which bone pieces are surgically exposed and joined by plates, screws, or pins.

Osteoporosis

Osteoporosis[20] is the most common of all bone diseases. It is a loss of bone tissue (especially spongy bone) to the point that the bones become unusually porous and brittle. It is usually associated with aging, with postmenopausal white women being at greatest risk; however, it can occur in either sex and at any age. Loss of bone mass is a significant problem among astronauts, who may spend time in microgravity conditions with too little bone-sustaining stress on the skeleton. Osteoporosis compromises a person's physical activity and presents a high risk of disabling bone fractures, often leading to death among the elderly. Fractures of the wrist and hip are common, as are crushing compression fractures of the vertebrae. Vertebral fractures commonly result in such spinal deformities as hyperkyphosis (see "The Vertebral Column" in section 6.3). The fracture of osteoporotic bones can result from stresses as slight as sitting down too quickly or supporting the body on one's wrists when rising from an armchair.

After menopause, women are especially at risk because the ovaries cease to secrete estrogen, which normally stimulates bone deposition. Other risk factors include smoking, insufficient exercise, inadequate calcium and protein intake, vitamin C deficiency, and diabetes mellitus.

Arthritis

Gradually declining bone density can be a silent, insidious menace to one's health, but a more noticeable effect of aging is the degenerative joint disease called *osteoarthritis (OA)*. Nicknamed "wear-and-tear arthritis," OA sets in as the synovial joints contain less lubricating synovial fluid and the articular cartilages soften and degenerate. As the cartilages become roughened by wear, joint movements may be accompanied by crunching or crackling sounds called *crepitus*. As OA progresses, exposed bone surfaces tend to develop spurs that grow into the

---

[20]*osteo* = bone; *porosis* = a state of being porous

joint cavity, restrict movement, and cause pain—especially in the fingers, spine, hips, and knees. About 85% of people over age 70 experience some degree of OA. It is seldom crippling, but in extreme cases it can immobilize the hip.

Rheumatoid arthritis (RA) is less common but more often crippling. It results from an autoimmune attack against the joint tissues. It begins when a misguided antibody called *rheumatoid factor* attacks the synovial membranes. Inflammatory cells accumulate in the joint and produce enzymes that degrade the articular cartilage. As the cartilage degenerates, the joint begins to ossify. The bones sometimes become solidly fused and immobilized, a condition called *ankylosis* (fig. 6.9). Rheumatoid arthritis is named for the fact that symptoms tend to flare up and subside periodically.[21] It affects more women than men, and because it often begins between the ages of 30 and 40, it can cause decades of pain and disability. There is no cure; it is possible only to slow its progression with anti-inflammatory drugs and physical therapy.

A hereditary arthritis more common in men is *gouty arthritis*. *Gout* is the accumulation of uric acid crystals in the joints, especially of the great toe. Uric acid irritates the articular cartilage and synovial membrane, causing swelling, tissue generation, and sometimes joint fusion.

Rheumatism is a broad term for any pain in the supportive and locomotor organs, including muscles, tendons, ligaments, and bones. It includes arthritis, bursitis, gout, and many other conditions. Physicians who specialize in the rheumatic diseases are called *rheumatologists*.

---

[21]*rheumat* = tending to change

(a)

(b)

Figure 6.9 Rheumatoid Arthritis (RA). (a) A severe case with ankylosis of the joints. (b) Colorized X-ray of hands with RA.
a: John Watney/Science Source b: CNRI/SPL/Science Source

6.3 The Axial Skeleton

Expected Learning Outcomes

When you have completed this section, you should be able to

a. define terms for the surface features and markings of bones;
b. name and define the two basic subdivisions of the skeleton;
c. name and describe the cavities within the skull;
d. identify each of the skull bones, name specific features, and name some of the main sutures that join them;
e. describe the general features of the vertebral column, general structure of a vertebra and intervertebral disc, and regional differences between the vertebrae; and
f. describe the anatomy of the ribs and sternum, regional differences between the ribs, and the articulation of the ribs with the vertebral column and sternum.

We turn our attention now to the anatomy of the individual bones that make up the adult skeleton. Descriptions of these bones include a variety of ridges, spines, bumps, depressions, holes, and joint surfaces, collectively called *bone markings*. It is important to know the names of these features because later descriptions of joints, muscle attachments, and the routes traveled by nerves and blood vessels are based on this terminology. The most common of these features are defined in **table 6.1**, and several are illustrated in **figure 6.10**. Other terms specific to one or a few bones are introduced when those bones are described. A few bone and joint features not

Table 6.1 Bone Markings

Term	Description and Example
Canal	A tubular passage or tunnel in a bone (auditory canal of the ear)
Condyle	A rounded knob (occipital condyles of the skull)
Crest	A narrow ridge (iliac crest of the hip bone)
Epicondyle	A flare superior to a condyle (medial epicondyle of the femur)
Facet	A smooth joint surface that is flat or only slightly concave or convex (articular facets of the vertebrae)
Fissure	A slit through a bone (orbital fissures behind the eye)
Foramen	A hole through a bone, usually round (foramen magnum of the skull)
Fossa	A shallow, broad, or elongated basin (infraspinous fossa of the scapula)
Process	Any bony prominence (mastoid process of the skull)
Sinus	A cavity within a bone (frontal sinus of the skull)
Spine	A sharp, slender, or narrow process (spine of the scapula)
Tubercle	A small, rounded process (greater tubercle of the humerus)
Tuberosity	A rough surface (tibial tuberosity of the tibia)

Figure 6.10 Bone Markings. Most of these features also occur on many other bones of the body.

discussed in the text are labeled in the art for context and comparison to other books you may consult.

There are 109 differently named bones of the skeleton but many of these are paired (left and right), so an adult skeleton typically has 206 individual bones. These are divided into two regions, axial and appendicular (fig. 6.11). (1) The **axial skeleton** forms the central supporting axis of the body and comprises 80 bones, including the skull, vertebral column, ribs, and sternum. (2) The **appendicular skeleton** comprises the remaining 126 bones—those of the upper limb and pectoral girdle, and of the lower limb and pelvic girdle.

Figure 6.11 The Adult Skeleton. The appendicular skeleton is colored green, and the rest is axial skeleton. APR

? *The appendicular skeleton is not entirely within the appendicular region of the body (see fig. 1.9). Explain.*

(a) Anterior view   (b) Posterior view

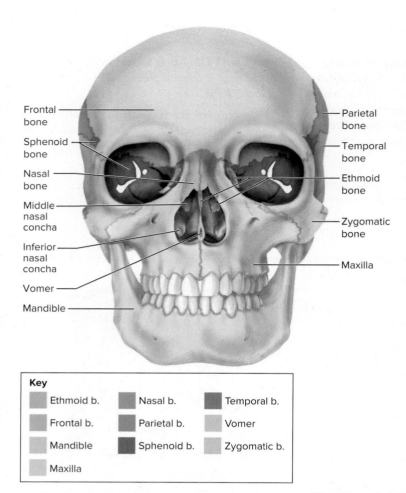

Figure 6.12 The Skull, Anterior View.

6.3a The Skull

The skull is the most complex part of the skeleton, a reflection of its multiple tasks including air and food intake, housing the brain and special sense organs, transmitting nerves that connect the brain to the rest of the body, and providing scaffolding for muscles of facial expressions, vocalization, and head movement. As shown in **figure 6.12** and the six figures thereafter, it consists of one freely movable bone, the mandible, and 21 relatively immobile bones connected by **sutures** (SOO-chers)—joints that appear as seams on the skull surface. Most bones of the skull are perforated by **foramina** (singular, *foramen*)—holes that allow passage for nerves and blood vessels to go to and from the brain.

The skull contains several prominent cavities—the **cranial cavity,** which encloses the brain; the **orbits** (eye sockets); the **nasal cavity;** the **oral (buccal) cavity;** the **middle-** and **inner-ear cavities;** and **paranasal sinuses,** air-filled spaces connected to the nasal cavity and named for the bones in which they occur—the **frontal, sphenoid, ethmoid,** and **maxillary sinuses (fig. 6.13)**.

The Cranial Bones

Eight bones of the skull are called *cranial bones* and the other 14 are called *facial bones*. The **cranial bones** form the **cranium.**[22] They are distinguished by being in direct contact with the fibrous membranes, the *meninges,* that enclose the brain. They are as follows:

1 frontal bone	1 occipital bone
2 parietal bones	1 sphenoid bone
2 temporal bones	1 ethmoid bone

The Frontal Bone

The frontal bone is the broad bone that forms the forehead, the roof of the orbit, and the anterior one-third of the roof of the cranial cavity (**figs. 6.14** and **6.15b;** also see fig. 6.12). It has a thick ridge in the region of the eyebrows and it contains the frontal sinus.

The Parietal Bones

The right and left **parietal bones** (pa-RY-eh-tul) form most of the cranial roof and part of its walls (figs. 6.14 and **6.16**). They extend from the **coronal**[23] **suture,** which separates them from the frontal bone anteriorly, to the **lambdoid**[24] **suture** (LAM-doyd), which separates them from the occipital bone posteriorly. A median **sagittal suture** separates the right and left parietal bones from each other.

The Occipital Bone

The **occipital bone** (oc-SIP-ih-tul) forms the rear of the skull and much of its base (figs. 6.14 and 6.15). Its most conspicuous feature is a

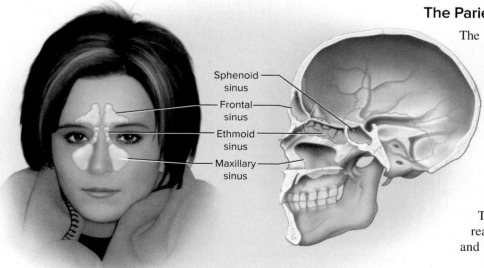

Figure 6.13 The Paranasal Sinuses.

---

[22]*crani* = helmet
[23]*corona* = crown
[24]Shaped like the Greek letter lambda (λ)

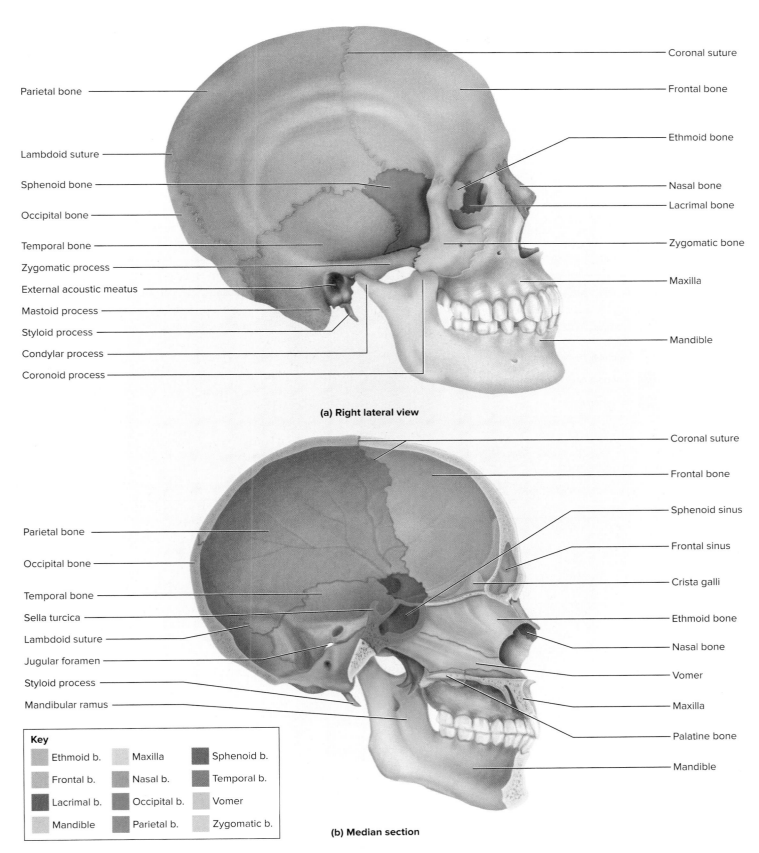

Figure 6.14 The Skull, Lateral Surface and Median Section. APR

? List all bones that join with the temporal bone.

160 PART 2 Support and Movement

(a) Inferior view

(b) Internal view of cranial floor

Figure 6.15 The Base of the Skull. (a) Inferior view. (b) Internal view of the cranial floor.

large opening, the **foramen magnum,**[25] which admits the spinal cord to the cranial cavity. On each side of the foramen magnum is a smooth knob, the **occipital condyle** (CON-dile), where the skull rests on the vertebral column.

The Temporal Bones

The temporal bones are the largest bones on the lateral surface of the cranium (fig. 6.14). They form the region around the ears (the *temples*) and part of the lower wall and floor of the cranial cavity. They derive their name from the fact that people often develop their first gray hairs on the temples with the passage of time.[26] They are more complex than the preceding bones, with the following major features.

- The **external acoustic meatus** (me-AY-tus), or **auditory** (ear) **canal.**
- Part of the **zygomatic arch.** The zygomatic arch is a prominent feature of the cheek formed by parts of the temporal and zygomatic bones and the maxilla.
- The **mastoid process,** a blunt downward growth that you can palpate as a prominent lump behind your earlobe.
- The **mandibular fossa,** a depression on the inferior surface where the mandible articulates with the cranium, forming the jaw joint (fig. 6.15a).

Also visible on the inferior surface of each temporal bone are two foramina, the **carotid canal** and **jugular foramen.** These are passages for the internal carotid artery and internal jugular vein, respectively, which supply blood to the brain and drain blood from it. Within the cranial cavity, the temporal bone forms a prominent ridge that contains the middle- and inner-ear cavities, which house the organs of hearing and balance (fig. 6.15b).

The Sphenoid Bone

The **sphenoid**[27] **bone** (SFEE-noyd) (figs. 6.14 and 6.15) has a complex shape with a thick median **body** and outstretched **wings,** which give the bone as a whole a ragged mothlike shape. The body contains a pair of sphenoid sinuses and has a saddlelike structure named the **sella turcica**[28] (SEL-la TUR-sih-ca), which houses the pituitary gland. The wings form part of the lateral surface of the cranium anterior to the temporal bone and part of the wall of the orbit. Each wing exhibits an **optic canal,** which permits passage of the optic nerve from the eye, and a gash, the **superior orbital fissure** (fig. 6.17), which allows passage of nerves that supply the muscles of eye movement.

The Ethmoid Bone APR

The **ethmoid**[29] **bone** (ETH-moyd) is located between the orbits and forms the roof of the nasal cavity. It is honeycombed with cavities that form the ethmoid sinus. A vertical plate of the ethmoid forms the superior part of the **nasal septum,** which divides the nasal cavity into right and left spaces. Three curled, scroll-like **nasal conchae**[30] (CON-kee) project into each fossa from the lateral wall (see fig. 15.2).

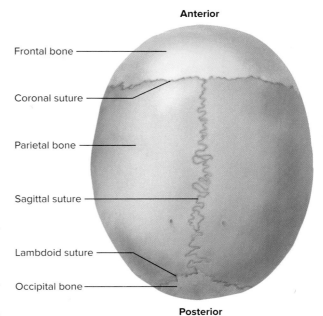

Figure 6.16 The Calvaria (Skullcap), Superior View. APR

---

[25]*foramen* = hole; *magnum* = large
[26]*tempor* = time
[27]*sphen* = wedge; *oid* = resembling
[28]*sella* = saddle; *turcica* = Turkish
[29]*ethmo* = sieve, strainer; *oid* = resembling
[30]*conchae* = conchs (large marine snails)

The superior and middle conchae are extensions of the ethmoid bone. The inferior concha is a separate bone. From within the cranial cavity, one can see a triangular ridge of the ethmoid called the **crista galli**,[31] which is surrounded by porous plates on either side that allow passage of nerves for smell.

The Facial Bones

The 14 **facial bones** shape the face and support internal structures of the oral and nasal cavities, but have no contact with the meninges or brain. They are as follows:

2 maxillae	2 nasal bones
2 palatine bones	2 inferior nasal conchae
2 zygomatic bones	1 vomer
2 lacrimal bones	1 mandible

The Maxillae

The two **maxillae** (mac-SILL-ee) form the upper jaw and support its teeth (see figs. 6.12 and 6.14). Even though the teeth are preserved with the skull, they are not bones. They are discussed in chapter 17. Each maxilla extends upward to form the floor and medial wall of the orbit. Within the orbit, it exhibits a gash, the **inferior orbital fissure** (fig. 6.17), which is a passage for blood vessels and sensory nerves from the face. Behind the teeth, the maxillae turn inward as a shelf that forms about four-fifths of the **hard palate,** the anterior part of the roof of the mouth (fig. 6.15a).

The Smaller Facial Bones

A number of small facial bones support the oral, nasal, and orbital cavities.

- Two small **palatine bones** form the posterior one-fifth of the hard palate, part of the wall of the nasal cavity, and part of the floor of the orbit (see fig. 6.15a). They are L-shaped, but this is apparent only if you see them dissected free of the rest of the skull.

---

[31]*crista* = crest; *galli* = of a rooster

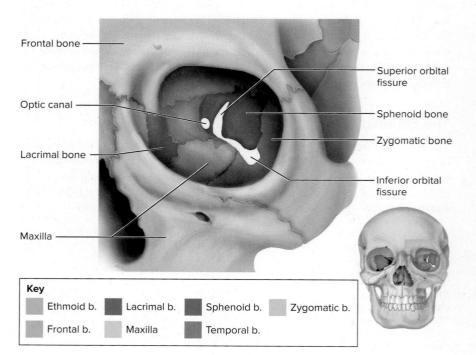

Figure 6.17 The Orbital (Eye) Region.

- The **zygomatic**[32] **bones** form the angles of the cheeks inferolateral to the eyes, and part of the lateral wall of each orbit (see fig. 6.12). The prominent zygomatic arch ("cheekbone") is partly formed by the zygomatic bones (see fig. 6.15a).
- The tiny **lacrimal**[33] **bones** (LACK-rih-mul) form part of the medial wall of each orbit (fig. 6.17). Each houses a membranous sac that collects tears from the eyes and drains them into the nasal cavity.
- The small nasal bones form the bridge of the nose and support cartilages that shape the lower portion of the nose (see figs. 6.12 and 6.14). The nasal bones are often fractured by blows to the nose.
- The three conchae of the nasal cavity were mentioned earlier. The superior and middle conchae are parts of the ethmoid bone. The **inferior nasal concha**, the largest of the three, is a separate bone (see fig. 6.12; see also fig. 15.2).
- The **vomer** forms the lower part of the nasal septum and joins the perpendicular plate of the ethmoid bone, which forms the upper part. It is a thin, flat, sharply pointed plate of bone (see fig. 6.14b).

The Mandible

The **mandible** is the lower jaw (see figs. 6.12 and 6.14); it supports the lower teeth, provides attachment for muscles of mastication, and forms a hinge joint with the temporal bone. The horizontal portion, carrying the teeth, is the **body** and the posterior part that rises to meet the floor of the cranium is the **ramus** (RAY-mus). The body and ramus meet at a corner called the **angle.** The ramus is divided by a U-shaped notch at its superior end into an anterior blade, the **coronoid process,** and the posterior **condylar process** (CON-dih-lur). The condylar process is capped by a knob, the **mandibular condyle,** which meets the temporal bone to form the hinge of the jaw, called the **temporomandibular joint (TMJ).** The coronoid process and angle are important attachment points for chewing muscles discussed in chapter 7.

Bones Associated with the Skull

Seven bones are closely associated with the skull but not considered part of it. These are the three **auditory ossicles**[34] in each middle ear (see chapter 10) and the **hyoid**[35] **bone** beneath the chin. The hyoid serves for the attachment of muscles that control the mandible, tongue, and larynx; the larynx is suspended from the hyoid by a ligament.

The Infant Skull

You probably know that a newborn baby (*neonate*) has "soft spots" in the skull. These are called *fontanels* ("little fountains"), and represent places where the intramembranous ossification of the cranial bones is not yet complete at birth. There are two on the superior midline of the cranium—the **anterior fontanel,** which is the largest and easiest to palpate, and the **posterior fontanel**—and there are two small fontanels on each lateral aspect of the skull, the **sphenoid** and **mastoid fontanels** (fig. 6.18). Being born before the fontanels fuse allows the fetal cranium to deform harmlessly as the head squeezes through the tight birth canal. The head of the neonate sometimes looks temporarily misshapen as a result, but soon assumes a normal shape. The fontanels fill in with bone as the infant ages—the anterior fontanel last of all, finally fusing as late as 18 to 24 months of age.

---

[32]*zygo* = to join, unite
[33]*lacrim* = tear, to cry
[34]*os* = bone; *icle* = little
[35]*hy* = the letter U; *oid* = resembling

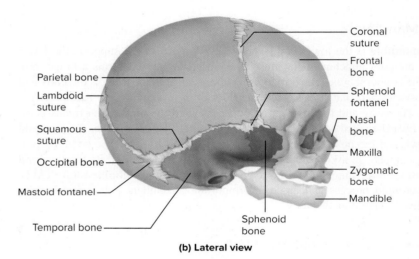

Figure 6.18 The Infant Skull at Birth. (a) Superior view showing anterior and posterior fontanels. (b) Lateral view showing mastoid and sphenoid fontanels.

6.3b The Vertebral Column

The **vertebral column (spine)** physically supports the skull and trunk; protects the spinal cord; absorbs stresses produced by walking, running, and lifting; and provides attachment for the limbs, thoracic cage, and postural muscles. Although commonly called the backbone, it is not a single bone but a flexible chain of 33 **vertebrae** and 23 cartilaginous **intervertebral discs** (fig. 6.19). The adult vertebral column averages about 71 cm (28 in.) long, with the discs accounting for about one-quarter of this.

The vertebrae are divided into five groups, usually numbering 7 *cervical vertebrae* in the neck, 12 *thoracic vertebrae* in the chest, 5 *lumbar vertebrae* in the lower back, 5 *sacral vertebrae* at the base of the spine, and 4 tiny *coccygeal vertebrae*. In adults, the sacral and coccygeal vertebrae are fused together into single bones, the *sacrum* and *coccyx;* the other three groups of vertebrae remain separate bones. To help remember that the first three groups are 7, 12, and 5 in number, think of a typical workday: Go to work at 7, get off for lunch at 12, and go home at 5.

The vertebral column is C-shaped at birth. Beyond the age of 3 years, it is slightly S-shaped, with four bends in the cervical, thoracic, lumbar, and pelvic regions. A bend that curves outward toward the back **kyphosis**. There are two of these, the *thoracic kyphosis* and the *pelvic kyphosis*. A bend toward the front is called a **lordosis,** and there are two of these, the *cervical lordosis* and the *lumbar lordosis* (fig. 6.20a). The curvatures make sustained bipedal walking possible because the trunk does not lean forward, as it does in primates such as a chimpanzee. The head is balanced over the body's center of gravity, and the eyes are directed straight forward.

CHAPTER 6 The Skeletal System 165

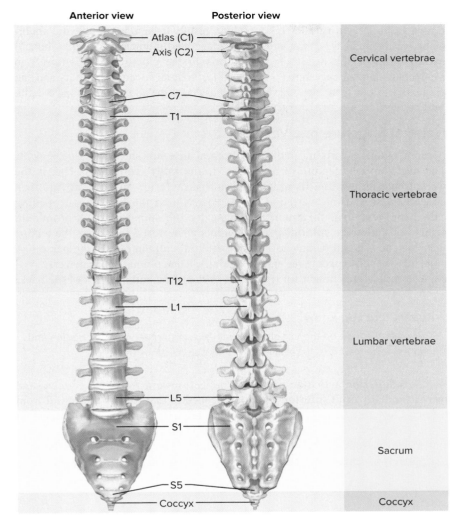

Figure 6.19 The Vertebral Column. **APR**

Figure 6.20 Adult Spinal Curvatures. (a) Normal curvatures. (b) Common abnormalities. Scoliosis often results from a developmental abnormality of the spine. Hyperkyphosis most often results from osteoporosis. Hyperlordosis commonly occurs in pregnancy and obesity.

Three common abnormalities of the spinal curvatures are shown in figure 6.20b. *Scoliosis,* a lateral curvature, results from a developmental abnormality of the spine. *Hyperkyphosis,* an exaggerated thoracic curvature ("hunchback"), often is the result of osteoporosis in the elderly. *Hyperlordosis,* an exaggerated lumbar curvature ("swayback"), commonly occurs in obesity and late pregnancy as one leans backward to compensate for the added abdominal weight.

General Structure of a Vertebra

A representative vertebra and intervertebral disc are shown in figure 6.21. The most obvious feature of a vertebra is the **body (centrum).** This is the weight-bearing portion. Its rough superior and inferior surfaces provide firm attachment to the intervertebral discs. Posterior to the body is an opening called the **vertebral foramen,** a passage for the spinal cord. A prominent projection called the **spinous process** points posteriorly, and forms the bumps that you can see and palpate along the vertebral column of a living person. A pair of **transverse processes** extend laterally. The spinous and transverse processes provide attachment for ligaments of the spine and muscles of the back.

> ***Apply What You Know***
>
> The lower we look on the vertebral column, the larger the vertebral bodies and intervertebral discs are. What is the functional significance of this?

Two **superior articular processes** project upward from one vertebra and meet a similar pair of **inferior articular processes** that project downward from the one above (see fig. 6.23b, c). Each process has a flat surface (facet) facing that of the adjacent vertebra. These processes restrict twisting of the vertebral column. Where two vertebrae meet, they exhibit a lateral gap between them called the **intervertebral foramen.** This allows passage for spinal nerves emerging from the spinal cord.

Figure 6.21 A Representative Vertebra and Intervertebral Disc, Superior Views. (a) A typical vertebra. (b) An intervertebral disc, oriented the same way as the vertebral body in part (a) for comparison. (c) A herniated disc, showing compression of the spinal nerve roots by the nucleus pulposus oozing from the disc.

Regional Characteristics of the Vertebrae

The vertebrae differ from one region of the vertebral column to another, as follows.

Cervical vertebrae (C1–C7) are the smallest and lightest. The first two (C1 and C2) have unique structures that allow for head movements (fig. 6.22). Vertebra C1 is called the **atlas** because it supports the head in a manner reminiscent of the giant of Greek mythology who carried the heavens on his shoulders. It is little more than a delicate ring surrounding a large vertebral foramen. On the upper surface is a pair of concave **superior articular facets** that meet the occipital condyles of the skull. When you nod your head "yes," the occipital condyles rock back and forth on these facets. Vertebra C2, the **axis,** allows rotation of the head as in gesturing "no." Its most distinctive feature is a prominent knob called the **dens**[36] (pronounced "denz"), which projects into the vertebral foramen of C1. C2 is the first vertebra that exhibits a spinous process.

In vertebrae C2 to C6, the spinous process is forked—a feature not seen in C7 or any lower vertebrae (fig. 6.23a). All seven cervical vertebrae have a prominent **transverse foramen** in each transverse process, another feature not found in any other class of vertebrae. These foramina provide passage for the *vertebral arteries,* which supply blood to the brain.

Thoracic vertebrae (T1–T12) are those that support the ribs (fig. 6.23b). They have relatively pointed spinous processes that angle sharply downward, and they exhibit small smooth depressions for attachment of the ribs. These depressions include slightly concave spots called **costal**[37] **facets** on the body of each vertebra and

---

[36]*dens* = tooth
[37]*costa* = rib; *al* = pertaining to

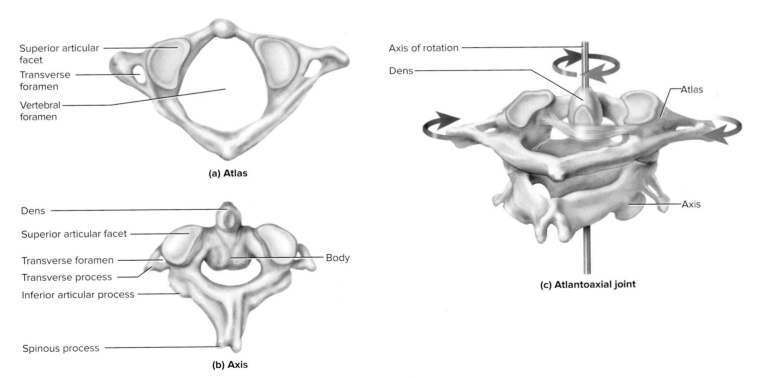

Figure 6.22 The Atlas and Axis, Vertebrae C1 and C2. (a) The atlas, C1, which directly supports the skull. (b) The axis, C2, which inserts into the vertebral foramen of the atlas. (c) Articulation of the two vertebrae and rotation of the atlas. This movement turns the head from side to side, as in gesturing "no." Note the ligament holding the dens of the axis in place. **APR**

? *What serious consequence could result from the rupture of the transverse ligament in an automobile accident?*

Figure 6.23 Typical Cervical, Thoracic, and Lumbar Vertebrae. Superior views of thoracic and lumbar vertebrae look like figure 6.21a. APR

? *What features unique to cervical vertebrae are shown in the superior view of part (a)?*

a **transverse costal facet** at the ends of each transverse process of vertebrae T1 to T10 (lacking on T11–T12). Most ribs insert into the gaps between adjacent thoracic vertebrae, so each vertebra contributes one-half of the articular surface. Vertebrae T1 and T10 to T12, however, have complete costal facets on the bodies; the corresponding ribs articulate on the vertebral bodies instead of between vertebrae.

Lumbar vertebrae (L1–L5) have a thick, stout body and a blunt, squarish spinous process (fig. 6.23c). For the most part, their superior articular processes face medially (like the palms of your hands about to clap), and the inferior processes face laterally, toward the superior processes of the next vertebra. This arrangement enables the lumbar region of the spine to resist twisting.

Sacral vertebrae (S1–S5) begin to fuse around age 16, and by age 26 they are fully fused into a bony plate, the **sacrum**[38] (SAY-krum) (**fig. 6.24**). The sacrum forms the posterior wall of the pelvic cavity. It was named for its prominence as the largest and most durable bone of the vertebral column. It has four pairs of large **sacral foramina** on the anterior and posterior surfaces, for emergence of spinal nerves to the pelvic organs and lower limbs. A **sacral canal** runs through the sacrum and contains spinal nerve roots.

Coccygeal vertebrae (coc-SIDJ-ee-ul) (Co1–Co4) fuse by the age of 20 to 30 into a single small triangular bone, the **coccyx** (COC-six) (fig. 6.24). The Roman anatomist Claudius Galen (c. 130–c. 200) named it this because he thought it resembled the beak of a cuckoo.[39] The coccyx provides attachment for muscles of the pelvic floor. It is commonly known as the "tailbone." It is a vestige of the bones that, in other mammals, continue as the vertebrae of the tail.

Intervertebral Discs

An **intervertebral disc** is a pad consisting of an inner gelatinous **nucleus pulposus** surrounded by a ring of fibrocartilage, the **anulus fibrosus** (see fig. 6.21b).

---

[38]*sacr* = great, prominent
[39]*coccyx* = cuckoo

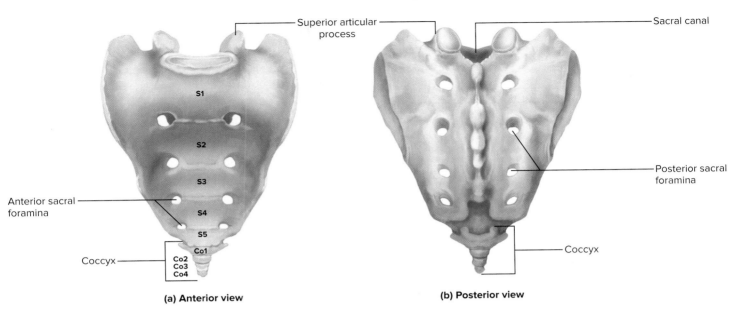

Figure 6.24 The Sacrum and Coccyx. APR

The discs bind adjacent vertebrae together, enhance spinal flexibility, support the weight of the body, and absorb shock. Under stress—for example, when you lift a heavy weight—the discs bulge laterally. Excessive stress can cause a *herniated disc*—a condition in which the anulus fibrosus cracks and the nucleus pulposus oozes out. The extruded gel may put painful pressure on the spinal cord or a spinal nerve (see fig. 6.21c).

6.3c The Thoracic Cage

The **thoracic cage** (fig. 6.25) consists of the thoracic vertebrae, sternum, and ribs. It encloses the heart and lungs and provides attachment for the pectoral girdle and upper limbs. The ribs protect not only the thoracic organs but also the spleen, most of the liver, and to some extent the kidneys. The thoracic cage is rhythmically expanded by the respiratory muscles to create a vacuum that draws air into the lungs.

Sternum

The **sternum** (breastbone) is a bony plate anterior to the heart. It consists of three parts: a short superior **manubrium**[40] (ma-NOO-bree-um) with a shape resembling the knot in a necktie; an elongated swordlike **body;** and at the inferior end, a small, daggerlike **xiphoid**[41] **process** (ZIF-oyd or ZY-foyd). The first pair of ribs articulates with the manubrium; the second pair articulates at the junction of the manubrium and body; and the next eight pairs articulate with the body only. The clavicles (collarbones) articulate with the manubrium.

Ribs

There are 12 pairs of **ribs.** Each attaches posteriorly to the vertebral column, and all but the last two arch around the chest and attach by way of a cartilaginous strip, the **costal cartilage,** to the sternum. Most of the ribs are curved, flattened blades. Rib 1, however, is a horizontal plate, and ribs 11 and 12 are pointed at the distal ends.

---

[40]*manubrium* = handle
[41]*xipho* = sword; *oid* = resembling

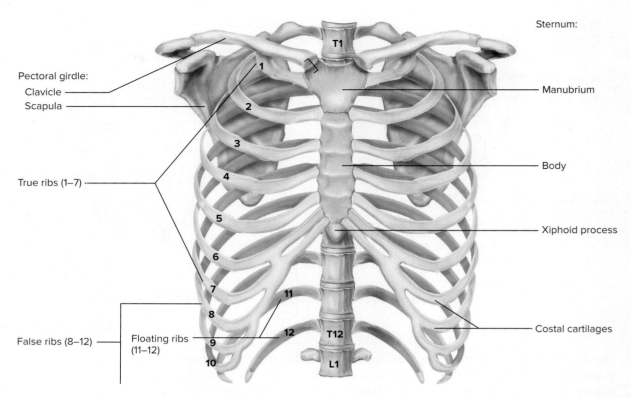

Figure 6.25 The Thoracic Cage and Pectoral Girdle, Anterior View. **APR**

Ribs 1 to 7 are called **true ribs** because each has its own costal cartilage. Ribs 8 to 10 attach to the costal cartilage of rib 7, and ribs 11 and 12 have no costal cartilages at all. Ribs 8 to 12 are therefore called **false ribs,** and ribs 11 and 12 are also called **floating ribs** because they lack of any connection to the sternum. The first 10 ribs articulate with the vertebrae at two points: the vertebral bodies and the transverse costal facets. Ribs 11 and 12 articulate with the vertebral bodies only.

Before You Go On

Answer these questions from memory to test your understanding of the previous section. Reread the material if you can't answer the questions.

12. Explain the distinction between cranial and facial bones; name six examples of each and locate them on the skull.
13. Prepare a list of all boldfaced key terms in the descriptions of the skull bones and try to identify each of those features on unlabeled models or illustrations of the skull.
14. Prepare a list of all boldfaced key terms in the descriptions of the sternum, ribs, and vertebrae and try to identify each of those features on unlabeled models or illustrations of the skeleton.
15. Summarize the features you would look for to identify a single isolated vertebra as cervical, thoracic, or lumbar.
16. What part of the sternum articulates with the clavicles? What parts articulate with the ribs? What part articulates with no other bones except for another part of the sternum?
17. Explain the distinction between true, false, and floating ribs. State which ribs, by number, belong in each category.

6.4 The Appendicular Skeleton

Expected Learning Outcomes

When you have completed this section, you should be able to

a. describe the bones of the pectoral girdle and upper limb, and major features of individual bones; and
b. do the same for the bones of the pelvic girdle and lower limb.

The *appendicular skeleton* consists of bones of the upper and lower limbs and the pectoral and pelvic girdles that attach the limbs to the axial skeleton. Injuries to the appendicular skeleton are common in the elderly, athletes, and as a result of recreational activities such as cycling, skateboarding, and skiing. They are quite disabling because we depend so much on the limbs for mobility and manipulating objects.

6.4a The Pectoral Girdle and Upper Limb

The **pectoral girdle** (shoulder girdle) supports the arm. It consists of a *clavicle* and a *scapula* on each side of the body.

The Clavicle

The **clavicle,**[42] or collar bone (fig. 6.26), is a slightly S-shaped bone easily seen and palpated on the upper thorax. Its superior surface is relatively rounded and smooth and its inferior surface flatter and slightly rough. The medial **sternal end** has a rounded, hammerlike head that articulates with the manubrium of the sternum. The lateral **acromial end** is markedly flattened and articulates with the acromion of the scapula. The clavicles brace the shoulders; without them, the pectoralis major muscles would pull the shoulders forward and medially.

The Scapula

The **scapula,** or shoulder blade (fig. 6.27), is a triangular plate that overlies ribs 2 to 7 on the upper back. The triangle is bounded by *superior, medial,* and *lateral*

Figure 6.26 The Right Clavicle (Collarbone). **APR**

[42]*clav* = hammer, club; *icle* = little

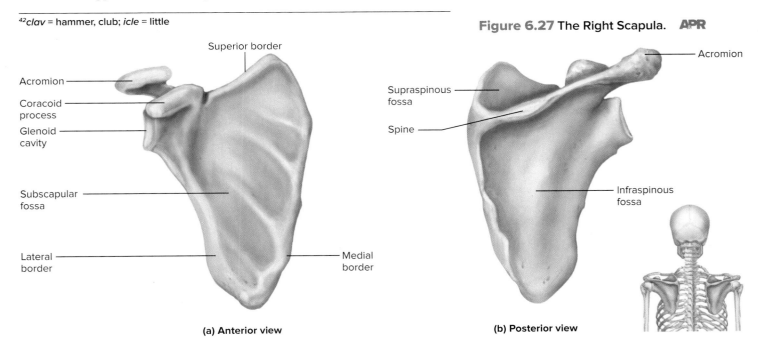

Figure 6.27 The Right Scapula. **APR**

borders; its broad anterior surface, called the **subscapular fossa,** is slightly concave and relatively featureless. The posterior surface has a prominent transverse ridge called the **spine,** a deep indentation superior to the spine called the **supraspinous fossa,** and a broad surface inferior to it called the **infraspinous fossa.**[43] The fossae are occupied mainly by muscles of the *rotator cuff,* which are concerned with movements of the arm (see fig. 7.18).

The most complex part of the scapula is the shoulder region, where it has three main features: (1) The **acromion**[44] (ah-CRO-me-on), a platelike extension of the scapular spine, forms the apex of the shoulder. It articulates with the clavicle. (2) The **coracoid**[45] **process** (COR-uh-coyd), shaped like a bent finger, provides attachment for the biceps brachii of the arm. (3) The **glenoid**[46] **cavity** (GLEN-oyd) is a shallow socket that articulates with the rounded head of the humerus.

The Humerus

There are 30 bones in each of the upper limbs: the *humerus* in the arm proper, the *radius* and *ulna* in the forearm, 8 *carpal bones* in the wrist, and 5 *metacarpal bones* and 14 *phalanges* in the hand.

The **humerus** is the only bone of the arm *(brachial region),* from shoulder to elbow, and is the largest bone of the upper limb. It has a hemispherical head that inserts into the glenoid cavity of the scapula (**fig. 6.28**). Proximally,

---

[43]*supra* = above; *infra* = below
[44]*acr* = extremity, point; *omi* = shoulder
[45]*corac* = crow; *oid* = resembling
[46]*glen* = pit, socket

Figure 6.28 The Right Humerus.   (a) Anterior view   (b) Posterior view

a distinctive groove, the *intertubercular sulcus,* accommodates a tendon of the biceps brachii muscle. The shaft has a rough area called the **deltoid tuberosity** on its lateral surface. This is the insertion for the deltoid muscle of the shoulder.

The distal end of the humerus has two smooth condyles where movement of the elbow occurs—the **capitulum**[47] (ca-PIT-you-lum) on the lateral side and the **trochlea**[48] (TROCK-lee-uh) on the medial side. Immediately proximal to these condyles, the humerus flares to form two bony processes, the **lateral** and **medial epicondyles,** which are easily palpated at the widest point of your elbow and anchor the bulk of the forearm muscles. The distal end also shows three deep pits or fossae. The **olecranon fossa** (oh-LEC-ruh-non) on the posterior side accommodates the ulna when the elbow is extended. On the anterior surface, a medial pit called the **coronoid fossa** and a lateral pit called the **radial fossa** accommodate the heads of the ulna and radius, respectively, when the elbow is flexed.

The Radius

The **radius** extends from elbow to wrist on the lateral side of the forearm, ending just proximal to the base of the thumb (fig. 6.29). Its proximal head is a distinctive disc that rotates freely on the capitulum of the humerus when the palm is turned forward and back. Just distal to the head is a process called the **radial tuberosity,** an attachment for the tendon of the biceps brachii muscle. The flared distal end of

---

[47]*capit* = head; *ulum* = little
[48]*troch* = wheel, pulley

Figure 6.29 The Right Radius and Ulna.

the radius articulates with three of the carpal bones, and has a small point called the **styloid process** that can be palpated on the lateral side of the wrist.

The Ulna

The **ulna**[49] is the medial bone of the forearm (fig. 6.29). Its proximal end (at the elbow) has a wrenchlike shape, with a deep C-shaped **trochlear notch** that wraps around the trochlea of the humerus. The posterior wall of the notch is formed by a prominent bony point called the **olecranon.** If you rest with your chin in your hands and your elbows on a table, it is the olecranon that contacts the table. The anterior wall of the notch is formed by a less prominent **coronoid process.** Medially, the **radial notch** accommodates the edge of the head of the radius and allows it to rotate as happens when you turn your palm upward. The distal end of the ulna flares into a small head and a pointed styloid process similar to that of the radius. It can be palpated on the medial side of the wrist.

A fibrous sheet called the **interosseous**[50] **membrane** (IN-tur-OSS-ee-us) loosely joins the radius and ulna along the length of their shafts. If you stand and lean forward and support your weight with your hands on a table, about 80% of the force is borne by the radius. The interosseous membrane pulls the ulna upward and transfers some of this force through the ulna to the humerus, thus distributing your weight more evenly across the elbow joint. Such weight distribution is also important in four-legged animals.

The Carpal Bones

The **carpal bones** in the base of the hand are arranged in two rows of four bones each (**fig. 6.30**). They allow movements of the hand from side to side and anterior to posterior. The carpal bones of the proximal row, starting at the lateral (thumb) side, are the **scaphoid, lunate, triquetrum** (tri-QUEE-trum), and **pisiform** (PY-sih-form). Translating from the Latin, these words loosely mean "boat-," "moon-," "triangle-," and "pea-shaped," respectively. The bones of the distal row, again starting on the lateral side, are the **trapezium,**[51] **trapezoid, capitate,**[52] and **hamate.**[53] The hamate can be recognized by a prominent hook, or *hamulus,* on the palmar side.

The Metacarpal Bones

The **metacarpal bones**[54] occupy the palmar region. They are numbered I through V with metacarpal I at the base of the thumb and metacarpal V at the base of the little finger (fig. 6.30). On a skeleton, the metacarpals look like extensions of the fingers, so the fingers seem much longer than they really are. Each is divided into a *base, body,* and *head;* the heads form the knuckles of a clenched fist.

The Phalanges

The **phalanges** (fah-LAN-jeez) are the bones of the fingers; the singular is *phalanx* (FAY-lanks). They are identified by roman numerals corresponding to the numbering of the metacarpals, preceded by *proximal, middle,* and *distal.* Digits II through V have three bones, called the proximal, middle, and

---

[49]*ulna* = elbow
[50]*inter* = between; *osse* = bones
[51]*trapez* = table, grinding surface
[52]*capit* = head; *ate* = possessing
[53]*ham* = hook; *ate* = possessing
[54]*meta* = beyond; *carp* = wrist

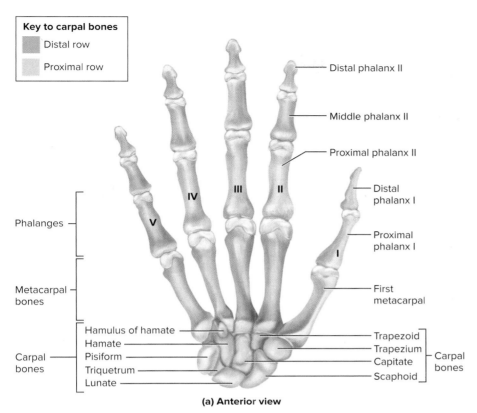

Figure 6.30 The Right Hand. (a) Anterior view with carpal bones color-coded to distinguish the proximal and distal rows. (b) X-ray of an adult hand. Sesamoid bones are small "extra" bones that often develop within tendons after birth. The pisiform bone and patella (kneecap) are also sesamoid bones. APR
b: RNHRD NHS Trust/The Image Bank/Getty Images

distal phalanges. Most people with wedding rings, for example, wear them encircling the left proximal phalanx IV. The thumb has only two phalanges, proximal and distal (fig. 6.30).

6.4b The Pelvic Girdle and Lower Limb

The adult **pelvic**[55] **girdle** is composed of three bones: the right and left *hip bones* and the *sacrum* (**fig. 6.31a**). The sacrum has already been described, because it is part of the vertebral column. Posteriorly, the hip bones are joined to the sacrum at the *sacroiliac joints;* anteriorly, they are joined to each other at the *pubic symphysis*.

The pelvic girdle plus the associated muscles and ligaments constitute the bowl-like **pelvis,** which has two regions. The broad superior region between the flare of the hips forms the wall of the lower abdominal cavity and contains mainly the lower intestines. A somewhat round opening called the **pelvic inlet** leads into a narrower inferior space, which forms the wall of the pelvic cavity and contains mainly the rectum; the urinary bladder; and in women, the uterus. The lower opening of the lesser pelvis is called the **pelvic outlet.** The inlet and outlet refer to the passage of an infant during birth.

The pelvis is the most *sexually dimorphic* part of the skeleton—that is, the one whose anatomy most differs between the sexes. The average male pelvis is heavier and thicker than the female's owing to the forces exerted on the bone by more robust muscles. The female pelvis is adapted to the needs of both locomotion and pregnancy and childbirth. It is wider and shallower and has a larger, rounder pelvic inlet

---

[55]*pelv* = basin, bowl

176 PART 2 Support and Movement

Figure 6.31 The Pelvic Girdle. (a) Anterosuperior view. The pelvic girdle consists of the two hip bones and the sacrum. (b) The right hip bone, lateral view. The three childhood bones that fuse to form the adult hip bone are identified by color. **APR**

and outlet for passage of the infant. These and other features are used to determine the sex of skeletal remains.

The Hip Bone

The **hip bone** is also known as the *os coxae*[56] (oss COC-see) or *coxal bone*. It forms by the fusion of three childhood bones called the *ilium, ischium* (ISS-kee-um), and *pubis* (PEW-biss), identified by color in **figure 6.31b**. The largest of these is the

---

[56]*os* = bone; *coxae* = of the hip

ilium[57] (ILL-ee-um). If you rest your hand on your hip, you are supporting it on the upper margin of the ilium, called the **iliac crest.** A prominent feature of the posterior margin of the ilium is the deep **greater sciatic notch** (sy-AT-ic), through which the sciatic nerve passes on its way down the posterior side of the thigh. The posterolateral surface of the ilium is relatively rough because it serves for attachment of several muscles of the buttocks and thighs. The anteromedial surface, by contrast, is the smooth, slightly concave **iliac fossa,** covered by the broad *iliacus muscle*.

The **ischium** forms the lower posterior portion of the hip bone. It is a roughly C-shaped bone that contributes most of the posterior wall of the acetabulum. Its lowermost part is the thick, rough-surfaced **ischial tuberosity,** which you are probably sitting on as you read this. It can be palpated by sitting on your fingers.

The **pubis (pubic bone)** is the most anterior portion of the hip bone. It contributes the anterior part of the acetabulum. The **pubic symphysis**[58] is the point where the anterior bodies of the two pubic bones are joined. The ischium and pubis encircle the **obturator**[59] **foramen,** a large round-to-triangular hole below the acetabulum, closed in life by a ligamentous sheet. Laterally, the most prominent feature of the hip bone is the **acetabulum**[60] (ASS-eh-TAB-you-lum), a deep socket for the head of the femur. As you can see from the color coding in the figure, the ilium, ischium, and pubis meet here and each contributes roughly one-third of this socket.

The number and arrangement of bones in the lower limb are similar to those of the upper limb, and again number 30 bones per limb: the femur in the thigh, the patella (kneecap), the tibia and fibula in the leg proper, 7 tarsal bones in the ankle, and 5 metatarsal bones and 14 phalanges in the foot. The bones of the lower limb, unlike those of the upper, are adapted for weight bearing and locomotion, and are therefore shaped and articulated differently, as we shall see.

The Femur

The **femur** (FEE-mur) is the body's longest and strongest bone, measuring about one-quarter of one's height (fig. 6.32). It has a hemispherical head that inserts into the acetabulum. Distal to the head is a constricted **neck** and two massive, rough processes of the upper femur called the **greater** and **lesser trochanters** (tro-CAN-turs); these are attachments for the powerful muscles of the hip. The anterior surface of the shaft is smooth and rounded, but the posterior surface is marked by a strongly developed ridge, the **linea aspera,** an attachment for strong thigh muscles. The distal end of the femur flares into **medial** and **lateral epicondyles,** which serve as sites of muscle and ligament attachment and which can be palpated at the widest point of the knee. Distal to these are two smooth wheel-like surfaces of the knee

Figure 6.32 The Right Femur and Patella.

(a) Anterior view (b) Posterior view

---

[57]*ilium* = flank, loin
[58]*sym* = together; *physis* = growth
[59]*obtur* = to close, stop up; *ator* = that which
[60]*acetabulum* = vinegar cup

joint, the **medial** and **lateral condyles.** On the anterior side of the femur, a smooth medial depression called the **patellar surface** accommodates the patella.

The Patella

The **patella**[61] (kneecap) is a roughly triangular bone that develops within the quadriceps tendon of the knee (fig. 6.32). It is cartilaginous at birth and ossifies between 3 and 6 years of age. It has a broad superior **base,** a pointed inferior **apex,** and a pair of shallow **articular facets** on its posterior surface where it articulates with the femur. The patella glides up and down on the patellar surface of the femur when the knee moves, and improves the efficiency of the quadriceps femoris muscle when it extends the knee. The power of a soccer player's kick hinges on this contribution from the patella.

The Tibia

The **tibia** is the thick, strong bone on the medial side of the leg proper (**fig. 6.33**). (Remember that in anatomical terminology, *leg* refers only to the *crural region* between the knee and ankle.) The tibia is the only weight-bearing bone of this region. Its broad superior head has two fairly flat surfaces, the **medial** and **lateral condyles.** The condyles of the femur rock on these surfaces when the knee is flexed. You can palpate the rough anterior surface of the tibia, the **tibial tuberosity,** just below your patella. This is where the patellar ligament inserts and the quadriceps muscle of the thigh exerts its pull when it extends the knee, as in kicking a football. Distal to this, the shaft has a sharply angular **anterior border,** which you can easily palpate in the shin. At the ankle, just above the brim of a standard dress shoe, you can palpate a prominent bony knob on each side. These knobs are the **medial malleolus**[62] (MAL-ee-OH-lus) of the tibia and the **lateral malleolus** of the fibula.

The Fibula

The **fibula** (fig. 6.33) is a slender lateral strut that helps to stabilize the ankle. It does not bear any weight. The distal expansion is the *lateral malleolus* just mentioned. The shafts of the tibia and fibula are connected through a flexible *interosseous membrane* similar to that of the forearm.

The Ankle and Foot

The **tarsal bones** of the ankle are arranged in proximal and distal groups somewhat like the carpal bones of the wrist (**fig. 6.34**). Because of the load-bearing role of the ankle, however, they are arranged quite differently from the carpal bones and are thoroughly integrated into the arches of the foot. The proximal group includes the **calcaneus**[63] (cal-CAY-nee-us) of the heel; the **talus,** a superior bone that articulates with the tibia; and the short, wide **navicular** bone anterior to the talus. The distal group forms a row of four bones. Proceeding

Figure 6.33 The Right Tibia and Fibula.

(a) Anterior view
(b) Posterior view

[61]*pat* = pan; *ella* = little
[62]*malle* = hammer; *olus* = little
[63]*calc* = stone, chalk

from medial to lateral, these are the **medial, intermediate,** and **lateral cuneiforms**[64] (cue-NEE-ih-forms) and the **cuboid.** The cuboid is the largest of the four.

The remaining bones of the foot are similar in arrangement and name to those of the hand. The **metatarsals**[65] are similar to the metacarpals. They are numbered I to V from medial to lateral, metatarsal I being proximal to the great toe. (Roman numeral I refers to the largest digit in both the hand and foot—the *thumb* and the *great toe,* respectively.) Bones of the toes, like those of the fingers, are called **phalanges.** The great toe

---

[64]*cunei* = wedge; *form* = in the shape of
[65]*meta* = beyond; *tars* = ankle

Figure 6.34 The Right Foot. (a) Superior (dorsal) view. (b) Inferior (plantar) view. (c) Medial view. (d) X-ray of lateral view. The white arrow represents the lateral longitudinal arch. **APR**
d: Medical Body Scans/Science Source

contains only two bones, the proximal and distal phalanx I. The other toes each contain a proximal, middle, and distal phalanx.

Three strong, springy foot arches absorb shock as the body jostles up and down during walking and running. The **medial longitudinal arch** extends from the heel to the great toe; the **lateral longitudinal arch** extends from the heel to the little toe (fig. 6.34d); and the **transverse arch** extends from side to side at the level where the metatarsal bones articulate with the distal row of tarsal bones (the cuboid and cuneiforms). These arches are held together by short, tough ligaments. Excessive weight or repetitious stress can weaken and stretch these ligaments, resulting in the condition commonly called *fallen arches* or *flat feet (pes planus*[66]*)*. This condition makes it harder to tolerate prolonged standing and walking.

Before You Go On

Answer these questions from memory to test your understanding of the previous section. Reread the material if you can't answer the questions.

18. Prepare a list of all boldfaced key terms in the descriptions of the upper and lower limbs and the two limb girdles; try to identify each of those features on unlabeled models or illustrations of the skeleton.
19. What structure is found in the intertubercular groove of the humerus? What structure attaches to its deltoid tuberosity? What bones articulate with its head, capitulum, and trochlea, respectively?
20. Describe the features you can use to distinguish a radius from an ulna, and how you can tell the proximal from the distal end of each bone.
21. How many carpal bones are there? How many tarsal bones? If there are 30 bones in both the upper and the lower limb, what makes up for the unequal number of carpal and tarsal bones?
22. Give the anatomical (osteological) name for each of the following bony structures that you can palpate on the body surface: the apex of the shoulder; the point of the elbow; the bumps on each side of the wrist; the bumps on each side of the ankle; the heel; and the "ball" of the foot at the base of the great toe.

6.5 Joints

Expected Learning Outcomes

When you have completed this section, you should be able to

a. define and describe three major classes of joints, the subclasses of each, and examples;
b. identify the anatomical components seen in most synovial joints;
c. define and demonstrate the types of movement that occur at synovial joints; and
d. describe the basic anatomy of the shoulder, elbow, hip, and knee joints.

---

[66]*pes* = foot; *planus* = flat

A **joint,** or **articulation,** is any point at which two bones meet, whether or not it is movable at that point. Your shoulder, for example, is a very movable joint, whereas the skull sutures are immovable joints.

6.5a Classification

Joints are classified according to the manner in which the adjacent bones are connected, with corresponding differences in how freely the bones can move. There are three major structural categories of joints: fibrous, cartilaginous, and synovial.

Fibrous Joints

A **fibrous joint,** or **synarthrosis**[67] (SIN-ar-THRO-sis), is a point at which adjacent bones are bound by collagen fibers that emerge from the matrix of one bone, cross the space between them, and penetrate into the matrix of the other. Most of these have very little mobility—for example, the skull sutures and the joints that bind a tooth to the jaw bone. (The latter is a minor exception to the general definition of a joint given above, because a tooth is not a bone.) The radius and ulna, however, are bound to each other by a relatively mobile fibrous joint formed by the interosseous membrane along their length (see fig. 6.29). It allows for such movements as rotation of the forearm from a palm-up to a palm-down position (see fig. 6.42). A similar membrane binds the tibia and fibula, but with less mobility than seen in the forearm.

Cartilaginous Joints

A **cartilaginous joint,** or **amphiarthrosis**[68] (AM-fee-ar-THRO-sis), is a point at which two bones are linked by cartilage. Examples include the pubic symphysis, where fibrocartilage joins the right and left bones to each other; intervertebral discs, where fibrocartilage joins consecutive vertebrae to each other; and the attachment of rib 1 to the sternum by a hyaline costal cartilage (see fig. 6.25). (Ribs 2 through 10, however, are joined to the sternum by synovial joints.)

Synovial Joints

A **synovial joint** (sih-NO-vee-ul), or **diarthrosis**[69] (DY-ar-THRO-sis), is what usually comes to mind when one thinks of a skeletal joint. Synovial joints include the most familiar and movable joints of the body, such as the shoulder, elbow, hip, and knee joints, as well as some less obvious examples among the carpal and tarsal bones. Synovial joints are the most structurally complex type of joint, and are the most likely to develop uncomfortable and crippling dysfunctions. They are therefore the most important joints for professionals such as physical and occupational therapists, nurses, fitness trainers, and athletic and dance coaches to understand well.

The relative mobility of synovial joints stems partly from the fact that the two bones are separated by a narrow space, the **joint (articular) cavity;** the space contains a slippery lubricant called **synovial**[70] **fluid** (fig. 6.35), similar

Figure 6.35 Structure of a Simple Synovial Joint. **APR**

? *Why is a meniscus unnecessary in an interphalangeal joint?*

---

[67]*syn* = together; *arthr* = joined; *osis* = condition
[68]*amphi* = on all sides; *arthr* = joined; *osis* = condition
[69]*dia* = separate, apart; *arthr* = joint; *osis* = condition
[70]*ovi* = egg

to raw egg white in consistency. The facing bone surfaces are covered with articular cartilages up to 2 mm thick. A connective tissue **joint (articular) capsule** encloses the cavity. It consists of a sturdy outer *fibrous capsule* that encloses the entire joint and an inner *synovial membrane* that secretes the fluid and retains it in the joint cavity.

In some synovial joints, especially those subjected to a lot of weight or pressure, fibrocartilage grows inward from the joint capsule and forms a pad between the bones. In the jaw, between the distal ends of the radius and ulna, and at both ends of the clavicle, the pad crosses the entire joint capsule and is called an **articular disc.** In the knee, two cartilages extend inward from the left and right but do not entirely cross the joint. Each is called a **meniscus**[71] because of its crescent shape. Such cartilages absorb shock and pressure and improve the fit between the bones, guide the bones across each other, and stabilize the joint, reducing the chance of dislocation.

Accessory structures associated with a synovial joint include tendons, ligaments, and bursae. Tendons are often the most important structures in stabilizing a joint. A **bursa**[72] is a fibrous sac filled with synovial fluid, located between adjacent muscles, between bone and skin, or where a tendon passes over a bone (see fig. 6.44). Bursae cushion muscles, help tendons slide more easily over the joints, and sometimes enhance the mechanical effect of a muscle by modifying the direction in which its tendon pulls. They are especially numerous in the hand, knee, and foot. Inflammation of a bursa is called *bursitis*. It is usually caused by overuse or repeated trauma to a joint, and it occurs most commonly in the shoulder. A *bunion* is bursitis of the great toe.

There are six classes of synovial joints, distinguished by patterns of motion determined by the shapes of the articulating bone surfaces (**fig. 6.36**). A bone's movement at a joint can be described with reference to three mutually perpendicular planes in space (x, y, and z). Some bones can move in only one plane, like a door on its hinges; others move in two planes, like the gear shift of a manual transmission; and still others move in three, like the shoulder. The six synovial joints are described here in descending order of mobility: one that moves in three planes (ball-and-socket), three that move in two planes (condylar, saddle, and gliding), and two that move in a single plane (hinge and pivot).

1. **Ball-and-socket joints** occur at the shoulder and hip, where one bone has a smooth hemispherical head that fits within a cuplike depression (socket) on the other. The head of the humerus fits into the glenoid cavity of the scapula, and the head of the femur fits into the acetabulum of the hip bone.
2. **Condylar (ellipsoid) joints** have an oval convex surface on one bone that fits into a complementary depression on the next—for example, the joints between the phalanges and metacarpal bones at the base of the fingers. These joints move in two planes. To demonstrate this, hold your hand with your palm facing you. Make a fist, and these joints flex in the sagittal plane. Fan your fingers apart, and they move in the frontal plane.
3. **Saddle joints** occur at the base of the thumb between the trapezium and metacarpal I, and between the clavicle and sternum. Each bone has a saddle-shaped surface—concave in one direction (like the front-to-rear curvature of a horse saddle) and convex in the other (like the left-to-right curvature of a saddle). These joints move in two planes. The thumb, for example, moves in a frontal plane when you spread your fingers and in a sagittal plane when you move it as if to grasp the handle

---

[71]*men* = moon, crescent; *iscus* = little
[72]*burs* = purse

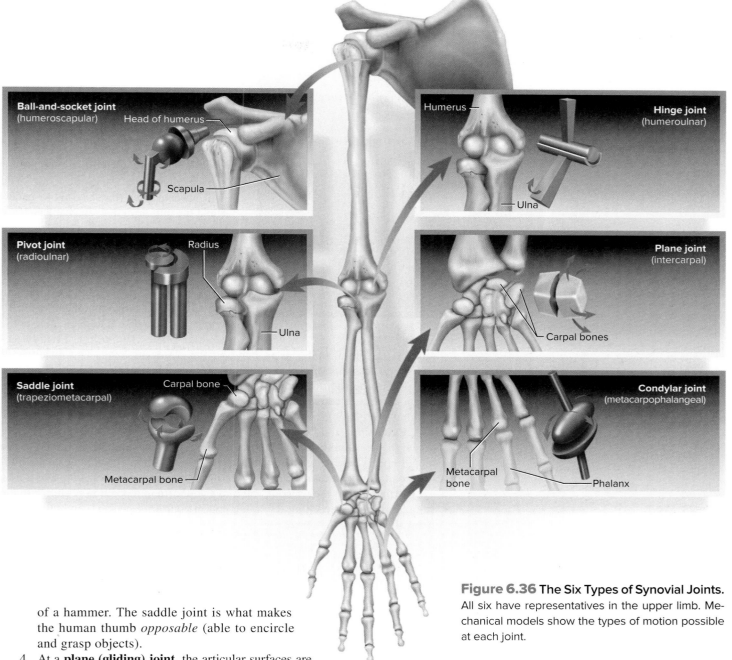

Figure 6.36 The Six Types of Synovial Joints. All six have representatives in the upper limb. Mechanical models show the types of motion possible at each joint.

of a hammer. The saddle joint is what makes the human thumb *opposable* (able to encircle and grasp objects).

4. At a **plane (gliding) joint,** the articular surfaces are flat or only slightly concave and convex. Plane joints occur between the wrist and ankle bones. The adjacent bones slide over each other and have rather limited movement. Their movements are complex but usually in two planes.
5. At a **hinge joint,** one bone has a convex surface that fits into a concave depression of the other one. Hinge joints can move freely in one plane, but they have very little movement in any other, similar to a door hinge. Examples include the elbow, knee, and interphalangeal (finger and toe) joints.
6. **Pivot joints** are joints in which a bone spins on its longitudinal axis. The principal examples are the joints between the first two vertebrae (see fig. 6.22) and between the radius and ulna at the elbow. The skull and atlas pivot on the axis when you shake your head "no," and the radius spins on its axis when you rotate your forearm.

Figure 6.37 Flexion and Extension. (a) Flexion and extension of the elbow. (b) Flexion, extension, and hyperextension of the wrist. (c) Flexion and hyperextension of the shoulder. (d) Hip and knee flexion and extension.
a–d: ©Ken Saladin/Timothy L. Vacula, photographer

Joints are not as easy to classify as this scheme may make it seem, however. The jaw (temporomandibular) joint, for example, has elements of condylar, hinge, and plane joints. Some other joints also show aspects of two or more of these six types.

6.5b Movement

Kinesiology, physical therapy, and related fields have a specific vocabulary for movements of the synovial joints. You will need to know these terms to understand the muscle actions in chapter 7.

Flexion, Extension, and Hyperextension

Most acts of flexion, extension, and hyperextension are movements that increase or decrease a joint angle in the sagittal plane (fig. 6.37). **Flexion** decreases the joint angle, as in bending the elbow or knee. The meaning of *flexion* is less obvious in the ball-and-socket joints of the shoulder and hip. Flexion of the shoulder consists of raising the arm in a sagittal plane, as if to point in front of you or toward the ceiling. Flexion of the hip entails raising the thigh, as in placing your foot on the next step when ascending stairs.

Extension straightens a joint—for example, straightening the elbow or knee. In stair climbing, you extend both the hip and knee to lift your body to the next step. **Hyperextension** extends a joint beyond where it would be in anatomical position. For example, raising the back of your hand as if admiring a new ring hyperextends the wrist. If you move your arm to a position posterior to the shoulder, such as reaching for something in your back pocket, you hyperextend your shoulder.

> **Apply What You Know**
> Some synovial joints have articular surfaces or ligaments that prevent them from being hyperextended. Try hyperextending some of your synovial joints and list a few for which this is impossible.

Abduction and Adduction

Abduction[73] (ab-DUC-shun) (fig. 6.38) is movement of a body part in the frontal plane away from the midline of the body—for example, raising the arm away from one side of the body or standing spread-legged. To abduct the fingers is to spread them apart. **Adduction**[74] (ah-DUC-shun) is movement toward the median plane.

---

[73]*ab* = away; *duc* = to carry, lead
[74]*ad* = toward; *duc* = to carry, lead

Elevation and Depression

Elevation is movement that raises a body part vertically. You elevate your mandible to bite off a piece of food, and elevate your clavicle and scapula when lifting a suitcase. The opposite of elevation is **depression,** such as lowering your mandible to open the mouth or lowering your shoulders.

Protraction and Retraction

Protraction[75] is anterior movement of a body part on a horizontal plane, and **retraction**[76] is posterior movement (fig. 6.39). You protract your shoulders when you reach out to hug someone or push open a door. You would retract them to stand at military attention. Such exercises as bench presses and push-ups involve repeated protraction and retraction of the shoulders. Most of us have some degree of overbite, so we protract the mandible to make the edges of the incisors meet when biting off a piece of food, then retract it to make the molars meet and grind food between them.

(a) Abduction

(b) Adduction

Figure 6.38 Abduction and Adduction.
a–b: ©Ken Saladin/Timothy L. Vacula, photographer

Circumduction

Circumduction[77] (fig. 6.40) is movement in which one end of an appendage remains relatively stationary while the other end makes a circular motion. The appendage as a whole thus describes a conical space. For example, if an artist standing at an easel reaches out and draws a circle on the canvas, the shoulder remains relatively stationary while the hand makes a circle. The extremity as a whole thus

---

[75]*pro* = forward; *trac* = pull, draw
[76]*re* = back; *trac* = pull, draw
[77]*circum* = around; *duc* = to carry, lead

(a) Protraction (b) Retraction

Figure 6.39 Protraction and Retraction of the Shoulder.
a–b: ©Ken Saladin/Timothy L. Vacula, photographer

Figure 6.40 Circumduction of the Upper Limb.
©Ken Saladin/Timothy L. Vacula, photographer

Figure 6.41 Medial and Lateral Rotation.
(a) Medial (internal) rotation of the humerus and femur. (b) Lateral (external) rotation of the same.
a–b: ©Ken Saladin/Timothy L. Vacula, photographer

(a) Medial (internal) rotation   (b) Lateral (external) rotation

exhibits circumduction. A baseball player winding up for the pitch circumducts the arm in a more extreme "windmill" fashion.

Rotation

Rotation is a movement in which a bone spins on its longitudinal axis like the axle of a bicycle wheel. Figure 6.41 shows limb movements that occur in **lateral (external)** and **medial (internal) rotation** of the femur and humerus. Turning the head from side to side is called **right** and **left rotation.** Powerful right and left rotation at the waist is important in such actions as baseball pitching and golf. Good examples of the lateral and medial rotation of the humerus are its movements in the forehand and backhand strokes of tennis.

Supination and Pronation

Supination[78] (SOO-pih-NAY-shun) is rotation of the forearm so that the palm faces forward or upward; in anatomical position, the forearm is supinated and the radius and ulna are parallel to each other (**fig. 6.42a**). Supination is the movement you would usually make with the right hand to turn a doorknob clockwise or drive a screw into a piece of wood. **Pronation**[79] is rotation of the forearm so that the palm faces toward the rear or downward (**fig. 6.42b**).

---

[78]*supin* = to lay back
[79]*pron* = to bend forward

As an aid to memory, think of it this way: You are *prone* to stand in the most comfortable position, which is with the palm *pronated*. If you were holding a bowl of *soup* in the palm of your hand, your forearm would have to be *supinated* to avoid spilling it. Supination and pronation should not be confused with *supine* (lying on one's back) and *prone* (lying face down).

Opposition and Reposition

Opposition[80] is movement of the thumb to approach or touch the fingertips (hence we say humans have an *opposable thumb*), and **reposition**[81] is its movement back to anatomical position, parallel to the index finger. Opposition enables the hand to grasp objects and is the single most important hand function.

Dorsiflexion and Plantar Flexion

Special names are given to vertical movements at the ankle (fig. 6.43a). **Dorsiflexion** (DOR-sih-FLEC-shun) is a movement in which the toes are raised (as one might do to apply toenail polish). Dorsiflexion occurs in each step you take as your foot comes forward. It prevents your toes from scraping on the ground and results in a *heel strike* when the foot touches down in front of you. **Plantar flexion** is a movement that points the toes downward, as in standing on tiptoe or pressing the gas pedal of a car. This motion also produces the *toe-off* in each step you take, as the heel of the foot behind you lifts off the ground before your toes do.

Inversion and Eversion

Inversion[82] is a foot movement that tilts the soles medially toward each other; **eversion**[83] tilts the soles away from each other (fig. 6.43b, c).

---

[80]*op* = against; *posit* = to place
[81]*re* = back; *posit* = to place
[82]*in* = inward; *version* = turning
[83]*e* = outward; *version* = turning

(a) Supination

(b) Pronation

Figure 6.42 Supination and Pronation of the Forearm. Note the way these forearm rotations affect the relationship of the radius and ulna.
a–b: ©Ken Saladin/Timothy L. Vacula, photographer

(a) Flexion of ankle

(b) Inversion

(c) Eversion

Figure 6.43 Movements of the Foot. (a) Dorsiflexion and plantar flexion. (b) Inversion. (c) Eversion.
a–c: ©Ken Saladin/Timothy L. Vacula, photographer

These movements are common in fast sports such as tennis and football, and sometimes result in ankle sprains. They are also important in walking across uneven surfaces, for example in mountain hiking on a rocky trail. These terms also refer to congenital deformities of the feet, which are often corrected by orthopedic shoes or braces.

6.5c Anatomy of Selected Synovial Joints

We now examine the gross anatomy of a few representative synovial joints at the shoulder, elbow, hip, and knee. These joints most often require medical attention, and they have a strong bearing on athletic performance.

The Shoulder

At the shoulder, the head of the humerus inserts into the shallow glenoid cavity of the scapula (fig. 6.44). The glenoid cavity is somewhat deeper than it appears on a dried skeleton because of a ring of fibrocartilage around its margin. This joint is stabilized mainly by a tendon of the biceps brachii muscle, which extends like a strap through the groove between the tubercles of the humerus, over the shoulder, and to the scapula. Four shoulder muscles at this joint contribute another stabilizing complex, the **rotator cuff** (see fig. 7.18).

This is the most mobile joint of the body, and serves with the elbow to position the hand for a given task. Its mobility, however, comes at considerable cost to stability. The shallowness of the glenoid cavity and looseness of the joint capsule make the shoulder highly subject to dislocation and other injuries.

The Elbow

The elbow is a hinge with three joints (fig. 6.45)—one where the trochlear notch of the ulna encircles the trochlea of the humerus, one where the disc-shaped head of the radius meets the capitulum of the humerus; and one where the edge of the radial head meets the ulna. Only the first two joints contribute to the elbow hinge, but the last two act in forearm supination and pronation. In these actions, the upper discoidal surface of the radial head swivels on the capitulum of the humerus, and the edge of the radial head spins like a

Figure 6.44 The Shoulder Joint. Bursae are shown in green. Gray bands not otherwise labeled are ligaments. The supraspinatus and subscapularis tendons are two tendons of the rotator cuff, a common site of athletic and workplace injuries. APR

(a) Anterior view (b) Frontal section

tire in a notch of the ulna. A ligament attaches to the ulna at both its ends and loops around the neck of the radius, holding the radius against the radial notch but allowing it to spin freely. On the posterior side of the elbow, there is a prominent bursa that eases the movement of tendons over the elbow. This is a common site of painful inflammation (*bursitis of the elbow,* or *olecranon bursitis*) resulting from impact injuries or repetitive motions on one's job or in sports such as tennis and golf.

The Hip

At the **hip joint,** the head of the femur inserts into the acetabulum of the hip bone (fig. 6.46). Because the hip joints bear much of the body's weight, they have deep sockets and are much more stable than the shoulder joint. The depth of the socket is somewhat greater than we see on dried bones because of a horseshoe-shaped ring of fibrocartilage attached to its rim. The joint is enclosed in three strong ligaments that twist when one stands, pulling the head of the femur tightly into the acetabulum and stabilizing the joint. This is important to human bipedalism because it allows one to stand with minimal continual muscle exertion at the joint. Because of the depth of the acetabulum and tautness of these ligaments, hip dislocations are rare. Hip fractures are common in the elderly, however, usually at the femoral neck just below the head, and hip replacement is one of the most common procedures of orthopedic surgery.

Figure 6.45 The Elbow Joint. **APR**

(a) Anterior view

(b) Lateral view, femoral head retracted from acetabulum

Figure 6.46 The Hip Joint. (a) The anterior view shows the ligaments that stabilze the joint. (b) The acetabulum with the femoral head retracted. **APR**

The Knee

The **knee joint** is the largest and most complex synovial joint (fig. 6.47). Anteriorly, it is partially covered by the patella. The tendon of the quadriceps muscle of the thigh extends from the muscle to the patella, and then continues as the patellar ligament to the tibial tuberosity below the knee (see fig. 7.20). The lateral and posterior sides of the knee joint are enclosed by a fibrous joint capsule. The knee is surrounded by about 13 bursae. It is stabilized mainly by the quadriceps tendon in front and the tendon of one of the hamstring muscles, the *semimembranosus*, on the rear of the thigh. Developing strength in these muscles therefore reduces the risk of knee injury.

Figure 6.47 The Knee Joint. Bursae are shown in green, articular cartilages in light blue, and tendons and ligaments in gray. APR

(a) Anterior view

(b) Posterior view

(c) Sagittal section

(d) Superior view of tibia and menisci

Clinical Application 6.2

KNEE INJURIES AND ARTHROSCOPIC SURGERY

The knee is highly vulnerable to rotational and horizontal stress, especially when it is flexed (as in skiing or running) and receives a blow from behind or from the side. The most common injuries are to a meniscus or the anterior cruciate ligament (ACL). Knee injuries heal slowly because ligaments and tendons have very little blood supply, and cartilage usually has no blood vessels at all. A damaged ACL is often replaced with a tissue graft from the patellar ligament or from a hamstring tendon.

The diagnosis and surgical treatment of knee injuries have been greatly improved by *arthroscopy*, a procedure in which a slender viewing instrument *(arthroscope)* and surgical instruments are inserted through small incisions in the knee. Shreds of torn menisci can be removed in this way. Arthroscopic surgery produces much less tissue damage than conventional surgery and enables patients to recover more quickly.

The joint cavity contains two crescent-shaped cartilages called the *lateral* and *medial menisci* (singular, *meniscus*) (fig. 6.47d). They absorb pressure from the weight of the body and stabilize the knee as described earlier. Deep within the joint cavity is a pair of ligaments called the **anterior cruciate**[84] **ligament (ACL)** (CROO-she-ate) and **posterior cruciate ligament (PCL).** These are called *cruciate* for the fact that they cross each other in the form of an X, and are called *anterior* and *posterior* for their attachments on the respective surfaces of the tibia. When the knee is extended, the ACL is pulled tight and prevents hyperextension. The PCL prevents the femur from sliding off the front of the tibia and prevents the tibia from being displaced backward. The most common knee injuries, often requiring surgical treatment, are to the ACL and menisci (see Clinical Application 6.2).

An important aspect of human bipedalism is the ability to "lock" the knees and stand erect without tiring the extensor muscles of the leg. When the knee is extended to the fullest degree allowed by the ACL, the femur rotates medially on the tibia. This action locks the knee, and in this state all the major knee ligaments are twisted and taut. To unlock the knee, a small posterior *popliteus muscle* rotates the femur laterally and untwists the ligaments.

Before You Go On

Answer these questions from memory to test your understanding of the previous section. Reread the material if you can't answer the questions.

23. Explain the distinction between fibrous and cartilaginous joints and give an example of each.
24. Explain what unique structural features of a synovial joint make some of these the most freely movable joints of the body.
25. Give one example of each of the following synovial joint types: ball-and-socket, condyloid, gliding, hinge, pivot, and saddle.
26. Explain why both the shoulder and hip sockets are deeper in living persons than they appear on a dried skeleton.
27. What function is served by the cruciate ligaments of the knee? What function is served by the menisci?

---

[84]*cruci* = cross; *ate* = characterized by

Aging of the Skeletal System

Nearly everyone experiences some degree of bone loss, called *osteopenia*,[85] with age. It begins as early as the 30s, as osteoblasts become less active than osteoclasts and bone deposition fails to keep pace with the rate of resorption. The aging of other organ systems also contributes to aging of the skeleton. As the skin ages, its role in vitamin D synthesis declines by as much as 75%, and declining levels of vitamin D mean we absorb less calcium from our food. As muscle mass and strength decline, the bones are subjected to less stress and are therefore less stimulated to deposit osseous tissue. One skeletal condition that becomes less common in old age, however, is herniated intervertebral discs. The discs become more fibrous in old age and have less nucleus pulposus, so they are less likely to rupture. Their degeneration, however, contributes to back pain and stiffness as the vertebral column becomes less supple. The thoracic cage also stiffens with age, because the costal cartilages tend to ossify. As a consequence, breathing becomes shallower and more laborious. We noted earlier that bone cells respond to mechanical stress by increasing calcium deposition, which is why regular physical exercise is essential for the maintenance of bone mass. Engaging in physical activity throughout life improves bone strength and helps to prevent osteoporosis.

---

[85]*osteo* = bone; *penia* = deficiency

CAREER SPOTLIGHT

Orthopedic Nurse

An orthopedic nurse is a specialist in the prevention and treatment of musculoskeletal disorders. He or she has achieved RN or LPN certification and then completed advanced training in orthopedic nursing. Many achieve certification as an orthopedic nurse certified (ONC), orthopedic nurse practitioner (ONP-C), or orthopedic clinical nurse specialist (OCNS-C); some go as far as master's and doctoral degrees. Orthopedic nurses must be skilled in such practices as traction, casting, external therapy, continuous passive motion therapy, and monitoring of neurological and circulatory status. Their work ranges from neonatal to geriatric care, in settings as diverse as hospitals, nursing homes, physician outpatient practices, industry, academics, and home health agencies. Another growing career in skeletal health is bone densitometrist, a radiologic technologist who specializes in assessment of bone density, especially in the evaluation and treatment of osteoporosis. See appendix B for links to further information on careers in orthopedic nursing and bone densitometry.

©ERproductions Ltd/Getty Images

CONNECTIVE ISSUES

Ways in Which the SKELETAL SYSTEM Affects Other Organ Systems

Integumentary System
Bones close to the body surface, such as skull bones, support and shape the skin.

Lymphatic and Immune Systems
The red bone marrow is the origin of all immune cells.

Muscular System
Bones provide the attachment sites for most muscles and furnish leverage, by means of which muscles exert most of their actions. Muscle contraction depends on calcium, for which the skeleton is the body's primary reservoir.

Respiratory System
Bones support the respiratory passages through the nasal cavity. The ribs protect the lungs, and rib movements are crucial to pulmonary ventilation.

Nervous System
The cranium and vertebral column protect the brain and spinal cord. Neural function is strongly affected by the skeleton's storage and release of calcium.

Urinary System
The skeleton supports and protects organs of the urinary system.

Endocrine System
Bones protect the endocrine glands of the head, chest, and pelvic cavity. Calcium is a necessary trigger for hormone secretion. Bone cells secrete hormones that target distant tissues.

Digestive System
The skeleton provides protective enclosure for some of the digestive organs.

Reproductive System
The skeleton provides protective enclosure for some reproductive organs. Pelvic anatomy contributes to the difficulty and complications of childbirth.

Circulatory System
The red bone marrow is the ultimate source of all blood cells. Calcium is needed for contraction of the heart. Osteoblasts and osteoclasts maintain blood calcium homeostasis.

Study Guide

Assess Your Learning Outcomes

To test your knowledge, discuss the following topics with a study partner or in writing, ideally from memory.

6.1 Skeletal Structure and Function

1. Bones and other components of the skeletal system
2. Five functions of the skeletal system
3. Differences in location and structure between compact and spongy bone
4. Four kinds of bone cells, their respective functions, and the relationship of osteocytes to the bone matrix and to each other
5. The components of the bone matrix and their functional significance
6. The histology of spongy and compact bone
7. Features of the gross anatomy of long and flat bones
8. Differences between the two types of bone marrow

6.2 Bone Development and Metabolism

1. The basic difference between intramembranous and endochondral ossification, and examples of bones that develop by each method
2. Stages of intramembranous and endochondral ossification; how each process accounts for the mature anatomy of a flat cranial bone and a long limb bone, respectively
3. The mechanisms by which bones grow in length and thickness
4. Terms for the addition of minerals to the skeleton and their removal from it; the cells that carry out these two processes, and how they do it
5. Terms for a deficiency and an excess of calcium in the blood, and the pathological consequences of such imbalances
6. The actions of calcitriol and parathyroid hormone on the skeleton
7. Other hormones and nutrients that affect skeletal metabolism

6.3 The Axial Skeleton

1. Meanings and examples of the bone surface markings listed in table 6.1
2. The usual number of named bones in the adult skeleton, and which groups of bones belong to the axial and appendicular skeleton
3. The cavities of the skull
4. The two categories of skull bones, how each category is defined, and the category to which any named bone of the skull belongs
5. Major features of any of the following cranial bones, including the ability to recognize and name these bones and their features in illustrations and laboratory materials: the frontal, parietal, temporal, occipital, sphenoid, and ethmoid bones
6. Major features of any of the following facial bones, including the ability to recognize and name these bones and their features in illustrations and laboratory materials: the maxillae, mandible, vomer, and inferior nasal conchae; and the palatine, zygomatic, lacrimal, and nasal bones
7. The locations and functions of the auditory ossicles and the hyoid bone
8. The usual number of vertebrae and intervertebral discs, the five groups of vertebrae, and the number of vertebrae in each group
9. The four curvatures of the adult spine, their functional significance, and the three main pathological deviations from normal
10. The general structural features seen in most cervical through lumbar vertebrae, and how their structure relates to the anatomy of the spinal cord and spinal nerves
11. Structural features that distinguish cervical, thoracic, and lumbar vertebrae from each other, and how one can identify a single isolated vertebra on the basis of these features
12. Unique features and names of the first two cervical vertebrae, and how their anatomy relates to movements of the head
13. Anatomy of the sacrum and coccyx, and how certain features of these can be seen as derived from features of the vertebrae higher on the spine
14. The structure of an intervertebral disc and how it relates to the condition known as a herniated disc
15. The three components of the thoracic cage
16. The three parts of the sternum, all bones that articulate with it, and the locations of these articulations
17. The distinction between true ribs, false ribs, and floating ribs, and which ribs belong to each category
18. How the anatomy of ribs 1, 11, and 12 differs from the more typical anatomy of ribs 2 through 10
19. How the ribs articulate with the vertebrae, and what features on the vertebrae mark these points of articulation

6.4 The Appendicular Skeleton

1. The names of the bones that form the pectoral and pelvic girdles
2. The shape and function of the clavicle and how one can distinguish the lateral from the medial end and superior from inferior surface of an isolated clavicle
3. Names and locations of the three fossae of the scapula; what landmark separates the two posterior fossae; and in general terms, what occupies these three fossae in a living person
4. Names and structures of the processes and joint cavity at the lateral end of the scapula
5. Names and locations of the 30 bones of the upper limb

6. Major anatomical features of the humerus, radius, and ulna, especially at the proximal and distal ends of each bone
7. The fibrous membrane between the radius and ulna, and its functional significance
8. Names of the four carpal bones of the proximal row, in order from lateral to medial, and the same for the four bones of the distal row
9. Structure of the metacarpal bones and phalanges, and the system for giving each of these 19 bones a unique name and number
10. Components of the pelvic girdle; the distinction between *pelvic girdle* and *pelvis;* and differences between the male and female pelvis
11. Anatomy of the hip bone; the childhood bones that fuse to form the adult hip bone; and articulations of the hip bone with the sacrum and femur
12. Anatomical features of the femur, patella, tibia, and fibula, especially at the proximal and distal ends of the long bones
13. Names and locations of the tarsal bones, and similarities and differences between the tarsal and carpal bones
14. The system for naming and numbering the metatarsal bones and phalanges of the foot
15. The three foot arches and their relationship to human bipedalism and locomotion

6.5 Joints

1. The differences between fibrous, cartilaginous, and synovial joints; synonyms for each; and examples of where each can be found
2. The general features of synovial joints and the modifications seen in synovial joints that bear unusual weight or pressure
3. Comparisons and contrasts between tendons and ligaments
4. The structure, function, and location of bursae
5. Examples of joints that move in one, two, and three planes
6. Six kinds of synovial joints, how they differ, and where each can be found
7. Definitions, examples of, and ability to demonstrate joint *flexion, extension,* and *hyperextension*
8. The same for *abduction* and *adduction*
9. The same for *elevation* and *depression*
10. The same for *protraction* and *retraction*
11. The same for *circumduction*
12. The same for *lateral* and *medial rotation,* and *right* and *left rotation*
13. The same for *supination* and *pronation*
14. The same for *opposition* and *reposition* of the thumb
15. The same for *dorsiflexion* and *plantar flexion* of the ankle
16. The same for *inversion* and *eversion* of the foot
17. Features of the articular bone surfaces at the shoulder joint; its relationship to the rotator cuff; and the advantage and disadvantage of its great mobility
18. Features of the articular bone surfaces at the elbow joint; how they relate to elbow and forearm movements; and the location and clinical significance of the olecranon bursa
19. Similarities and differences between the hip joint and shoulder joint; the role of cartilages in stabilizing these two joints; and the effects of standing on the hip joint
20. Features of the articular bone surfaces at the knee joint; the muscles most responsible for stabilizing the knee joint; and the structure, importance, and clinical significance of the menisci and cruciate ligaments

Testing Your Recall

1. Which cells secrete hydrochloric acid, resorb osseous tissue, and raise the blood calcium concentration?
   a. chondrocytes
   b. osteocytes
   c. osteogenic cells
   d. osteoblasts
   e. osteoclasts

2. The medullary cavity of an adult bone may contain
   a. adipose tissue.
   b. hyaline cartilage.
   c. periosteum.
   d. osteocytes.
   e. articular cartilages.

3. Which of the following movements are unique to the foot?
   a. dorsiflexion and inversion
   b. elevation and depression
   c. circumduction and rotation
   d. abduction and adduction
   e. opposition and reposition

4. Which of the following joints has anterior and posterior cruciate ligaments?
   a. the shoulder
   b. the elbow
   c. the hip
   d. the knee
   e. the ankle

5. Which of these is the bone of the heel?
   a. cuboid
   b. calcaneus
   c. navicular
   d. trochlear
   e. talus

6. _____ is secreted in response to a falling blood calcium level, and raises the level by promoting osteoclast activity.
   a. Parathyroid hormone
   b. Calcitonin
   c. Growth hormone
   d. Estrogen
   e. Calcitriol

7. A child jumps to the ground from the top of a playground "jungle gym." His leg bones do not shatter mainly because they have
   a. more osteons than older bone.
   b. young, resilient osteocytes.
   c. an abundance of calcium phosphate.
   d. collagen fibers.
   e. a strong periosteum.

8. One long bone meets another at its
   a. diaphysis.
   b. epiphyseal plate.
   c. periosteum.
   d. metaphysis.
   e. epiphysis.

9. The spinal cord passes through the _____ of the vertebrae.
   a. vertebral foramina
   b. intervertebral foramina
   c. transverse foramina
   d. obturator foramina
   e. bodies

10. The principal facial bone between the orbit and upper teeth is the
    a. frontal bone.
    b. zygomatic bone.
    c. maxilla.
    d. mandible.
    e. parietal bone.

11. The mastoid process and external acoustic meatus are parts of the _____ bone.

12. Osteocytes contact each other through channels called _____ in the bone matrix.

13. A tendon joins a muscle to a bone whereas a _____ joins one bone to another.

14. The lubricant in the shoulder and hip joints is called _____.

15. As people age they experience _____, a gradual loss of bone mass.

16. _____ are bone cells that secrete collagen and stimulate calcium phosphate deposition.

17. _____ is the science of body movement.

18. The femur is prevented from slipping sideways off the tibia in part by a pair of crescent-shaped cartilages called the lateral and medial _____.

19. Bones of the skull are joined along lines called _____.

20. A herniated disc occurs when a ring called the _____ cracks.

Answers in Appendix A

What's Wrong with These Statements?

Briefly explain why each of the following statements is false, or reword it to make it true.

1. There are equal numbers of carpal and tarsal bones.

2. The growth zone of a child's long bones is the articular cartilage.

3. Bones that articulate with the parietal bone include the frontal, temporal, occipital, and zygomatic bones.

4. On a living person, it would be possible to palpate the muscles in the subscapular fossa but not those of the supraspinous fossa.

5. Reaching behind you to take something out of your hip pocket involves hyperextension of the elbow.

6. The bones that enclose the brain develop from fetal cartilage.

7. Synovial fluid is secreted by the bursae.

8. In strict anatomical terminology, the words *arm* and *leg* both refer to regions with only one bone.

9. Parathyroid hormone is an important stimulus for bone deposition.

10. Climbing a flight of stairs requires strong, alternating adduction and abduction of the hip joint.

Answers in Appendix A

Testing Your Comprehension

1. Most osteocytes of an osteon are far removed from blood vessels, but are still able to respond to hormones in the blood. Explain how it is possible for hormones to reach and stimulate these cells.

2. How does the regulation of blood calcium concentration exemplify negative feedback and homeostasis?

3. Name the action that would occur at each of the following joints in the indicated situation. (For example, the shoulder in picking up a suitcase. Answer: elevation.) (a) The arm when you raise it to rest your hand on the back of a sofa on which you're sitting. (b) Your neck when you look up at a plane in the sky. (c) Your tibia when you turn the toes of one foot to touch the heel of the other foot. (d) Your humerus when you reach up to scratch the back of your head. (e) A bowler's backswing. (f) A basketball player's foot as she makes a jump shot. (g) Your shoulder when you pull back on the oars of a rowboat. (h) Your elbow when lifting a barbell. (i) A soccer player's knee when kicking the ball. (j) Your index finger when dialing an old rotary telephone. (k) Your thumb when you pick up a tiny bead between your thumb and index finger.

Chapter 7

The Muscular System

Chapter Outline

7.1 Muscular Tissue and Cells
- 7.1a The Functions of Muscles
- 7.1b Skeletal Muscle Fibers
- 7.1c The Nerve–Muscle Relationship

7.2 Physiology of Skeletal Muscle
- 7.2a Muscle Excitation, Contraction, and Relaxation
- 7.2b Whole-Muscle Contraction
- 7.2c Muscle Metabolism
- 7.2d Muscular Strength and Conditioning

7.3 Cardiac and Smooth Muscle
- 7.3a Cardiac Muscle
- 7.3b Smooth Muscle

7.4 Anatomy of the Muscular System
- 7.4a General Aspects of Muscle Anatomy
- 7.4b Muscles of the Face
- 7.4c Muscles of Chewing
- 7.4d Muscles of Head and Neck Movement
- 7.4e Muscles of the Trunk
- 7.4f Muscles of the Pectoral Girdle and Upper Limb
- 7.4g Muscles of the Pelvic Girdle and Lower Limb

Clinical Applications/Perspectives on Health
- Clinical Application 7.1: Neuromuscular Toxins and Paralysis
- Clinical Application 7.2: Rigor Mortis
- Clinical Application 7.3: Carpal Tunnel Syndrome
- Perspectives on Health

End of Chapter
- Aging of the Muscular System
- Career Spotlight: Massage Therapist
- Connective Issues
- Study Guide

A colorized frontal MRI scan, showing muscles of the lumbar, pelvic, and upper femoral regions. After reading this chapter, return to this photo and identify as many muscles as you can.
Simon Fraser/Science Source

BASE CAMP

Before ascending to the next level, be sure you're properly equipped with a knowledge of these concepts from earlier chapters.

- The functions of membrane proteins, especially receptors and ion gates (see "Membrane Proteins" in section 3.2a)
- Desmosomes and gap junctions (see "Cell Junctions" in section 3.2d)
- Gross anatomy of the skeleton (see sections 6.3 and 6.4)
- Terminology of joint movements (see "Movement" in section 6.5b)

Module 6
Muscular System

The muscular system is fundamental to nearly every aspect of our day-to-day lives. It is the basis of every move we make, every breath we take. We sometimes take for granted our ability to walk, chew and swallow, and lift and carry a child, yet all of these common activities require precise coordination of a symphony of muscles. Many of us pursue fitness training to improve muscular strength and endurance. We might do exercises that target our "lats," "abs," or "biceps," and to hone strength, speed, and endurance when we swim, bike, run, play soccer, or do other activities for recreation or competition. People are motivated to engage in physical activity not only for the joy of movement, but because most of us know that it promotes health. At every age, maintaining muscle mass contributes to quality of life and good health. In fact, research shows that lack of exercise contributes to even seemingly unrelated conditions such as diabetes and Alzheimer disease. The modern pattern of chronic diseases is associated with our increasingly sedentary lifestyle. Many of you reading this book may aspire to careers in fitness training, coaching, or rehabilitation training, and a knowledge of the muscular system will be central to your profession.

In this chapter, we'll survey the anatomy and function of the body's major skeletal muscles. We'll also examine how muscles work, right down to the cellular and molecular level, including the factors that underlie variations in strength, effects of exercise, and causes of fatigue. More briefly, we'll also compare the properties of skeletal muscle with the cardiac muscle of the heart and the smooth muscle of our internal organs, which underlie all manner of processes from digestive contractions to childbirth.

7.1 Muscular Tissue and Cells

Expected Learning Outcomes

When you have completed this section, you should be able to

a. identify diverse functions of muscular tissue;
b. describe the structure of a skeletal muscle fiber and relate this to its function; and
c. describe the nerve–muscle relationship in skeletal muscle.

There are three types of muscular tissue: skeletal, cardiac, and smooth, compared in table 4.11. The term **muscular system** refers only to the skeletal muscles, which are the primary focus of this chapter. However, we will begin with some characteristics that the three types have in common, and look more closely at their differences later in the chapter.

7.1a The Functions of Muscles

Collectively, the three types of muscle tissue have the following functions:

- **Movement.** This includes externally visible movements of the head, trunk, and limbs and less conspicuous actions such as breathing, propelling contents through the digestive tract, pumping blood, and the dilation and constriction of blood vessels to regulate blood pressure and flow. Muscular movements also play important roles in communication: speech, writing, gestures, and facial expressions.
- **Stability.** It is also important that the muscles prevent unwanted movement, as in maintaining posture and holding one bone still while another one moves.
- **Control of body openings and passages.** Ring-shaped **sphincter muscles** around the eyelids and pupils control admission of light to the eye; others regulate waste elimination and control the movement of food, bile, and other materials through the body.
- **Heat generation.** Body heat is necessary for enzymes to function and regulate one's metabolism. The skeletal muscles produce up to 30% of our body heat at rest and up to 40 times as much during exercise.
- **Glycemic control.** This means the regulation of blood glucose within normal limits. The skeletal muscles play a significant role in stabilizing blood sugar levels by absorbing a large share of it. In old age, in obesity, and when muscles become deconditioned and weakened, people suffer an increased risk of diabetes mellitus because of the decline in this glucose-buffering function.

7.1b Skeletal Muscle Fibers

Skeletal muscle may be defined as voluntary striated muscle that is usually attached to one or more bones. It is called **voluntary** because it is usually subject to conscious control, and **striated** because it has alternating light and dark transverse bands, or **striations** (fig. 7.1), that reflect the overlapping arrangement of the internal proteins that enable it to contract. Skeletal muscle cells are perhaps the most internally complex, tightly organized of all human cells. In order to understand muscle function, one must know how the organelles and contractile proteins of a muscle fiber are arranged.

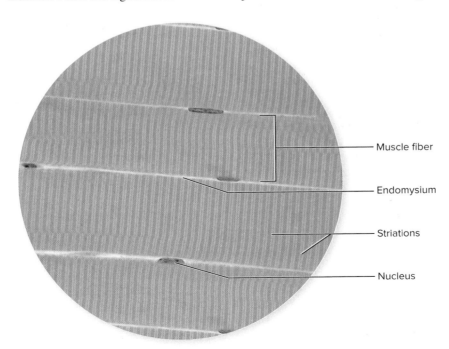

Figure 7.1 Skeletal Muscle Fibers (×1,000).
©Ed Reschke

❓ *Why are skeletal muscle cells called fibers, but cardiac and smooth muscle cells are not?*
APR

Structure of the Muscle Fiber

Skeletal muscle cells are called **muscle fibers** because of their exceptionally long slender shape: typically about 100 μm (0.1 mm) in diameter and 3 cm long, but sometimes up to 500 μm thick and 30 cm long. Each muscle fiber has multiple nuclei pressed against the inside of the plasma membrane, reserving the deeper part of the cell mainly for thick bundles of contractile protein (fig. 7.2).

These bundles, called **myofibrils,** number from several dozen to a thousand or more. Packed between them are numerous mitochondria, a network of smooth endoplasmic reticulum, deposits of the high-energy carbohydrate **glycogen,** and a red oxygen-binding pigment, **myoglobin.**

The plasma membrane, called the **sarcolemma,**[1] has tunnel-like infoldings called **transverse (T) tubules** that penetrate through the fiber and emerge on the other side. The function of a T tubule is to carry an electrical current from the surface of the cell to the interior when the cell is stimulated.

The smooth endoplasmic reticulum of a muscle fiber is called **sarcoplasmic reticulum (SR).** It forms a web around each myofibril, and alongside the T tubules

---

[1]*sarco* = flesh, muscle; *lemma* = husk

Figure 7.2 Structure of a Skeletal Muscle Fiber. This is a single cell containing 11 myofibrils (9 shown at the left end and 2 cut off at midfiber). A few myofilaments are shown projecting from the myofibril at the left. Their finer structure is shown in figure 7.3.

it exhibits dilated sacs called **terminal cisterns.** The SR is a reservoir for calcium ions; when the muscle fiber is stimulated, the SR releases a flood of calcium into the cytosol to activate the contraction process.

Myofilaments and Striations

Understanding muscle contraction requires a closer look at the myofibrils that pack the muscle fiber. Each of these is a bundle of parallel protein microfilaments called **myofilaments** (fig. 7.3). There are two main kinds of myofilaments.

Figure 7.3 Molecular Structure of Thick and Thin Filaments. (a) A single myosin molecule consists of two intertwined polypeptides forming a threadlike tail and a double globular head. (b) A thick filament consists of 200 to 500 myosin molecules bundled together with the heads projecting outward. (c) A thin filament consists of two intertwined chains of globular actin molecules, smaller filamentous tropomyosin molecules, and a calcium-binding protein called troponin associated with the tropomyosin. (d) A region of overlap between the thick and thin myofilaments.

1. **Thick filaments** are made of several hundred molecules of a protein called myosin. A **myosin** molecule is shaped like a golf club, with two polypeptides intertwined to form a shaftlike *tail,* and a double globular *head* projecting from it at an angle. A thick filament may be likened to a bundle of 200 to 500 such "golf clubs," with their heads directed outward in a helical array around the bundle.
2. **Thin filaments,** about half as wide as the thick filaments, are composed mainly of two intertwined strands of a protein called **actin.** Each actin is like a bead necklace—a string of globular subunits. A thin filament also has two proteins called **tropomyosin** and **troponin** that act together as a molecular "switch" that either allows or prohibits muscle contraction, as we shall see.

Myosin and actin are not unique to muscle; they occur in nearly all cells, where they function in moving and changing shape, mitosis, and transport of intracellular materials. In skeletal and cardiac muscle they are especially abundant, however, and are organized into a precise array that accounts for the striations of these two muscle types. Springy *elastic filaments* help to hold the myofilaments in proper alignment.

The highly regular pattern of myofilament overlap gives striated muscle an appearance of dark **A bands** alternating with lighter **I bands** (fig. 7.4). To help remember which band is which, think "dArk" and "lIght." Each I band consists only of thin filaments. The I band is bisected by a thin dark line

Figure 7.4 Muscle Striations and Myofilament Overlap. (a) Five myofibrils of a single muscle fiber, showing the striations in the relaxed state (electron micrograph). (b) The overlapping pattern of thick and thin myofilaments that accounts for the striations seen in part (a). **APR**

a: Don W. Fawcett/Science Source

? Why are A bands darker than I bands in the photo?

called the **Z disc**,[2] a dense plaque of protein that anchors the thin filaments. The darker A bands are regions in which thick and thin filaments overlap, accounting for the darker appearance of these bands. The middle of an A band is a little lighter than the rest, however, because the thin filaments do not reach that far.

Each segment of a myofibril from one Z disc to the next is called a **sarcomere**[3] (SAR-co-meer). A muscle shortens because its sarcomeres shorten and pull the Z discs closer to each other.

7.1c The Nerve–Muscle Relationship

Skeletal muscle cannot contract unless it is stimulated by a nerve. If its nerve connections are severed or poisoned, a muscle is paralyzed. Thus, muscle contraction cannot be understood without first understanding the relationship between nerve and muscle cells.

The nerve cells that stimulate skeletal muscles, called **motor neurons,** are located in the brain and spinal cord. Their axons lead to the muscles. At its distal end, each axon branches to multiple muscle fibers, but each muscle fiber receives only one nerve fiber. Each motor neuron stimulates all the muscle fibers of its group to contract at once, so one motor neuron and all muscle fibers supplied by it are called a **motor unit** (fig. 7.5). The muscle fibers of a given motor unit are dispersed through the muscle, not all concentrated in one place, so they contribute to a uniform contraction rather than producing an ineffective local twitch. Large motor neurons also require a stronger stimulus to excite them than small motor neurons, and are involved in stronger muscle contractions. Large motor neurons also require a stronger stimulus to excite them—and are involved in stronger muscle contractions—than small motor neurons.

The point where the end of an axon meets another cell is called a **synapse** (SIN-aps). When the second cell is a skeletal muscle fiber, the synapse is also called a **neuromuscular junction (NMJ)** (fig. 7.6). Each branch of the neuron ends in a bulbous swelling called an **axon terminal,** which is nestled in a depression on the muscle fiber. The two cells don't actually touch each other but are separated by a tiny gap, the **synaptic cleft.** A protein *basal lamina* encloses the entire NMJ and isolates it from the surrounding tissue fluid. Terminal Schwann cells cover the ends of the axon terminals and are important in repair and maintenance of the NMJ.

The axon terminal contains membrane-bounded sacs called **synaptic vesicles,** which contain a signaling chemical called **acetylcholine (ACh)** (ASS-eh-till-CO-leen). ACh is one of many signaling chemicals called neurotransmitters employed by the nervous system; others are introduced in section 8.2d. When the neuron releases ACh, the ACh diffuses quickly across the synaptic cleft and binds to proteins called **ACh receptors** on the surface of the muscle fiber. This binding stimulates the muscle fiber to contract. Also found on the muscle fiber and in the synaptic cleft is an enzyme called **acetylcholinesterase (AChE)** (ASS-eh-till-CO-lin-ESS-ter-ase). It quickly breaks down ACh to stop stimulation of the muscle and allow it to relax. Interference with ACh action can weaken or paralyze the muscles (see Clinical Application 7.1).

Figure 7.5 Motor Units. (a) A slice of the spinal cord depicting two motor neurons. Some motor neurons (green) are small, relatively sensitive, and easily activated. Others (violet) are large, less sensitive, and activated only when greater muscular strength is needed. (b) In the muscle, the muscle fibers of small motor units are relatively small (green). Large motor units have larger and more numerous muscle fibers (violet). Note that the muscle fibers of any given motor unit are distributed throughout the muscle and commingled with the fibers of other motor units (red), not clustered in one place.

---

[2] *Z = Zwischenscheibe* (German) = "between disc"
[3] *sarco* = muscle; *mere* = part, segment

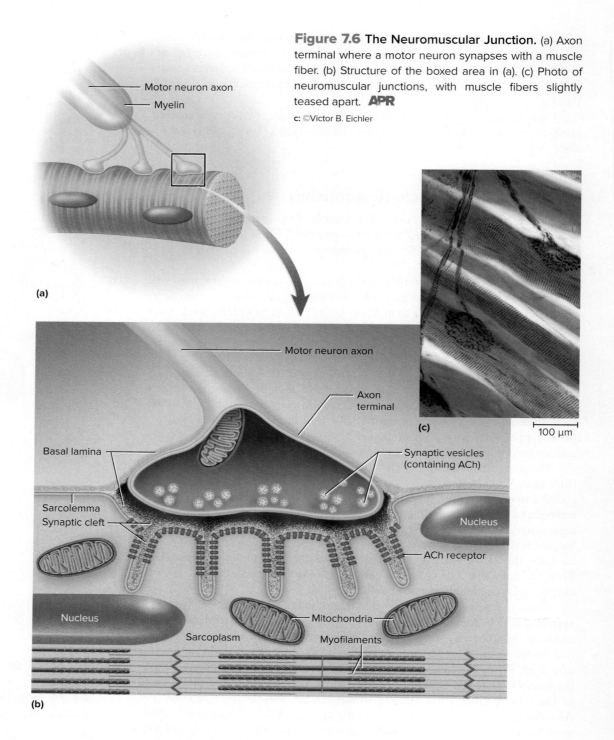

Figure 7.6 The Neuromuscular Junction. (a) Axon terminal where a motor neuron synapses with a muscle fiber. (b) Structure of the boxed area in (a). (c) Photo of neuromuscular junctions, with muscle fibers slightly teased apart. **APR**

c: ©Victor B. Eichler

Clinical Application 7.1

NEUROMUSCULAR TOXINS AND PARALYSIS

Some forms of muscle paralysis are caused by toxins that interfere with functions at the neuromuscular junction. Some pesticides, for example, contain *cholinesterase inhibitors,* which bind to AChE and prevent it from degrading ACh. As a result, ACh lingers in the synapse and overstimulates the muscle, causing *spastic paralysis*—a state of continual contraction. This can suffocate a person if the laryngeal muscles contract spasmodically and close the airway, or if the respiratory muscles cannot relax and allow free breathing. *Tetanus* (lockjaw) is a form of spastic paralysis caused by a bacterial toxin. The toxin acts in the spinal cord to cause overstimulation of the motor neurons, producing unwanted muscle contractions.

Flaccid paralysis is a state in which the muscles are limp and cannot contract. *Curare* (cue-RAH-ree), a poison extracted from certain plants in South America, binds to the ACh receptors but fails to stimulate the muscle. ACh is prevented from acting, and muscles can't contract. Curare was discovered by indigenous people in South America, and some tribes still use blowgun darts dipped in curare to hunt animals. Medically, curare has been used to treat muscle spasms in some neurological disorders.

Before You Go On

Answer these questions from memory to test your understanding of the previous section. Reread the material if you can't answer the questions.

1. Describe some functions of the muscular system other than external movements of the body.
2. Explain, in terms of thick and thin myofilaments, why skeletal muscle appears striated under the microscope. Identify the bands of a sarcomere and which contractile proteins are in each band.
3. What are the functions of synaptic vesicles, acetylcholine, and acetylcholinesterase?

7.2 Physiology of Skeletal Muscle

Expected Learning Outcomes

When you have completed this section, you should be able to

a. describe how a neuron stimulates a muscle fiber and initiates contraction;
b. explain the mechanisms of contraction and relaxation;
c. describe and explain twitch, summation, and other aspects of muscle behavior;
d. contrast isometric and isotonic contraction;
e. describe two ways in which muscle meets the energy demands of exercise;
f. discuss the factors that cause muscle fatigue and limit endurance;
g. distinguish between fast and slow types of muscle fibers; and
h. identify some variables that determine muscular strength.

7.2a Muscle Excitation, Contraction, and Relaxation

Movement results from a series of events that begin at the neuromuscular junction and continue in the thick and thin filaments within the muscle fiber. We can view the cycle of muscle contraction and relaxation as occurring in stages: excitation, contraction, and relaxation.

Excitation

Excitation is the process of converting an electrical nerve signal to an electrical signal in the muscle fiber. Steps 1 to 3 below, numbered to match **figure 7.7**, constitute the excitation process.

1. A nerve signal arrives at the synapse and stimulates synaptic vesicles to release acetylcholine (ACh) into the synaptic cleft.
2. ACh binds to ACh receptors in the sarcolemma. Each receptor is a gated channel that opens in response to ACh. Driven by concentration and electrical gradients across the sarcolemma, sodium ions (Na^+) quickly enter the muscle fiber and potassium ions (K^+) exit through these open gates.
3. These ion movements electrically excite the sarcolemma and initiate a wave of electrical changes called *action potentials* (discussed in more depth in chapter 8). Action potentials spread in all directions away from the neuromuscular junction—like ripples spreading out in a pond when you drop a stone into it—and pass down into the T tubules.

The muscle fiber is now excited, but three other events must happen before it can start to contract.

Muscle contraction is possible because of rapid, cyclic interactions between the myosin and actin of the thick and thin filaments. In a relaxed muscle, regulatory proteins block the myosin from binding to actin. Excitation initiates a chain of events that allows the myosin and actin to interact.

4. Excitation of a T tubule opens calcium (Ca^{2+}) channels in the adjacent sarcoplasmic reticulum. Calcium escapes the sarcoplasmic reticulum and floods the cytosol of the muscle fiber.

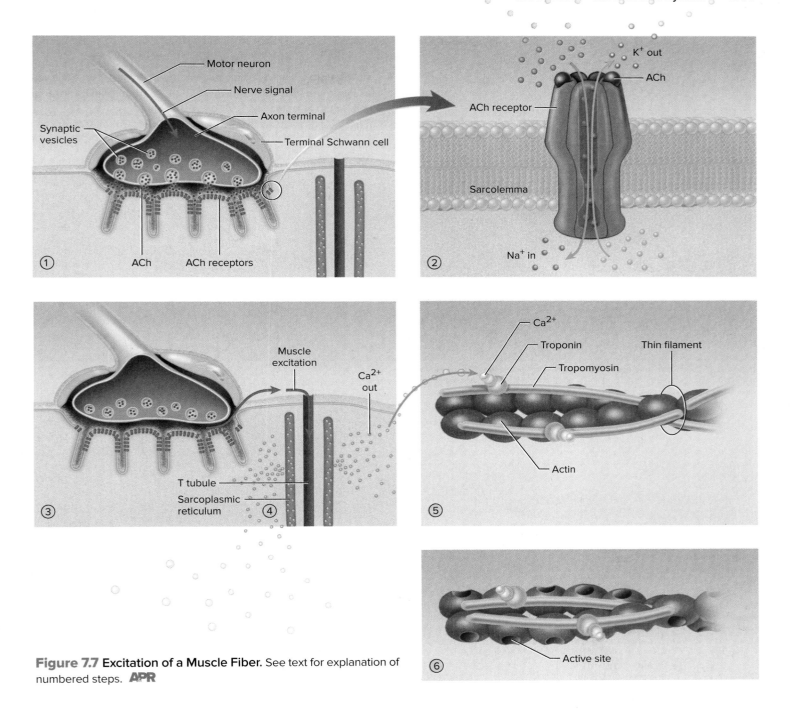

Figure 7.7 Excitation of a Muscle Fiber. See text for explanation of numbered steps. **APR**

⑤ Ca^{2+} binds to the troponin molecules attached to the thin filaments.
⑥ This induces the associated tropomyosin molecules to shift position, exposing the myosin-binding active sites on the actin.

Contraction

Contraction is the step in which the muscle fiber develops tension and may shorten. (Muscles often "contract," or develop tension, without shortening, as we see later.) It was once thought that a muscle cell contracted because its thick or thin filaments became shorter. Now it is understood that these filaments do not shorten; they slide across each other (fig. 7.8). The mechanism of contraction is therefore called the *sliding filament model*.

7. Each myosin head binds an ATP molecule and hydrolyzes it into ADP and a phosphate (P_i) group. This causes the head to move from a flexed position (like a bent elbow) to an extended, high-energy position.
8. The myosin head binds to an active site on the thin filament, like straightening your elbow to reach out and grasp a rope in your hand. The link between a myosin head and actin filament is called a **cross-bridge.**

Figure 7.8 The Sliding Filament Mechanism of Contraction. This is a cycle of repetitive events that cause a thin filament to slide over a thick filament and generate tension in the muscle. See text for explanation of numbered steps. **APR**

⑨ Myosin then releases the ADP and P_i and flexes into its original bent, low-energy position, tugging the thin filament along with it (like bending your elbow and tugging on the rope). This is called the **power stroke.** The head remains bound to actin until it binds to a new ATP.

⑩ Upon binding more ATP, the myosin head releases the actin. (Many other myosin heads retain their hold on the actin while some release it, so the actin doesn't slip back to the starting point.) The head is now prepared to repeat the whole process—it will split the ATP, return to its extended position (the **recovery stroke**), attach to a new site farther down the thin filament, and produce another power stroke.

This cycle repeats at a rate of about five strokes per second, with each stroke consuming one ATP molecule. Hundreds of myosin heads "crawl" along each thin filament at once, each head taking tiny jerky "steps" but with the net effect being a smooth motion, much like the smooth crawling of a millipede resulting from the tiny steps taken by hundreds of legs in succession. As the thin filament is pulled along the thick filament, it pulls the Z disc along behind it. The cumulative effect is that all the sarcomeres shorten. Through their linkage to the sarcolemma, they shorten the entire cell.

Figure 7.9 compares the appearance of sarcomeres in relaxed and contracted states. Note that in the contracted state, the sarcomere is shorter, but the thick and thin filaments are the same length as they are at rest. The sarcomere is shorter not because the filaments are shorter, but because the filaments overlap more extensively.

Relaxation

When its work is done and the nerve stops stimulating it, a muscle fiber relaxes and returns to its resting length. This is achieved by the following steps (not illustrated):

⑪ When nerve signals cease, the axon terminal stops releasing ACh.

⑫ Acetylcholinesterase breaks down the ACh that is already in the synapse, so stimulation of the muscle fiber ceases.

⑬ The sarcoplasmic reticulum reabsorbs Ca^{2+}, taking it back into storage. This is done by active-transport pumps, which require ATP; therefore, ATP is required not only for muscle contraction but also for its relaxation.

⑭ Deprived of Ca^{2+}, the troponin–tropomyosin complex shifts back into the resting position shown in figure 7.7 (step 5), where it blocks myosin from binding to actin. The muscle fiber ceases to produce or maintain tension and relaxes.

The return of a muscle to its resting length can be achieved by two means. One is simply the action of gravity. Suppose, for example, you contract your biceps brachii to flex your elbow and raise your hand. When you relax the biceps, gravity alone pulls your forearm downward and stretches the biceps back to its resting length. The other means of lengthening a muscle is the action of an antagonistic muscle operating at the same joint. Your *triceps brachii* on the posterior side of your arm is an *antagonist* of the biceps (see fig. 7.15). When it contracts, it extends the elbow and stretches the biceps.

Muscles normally do not relax to their *maximum* lengths. A low level of stimulation from the nervous system keeps even relaxed muscles in a state of partial contraction called **muscle tone (tonus).** By keeping the muscles firm even at rest, muscle tone helps to stabilize the joints. It also ensures that the thick and thin filaments always overlap enough so that when the muscle is called into action, a sufficient number of myosin heads can bind to actin and initiate contraction.

> ### Apply What You Know
> Chapter 2 noted that one of the most important properties of proteins is their ability to change shape (conformation) when other chemicals bind to them. Explain the relevance of this property to at least two proteins involved in muscle contraction.

(a) Relaxed sarcomere

(b) Fully contracted sarcomere

Figure 7.9 Relaxed and Contracted Sarcomeres. (a) A relaxed sarcomere, with partially overlapping thick and thin filaments and wide I bands. This is the state typical of resting muscle tone. (b) Fully contracted sarcomeres. Thick filaments abut the Z discs, and the sarcomeres are not capable of contracting any further.

❓ *If you looked at an electron micrograph of muscle fibers in the state depicted in part (b), which band of the striations would be absent? Explain.*

Clinical Application 7.2

RIGOR MORTIS

Rigor mortis[4] is the hardening of the muscles and stiffening of the body that begins 3 to 4 hours after death. It occurs partly because the sarcoplasmic reticulum deteriorates upon death, releasing calcium into the cytosol. Also, calcium leaks into the cell from the extracellular fluid through the deteriorating sarcolemma. The calcium ions activate myosin–actin cross-bridging. Once bound to actin, myosin cannot release it without first binding an ATP molecule, and of course ATP is unavailable in a dead body. The thick and thin filaments remain rigidly cross-linked until the myofilaments begin to decay. Rigor mortis peaks about 12 hours after death and then diminishes over the next 48 to 60 hours.

7.2b Whole-Muscle Contraction

Now you know how an individual muscle cell shortens. Our next objective is to consider how this relates to the action of the muscle as a whole.

Muscle Twitch and Tetanus

The minimum contraction exhibited by a muscle is called a **muscle twitch,** a single cycle of contraction and relaxation (**fig. 7.10a**). A twitch is very brief, lasting as little as 7 milliseconds (ms) in the fastest muscles and no more than 100 ms (0.1 s) even in the slowest ones. It is too brief and too weak to do useful muscular work such as move a joint. Useful work depends on **summation,** or the addition of multiple twitches that occurs when multiple nervous stimuli arrive in rapid succession.

When stimuli arrive in quick enough succession, the muscle relaxes only partially between twitches. Each new twitch "rides piggyback" on the previous one and generates higher tension. This results in a state of smooth contraction called **incomplete tetanus** (**fig. 7.10b**). The reason for the smoothness of overall muscle contraction despite the "flutter" in an individual motor unit is that motor units work in shifts; when one motor unit relaxes, another contracts and takes over so that the muscle doesn't lose tension. *Complete tetanus,* a state of constant spasmodic tension with no relaxation at all (**fig. 7.10c**), is achieved only by artificial high-frequency stimulation in the laboratory or by toxins (see Clinical Application 7.1). Such strong contraction can severely injure muscles, tendons, and other soft tissues and is usually prevented in the living body by a spinal negative feedback mechanism.

We can see, then, that variation in muscle tension can result from differences in the frequency of stimulation by a nerve fiber. Another factor in muscle tension is the number of motor units activated at once. If a gentle contraction is needed, as in lifting a teacup, the brain and spinal cord activate relatively few motor units in the muscle. For a stronger contraction, as in lifting a barbell, the nervous system activates a greater number of motor units, and activates larger ones (with more muscle fibers per motor neuron; see fig. 7.5).

Isometric and Isotonic Contraction

In muscle physiology, "contraction" doesn't always mean the shortening of a muscle—it may mean only that the muscle is producing tension while an external resistance causes it to stay the same length or even to become longer. Suppose you

---

[4]*rigor* = rigidity; *mortis* = of death

Figure 7.10 Stimulus Frequency and Muscle Tension. (a) Three complete muscle twitches produced at an unnaturally low stimulus frequency. (b) Summation of twitches, and progressively stronger muscle contraction, that occur naturally in the body at normal stimulus frequencies. (c) Fusion of twitches (complete tetanus) seen at high stimulus frequency in the laboratory, but almost never in the living body.

lift a heavy box of books from a table. When you first contract the muscles of your arms, you can feel the tension building in them even though the box is not yet moving. At this point, your muscles are contracting at a cellular level, but their tension is resisted by the weight of the load and the muscle produces no external movement. This phase is called **isometric**[5] **contraction**—contraction without a change in length. Once muscle tension rises enough to move the box the muscle begins to shorten while maintaining constant tension. This phase is called **isotonic**[6] **contraction.** Isometric and isotonic contraction are both phases of normal muscular action. Isometric contraction is also important in muscles that stabilize joints and maintain posture.

Isotonic contraction is further divided into two forms: concentric and eccentric. In **concentric contraction,** a muscle shortens as it maintains tension—for example, when the biceps brachii contracts and flexes the elbow such as during biceps curls. In **eccentric contraction,** a muscle maintains tension as it lengthens, such as the biceps when you lower a weight to the floor but maintain enough control not to simply let it drop. When people suffer muscle injuries in weight lifting, it is usually during the eccentric phase, because the sarcomeres and connective tissues of the muscle are pulling in one direction while the weight is pulling the muscle in the opposite direction.

7.2c Muscle Metabolism

All muscle contraction depends on ATP; this is the only energy source that myosin can use, so no other molecule can serve in its place. The question of providing energy to muscles during exercise therefore comes down to alternative ways of providing ATP.

There are two mechanisms for generating ATP: *anaerobic fermentation* and *aerobic respiration*. Anaerobic fermentation is a pathway in which glucose is ultimately converted to lactate. For each glucose consumed, this pathway produces a net yield of 2 ATP. Anaerobic fermentation has the advantage that it does not use oxygen. Therefore it is a way for a muscle to produce ATP during times when demand is high (in short bursts of exercise) but oxygen cannot be delivered fast enough to meet the needs of aerobic respiration. Its disadvantage is that the ATP yield is low. The lactate generated by anaerobic fermentation was long thought to be a significant factor in muscle fatigue, but recent research has cast doubt on this; the bloodstream washes away lactate as fast as it is produced and lactate level does not rise in an exercising muscle. Fatigue is due to other causes discussed later.

---

[5]*iso* = same, uniform; *metr* = length
[6]*iso* = same, uniform; *ton* = tension

Figure 7.11 Modes of ATP Production During Exercise.

In the alternative pathway, aerobic respiration, an intermediate product of glucose metabolism called *pyruvate* is oxidized in the mitochondria to carbon dioxide and water. In skeletal muscle, this method yields another 30 ATP molecules for each original glucose. It is far more efficient than anaerobic fermentation. Its disadvantage is that it requires oxygen, so aerobic respiration cannot function adequately during periods when exercise is so intense that the respiratory and cardiovascular system cannot keep pace with the needs of the muscular system.

Figure 7.11 shows how muscle shifts from one mode of ATP production to another over the course of exercise. In intense but brief exercise such as a 100-meter dash or a basketball play, muscle uses oxygen from its myoglobin to make ATP aerobically. As this ATP is consumed, it breaks down to ADP. The muscle fiber converts some of this ADP back to ATP by borrowing a phosphate group, either from a phosphate-storage molecule called *creatine phosphate,* or from other ADP molecules. This mechanism, called the *phosphagen system,* can supply enough ATP for about 6 seconds of sprinting or fast swimming, or about 1 minute of brisk walking.

When these sources of ATP are exhausted, the muscle gradually shifts to anaerobic fermentation (the *glycogen–lactate system*), which provides enough ATP for about 30 to 40 seconds of maximum activity, as in running completely around a baseball diamond. In time, a faster heartbeat and faster, deeper respiration catch up with the demands of the muscle, providing enough oxygen for it to shift back to aerobic respiration. Aerobic respiration usually takes over after about 40 seconds. In exercises lasting more than 10 minutes, such as bicycling or marathon running, more than 90% of the ATP is produced aerobically.

Fatigue and Endurance

Muscle fatigue is the progressive weakness that results from prolonged use of the muscles. This is what sets a limit, for example, on how many chin-ups or bench presses a person can perform. Rapidly squeezing a rubber ball or trying to take lecture notes from a fast-talking professor results in fatigue in hand muscles. There are numerous causes of fatigue, including depletion of glycogen and blood glucose, leakage of calcium from the sarcoplasmic reticulum, and accumulation of potassium in the extracellular fluid, which reduces the excitability of the muscle fiber.

Endurance, or tolerance of prolonged exercise, depends on several factors: a muscle's supply of myoglobin and glycogen; its density of blood capillaries; the supply of organic nutrients such as fatty acids and amino acids; the number of mitochondria; one's maximal rate of oxygen uptake (which depends on physical conditioning); and simple willpower and other psychological factors.

Recovery Period

When you have just completed a strenuous exercise such as a run, you don't return immediately to your normal physiological state, but experience a recovery period in which you continue to breathe more heavily than usual (as well as exhibit an elevated heart rate and blood pressure). The heavy breathing reflects the body's need for extra oxygen, called **excess postexercise oxygen consumption (EPOC)** (also known by the older expression *oxygen debt*). EPOC is the difference between the resting rate of oxygen consumption and the elevated rate following an exercise. This oxygen serves to replenish the resting levels of ATP and creatine phosphate in the muscles; to oxidize and dispose of lactate; to meet the elevated overall metabolic demand of the body heated by exercise; and to rebuild oxygen levels in the muscle myoglobin, blood hemoglobin, and blood plasma.

Physiological Types of Muscle Fibers

Not all skeletal muscle fibers are metabolically alike or adapted to perform the same task. There are two primary fiber types: some that respond slowly but are relatively resistant to fatigue, and some that respond more quickly but also fatigue quickly. Most muscles are composed of a mixture of the two.

- **Slow-twitch fibers** exhibit relatively long, slow twitches (up to 100 ms), so they aren't well adapted for quick responses. On the other hand, they are well adapted for aerobic respiration and do not fatigue easily. Their aerobic adaptations include abundant mitochondria, myoglobin, and blood capillaries. Myoglobin and blood capillaries give the tissue a deep red color, so muscles made predominantly of slow-twitch fibers are sometimes called *red muscles*. Examples include postural muscles of the back and a deep calf muscle called the *soleus*.
- **Fast-twitch fibers** are well adapted for quick responses (twitches as fast as 7.5 ms), but not for fatigue resistance. They are especially important in sports such as basketball that require stop-and-go activity and frequent changes of pace. They are rich in enzymes for anaerobic fermentation and for regenerating ATP from creatine phosphate. Their sarcoplasmic reticulum releases and reabsorbs Ca^{2+} quickly, which partially accounts for their quick, forceful contractions. They have more glycogen but less myoglobin and fewer mitochondria and blood capillaries than slow-twitch fibers, so they are relatively pale. Muscles composed predominantly of fast-twitch fibers are sometimes called *white muscles*. Examples include the *biceps brachii* of the arm, muscles of eye movement, and a superficial calf muscle called the *gastrocnemius*.

People with different types and levels of physical activity differ in the proportion of one fiber type to another even in the same muscle. For example, one study of male athletes found the quadriceps muscle of the thigh in marathon runners to be 82% slow-twitch and 18% fast-twitch fibers, but quite the opposite in sprinters and jumpers: 37% slow- and 63% fast-twitch. People are born with a genetic predisposition for a certain ratio of fiber types; one person might be a "born sprinter" and another a "born marathoner." One who is born with a genetic predisposition for predominantly slow-twitch fibers in the powerful muscles of the thighs might become a world-class marathoner, but be unable to cross over to a competitive level of sprinting, and vice versa.

Sometimes two or more muscles act across the same joint and superficially seem to have the same function. For example, the gastrocnemius and soleus muscles both insert on the heel through the same *calcaneal (Achilles) tendon*, so they have the same action—to lift the heel and raise the body, as in walking, running, and jumping. But they're not redundant. The gastrocnemius is a white muscle adapted for quick, powerful movements such as jumping, whereas the soleus is a red muscle that does most of the work in endurance exercises such as running, skiing, and climbing.

7.2d Muscular Strength and Conditioning

Muscular strength depends on a variety of anatomical and physiological factors. We have just seen that sustained or repetitive contraction can cause muscle fatigue, a gradual loss of strength; muscle strength therefore depends in part on fatigue resistance. Strength is also proportional to the diameter of a muscle at its thickest point (such as the bulge at the middle of the biceps brachii), which is why body building increases strength. There are two types of exercise that improve muscle performance.

Resistance exercise, such as weight lifting and chin-ups, is the contraction of muscles against a load that resists movement. A goal of weight lifting is to build muscle mass. Because muscle fibers are incapable of mitosis, the growth results not from an increase in the number of muscle cells but from an increase in the size of preexisting cells.

Endurance (aerobic) exercise, such as running and swimming, improves fatigue resistance. In slow-twitch fibers, especially, such activities increase glycogen stores, numbers of mitochondria, and density of blood capillaries—all of which promote efficient production of ATP. Endurance exercise also increases the oxygen-transport capacity of the blood and enhances the efficiency of the cardiovascular, respiratory, and nervous systems.

Resistance exercise doesn't significantly improve fatigue resistance, and endurance exercise doesn't significantly increase strength. Optimal performance comes from *cross-training*, which incorporates elements of both types.

> **Apply What You Know**
>
> Review hyperplasia and hypertrophy in section 4.6a. To which of these processes would you attribute the growth of a weight lifter's muscles? Explain.

Before You Go On

Answer these questions from memory to test your understanding of the previous section. Reread the material if you can't answer the questions.

4. What is the role of calcium in muscle contraction? Where does it come from and what triggers its release from that source?
5. What process makes the thin myofilaments "slide" over the thick ones during muscle contraction? What is the role of ATP in that process?
6. Explain why high-frequency stimulation makes a muscle contract with more tension than a single twitch produces.
7. Describe what roles isometric and isotonic contraction play when you lift a heavy object.
8. What is the difference between anaerobic fermentation and aerobic respiration? Identify a physical activity that would depend mainly on each type of ATP production.
9. Describe some tasks for which slow-twitch fibers are more effective than fast-twitch fibers, and vice versa.
10. Identify the benefits and shortcomings of both resistance and endurance exercise.

7.3 Cardiac and Smooth Muscle

Expected Learning Outcomes

When you have completed this section, you should be able to

a. describe the special functional roles of cardiac and smooth muscle; and
b. explain how the structure of these two forms of muscle accounts for their functional properties.

Until now we have been concerned with skeletal muscle. The other two types of muscular tissue are cardiac and smooth.

7.3a Cardiac Muscle

Cardiac muscle (fig. 7.12) is limited to the heart, where its function is to pump blood. Knowing that, we can predict the properties that it must have: (1) It must contract with a regular rhythm; (2) the muscle cells of a given heart chamber must be well synchronized so the chamber can effectively eject blood; (3) each contraction must last long enough to expel blood from the chamber; (4) it must function in sleep and wakefulness, without fail and without need of conscious attention; and

Figure 7.12 Cardiac Muscle. (a) Light micrograph (×400). (b) Structure of an individual cardiomyocyte. The portion in color is a single cell; adjacent cells are shown in gray. Intercalated discs connect cells together end to end through electrical and mechanical junctions. APR

a: ©Ed Reschke

(5) it must be highly resistant to fatigue. These functional properties are keys to understanding how cardiac muscle differs structurally and physiologically from skeletal muscle.

Cardiac muscle is striated like skeletal muscle, but it is **involuntary** (it contracts automatically, without requiring one's conscious control and attention), and the heart is **autorhythmic**[7] (it has a self-maintained rhythm of contraction). The rhythm is set by a pacemaker in the wall of the upper heart. Although the heart receives nerves from the brainstem and spinal cord, they only modify the rate and force of the heartbeat; they do not initiate each beat.

Cardiac muscle cells are not long fibers like those of skeletal muscle, but are shorter and appear roughly rectangular in tissue sections. They are slightly forked or notched at the ends, and each fork links one cell to the next. The linkages, called **intercalated discs** (in-TUR-kuh-LAY-ted), appear in stained tissue sections as dark lines thicker than the striations. An intercalated disc has electrical *gap junctions* that allow each cell to directly stimulate its neighbors, and has *desmosomes* and other mechanical junctions that keep the cells from pulling apart when the heart contracts.

Cardiac muscle doesn't exhibit quick twitches like skeletal muscle; rather, its contractions are prolonged for as much as a quarter of a second (250 ms). This means that each heart chamber sustains its contraction long enough to effectively expel blood.

Cardiac muscle uses aerobic respiration almost exclusively. It is very rich in myoglobin and glycogen, and it has especially large mitochondria that fill about 25% of the cell, compared to smaller mitochondria occupying about 2% of a skeletal muscle fiber. Because it makes little use of anaerobic fermentation, cardiac muscle is very resistant to fatigue. The skeletal muscles in your hand would quickly tire if you squeezed every 0.8 s, but the heart contracts at this rate, and often faster, for a lifetime without fatigue.

Chapter 13 explores cardiac muscle in greater depth.

7.3b Smooth Muscle

Smooth muscle (fig. 7.13) is involuntary, like cardiac, but it lacks striations; the latter quality is the reason it is called smooth. Smooth muscle cells are much

---

[7]*auto* = self

Figure 7.13 Smooth Muscle. (a) Light micrograph (×400). (b) Layered arrangement of smooth muscle in a cross section of the esophagus. Many hollow organs have alternating layers of smooth muscle like this. **APR**

a: Dennis Strete/McGraw-Hill

? *For what purpose does the esophagus require such a muscular wall?*

shorter than skeletal muscle fibers; have only one nucleus located near the middle of the cell; and are fusiform in shape—thick in the middle and tapered at the ends. Smooth muscle occurs in the walls of the blood vessels and many body-cavity organs (viscera)—such as in the respiratory, urinary, digestive, and reproductive tracts—where it is also called **visceral muscle.** A slightly different form of smooth muscle occurs in the iris of the eye and is associated with the hair follicles of the skin.

Some functions of smooth muscle, and its special properties, can be inferred by reflecting on organs such as the stomach, urinary bladder, and uterus, whose walls are composed primarily of this tissue. Such organs relax and stretch a great deal as they fill with food, urine, or a growing fetus. On the other hand, when such organs are empty, they don't collapse into flaccid bags. Their smooth muscle maintains a state of contraction called **smooth muscle tone,** which keeps the walls of the organs firm.

Special structural characteristics of smooth muscle cells underlie their distinctive functional properties. Thick and thin myofilaments are present, but they do not overlap each other with the regularity that produces the striations of skeletal and cardiac muscle. Sarcomeres and Z discs are absent. The thin filaments slide over thick filaments during contraction, but instead of pulling on Z discs and shortening a sarcomere, they pull on dense plaques of protein scattered throughout the cytoplasm and on the inner surface of the sarcolemma. In doing so, they cause the myocyte to contract with a twisting motion, like wringing out a wet towel.

Smooth muscle contracts in response to nervous stimulation as well as such stimuli as stretch (as in a full bladder), hormones (for example, during labor contractions), and blood levels of CO_2, O_2, and pH. As in the other muscle types, contraction is triggered by calcium ions (Ca^{2+}). However, smooth muscle contains very little sarcoplasmic reticulum and obtains its calcium from the extracellular fluid by way of calcium channels in the sarcolemma. During relaxation, calcium is pumped back out of the cell by active transport.

Smooth muscle contracts slowly, with tension peaking about 0.5 second after stimulation and then declining over a period of 1 to 2 seconds. Its myosin doesn't always detach from actin immediately; it can remain attached for prolonged periods without consuming more ATP. Smooth muscle therefore maintains tension while consuming only 1/10 to 1/300 as much ATP as skeletal muscle does; this makes it very resistant to fatigue.

Unlike skeletal and cardiac muscle, smooth muscle is fully capable of mitosis. An organ such as the pregnant uterus can grow by adding more muscle cells, and injured smooth muscle regenerates much better than the other two muscle types.

Before You Go On

Answer these questions from memory to test your understanding of the previous section. Reread the material if you can't answer the questions.

11. Explain how both cardiac muscle and smooth muscle are able to resist fatigue better than skeletal muscle does.
12. What are the functions of intercalated discs?
13. Explain how smooth muscle can maintain tension while consuming less energy (ATP) than skeletal muscle does. Why is this property important to the functions of blood vessels and viscera?

7.4 Anatomy of the Muscular System

Expected Learning Outcomes

When you have completed this section, you should be able to

a. describe the relationship of muscle fibers to connective tissues in a muscle;
b. distinguish between the attachments and belly of a muscle;
c. explain how muscles act in groups to govern the movements of a joint;
d. interpret some Latin terms commonly used in the names of muscles; and
e. identify or describe the locations and functions of several major skeletal muscles of the body.

7.4a General Aspects of Muscle Anatomy

The rest of this chapter is concerned with the anatomy of skeletal muscles. Before we examine individual muscles, we will survey general features of gross anatomy that muscles have in common, the functional relationships of different muscles to each other, and principles of naming skeletal muscles.

Connective Tissues and Muscle Attachments

A skeletal muscle is composed of more than muscular tissue. It also includes fibrous connective tissue, blood vessels, and nerves (**fig. 7.14**). Each muscle fiber is enclosed in a thin sleeve of loose connective tissue called the **endomysium**[8] (EN-doe-MIZ-ee-um). This tissue slightly separates the cells and allows room for capillaries and neurons to reach every muscle fiber.

Muscle fibers are arranged in bundles called **fascicles**[9] (FASS-ih-culs) (the grain visible in a cut of meat such as roast beef). Each fascicle is surrounded and set off from neighboring ones by a connective tissue sheath called the **perimysium**,[10] thicker than the endomysium. The muscle as a whole is surrounded by still another connective tissue layer, the **epimysium**.[11] Adjacent muscles are separated from each other and from the skin by fibrous sheets called **fasciae** (FASH-ee-ee) (singular, *fascia*).

Some fasciae separate groups of functionally related muscles into **muscle compartments**. A muscle compartment also contains the nerves and blood vessels that supply the muscle group. The tight binding of muscles by these fasciae contributes to a clinical problem called *compartment syndrome*, a common athletic injury (see Perspectives on Health).

---

[8]*endo* = within; *mys* = muscle
[9]*fasc* = bundle; *icle* = little
[10]*peri* = around; *mys* = muscle
[11]*epi* = upon, above; *mys* = muscle

Figure 7.14 Muscle–Connective Tissue Relationships. (a) The muscle–bone attachment. (b) Muscle fascicles in the tongue. Vertical fascicles passing between the superior and inferior surfaces of the tongue are seen alternating with cross-sectioned horizontal fascicles that pass from the tip to the rear of the tongue. A fibrous perimysium can be seen between the fascicles, and endomysium can be seen between the muscle fibers within each fascicle (c.s. = cross section; l.s. = longitudinal section). (c) Muscle compartments in the leg. Fibrous fasciae separate muscle groups into tight-fitting compartments along with their blood vessels and nerves.
b: Victor Eroschenko

PERSPECTIVES ON HEALTH

The muscular system suffers fewer disorders than any other organ system, but muscle soreness, cramps, and injuries are common, and some people suffer muscular dystrophy and other serious muscle diseases.

Muscle Pain

The most common form of muscle pain is *delayed onset muscle soreness (DOMS)*—the pain and stiffness commonly felt from hours to a day after strenuous muscular work. DOMS is associated with microtrauma to the muscles, including disrupted plasma membranes and sarcomere structure. *Cramps* are painful muscle spasms triggered by heavy exercise, extreme cold, dehydration, electrolyte loss, low blood glucose, or inadequate blood flow to the muscles.

Muscle Injuries

Every year, thousands of competitive athletes from the high school to professional levels sustain some type of injury to their muscles, as do many people who run, weight lift, do cross fit, and engage in other types of exercise. Long, hard running and repetitive kicking (as in soccer and football), for example, sometimes result in *pulled hamstrings*—strained hamstring muscles (see fig. 7.21) or tears in the hamstring tendons, often with a hematoma (blood clot) in the fascia around the muscles. *Shinsplints* are various injuries producing pain in the leg, such as inflammation of posterior leg muscle tendons (tendinitis) or periosteum of the tibia. Shinsplints commonly result from unaccustomed jogging or other vigorous activity of the legs after a period of relative inactivity.

The *rotator cuff* (see figs. 6.44 and 7.21) is a site of frequent injuries among baseball pitchers, bowlers, swimmers, weight lifters, and those who play racquet sports, among other people. Injuries here are usually caused by strenuous circumduction of the shoulder, shoulder dislocation, or persistent use of the arm in a position above horizontal. Such actions can tear the tendons of any of the rotator cuff muscles, most often the tendon of the supraspinatus. Recurrent inflammation of a rotator cuff tendon can cause the tendon to degenerate and eventually rupture in response to moderate stress. The injury is not only painful but also makes the shoulder joint unstable and subject to dislocation.

Compartment syndrome is a disorder in which overuse, muscle strain, or trauma damages the blood vessels in a muscle compartment. Because the fasciae cannot stretch, a blood clot *(contusion)* or tissue fluid in the compartment compresses the muscles, nerves, and vessels. The reduced blood flow can lead to nerve destruction within 2 to 4 hours and death of muscular tissue if the condition remains untreated for 6 hours or more. While nerves can regenerate if blood flow is restored, muscle death is irreversible. Depending on its severity, compartment syndrome may be treated with immobilization of the limb, rest, or an incision to drain fluid and relieve compartment pressure.

Muscular Dystrophy

Duchenne muscular dystrophy (DMD) is hereditary. It involves progressive replacement of muscular tissue with adipose and fibrous tissue. Most cases occur in males and are diagnosed between the ages of 2 and 10 years. When a child with DMD begins to walk, he may fall frequently and have difficulty standing up again. DMD tends to affect the hips first, then the lower limbs, and then progresses to the abdominal and spinal muscles. Patients are usually confined to a wheelchair by early adolescence and rarely live past the age of 20. Genetic screening can determine whether a prospective parent is a DMD carrier, and can serve as a basis for determining the risk of having a child with DMD.

Most skeletal muscles attach to bones by way of fibrous cords or sheets called **tendons.** Collagen fibers of the tendon continue into the periosteum and bone matrix. You can easily palpate tendons at the heel and wrist to feel their texture. In some cases, the tendon is a broad sheet called an **aponeurosis**[12] (AP-oh-new-RO-sis). This term originally referred to the tendon located beneath the scalp, but now it also refers to similar tendons associated with certain abdominal, lumbar, hand, and foot muscles (see fig. 7.19a).

---

[12]*apo* = upon, above; *neuro* = nerve

Most muscles attach to a different bone at each end, so either the muscle or its tendon spans at least one joint. The thick middle region of a muscle between its attachments is called the **belly** of the muscle (fig. 7.15). When a muscle contracts, it moves one bone relative to the other. In older literature, the stationery end of the muscle was called the *origin* and the moving end was called the *insertion*. Authorities have lately abandoned this terminology, however, because the end that moves may differ from one joint action to another. For example, the quadriceps femoris arises mainly on the femur and connects at its distal end to the tibia just below the knee. If you kick a soccer ball, this muscle moves the tibia more than the femur, so the femur would be considered its origin and the tibia its insertion. But when you stand up from a chair, the tibia remains stationary and the femur moves, so by the older principle, we would now have to regard the tibia as the origin and the femur as the insertion.

Coordinated Action of Muscle Groups

The movement produced by a muscle, or its prevention of an unwanted movement, is called its **action.** Skeletal muscles seldom act alone; instead, they function in groups whose combined actions produce the coordinated motion of a joint (fig. 7.15). Muscles can be classified into four categories according to their actions, but it must be stressed that a particular muscle can act in a certain way during one joint action and in a different way during other actions of the same joint.

1. The **prime mover (agonist)** is the muscle that produces most of the force during a particular joint action. In flexing the elbow, for example, the prime mover is the *brachialis,* which lies on the anterior side of the humerus deep to the biceps brachii.
2. A **synergist**[13] (SIN-ur-jist) is a muscle that aids the prime mover. The biceps brachii, for example, works as a synergist in elbow flexion. It is not

---

[13]*syn* = together; *erg* = work

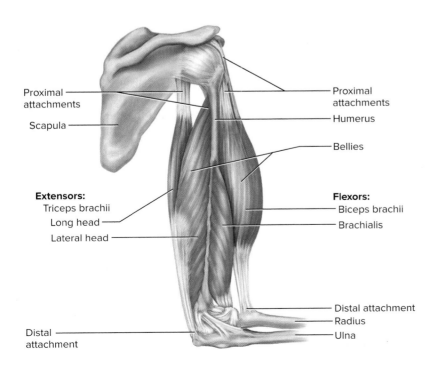

Figure 7.15 Synergistic and Antagonistic Muscle Pairs. The biceps brachii and brachialis muscles are synergists in elbow flexion. The triceps brachii is an antagonist of those two muscles and is the prime mover in elbow extension.

? *Suppose you threw a dart at a dart board. Which of these muscles would be the prime mover in that action, and which would act as a brake or antagonist?*

as strong as the brachialis. The actions of a prime mover and its synergist are not necessarily identical and redundant. A synergist may stabilize a joint and restrict unwanted movements, or modify the direction of a movement so that the action of the prime mover is more coordinated and specific.

3. An **antagonist**[14] is a muscle that opposes the prime mover. The brachialis and triceps brachii, for example, represent an **antagonistic pair** of muscles that act on opposite sides of the elbow. The brachialis flexes the elbow and the triceps extends it. Moreover, when either of these muscles contracts, the other one exerts a braking action that moderates the joint movement and makes it smoother and better coordinated. For example, if you extend your arm to reach out and pick up a cup of tea, your triceps acts as the prime mover of elbow extension and your brachialis and biceps act as antagonists to slow the extension and stop it at the appropriate point. Which member of the group acts as the agonist depends on the motion under consideration. In flexion of the elbow, the brachialis is the prime mover and the triceps is the antagonist; when the elbow is extended, their roles are reversed.

4. A **fixator** is a muscle that restricts a bone from moving. To *fix* a bone means to hold it steady, allowing another muscle attached to it to pull on something else. For example, if no fixators were involved, contraction of the biceps brachii would tend not only to flex the elbow but also to pull the scapula forward. There are fixators (the *rhomboids*), however, that attach the scapula to the vertebral column and hold it firmly in place so that the force of the biceps acts almost entirely on the elbow joint.

How Muscles Are Named

The remainder of this chapter is mostly a description of major skeletal muscles. Long muscle names such as *flexor carpi radialis brevis* may seem intimidating at first, but they are really more of a help than an obstacle to understanding if you gain a little familiarity with the words most commonly used in naming muscles. This name means "a short *(brevis)* flexor of the wrist *(carpi)* that lies alongside the radius *(radialis)*." Muscles are typically named for such characteristics as their action, location, shape, size, orientation, and number of heads. Footnotes throughout the following pages will help give you insight into the muscle names.

Many people who exercise know some of the skeletal muscles by common, slang terms—the "glutes," "abs," and "pecs," for example. Table 7.1 relates these common terms to the formal anatomical names of the muscles.

In the following pages, we will examine many of the most prominent muscles (fig. 7.16). We begin with those of the axial region—the head, neck, abdomen, and back. The general location and everyday actions of these muscles will be described first. Table 7.2 then summarizes the muscles with more specific information on their attachments and actions.

Table 7.1 Slang Terms for Major Skeletal Muscles

Slang Term	Anatomical Term
Abs	Abdominal obliques, rectus abdominis, transverse abdominal
Calf muscles	Gastrocnemius, soleus
Delts	Deltoids
Glutes	Gluteus maximus, medius, and minimus
Hamstrings	Biceps femoris, semimembranosus, semitendinosus
Lats	Latissimus dorsi
Pecs	Pectoralis major
Quads	Quadriceps femoris
Six-pack	Rectus abdominis

---

[14]*ant* = against; *agonist* = competitor

Figure 7.16 The Muscular System. Part (a) shows anterior superficial muscles on the anatomical right and deeper muscles on the left. Part (b) shows the posterior superficial and deeper muscles.

(a) Anterior view

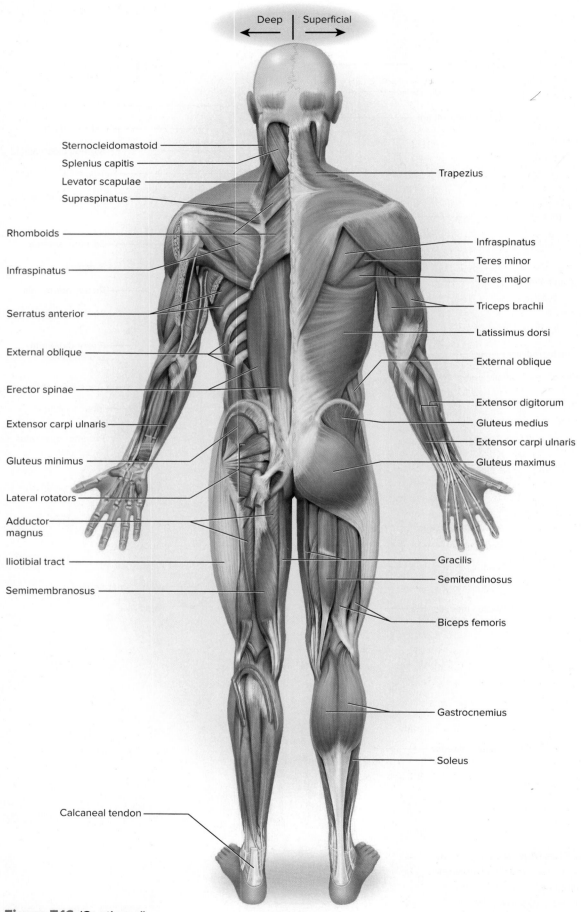

Figure 7.16 (Continued) (b) Posterior view

7.4b Muscles of the Face

One of the most striking contrasts between a human face and that of a rat, horse, or dog, for example, is the variety and subtlety of human facial expression. This is made possible by a complex array of small muscles that insert in the dermis and tense the skin when they contract. These muscles produce effects as diverse as a smile, a scowl, a frown, and a wink. They add subtle shades of meaning to our spoken words and are enormously important in nonverbal communication. The following are just a few of the most prominent facial muscles.

Frontalis (frun-TAY-lis) (fig. 7.17a). The frontalis is a muscle of the forehead that extends from a tendon beneath the scalp to the skin beneath the eyebrows. It expressively elevates the eyebrows and wrinkles the skin of the forehead.

Orbicularis oculi[15] (or-BIC-you-LERR-is OC-you-lye) (fig. 7.17a). This is a circular muscle (sphincter) located in and beyond the eyelid. Arising largely from the lacrimal bone and adjacent regions, it inserts in both eyelids and serves to blink and close the eyes. Blinking promotes the flow of tears and their drainage into the nasal cavity.

Orbicularis oris[16] (or-BIC-you-LERR-is OR-is) (fig. 7.17a). This is an array of small interlacing muscles located in the lips, encircling the mouth. It closes the lips (but not the jaw) and, with harder contraction, protrudes them as in kissing. It is important in speech.

Zygomaticus[17] (ZY-go-MAT-ih-cus) (fig. 7.17a). Several muscles converge on the lips and make the mouth the most expressive part of the face. The *zygomaticus major* and *minor* arise on the zygomatic bone and end on the orbicularis oris at the corner of the mouth. They draw the corner of the mouth laterally and upward as in smiling and laughing.

Buccinator[18] (BUCK-sin-AY-tur) (fig. 7.17a). This muscle constitutes the fleshy mass of the cheek. It arises from the mandible and maxilla and ends deep in the mucous membranes of the cheeks and lips. It has multiple functions in airflow, drinking, and chewing. It is well developed in infants, who use it to nurse, and it is used when blowing out air or spitting out liquid, playing a wind instrument, helping to control airflow in speech, and pushing and retaining food between the teeth during chewing. When one closes the mouth, the buccinator retracts the cheek to protect it from being bitten.

7.4c Muscles of Chewing

There are four paired muscles of mastication: the temporalis, masseter, and medial and lateral pterygoids. Collectively, they have four actions on the mandible: *elevation,* to bite; *protraction,* a forward movement of the mandible to align the incisors for biting; *retraction,* to draw the mandible posteriorly and align the molars for chewing; and *lateral* and *medial excursion,* left–right movements that grind food between the molars.

Temporalis[19] (TEM-po-RAY-liss) (fig. 7.17a). This is a broad fan-shaped muscle on the side of the cranium. It originates on faint ridges along the side of the frontal, parietal, and occipital bones. It converges toward the zygomatic arch, passes behind (medial to) it, and ends on the coronoid process of the mandible. Its main action is to elevate and retract the mandible.

---

[15] *orb* = circle; *ocul* = eye
[16] *orb* = circle; *oris* = of the mouth
[17] refers to the zygomatic arch
[18] *bucc* = cheek
[19] *temporalis* = of the temporal region of the head

Figure 7.17 Muscles of Facial Expression and Chewing. (a) Right lateral view of the head. (b) A view into the skull from behind, showing the relationship of the pterygoid muscles to the sphenoid bone and mandible. APR

? *Why are the pterygoids better able than the temporalis to produce the side-to-side action of grinding food between the molars?*

Masseter[20] (ma-SEE-tur) (fig. 7.17a). The masseter is shorter and superficial to the temporalis, arising from the zygomatic arch and inserting laterally on the angle of the mandible. It is a thick muscle easily palpated at the rear of your cheek. It is a synergist of the temporalis in elevating the mandible, but also produces some degree of protraction and excursion.

Pterygoids[21] (TERR-ih-goyds) (**fig. 7.17b**). The *lateral* and *medial pterygoid* muscles arise from the *pterygoid plates* of the sphenoid bone on the inferior

---
[20]*masset* = chew
[21]*pteryg* = wing; *oid* = shaped

surface of the cranium. They pass obliquely to the inner (medial) surface of the mandible. They elevate and protract the mandible and move it from side to side for grinding food.

7.4d Muscles of Head and Neck Movement

Muscles that move the head arise from the vertebral column, thoracic cage, and pectoral girdle and insert on the cranial bones. The principal flexors of the neck are the *sternocleidomastoids,* which converge like a V toward the sternum. Extensors include the *trapezius* and *splenius capitis,* located on the rear of the neck.

Sternocleidomastoid[22] (STIR-no-CLY-do-MAST-oyd) (figs. 7.16a, 7.17a). This is the prime mover of neck flexion. It originates on the sternum and medial half of the clavicle and ends on the mastoid process behind the ear. It is a thick, cordlike muscle often visible on the side of the neck, especially when one turns the head to one side and tips the chin up. It rotates the head to the left or right or tilts it upward and toward the opposite side, as in looking over one's shoulder, depending on what other muscles do at the same time. When both sternocleidomastoids contract at once, the neck flexes forward, for example, when you look down to eat or read.

Trapezius (tra-PEE-zee-us) (fig. 7.16b). This is a vast triangular muscle of the neck and upper back; together, the right and left trapezius muscles form a trapezoidal shape. The trapezius has a broad origin extending from the *occipital protuberance*—a median bump at the rear of the skull—to vertebra T12. The muscle converges toward the shoulder, where it terminates on the spine and acromion of the scapula and the lateral end of the clavicle. Either trapezius working alone flexes the neck to one side. When they contract together, the trapezius muscles extend the neck, as in looking straight upward. The trapezius also has actions on the scapula and arm described later, under limb muscles.

7.4e Muscles of the Trunk APR

Here we consider muscles of the abdominal wall as well as back muscles that act on the vertebral column. The trunk muscles also include muscles of the pelvic floor and muscles of respiration—the *intercostal muscles* between the ribs and the *diaphragm* between the thoracic and abdominal cavities, which are described in chapter 15. Some prominent muscles seen on the trunk, such as the *pectoralis major* of the chest, act on the arm and are therefore considered later as appendicular muscles.

Muscles of the Abdomen APR

Four pairs of sheetlike muscles reinforce the anterior and lateral abdominal walls. They support the intestines and other viscera against the pull of gravity; stabilize the vertebral column during heavy lifting; and as abdominal compressors, aid in expelling organ contents during respiration, urination, defecation, vomiting, and childbirth. They are the *rectus abdominis, external* and *internal obliques,* and *transverse abdominal muscle.*

Unlike the thoracic cavity, the abdominal cavity lacks a protective bony enclosure. However, the fascicles of these three muscle layers run in different directions

---

[22]*sterno* = sternum; *cleido* = clavicle; *mastoid* = mastoid process of skull

(horizontally, vertically, and obliquely), lending strength to the abdominal wall much like a sheet of plywood, in which wood fibers run in different directions from layer to layer. The tendons of the abdominal muscles are aponeuroses. At its inferior margin, the aponeurosis of the external oblique forms a strong, cordlike *inguinal ligament.*

Rectus[23] abdominis (REC-tus ab-DOM-ih-nis) (fig. 7.16a). This is a median muscular strap extending vertically from the pubis to the sternum and lower ribs. It flexes the waist, as in bending forward or doing sit-ups; stabilizes the pelvic region during walking; and compresses the abdominal viscera. It is separated into four segments by fibrous *tendinous intersections* that give the abdomen a segmented appearance in well-muscled individuals. The rectus abdominis is enclosed in a fibrous sleeve called the *rectus sheath,* and the right and left rectus muscles are separated by a median fibrous strip called the *linea alba.*[24]

Abdominal obliques (fig. 7.16a). The **external oblique** is the most superficial muscle of the lateral abdominal wall. Its fascicles arise from the lower eight ribs, run anteriorly and downward along the lateral abdomen, and end on the pubic bone and ilium. Deep to it is the **internal oblique.** Its fascicles arise from the ilium and from inguinal and lumbar connective tissues and run anteriorly and upward to terminate mainly on the lower ribs and their cartilages. These muscles maintain posture; flex and rotate the vertebral column, as in twisting at the waist; support the viscera; aid in deep breathing and loud vocalizations such as singing and public speaking; and compress the abdominal viscera.

Transverse abdominal (fig. 7.16a). This is the deepest of the abdominal group. It crosses the abdomen horizontally, like the cummerbund of a tuxedo, and acts primarily as an abdominal compressor. It arises largely from the same places as the internal abdominal oblique and ends on the abdominal midline from the xiphoid process above to the pubis below.

Muscles of the Back APR

The most superficial muscles of the back are the trapezius of the upper back and latissimus dorsi of the lower back (see fig. 7.16b). These act on the arm and are therefore discussed later under the appendicular musculature. The strongest of the deeper back muscles is the following.

Erector spinae (eh-RECK-tur SPY-nee) (fig. 7.16b). This is a multipart muscle group with cervical, thoracic, and lumbar portions and three parallel, vertical columns on each side of the spine. It has complex, extensive attachments to the ribs, vertebrae, and iliac crest. It maintains posture, straightens the spine after one bends at the waist, and is employed in arching the back. As the name implies, it is important in standing and sitting erect. Unilateral contraction flexes the waist laterally, and its superior (cervical) portion contributes to rotation of the head.

Table 7.2 summarizes the axial muscles and provides more detail as to their attachments and actions. In this table and the next, we use the term muscle *attachments.* However, the first bulleted attachment note for each muscle is the one historically called the origin, and the second bullet point for each is the insertion. Bear in mind the shortcomings of those terms explained earlier in this section.

---

[23]*rect* = straight
[24]*linea* = line; *alb* = white

Table 7.2 Axial Muscles

Name	Attachments	Principal Actions
Facial Muscles		
Buccinator (fig. 7.17a)	• Lateral aspect of maxilla and mandible • Orbicularis oris muscle; submucosa of cheek and lips	Compresses cheek; pushes food between teeth; creates suction; expels air or liquid from mouth; retracts cheek when mouth is closing so it is not bitten
Frontalis (fig. 7.17a)	• Subcutaneous tissue of scalp • Subcutaneous tissue of eyebrows	Raises eyebrows; wrinkles forehead
Orbicularis oculi (fig. 7.17a)	• Lacrimal bone; adjacent regions of frontal bone and maxilla; median angle of eyelid • Upper and lower eyelids; skin adjacent to orbit	Closes eye; promotes flow of tears
Orbicularis oris (fig. 7.17a)	• Connective tissue near angle of lips • Mucosa of lips	Closes lips; protrudes lips as in kissing; aids in speech
Zygomaticus major and minor (fig. 7.17a)	• Zygomatic bone • Superolateral corner of mouth	Draw corner of mouth laterally and upward, as in smiling
Muscles of Chewing		
Masseter (fig. 7.17a)	• Zygomatic arch • Lateral surface of mandibular ramus and angle	Elevates and protracts mandible in biting; aids in mandibular excursion (grinding of food)
Pterygoids (fig. 7.17b)	• Pterygoid plates of sphenoid bone; palatine bone; maxilla • Medial surface of mandible near angle and mandibular condyle	Protract mandible in biting and produce mandibular excursion in chewing
Temporalis (fig. 7.17a)	• Lateral surface of occipital, temporal, and frontal bones • Mandible at coronoid process and anterior border of ramus	Elevates and retracts mandible; aids in excursion
Muscles of Head–Neck Movement		
Sternocleidomastoid (figs. 7.16a, 7.17a)	• Clavicle and manubrium of sternum • Mastoid process of skull	Unilateral contraction rotates head left or right and directs gaze over opposite shoulder; bilateral contraction tilts head downward
Trapezius (fig. 7.16b)	• External occipital protuberance of skull; spinous processes of vertebrae C7–T12 • Clavicle; acromion and spine of scapula	Flexes neck to one side; extends neck (as in looking upward). See other functions in table 7.3
Muscles of the Abdomen		
Abdominal obliques (external and internal) (fig. 7.16a)	• Ribs 5–12; iliac crest; inguinal and lumbar connective tissues • Ribs 10–12; abdominal midline from xiphoid process to pubis	Flexes abdomen in sitting up; rotates trunk at waist; compresses abdomen
Rectus abdominis (fig. 7.16a)	• Pubis • Xiphoid process; costal cartilages of lower ribs	Flexes waist in sitting up; stabilizes pelvis in walking; compresses abdomen
Transverse abdominal (fig. 7.16a)	• Iliac crest; costal cartilages; inguinal and lumbar connective tissues • Abdominal midline from xiphoid process to pubis	Compresses abdomen
Muscles of the Back		
Erector spinae (fig. 7.16b)	• Extensive attachments to ribs, vertebrae, sacrum, iliac crest • Vertebrae and ribs	Extends vertebral column as in arching back; laterally flexes vertebral column

7.4f Muscles of the Pectoral Girdle and Upper Limb

The rest of this chapter concerns muscles that act on the limbs and limb girdles—that is, the appendicular skeleton. We begin with muscles that originate on the axial skeleton and act on the scapula. Table 7.3 summarizes attachments and actions of the appendicular muscles.

Pectoralis minor (PECK-toe-RAY-lis) (fig. 7.16a). This is a deep muscle of the mammary region. It arises from ribs 3 through 5 and converges on the coracoid process of the scapula. It draws the scapula laterally and forward, protracting the shoulder as in reaching out for a door handle. Acting with other muscles, it depresses the shoulder, as in reaching down to pick up a suitcase.

Serratus anterior (serr-AY-tus) (fig. 7.16a). This muscle originates on all or nearly all of the ribs, wraps around the chest between the rib cage and scapula, and ends on the medial border of the scapula. It holds the scapula against the rib cage, protracts it, and is the prime mover in forward thrusting, throwing, and pushing; it is nicknamed the "boxer's muscle" for these actions. It also elevates the shoulder, as in lifting a suitcase.

Trapezius (tra-PEE-zee-us) (fig. 7.16b). The trapezius was discussed earlier for its actions on the head, but it also has important actions on the scapula. Its action depends on whether its superior, middle, or inferior portions contract and whether it acts alone or with other muscles. It can either elevate or depress the shoulder; it retracts the scapula when one draws back the shoulders, as in pulling on a rope; and it can stabilize the scapula and shoulder during arm movements. Normal depression of the scapula occurs mainly by gravitational pull, but the trapezius and serratus anterior can cause faster, more forcible depression, as in swimming, hammering, and rowing a boat.

Levator scapulae (leh-VAY-tur SCAP-you-lee) (figs. 7.16b, 7.18a). The levator scapulae extends from vertebrae C1 through C4 to the superior angle and medial border of the scapula. It works with the superior portion of the trapezius to elevate the scapula, for example when you shrug your shoulders, lift a suitcase, or carry a

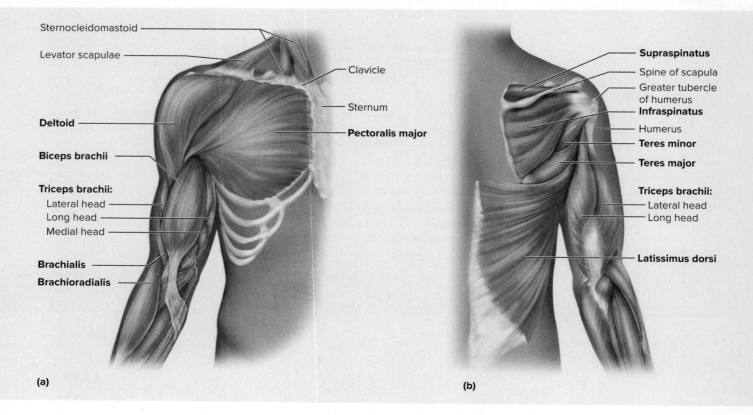

Figure 7.18 Muscles of the Pectoral, Scapular, and Brachial Regions. (a) Anterior view of superficial pectoral and brachial muscles. (b) Posterior view of muscles that act on the shoulder and elbow joints. (c) Anterior view of scapular and brachial muscles, with rib cage removed to see facing surface of scapula behind it. (d) Left lateral view of rotator cuff muscles in relation to scapula. The rotator cuff itself consists of the tendons of these muscles, and is not shown here; some of the tendons are illustrated in figure 6.44. **APR**

heavy weight on your shoulder. Depending on the action of other muscles, it can also flex the neck laterally, retract the scapula and brace the shoulder, or rotate the scapula to lower the apex of the shoulder.

Rhomboids (ROM-boyds) (see fig. 7.16b). The rhomboids extend from the lower cervical and upper thoracic vertebrae to the medial border of the scapula. They retract and elevate the scapula by drawing it toward the spine and, when the biceps brachii contracts, the rhomboids fix the scapula so that the biceps flexes the elbow. Otherwise the biceps would tend to pull the scapula forward.

Muscles Acting on the Humerus

Nine muscles cross the shoulder joint and insert on the humerus. The first two of these, the *pectoralis major* and *latissimus dorsi,* are considered *axial muscles* because they arise mainly on the axial skeleton. They bear the primary responsibility for attaching the arm to the trunk. The axilla (armpit) is the depression between these two muscles.

Pectoralis major (PECK-toe-RAY-lis) (fig. 7.18a). This is the thick, superficial, fleshy muscle of the chest. It extends mainly from the clavicle and sternum to the upper humerus. It adducts the humerus and rotates it medially; flexes the shoulder as in reaching out to hug someone; and aids in climbing, pushing, and throwing.

Latissimus dorsi[25] (la-TISS-ih-mus DOR-sye) (**fig. 7.18b**). This is a broad muscle of the back that extends from vertebrae and fascia of the waist and lower back to the humerus and forms the posterior border of the axilla (armpit). It extends the shoulder as in reaching behind you to clasp your hands behind your back; thus, it is an antagonist of the pectoralis major. It pulls the body upward in climbing, and produces strong downward strokes of the arm as in hammering or swimming; it is sometimes nicknamed the "swimmer's muscle."

---

[25]*latissimus* = broadest; *dorsi* = of the back

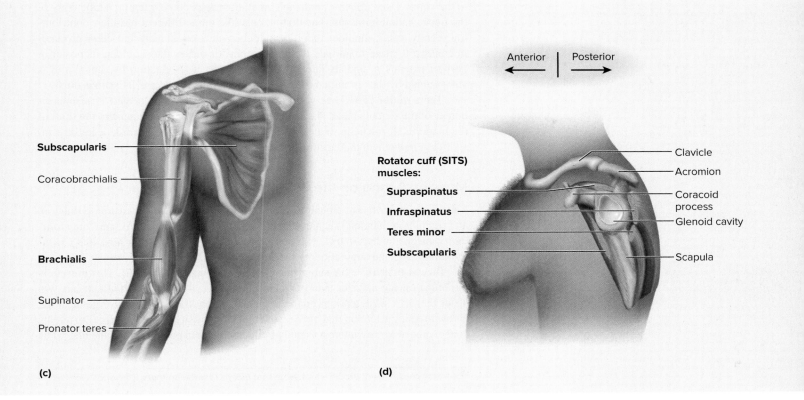

The rest of these are called *scapular muscles* because they arise from the scapula. The first four provide the tendons that form the **rotator cuff** of the shoulder (**fig. 7.18d**). They are nicknamed the "SITS muscles" after the first letters of their formal names—the *supraspinatus, infraspinatus, teres minor,* and *subscapularis.* The subscapularis fills most of the subscapular fossa on the anterior surface of the scapula. The other three lie on the posterior surface. The tendons of these muscles merge with the joint capsule of the shoulder as they pass to the humerus. They end on the proximal end of the humerus, forming a partial sleeve around it. The rotator cuff reinforces the joint capsule and holds the head of the humerus in the glenoid cavity. The rotator cuff, especially the tendon of the supraspinatus, is easily damaged by strenuous circumduction, as in bowling and baseball pitching (see Perspectives on Health earlier in this section).

Supraspinatus (SOO-pra-spy-NAY-tus) (fig. 7.18b). This muscle occupies the *supraspinous fossa* above the scapular spine on the posterior side of the scapula. It abducts the humerus and resists downward slippage of the humerus when one is carrying weight.

Infraspinatus (IN-fra-spy-NAY-tus) (fig. 7.18b). This muscle occupies most of the *infraspinous fossa* below the scapular spine. It extends and laterally rotates the humerus and prevents the humeral head from slipping upward out of the glenoid cavity.

Teres minor (TARE-eez) (fig. 7.18b). This muscle lies on the posterior side of the scapula near its inferior border, between the infraspinatus and teres major. It adducts and laterally rotates the humerus, and prevents it from slipping out of the glenoid cavity when the arm is abducted.

Subscapularis (SUB-SCAP-you-LARE-is) (**fig. 7.18c**). This muscle lies on the anterior side of the scapula, filling the subscapular fossa. It medially rotates the humerus and otherwise acts similarly to the teres minor.

The following two scapular muscles do not contribute to the rotator cuff but are also important in movements of the arm at the shoulder.

Deltoid (DEL-toyd) (fig. 7.18a). This is the thick muscle that caps the shoulder; it is a commonly used site for intramuscular drug injections. It is named for its triangular shape (after the Greek delta, Δ), having a broad superior base attached to the clavicle and upper scapula and an inferior apex converging on a tuberosity of the upper lateral humerus. The deltoid acts like three different muscles, contributing greatly to the shoulder mobility that characterizes humans and other primates in contrast to other mammals. Its anterior portion flexes the shoulder, its posterior portion extends it, and its lateral portion abducts it. Abduction by the deltoid is antagonized by the combined action of the pectoralis major and latissimus dorsi.

Teres major (TARE-eez) (fig. 7.18b). The teres major is located at the inferior border of the scapula on its posterior side. It extends from the scapula to the shaft of the humerus. It medially rotates the humerus and draws it backward, as in the arm-swinging movement that accompanies walking.

Muscles Acting on the Forearm

The elbow and forearm are capable of four motions: flexion, extension, pronation, and supination. Flexion is achieved by the *biceps brachii, brachialis,* and *brachioradialis;* extension by the *triceps brachii;* pronation by the *pronator teres* and *pronator quadratus;* and supination by the *supinator* and biceps brachii.

Biceps brachii[26] (BY-seps BRAY-kee-eye) (figs. 7.15, 7.18a). This muscle is the bulge on the anterior side of the arm, superficial to the brachialis. It has two heads that arise from tendons on the scapula. Distally, the muscle converges near the elbow to a single tendon that ends on the radial tuberosity of the forearm. Despite its conspicuousness and interest to body builders, it is less powerful than the brachialis

---

[26]*bi* = two; *ceps* = head; *brachi* = arm. Note that *biceps* is singular; there is no such word as *bicep.*

and acts as a synergist in flexing the elbow, not the prime mover. It also contributes to supination of the forearm.

Brachialis (BRAY-kee-AL-is) (figs. 7.15, 7.18c). This is a deep muscle lying along the anterior surface of the lower half of the humerus. It extends from mid-humerus to the superior end of the ulna and is the prime mover of elbow flexion.

Brachioradialis (BRAY-kee-oh-RAY-dee-AL-is) (fig. 7.18a). This muscle flexes the elbow although it is located in the forearm. It is the thick muscular mass on the radial (lateral) side of the forearm, near the elbow; it extends from the distal end of the humerus to the distal end of the radius. It is much less powerful than the biceps and brachialis.

Triceps brachii (TRY-seps BRAY-kee-eye) (figs. 7.15, 7.18b). This is a three-headed muscle on the posterior side of the humerus. It extends from the scapula and humerus to the olecranon of the ulna, and is the prime mover of elbow extension.

Pronators (figs. 7.18c, 7.19b). Pronation of the forearm is achieved by two muscles: The **pronator teres** (PRO-nay-tur TARE-eez), near the elbow, runs obliquely from the lower humerus and upper ulna to the middle of the radius. The **pronator quadratus** (kwa-DRAY-tus), a more squarish muscle, connects the anterior surfaces of the radius and ulna just above the wrist.

Supinator (SOO-pih-NAY-tur) (figs. 7.16a, 7.18c). Supination of the forearm is achieved by the biceps brachii, as noted earlier, and the supinator muscle slightly distal to the elbow. The supinator extends from the humerus and ulna near the elbow to the upper radius.

Muscles Acting on the Wrist and Hand

The hand carries out more subtle and sophisticated movements than any other part of the body; the wrist and hand are therefore acted upon by at least 15 muscles in the forearm and 18 muscles in the hand itself. Relatively few of them can be considered here.

The bellies of the forearm muscles form its fleshy roundness. Their actions are mainly flexion and extension of the hand and fingers. Flexors are on the anterior side. Their tendons pass through the wrist and end mainly on carpal and metacarpal bones and the phalanges. Most of these tendons pass under a transverse ligament that crosses the wrist like a bracelet, the *flexor retinaculum*[27] (**fig. 7.19a**). One exception is the *palmaris longus* muscle, which is relatively weak and even absent from some people; its tendon passes over the retinaculum. The extensor muscles are on the posterior side of the forearm, and their tendons pass under a similar *extensor retinaculum* on that side of the wrist (fig. 7.16b). Although the forearm muscles are numerous and complex, most of their names suggest their actions, and from their actions, their approximate locations in the forearm can generally be deduced.

Following are three important flexors of the wrist and fingers. All three arise primarily from a single shared tendon on the medial epicondyle of the humerus. These are especially important to the strength of your grip.

Flexor carpi radialis (FLEX-ur CAR-pye RAY-dee-AL-is) (fig. 7.19a). This is a superficial muscle on the radial (lateral) side of the forearm, bordered laterally by the brachioradialis and medially by (in most people) the palmaris longus. It inserts on metacarpals II and III and flexes the wrist anteriorly and laterally (toward the thumb).

Flexor carpi ulnaris (FLEX-ur CAR-pye ul-NAIR-is) (fig. 7.19a). This is a superficial muscle on the ulnar (medial) side of the forearm. It inserts on some of the wrist bones and metacarpal V, and also flexes the wrist anteriorly and medially (toward the little finger).

Flexor digitorum superficialis (FLEX-ur DIDJ-ih-TORE-um SOO-per-FISH-ee-AY-lis) (**fig. 7.19b**). This muscle lies deep to the previous two. It has four distal tendons leading to the middle phalanges of fingers II through V. Depending

---

[27]*retinac* = retainer, bracelet; *cul* = little

Figure 7.19 Muscles of the Forearm. (a) Anterior view of superficial flexors. (b) Anterior view of deeper flexors and pronators. (c) Posterior view of superficial extensors. **APR**

(a) Superficial flexors

on the action of other muscles, it flexes the fingers as in making a fist, or flexes just the wrist. Beneath it, the *flexor digitorum profundus* (fig. 7.16a) performs a similar action but it attaches to the distal phalanges of the same four fingers and is therefore the only muscle that can independently flex the distal segments of those fingers.

A frequent site of discomfort and disability is the **carpal tunnel,** a tight space between the carpal bones and flexor retinaculum. The flexor tendons passing through the tunnel are enclosed in tubular bursae called *tendon sheaths* that enable them to slide back and forth quite easily, although this region is very subject to injury (see Clinical Application 7.3).

Clinical Application 7.3

CARPAL TUNNEL SYNDROME

Prolonged, repetitive motion of the wrist and fingers can cause tissues in the carpal tunnel to become inflamed, swollen, or fibrotic. Because the carpal tunnel cannot expand, swelling compresses the *median nerve* of the wrist, which passes through the tunnel with the flexor tendons. This pressure causes tingling and muscular weakness in the palm and lateral side of the hand and pain that may spread to the arm and shoulder. This condition, called **carpal tunnel syndrome**, is common among pianists, meat cutters, and others who spend long hours making repetitive wrist motions. It is treated with aspirin and other anti-inflammatory drugs, immobilization of the wrist, and sometimes surgical removal of part or all of the flexor retinaculum to relieve pressure on the nerve.

The extensors on the posterior side of the forearm share a single tendon arising from the humerus. We will consider only the four most prominent ones.

Extensor digitorum (ex-TEN-sur DIDJ-ih-TORE-um) (fig. 7.19c). This broad, superficial extensor on the midline of the forearm has four distal tendons leading to the phalanges of fingers II through V. Its tendons are easily seen and palpated on the back of the hand when the fingers are extended and the wrist is hyperextended. It extends the wrist and abducts the fingers (spreads them apart). The little finger and the index finger also have their own extensors, respectively called the *extensor digiti minimi* and *extensor indicis,* and there are several muscles for thumb movement alone. Loss of the thumb is the most disabling of all hand injuries.

Extensor carpi ulnaris (ex-TEN-sur CAR-pye ul-NARE-is) (fig. 7.19c). This muscle lies medial to the extensor digitorum (on the ulnar side) and attaches distally to the base of metacarpal V. It extends the wrist and flexes it medially, and it can hold the wrist straight when the fist is clenched or the hand grips an object.

Extensor carpi radialis longus and **brevis** (ex-TEN-sur CAR-pye RAY-dee-AY-lis LAWN-gus and BREV-is) (fig. 7.19c). These muscles lie side by side, lateral to the extensor digitorum, and insert on metacarpal bones II and III. Both of them extend the wrist and flex it laterally.

Apply What You Know

Andre is a renowned symphony pianist. Explain why his career would result in more than average strength in his extensor digitorum and flexor digitorum profundus muscles.

7.4g Muscles of the Pelvic Girdle and Lower Limb

The largest and strongest muscles in the body are found in the lower limb. Unlike those of the upper limb, they are adapted less for precision than for the strength needed to stand, maintain balance, walk, and run. Several of them cross and act upon two or more joints, such as the hip and knee.

Muscles Acting on the Hip and Femur

There are four principal groups of muscles that act on the femur, producing movement at the hip joint: anterior muscles that flex the hip; gluteal muscles of the buttocks, which extend the hip and abduct the thigh; additional abductors on the lateral side of the femur; and adductors on the medial side.

Iliopsoas (IL-ee-oh-SO-us) (fig. 7.20). The principal anterior hip muscle is the powerful iliopsoas, which is actually composed of two smaller muscles—the **iliacus** (ih-LY-uh-cus), a broad muscle that fills the iliac fossa of the pelvis, and the **psoas major** (SO-us), which arises mainly from the lumbar vertebrae. They converge on a single tendon that terminates on the femur, so they have the same action. Together, these muscles flex the hip and swing the leg forward when walking; they flex the trunk if the femur is fixed, as in bending forward to pick up something from the floor or doing abdominal crunches (sit-ups); and they balance the trunk during sitting and standing.

Gluteal muscles (fig. 7.21). The gluteus maximus, medius, and minimus are the muscles of the buttocks. They arise from the lateral surfaces of the hip bone and sacrum and lead to the greater trochanter and upper, lateral femur. The **gluteus maximus** is the largest. It is a hip extensor that produces the backswing of the lower limb in walking; it helps you stand straight after bending forward at the waist; and it provides most of the lift when you climb stairs and most of the thrust in pushing on a bicycle pedal. The **gluteus medius** and **minimus** abduct the femur, for example when moving the feet apart to stand in a spread-legged position. They are also important in walking. When one leg is lifted from the ground, balance is precarious and the upper body tends to lean toward the lifted, unsupported leg. The gluteus medius and minimus counteract this, shifting the trunk back over the supporting leg, preventing the swaying of the torso from side to side and keeping you from falling over.

Thigh adductors (fig. 7.20). The medial side of the thigh has several adductor muscles—the *adductor longus, adductor brevis, adductor magnus,* and *gracilis.* Collectively, these muscles begin on the pubis and ischium of the hip bone and mostly attach to the shaft of the femur, except for the gracilis that ends medially on the tibia just below the knee. In addition to adducting

Figure 7.20 Anterior Muscles of the Hip and Thigh. APR

? *What two joints are acted upon by the rectus femoris?*

Figure 7.21 Gluteal, Hamstring, and Lateral Rotator Muscles (Posterior View). Each figure shows superficial muscles on the anatomical right and deeper muscles on the left. **APR**

the thigh, most of these muscles aid in flexing the hip, and the gracilis aids in flexing the knee. You can imagine the importance of the adductors in riding a horse.

Thigh rotators (fig. 7.21). The lateral side of the hip has a group of short muscles collectively called the **lateral rotators.** They arise from the hip bone and end, along with the gluteus medius and minimus, on or near the greater trochanter of the femur. They rotate the femur laterally, as in standing with the toes pointed outward, and like those two gluteal muscles, they aid in shifting the body weight and maintaining balance when one is walking.

Muscles Acting on the Knee

The bulk of the anterior thigh is formed by the four-part *quadriceps femoris*. On the posterior side, we find the *hamstring* group, consisting of three muscles. Also crossing the thigh anteriorly is the *sartorius*.

Quadriceps femoris (QUAD-rih-seps FEM-oh-ris) (fig. 7.20). This large muscle consists of four heads that arise on the ilium and the shaft of the femur and converge on a single tendon at the knee. The heads are the **rectus femoris** in the middle; flanked by the **vastus lateralis** and **vastus medialis** on the lateral and medial sides; and the **vastus intermedius,** which is located between these and deep to the rectus femoris (therefore not illustrated). The **quadriceps (patellar) tendon** extends from the distal end of this muscle group to arch over the knee and enclose the patella; it continues beyond the patella as the **patellar ligament,** ending on the anterior surface of the tibia just below the knee. This change in terminology reflects the definition of a tendon as connecting muscle to bone (quadriceps to patella) and a ligament as connecting bone to bone (patella to tibia). All quadriceps heads extend the knee, as in kicking a football or standing up from a chair, and the rectus femoris helps to flex the hip.

Sartorius[28] (sar-TORE-ee-us) (fig. 7.20). This straplike muscle is the longest in the body. It extends from the lateral side of the hip to the medial side of the tibia, crossing diagonally over the quadriceps femoris. It is used to cross the legs and aids in flexing the hip and knee.

Hamstrings (fig. 7.21). *Hamstrings* is a colloquial name for the three major muscles on the posterior side of the thigh—the **biceps femoris** (BY-seps FEM-oh-ris), **semimembranosus** (SEM-ee-MEM-bra-NO-sus), and **semitendinosus** (SEM-ee-TEN-dih-NO-sus). They arise mainly from the ischial tuberosity of the hip bone and lead to the upper tibia and fibula. The pit at the rear of the knee *(popliteal fossa)* is bordered laterally by the biceps tendon and medially by the tendons of the other two. The hamstrings flex the knee, and aided by the gluteus maximus, they extend the hip during walking and running. Hamstring injuries are common among sprinters, soccer players, and other athletes who rely on quick acceleration (see Perspectives on Health earlier in this section). The hamstrings are important in supporting the body weight. Chronic low-back pain can be relieved by exercises that strengthen the hamstring and abdominal muscles.

Muscles Acting on the Foot

The fleshy mass of the leg proper (below the knee) is formed by muscles that act on the ankle and foot.

Anterior group (fig. 7.22a). Two principal muscles on the anterior side of the leg are the **extensor digitorum longus** (ex-TEN-sur DIDJ-ih-TORE-um LAWN-gus) and **tibialis anterior** (TIB-ee-AY-lis). They arise from the anterior tibia, fibula, and interosseous membrane between those bones. Their tendons terminate on the phalanges of toes II to V, metatarsal bone I, and the medial cuneiform bone of the tarsal group. These muscles dorsiflex the foot, and the first of them extends all of the toes except the great toe, which has its own extensor. Dorsiflexion lifts the toes and prevents them from scuffing on the ground during walking.

---

[28]*sartor* = tailor

Figure 7.22 Muscles of the Leg. (a) Superficial anterior group. (b) Superficial posterior muscles. (c) Gastrocnemius removed to show the deeper posterior soleus muscle. APR

Posterior group (fig. 7.22b, c). The major posterior (calf) muscles are the superficial **gastrocnemius**[29] (GAS-trock-NEE-me-us), whose two heads form the prominent bulge of the upper calf, and the deeper **soleus.**[30] The gastrocnemius arises from the lower femur and knee joint and the soleus from the upper fibula and middle tibia. Both of these muscles converge on the same tendon, the **calcaneal (Achilles) tendon,** which inserts on the calcaneus (heel bone). They plantar flex the foot, as in standing on tiptoe or jumping. Plantar flexion also provides lift and forward thrust in walking.

Like the hand, the foot has 19 small muscles that support the foot arches and flex, extend, abduct, and adduct the toes.

Table 7.3 summarizes the appendicular muscles and provides more detailed information on their bone attachments.

---

[29]*gastro* = belly; *cnem* = leg
[30]named for its resemblance to a flatfish, the sole

Table 7.3 Appendicular Muscles

Name	Attachments	Action
Muscles Acting on the Scapula		
Levator scapulae (fig. 7.16b, 7.18a)	• Vertebrae C1–C4 • Medial border and superior angle of scapula	Elevates scapula; tilts apex of shoulder downward; braces shoulder; flexes neck laterally
Pectoralis minor (fig. 7.16a)	• Ribs 3–5 • Coracoid process of scapula	Draws scapula laterally and forward around chest; lowers apex of shoulder
Rhomboids (fig. 7.16b)	• Vertebrae C7–T5 • Medial border of scapula	Retract and elevate scapula; rotate apex of shoulder downward; fix scapula during arm movements; brace shoulder
Serratus anterior (fig. 7.16a)	• Most or all ribs • Medial border of scapula	Draws scapula laterally and forward around chest; raises apex of shoulder; prime mover in forward thrusting, throwing, and pushing
Trapezius (fig. 7.16b)	• External occipital protuberance of skull; ligament of nuchal region; spinous processes of vertebrae C7–T12 • Clavicle; acromion and spine of scapula	Different parts of trapezius raise or lower apex of shoulder, retract scapula, and stabilize scapula during arm movements. See other functions in table 7.2
Rotator Cuff Muscles Acting on the Humerus		
Infraspinatus (fig. 7.18b)	• Infraspinous fossa of scapula • Greater tubercle of humerus	Extends and laterally rotates humerus and prevents it from sliding upward
Subscapularis (fig. 7.18c)	• Subscapular fossa of scapula • Lesser tubercle of humerus, anterior surface of joint capsule	Medially rotates humerus and prevents it from sliding upward
Supraspinatus (fig. 7.18b)	• Supraspinous fossa of scapula • Greater tubercle of humerus	Abducts humerus and resists its downward displacement when carrying heavy weight
Teres minor (fig. 7.18b)	• Lateral border of scapula • Greater tubercle of humerus, posterior surface of joint capsule	Adducts and laterally rotates humerus
Other Muscles Acting on the Humerus		
Deltoid (fig. 7.18a)	• Clavicle; spine and acromion of scapula • Deltoid tuberosity of humerus	Different parts abduct, flex, and medially rotate humerus, and extend and laterally rotate it
Latissimus dorsi (fig. 7.18b)	• Vertebrae T7–L5; ribs 9 or 10 through 12; iliac crest; lumbar connective tissues • Floor of intertubercular sulcus of humerus	Extends shoulder; pulls body upward in climbing; produces strong downward thrusts as in hammering and swimming; adducts and medially rotates humerus
Pectoralis major (fig. 7.18a)	• Clavicle, sternum, costal cartilages 1–6 • Lateral margin of intertubercular sulcus of humerus	Adducts and medially rotates humerus; flexes shoulder as in reaching forward; aids in climbing, pushing, and throwing
Teres major (fig. 7.18b)	• Inferior angle of scapula • Medial margin of intertubercular sulcus of humerus	Adducts and medially rotates humerus; extends shoulder
Muscles Acting on the Forearm		
Biceps brachii (fig. 7.15, 7.18a)	• On and near coracoid process of scapula • Radial tuberosity	Strongly supinates forearm; synergizes elbow flexion; secures head of humerus in glenoid cavity
Brachialis (fig. 7.15, 7.18c)	• Anterior aspect of lower humerus • Coronoid process and tuberosity of ulna	Flexes elbow
Brachioradialis (fig. 7.18a)	• Lateral supracondylar ridge of humerus • Distal radius near styloid process	Flexes elbow
Pronator quadratus (fig. 7.19b)	• Anterior aspect of lower ulna • Anterior aspect of lower radius	Pronates forearm

Table 7.3 Appendicular Muscles *(continued)*

Name	Attachments	Action
Pronator teres (fig. 7.18c, 7.19b)	• Near medial epicondyle of humerus; coronoid process of ulna • Lateral midshaft of radius	Pronates forearm
Supinator (fig. 7.18c, 7.19b)	• Lateral epicondyle of humerus; shaft of upper ulna • Shaft of upper radius	Supinates forearm
Triceps brachii (fig. 7.15, 7.18b)	• Inferior margin of glenoid cavity; posterior aspect of humerus • Olecranon of ulna	Extends elbow; long head adducts humerus
Flexors Acting on the Wrist and Hand		
Flexor carpi radialis (fig. 7.19a)	• Medial epicondyle of humerus • Base of metacarpals II and III	Flexes wrist anteriorly and laterally (radial flexion)
Flexor carpi ulnaris (fig. 7.19a)	• Medial epicondyle of humerus • Pisiform, hamate, and metacarpal V	Flexes wrist anteriorly and medially (ulnar flexion)
Flexor digitorum superficialis (fig. 7.19b)	• Medial epicondyle of humerus; radius; coronoid process of ulna • Middle phalanges II–V	Flexes wrist and knuckles II–V
Extensors Acting on the Wrist and Hand		
Extensor carpi radialis longus and brevis (fig. 7.19c)	• Lateral supracondylar ridge and lateral epicondyle of humerus, respectively • Base of metacarpals II and III, respectively	Extend wrist and aid in radial flexion
Extensor carpi ulnaris (fig. 7.19c)	• Lateral epicondyle of humerus; posterior side of ulnar shaft • Base of metacarpal V	Extends wrist and aids in ulnar flexion
Extensor digitorum (fig. 7.19c)	• Lateral epicondyle of humerus • Posterior (dorsal) side of phalanges II–V	Extends wrist and fingers II–V
Muscles Acting on the Hip and Femur		
Adductor longus and brevis (fig. 7.20)	• Pubis • Posterior shaft of femur	Adduct and medially rotate femur; flex hip
Adductor magnus (fig. 7.20)	• Ischium and pubis • Posterior shaft of femur	Adducts and medially rotates femur; posterior part extends hip
Gluteus maximus (fig. 7.21)	• Posterior surface of ilium, sacrum, and coccyx • Upper lateral femur	Extends hip joint, important in backswing of stride, abducts and laterally rotates femur
Gluteus medius and minimis (fig. 7.21)	• Posterior surface of ilium • Greater trochanter of femur	Abduct and medially rotate femur, maintain balance by shifting body weight during walking
Gracilis (fig. 7.20)	• Pubis and ischium • Medial surface of upper tibia	Adducts femur; flexes knee
Iliopsoas (fig. 7.20)	• Iliac crest and fossa; sacrum; vertebral bodies T12–L5 • Lesser trochanter and adjacent shaft of femur	Flexes hip as in bending at waist or raising thigh
Muscles Acting on the Knee		
Biceps femoris (fig. 7.21)	• Ischial tuberosity and posterior surface of shaft of femur • Head of fibula	Flexes knee; extends hip; laterally rotates leg
Quadriceps femoris (fig. 7.20)	• Ilium; posterior and anterior aspects of shaft of femur • Tibial tuberosity	Extends knee; flexes hip

Table 7.3 Appendicular Muscles *(continued)*

Name	Attachments	Action
Sartorius (fig. 7.20)	• Anterior superior spine of ilium • Medial aspect of tibial tuberosity	Flexes hip and knee; used in crossing legs
Semimembranosus (fig. 7.21)	• Ischial tuberosity • Medial condyle of tibia; lateral condyle of femur	Flexes knee; extends hip
Semitendinosus (fig. 7.21)	• Ischial tuberosity • Medial surface of upper tibia	Same as semimembranosus
Muscles Acting on the Foot		
Extensor digitorum longus (fig. 7.22a)	• Lateral condyle of tibia, shaft of fibula, interosseous membrane • Middle and distal phalanges II–V	Extends toes II–V; dorsiflexes foot
Gastrocnemius (fig. 7.22b)	• Condyles and popliteal surface of femur; capsule of knee joint • Calcaneus	Flexes knee; plantar flexes foot
Soleus (fig. 7.22c)	• Posterior surface of proximal one-fourth of fibula and middle one-third of tibia • Calcaneus	Plantar flexes foot
Tibialis anterior (fig. 7.22a)	• Lateral margin of tibia; interosseous membrane • Medial cuneiform; metatarsal I	Dorsiflexes and inverts foot

Before You Go On

Answer these questions from memory to test your understanding of the previous section. Reread the material if you can't answer the questions.

14. In what ways does the function of the endomysium differ from that of the epimysium of a muscle?
15. How does a synergist modify the action of a prime mover? How does an antagonist modify it? What role does a fixator play?
16. Name a muscle employed in each of the following: a wink, a kiss, a smile, and a frown.
17. Name the three muscle layers of the lateral abdominal region from superficial to deep, and describe the actions of these muscles.
18. Name the muscles of the rotator cuff and describe their locations.
19. Identify three antagonists of the triceps brachii and describe what actions the triceps and these antagonists perform.
20. Describe the functions of the gluteal muscles, hamstrings, and quadriceps femoris.
21. List two muscles and their actions involved in each of the following activities: sit-ups, dribbling a basketball, kicking a soccer ball, and rising from a chair.

Aging of the Muscular System

A common effect of aging is loss of muscle mass, and while the degree varies among individuals, everyone experiences some muscle atrophy and loss of strength with age. Muscular strength and mass peak in the 20s, and by the age of 80, most people have only half as much strength and endurance. Many people over age 75 cannot lift a 4.5 kg (10 lb) weight, making such simple tasks as carrying a bag of groceries very difficult. Loss of muscle mass not only reduces mobility and ability to carry out normal daily tasks, but also increases the risk of obesity, cardiovascular disease, and type 2 diabetes. Fast-twitch muscle fibers show the earliest and greatest atrophy, resulting in increased reaction time, slower reflexes, and reduced coordination, meaning that tasks such as buttoning the clothes take more time and effort.

There are multiple reasons for the loss of strength. Aged muscle has fewer myofibrils; more disorganized sarcomeres; smaller mitochondria; and reduced amounts of ATP, myoglobin, glycogen, and creatine phosphate. Increased adipose and fibrous tissue in the muscles limits their movement and blood circulation. In addition, there are fewer motor neurons in the spinal cord, so some muscle atrophy may result from reduced nerve supply. The neurons that do remain produce less acetylcholine and stimulate the muscles less effectively.

Even though people typically lose muscle mass and function as they age, these effects are noticeably less in people who continue to exercise throughout life. Statistics indicate that less than a quarter of adults in the U.S. get the recommended amount of aerobic and strength-training exercise. The rising proportion of inactive adults and the associated high rates of obesity contribute to increased risk for several chronic, deadly diseases, including Alzheimer disease. Regular exercise, even if one starts late in life, counteracts these age-related diseases and improves overall quality of life.

CAREER SPOTLIGHT

Massage Therapist

A massage therapist manipulates muscles and other soft tissues by kneading, gliding, and compression in order to promote relaxation; relieve pain; stimulate blood and lymph flow; and address issues of stress, anxiety, and tension. Massage therapists work with clients as diverse as athletes, people with soft-tissue injuries, pregnant women, and children, but most often, simply with persons seeking relaxation and a sense of wellness. There are more than 80 types of massage therapy, such as Swedish massage, deep pressure, and acupressure, and spas often combine massage with hydrotherapy, aromatherapy, and other methods. Massage therapists work in a variety of settings including hospitals, long-term care facilities, chiropractic and physical therapy clinics, wellness centers, fitness centers, and hotels, airports, and spas. Most states require one to be a Certified Massage Therapist (CMT) or equivalent before granting a license to practice. Candidates must complete a training program of usually 500 to 1,000 hours of instruction, entailing both hands-on practice and classroom instruction in anatomy and physiology, kinesiology, first aid, CPR, business practices, ethics, and legal issues. One must also pass a licensing examination with both written and demonstration components. See appendix B for further information.

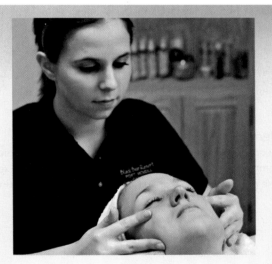
Rolf Hicker/All Canada Photos/Getty Images

CONNECTIVE ISSUES

Ways in Which the MUSCULAR SYSTEM AND OTHER MUSCULAR TISSUES Affect Other Organ Systems

Integumentary System
Superficial muscles support and shape the skin; facial muscles act on the skin to produce facial expressions.

Skeletal System
Muscles move the bones, stabilize the joints, and produce stress that stimulates bone ossification and remodeling; the shapes of mature bones are determined partly by the muscular stresses applied to them.

Nervous System
Muscles give expression to thoughts, emotions, and motor commands that arise in the central nervous system. Exercise positively affects brain health and cognitive function.

Endocrine System
Exercise stimulates the secretion of stress hormones; skeletal muscles protect some endocrine organs; muscle mass affects the body's insulin sensitivity.

Circulatory System
Skeletal muscle contractions help to move blood through veins; exercise stimulates the growth of new blood vessels; cardiac muscle pumps blood, and smooth muscle governs blood vessel diameter and blood and flow.

Lymphatic and Immune Systems
Muscle contractions promote lymph flow; exercise elevates levels of immune cells and antibodies; excessive exercise can inhibit immune responses.

Respiratory System
Muscle contractions ventilate the lungs; skeletal muscles of the larynx and pharynx and smooth muscle of the trachea and bronchial tree regulate airflow; CO_2 generated by exercising muscles stimulates heavier breathing.

Urinary System
Skeletal and smooth muscles control urination; muscles of the pelvic floor support the bladder.

Digestive System
Skeletal muscles enable chewing and swallowing and control defecation; smooth muscle propels material through the digestive tract and also controls defecation; abdominal and lumbar muscles protect the lower digestive organs.

Reproductive System
Muscles contribute to erection and orgasm; abdominal and pelvic muscles aid in childbirth.

Study Guide

Assess Your Learning Outcomes

To test your knowledge, discuss the following topics with a study partner or in writing, ideally from memory.

7.1 Muscular Tissue and Cells

1. Five functions of muscular tissue
2. Why skeletal muscle is described as striated and voluntary
3. General appearance and typical dimensions of muscle fibers
4. Description of the sarcolemma, transverse tubules, sarcoplasmic reticulum, myofibrils, and myofilaments and their locations in a muscle fiber
5. The roles of glycogen and myoglobin in muscle fibers
6. The composition, molecular arrangement, and functions of thick and thin myofilaments
7. The role of the proteins myosin, actin, tropomyosin, and troponin in muscle fibers
8. Structural relationships of the A bands, I bands, Z discs, and sarcomeres; how the arrangement of the myofilaments accounts for the A and I bands
9. Why the mechanism of contraction is called the sliding filament model
10. Motor neurons and their relationship with skeletal muscle fibers
11. What a motor unit is, why it is called that, and the functional advantages of a muscle being composed of many motor units
12. Structural components of a neuromuscular junction and the function of each
13. The locations and roles of acetylcholine and acetylcholinesterase

7.2 Physiology of Skeletal Muscle

1. The stages of muscle contraction and relaxation
2. How an arriving nerve signal excites a muscle fiber
3. Events that link muscle excitation to the onset of contraction
4. How a muscle fiber shortens; the roles of myosin, actin, tropomyosin, calcium, troponin, and ATP in this process
5. How a muscle relaxes; the roles of acetylcholinesterase, the sarcoplasmic reticulum, ATP, and tropomyosin in this process
6. Why a single muscle twitch cannot perform useful muscular work; the effect of summation of twitches on muscle performance
7. How muscle tone is maintained and why this is important
8. The difference between isometric and isotonic muscle contraction, and how these relate to ordinary muscle actions; the roles of concentric and eccentric contraction in weight lifting
9. The two metabolic pathways for producing the ATP needed for muscle contraction, and the advantage and disadvantage of each
10. How a muscle shifts from one mode of ATP production to another over the course of a prolonged exercise
11. Causes of muscle fatigue and variables that improve endurance
12. Why a person continues to breathe heavily after the end of a strenuous exercise
13. How slow- and fast-twitch muscle fibers differ metabolically; the advantage and limitation of each; and examples of muscles in which each type is abundant
14. The cellular effects and respective benefits of resistance and endurance exercise

7.3 Cardiac and Smooth Muscle

1. The properties of cardiac muscle that enable it to reliably pump blood
2. The structure of cardiac muscle cells and how they differ from skeletal muscle fibers
3. Significance of the prolonged contraction of cardiac muscle cells
4. The significance of aerobic respiration in cardiac muscle, and how this is reflected in the internal structure of the muscle cell
5. Locations of smooth muscle
6. Similarities and differences between smooth muscle myocytes, cardiomyocytes, and skeletal muscle fibers
7. Why smooth muscle tone is important in hollow organs
8. How smooth muscle tissue differs from skeletal and cardiac muscle in its capacity for repair

7.4 Anatomy of the Muscular System

1. The connective tissue components of a skeletal muscle and their spatial relationship to the muscle fibers
2. Attachments between skeletal muscle and bone; the difference between an aponeurosis and other tendons
3. The four functional categories of muscles that may be found at a joint, and how these four types modify each other's actions
4. Why the Latin names of muscles can be helpful in understanding their locations and functions once one becomes accustomed to them
5. The proper anatomical names of several muscles that have popular nicknames commonly used by those interested in physical training
6. The locations, attachments, and actions of the following facial muscles—the frontalis, orbicularis oculi, orbicularis oris, zygomaticus, and buccinator—and some ordinary activities in which these muscles are involved
7. The same for the following muscles of mastication: temporalis, masseter, and pterygoids
8. The same for the following muscles of head and neck movement: sternocleidomastoid and trapezius

9. The same for the following abdominal muscles: rectus abdominis, external and internal obliques, and transverse abdominal
10. The same for the erector spinae
11. The same for the following muscles acting on the scapula: pectoralis minor, serratus anterior, trapezius, levator scapulae, and rhomboids
12. The same for additional muscles that act on the humerus: pectoralis major, latissimus dorsi, deltoid, and teres major
13. The same for the four rotator cuff muscles: the supraspinatus, infraspinatus, teres minor, and subscapularis; why their tendons are called the rotator cuff, and some actions that often cause rotator cuff injuries
14. The same for the following muscles acting on the elbow or forearm: biceps brachii, brachialis, brachioradialis, triceps brachii, pronator teres, pronator quadratus, and supinator
15. The locations of muscles that move the hand, and why there are so many hand muscles
16. The locations and functional significance of the flexor retinaculum, extensor retinaculum, and carpal tunnel
17. The locations, attachments, and actions of the following wrist and finger flexors—flexor carpi radialis, flexor carpi ulnaris, flexor digitorum superficialis, and flexor digitorum profundus—and some ordinary activities in which these muscles are involved
18. The same for the following wrist and finger extensors: extensor digitorum, extensor carpi ulnaris, extensor carpi radialis longus, and extensor carpi radialis brevis
19. The same for the two components of the iliopsoas muscle complex: the iliacus and psoas major
20. The same for the three gluteal muscles: gluteus maximus, medius, and minimus
21. The same for the adductor and abductor groups of thigh muscles, considering each of the two groups collectively
22. The same for the four heads of the quadriceps femoris; the sartorius; and the three hamstring muscles
23. The same for the anterior muscles of the leg: extensor digitorum longus and tibialis anterior
24. The same for the posterior muscles of the leg: gastrocnemius and soleus
25. Location of muscles that act on the foot, and their major functions

Testing Your Recall

1. The relatively light middle portion of the A band of skeletal muscle
   a. has myosin but no actin.
   b. is the site of the Z disc.
   c. is light because of the glycogen stored there.
   d. has thin myofilaments but no thick myofilaments.
   e. marks the boundary between one sarcomere and the next.

2. Before a muscle fiber can contract, ATP must bind to
   a. a Z disc.
   b. the myosin head.
   c. tropomyosin.
   d. troponin.
   e. actin.

3. ACh receptors are found in
   a. the synaptic vesicles.
   b. the terminal cisterns.
   c. the thick filaments.
   d. the thin filaments.
   e. the sarcolemma.

4. In the activation of muscle contraction, acetylcholine is released from
   a. the terminal cisterns.
   b. the sarcolemma.
   c. synaptic vesicles.
   d. the synaptic cleft.
   e. intercalated discs.

5. Slow-twitch fibers have all of the following *except*
   a. an abundance of myoglobin.
   b. an abundance of glycogen.
   c. high fatigue resistance.
   d. a deep red color.
   e. a high capacity to synthesize ATP aerobically.

6. Which of these muscles is an extensor of the neck?
   a. external oblique
   b. sternocleidomastoid
   c. trapezius
   d. temporalis
   e. latissimus dorsi

7. Which of the following muscles does *not* extend the hip joint?
   a. quadriceps femoris
   b. gluteus maximus
   c. biceps femoris
   d. semitendinosus
   e. semimembranosus

8. Both the gastrocnemius and _____ muscles insert on the heel by way of the calcaneal tendon.
   a. semimembranosus
   b. extensor digitorum longus
   c. tibialis anterior
   d. soleus
   e. quadriceps femoris

9. Which of the following muscles contributes the most to smiling?
   a. frontalis
   b. orbicularis oris
   c. zygomaticus major
   d. masseter
   e. mentalis

10. Four scapular muscles contribute tendons to the rotator cuff, including all of these *except*
    a. the supraspinatus.
    b. the teres minor.
    c. the subscapularis.
    d. the levator scapulae.
    e. the infraspinatus.

11. Thick myofilaments consist mainly of the protein myosin.

12. The neurotransmitter that stimulates skeletal muscle is acetylcholine.

13. A state of continual partial muscle contraction is called muscle tone.

14. A bundle of muscle fibers surrounded by perimysium and seen as the grain in a cut of meat is called a/an fascicle.

15. Skeletal muscle contraction requires the release of calcium ions from the sarcoplasmic reticulum.

16. Cardiac and smooth muscle are called involuntary because they can contract without nervous stimulation and are usually not subject to conscious control.

17. A circular muscle that regulates a body opening or passage is called a/an sphincter.

18. The knee is extended by a four-part muscle called the quadriceps femoris on the anterior side of the thigh.

19. The largest muscle of the back, extending from the midthoracic to lumbar region and converging on the humerus, nicknamed the swimmer's muscle, is the latissimus dorsi.

20. The large bulge of the calf, just below the popliteal region, is a two-headed muscle called the gastrocnemius.

Answers in Appendix A

What's Wrong with These Statements?

Briefly explain why each of the following statements is false, or reword it to make it true.

1. Each motor neuron supplies just one muscle fiber.

2. To initiate muscle contraction, calcium ions must bind to the myosin heads.

3. Slow-twitch fibers fatigue relatively quickly.

4. Thin filaments get shorter when a muscle contracts.

5. Of the three muscle types, skeletal muscle has the largest and most numerous mitochondria.

6. The zygomaticus major and minor muscles attach to the zygomatic bone and mandible.

7. To push someone away from you, you would use the trapezius more than the serratus anterior.

8. In isometric contraction, a muscle maintains a constant level of tension as it shortens or lengthens.

9. The iliopsoas is the main muscle of hip extension.

10. In lifting barbells by flexion of the elbow, the biceps brachii muscles provide more power than the brachialis muscles do.

Answers in Appendix A

Testing Your Comprehension

1. Without ATP, relaxed muscle cannot contract and a contracted muscle cannot relax. Explain why.

2. Consider the relaxed sarcomere in figure 7.9a. Suppose a relaxed skeletal muscle were stretched so much that the thick filaments barely overlapped with the tips of the thin filaments or did not overlap at all. Predict what would happen when a motor nerve fiber stimulated such a muscle, and explain why.

3. Radical mastectomy, once a common treatment for breast cancer, involved removal of the pectoralis major along with the breast. What functional impairments would result from this? What synergists could a physical therapist train a patient to use to recover some lost function?

PART 3 Internal Coordination

Chapter 8

The Nervous System I

NERVE CELLS, THE SPINAL CORD, AND REFLEXES

Nerve cells growing in laboratory culture. These were produced in an experiment to culture nervous tissue and study regeneration for the treatment of spinal paralysis.
Science Source

Chapter Outline

8.1 Cells and Tissues of the Nervous System
- 8.1a Overview of the Nervous System
- 8.1b Neurons (Nerve Cells)
- 8.1c Neuroglia
- 8.1d The Myelin Sheath
- 8.1e Forms of Nervous Tissue

8.2 The Physiology of Neurons
- 8.2a Producing a Resting Membrane Potential
- 8.2b Generating an Action Potential
- 8.2c Conducting a Nerve Signal to Its Destination
- 8.2d Transmitting a Nerve Signal to the Next Cell

8.3 The Spinal Cord, Spinal Nerves, and Reflexes
- 8.3a The Spinal Cord—Functional Anatomy
- 8.3b The Spinal Meninges—Protective Membranes
- 8.3c Spinal Nerves—Communicating with the Rest of the Body
- 8.3d Somatic Reflexes

Clinical Applications/Perspectives on Health
- Clinical Application 8.1: Multiple Sclerosis
- Clinical Application 8.2: Spinal Cord Injury and Paralysis
- Perspectives on Health

End of Chapter
- Career Spotlight: Occupational Therapist
- Study Guide

BASE CAMP

Before ascending to the next level, be sure you're properly equipped with a knowledge of these concepts from earlier chapters.
- Ions and electrolytes (see section 2.1d)
- Plasma membrane structure (see section 3.2a)
- Voltage-gated and ligand-gated membrane channels (see "Membrane Proteins" in section 3.2a)
- Anatomy of the vertebral column in relationship to the spinal cord (see "The Vertebral Column" in section 6.3b)

Anatomy & Physiology Revealed 4.0
Module 7
Nervous System

We now turn our attention to the nervous system, perhaps the most complex and mysterious body system. It forms the foundation for our sensory awareness, consciousness, personality, and movement. Although the nervous system has been the subject of intense research for well over a hundred years, there is still much that is unknown. Ultimately, the nervous system is a system of internal communication that coordinates the activities of the some 30 trillion cells of the body. If the cells acted independently, the result would be physiological chaos and death. Two systems of internal communication coordinate cellular activities: (1) the endocrine system, which uses hormones as intercellular signals (this will be considered in chapter 11); and (2) the nervous system, which uses electrical and chemical signals, and is the subject of the next three chapters. In this chapter, we will study nerve cells, the spinal cord and its nerves, and spinal reflexes. Chapter 9 will survey the brain, its associated nerves, and the autonomic nervous system, and chapter 10 concerns the sense organs.

8.1 Cells and Tissues of the Nervous System

Expected Learning Outcomes

When you have completed this section, you should be able to

a. state three fundamental functions of the nervous system;
b. define the two major anatomical subdivisions of the nervous system;
c. define three functional categories of neurons;
d. describe the structure of a generalized neuron and the common variations on this structure;
e. discuss the neuroglial cells of nervous tissue and their functions;
f. discuss the myelin sheath that envelops many nerve fibers, including its function and how it is produced; and
g. distinguish between the gray and white matter of the central nervous system.

8.1a Overview of the Nervous System

The nervous system coordinates diverse cellular functions in three basic steps: (1) It senses information about changes within the body and in the external environment and sends messages to the *central nervous system (CNS)*. (2) The CNS receives and processes (integrates) the information and determines what response, if any, is appropriate. (3) The CNS issues commands to the skeletal muscles and other organs to carry out the necessary response.

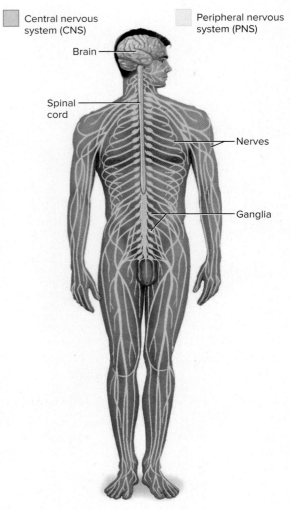

Figure 8.1 The Nervous System.

Figure 8.2 The Organization of the Nervous System. The flow chart shows the subdivisions of the nervous system.

The nervous system has two main anatomical subdivisions (fig. 8.1).

1. The **central nervous system (CNS)** is composed of the brain and spinal cord and is enclosed and protected by bone—the cranium and vertebral column. It carries out the integrative functions of the nervous system.
2. The **peripheral nervous system (PNS)** is composed of nerves leading to and from the CNS. The PNS provides pathways of sensory input and motor output to structures that carry out its commands.

The peripheral nervous system is functionally divided into *sensory* and *motor* divisions (fig. 8.2).

- The **sensory (afferent**[1]**) division** carries signals from various **receptors** (sense organs and simple sensory nerve endings) to the CNS to provide information about environmental changes (stimuli) within and outside the body.
- The **motor (efferent**[2]**) division** carries signals from the CNS to structures such as muscles and glands that carry out the body's responses. Cells and organs that respond to these commands are called **effectors.** The motor division has two subdivisions.
  - The **somatic motor division** carries signals to skeletal muscles. This produces voluntary movements, such as when you decide to reach for a book or throw a ball.
  - The **autonomic nervous system (ANS; visceral motor division)** carries signals to cardiac muscle, smooth muscle, and glands. We usually have little voluntary control over these effectors. The ANS is further divided into:
    - The **sympathetic division,** which prepares the body for action—for example, by accelerating the heartbeat and increasing respiratory airflow. It is often called the "flight or fight" system and is activated by fear or when we need extra energy for exercise.
    - The **parasympathetic division,** which tends to have the opposite effect of the sympathetic division—calming the body and slowing the heartbeat, for example. It also stimulates digestion.

8.1b Neurons (Nerve Cells)

The **neurons (nerve cells)** carry out the primary task of the nervous system, communication. There are about a trillion neurons in the nervous system—10 times as many neurons in your body as there are stars in our galaxy! A neuron usually consists of a more or less globular or stellate cell body where its nucleus is located, and two or more long fibrous processes that reach out to other cells.

Basic Structure of Neurons

Neurons are highly variable in shape, but a good starting point for understanding their structure is a *spinal motor neuron,* a cell that originates in the spinal cord and leads to a skeletal muscle—the type of neuron that enables us to move the body (fig. 8.3). The control center of the neuron is its **neurosoma**[3] **(soma or cell body).** It contains the nucleus and is therefore the cell's site of genetic control. It also has most of the usual organelles found in other cells: mitochondria, lysosomes, a Golgi complex, and an extensive rough endoplasmic reticulum. There are no centrioles, however, and for this and other reasons, neurons cannot undergo mitosis. Neurons that die are therefore usually irreplaceable; surviving neurons cannot multiply to replace those that are lost.

---

[1] *af,* from *ad* = toward; *fer* = to carry
[2] *ef,* from *ex* = out; *fer* = to carry
[3] *soma* = body

Figure 8.3 A Representative Neuron. (a) A multipolar neuron such as a spinal motor neuron. (b) Detail of the myelin sheath. The Schwann cells and myelin are explained later in this chapter. (c) Photograph of this neuron type. **APR**

C: ©Ed Reschke

(a) Multipolar neurons

(b) Bipolar neurons

(c) Unipolar neuron

Figure 8.4 Variation in Neuron Structure. (a) Two multipolar neurons of the brain—a pyramidal cell (left) and a Purkinje cell. (b) Two bipolar neurons—a bipolar cell of the retina (left) and an olfactory neuron of the nose. (c) A unipolar neuron of the type involved in the senses of touch and pain.

Arising from the neurosoma, there are several thick arms that divide repeatedly into fine branches. They are named **dendrites**[4] after their striking similarity to the branches and twigs of a leafless tree in winter. They are the "receiving end" of a neuron, the primary route by which it receives signals from other neurons. Some neurons have only one dendrite, and some have thousands. The more dendrites a neuron has, the more information it can receive and incorporate into its decision making. Dendrites have tiny protrusions called *dendritic spines* that increase the connectivity between neurons. Dendritic spines dynamically change their density and structure in response to internal and external stimuli. Their plasticity is related to learning and memory as well as other cognitive processes.

On one side of the neurosoma is a conical mound called the *axon hillock,* which gives rise to a long process called the **axon (nerve fiber)**—the neuron's output pathway for signals that it sends to other cells. An axon is roughly cylindrical and relatively unbranched except at its distal end. A neuron never has more than one axon, and some neurons in the brain and retina have none. At the distal end, an axon usually exhibits a profusion of fine branches, with each twig ending with a bulb called the *axon terminal.* This will be the focus of our attention when we consider how a neuron transmits a signal to the next cell across a junction called the *synapse.*

Simplified textbook drawings cannot do justice to the structure of a neuron, but an analogy may create a useful impression of its dimensions. A typical neuron has a neurosoma from 5 to 135 μm in diameter, whereas the axon is about 1 to 20 μm in diameter and is sometimes more than 1.5 m (5 ft) long, yet too slender to see without a microscope. If we scale the neurosoma up to the size of a tennis ball, such an axon would be up to 1.7 km (1 mile) long and a little narrower than a garden hose, and the dendrites could densely fill a 30-seat classroom from wall to wall and ceiling to floor, like an impenetrable briar thicket. Such proportions are all the more impressive when we consider that the neuron must assemble molecules and organelles in that "tennis ball" neurosoma and deliver them through its "mile-long hose" to the end of the axon. Neurons use mobile *motor proteins* to "walk" up and down the axon carrying substances to and from the neurosoma.

Apply What You Know

The axon of a neuron has a dense cytoskeleton. Considering the functions of a cytoskeleton you studied in section 3.3a, suggest at least two reasons that this cytoskeleton is so necessary to an axon.

Structural and Functional Classes of Neurons

Not all neurons fit the preceding basic description. They vary in shape depending on their function and location. Neurons are classified structurally according to the number of processes that arise from the neurosoma (fig. 8.4):

1. **Multipolar** neurons, like the one just described, have one axon and multiple dendrites. This is the most common type.
2. **Bipolar** neurons have one axon and one dendrite. These include olfactory cells of the nose, neurons in the retina associated with vision, and sensory neurons of the ear.
3. **Unipolar** neurons have only one process leading away from the neurosoma. It branches like a T a short distance away, with one branch bringing signals from sources such as the skin and joints, and the other branch leading to the

---

[4]*dendr* = tree, branch; *ite* = little

spinal cord. The dendrites are the branching receptive tips at the end of the former branch, and the rest of the fiber is considered to be the axon.

Functionally, neurons fit into three basic classes corresponding to the sensory, integrative, and motor functions of the nervous system (fig. 8.5):

1. **Sensory (afferent) neurons** are specialized to detect stimuli and transmit information about them to the CNS. Afferent neurons originate in the eyes, ears, skin, joints, and other organs and carry information about light, sound, touch, pain, and other stimuli to the brain or spinal cord. The word *afferent* refers to the fact that signals in these cells travel *toward* the CNS.
2. **Interneurons** perform the integrative functions of the nervous system—they process, store, and retrieve information and "make decisions" about how the body should respond to a stimulus. They are contained entirely within the CNS and are the most abundant of all neurons.
3. **Motor (efferent) neurons** are specialized to carry outgoing signals from the CNS to the cells and organs that carry out its commands. They begin in the CNS and extend to muscle and gland cells. The word *efferent* refers to the fact that these cells carry nerve signals away from the CNS.

Most sensory neurons are unipolar or bipolar; interneurons and motor neurons are generally multipolar.

8.1c Neuroglia

Although neurons carry out the communicative function of the nervous system, they are outnumbered at least 10 to 1 by cells called **neuroglia**[5] (noor-OG-lee-uh), or **glial cells** (GLEE-ul). Traditionally, these cells were seen as passive players that merely support neurons, but we now know that neuroglial cells have central roles in the healthy functioning of the nervous system (see Table 8.1). We still know less about them than neurons, and their complete range of functions remains to be discovered. There are four kinds of glial cells in the central nervous system (fig. 8.6).

1. **Oligodendrocytes**[6] (OL-ih-go-DEN-dro-sites) are large bulbous cells with as many as 15 armlike processes, giving them an appearance somewhat like an octopus. Each arm reaches out to a nearby nerve fiber and spirals around it to form a layer of insulation called the *myelin sheath,* which we'll examine later.
2. **Ependymal**[7] **cells** (ep-EN-dih-mul) are cuboidal cells that line the internal, fluid-filled cavities of the brain and spinal cord. They produce the *cerebrospinal fluid (CSF)* that fills these cavities and bathes the CNS surface. They have cilia on their surfaces that help to circulate the CSF. Functions of the CSF are discussed in section 9.1c.
3. **Microglia** are small phagocytic cells that wander through the CNS tissue searching out and destroying tissue debris, infectious microorganisms, and other foreign matter. They become concentrated in areas damaged by infection, trauma, or stroke.

Figure 8.5 Functional Classes of Neurons. All neurons can be classified as sensory, motor, or interneurons based on their location and the direction of signal conduction.

---

[5]*glia* = glue
[6]*oligo* = a few; *dendro* = branches; *cyte* = cell
[7]*ep* = above, upper; *endym* = garment, wrap

Table 8.1 Neuroglial Cells and Their Functions APR

Neuroglia of CNS	
Oligodendrocytes	Form myelin in the brain and spinal cord
Ependymal cells	Line cavities of brain and spinal cord; secrete and circulate cerebrospinal fluid
Microglia	Phagocytize and destroy microorganisms, foreign matter, and dead nervous tissue
Astrocytes	Cover brain surface and parts of individual neurons; form supportive framework in the CNS; contribute to the blood–brain barrier; nourish neurons; produce growth factors that stimulate neurons; promote the formation of synapses; communicate electrically with neurons and influence synaptic signaling; remove K^+ and other chemicals from the ECF of the brain and spinal cord; help to regulate composition of ECF; form scar tissue to replace damaged nervous tissue
Neuroglia of PNS	
Schwann cells	Wrap around nerve fibers in the PNS and form myelin sheath in the PNS; aid in regeneration of damaged nerve fibers
Satellite cells	Surround neurosomas in the ganglia; provide electrical insulation and regulation of chemical environment of neurons

4. **Astrocytes,**[8] the most abundant glial cells in the CNS, constitute over 90% of the brain tissue in some areas. They have the most diverse roles of all glial cells.

- They form a supportive structural framework for nervous tissue.
- Their extensions, *perivascular feet,* contact blood vessels and maintain the *blood–brain barrier* that separates the CNS from the rest of the body. The blood–brain barrier is further described in section 9.1e.

[8]*astro* = starlike; *cyte* = cell

Figure 8.6 Neuroglia of the Central Nervous System.

❓ *What cells in the peripheral nervous system perform the same function as the oligodendrocytes seen in this figure?*

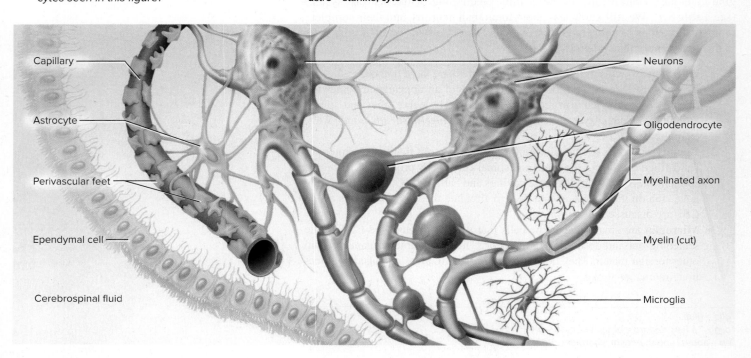

- They monitor neuron activity and adjust local blood flow, thereby supporting energetic demands of the neurons.
- They secrete chemicals called *growth factors* that promote neuron growth and synapse formation.
- They maintain a stable extracellular environment, regulating the amount of potassium, for example.
- They convert glucose to lactate and supply this to neurons for nourishment.
- They form scar tissue in damaged regions of the CNS.

Two more glial cell types occur in the peripheral nervous system.

1. **Satellite cells** surround the cell bodies of peripheral neurons. They insulate them electrically and regulate their chemical environment.
2. **Schwann**[9] **cells** (pronounced "shwon") wrap around nerve fibers of the PNS, enclosing each in a sleeve called the **neurilemma**[10] (NOOR-ih-LEM-ah). In most cases, a Schwann cell also spirals repeatedly around the axon, depositing layer after layer of its own membrane between the neurilemma and nerve fiber. This wrapping forms the myelin sheath of peripheral nerve fibers, similar to the one produced by oligodendrocytes in the CNS.

Schwann cells are also critical for the regeneration and healing of damaged nerve fibers. Unlike Schwann cells, oligodendrocytes cannot perform this regenerative role, and damaged fibers in the CNS are lost forever. Being enclosed in bone, however, nervous tissue of the CNS is injured far less frequently than nerves of the PNS.

8.1d The Myelin Sheath

The **myelin sheath** (MY-eh-lin) bears closer examination because it is so important in signal conduction by nerve fibers. In fact, it is absent in invertebrates, and was an important feature in the evolution of the complexity of the vertebrate nervous system. It is formed by oligodendrocytes in the brain and spinal cord and Schwann cells in the peripheral nerves (**fig. 8.7**). The sheath consists of layers of plasma membranes that encircle the axons, wrapping them in an insulating lipid blanket that ensures rapid, efficient signal conductance. In this way, it is similar to the rubber or plastic covering of an electrical cord. In the CNS, each oligodendrocyte reaches out to several nearby nerve fibers and contributes a myelin sheath to each of them. In the PNS, one Schwann cell wraps around only one nerve fiber.

In both cases, a nerve fiber is much longer than the reach of a single glial cell, so it requires many Schwann cells or oligodendrocytes to myelinate one nerve fiber. Consequently, the myelin sheath is segmented. The gaps between segments are called **nodes of Ranvier**[11] (RON-vee-AY), and the myelin-covered segments are called **internodes** (see fig. 8.3). The internodes are about 0.2 to 1.0 mm long. Myelinated fibers conduct signals much faster than unmyelinated fibers because of the arrangement of the nodes and internodes, as will be explained later in the chapter. Not all nerve fibers are myelinated—slower signals carried by unmyelinated fibers are adequate for processes such as digestive secretion. In contrast, speedy signals carried along myelinated fibers provide vertebrates with the ability to respond swiftly—for example, a cheetah's speed or the rapid, graceful movements of a soccer player are possible because of myelin.

8.1e Forms of Nervous Tissue

In the central nervous system, we find two fundamental types of nervous tissue called white and gray matter. **White matter** consists of bundles of nerve fibers

---

[9]Theodor Schwann (1810–82), German histologist
[10]*neuri* = nerve; *lemma* = sheath, husk
[11]L. A. Ranvier (1835-19-22), French histologist and pathologist

called **tracts** that travel up and down the spinal cord, between one region of the brain and another, and between brain and cord. Many fibers of these tracts are myelinated; myelin gives the white matter a glistening, pearly white color. All fibers in one tract have a similar origin, destination, and function. We'll take a closer look at tracts of the spinal cord in this chapter and tracts of the brain in chapter 9. There are no nerves in the brain or spinal cord. The body's bundles of nerve fibers are called *nerves* in the PNS and *tracts* in the CNS.

Gray matter is where the neurosomas, dendrites, and synapses are located. There is relatively little myelin here, so this tissue has a duller color in fresh nervous tissue. It is the information-processing part of the CNS, whereas the white matter is more like a telephone cable—a bundle of "wires" (nerve fibers) carrying signals from place to place but not processing signals along the way.

Figure 8.7 Myelination. (a) A Schwann cell of the PNS, wrapping repeatedly around an axon to form the multilayered myelin sheath. The myelin spirals outward away from the axon as it is laid down. The outermost coil of the Schwann cell constitutes the neurilemma. (b) An oligodendrocyte of the CNS wrapping around the axons of multiple neurons. Here, the myelin spirals inward toward the axon as it is laid down. (c) A myelinated axon (top) and unmyelinated axon (bottom) (electron micrograph). **APR**

c: Dr. Dennis Emery, Iowa State University/McGraw-Hill

Clinical Application 8.1

MULTIPLE SCLEROSIS

Multiple sclerosis[12] (MS) is a degenerative disorder of the myelin sheath. It is widely thought to be caused by an autoimmune reaction triggered by a virus in genetically susceptible individuals. It is characterized by deterioration of the oligodendrocytes and myelin sheaths of the CNS, especially in individuals between the ages of 20 and 40. MS is named for the hardened scar tissue that replaces the myelin. This degeneration and scarring disrupts nerve conduction, with effects that depend on what part of the CNS is involved—double vision, blindness, speech defects, neurosis, tremors, or numbness, for example. Patients experience variable cycles of milder and worse symptoms until they may eventually become bedridden. As yet there is no cure, but there is conflicting evidence of how much it shortens a person's life expectancy, if at all. Some MS victims die within 1 year of diagnosis, but many live with it for 25 or 30 years.

In the spinal cord, white matter forms the surface tissue and the gray matter forms the inner core. In the brain, by contrast, white matter forms most of the deep tissue, whereas gray matter forms a surface layer as well as a few deep masses embedded in the white matter. In the CNS, these deep gray matter masses surrounded by white matter are called **nuclei** (not to be confused with cell nuclei). A given nucleus of CNS gray matter typically has a specific function such as regulating your respiratory rhythm or controlling your appetite.

Before You Go On

Answer these questions from memory to test your understanding of the previous section. Reread the material if you can't answer the questions.

1. Which nervous system function pertains to information input to the CNS? Which function pertains to its output to muscles and glands? What are the terms for the input and output neurons that correspond to these two functions?
2. Distinguish between the central and peripheral nervous systems and identify the components of each.
3. Make a sketch of a generalized multipolar neuron and label its dendrites, soma, axon, and axon terminals.
4. What types of neuroglia occur in the central nervous system? What are their functions? Name the types in the peripheral nervous system and state their functions.
5. Describe the structure of the myelin sheath found around many nerve fibers.
6. What is the difference between the gray and white matter of the central nervous system? Which of these is the seat of the CNS's integrative function?

---

[12]*scler* = hard, tough; *osis* = condition

8.2 The Physiology of Neurons

Expected Learning Outcomes

When you have completed this section, you should be able to

a. explain how a neuron maintains a resting potential across its plasma membrane;
b. describe the mechanism behind the voltage changes that occur when a neuron is stimulated;
c. explain how a neuron conducts a signal from the neurosoma to the end of its axon; and
d. explain how a neuron transmits information to the next cell.

8.2a Producing a Resting Membrane Potential

Nerve signals are essentially electrical events created by the movement of sodium and potassium ions. Every living cell is like a little biological battery. Just as a battery has a positive and a negative pole, a cell has a positive charge on the outer surface of the plasma membrane and a negative charge on the inner surface. In a resting, unstimulated neuron, this charge difference is called the **resting membrane potential (RMP).** It is a readiness for action, a starting point for generating a nerve signal. To understand how nerve cells communicate, the first thing we must address is how neurons produce the RMP; how does the battery get "charged up"?

In a battery, the charge difference results from a difference in the concentration of electrons at the two poles. In a cell, by contrast, it results from a difference in the concentrations of sodium, potassium, and other ions on the two sides of the membrane. Sodium ions (Na^+) are about 12 times as concentrated on the outside of the membrane as on the inside, and potassium ions (K^+) are almost 38 times as concentrated on the inside (fig. 8.8). In addition, the cell contains larger negative ions (anions) such as phosphates, proteins, and nucleic acids that cannot pass out through the membrane. All of these inequalities of ion distribution collectively make the interior of the cell negatively charged relative to the exterior.

When a battery has such an unequal charge distribution, we say it is charged; when a cell membrane does so, we say it is **polarized.** Electrical charges are measured in volts (V) or fractions of a volt (millivolts, or mV). For example, an automobile battery typically has a charge of 12 V and a flashlight battery, 1.5 V. The charges across a cell membrane are much lower—typically an RMP of about –70 mV in neurons. The minus sign indicates the negative internal side of the membrane.

Figure 8.8 Ionic Basis of the Resting Membrane Potential. Note that sodium ions are much more concentrated in the extracellular fluid (ECF) than in the intracellular fluid (ICF), whereas potassium ions are more concentrated in the ICF. Large anions unable to penetrate the plasma membrane give the cytoplasm a negative charge relative to the ECF. Milliequivalents per liter (mEq/L) is a unit of concentration that reflects the number of particles per given volume of fluid and the electrical charge on each particle (see appendix C).

- Na^+ 145 mEq/L
- K^+ 4 mEq/L
- Na^+ 12 mEq/L
- K^+ 150 mEq/L
- Large anions that cannot escape cell

The plasma membrane includes proteins that act as channels for Na⁺ and K⁺, and the slow constant leakage of these ions through the membrane always threatens to abolish the RMP. However, the sodium–potassium (Na⁺–K⁺) pump in the membrane continually ejects 3 Na⁺ and pumps in 2 K⁺ for each cycle of the pump. Since it ejects three positive charges for every two it brings in, the pump maintains the cell's negative internal charge (see "Active Transport" in section 3.2). The Na⁺–K⁺ pump requires ATP to do this, and accounts for about 70% of the energy requirement of the nervous system.

8.2b Generating an Action Potential

Nerve signaling is based on ion channels in the plasma membrane called ligand-gated and voltage-gated channels. A *ligand* is any chemical that binds to a protein. A **ligand-gated channel** is a membrane protein that opens or closes in response to a ligand binding to it. A **voltage-gated channel** is one that opens or closes in response to changes in membrane potential (see "Membrane Proteins" in section 3.2a).

The first step in nerve signaling is excitation of a neuron, usually at a dendrite. Often, the stimulus is a ligand that binds to a channel protein. In response, the channel opens and allows sodium ions to flow into the cell like water through a faucet that has just been opened (fig. 8.9). The influx of positive Na⁺ ions cancels some of the negative charge inside the cell and the membrane potential shifts from the −70 mV resting voltage toward zero. Such a change to a less negative voltage is called **depolarization.** Any such voltage change occurring near the point of stimulation is called a **local potential.** The Na⁺ diffuses under the plasma membrane to adjacent regions of the cell, changing the voltage there as well. A wave of excitation thereby spreads out from the point of stimulation like ripples in a pond.

A local potential depolarizes a short section of the membrane. If the stimulus is strong enough, the signal will flow all the way to the axon hillock, where it may produce an **action potential,** a more dramatic shift in membrane voltage in the axon. A neuron is said to "fire" when it produces an action potential. The steps of the action potential follow and are numbered to match figure 8.10.

① Upon stimulation, Na⁺ enters a neuron and the local potential spreads toward the cell's axon hillock.

② The local potential reaches the axon hillock, a region with a high density of voltage-gated channels for sodium and potassium. If the local potential attains a minimum voltage called **threshold,** it opens the channels in the axon hillock and produces an action potential.

③ Sodium ions rush into the cell and neutralize the negative charges near the inside of the membrane, causing the membrane voltage to shoot rapidly upward. As the voltage passes 0 mV, the Na⁺ channels start to close, but by the time all of them do so, the voltage generally has risen to about +35 mV.

④ Voltage-gated potassium channels in the hillock also open, but more slowly. Potassium ions rapidly leave the cell when these gates open. The outflow of K⁺ makes the interior of the cell turn negative again. The shift of the voltage back into the negative numbers is called **repolarization.** Slightly more potassium leaves the cell than is necessary to reach the resting membrane potential, which produces a slight negative overshoot called *hyperpolarization.*

8.2c Conducting a Nerve Signal to Its Destination

We have seen how a nerve signal is initiated; now we examine how it travels to its final destination. The action potential is a voltage spike over a limited area of plasma membrane. However, it triggers another

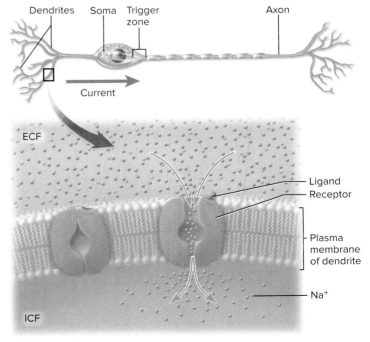

Figure 8.9 Excitation of a Neuron by a Chemical Stimulus. When the chemical (ligand) binds to a receptor on the neuron, the receptor acts as a ligand-gated channel that opens and allows Na⁺ to diffuse into the cell. This depolarizes the plasma membrane.

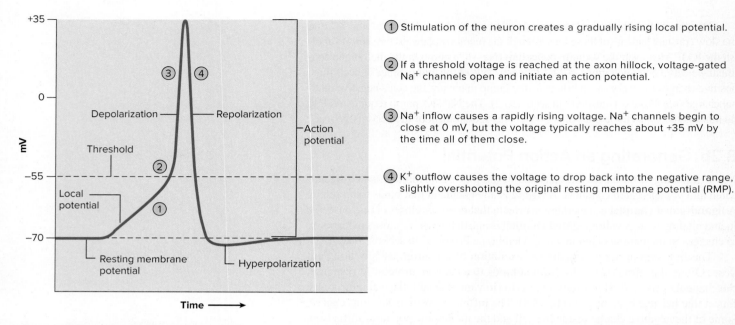

Figure 8.10 An Action Potential. APR

action potential in the membrane immediately ahead of it, and that action potential triggers another, and so forth. Thus, we get a chain reaction of one action potential after another along the length of a nerve fiber. This chain reaction constitutes the **nerve signal.** An illuminating analogy to this is standing up a long row of dominoes and pushing the first one over. When that domino falls, it pushes over the second, and so forth—and the chain reaction produces a wave of energy traveling to the end of the line. No one domino moves to the other end of the line; a falling domino is a local event. Similarly, an action potential is a local event, but it triggers the next one and, like the row of falling dominoes, we get a wave of energy traveling from one end of the axon to the other. That traveling wave is the nerve signal (fig. 8.11). Action potentials do not travel; nerve signals do.

Figure 8.11 Continuous Conduction of a Signal in an Unmyelinated Nerve Fiber. Action potentials are occurring in areas colored red; note the reversal of membrane polarity in those areas. Yellow areas are in a refractory period of reduced excitability, and green areas of membrane are fully excitable, ready to respond. APR

? What prevents a nerve signal from backing up and returning to the cell body?

For a short time after an action potential has occurred, that patch of membrane cannot be stimulated to produce another; it is said to be in a **refractory period.** Therefore, a zone of refractory membrane trails behind the nerve signal (the yellow zone in the figure). This prevents a nerve signal from backing up and returning to the neurosoma.

An unmyelinated nerve fiber has a high density of voltage-gated Na^+ and K^+ channels along its entire length, so everywhere an action potential occurs, it can excite another one immediately ahead of it until the signal arrives at the end of the fiber. This type of signal conduction, with action potentials occurring at every point along the axon, is called **continuous conduction.**

Things are different in a myelinated fiber (fig. 8.12). There are very few voltage-gated channels in the myelin-covered internodes—not enough to generate an action potential. No action potentials could occur in the internodes anyway, since myelin insulates the nerve fiber from the surrounding extracellular fluid (ECF) and no significant amount of Na^+ can get into the nerve fiber. At each node of Ranvier, however, the nerve fiber is exposed to the ECF, and here is where we find the greatest concentration of ion channels. This is the only place a myelinated fiber can generate action potentials.

Figure 8.12 Saltatory Conduction of a Signal in a Myelinated Nerve Fiber. (a) Sodium inflow at a node of Ranvier initiates a force that is quickly transferred from ion to ion along the internode to the next node, getting weaker with distance. At the next node, Na^+ ion (charge) density is just sufficient to open new voltage-gated channels and repeat the process. (b) Action potentials can occur only at nodes of Ranvier, so the nerve signal appears to jump from node to node. **APR**

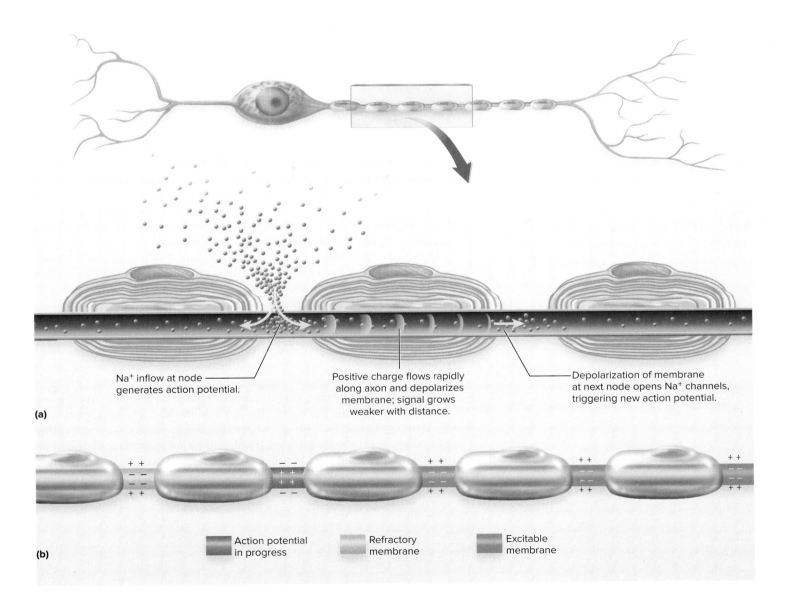

From one node to the next, the signal travels by a very rapid energy transfer from ion to ion, somewhat like the opposing north poles of two magnets pushing each other apart. Since action potentials occur only at the nodes, it appears as if the nerve signal were jumping from node to node; this mode of conduction is called **saltatory**[13] **conduction.**

Nerve signals travel as fast as 120 m/s in large myelinated fibers, compared to no more than 2 m/s in unmyelinated ones. This is because the energy transfer along the internodes of myelinated fibers is faster than the generation of action potentials at the nodes, and for most of the distance that it travels, a nerve signal passes through internodes. It slows down slightly at each node of Ranvier, but each node reamplifies the signal back to its original strength. Even if a nerve signal travels a meter or more from neurosoma to the end of the axon, it is just as strong at the end as it was at the beginning—much like the last domino in a line falling just as forcefully as the first one did.

8.2d Transmitting a Nerve Signal to the Next Cell

All good things must come to an end, and so it is with the wave of action potentials; they reach the end of the axon, where there is a narrow gap between it and the next cell. How is the neuron to communicate with the cell across the gap? In the early twentieth century, many biologists refused to believe such gaps existed. They thought a space would stop the signal, like a cut in an electrical wire.

But this is where the chemical part of nervous communication comes in. The point where the two cells meet is called a **synapse** (fig. 8.13). The nerve signal arrives at the synapse by way of the **presynaptic neuron,** and the desirable effect is to stimulate the next cell, the **postsynaptic neuron.** The gap between the two is called the **synaptic cleft.**

The axon terminal of the presynaptic neuron contains numerous spheroidal **synaptic vesicles** loaded with chemicals called **neurotransmitters.** The arrival of a nerve signal at the end of the presynaptic fiber triggers several synaptic vesicles to undergo exocytosis, releasing their neurotransmitter into the synaptic cleft. Neurotransmitter molecules diffuse across the cleft and bind to receptors on the membrane of the postsynaptic neuron.

One such neurotransmitter is **acetylcholine (ACh).** In the previous chapter (see "The Nerve–Muscle Relationship" in section 7.1c), we studied its binding to receptors on the surface of a muscle fiber. The receptors are ligand-gated channels that open in response to ACh and allow Na^+ and K^+ to flow through, producing a local potential in the postsynaptic cell. If local potentials in the postsynaptic cell are great enough to depolarize it to its threshold, that cell fires and forwards the signal to another cell farther down the line.

Neurotransmitters stimulate the postsynaptic cell for very brief periods of time. If this were not the case, the postsynaptic cell would continue firing and we would not have the ability to control the desired functions. The neurotransmitter in the cleft may be chemically broken down by enzymes in the cleft or reabsorbed by the presynaptic neuron, or it may simply diffuse away from the cleft into the surrounding tissue fluid. A failure to remove the neurotransmitter and stop synaptic transmission can result in consequences as severe as muscle spasms, paralysis, and suffocation (see Clinical Application 7.1, section 7.1c).

---

[13]from *saltare* = to leap or dance

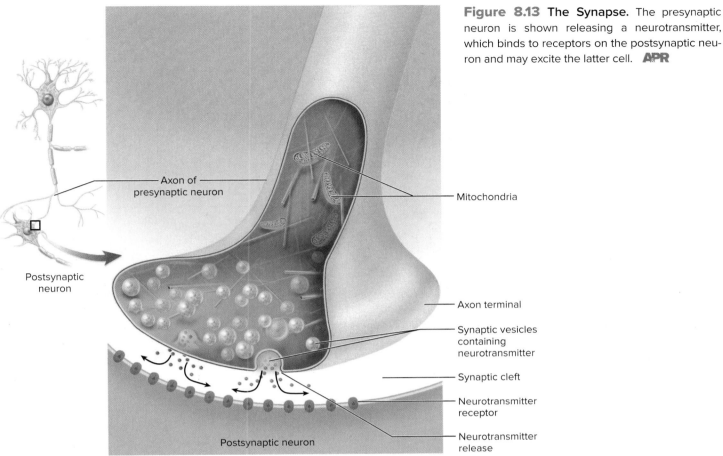

Figure 8.13 The Synapse. The presynaptic neuron is shown releasing a neurotransmitter, which binds to receptors on the postsynaptic neuron and may excite the latter cell. **APR**

Acetylcholine is but one neurotransmitter. More than a hundred others are known, including norepinephrine, serotonin, and dopamine, which play key roles in mood, alertness, muscular control, glandular secretion, and other functions. The importance of dopamine in motor control and Parkinson disease is discussed in Perspectives on Health at the end of this section.

Before You Go On

Answer these questions from memory to test your understanding of the previous section. Reread the material if you can't answer the questions.

7. What is the difference between a resting membrane potential and an action potential? Explain the role of sodium and potassium ions in each of these.
8. Explain how an unmyelinated nerve fiber conducts a signal from the beginning to the end of an axon. How does the process differ in a myelinated nerve fiber? Which one is faster?
9. Describe the structure of a synapse.
10. What is the role of a neurotransmitter in communication from one neuron to the next? Name some neurotransmitters.

PERSPECTIVES ON HEALTH

Alzheimer and Parkinson diseases are the two most common degenerative nervous system disorders. Both are associated with neurotransmitter deficiencies.

Alzheimer Disease

Alzheimer[14] disease (AD) may begin before the age of 50 with signs so slight and ambiguous that early diagnosis is difficult. A common early sign is memory loss, especially for recent events. A person with AD may ask the same questions repeatedly, show a reduced attention span, and become disoriented and lost in previously familiar places. Family members often feel helpless as they watch their loved one's personality gradually deteriorate beyond recognition. The AD patient may become moody, confused, or even paranoid, combative, and hallucinatory. Some lose the ability to read, write, talk, walk, and eat. Death ensues from complications of confinement and immobility, such as pneumonia.

AD affects about 11% of the U.S. population over the age of 65; the incidence rises to 47% by age 85. It accounts for nearly half of all nursing home admissions and is a leading cause of death among the elderly. AD claims about 100,000 lives per year in the United States.

Diagnosis can be confirmed by autopsy. There is atrophy of some of the gyri (folds) of the cerebral cortex and the hippocampus, an important center of memory. Nerve cells exhibit *neurofibrillary tangles*—dense masses of broken and twisted cytoskeleton (fig. 8.14). The number of tangles is proportional to the severity of the disease. In the intercellular spaces, there are *senile plaques* consisting of aggregations of cells, altered nerve fibers, and a core of *β-amyloid protein*—the breakdown product of a glycoprotein of plasma membranes. Oxidative stress (free radical injury) is now widely believed to underlie the formation of amyloid protein and neurofibrillary tangles, as well as all the other aspects of AD pathology. This has led to interest in whether dietary antioxidants or antioxidant therapy can reduce the incidence or progression of AD.

Intense biomedical research efforts are currently geared toward identifying the causes of AD and developing treatment strategies. Three genes on chromosomes 1, 14, and 21 have been implicated in various forms of early- and late-onset AD. Interestingly, persons with Down syndrome (trisomy 21), who have three copies of chromosome 21 instead of the usual two, tend to show early-onset

Figure 8.14 Alzheimer Disease (AD). (a) Brain of a person who died of AD. Note the shrunken folds of cerebral tissue (gyri) and wide gaps (sulci) between them, indicating atrophy of the brain. (b) Cerebral tissue from a person with AD. Neurofibrillary tangles are present within the neurons, and a senile plaque is evident in the extracellular matrix
a: Science Source b: Simon Fraser/Science Source

Alzheimer disease. But genes don't tell the whole story. Nongenetic (environmental) factors are also involved. Among its other benefits, consistent physical exercise decreases the risk of developing AD.

As for treatment, there is no cure, but researchers are developing drugs that focus on β-amyloid and a misfolded *tau protein* that underlies the neurofibrillary tangles. AD patients show deficiencies of acetylcholine (ACh) and nerve growth factor (NGF). Some patients show improvement when treated with NGF or cholinesterase inhibitors, but results so far have been modest.

---

[14]Alois Alzheimer (1864–1915), German neurologist

Parkinson Disease

Parkinson[15] disease (PD), or *parkinsonism,* is a progressive loss of motor function beginning in a person's 50s or 60s. It is due to degeneration of dopamine-releasing neurons in a brainstem region called the *substantia nigra.* A gene has been identified for a hereditary form of PD, but most cases are nonhereditary and of little-known cause.

Dopamine is an inhibitory neurotransmitter that normally prevents excessive activity in motor centers of the brain called the *basal nuclei.* Hyperactivity of the basal nuclei results in involuntary muscle contractions. Shaking of the hands (tremor) and compulsive "pill-rolling" motions of the thumb and fingers are common manifestations. In addition, the facial muscles may become rigid and produce a staring, expressionless face with a slightly open mouth. The patient's range of motion diminishes. He or she takes smaller steps and develops a slow, shuffling gait with a forward-bent posture and a tendency to fall forward. Speech becomes slurred and handwriting becomes cramped and eventually illegible. Tasks such as buttoning clothes and preparing food become increasingly laborious.

Patients cannot be expected to recover from PD, but its effects can be alleviated with drugs and physical therapy. Treatment with dopamine is ineffective, but drugs that enhance the action of the brain's self-produced dopamine are of some benefit. The dopamine precursor, levodopa (L-dopa), affords some relief from symptoms, but it does not slow progression of the disease and it has undesirable side effects on the liver and heart. It is effective for only 5 to 10 years of treatment. In severe cases that are unresponsive to medication, certain surgical treatments can help to quell the muscle tremors.

---

[15]James Parkinson (1755–1824), British physician

8.3 The Spinal Cord, Spinal Nerves, and Reflexes

Expected Learning Outcomes

When you have completed this section, you should be able to

a. describe the functions of the spinal cord;
b. name and describe the three protective membranes that envelop the spinal cord;
c. describe the basic anatomy of nerves and ganglia, and anatomy of the spinal nerves in particular;
d. state the universal characteristics of reflexes and distinguish between somatic and visceral reflexes;
e. identify the basic components of a somatic reflex arc; and
f. explain the mechanisms of stretch and withdrawal reflexes.

8.3a The Spinal Cord—Functional Anatomy

The **spinal cord** (fig. 8.15) is a cylinder of nervous tissue enclosed in the vertebral canal. Its functions include:

- **Conduction.** It contains bundles of nerve fibers that conduct information up and down the cord—sensory signals from the lower body up to the brain and motor signals down to the muscles and glands below the head.
- **Neural integration.** Groups of spinal neurons receive input from a variety of sources, integrate the information, and execute appropriate output. For example, the spinal cord can integrate the stretch sensation from a full bladder with cerebral input concerning the appropriate time and place to urinate and issue messages to the urinary sphincters to either allow or delay urination.
- **Locomotion.** The decision to walk is generated in the brain, but the simple, repetitive muscle contractions that put one foot in front of another, over and over, are coordinated by groups of spinal neurons called *central pattern generators.*

- **Reflexes.** Spinal reflexes play vital roles in posture, motor coordination, and protective responses to pain or injury.

The spinal cord extends for about 45 cm (18 in.) from the foramen magnum of the skull to the level of the first or second lumbar vertebra, just below the ribs (farther in infants and children). It is divided into **cervical, thoracic, lumbar,** and **sacral regions.** The last of these is named for the fact that its nerves emerge from the vertebral canal in the sacral region, not because the spinal cord itself extends that far. The cord averages about 1.8 cm wide (comparable to the little finger), but it widens slightly at the *cervical* and *lumbar enlargements* because of the greater mass of neurons here to control the complex movements of the upper and lower limbs.

At its lower end, the spinal cord tapers to a blunt point, the *medullary cone,* and gives off a bundle of nerves that occupy the vertebral canal from there through the sacrum. This bundle somewhat resembles a horse's tail and is therefore named the *cauda equina*[16] (CAW-duh ee-KWY-nuh).

---

[16]*cauda* = tail; *equin* = horse

Figure 8.15 The Spinal Cord, Posterior View. (a) Overview of spinal cord structure. (b) Detail of the spinal cord and associated structures. APR

Apply What You Know

Spinal cord injuries commonly result from fractures of vertebrae C5 to C6 (a "broken neck"), but never from fractures of L3 to L5. Explain these contrasting observations in light of the anatomy you have just read.

In cross section, the cord is elliptical in shape and exhibits a central butterfly- or H-shaped core of gray matter surrounded by white matter (fig. 8.16). A pair of median grooves almost divides the cord in half, but they stop short of a gray matter bridge *(gray commissure)* that connects the right and left halves of the cord. In the

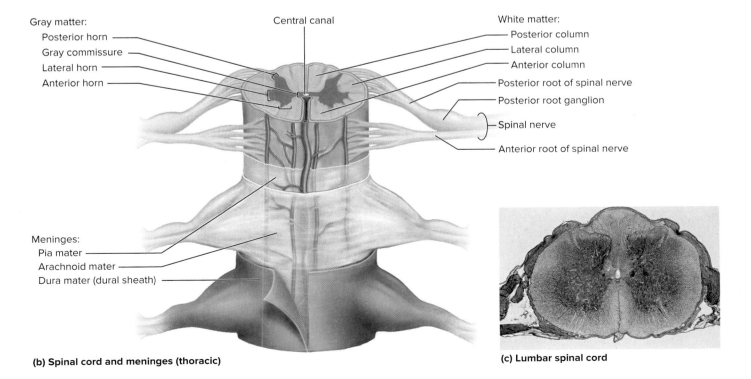

Figure 8.16 Cross-Sectional Anatomy of the Spinal Cord. (a) Relationship to the vertebra, meninges, and spinal nerve. (b) Detail of the spinal cord, meninges, and spinal nerves. (c) Cross section of the lumbar spinal cord with spinal nerves.

c: Ed Reschke/Getty Images

middle of the gray commissure, there is often a *central canal* filled with cerebrospinal fluid, but in most regions of the adult spinal cord, the canal is collapsed and closed.

The Spinal Gray Matter

In the gray matter, the two posterior "wings of the butterfly" are called the **posterior (dorsal) horns.** They extend to the surface of the cord and serve as receiving points for incoming sensory information. The two anterior wings are called the **anterior (ventral) horns.** They are wider than the posterior horns and extend only partially toward the spinal cord surface. They are the locations of the neurosomas of the motor neurons depicted in figure 8.3. The axons of these neurons extend to the cord surface and emerge there to become part of a spinal nerve. Between the incoming nerve fibers of the posterior horn and the neurosomas of the motor neurons in the anterior horn, there are usually many interneurons forming complex networks in the gray matter of the cord, including some whose axons cross through the commissure and connect the left and right sides of the cord. From the second thoracic to the first lumbar segment of the cord, there is also a lateral projection of gray matter called the **lateral horn,** the origin of neurons of the sympathetic nervous system described in section 9.5.

The Spinal White Matter

The white matter of the spinal cord consists of bundles of nerve fibers traveling up and down the cord, serving as communication pathways between different levels of the body. Some of these fibers originate from brain neurons, and others from neurons in the gray matter of the cord. These bundles of nerve fibers are arranged in three groups on each side of the cord—*posterior, lateral,* and *anterior columns.*

A functional knowledge of these columns is crucial in understanding spinal cord injuries, because often a patient loses specific sensory or motor abilities depending on which column, and which of its subdivisions, has been injured. Each column is divided into functionally distinct tracts of white matter. Some of them are **ascending tracts** that carry sensory information to higher levels of the cord and the brain. Others are **descending tracts** that carry motor signals from the brain down the cord to meet the motor neurons that control muscles and glands.

A peculiar and clinically important point about these spinal tracts is that most of them cross over from the left side of the cord or brainstem to the right, or vice versa, along the way. Therefore, sensory signals originating on the left side of the body are transmitted to the right side of the brain, and motor signals originating in the right side of the brain command muscles on the left side of the body, and vice versa. Such crossing-over is called **decussation.**[17] It explains why a person who has had a stroke that kills brain tissue on one side of the brain often experiences paralysis or sensory loss on the opposite side of the body.

---

[17]*decuss* = to cross, form an X

Clinical Application 8.2

SPINAL CORD INJURY AND PARALYSIS

In the United States, more than 17,000 people per year suffer spinal cord injuries, the most severe resulting in paralysis. Most such cases result from vertebral fractures incurred in automobile and motorcycle accidents, falls, sports, and gunshot and stab wounds. Males have a higher incidence of paralysis than females because of their high-risk behaviors; elderly people are at increased risk because of falls; and in wartime, many such injuries are incurred by combat soldiers and are thus a common issue in the medical care of veterans.

If the spinal cord is completely severed, one loses all sensation and motor control at and below the point of injury. A patient initially exhibits *spinal shock,* characterized by *flaccid paralysis*—muscle slackness and inability to contract, along with loss of bladder and bowel control, dangerously erratic changes in blood pressure, and other problems. As spinal shock subsides and reflexes are regained, the muscles commonly exhibit *spastic paralysis*—exaggerated contraction with an inability to relax owing to a lack of inhibitory signals from the brain. Spastic paralysis commonly progresses to a state of chronic flexion of the hips and knees (flexor spasms) to a state in which the limbs are extended and rigid (extensor spasms).

There are three forms of limb paralysis. *Paraplegia* is paralysis of both lower limbs resulting from spinal cord lesions at levels T1 to L1. *Quadriplegia* is paralysis of all four limbs resulting from lesions above level C5, such as a "broken neck." *Hemiplegia* is paralysis of either the right or left side of the body, usually resulting not from spinal cord injuries but from a stroke or other brain lesion. Because of the decussation (crossover) of motor nerve fibers in the brainstem, the hemiplegia occurs on the side of the body opposite from the side where the stroke occurred.

8.3b The Spinal Meninges—Protective Membranes APR

The brain and spinal cord are covered by three membranes that lie between the nervous tissue and bone (fig. 8.16b). They help to protect the delicate nervous tissue from abrasion and other trauma. Collectively, they are called **meninges** (men-IN-jeez) (singular, *meninx*). Individually, they are

1. The **dura mater**[18] (DOOR-uh MAH-tur), a tough collagenous membrane about as thick as a rubber kitchen glove. This is the outermost membrane. It lies against the inner surface of the bone in the cranial cavity, but in the vertebral canal it forms a loose-fitting *dural sheath* around the spinal cord and is separated from the bone by a fat-filled *epidural space.* Anesthetics are often introduced into this space to deaden the pain of childbirth or surgery *(epidural anesthesia).*

---

[18]*dura* = tough; *mater* = mother, womb

2. The **arachnoid**[19] **mater** (ah-RACK-noyd), a delicate middle layer named for an appearance suggestive of a spider's web. It consists of a thin epithelium adhering to the inside of the dura and a loose web of fibers extending inward to the pia.
3. The **pia**[20] **mater** (PEE-uh MAH-tur), the innermost layer, a transparent tissue of microscopic thinness that closely follows the surface contours of the brain and spinal cord.

8.3c Spinal Nerves—Communicating with the Rest of the Body

The peripheral nervous system is composed of two kinds of structures: nerves and ganglia. A **nerve** is a bundle of nerve fibers, blood vessels, and connective tissue wrappings. It resembles a string that frays into progressively finer branches the farther away from the CNS we look. Nerves have a pearly white color due mainly to their fibrous connective tissues.

Each nerve fiber in a nerve is wrapped in a thin layer of loose connective tissue called the **endoneurium**[21] (**fig. 8.17**). In most nerves, the fibers are gathered in bundles called **fascicles,**[22] each wrapped in a sheath of flat cells called the **perineurium.**[23] Several fascicles are then bundled together and wrapped in an outer, fibrous sleeve called the **epineurium**[24] to compose the nerve as a whole. The tough epineurium protects the nerve from stretching and injury. Nerves have a high metabolic rate and need a plentiful blood supply. These three layers of connective tissues ensure that blood vessels can reach every nerve fiber.

If we compare a nerve to a string, then a **ganglion**[25] is like a knot in the string—a swelling, usually near one end of the nerve, that contains the cell bodies of the peripheral neurons. In some ganglia, the neurons form synaptic contacts with each other, and the ganglion is therefore an information-processing center in the PNS.

The spinal cord gives off 31 pairs of **spinal nerves,** emerging with ladderlike regularity from the intervertebral foramina between the vertebrae. They provide a means for the cord, and therefore the whole CNS, to communicate with the rest of the body. In this section, we take a closer look at these nerves and their relationship to spinal cord anatomy.

Each spinal nerve originates from the cord as six to eight *nerve rootlets* emerging from the posterior horn of the cord and a similar number arising from the anterior horn (see figs. 8.15 and 8.16). The posterior rootlets converge to form a single **posterior (dorsal) root** of the nerve, which immediately expands into a swelling called the **posterior root ganglion;** it narrows again as it approaches the intervertebral foramen. The posterior root is composed of sensory (afferent) nerve fibers, carrying signals from sensory nerve endings to the spinal cord. The anterior rootlets also converge into a single **anterior (ventral) root,** but with no ganglion. The anterior root is composed of motor (efferent) nerve fibers, carrying signals leaving the spinal cord and destined for muscles and glands. The posterior and anterior roots unite into a single spinal nerve, which then passes through the intervertebral foramen.

---

[19]*arachn* = spider, spider web; *oid* = like
[20]*pia* = through mistranslation, now construed as tender, thin, or soft
[21]*endo* = within; *neuri* = nerve
[22]*fasc* = bundle; *icle* = little
[23]*peri* = around
[24]*epi* = upon
[25]*gangli* = knot

Figure 8.17 Anatomy of a Nerve. (a) A spinal nerve and its association with the spinal cord. (b) Cross section of a nerve (light micrograph). Individual nerve fibers show as tiny red dots, each surrounded by a light ring of myelin. **APR**
b: PASIEKA/Science Photo Library/Getty Images

Nerve roots are shortest in the cervical region and pass almost horizontally to the nearest intervertebral foramen. From infancy to adulthood, however, the spine grows faster and farther than the spinal cord does, and the nerve roots must grow and sweep downward to stay with their corresponding intervertebral foramen. Therefore, the roots become progressively longer as we descend the cord. The roots from spinal segments L2 to Co1 form the cauda equina and emerge from the vertebral canal below the lowest part of the spinal cord.

After emerging from the intervertebral foramen, each spinal nerve branches into a *posterior ramus*[26] that supplies the skin, muscles, and joints of the back; an *anterior ramus* that supplies the anterior and lateral skin and muscles as well as the limbs; and a small *meningeal branch* that reenters the vertebral canal to supply the meninges and vertebrae.

---

[26]*ramus* = branch

The anterior ramus differs from one region of the trunk to another. In the thoracic region, it forms an *intercostal nerve* that travels between the ribs, innervating the intercostal muscles, skin, and abdominal muscles. Everywhere else, anterior rami from multiple spinal nerves join each other to form a web of nerves called a **plexus.** There are five of these: a small **cervical plexus** deep in the neck; a large **brachial plexus** near the shoulder; a **lumbar plexus** in the lower back; below that, a **sacral plexus;** and finally a tiny **coccygeal plexus.** The nerves that emerge from these plexuses supply the skin, muscles, and other structures at their respective regions of the trunk and issue nerves to the limbs. Figure 8.18 shows a general view of these plexuses, and table 8.2 summarizes the structures innervated by each one.

Figure 8.18 The Spinal Nerve Roots and Plexuses, Posterior View. Boldface terms on the left identify the five nerve plexuses; boldface terms on the right identify the five regional nerve groups.

Table 8.2 Structures Innervated by the Spinal Nerve Plexuses APR

Plexus	Structures Innervated
Cervical plexus	**Sensory:** Skin of external ear and from surrounding region of head and neck, underside of chin, and shoulder; anterior chest, diaphragm, pleurae, and pericardium **Motor:** Diaphragm and a few muscles associated with larynx and hyoid bone
Brachial plexus	**Sensory:** Skin, muscles, and joints of shoulder and upper limb **Motor:** Muscles of upper limb
Lumbar plexus	**Sensory:** Skin of lower abdominal, gluteal, and genital regions; skin, muscles, and joints of lower limb **Motor:** Muscles of abdominal and pelvic regions and thigh
Sacral and coccygeal plexuses	**Sensory:** Skin of gluteal and perineal regions; genitals; skin, muscles, and joints of gluteal region and lower limb **Motor:** Muscles of gluteal and perineal regions and lower limb

8.3d Somatic Reflexes

A **reflex** may be defined as a quick, involuntary, stereotyped reaction of a gland or muscle to stimulation. This definition sums up four important properties: (1) Reflexes are *quick,* usually involving simple neural pathways with relatively few interneurons. (2) They are *involuntary,* meaning they occur without our intent, often without our awareness, and they are difficult or impossible to suppress. (3) They are *stereotyped,* meaning they occur in essentially the same way every time, unlike our more variable voluntary and learned behaviors. (4) They require *stimulation;* they are not spontaneous actions like muscle tics but responses to sensory input.

Some reflexes involve cardiac muscle, smooth muscle, or glands, such as acceleration of the heartbeat in fear, contraction of the esophagus in swallowing, or the secretion of tears in response to irritation of the cornea. These are called **visceral reflexes** and are discussed in section 9.5. Other reflexes involve skeletal muscles, such as pulling your hand back from a hot stove or the knee-jerk reflex so often evoked in a medical examination. These are called **somatic reflexes** and are the type we'll discuss here. They are sometimes called *spinal reflexes,* but this is a misleading expression because (1) many visceral reflexes also involve the spinal cord, and (2) many somatic reflexes are mediated more by the brain than by the spinal cord.

The Reflex Arc

Somatic reflexes usually involve a relatively simple neural pathway called a **reflex arc.** Figure 8.19 illustrates the essential components of an arc:

- A **receptor,** a sensory nerve ending or simple sense organ in the skin, a muscle, or a tendon.
- An **afferent (sensory) nerve fiber,** the axon of a unipolar neuron that carries signals from the receptor to the spinal cord or brainstem.
- An **integrating center,** a point of synaptic contact between neurons in the gray matter of the spinal cord or brainstem. In most reflex arcs, this involves one or more interneurons, but in the simplest arcs, the afferent fiber may synapse directly with the efferent fiber. These simple arcs allow for quicker but less complex or refined responses.

- An **efferent (motor) nerve fiber,** which carries motor signals to a skeletal muscle.
- An **effector,** which in this case is the skeletal muscle. In reflexes in general, *effector* means any organ or cell that carries out the final response.

The Stretch Reflex

The workings of a reflex arc can be exemplified most simply by considering the test so commonly performed in routine medical examinations, the *patellar tendon ("knee-jerk") reflex* (fig. 8.19). The doctor taps on the patellar ligament just below the kneecap with a rubber reflex hammer. This stretches the ligament and its continuation, the quadriceps tendon leading to the quadriceps femoris muscle of the thigh. At the junction between the muscle and tendon, and less abundantly elsewhere in the muscle, are specialized stretch receptors called **muscle spindles.** The sudden stretch produced by the hammer excites the spindle's nerve fibers, which conduct signals to the spinal cord.

In the cord, these nerve fibers synapse directly with motor neurons. Thus, such an arc is called a *monosynaptic reflex arc* because there is only one synapse between the sensory and motor neurons. Those neurons transmit signals immediately back to the muscle fibers of the quadriceps, which contract abruptly, pulling on the tibia and producing the twitch of the knee. By observing the strength of the response, a physician can get clues to the health of the nervous system as well

Figure 8.19 A Representative Reflex Arc. The monosynaptic reflex arc of the patellar tendon stretch reflex. **APR**

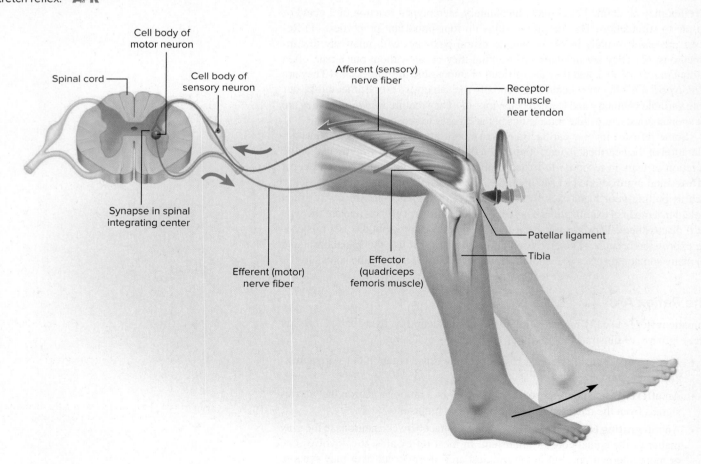

as to disorders such as hormone and electrolyte imbalances, diabetes, alcoholism, and some infectious diseases.

In order for the quadriceps to produce the knee jerk, the hamstrings must be simultaneously inhibited from contracting and fighting the quadriceps. This is achieved through a *polysynaptic reflex arc* in which input from the muscle spindle also inhibits the spinal motor neurons that would otherwise stimulate the hamstrings. It is called polysynaptic because at least one interneuron and multiple synapses intervene between the afferent and efferent neurons.

Some natural functions of stretch reflexes are to respond to joint movements and produce corrections in muscle tension that make our actions better coordinated, to maintain posture, and to prevent your head from tipping forward or your jaw from hanging open (unless you are nodding off in class). The vigorous and precisely coordinated movements of dance and athletics would be impossible without stretch reflexes.

The Flexor Reflex

Another well-known reflex is the **flexor (withdrawal) reflex,** which occurs, for example, when you pull back your hand upon burning your fingers, or your leg when you've stepped on something sharp (fig. 8.20). The protective function of this reflex requires more than the quick jerk and return to normal that we see in the stretch reflex; it requires the coordinated action of multiple muscles and a more sustained contraction so you don't put your hand or foot right back in harm's way. The flexor reflex also involves a polysynaptic reflex arc. Some signals follow pathways with relatively few synapses and thus head out to the muscles relatively quickly. Others follow pathways with more synapses and delays, and therefore follow a little behind the first. Consequently, the muscles receive a more prolonged output from the spinal cord and produce a more sustained contraction. In the case of a leg withdrawal as in the figure, note that the knee flexors (hamstrings) are excited, whereas the extensor (quadriceps) is inhibited.

There are other kinds of somatic reflexes that we will not explore in detail. For example: (1) A *crossed-extension reflex* causes you to stiffen one leg when you lift the other from the ground, so you don't fall over when withdrawing a limb from pain. (2) An *intersegmental reflex arc* travels up and down the spinal cord, from one segment to another, to produce reactions at a level different from the one where the stimulus is received; an example of such a reaction is tilting your torso to the left when you raise the right leg, in order to keep your balance. (3) A *tendon reflex* inhibits excessive muscle contractions to prevent the rupturing of tendons.

Figure 8.20 The Flexor Reflex. A pain stimulus triggers a flexor reflex, which results in contraction of flexor muscles of the injured limb and withdrawal of the limb from danger. The polysynaptic reflex arc produces a more sustained, protective response than the quick jerk of the stretch reflex in figure 8.19.

⊕ Excitatory effect
⊖ Inhibitory effect

① Stepping on glass stimulates pain receptors in foot.

② Pain signals travel to spinal cord via sensory nerve fibers and posterior root.

③ Sensory neuron excites multiple interneurons in spinal cord.

④ Motor nerve fiber stimulates hamstrings to contract, flexing knee and withdrawing foot.

⑤ Motor neuron to quadriceps is inhibited.

⑥ Quadriceps relaxes so it does not oppose knee flexion.

⑦ Signals cross to opposite side of spinal cord, extending and stiffening the left leg to maintain balance.

Before You Go On

Answer these questions from memory to test your understanding of the previous section. Reread the material if you can't answer the questions.

11. What are the four regions of the spinal cord? Which two of them have enlargements, and why?
12. At what level of the vertebral column does the adult spinal cord end? What exists in the vertebral canal below that point?
13. Describe the butterfly-shaped components of the spinal gray matter and the columns of white matter that surround the gray. How do the spinal gray matter and white matter differ in function?
14. From superficial to deep, name and describe the three spinal meninges.
15. What are the fascicles of a nerve? How do they relate to the endoneurium, perineurium, and epineurium?
16. Describe the roots and rami of a spinal nerve and how these relate to the vertebrae.
17. What are the five basic components of a reflex arc?
18. How do a stretch reflex and a withdrawal reflex differ in their synaptic connections and functions?

CAREER SPOTLIGHT

Occupational Therapist

David Grossman/Science Source

An occupational therapist (OT) is a health-care professional who assists people who are impaired in various ways to restore or sustain the skills needed for independent, productive, and satisfying lives. OTs work with a highly diverse range of clients—persons with birth defects, spinal cord injuries, hip replacements or artificial limbs, stroke, visual impairments, alcohol or drug dependence, mental retardation, emotional disturbances, and other conditions. The objectives of occupational therapy range from the most basic skills—dressing, grooming, bathing, cooking, eating—to logical reasoning, social poise, computer use, driving a car, strength conditioning, use of adaptive equipment such as wheelchairs, caring for a home, advancing one's education, or holding a job. Many occupational therapists specialize in certain types of disability or in certain age groups, ranging from premature infants to the elderly. OTs work in hospitals, outpatient clinics, rehabilitation centers, fitness centers, hospices, schools, and clients' homes.

To qualify for a career in occupational therapy, one must complete a graduate degree and a period of supervised fieldwork, and pass a national board exam. Many colleges and universities offer master's and doctoral programs in OT. Prerequisites for admission vary from school to school, but may require biology, anatomy and physiology, kinesiology, physics, psychology, sociology, and other courses, as well as volunteer or paid work experience with disabled persons. See appendix B for further career information.

Study Guide

Assess Your Learning Outcomes

To test your knowledge, discuss the following topics with a study partner or in writing, ideally from memory.

8.1 Cells and Tissues of the Nervous System

1. The three fundamental functions of the nervous system
2. Distinction between the central and peripheral nervous system, and the components of each
3. The organization of the nervous system, including the two main divisions of the PNS and their subdivisions
4. The structure of a generalized neuron, and the most common variations on that basic structure
5. Three functional categories of neurons—sensory neurons, interneurons, and motor neurons
6. Six kinds of neuroglia, whether each kind is located in the CNS or PNS, and the functions performed by each
7. Structure and function of the myelin sheath, and how it is produced in the CNS and PNS
8. The distinction between the gray and white matter of the CNS

8.2 The Physiology of Neurons

1. What is meant by saying a cell has a resting membrane potential and that its plasma membrane is electrically polarized
2. Distinctions between a resting membrane potential, a local potential, and an action potential
3. How the resting membrane potential is produced by the distribution of cations and anions on the two sides of the plasma membrane, and how the Na$^+$–K$^+$ pump maintains this potential
4. The mechanism by which stimulation of a neuron produces a local potential
5. The mechanism by which a local potential can lead to an action potential; the role of voltage-gated ion channels in producing action potentials
6. The phases of an action potential and the ion movements that account for each
7. The difference between an action potential and a nerve signal
8. The mechanisms of continuous and saltatory conduction and how they differ from each other
9. Why myelinated nerve fibers conduct signals faster than unmyelinated ones
10. The structure of a synapse where neurons meet
11. How a neuron transmits information across a synapse to the next cell
12. The definition of *neurotransmitter* and some examples

8.3 The Spinal Cord, Spinal Nerves, and Reflexes

1. Gross anatomy of the spinal cord
2. The relationship between the spinal cord and vertebral canal, and the contents of the adult vertebral canal below the termination of the spinal cord
3. The shape of the spinal gray matter and the terminology and functional significance of its horns
4. Three zones of white matter on each side of the spinal cord; how it is organized into columns and tracts; and the two fundamental types of spinal tracts
5. Differences between ascending and descending spinal cord tracts, and an example of each
6. The meaning of *decussation* and its significance in the relationship between CNS function and the sensory and motor functions of the body outside of the CNS
7. Names, composition, and structural relationships of the three meninges of the spinal cord
8. Names, composition, and structural relationships of the three connective tissues of a nerve, and how these relate to nerve fibers and fascicles
9. Where ganglia occur, what composes them, and how they relate structurally and functionally to nerves
10. Anatomy of a spinal nerve, including its two roots, its rootlets, the posterior root ganglion, its rami, and its spatial relationship to the spinal cord and vertebrae
11. The five plexuses of spinal nerves, where they are located, and the general destinations or origins of nerve fibers that lead to or exit from each plexus
12. General properties of a reflex and the distinctions between visceral and somatic reflexes
13. Fundamental components of a somatic reflex arc
14. The mechanism of a stretch reflex, the role of muscle spindles, and why it is advantageous for stretch reflexes to be monosynaptic
15. The mechanism of a flexor reflex and why it is advantageous for flexor reflexes to be polysynaptic
16. Some general purposes of crossed-extension, intersegmental, and tendon reflexes

Testing Your Recall

1. Myelinated nerve fibers produce action potentials only at their
   a. nodes of Ranvier.
   b. internodes.
   c. synapses.
   d. dendrites.
   e. neurosomas.

2. The posterior root ganglion next to the spinal cord contains the somas of
   a. efferent neurons.
   b. interneurons.
   c. unipolar neurons.
   d. multipolar neurons.
   e. motor neurons.

3. An oligodendrocyte serves a similar function in the CNS that a/an _____ serves in the PNS.
   a. Schwann cell
   b. microglia cell
   c. endoneurium
   d. ependymal cell
   e. satellite cell

4. An action potential begins with
   a. a flow of K^+ out of a neuron.
   b. a flow of Ca^{2+} into a neuron.
   c. the release of a neurotransmitter.
   d. a flow of Na^+ out of a neuron.
   e. a flow of Na^+ into a neuron.

5. A synaptic vesicle is a storage site for
   a. calcium ions.
   b. sodium ions.
   c. neurotransmitter receptors.
   d. neurotransmitter.
   e. acetylcholinesterase.

6. The spinal cord is divided into all of the following regions *except*
   a. sacral.
   b. coccygeal.
   c. lumbar.
   d. cervical.
   e. thoracic.

7. Neurons that carry sensory information to the spinal cord are called
   a. efferent neurons.
   b. effluent neurons.
   c. interneurons.
   d. affluent neurons.
   e. afferent neurons.

8. The nerve fibers that stimulate skeletal muscles below the neck arise from neurosomas in
   a. the posterior horn.
   b. the anterior horn.
   c. the lateral horn.
   d. the posterior root ganglion.
   e. the brain.

9. A bundle of nerve fibers within a nerve is called
   a. the gray matter of the nerve.
   b. an endoneurium.
   c. a fascicle.
   d. the white matter of the nerve.
   e. a perineurium.

10. Jerking your hand back when you are burned is an example of
    a. an alarm reflex.
    b. a stretch reflex.
    c. decussation.
    d. a visceral reflex.
    e. a monosynaptic reflex.

11. The outermost and toughest of the meninges is the _____.

12. The brain and spinal cord compose the _____ nervous system.

13. Stretch reflexes are activated when stretching of a muscle stimulates encapsulated receptors called _____ in the muscle tissue.

14. Muscles of the forearm are controlled by nerves arising from the _____ plexus.

15. In a myelinated peripheral nerve fiber, the gap between one Schwann cell and the next is called a/an _____.

16. The sensory neurons for smell have the _____ shape, meaning they possess only one dendrite and one axon.

17. Neuroglia called _____ cells line the internal cavities of the brain and spinal cord and secrete cerebrospinal fluid.

18. A nerve signal is a chain reaction of one _____ after another, analogous to a row of falling dominoes.

19. Neurons communicate with each other by means of chemical secretions called _____.

20. The posterior and anterior horns of the spinal cord are composed of a type of nervous tissue called _____.

Answers in Appendix A

What's Wrong with These Statements?

Briefly explain why each of the following statements is false, or reword it to make it true.

1. Motor signals travel the spinal cord by way of its ascending tracts.
2. The arachnoid mater lies between the pia mater and the nervous tissue of the spinal cord.
3. The outermost layer of a nerve is a fibrous covering called the perineurium.
4. Spinal nerves of the thoracic region form a weblike nerve plexus adjacent to the vertebral column.
5. All somatic reflex arcs pass through the spinal cord.
6. The nodes of Ranvier speed up signal conduction in myelinated axons.
7. A neuron at rest has a positive internal charge relative to the charge outside its plasma membrane.
8. *Neuron* is another name for *nerve fiber*.
9. The myelin sheath of brain and spinal cord neurons is produced by Schwann cells.
10. A bipolar neuron has two axons.

Answers in Appendix A

Testing Your Comprehension

1. The local anesthetics tetracaine and procaine (Novocain) prevent voltage-gated Na^+ channels from opening. Explain why this would block the conduction of pain signals in a sensory nerve.
2. Mr. Richards has had a stroke that damaged some of the tissue on the left side of his brain and paralyzed his right arm. He doesn't understand why his *right* arm is paralyzed when the neurologist told him the stroke was on the *left* side of his brain, and he feels sure his neurologist is wrong. If you were his nurse or physical therapist, how would you explain this to him in simple, lay terms?
3. In a spinal tap, a needle is inserted between two of the lower lumbar vertebrae into the vertebral canal in order to withdraw a sample of cerebrospinal fluid. From what you know of spinal cord anatomy, deduce why it is safer to perform the procedure in this region rather than higher on the vertebral column.

Chapter 9

The Nervous System II
THE BRAIN, CRANIAL NERVES, AND AUTONOMIC NERVOUS SYSTEM

White matter tracts in the brain visualized using diffusion tensor imaging.
© Pasieka/Science Source

Chapter Outline

9.1 Overview of the Brain
- 9.1a Major Landmarks
- 9.1b Meninges
- 9.1c Gray and White Matter
- 9.1d Ventricles and Cerebrospinal Fluid
- 9.1e Blood Supply and the Blood–Brain Barrier

9.2 Principal Divisions of the Brain
- 9.2a The Brainstem and Diencephalon
- 9.2b The Cerebellum
- 9.2c The Cerebrum

9.3 Multiregional Brain Functions
- 9.3a Sensation
- 9.3b Cognition
- 9.3c Voluntary Movement
- 9.3d Language
- 9.3e Cerebral Lateralization
- 9.3f The Electroencephalogram

9.4 The Cranial Nerves

9.5 The Autonomic Nervous System
- 9.5a Visceral Reflexes
- 9.5b Neural Pathways
- 9.5c Subdivisions of the ANS
- 9.5d Neurotransmitters and Receptors
- 9.5e Dual Innervation
- 9.5f Higher-Level Regulation of the ANS

Clinical Applications/Perspectives on Health
- Clinical Application 9.1: Meningitis
- Clinical Application 9.2: Aphasia
- Perspectives on Health

End of Chapter
- Aging of the Nervous System
- Career Spotlight: Electroneurodiagnostic Technologist
- Connective Issues
- Study Guide

BASE CAMP

Before ascending to the next level, be sure you're properly equipped with a knowledge of these concepts from earlier chapters.

- Anatomy of the cranium in relation to the brain and cranial nerves (see "The Skull" in section 6.3a)
- Structure and function of neurons (see section 8.1b)
- Neuroglia in section 8.1c
- Meninges of the CNS in section 8.3b
- Anatomy of nerves (see section 8.3d)

Anatomy & Physiology Revealed 4.0
Module 7
Nervous System

Perhaps the most fascinating scientific pursuit in the 21st century is understanding the brain. The human brain has some 80 billion neurons. The trillions of connections between these cells underlie sensory processing and perception, memory, emotions, and control of movement. The aspects of humanity that we hold most dear—creativity, compassion, intelligence, indeed the very curiosity that makes us strive to understand it—all stem from the brain itself. But how? Even after more than a hundred years of intense study, we are seeking to understand how cellular structures and processes give rise to the complexities of the human mind. Advances in molecular biology and the development of imaging techniques such as PET scans and MRIs that reveal the workings of the living brain have advanced our knowledge considerably in recent years. Still, there is much more to be discovered.

In this chapter, we will explore major anatomical features of the brain and some of its complex functions. We also will explore how the brain connects to the rest of the body through 12 pairs of cranial nerves, which arise from the floor of the brain and emerge from the skull; and we'll examine the sympathetic and parasympathetic nervous systems, which regulate many of our unconscious physiological processes such as heart rate and digestion.

9.1 Overview of the Brain

Expected Learning Outcomes

When you have completed this section, you should be able to

a. describe the major subdivisions and anatomical landmarks of the brain;
b. describe the locations of the gray and white matter of the brain;
c. name and describe the three meninges that enclose the brain;
d. describe the system of chambers and channels in the brain and the flow of cerebrospinal fluid through this system; and
e. describe the brain's blood supply and the selective barrier between the blood and brain tissue.

9.1a Major Landmarks

A broad overview of the brain's gross anatomy will provide important points of reference as we progress through a more detailed study. The largest, most conspicuous part of the brain is the **cerebrum** (seh-REE-brum or SER-eh-brum), composed of a pair of half-globes called the **cerebral hemispheres** (fig. 9.1). The cerebral surface is marked by thick folds called **gyri**[1] (JY-rye; singular, *gyrus*) separated by shallow

---

[1]*gyr* = turn, twist

282 PART 3 Internal Coordination

grooves called **sulci**[2] (SUL-sye; singular, *sulcus*). The left and right hemispheres are separated by a deep groove, the **longitudinal fissure.** At the bottom of this fissure, the hemispheres are connected by a prominent bundle of nerve fibers called the **corpus callosum**[3]—a conspicuous landmark in the bisected brain (fig. 9.2).

---

[2]*sulc* = furrow, groove
[3]*corpus* = body; *call* = thick

Figure 9.1 Surface Anatomy of the Brain.
(a) Superior view of the cerebral hemispheres. (b) Left lateral view. (c) The brain of a cadaver.
APR
c: © Rebecca Gray/McGraw-Hill Education

(a) Superior view

(b) Lateral view

(c) Lateral view

Figure 9.2 Medial Aspect of the Brain.
(a) Major anatomical landmarks of the medial surface. (b) Median section of the cadaver brain. **APR**
b: © Dennis Strete/McGraw-Hill Education

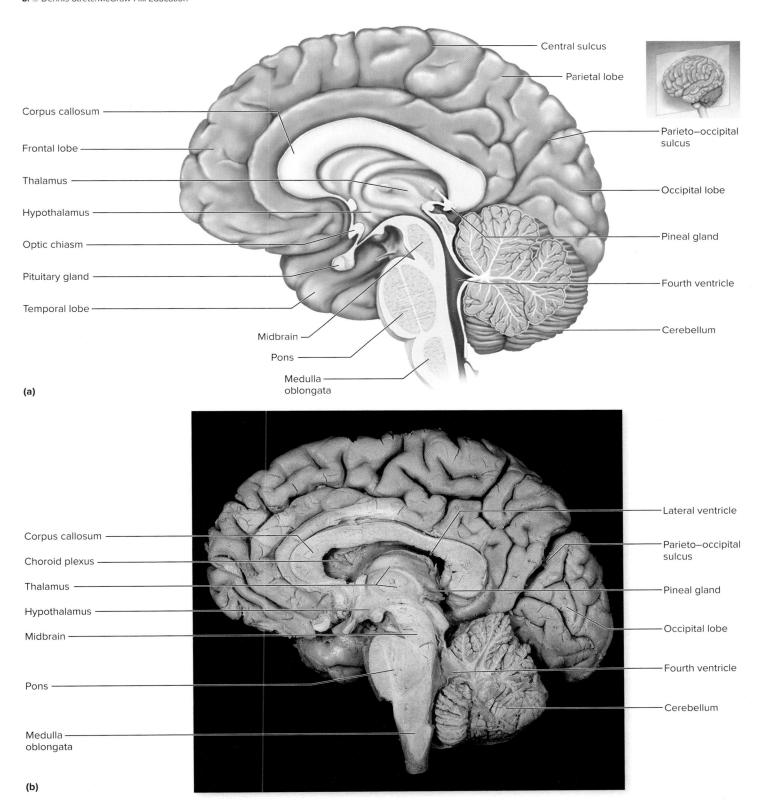

The **cerebellum**[4] (SER-eh-BEL-um) is the second-largest part of the brain. It lies at the rear of the head beneath the cerebrum. Like the cerebrum, it is composed of right and left hemispheres. It is marked by slender gyri, here called **folia,**[5] separated by shallow sulci.

The third major part of the brain is the **brainstem** (fig. 9.1b), the smallest of all and yet least dispensable to survival. Strokes and injuries here are more likely to be fatal than comparable injuries to the cerebrum or cerebellum. We can liken the brain to a mushroom, with the cerebrum being the cap and the brainstem being the stalk. At the foramen magnum of the skull, the brainstem connects to the spinal cord.

9.1b Meninges

Like the spinal cord, the brain is enveloped in three connective tissue membranes, the **meninges,** which lie between the nervous tissue and bone (fig. 9.3). The thickest and toughest of these is the **dura mater,** which lines the inside of the cranium and forms a fibrous sac around the brain. In the cranium, it consists of two layers that are separated in some places to form blood-filled spaces called **dural sinuses.** These collect blood that has circulated through the brain tissue and ultimately empty into the internal jugular veins of the neck. The dura mater folds inward to form a tough fibrous wall between the two cerebral hemispheres, another between the two cerebellar hemispheres, and a horizontal shelf between the cerebrum and cerebellum. These folds limit potentially injurious movements of the brain when the head is jarred. Severe blows to the head, however, can nevertheless cause *traumatic brain injury (TBI),* as in car and motorcycle accidents, football injuries, and falls.

Beneath the dura is an **arachnoid mater** similar to that of the spinal cord; the two membranes are separated in some places by a *subdural space.* Beneath the arachnoid mater, a *subarachnoid space* separates it from the innermost membrane, the **pia mater.** The pia mater closely follows all the contours of the brain surface, and is so thin and transparent that in most places it is visible only with a microscope. Delicate filaments extend from the arachnoid mater to the pia mater and suspend the brain in its bath of cerebrospinal fluid.

Clinical Application 9.1

MENINGITIS

Meningitis—inflammation of the meninges—is one of the most serious diseases of infancy and childhood. It is caused by a variety of bacterial and viral infections of the CNS. These usually invade by way of the nose and throat, often following respiratory, throat, or ear infections. Meningitis can cause swelling of the brain, cerebral hemorrhaging, and sometimes death within mere hours of onset. Signs and symptoms include high fever, stiff neck, drowsiness, intense headache, and vomiting. Death can occur so suddenly that infants and children with a high fever should receive immediate medical attention. Freshman college students show a slightly elevated incidence of meningitis, especially those living in crowded dormitories rather than off campus.

---

[4]*cereb* = brain; *ellum* = little
[5]*foli* = leaf

Figure 9.3 The Meninges of the Brain. Frontal section of the head. **APR**

❓ *In what ways does the dura mater here differ from the dura mater enclosing the spinal cord?*

9.1c Gray and White Matter

The tissues of the brain are divided into gray and white matter, like those of the spinal cord. In the brain, however, most of the gray matter is on the surface, where it forms a layer called the **cortex** of the cerebrum and cerebellum (fig. 9.4c). Smaller masses of gray matter called **nuclei** occur more deeply, surrounded by white matter. The gray matter contains the cell bodies, dendrites, and synapses of the neurons, so it is the site of all information processing, memory, thought, and decision making by the brain. The white matter is composed of **tracts** (bundles) of myelinated nerve fibers connecting one region of the CNS to another. The aforementioned corpus callosum is the largest tract. Tracts get their glistening white color from the myelin.

9.1d Ventricles and Cerebrospinal Fluid APR

Sections through the brain reveal four internal chambers called **ventricles** (fig. 9.4)—two *lateral ventricles,* one in each cerebral hemisphere; a single *third ventricle* in a narrow space between the hemispheres; and a *fourth ventricle* between the cerebellum and brainstem. A small pore connects each lateral ventricle to the third, and a canal through the brainstem connects the third ventricle to the fourth. The system then continues into the spinal cord as the central canal. These chambers and canals are filled with **cerebrospinal fluid (CSF),** which also fills the subarachnoid space and bathes the external surface of the brain and spinal cord.

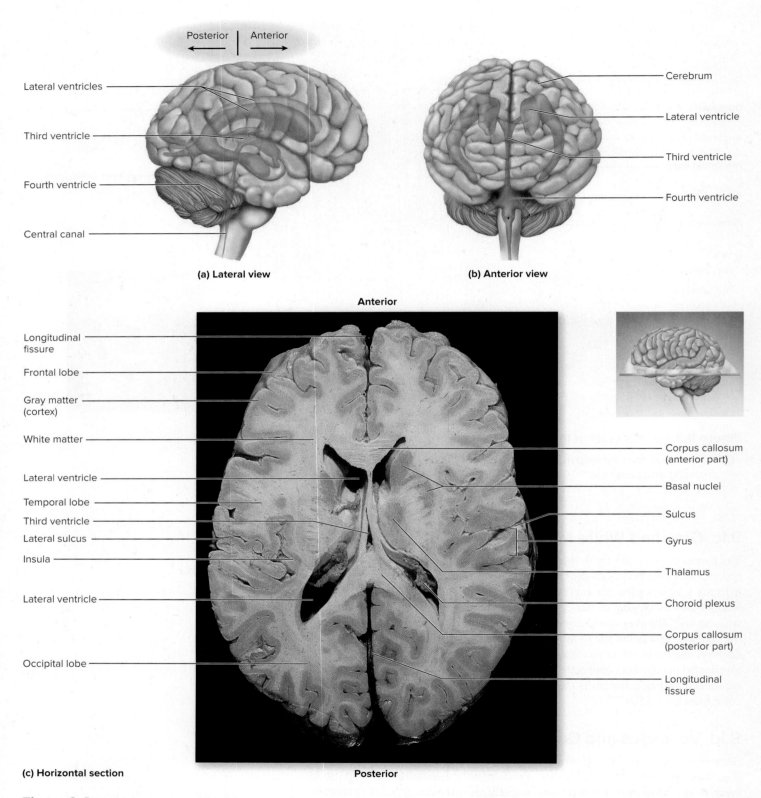

Figure 9.4 Ventricles of the Brain. (a) Right lateral view. (b) Anterior view. (c) Superior view of a horizontal section of the cadaver brain, showing the lateral ventricles and some other features of the cerebrum. Note the thin layer of gray matter (cerebral cortex) overlying the cerebral white matter. APR

c: © Rebecca Gray/Don Kincaid/McGraw-Hill Education

CSF is produced by the filtration of blood plasma through the brain surface into the subarachnoid space; by the layer of *ependymal cells* that lines the ventricles; and by a spongy mass of blood capillaries, the **choroid plexus** (CORE-oyd), in each of the four ventricles. The CSF produced in the ventricles circulates from the lateral ventricles to the third, and down the brainstem canal to the fourth ventricle and spinal cord. The fourth ventricle has three pores through which CSF exits the internal spaces and enters the subarachnoid space. Here, it bathes the outer surfaces of the brain and spinal cord. It is eventually reabsorbed into the bloodstream through outgrowths of the arachnoid mater called **arachnoid granulations.**

The CSF provides a carefully regulated chemical environment for the brain and continually rinses metabolic wastes from the brain and spinal cord. It also provides buoyancy for the brain so it does not rest heavily in the floor of the cranium, and cushions the brain to protect it from minor jolts or blows to the head.

9.1e Blood Supply and the Blood–Brain Barrier

The brain is only 2% of the body weight, yet it has such a high metabolic rate that it demands 15% of the blood supply. A mere 4-minute cessation of cerebral blood flow can cause irreversible brain damage. It receives its supply through two **internal carotid arteries** that begin near the angle of the mandible and enter the cranial cavity through the *carotid canals* in the base of the skull, and two **vertebral arteries** that pass up the back of the neck through the transverse foramina of the cervical vertebrae and enter the cranial cavity through the foramen magnum. The cerebral arteries and veins are further described at "The Systemic Circuit" in chapter 13.

Despite its critical importance to the energy-hungry brain, the blood is also a source of antibodies, toxins, and other agents that are potentially harmful to the irreplaceable brain tissue. Consequently, there is a **blood–brain barrier (BBB)** that strictly regulates what substances can get from the bloodstream into the tissue fluid of the brain. The BBB consists mainly of tight junctions between the epithelial cells of the cerebral blood capillaries, and is maintained by astrocytes (see section 8.1c). The BBB is freely permeable to oxygen and glucose (and some drugs such as alcohol and nicotine), but it prevents many other substances from passing through. Unfortunately, it makes it very difficult to deliver drugs for the treatment of brain diseases. For example, the BBB is an obstacle to treating brain cancer with chemotherapy.

Before You Go On

Answer these questions from memory to test your understanding of the previous section. Reread the material if you can't answer the questions.

1. What are the three principal subdivisions of the brain? Which ones are divided into hemispheres and have gyri and sulci? What do the terms gyri and sulci mean?
2. Where are the gray matter and white matter of the brain located?
3. Describe the source of CSF, the route that it takes through and around the CNS, and its functions.
4. What is the blood–brain barrier, and why is it important?

9.2 Principal Divisions of the Brain

Expected Learning Outcomes

When you have completed this section, you should be able to

a. list the parts of the brainstem and describe their structure and function;
b. name the three components of the diencephalon and describe their locations and functions;
c. describe the structure and function of the cerebellum;
d. describe the locations of the gray matter and tracts of white matter in the cerebrum;
e. list the five lobes of the cerebrum, identify their anatomical boundaries, and state their functions; and
f. describe the limbic system and identify its major functions.

In our regional survey of the brain, we will start at the rear and work forward, from brainstem to cerebrum. This has the advantage of beginning with regions that are relatively simple, anatomically and functionally, and working our way up to the most complex and mysterious part of the brain, the cerebrum—the seat of such complex functions as thought, memory, and emotion. **APR**

9.2a The Brainstem and Diencephalon

The brainstem (fig. 9.5) is an elongated pillar extending vertically from the foramen magnum to the diencephalon, part of the embryonic forebrain that sits beneath the cerebrum. From inferior to superior, the brainstem consists of the medulla oblongata, pons, and midbrain. The brainstem serves in part as the "information highway" between the cerebrum and the lower body; it contains prominent tracts of nerve fibers carrying signals coming up from the spinal cord and down from the cerebrum.

Embedded among these tracts are at least 100 pairs of nuclei—small masses of gray matter that control fundamental physiological functions of the body. Many of these nuclei form a loosely organized web of gray matter called the **reticular formation** with functions as varied as regulation of sleep and consciousness, cardiovascular and respiratory control, and pain modulation. Our survey of the brainstem will proceed from bottom to top, and will emphasize functions of nuclei associated with the reticular formation. We will then discuss the parts and functions of the diencephalon.

The Medulla Oblongata

The **medulla oblongata**[6] looks superficially like a slightly wider continuation of the spinal cord. It contains fiber tracts that carry sensory information from the spinal cord to higher levels of the brain, and motor commands from higher brain centers to the spinal cord. The motor signals travel predominantly through a pair of anterior ridges called **pyramids,** shaped somewhat like two parallel baseball bats. When you consciously will your muscles to contract, the motor signals pass through here. Ridges on the posterior surface of the medulla carry sensory signals destined for the cerebrum.

Amid these tracts of white matter are gray matter nuclei concerned with controlling the heart rate and blood pressure; the respiratory rhythm, coughing, and sneezing; the digestive tract functions of swallowing, salivation, gastrointestinal secretion, gagging, and vomiting; and roles in speech, sweating, and tongue and head movements.

---

[6]*medulla* = pith, inner core; *oblongata* = elongated

Regions of the brainstem
- Diencephalon
- Midbrain
- Pons
- Medulla oblongata

Figure 9.5 The Brainstem and Diencephalon. (a) Anterior view. (b) Posterolateral view. The cerebellum is cut off to open the fourth ventricle.

The Pons

The **pons**[7] (pronounced "ponz") is a distinctly wider segment of the brainstem superior to the medulla. It is the most significant relay center for signals going to the cerebellum. It also contains nuclei concerned with sleep, hearing, equilibrium, taste, eye movements, facial expressions, facial sensation, respiration, swallowing, bladder control, and posture.

The Midbrain

The **midbrain** is a relatively short bridge between the pons and higher brainstem centers. It has a pair of stalks called *cerebral peduncles*[8] (peh-DUN-culs) and another pair called *cerebellar peduncles* that anchor the cerebral hemispheres and cerebellum to the brainstem; these are major communication pathways between these brain regions. The midbrain filters pain stimuli and determines whether we become consciously aware of pain and how intensely we feel it. It also has a darkly pigmented tissue called the *substantia nigra*[9] (sub-STAN-she-uh NY-gruh), a dark gray to black nucleus pigmented with melanin. The substantia nigra produces the neurotransmitter *dopamine,* which acts on the *basal nuclei* of the cerebrum (described later) to inhibit unwanted muscular movements. Degeneration of the substantia nigra results in the tremors and other motor deficiencies of Parkinson disease (see Perspectives on Health in section 8.2). The posterior surface of the midbrain has four conspicuous bumps called *colliculi*[10] concerned with visual and auditory reflexes and other functions related to vision and hearing.

The Diencephalon

The **diencephalon** (DI-en-SEF-a-lon) consists of the *thalamus,* the funnel-shaped *hypothalamus,* and the *epithalamus* (fig. 9.5).

Each side of the brain has a large, ovoid **thalamus**[11] perched at the superior end of the brainstem (the large blue bodies in fig. 9.5). In about 70% of people, the right and left thalami are joined medially by a narrow bridge of tissue. The thalamus plays various roles in memory, emotion, motor control, and sensation. Its main function is to process incoming sensory information and relay it to the appropriate places in the cerebrum. For example, it receives visual signals from the optic nerve, processes them, and relays information to the posterior region of the cerebral cortex, where we first become aware of what we see. Similarly, the thalamus processes and relays signals for taste, hearing, equilibrium, vision, touch, pain, pressure, heat, and cold to appropriate processing centers in the cerebrum. Thus, the thalamus is often nicknamed "the gateway to the cerebral cortex."

A single **hypothalamus,**[12] lies medially beneath the thalami, as its name implies. It is shaped somewhat like a flattened funnel, forming the floor and the lateral walls of the third ventricle (see fig. 9.2). The hypothalamus is a major control center for many instinctive and automatic functions of the body including water balance, blood pressure, hunger and satiety (feeling full), thirst, digestive functions, metabolism, stress responses, shivering and sweating, sleep, memory, growth, emotion, reproductive cycles, sexual behavior, childbirth, and lactation. The pituitary gland attaches by a stalk to the base of the hypothalamus, and the hypothalamus carries out several of its functions by regulating the secretion

---

[7] *pons* = bridge
[8] *ped* = foot; *uncle* = little
[9] *substantia* = substance; *nigra* = black
[10] *coll* = hill; *cul* = little
[11] *thalamus* = chamber, inner room
[12] *hypo* = below

of pituitary hormones (see section 11.3). The hypothalamus may be dwarfed in size by other brain structures, but it is mighty in terms of function, playing an essential role in the homeostatic regulation of nearly all organs of the body.

The **epithalamus** is a small median sliver of tissue that arches over the midline between the thalami. Its most prominent component is the **pineal gland** (see fig. 9.2a), an endocrine gland that synthesizes melatonin (see section 11.4a).

9.2b The Cerebellum

The cerebellum is the second-largest part of the brain (fig. 9.6). Even though it is only 10% of the brain mass, it contains over 50% of its neurons. As noted earlier, the cerebellum consists of right and left hemispheres connected by a narrow bridge, and is marked by slender folia and sulci. The cerebellum has a surface cortex of gray matter and a deeper layer of white matter. In a sagittal section, the white matter exhibits a branching, fernlike pattern called the *arbor vitae*[13] (AR-bur VEE-tee).

Until recently, the function of the cerebellum was thought to be largely limited to motor coordination, as described later in this chapter. However, with modern brain-imaging techniques such as PET scans and functional MRI (fMRI), it has become clear that this is only one part of a much broader cerebellar role. Its more general function seems to be the evaluation of various forms of sensory input. It is highly engaged in fingertip sensation; strategizing movements (such as knowing just when and where to reach to catch a flying baseball); judging the three-dimensional shape of objects (spatial perception); judging the passage of time; and predicting how much the eyes must move to compensate for head movements and remain visually fixed on an object. Cerebellar lesions impair a person's ability to distinguish differences in pitch between two sounds and to distinguish between similar-sounding words such as *rabbit* and *rapid*. Some people with cerebellar lesions have serious motor deficits. Others have difficulty planning and scheduling tasks or controlling their impulses and emotions. Many children with attention-deficit/hyperactivity disorder (ADHD) have abnormally small cerebellums.

Figure 9.6 The Cerebellum. (a) Median section, showing relationship to the brainstem. (b) Superior view. APR

Apply What You Know
Discuss the multifaceted importance of the cerebellum to a tennis player, a concert pianist, and a student riding a bicycle to class in city traffic.

---

[13]*arbor* = tree; *vitae* = of life

9.2c The Cerebrum

The cerebrum is the largest and most conspicuous part of the brain. It enables you to turn pages in a book (or scroll down a computer screen), read and comprehend the words, remember ideas, talk about them, and take an examination. It is the seat of your sensory perception, memory, thought, judgment, and voluntary movement, among other functions.

Cerebral Tissues

As already described, the cerebrum consists of two half-globes, the *cerebral hemispheres,* separated by the deep *longitudinal fissure* (see fig. 9.1a). The surface of each hemisphere is marked by conspicuous gyri and sulci. Its surface tissue is a layer of gray matter called the **cerebral cortex,**[14] which contains the cell bodies of the cerebral neurons, their dendrites, and their synapses. Even though it is only 2 to 3 mm thick, the cortex constitutes about 40% of the mass of the brain and contains 14 to 16 billion neurons. Elaborate folding allows the cerebrum to contain three times as much cortex as we would have if it had a smooth surface. Animals with less behavioral sophistication, variability, and adaptability have relatively smooth-surfaced cerebrums. Only whales, dolphins, and apes rival humans in the extent of their cerebral convolutions.

The deeper white matter (see fig. 9.4c) of the cerebrum is composed of neuroglia and myelinated nerve fibers (axons) leading to and from the cortex. These fibers are bundled into tracts that connect regions within a hemisphere, allow communications between the cerebrum and lower brain regions and the spinal cord, and connect the right and left hemispheres through the thick C-shaped corpus callosum. It is chiefly through the corpus callosum that the left and right hemispheres coordinate their activities with each other.

Lobes of the Cerebrum

Some cerebral gyri vary from one person's brain to another and between the right and left hemispheres of the same person. Others, however, are quite consistent and serve as landmarks that universally define the functional *lobes* of any human cerebrum. There are five lobes in each hemisphere, mostly named for the overlying cranial bones (fig. 9.7):

1. The **frontal lobe** begins immediately behind the frontal bone, superior to the eyes. Its posterior boundary is a wavy vertical groove, the *central sulcus,* that extends from the top of the head to the margin of the lateral sulcus, which is described shortly. A wide region of the frontal lobe just anterior to the central sulcus is concerned with voluntary motor functions. Farther anterior to this, between the motor region and the forehead, is the **prefrontal cortex.** This is little developed in other mammals but very prominent in humans, and is concerned with *explicit memory* (memory of facts that you can verbalize), mood, emotion, motivation, and abstract thought such as foresight, planning, problem solving, and social judgment. On the underside of each frontal lobe is a club-shaped *olfactory bulb,* which receives nerve fibers from the nose for the sense of smell and relays these signals to more posterior processing centers.
2. The **parietal lobe** forms the uppermost part of the brain and underlies the parietal bone. Its anterior boundary is the central sulcus and its posterior boundary is the *parieto–occipital sulcus,* visible on the medial surface of each hemisphere (see fig. 9.2). This lobe is concerned with the sensory perception and integration of taste; processing visual information; spatial

---

[14]*cortex* = bark, rind

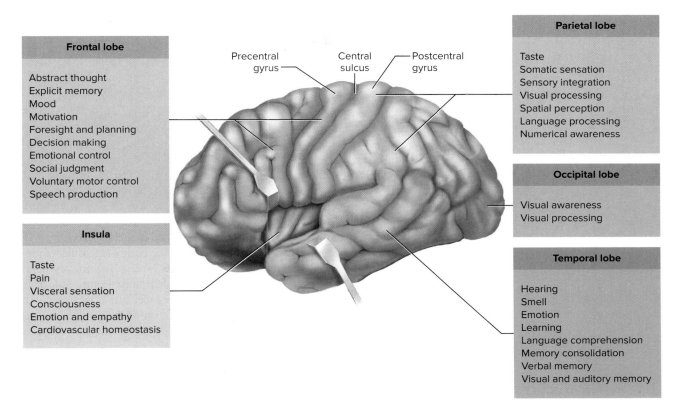

Figure 9.7 Lobes of the Cerebrum. The frontal and temporal lobes are retracted slightly to reveal the insula. APR

perception; language processing; numerical awareness (such as how many objects one is seeing); and *somatic* (body-wall and musculoskeletal) sensations such as touch, stretch, and pain.

3. The **occipital lobe** is at the rear of the head underlying the occipital bone. It is the principal visual center of the brain.
4. The **temporal lobe** is a lateral, horizontal lobe deep to the temporal bone, separated from the parietal lobe above it by a deep *lateral sulcus*. It is concerned with hearing, smell, emotional behavior, learning, various aspects of memory, visual recognition, and language comprehension.
5. The **insula**[15] is a small mass of cortex deep to the lateral sulcus, made visible only by retracting or cutting away some of the overlying cerebrum (see figs. 9.1c, 9.4c, and 9.7). It plays roles in taste, pain, visceral sensations (such as a stomachache), consciousness, emotion, empathy, and cardiovascular homeostasis.

The Basal Nuclei

The **basal nuclei** (sometimes called the **basal ganglia**) are masses of cerebral gray matter buried deep in the white matter, lateral to the thalamus (see fig. 9.4). They receive input from the substantia nigra of the midbrain and motor areas of the cerebral cortex and send signals back to both of these locations. They are part of a feedback loop that serves in motor control and are further discussed in a later section on that topic. The classical view of the function of the basal nuclei stemmed from

---

[15]*insula* = island

20th-century observations that injury to these areas resulted in movement disorders. Recently, it has been recognized that the basal nuclei have a range of functions, including emotion, learning, and motor skills.

The Limbic System

The **limbic**[16] **system** is a group of brain structures concerned primarily with learning, memory, and emotion. It forms a ring of tissue on the medial side of each hemisphere, encircling the corpus callosum and thalamus (fig. 9.8). It includes connections to other parts of the brain, including portions of the frontal and temporal lobes, hypothalamus, thalamus, and the olfactory apparatus. The last of these may explain why odors are associated with powerful memories and emotional responses. Two particularly important components of the limbic system are the **amygdala**[17] (ah-MIG-da-luh) and **hippocampus**[18]—both of which are parts of the temporal lobe. The hippocampus is central to memory, as described in secion 9.3b.

The amygdala receives sensory information from the cortex regarding vision, hearing, taste, and smell, as well as general senses such as pain. The sensory input enables it to mediate emotional responses to stimuli that range from a foul odor to pleasant music to a stomachache. The amygdala is especially associated with feelings of fear and anxiety, but also plays a part in motivation for many behaviors including food intake and sexual behavior, and is important in drawing our attention to novel stimuli.

The limbic system was originally thought to be associated mainly with smell, but later experiments revealed more significant roles in emotion and memory. It is involved in feelings that reinforce behaviors important to survival, such as the sense of reward associated with eating and sex, fear upon seeing a snake, and aversion to a food that has made one sick in the past. Most limbic structures have centers for both reward and aversion. Stimulation of a reward center produces a sense of pleasure; stimulation of an aversion center produces feelings such as sorrow, fear, or avoidance. Aversion centers dominate some limbic structures, such as the amygdala, whereas reward centers dominate others.

---

[16]*limb* = border; *ic* = pertaining to
[17]*amygdala* = almond
[18]*hippocampus* = sea horse, named for its shape

Figure 9.8 The Limbic System. This ring of structures (shown in violet) includes important centers of learning and emotion. In the frontal lobe, there is no sharp anterior boundary to limbic system components and there is no universal agreement among anatomists on how much to include in the limbic system. **APR**

❓ *What component of this system is especially associated with the sense of fear?*

Before You Go On

Answer these questions from memory to test your understanding of the previous section. Reread the material if you can't answer the questions.

5. List the four regions of the brainstem from inferior to superior, and list several functions of each.
6. Where are the thalami located, and what is their primary function?
7. Describe the location and functions of the hypothalamus.
8. State several roles of the cerebellum in addition to motor coordination.
9. List the five lobes of the cerebrum and some functions of each one.
10. Emotion is a very complex human function. Name five brain structures described in this section that play various roles in emotion.
11. State a location in the brain for each of the following functions: regulation of heart rate; formation of memories; visual awareness; blocking of pain signals; hearing; perception of touch; and coordination of movements.

9.3 Multiregional Brain Functions

Expected Learning Outcomes

When you have completed this section, you should be able to

a. identify the destinations of different types of sensory signals going to the brain;
b. identify some of the conscious, thinking (cognitive) areas of the brain;
c. identify areas of the brain involved in the creation of memories and in memory storage and retrieval;
d. describe how the brain controls the skeletal muscles;
e. describe the locations and functions of the language centers of the brain;
f. discuss the functional relationship between the right and left cerebral hemispheres; and
g. describe how electroencephalograms are recorded and why they are clinically useful.

Such brain functions as sensation, memory, language, emotion, and motor control are best considered as *multiregional* brain functions because they are not localized to one specific part of the brain, but involve many. There is no simple one-to-one relationship between these functions and the anatomical parts described up to this point. They are associated especially with the cerebral cortex, but not exclusively; they also involve such areas as the cerebellum, basal nuclei, and others. Some of these functions present the most difficult challenges for neurobiology, but they are also the most intriguing.

9.3a Sensation

The human senses fall into two categories: (1) The **special senses** employ sense organs limited to the head and include vision, hearing, equilibrium, taste, and smell. (2) The **somatosensory**[19] or **general senses** are distributed over the entire body and include touch, itch, pain, heat, cold, pressure, stretch, and movement. The regions of the brain that receive input for the special senses are (**fig. 9.9a**)

- vision—the posterior occipital lobe;
- hearing—the superior temporal lobe;
- equilibrium—mainly the cerebellum;
- taste—parts of the insula and lower parietal lobe; and
- smell—the medial surface of the temporal lobe and inferior surface of the frontal lobe.

These receiving areas are called **primary sensory areas.** They merely make one aware of a stimulus, but identifying and making sense of it are carried out by adjacent **sensory association areas** of the cerebrum. For example, a *visual association area* immediately anterior to the primary visual area enables one to identify what one sees and to relate effectively to one's visual world.

Signals for the general senses arrive in the **postcentral gyrus,** a vertical fold of the parietal lobe immediately posterior to the central sulcus. Its cortex is called the **primary somatosensory cortex.** Signals coming from the spinal cord decussate (cross to the opposite side) along the way (see "The Spinal White Matter" in section 8.3a), so those arising on the right side of the body end in the left postcentral gyrus, and vice versa. Each half of the cerebrum therefore senses the opposite side of the body. Furthermore, each postcentral gyrus is like an upside-down sensory map of the body (**fig. 9.9b**). This map, often called the *sensory homunculus,*[20] has an oddly distorted look because it shows body regions disproportionately large or small according to the relative amounts of cerebral cortex dedicated to them. As you can see in the figure, much more cortex is concerned with sensations from the face and hands, for example, than that concerned with the entire trunk of the body.

9.3b Cognition

Cognition[21] refers to mental processes such as awareness, perception, thinking, knowledge, and memory. It is an integration of sensory information and motor output. Cognitive association areas of the cerebrum account for 75% of our brain tissue. They include the sensory association areas just described, but there are other, diverse cognitive functions in all five lobes of the cerebrum and even in the cerebellum. Lesions (injuries) in these areas can produce a broad spectrum of effects such as inability to recognize familiar faces or to navigate within a familiar building; a variety of language deficits (aphasias) described later; or profound personality disorders and socially inappropriate behaviors.

Memory is one of the cognitive abilities we value most highly. It is a widely distributed function, not localized to any one area of the brain. The creation of new memories from ongoing sensory experience, called *memory consolidation,* is a function of the hippocampus, amygdala, and cerebellum, but these are not memory-storage depots. Different forms of memory are stored in very different regions of the cerebrum—for example, the memory of motor skills such as tying one's shoes or driving a car lies in the basal nuclei; a memory for faces and

---

[19]*somato* = body

[20]*hom* = man; *uncul* = little

[21]*cogni* = to know; *tion* = process

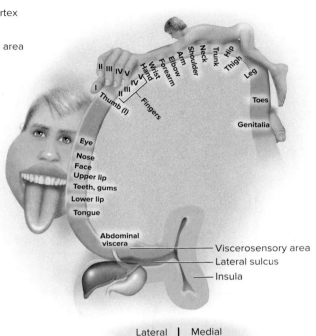

Figure 9.9 Sensory Areas of the Cerebrum.
(a) Primary sensory areas and association areas. (b) Sensory map of the postcentral gyrus. The human figure is distorted to represent the fact that larger areas of cortex are concerned with facial sensation, for example, than those concerned with the thigh and leg. Notice that higher regions of the gyrus, in general, are associated with lower regions of the body. **APR**

❓ How do you predict a person would be affected by a stroke that destroyed much of the postcentral gyrus within the longitudinal fissure?

voices resides in the temporal lobe; a memory of languages resides in the posterior temporal to anterior occipital lobe; and memories of odors, tastes, and one's social position are associated with the frontal lobe.

9.3c Voluntary Movement

Skeletal muscle movement is such an essential part of our lives that we take it for granted and don't think about the fact that many parts of the nervous system are involved in everything from rolling your eyes to moving a fork to your mouth. Skeletal muscles are activated by the **somatic motor division,** part of the peripheral nervous system (PNS) (see section 8.1a). This is distinguished from the *visceral motor (autonomic) division,* discussed later, which controls cardiac and smooth muscle and glands. The somatic motor division is mostly under voluntary control—we contract most skeletal muscles at will—whereas the visceral motor division is largely involuntary.

The decision to make a voluntary muscular movement originates in the **motor association area** of the frontal lobe (fig. 9.10a). A "plan of action" is then relayed to neurons in the **precentral gyrus**—the most posterior gyrus of the frontal lobe, immediately in front of the central sulcus. Its cortex is called the **primary motor cortex.** Axons from neurons of this gyrus descend through the brainstem and spinal cord to synapse with motor (efferent) neurons whose axons exit the CNS and supply the skeletal muscles. Most of the axons that originate in the primary motor cortex

Figure 9.10 Motor Areas of the Cerebrum. (a) Primary motor and motor association areas. (b) A frontal section through the precentral gyrus, mapping out areas of motor function. Like the sensory map in figure 9.9, the motor map shows an inverted arrangement with lower parts of the gyrus controlling muscles in higher areas of the body. More cerebral tissue is dedicated to fine motor control of the hands, face, and speech apparatus than to the trunk and lower limb. **APR**

cross over to the other side of the body as they pass through the medulla oblongata on their way to synapse with motor neurons in the spinal cord. Consequently, the right cerebrum controls movements on the left side of the body and vice versa. This is not true of most head–neck muscles, however, which are controlled by the cerebral hemisphere on the same side of the body.

The precentral gyrus, like the postcentral gyrus described earlier, has an upside-down map of the opposite side of the body (**fig. 9.10b**). This is depicted as a *motor homunculus* distorted to represent relative amounts of cerebrum dedicated to each region of the body. The amount of cerebral tissue dedicated to hands, face, and tongue reflects the importance of fine motor control in speech, facial expressions, and use of the hands.

Other areas of the brain important in muscle control are the basal nuclei (see fig. 9.4b) and cerebellum. Among other functions, the basal nuclei assume control of highly practiced behaviors that one carries out with little thought—riding a bicycle, typing, driving a car, or tying one's shoes, for example. They also control the onset and cessation of planned movements, and the repetitive movements at the shoulder and hip that occur during walking.

The cerebellum aids in learning motor skills, maintains muscle tone and posture, smooths muscle contractions, coordinates eye and body movements, and coordinates the motions of different joints with each other (such as the shoulder and elbow in pitching a baseball). The cerebellum monitors the body's position and movements by means of input from the inner ears and sense organs called **proprioceptors**[22] (PRO-pree-o-SEP-turs) in the muscles and joints. It compares the

---

[22]*proprio* = one's own; *ceptor* = receiver

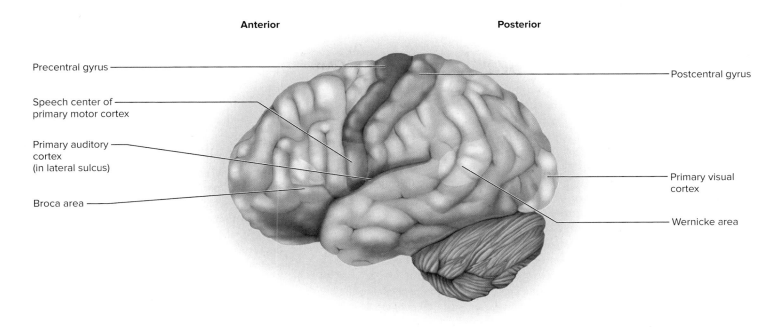

Figure 9.11 Language Centers of the Left Hemisphere.

body's actual movements with our cerebral intentions and issues output that adjusts and smooths out our movements.

9.3d Language

Language includes several cognitive and motor abilities—listening, understanding words, speaking, and reading and writing (fig. 9.11). The control of these tasks is very complex and involves multiple areas of the cerebrum; we can consider only a few key centers here. One is the **Wernicke**[23] **area** (WUR-ni-keh), usually found in the left cerebral hemisphere just posterior to the lateral sulcus. It lies at a crossroad between the visual cortex of the occipital lobe and the auditory cortex of the temporal lobe. It receives input from both and interprets visual images and sounds in terms of learned rules of grammar and vocabulary. That is, it creates linguistic meaning from what we see and hear, and begins to formulate the phrases of what we will speak.

The Wernicke area transmits a plan of speech forward to the **Broca**[24] **area,** which lies low in the prefrontal cortex of the same hemisphere (usually the left). The Broca area plays a part in constructing grammatically sensible sentences, and it compiles a "plan" of motor neuron action for the muscle contractions of the larynx, tongue, cheeks, and lips. It transmits this plan by way of the insula to the primary motor cortex (precentral gyrus), which then issues commands to the relevant muscles to carry out the speech. Injury to the Wernicke or Broca area from stroke or trauma can leave a person with language impairments called **aphasia**[25] (ah-FAY-zhee-uh) (see Clinical Application 9.2).

9.3e Cerebral Lateralization

The two cerebral hemispheres look identical at a glance, but close examination reveals a number of differences. For example, in women the left temporal lobe is often longer than the right. In left-handed people, the left frontal, parietal, and occipital lobes are usually wider than those on the right. The two hemispheres also differ

---

[23]Karl Wernicke (1848–1905), German neurologist
[24]Pierre Paul Broca (1824–80), French surgeon and anthropologist
[25]*a* = without; *phas* = speech

Clinical Application 9.2

APHASIA

Aphasia takes many forms depending on which regions of cerebral tissue, or the white-matter tracts between them, are damaged. Injury to the Wernicke area can make a person unable to recognize spoken or written words or remember the names of familiar objects *(anomic[27] aphasia)*. It often results in babbling incomprehensible speech filled with made-up words and nonsensical syntax *(fluent aphasia)*. Lesions of the Broca area, by contrast, leave a person knowing what to say but having great difficulty commanding the muscles to say it *(nonfluent aphasia)*. The result is typically a stammering, labored speech with partial sentences, or a frustrated tight-lipped reluctance to speak at all.

Some strokes reduce a person's vocabulary to only a few words, sometimes the last few that he or she spoke before the stroke occurred. Stranger still, damage to certain small areas of cerebral cortex can leave a person with impaired mathematical ability, a selective absence of vowels from one's writing, or difficulty understanding the second half of each word the person reads.

Speech therapy can help some individuals regain lost language skills, depending on the location and extent of tissue damage. Even sex can make a difference in recovery; men are three times as likely as women to suffer language deficits from left-sided cerebral damage, and women recover more easily than men.

in some of their functions. Brain imaging studies show that certain tasks activate regions of cerebral cortex localized more in one cerebral hemisphere than the other (fig. 9.12). However, neither hemisphere is "dominant," and the idea that a person is "left-brained" or "right-brained" is a pseudoscientific idea with no basis in neuroscience.

9.3f The Electroencephalogram

Brain functions and disorders are often studied or diagnosed by placing an array of recording electrodes on the scalp and recording the subject's **brain waves**—rhythmic voltage changes resulting from the synchronized activity of cerebral neurons close to the surface. The recording is called an **electroencephalogram**[26] **(EEG)**. It exhibits four types of brain waves, each of which dominates a different state of consciousness, wakefulness, or sleep (fig. 9.13). Normal EEGs are useful in studying brain functions such as sleep and consciousness, and abnormalities can help diagnose epilepsy, degenerative brain diseases, brain tumors, metabolic abnormalities, and cerebral trauma. The EEG changes with age and with states of consciousness ranging from high alert to deep sleep and coma. The complete and persistent absence of brain waves is often used as a clinical and legal criterion of brain death.

---

[26]*electro* = electricity; *encephalo* = brain; *gram* = record
[27]*a* = without; *nom* = names

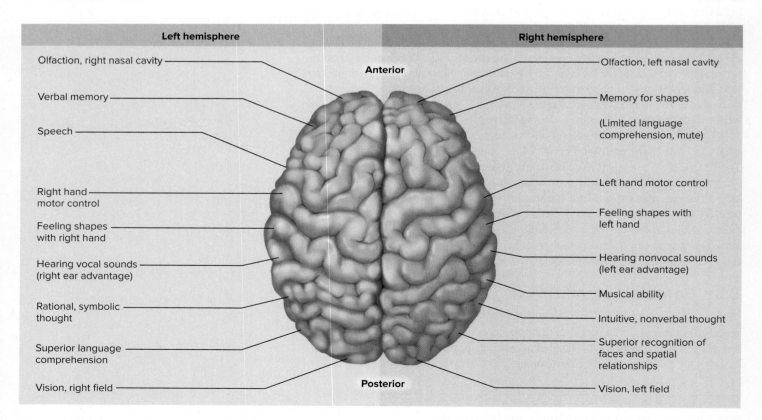

Figure 9.12 Lateralization of Cerebral Functions. The two cerebral hemispheres are not functionally identical.

Figure 9.13 The Electroencephalogram (EEG).
© Bob Coyle/McGraw-Hill Education

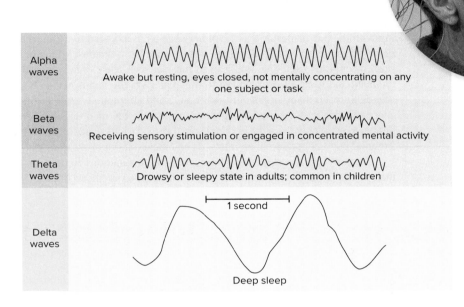

Before You Go On

Answer these questions from memory to test your understanding of the previous section. Reread the material if you can't answer the questions.

12. Describe the locations of the primary sensory areas for vision, hearing, taste, touch, and pain. How would a pain in the left foot differ from a pain in the right hand with respect to the area of cerebral cortex where these sensations are registered?

13. The hippocampus, amygdala, and cerebellum are all involved in memory, but how do their memory functions differ?

14. In what gyrus do motor commands to the skeletal muscles originate? How does this gyrus resemble and differ from the one that receives somatosensory information?

15. What are the roles of the Wernicke area, Broca area, and precentral gyrus in language?

9.4 The Cranial Nerves

Expected Learning Outcomes

When you have completed this section, you should be able to

a. list the 12 cranial nerve pairs by name and number;
b. identify where each cranial nerve originates and terminates; and
c. state the functions of each cranial nerve.

The brain communicates with the rest of the body through two routes: the spinal cord and 12 pairs of **cranial nerves.** The cranial nerves are numbered I to XII starting with the most anterior pair. Each nerve also has a descriptive name such as *optic nerve* or *vagus nerve.* Most of these nerves arise from the base of the brain, exit the skull through its foramina, and lead to muscles and sense organs located mainly in the head and neck. One exception to this pattern is the vagus nerve (X), which descends to reach many organs in the thoracic and abdominal cavities. Another is the accessory nerve (XI), once thought to originate in the medulla oblongata but now known to arise only from the cervical spinal cord. Strictly speaking, it is therefore not a cranial nerve, but neuroanatomists haven't kicked it out of the club yet.

Table 9.1 itemizes the cranial nerves and their functions, and **figure 9.14** shows their points of origin and termination in the brain and peripheral organs. Two of these

Table 9.1 The Cranial Nerves

Number and Name	Function	Anatomical Course
I. Olfactory	*Sensory:* Smell	From nasal cavity to olfactory bulbs of brain
II. Optic	*Sensory:* Vision	From eye to thalamus and midbrain
III. Oculomotor	*Motor:* Eye movements	From midbrain to all muscles of eye movement except as noted for cranial nerves IV and VI; muscle that opens eyelid; and internal eye muscles that control lens and iris
IV. Trochlear	*Motor:* Eye movements	From midbrain to superior oblique muscle of eye movement
V. Trigeminal (three branches, V_1, V_2, V_3)	*Motor:* Chewing *Sensory:* Touch, temperature, and pain from face	*Motor:* From pons to masseter, temporalis, and pteryoid muscles of chewing; tensor tympani muscle of middle ear *Sensory:* From facial skin, surface of eye, nasal mucosa, paranasal sinuses, palate, teeth and gums, tongue (but not taste buds), floor of mouth, and dura mater to pons
VI. Abducens	*Motor:* Eye movements	From pons to lateral rectus muscle of eye movement
VII. Facial	*Motor:* Facial expression; secretion of tears, saliva, and nasal and oral mucus *Sensory:* Taste	*Motor:* From pons to muscles of facial expression, salivary glands, tear glands, and glands of nasal cavity and palate *Sensory:* From taste buds of anterior tongue to thalamus
VIII. Vestibulocochlear	*Motor:* Cochlear tuning *Sensory:* Hearing and equilibrium	*Motor:* From pons to cochlea of inner ear *Sensory:* From cochlea, semicircular ducts, and vestibule of inner ear, to medulla oblongata and medulla–pons junction
IX. Glossopharyngeal	*Motor:* Salivation, swallowing, gagging *Sensory:* Taste; sensations of touch, pressure, pain, and temperature in tongue and outer ear; regulation of blood pressure and respiration	*Motor:* From medulla oblongata to salivary glands, glands of tongue, and muscle of pharynx *Sensory:* From middle and outer ear, posterior one-third of tongue (including taste buds), and pharynx, to medulla oblongata
X. Vagus	*Motor:* Swallowing, speech, regulation of heart rate and bronchial airflow, gastrointestinal secretion and motility *Sensory:* Taste; sensations of hunger, fullness, and gastrointestinal discomfort	*Motor:* From medulla oblongata to tongue, palate, larynx, lungs, heart, liver, spleen, digestive tract, kidney, and ureter *Sensory:* From root of tongue, pharynx, larynx, epiglottis, outer ear, and several thoracic and abdominal viscera, to medulla oblongata
XI. Accessory	*Motor:* Swallowing; head, neck, and shoulder movements	From spinal cord segments C1 to C6, into the cranial cavity and back out to palate, pharynx, trapezius, and sternocleidomastoid muscles
XII. Hypoglossal	*Motor:* Tongue movements	From medulla oblongata to muscles of tongue

Figure 9.14 The Cranial Nerves. Pathways concerned with sensory function are shown in green, and those concerned with motor function are shown in red. Arrows indicate the direction in which the nerve signals travel. All cranial nerves except I and II carry sensory fibers, but these functions are omitted in cases where the nerve is essentially motor and its sensory function is limited to proprioception (muscle feedback). **APR**

nerves are purely sensory—the olfactory (I) and optic (II) nerves. The rest are *mixed nerves*, carrying a two-way traffic of sensory signals to the brain and motor signals to muscles and other organs. In cases where only the motor function of a nerve is described in table 9.1, the sensory function is limited to proprioception—monitoring the contraction of a muscle and sending information about its performance back to the brain.

For centuries, students have invented rhymes and other phrases to help them remember the names of the cranial nerves in order. One of the senior author's students, now a neurologist,[28] created the following memory aid (mnemonic) in 1993, which has since been widely disseminated on the Internet. The first one to four letters of each word (boldfaced) match the initial letters of the cranial nerves: **Ol**d **Op**ie **oc**casionally **tr**ies **trig**onometry, **a**nd **f**eels **ve**ry **glo**omy, **vagu**e, and **hypo**active.

Before You Go On

Answer these questions from memory to test your understanding of the previous section. Reread the material if you can't answer the questions.

16. List the purely sensory cranial nerves and state the function of each.
17. What is the only cranial nerve that extends beyond the head–neck region? Name some organs in which it terminates.
18. If the oculomotor, trochlear, or abducens nerve was damaged, the effect would be similar in all three cases. What would that effect be?
19. Name two cranial nerves involved in the sense of taste and explain where their sensory fibers originate.

9.5 The Autonomic Nervous System

Expected Learning Outcomes

When you have completed this section, you should be able to

a. distinguish between the autonomic nervous system (ANS) and the somatic motor nervous system;
b. identify some visceral reflexes;
c. discuss the relevance of visceral reflexes to homeostasis;
d. name and compare the two subdivisions of the ANS;
e. explain the relationship of the adrenal medulla and enteric nervous system to the ANS;
f. identify the two principal neurotransmitters used by the ANS and where each of them is employed in the system; and
g. explain how the sympathetic and parasympathetic divisions can have cooperative or antagonistic effects on various organs.

---

[28]Marti Haykin, M.D.

The **autonomic nervous system (ANS)** is the part of the peripheral nervous system that controls glands, cardiac muscle, and smooth muscle. It is also called the **visceral motor division** to distinguish it from the somatic motor division that controls the skeletal muscles. The primary targets of the ANS are organs of the thoracic and abdominal cavities, such as the heart, lungs, digestive tract, and urinary tract, but it also innervates some structures of the body wall, including cutaneous blood vessels, sweat glands, and *arrector pili* muscles.

The word *autonomic* literally means "self-governed." It refers to the fact that the ANS usually carries out its actions without one's conscious intent, awareness, or ability to control it at will. It is for this reason that autonomic responses are used as a basis for polygraph ("lie detector") tests. Organs depend on the autonomic nervous system to adjust their activity to meet the body's changing needs. The ANS adjusts the heart rate to accommodate rest or exercise, for example.

9.5a Visceral Reflexes

The ANS works through **visceral reflexes,** which regulate primitive functions such as blood pressure, heart rate, body temperature, digestion, and many others. In short, the ANS quietly manages a myriad of unconscious processes responsible for homeostasis; these basic functions are not unique to humans but are shared by all animals. From a clinical perspective, one reason that it is important to understand the ANS is that many drug therapies are based on manipulation of autonomic functions (see Perspectives on Health at the end of this section).

A visceral reflex arc includes the same basic components as a somatic reflex arc (see section 8.3d)—receptors, afferent neurons, interneurons, efferent neurons, and effectors. For example, high blood pressure activates a **baroreflex**[29] involving the following components (fig. 9.15): ① It stimulates stretch receptors called *baroreceptors* in the internal carotid arteries and aorta. ② They issue signals through afferent fibers in cranial nerves to the medulla oblongata. ③ Interneurons of the medulla integrate this input with other information and issue signals back to the heart by way of efferent fibers in the vagus nerves. ④ The vagus nerves slow down the heart (the effector) and reduce blood pressure. This completes a homeostatic negative feedback loop and ideally restores blood pressure to normal.

9.5b Neural Pathways

The ANS is usually defined as a part of the peripheral nervous system, but it is controlled by autonomic control centers in the CNS, including the hypothalamus and other regions of the brainstem. Motor neurons originate in the brain and in the spinal cord, and in ganglia in the PNS. Nerve fibers travel in the cranial and spinal nerves on their paths to effectors.

The autonomic pathway to a target organ differs significantly from somatic motor pathways. In somatic pathways, a motor neuron in the brainstem or spinal cord issues an axon that reaches all the way to a skeletal muscle (fig. 9.16a). In autonomic pathways, the signal must travel across two neurons to get to the effectors, and it must cross a synapse where these neurons meet in an autonomic ganglion (fig. 9.16b, c). The first neuron, called the **preganglionic neuron,** has a neurosoma in the brainstem or spinal cord; its axon terminates in a ganglion outside the CNS. It synapses there with a **postganglionic neuron** whose axon extends the rest of the way to the target cells. The axons of these neurons are called the pre- and postganglionic fibers.

Figure 9.15 A Visceral Reflex Arc. In the arterial baroreflex, a rise in blood pressure is detected by baroreceptors in the carotid artery. Cranial nerves (glossopharyngeal and vagus) transmit signals to the medulla oblongata, resulting in parasympathetic output through the vagus nerve that reduces the heart rate and lowers blood pressure.

---

[29]*baro* = pressure

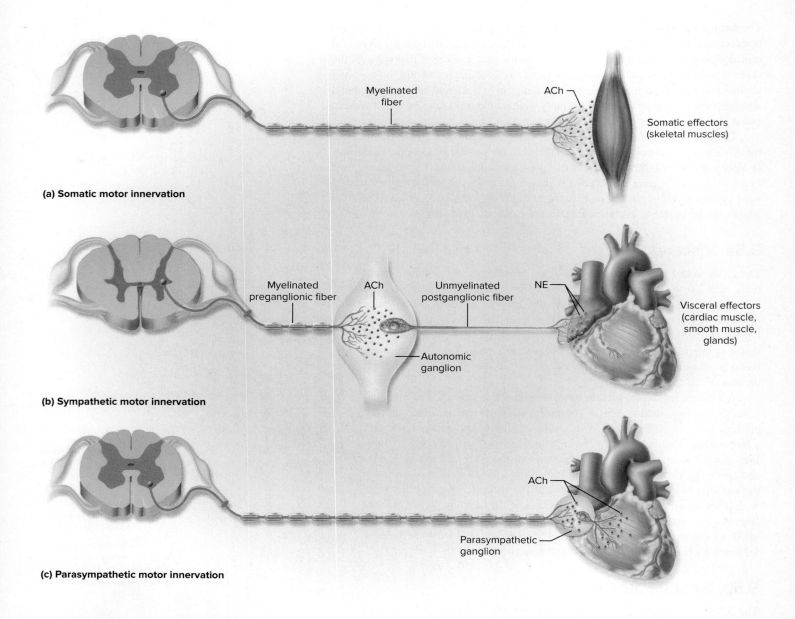

Figure 9.16 Comparison of Somatic and Autonomic Efferent Pathways. (a) In the somatic nervous system, each motor nerve fiber extends all the way from the CNS to a skeletal muscle. (b) In the sympathetic division, a short preganglionic fiber extends from the spinal cord to a sympathetic ganglion near the vertebral column. A longer postganglionic fiber extends from there to the target organ. (c) In the parasympathetic division, a long preganglionic fiber extends from the CNS to a ganglion in or near the target organ, and a short postganglionic fiber completes the path to specific effector cells in that organ (smooth or cardiac muscle, or gland cells). (ACh = acetylcholine; NE = norepinephrine)

9.5c Subdivisions of the ANS

The ANS has two subdivisions: sympathetic and parasympathetic. The **sympathetic division** adapts the body in many ways for physical activity—it increases alertness, heart rate, blood pressure and flow, blood glucose concentration, and pulmonary airflow. Extreme sympathetic responses are often called the "fight-or-flight" reaction because they come into play when an animal must attack, defend itself, or flee from danger. In our own lives, this reaction occurs in many situations involving arousal, competition, stress, danger, anger, or fear—ranging from giving a public speech to playing a game of soccer to defending oneself from an attacker. Ordinarily, however, the sympathetic division has more subtle effects that we barely notice, if at all.

The **parasympathetic division,** by comparison, has a calming effect on many body functions. It is associated with reduced energy expenditure and normal bodily maintenance, including such functions as digestion and waste elimination. This is sometimes called the "resting-and-digesting" state.

This doesn't mean that the body alternates between states where one division or the other is active and the opposite system is turned off. Normally, both divisions are active simultaneously, but the balance between them shifts with the body's changing needs. Neither division has universally excitatory or calming effects. The sympathetic division, for example, excites the heart but inhibits digestive and urinary functions, whereas the parasympathetic division has the opposite effects.

Anatomy of the Sympathetic Division

All efferent, preganglionic fibers of the sympathetic division arise from the thoracic and first two lumbar segments of the spinal cord. Most of these fibers travel a short distance to ganglia that lie alongside the vertebral column. Here, some of the fibers turn and travel up or down to higher and lower ganglia, uniting the ganglia into a string called the **sympathetic chain.** The chain extends upward into the cervical region and downward to the sacral and coccygeal levels (fig. 9.17a).

Usually, the preganglionic nerve fiber synapses with a postganglionic neuron somewhere in the sympathetic chain. Some of the fibers pass through the chain without synapsing, however, and go to ganglia wrapped around the abdominal aorta, the large artery on the posterior abdominal wall. From either the sympathetic chain or these ganglia, postganglionic nerve fibers then complete the path to the target organs. The general pattern in the sympathetic division is to have short preganglionic fibers and long postganglionic ones (fig. 9.16b).

Most preganglionic fibers branch before reaching their synapses. Typically, each preganglionic neuron synapses with 10 to 20 postganglionic neurons. This means that a localized output from the spinal cord can branch out and reach several target organs at once, such as the eyes, sweat glands, heart, and lungs, immediately creating multiple effects of the fight-or-flight response (pupillary dilation, sweating, faster heart rate, and increased airflow).

> *Apply What You Know*
> One of the targets of the sympathetic nervous system is the liver, where the effect is to release glucose from the glycogen that is stored there. How does this relate to the "fight-or-flight" response?

Another component of the sympathetic division is found in the **adrenal[30] glands,** which lie superior to the kidneys (see fig. 11.12). The inner core of each gland, the **adrenal medulla,** is essentially a sympathetic ganglion. Preganglionic fibers terminate on cells of the medulla, which secrete epinephrine (adrenaline) and norepinephrine (noradrenaline) into the blood. These hormones accentuate the effects of the sympathetic division.

Anatomy of the Parasympathetic Division

Preganglionic fibers of the parasympathetic division leave the CNS by way of four cranial nerves—the oculomotor (III), facial (VII), glossopharyngeal (IX), and vagus (X) nerves—and by way of spinal nerves from the sacral segments of the

---

[30]*ad* = near; *ren* = kidney

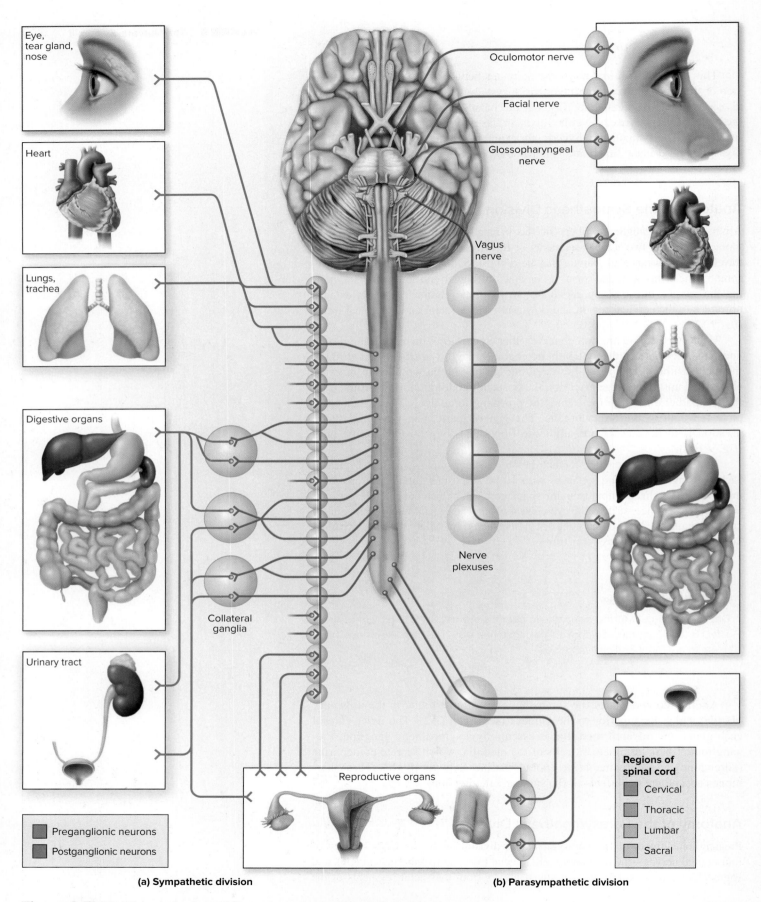

Figure 9.17 The Sympathetic and Parasympathetic Nerve Pathways. (a) Sympathetic pathways. (b) Parasympathetic pathways. The direction of signal conduction is away from the CNS in all these pathways.

❓ *Suppose a person had a spinal cord injury that severed the cord at the cervical level. Which of these functions do you think would be lost, and why—acceleration of the heart in response to fear, or the secretion of stomach juices in response to the smell of good food?*

spinal cord (fig. 9.17b). In comparison to sympathetic preganglionic fibers, parasympathetic preganglionics travel relatively long distances to *terminal ganglia* in or near their target organs, and synapse there with short postganglionic fibers that complete the path to the target cells (figs. 9.16c and 9.17b). Parasympathetic fibers in cranial nerves III, VII, and IX supply target organs in the head, such as tear glands, salivary glands, nasal glands, and internal muscles of the eyeball (for control of the iris and lens). Cranial nerve X (the vagus nerve), however, carries about 90% of all parasympathetic fibers and descends through the thorax and abdomen to reach numerous organs—heart, lungs, esophagus, stomach, intestines, liver, pancreas, and urinary tract. Parasympathetic fibers from the sacral region of the spinal cord supply the colon and rectum, urinary bladder, and reproductive organs. Fibers from the vagus and sacral nerves pass through various nerve plexuses (webs) on their way to the target organs.

9.5d Neurotransmitters and Receptors

We have already seen a few cases in which the two divisions of the ANS have contrasting effects on the same organs. The sympathetic division accelerates the heartbeat, for example, and the parasympathetic division slows it down; the sympathetic division inhibits digestion and the parasympathetic division stimulates it. The key to such contrasting effects, and to many drug actions, depends on differences in the neurotransmitters employed by the two divisions and in the types of neurotransmitter receptors found in the target cells. A single neurotransmitter can excite some organs and inhibit others because of differences in the type of receptor those organs have for it.

The two main neurotransmitters employed in the ANS are acetylcholine (ACh) and norepinephrine (NE). A *cholinergic*[31] nerve fiber or receptor is one that, respectively, secretes or binds ACh. An *adrenergic*[32] nerve fiber or receptor secretes or binds NE; it is named for an older synonym for NE, *adrenaline*.

All preganglionic fibers of the ANS are cholinergic; they release ACh to stimulate the postganglionic neurons. Parasympathetic postganglionic fibers also are cholinergic, whereas most sympathetic postganglionic fibers are adrenergic (table 9.2).

9.5e Dual Innervation

Many organs receive both sympathetic and parasympathetic input. In the eye, for example, the iris receives fibers of both types (fig. 9.18). Sympathetic fibers dilate the pupil, a well-known sympathetic effect, whereas parasympathetic fibers constrict it. When the two divisions of the ANS have opposite effects on the same organ, they are said to have **antagonistic effects.**

Table 9.2 Locations of Cholinergic and Adrenergic Fibers in the ANS

Division	Preganglionic Fibers	Postganglionic Fibers
Sympathetic	Always cholinergic	Mostly adrenergic; a few cholinergic
Parasympathetic	Always cholinergic	Always cholinergic

---

[31]*cholin*, from *acetylcholine*; *erg* = work, action
[32]after *noradrenaline*, a synonym for *norepinephrine*

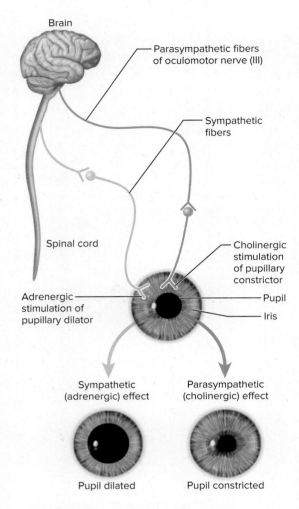

Figure 9.18 Dual Innervation of the Iris. Shows antagonistic effects of the sympathetic (yellow) and parasympathetic (blue) divisions on the iris.

In other cases, the two divisions have **cooperative effects.** Saliva, for example, consists of a watery solution of digestive enzymes plus slippery mucus that makes food easier to swallow. Sympathetic fibers stimulate mucus secretion, and parasympathetic fibers stimulate enzyme secretion.

It is not always necessary to have dual innervation, however, for the ANS to produce opposite effects on an organ. Most blood vessels, for example, receive only sympathetic nerves. When the nerves increase their firing rate, they constrict the vessels, which raises the blood pressure. When the nerves decrease their firing rate, they allow the blood vessels to relax and dilate, which lowers the blood pressure. In some cases, such as sudden emotional shock, the sympathetic firing rate decreases so much that the blood pressure falls too low and a person faints.

Table 9.3 presents several examples of the sympathetic and parasympathetic effects on various target organs and bodily processes.

9.5f Higher-Level Regulation of the ANS

Even though *autonomic* means "self-governed," the ANS is not an independent nervous system. All of its output originates in the CNS, and it is strongly influenced by the cerebral cortex, hypothalamus, limbic system, medulla oblongata, and spinal cord. Consequently, our emotions affect such autonomic functions as blood pressure, heart rate, digestion, and sexual function. Other autonomic responses originate in the brainstem or are at least influenced by it, such as thermoregulation, salivation, digestive secretion, bladder and bowel control, and pupillary light reflexes.

The spinal cord also contains integrating centers for such autonomic reflexes as defecation, urination, erection, and ejaculation. Fortunately, the brain is able to inhibit defecation and urination consciously, but when injuries sever the spinal cord from the brain, the autonomic spinal reflexes alone control the elimination of urine and feces. The lack of voluntary control over these functions is called urinary or fecal *incontinence.* Erection and ejaculation can occur through autonomic spinal reflexes alone even in men with spinal cord injuries who cannot feel the associated sensations.

Table 9.3 Effects of the Sympathetic and Parasympathetic Divisions

Target	Sympathetic Effect	Parasympathetic Effect
Eye		
Iris	Pupillary dilation	Pupillary constriction
Lens	Relaxation for distant vision	Contraction for near vision
Integumentary System		
Merocrine sweat glands	Secretion	No effect
Arrector pili muscles	Hair erection	No effect
Circulatory System		
Heart rate and force	Increase	Decrease
Cutaneous blood vessels	Constriction	Dilation, blushing
Respiratory System		
Bronchioles	Dilation, increased airflow	Constriction, reduced airflow
Urinary System		
Kidneys	Reduced urine output	No effect
Bladder wall	No effect	Contraction, urine release
Digestive System		
Salivary glands	Thick mucous secretion	Thin serous secretion
Gastrointestinal motility	Reduced	Increased
Liver	Glycogen breakdown	Glycogen synthesis
Reproductive System		
Penis and clitoris	No effect	Erection
Glandular secretions	No effect	Stimulation
Orgasm, smooth muscle roles	Stimulation	No effect

Before You Go On

Answer these questions from memory to test your understanding of the previous section. Reread the material if you can't answer the questions.

20. Trace the pathway of a parasympathetic nerve fiber from the medulla oblongata to the small intestine.
21. What neurotransmitters are secreted by adrenergic and cholinergic fibers?
22. What are the two ways in which the sympathetic and parasympathetic divisions can affect each other when they both innervate the same target organ? Give examples.
23. How can the sympathetic division have contrasting effects in a target organ without dual innervation?
24. Give examples of how the CNS influences the autonomic nervous system, which is regarded as part of the PNS.

PERSPECTIVES ON HEALTH

Drugs and the Nervous System

An understanding of neurotransmitter and receptor types in the brain and autonomic nervous system has given rise to the large field of neuropharmacology—the branch of science that deals with the design of drugs that mimic, enhance, or inhibit the action of these neurotransmitters.

A number of drugs work by stimulating adrenergic and cholinergic neurons or receptors. For example, phenylephrine, found in such cold medicines as Dimetapp and Sudafed PE, aids breathing by stimulating a subclass of adrenergic receptors called α-adrenergic (alpha-adrenergic) receptors. It dilates the bronchioles and constricts nasal blood vessels, thus reducing swelling in the nasal mucosa. Some drugs work by binding to adrenergic receptors without stimulating them, thereby blocking the action of epinephrine and norepinephrine. Propanolol, for example, reduces high blood pressure by blocking β-adrenergic (beta-adrenergic) receptors; it is in a family of drugs called β-blockers.

Other drugs target parasympathetic effects. Pilocarpine relieves glaucoma (excessive pressure in the eyeball) by mimicking an action of acetylcholine, dilating a vessel that drains fluid from the eye. Some drugs block ACh effects. For example, ACh normally stimulates constriction of the pupil and secretion of respiratory mucus. Atropine blocks certain ACh receptors, so it is useful for dilating the pupils for eye examinations and drying the mucous membranes of the respiratory tract in preparation for inhalation anesthesia. It is an extract of the deadly nightshade plant, *Atropa belladonna*.[33] Women of the Middle Ages used nightshade to dilate their pupils, which was regarded as a beauty enhancement.

Many other drugs are designed to act on the central nervous system. Sigmund Freud predicted that psychiatry would eventually draw upon biology and chemistry to deal with emotional problems once treated only by counseling and psychoanalysis. A branch of neuropharmacology called *psychopharmacology* has fulfilled his prediction. This field dates to the 1950s when chlorpromazine, an antihistamine, was incidentally found to relieve the symptoms of schizophrenia.

The management of clinical depression is one example of how psychopharmacology has supplemented counseling approaches. A person's mood is strongly influenced by neurotransmitters in a family called the *monoamines*, including norepinephrine and serotonin. Some forms of depression result from monoamine deficiencies, and therefore yield to drugs that prolong the effects of the monoamines already present at the synapses. Fluoxetine (Prozac), for example, blocks the reabsorption of serotonin by the presynaptic neurons that release it. This causes serotonin to linger longer in the synapse, producing a mood-elevating effect. Fluoxetine is called a *selective serotonin reuptake inhibitor (SSRI)*.

Our growing understanding of neurochemistry also gives us deeper insight into the action of addictive drugs of abuse such as amphetamines and cocaine. Amphetamines ("speed") chemically resemble norepinephrine and dopamine, two neurotransmitters associated with elevated mood. Dopamine is especially important in sensations of pleasure. Cocaine blocks dopamine reuptake and thus produces a brief rush of good feelings. But when dopamine is not reabsorbed by the neurons, it diffuses out of the synaptic cleft and is degraded elsewhere. Cocaine thus depletes the neurons of dopamine faster than they can synthesize it, so that finally there is no longer an adequate supply to maintain normal mood and the user becomes dependent on the cocaine. The postsynaptic neurons make new dopamine receptors as if "searching" for the neurotransmitter—all of which leads ultimately to anxiety, depression, and the inability to experience pleasure without the drug.

Another drug from which many derive pleasure, but with far less harmful consequences, is caffeine. Caffeine works because of its similarity to adenosine (fig. 9.19). Adenosine, which you know as a component of DNA, RNA, and ATP, also acts in the brain to inhibit ACh release. One theory of sleepiness is that it results when prolonged metabolic activity breaks down so much ATP that the accumulated adenosine has a noticeably inhibitory effect. Caffeine resembles adenosine closely enough to bind to its receptors, but it does not produce the inhibitory effect. Thus, it blocks adenosine action, enhances ACh secretion, and makes a person feel more alert.

---

[33]*bella* = beautiful, fine; *donna* = woman

Figure 9.19 Adenosine and Caffeine. Adenosine, a breakdown product of ATP and other chemicals, inhibits ACh release and produces a sense of sleepiness. Caffeine is similar enough to block the action of adenosine by binding to its receptors. This results in increased ACh release and heightened arousal.
left: © Stockbyte/Getty Images **right:** © Foodcollection

Aging of the Nervous System

The nervous system reaches its peak development around age 30. After age 40, the brain loses about 5% of its mass every 10 years, and possibly faster after age 70. In old age, the cerebral gyri are narrower, the sulci are wider, the cortex is thinner, and there is more space between the brain and meninges. The remaining cortical neurons have fewer synapses, and for multiple reasons, synaptic transmission is less efficient: The neurons produce less neurotransmitter, they have fewer receptors, and the neuroglia around the synapses are more leaky and allow neurotransmitter to escape from the synapses. The degeneration of myelin sheaths with age also slows down nerve conduction.

Neurons exhibit less rough endoplasmic reticulum and Golgi complex with age, which indicates that their metabolism is slowing down. Old neurons accumulate more neurofibrillary tangles—dense mats of cytoskeletal elements in their cytoplasm. In the extracellular material, plaques of amyloid protein appear, especially in people with Down syndrome and Alzheimer disease (AD). AD is the most common nervous disability of old age (see Perspectives on Health in section 8.2).

Not all functions of the central nervous system are equally affected by senescence. Motor coordination, intellectual function, and short-term memory decline more than language skills and long-term memory. Elderly people are often better at remembering things in the distant past than remembering recent events.

The sympathetic nervous system loses adrenergic receptors with age and becomes less sensitive to norepinephrine. This contributes to a decline in homeostatic control of such variables as body temperature and blood pressure. Many elderly people experience orthostatic hypotension—a drop in blood pressure when they stand, which sometimes results in dizziness, loss of balance, or fainting.

CAREER SPOTLIGHT

Electroneurodiagnostic Technologist

Electroneurodiagnostic (END) technologists are also called EEG technologists, although their duties are broader than recording electroencephalograms. END technologists work primarily in hospital neurology departments, operating equipment that records electrical activity of the brain, spinal cord, and peripheral nerves, thus obtaining data used by neurologists in diagnosing and monitoring a broad spectrum of neurological diseases: epilepsy, stroke, encephalitis, head trauma, and Alzheimer disease. END technologists are also involved in sleep studies and assessment of brain death.

There are more openings for END technologists than there are qualified people to fill them, so job opportunities are excellent. One must have good manual dexterity, excellent oral and written communication skills, and a basic ability to use computers and electronic equipment. Training programs require only a high school diploma for entry, and extend for 12 to 24 months leading to an associate degree or certification. The training includes electronics, computer skills, neuroanatomy, neuropathology, psychology, and other topics. See appendix B for further career information.

© Larry Mulvehill/Science Source

CONNECTIVE ISSUES

Ways in Which the NERVOUS SYSTEM Affects Other Organ Systems

Integumentary System
The autonomic nervous system regulates piloerection, sweating, and cutaneous blood flow, thereby adjusting heat loss through the skin. Cutaneous nerves provide sensations such as touch, itch, tickle, pressure, and temperature.

Skeletal System
Nervous stimulation affects bone growth and remodeling by regulating the muscle tension exerted on the bones; nerves in bones respond to strains and fractures.

Muscular System
Skeletal muscles can't contract without nervous system stimulation; the nervous system controls all body movements.

Endocrine System
The hypothalamus controls the pituitary gland; the sympathetic nervous system controls the adrenal medulla; the brain synthesizes several hormones.

Circulatory System
The autonomic nervous system regulates heartbeat, blood vessel diameters, blood pressure, and routing of blood; sympathetic stimulation enhances blood clotting.

Lymphatic and Immune Systems
Nerves to lymphatic organs influence production of immune cells; the nervous system regulates immune responses; emotional states influence disease susceptibility and severity.

Respiratory System
The brain regulates rate and depth of breathing, and the autonomic nervous system influences ease of pulmonary airflow.

Urinary System
The sympathetic division regulates rate of urine production by adjusting renal blood flow and pressure; nervous reflexes control emptying of bladder.

Digestive System
The nervous system regulates appetite, feeding behavior, digestive tract motility and secretion, and defecation.

Reproductive System
The nervous system regulates sex drive, arousal, and orgasm; the brain secretes a hormone (oxytocin) involved in sexual pair-bonding, childbirth, and lactation; the brain stimulates pituitary hormones that control sperm and egg production, the menstrual cycle, pregnancy, and lactation.

Study Guide

Assess Your Learning Outcomes

To test your knowledge, discuss the following topics with a study partner or in writing, ideally from memory.

9.1 Overview of the Brain

1. Principal portions of the brain—cerebrum, cerebellum, and brainstem—and major features of their gross anatomy
2. Meninges of the brain and ways in which they differ from the meninges of the spinal cord
3. The distribution of gray and white matter in the brain
4. Four ventricles of the brain
5. The functions, sources, flow, and reabsorption of cerebrospinal fluid
6. Arteries that supply the brain
7. The blood–brain barrier, how it protects the brain, and how it presents an obstacle to certain forms of medical treatment

9.2 Principal Divisions of the Brain

1. Locations, anatomical features, and functions of the medulla oblongata, pons, midbrain, thalamus, and hypothalamus
2. Gross anatomy, histology, and functions of the cerebellum
3. Gross anatomy of the cerebrum, including the names and boundaries of its five lobes
4. The distribution of gray and white matter in the cerebrum and the three types of tracts of white matter
5. Functions of the five lobes of the cerebrum
6. The location and basic functions of the basal nuclei of the cerebrum
7. Major components, locations, and functions of the limbic system, especially the amygdala and hippocampus

9.3 Multiregional Brain Functions

1. Areas of cerebrum that receive input for the special and general senses
2. The functional difference between primary sensory areas and sensory association areas
3. Sensory organization of the postcentral gyrus
4. Locations of some of the cognitive functions of the brain
5. Areas of the brain concerned with the creation and storage of memories
6. Areas of the brain concerned with voluntary motor control, especially organization of the precentral gyrus and the role of the cerebellum in motor coordination
7. Locations and functions of the language centers of the brain
8. Functional differences between the right and left cerebral hemispheres
9. Brain waves seen in the electroencephalogram, and the clinical uses of the EEG

9.4 The Cranial Nerves

1. Numbers and names of the cranial nerves
2. The origin, termination, and functions of each nerve

9.5 The Autonomic Nervous System

1. Differences between the autonomic and somatic motor divisions of the nervous system
2. Targets (effectors) of the autonomic nervous system (ANS)
3. Examples of visceral reflexes and a visceral reflex arc
4. Anatomical and functional differences between the sympathetic and parasympathetic divisions
5. Roles of the sympathetic chain ganglia and adrenal medulla in the sympathetic division
6. Which cranial and spinal nerves carry fibers of the parasympathetic nervous system, and the destinations of each
7. Two primary neurotransmitters of the ANS, and the distinction between and locations of adrenergic and cholinergic nerve fibers and receptors
8. Antagonistic and cooperative effects in the ANS
9. Examples of ANS regulation by the cerebrum, hypothalamus, limbic system, lower brainstem, and spinal cord

Testing Your Recall

1. As a result of a brain tumor, a person experiences intense hunger and never feels satiated (full). He gains a great deal of weight in a short time. What part of the brain has probably been damaged?
   a. the thalamus
   b. the amygdala
   c. the hypothalamus
   d. the pituitary gland
   e. the medulla oblongata

2. The blood–brain barrier is formed by
   a. Schwann cells.
   b. ependymal cells.
   c. epithelial cells of the capillaries.
   d. the pia mater.
   e. the arachnoid villi.

3. The right precentral gyrus controls muscles on the left side of the body because the motor nerve fibers decussate as they pass through
   a. the pyramids of the medulla oblongata.
   b. the posterior horn of the spinal cord.
   c. the corpus callosum between cerebral hemispheres.
   d. the white matter of the cerebral hemisphere.
   e. the tracts of the spinal cord.

4. The medulla oblongata plays a direct role in all of the following *except*
   a. speech.
   b. respiration.
   c. swallowing.
   d. the heartbeat.
   e. memory.

5. Nearly all sensory signals pass through a "gateway to the cerebral cortex" called the _____ on their way to centers of consciousness in the brain.
   a. thalamus
   b. hypothalamus
   c. midbrain
   d. pons
   e. cerebellum

6. Movement disorders are most likely to result from injury to
   a. the Wernicke area.
   b. the cerebellum.
   c. the prefrontal cortex.
   d. the temporal lobe.
   e. the postcentral gyrus.

7. To smile, frown, or blow a kiss depends most on which of the following cranial nerves?
   a. accessory
   b. glossopharyngeal
   c. trigeminal
   d. facial
   e. trochlear

8. Most parasympathetic nerve fibers travel through
   a. the corpus callosum.
   b. thoracic spinal nerves.
   c. sacral spinal nerves.
   d. the vagus nerve.
   e. the trigeminal nerve.

9. One of the hallmarks of Alzheimer disease is memory loss. Which of these structures is especially associated with this symptom?
   a. cerebellum
   b. corpus callosum
   c. precentral gyrus
   d. Broca area
   e. hippocampus

10. The autonomic nervous system targets all of these effectors *except*
    a. cardiac muscle.
    b. salivary glands.
    c. blood vessels.
    d. smooth muscles.
    e. skeletal muscles.

11. The right and left cerebral hemispheres are connected to each other by a thick C-shaped tract of fibers called the _____.

12. The brain has four chambers called _____ filled with _____ fluid.

13. On a sagittal plane, the white matter of the cerebellum exhibits a branching pattern called the _____.

14. Acetylcholine binds to a category of neurotransmitter receptors called _____ receptors.

15. Cerebrospinal fluid is secreted partly by a mass of blood capillaries called the _____ in each ventricle.

16. The primary motor area of the cerebrum is the _____ gyrus of the frontal lobe.

17. Sense organs called _____ monitor stretch, movement, and tension in the muscles and joints, giving one a nonvisual awareness of the body's position and movement.

18. Any area of cerebral cortex that identifies or interprets sensory information is called a/an _____ area.

19. The awareness of visual input lies in the _____ lobe of the cerebrum.

20. The motor pattern for speech is generated in an area of cortex called the _____ and then transmitted to the primary motor cortex to be carried out.

Answers in appendix A

What's Wrong with These Statements?

Briefly explain why each of the following statements is false, or reword it to make it true.

1. The two hemispheres of the cerebellum are separated by the longitudinal fissure.
2. All cerebrospinal fluid is produced by the choroid plexuses of the brain ventricles.
3. The functions of the limbic system include regulation of the respiratory and cardiac rhythms.
4. Each half of the brain has its own thalamus, hypothalamus, and epithalamus.
5. The sympathetic nervous system stimulates digestion.
6. Hearing is a function of the occipital lobe.
7. The gray matter of the brain is organized into functional regions called tracts.
8. In the parasympathetic nervous system, the preganglionic fibers are adrenergic and usually shorter than the postganglionic fibers.
9. All of the cranial nerves innervate structures limited to the head–neck region; everything below the neck is innervated by spinal nerves.
10. The optic nerve controls movements of the eye.

Answers in appendix A

Testing Your Comprehension

1. Which cranial nerve conveys pain signals to the brain in each of the following situations? (a) Sand blows into your eye. (b) You bite the rear of your tongue. (c) Your stomach hurts from eating too much.
2. A person can survive destruction of an entire cerebral hemisphere but cannot survive destruction of the hypothalamus, which is a much smaller mass of brain tissue. Explain this difference and describe some ways that destruction of a cerebral hemisphere would affect one's quality of life.
3. Suppose you were walking alone at night and suddenly heard a dog growling close behind you. Describe several ways in which your sympathetic nervous system would prepare you to deal with this situation.

The Sense Organs

Chapter 10

Chapter Outline

10.1 Receptors and Sensations
- 10.1a Types of Sensory Information
- 10.1b General and Special Senses

10.2 The General Senses
- 10.2a Simple Nerve Endings
- 10.2b Pain

10.3 The Chemical Senses—Taste and Smell
- 10.3a Gustation—The Sense of Taste
- 10.3b Olfaction—The Sense of Smell

10.4 The Ear—Equilibrium and Hearing
- 10.4a General Anatomy of the Ear
- 10.4b Equilibrium
- 10.4c The Cochlea and Hearing
- 10.4d The Auditory Projection Pathway

10.5 The Eye and Vision
- 10.5a Anatomy of the Orbital Region
- 10.5b Anatomy of the Eyeball
- 10.5c Forming the Visual Image
- 10.5d Generating the Visual Nerve Signal
- 10.5e The Visual Projection Pathway

Clinical Applications/Perspectives on Health
- Clinical Application 10.1: Migraine Headache
- Clinical Application 10.2: Middle-Ear Infection
- Clinical Application 10.3: Glaucoma
- Perspectives on Health

End of Chapter
- Aging of the Sense Organs
- Career Spotlight: Optician
- Study Guide

Iris camera technology reveals the beautiful intricacy of the human iris, as unique to each individual as fingerprints. Iris recognition systems are increasingly used for security clearances and in place of passports at some airports.
© Robert Mora/Alamy

BASE CAMP

Before ascending to the next level, be sure you're properly equipped with a knowledge of these concepts from earlier chapters.

- "Basic Structure of Neurons" in section 8.1b
- Action potentials in nerve fibers (see "Conducting a Nerve Signal to Its Destination" in section 8.2c)
- "Lobes of the Cerebrum" in section 9.2c
- Areas of primary sensory cortex (see "Sensation" in section 9.3a)

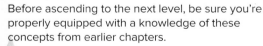

Anatomy & Physiology Revealed 4.0 | Module 7 Nervous System

Anyone who enjoys music, art, delicious food, or a good conversation knows the value of the human senses. Indeed, sensory input is necessary to the integrity of one's personality, intellectual function, and even sanity; extreme sensory deprivation can lead to hallucinations, incoherent thought patterns, deterioration of intellectual function, and sometimes morbid fear or panic. Such effects sometimes occur in burn patients who are immobilized and extensively bandaged (including the eyes). Sensory deprivation has been used by oppressive governments for the torture of political prisoners.

Yet much of the information communicated by the sense organs never comes to our conscious attention—for example, information about blood pressure, body temperature, and muscle tension. Despite our lack of awareness, such information is necessary for internal communication and control. The sense organs are therefore vital not just to our conscious awareness or enjoyment of the environment but also to homeostasis and our very survival in a ceaselessly changing and challenging world.

10.1 Receptors and Sensations

Expected Learning Outcomes

When you have completed this section, you should be able to

a. define *receptor* and *sense organ;*
b. describe the types of information obtained from sensory receptors; and
c. classify the human sense organs into broad functional categories.

A sensory **receptor** is any structure specialized to detect a stimulus. Some receptors are simple, bare nerve endings, such as the receptors for heat and pain, while others are true sense organs. A **sense organ** is a structure that combines nervous tissue with muscular, epithelial, or connective tissues that enhance its response to a specific stimulus. Sense organs can be as complex as the eye and ear or as microscopic and simple as a dendrite wrapped in a little connective tissue.

A **sensation** is produced when a receptor is excited by a stimulus. Sensory signals sent to the brain are processed in the cerebral cortex, and we may become aware of the sensation. Conscious awareness, or **perception,** of senses such as temperature, pain, and hunger allows us to respond appropriately to stimuli—putting on a coat if we're cold, avoiding putting weight on a hurt ankle, or eating, for example. However, most sensory signals delivered to the CNS produce no conscious sensation at all. Some are filtered out in the brainstem before reaching the cerebral cortex, a valuable function that keeps us from being distracted by innumerable unimportant stimuli detected by the sense organs. Other nerve signals concern functions that do not require our conscious attention, such as monitoring blood pressure and pH.

10.1a Types of Sensory Information

Sensory receptors provide the CNS with four kinds of information about a stimulus—*type, location, intensity,* and *duration:*

1. **Type (modality)** refers to kinds of environmental stimuli. Vision, hearing, taste, and pain are examples of sensory modalities, as are subcategories of these, such as red or blue color, or bitter or sweet taste. The action potentials for vision are identical to the action potentials for taste or any other sensory type. So how can the brain tell a visual signal from a taste signal? Modality is determined by where the sensory signals end in the brain. Hypothetically, if we could rewire the brain so signals from the ear were routed to the visual cortex of the occipital lobe, we would perceive acoustic signals as light rather than sound.

2. **Location** is also determined by which nerve fibers issue signals to the brain. Any sensory neuron detects stimuli within an area called its **receptive field.** In the sense of touch, for example, a single sensory neuron may receive information from an area of skin as great as 7 cm in diameter. No matter where the skin is touched within that field, the brain receives signals from that neuron and a person can't tell whether the skin was touched at point A or at some other point 4 or 5 cm away (fig. 10.1a). In some areas of the body, such as the back, it isn't necessary to make finer distinctions. On the other hand, we need to be able to more precisely localize touch sensations from the fingertips. Here, each sensory neuron may cover a receptive field as small as 1 mm in diameter, so two points of contact just 2 mm apart will be felt separately (fig. 10.1b). You can imagine the importance of this to reading braille, appreciating the fine texture of a fabric, or manipulating a small object such as a sesame seed.

Figure 10.1 The Receptive Field of a Nerve Ending. (a) A neuron with a large receptive field, as found in the skin of the back. If the skin is touched simultaneously at two closely spaced locations within the same receptive field, both touches stimulate the same nerve fiber and the brain perceives this as a single touch. (b) Neurons with small receptive fields, as found on the fingertips. Two closely spaced touches here are likely to stimulate different nerve fibers, enabling the brain to perceive them as separate touches.

? *Which of these do you think would be characteristic of your lips?*

(a) One large receptive field (arrow)

(b) Three small receptive fields (arrows)

3. **Intensity** refers to whether a sound is loud or soft, a light is bright or dim, or a pain is mild or excruciating. One way of communicating stimulus intensity is for a nerve fiber to fire more rapidly in response to a high-intensity stimulus and more slowly in response to a low-intensity stimulus. Another is that intense stimuli stimulate a larger number of nerve fibers to fire than weak stimuli do. Yet a third is that weak stimuli activate only the most sensitive neurons, whereas strong stimuli also activate less sensitive ones. Therefore, the brain can discern stimulus intensity by monitoring which nerve fibers are sending it signals, how many nerve fibers are doing so, and how fast those nerve fibers are firing.

4. **Duration,** or how long a stimulus lasts, is encoded by changes in the firing frequency of a nerve fiber with the passage of time. All receptors exhibit the property of **sensory adaptation.** If the stimulus is prolonged, the firing of the neuron gets slower over time; with it, so does our conscious sensation. Adapting to hot bathwater is an example of this. Sensory adaptation also explains why we may notice a suspicious odor (such as a gas leak) for several seconds and then the sensation fades in intensity even if the stimulus is still there. Some receptors adapt much more slowly, such as those for muscle tension and body position. The brain must always be aware of the body's position and movements, so it would be problematic if these receptors adapted too quickly and ceased to inform the brain.

Interestingly, regardless of the type of environmental stimulus, sensory information comes to the brain in one simple form—a chain of action potentials. We can think of all sense organs as *transducers*—devices that convert one form of energy (light, heat, chemicals, touch, sound) into another (action potentials). The process of converting stimulus energy to nerve energy is called *sensory transduction.* This chapter will explain how receptors achieve that.

10.1b General and Special Senses

There are multiple ways of classifying the senses and receptors. The **general senses** are distributed over much or all of the body—in the skin, muscles, tendons, joints, and viscera. They include the senses of touch, pressure, pain, heat and cold, stretch, and others. The **special senses** are limited to the head and employ receptors that are innervated by the cranial nerves. They include taste, smell, hearing, equilibrium, and vision.

Before You Go On

Answer these questions from memory to test your understanding of the previous section. Reread the material if you can't answer the questions.

1. Not every sensory receptor is a sense organ. Explain.
2. Not every sensory signal results in a person's conscious awareness of a stimulus. Explain.
3. What are the four important qualities of a stimulus that sense organs communicate to the brain? Give some examples.
4. State two examples of general senses and two examples of special senses.

10.2 The General Senses

Expected Learning Outcomes

When you have completed this section, you should be able to

a. describe some relatively simple and widespread sensory nerve endings and the sensations associated with them;
b. define and distinguish some types of pain;
c. identify some chemicals that stimulate pain receptors; and
d. explain the mysterious-seeming phenomena of referred pain, phantom pain, and the relative painlessness of some major injuries.

Receptors for the general senses are relatively simple in structure and physiology. They consist of one or a few sensory nerve fibers and, usually, a sparse amount of connective tissue. A few examples are illustrated in figure 10.2.

10.2a Simple Nerve Endings

The simplest receptors are bare dendrites with no connective tissue and include the following:

- **Free nerve endings** include *warm receptors* that respond to rising temperatures, *cold receptors* that respond to falling temperatures, and *nociceptors*[1] (NO-sih-sep-turs) that produce pain sensations in response to tissue injury. Free nerve endings are especially abundant in the skin and mucous membranes.
- **Tactile (Merkel) discs** are receptors for light touch that detect textures, edges, and shapes. They are flattened nerve endings that terminate adjacent to a specialized *tactile cell* at the base of the epidermis. Compression of the tactile cell releases a chemical signal that excites the associated nerve fiber.
- **Hair receptors** are dendrites that coil around a hair follicle and respond to movements of the hair. They are stimulated when, for example, an ant walks across one's skin, bending one hair after another. However, they adapt quickly, so we are not constantly irritated by the feel of clothing against the skin. Hair receptors are particularly important in the eyelashes, where the slightest touch triggers a protective blink reflex.

Some other simple nerve endings are encapsulated by glial cells or connective tissue that wraps around the nerve fiber. Most of these are receptors for touch, pressure, or stretch. Some of them, such as the muscle spindles involved in somatic reflexes (see "The Stretch Reflex" in section 8.3), are **proprioceptors**[2] (PRO-pree-oh-SEP-turs)—receptors specialized to detect the position and movement of the body and its parts. Two other encapsulated nerve endings are as follows (several additional types are not considered here):

- **Tactile (Meissner) corpuscles,** like tactile discs, are receptors for light touch and texture and are found in the dermal papillae at the boundary between the dermis and epidermis. Each consists of two or three nerve fibers meandering upward through a tall mass of flattened Schwann cells. They are especially concentrated in sensitive hairless areas such as the fingertips, palms, eyelids, lips, nipples, and genitals. Drag a fingernail lightly across the back of your hand, then across your palm. The difference in sensation that you feel is due to the high concentration of tactile corpuscles in the palmar skin. Tactile corpuscles enable you to tell the difference between satin and sandpaper, for example, by light strokes of your fingertips.

Free nerve endings

Tactile cell — Nerve ending
Tactile disc

Hair receptor

Tactile corpuscle

Lamellar corpuscle

Figure 10.2 Receptors for the General (Somatosensory) Senses.

---

[1]*noci* = pain
[2]*proprio* = of one's own; *ceptor* = sensory receptor

- **Lamellar (pacinian) corpuscles** are sensitive to deep pressure, stretch, tickle, and vibration. They are 1 or 2 mm long and visible to the naked eye. In cross section, they look like an onion slice, with the nerve fiber at the core surrounded by layers of flattened Schwann cells and fibroblasts. They are especially abundant in the periosteum of bone, joint capsules, the pancreas and some other viscera, and deep in the dermis, especially in the hands, feet, breasts, and genitals.

10.2b Pain

Pain is discomfort caused by tissue injury or noxious stimulation. While few of us enjoy pain and we may wish that no such thing existed, it is necessary and adaptive, providing powerful motivation to avoid or minimize tissue damage or to escape danger. The importance of pain is evident in conditions like leprosy and diabetes mellitus, where nerve damage often diminishes pain sensation and makes people unaware of minor injuries that, if not cared for, can become infected. Victims may lose fingers, toes, or entire limbs. In short, pain is an adaptive sensation that enhances survival and well-being.

Pain has its own specialized nerve endings, the **nociceptors,** which respond to chemicals released by injured tissues. They are especially dense in the skin and mucous membranes and occur in nearly all organs, but are absent from the brain. During brain surgery, patients sometimes must be awake to communicate with the surgeon but need only a local anesthetic for the skin and meninges; operation on the brain itself produces no pain.

Pain from the skin, muscles, and joints is called **somatic pain.** Pain from the internal organs of the body cavities is called **visceral pain.** The latter often results from stretch, chemical irritants such as alcohol and bacterial toxins, or a drop in blood flow, as in menstrual cramps. It is often accompanied by nausea. The precise location of visceral pain is more difficult to pinpoint than that of somatic pain.

Three fascinating aspects of pain are referred pain, phantom pain, and spinal gating of pain. **Referred pain** is a phenomenon in which pain from the viscera is mistakenly thought to come from the skin or other superficial sites—for example, when the pain of a heart attack is felt "radiating" along the left shoulder and arm. This results from the fact that pain fibers from the heart and skin converge on the same neurons in the central nervous system, so by the time these signals get to the pain centers of the brain, the brain cannot identify their source. Of these two, the skin is more frequently injured, so the brain acts as if it "assumes" that the pain most likely comes from the skin. Knowledge of the areas of referred pain (fig. 10.3) is important to a physician in diagnosing the true source of discomfort reported by a patient.

Figure 10.3 Referred Pain. (a) Basis of referred pain in a heart attack. The brain cannot distinguish the source of pain if two or more origins feed into the same interneurons and CNS pathways to the sensory cortex. (b) Pain from the viscera is often felt in specific regions of the skin.

(a)

(b)

Clinical Application 10.1

MIGRAINE HEADACHE

About 5% of men and twice as many women in the United States suffer from *migraine headaches*. These differ from ordinary *tension headaches* and the *symptomatic headaches* that stem from underlying disorders such as brain tumors and infections. Migraines are felt as throbbing, moderate to severe pain, often behind one eye or ear. In slightly over half of migraine sufferers, these headaches are limited to one side of the head (the word *migraine* is a French corruption of the Latin *hemicrania,* "half-cranium"). They are commonly preceded by an *aura,* or subjective sensation that a headache is coming on, that may include visual disturbances such as light flashes and blind spots, and sometimes dizziness or ringing in the ears. The aura often lasts for 20 to 30 minutes before the onset of pain itself, and is associated with a wave of reduced cerebral blood flow starting in the occipital region and progressing forward in one hemisphere of the brain. A migraine headache commonly lasts from 4 hours to 3 days and may be accompanied by photophobia (aversion to light), scalp tenderness, nausea, and vomiting. The pain arises not from the brain tissue but from fibers of the trigeminal nerve that innervate the dura mater and cerebral arteries. The cause of migraine headaches is still not well known; possible culprits of special interest to medical physiologists are serotonin and other neurotransmitters that stimulate blood vessel dilation and constriction.

Phantom pain is the eerie sensation of pain coming from a limb that has been amputated. It may be associated with other illusory sensations as well, such as the bulk and weight of the absent limb or feelings of itching or movement. Any irritation of the stump of a limb can set off nerve impulses along the corresponding spinal pathway, and the brain does not perceive that the distal part of the limb is absent.

People are sometimes severely injured and yet feel little or no pain—for example, soldiers mortally wounded on a battlefield. The absence or minimal sense of pain results from a mechanism called **spinal gating.** Certain spinal and brainstem neurons secrete pain-blocking substances called *endorphins* and *enkephalins*, which stop pain signals at synapses in the cord or brainstem before they reach conscious levels of the brain. We make more everyday use of this phenomenon when, for example, we painfully bump our elbow and rub the area to ease the pain. The rubbing action activates the spinal gating mechanism and lessens the pain. Unfortunately, efforts to employ endorphins or enkephalins for the clinical management of pain have been disappointing.

Before You Go On

Answer these questions from memory to test your understanding of the previous section. Reread the material if you can't answer the questions.

5. Name some simple sensory nerve endings and identify sensations produced by each.
6. What is the difference between somatic and visceral pain?
7. Why does the brain perceive the pain of a heart attack as coming from the shoulder and arm?
8. Explain how a major, fatal injury can feel less painful than a relatively minor cut or burn.

10.3 The Chemical Senses—Taste and Smell

Expected Learning Outcomes

When you have completed this section, you should be able to

a. describe the receptor cells for taste and smell and identify their anatomical locations;
b. identify the five primary taste sensations and the chemicals that produce them;
c. discuss factors other than taste that contribute to the flavor of food; and
d. describe gustatory and olfactory sensory pathways to the brain and identify the brain regions that process the information.

Our world is made of chemicals, and many of these play an essential survival role in the selection of appropriate foods, avoidance of danger, and so forth. We chemically monitor our food and drink through the sense of taste and monitor the air through the sense of smell. These are called the **chemical senses**. **Chemoreceptors** are organs and nerve endings that detect chemicals; they include receptors in some arteries for monitoring blood chemistry, and our organs of taste and smell.

10.3a Gustation—The Sense of Taste

Gustation begins with the chemical stimulation of sensory cells clustered in about 4,000 **taste buds.** Most of these are on the tongue, but some occur inside the cheeks and on the palate, pharynx, and epiglottis, especially in infants and children. The visible bumps on the tongue, called **lingual papillae** (fig. 10.4), include *vallate, foliate, fungiform,* and *filiform* types. They are not themselves taste buds. Up to half of our taste buds are found in just 7 to 12 vallate papillae that form a V at the rear of the tongue. The spiky filiform papillae (responsible for the roughness of a cat's tongue) have no taste buds in humans but are employed in perception of food texture, or what food technologists call *mouthfeel*.

Regardless of location, all taste buds look alike—ovoid clusters of banana-shaped **taste cells** mixed with nonsensory *supporting cells* and unspecialized *basal cells* that replace dead taste cells (fig. 10.4d). Taste cells have microvilli called *taste hairs* that project into a pit called the *taste pore* at the apex of the taste bud. To be tasted, food chemicals *(tastants)* must dissolve in the saliva and flood the taste pore to reach the taste hairs. Taste cells are epithelial cells, not neurons, but they have synaptic vesicles at the base of the cell that release a neurotransmitter and stimulate an adjacent nerve fiber.

The five **primary taste sensations** are:

1. **Salty,** produced by metal ions such as sodium and potassium. Electrolyte deficiencies can cause a person or animal to crave salty food.
2. **Sweet,** produced by sugars and many other organic compounds and usually associated with carbohydrates and high-calorie foods.
3. **Sour,** produced by the acids in such foods as citrus fruit.
4. **Bitter,** produced by organic alkaloid compounds such as nicotine, caffeine, and quinine, and associated with plant toxins and spoiled foods. The bitter sensation probably served in evolution as a warning not to ingest a substance, but many people cultivate a liking for certain bitter foods and drinks, such as capers and tonic water.
5. **Umami**[3] (pronounced "ooh-mommy"), a meaty taste produced by amino acids such as aspartic and glutamic acid and by monosodium glutamate (MSG), a glutamic acid salt widely used as a flavor-enhancing food additive.

---

[3]*umami* = Japanese slang loosely meaning delicious or yummy

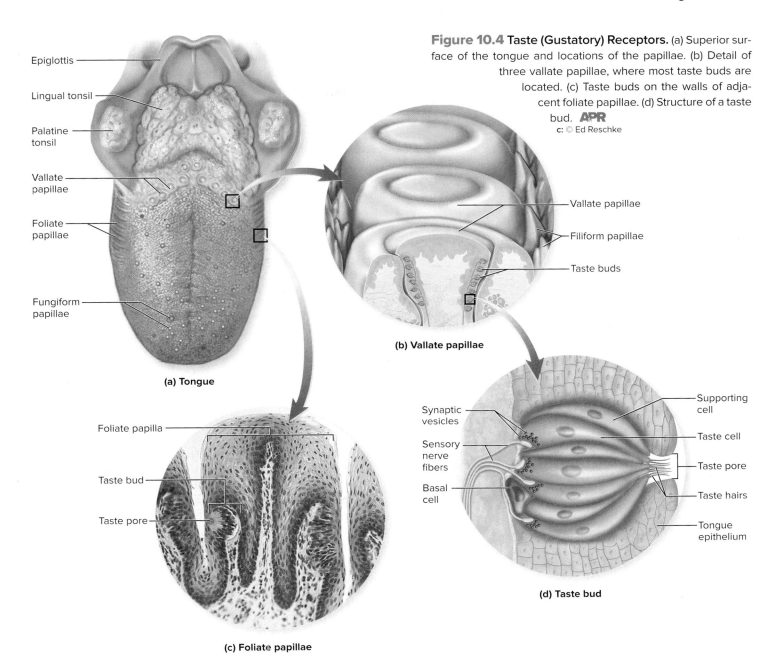

Figure 10.4 Taste (Gustatory) Receptors. (a) Superior surface of the tongue and locations of the papillae. (b) Detail of three vallate papillae, where most taste buds are located. (c) Taste buds on the walls of adjacent foliate papillae. (d) Structure of a taste bud. APR
c: © Ed Reschke

Recent research suggests that two more primary tastes may warrant inclusion in this list, as they each have their own unique oropharyngeal receptors: *water* (if one considers that a "taste") and *oleogustus*,[4] the proposed name for the taste of fats.

The *flavor* of a food involves much more than its taste; it is also affected by aroma, texture, temperature, appearance, and state of mind. Without their aromas, cinnamon merely tastes faintly sweet and coffee and peppermint are bitter. The hot flavor of peppers and some other foods stems from stimulation of free sensory endings of the trigeminal nerve rather than stimulation of the taste buds.

The facial, glossopharyngeal, and vagus nerves (cranial nerves VII, IX, and X) convey taste signals to the medulla oblongata of the brain. Signals are relayed

---

[4]*oleo* = oil; *gust* = taste

from here to regions of the insula and postcentral gyrus of the cerebrum, where we become conscious of a taste and integrate it with sensations of smell, vision, texture, and temperature to form our overall impression of the flavor and palatability of a food.

Why do humans prefer certain tastes? Evolutionarily, seeking foods with high nutrient quality and avoiding food that may be toxic is adaptive for survival. For our hunter–gatherer ancestors who rarely had access to sugary, salty, fatty foods, it was advantageous to seek sweet, salty, and umami tastes. In industrialized society where most people are relatively sedentary and where we have access to cheap, abundant sugary and high-fat foods, the motivation to seek out such tastes is no longer adaptive, but instead, is associated with obesity and the diseases that plague our modern lives.

10.3b Olfaction—The Sense of Smell

Olfaction, the sense of smell, is a response to airborne chemicals *(odorants)*. These are detected by a patch of sensory epithelium called the **olfactory mucosa** in the roof of the nasal cavity (**fig. 10.5**). The mucosa measures about 5 cm^2 and consists of 10 to 20 million **olfactory cells,** nonsensory supporting cells and basal cells, as well as mucus-secreting *olfactory glands.* Olfactory cells are true neurons, unlike taste cells. They have a swollen tip bearing 10 to 20 cilia called **olfactory hairs,** a bulbous body containing the nucleus, and a basal end that tapers to a thin axon leading to the brain. Being the only neurons directly exposed to the external environment, they live only about 60 days and are continually replaced by dividing basal cells.

It is a myth that humans have a poor sense of smell relative to other mammals. Although other animals may have more olfactory neurons (over 200 million in dogs for, example), humans are more sensitive to some odors than even dogs and rodents. Different species are able to detect different odorants, probably due to their unique sensory environment and ecological adaptation. Most people can distinguish 2,000 to 4,000 distinct odors; some, such as world-class food and wine experts, can distinguish up to 10,000. Human behavior is influenced by smell; for example, odor plays a role in unconscious communication between individuals and influences mate choice. Smell is closely tied to memories and emotion. Decreased ability to smell is an early predictor of Alzheimer disease and olfactory dysfunction due to disease or injury is associated with depression.

The first step in olfaction is usually that an odor molecule diffuses through the mucus on the surface of the olfactory epithelium and binds to one of the hairs on a receptor cell. This leads to depolarization of the cell, generating action potentials in its axon. Some chemicals, however, act on nociceptors of the trigeminal nerve rather than on olfactory cells—for example, ammonia, menthol, chlorine, and the capsaicin of hot peppers. "Smelling salts" revive unconscious persons by strongly stimulating the trigeminal nerve endings with ammonia fumes.

The axons from multiple olfactory cells converge to form little bundles that collectively constitute the olfactory nerve (cranial nerve I). These penetrate through several pores in the cribriform plate of the ethmoid bone. Directly above this plate, the nerve fibers end in a pair of *olfactory bulbs* of the brain (see fig. 9.14). From here, other neurons form bundles of fibers called *olfactory tracts* that relay signals to multiple destinations in the cerebrum and brainstem, especially the **primary olfactory cortex** in the temporal lobe. Higher brain centers interpret complex odors such as chocolate, coffee, and perfume by decoding a mixture of signals, like combining different primary colors to get all the colors of the light spectrum.

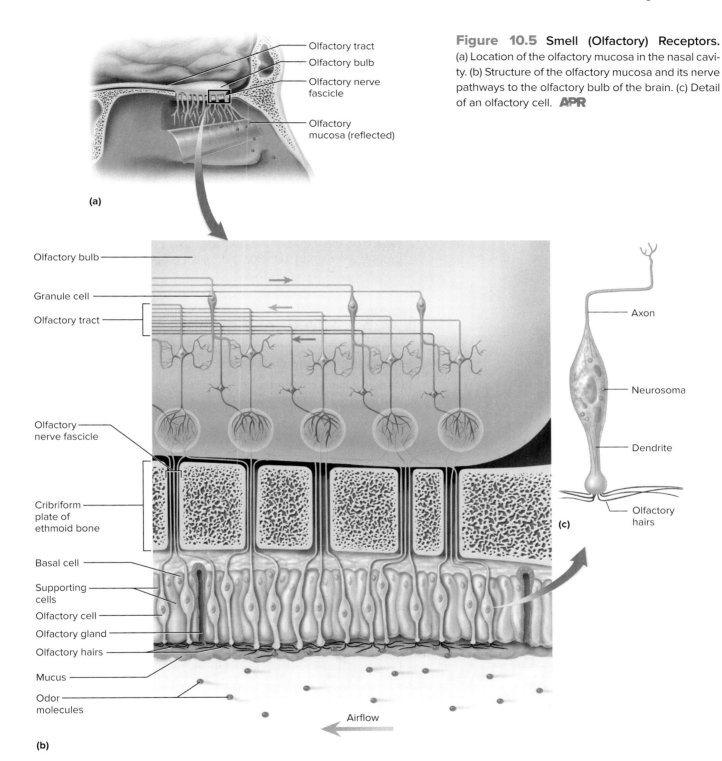

Figure 10.5 Smell (Olfactory) Receptors. (a) Location of the olfactory mucosa in the nasal cavity. (b) Structure of the olfactory mucosa and its nerve pathways to the olfactory bulb of the brain. (c) Detail of an olfactory cell. APR

Apply What You Know

Look up the meaning of *anosmia*. Explain why anosmia could endanger a person's health or life. Also describe a couple of your favorite foods or beverages and how your enjoyment of them would be affected by anosmia.

Before You Go On

Answer these questions from memory to test your understanding of the previous section. Reread the material if you can't answer the questions.

9. Other than the tongue, where are taste buds located? Are taste buds visible with the naked eye? Explain.
10. Which of these are neurons—taste cells or olfactory cells? Which cell type binds stimulant molecules with microvilli? Which one employs cilia?
11. What are the five primary taste sensations? Explain why humans are adapted to have certain taste preferences.
12. Where do the axons of the olfactory cells end? Where in the brain is the primary olfactory cortex located?

10.4 The Ear—Equilibrium and Hearing

Expected Learning Outcomes

When you have completed this section, you should be able to

a. identify the anatomical components of the outer, middle, and inner ear;
b. explain how the vestibular system of the inner ear functions to give us a sense of the body's position and movements; and
c. explain how the cochlea of the inner ear converts the vibrations of a sound wave into a nerve signal, and how the brain interprets such signals.

The ear serves two very different human senses—equilibrium and hearing. The original evolutionary function of the ear was equilibrium. Its role in hearing emerged only as vertebrates colonized land and benefited from sensitivity to airborne vibrations. This was when animals evolved the *outer* and *middle ears,* including an eardrum *(tympanic membrane),* and a new inner-ear structure, the *cochlea,* specialized for hearing. As you will soon see, however, the cochlea and organs of balance work by surprisingly similar means—the movement of inner-ear fluids and gelatinous membranes relative to a type of sensory cells called *hair cells.* The ears are **mechanoreceptors**—organs that generate nerve signals in response to physical distortion of the plasma membranes of receptor cells. Touch, pressure, and some of our other senses also fall in this category.

10.4a General Anatomy of the Ear

The ear consists of three sections called the *outer, middle,* and *inner ear* (fig. 10.6). The outer ear is essentially a funnel that collects sound waves; the middle ear relays them to the inner ear and protects the ear from loud sounds; and the inner ear converts vibrations and body movements into nerve signals.

The Outer Ear

The most prominent feature of the **outer ear** is the fleshy **auricle (pinna)** on the side of the head. The auricle is shaped and supported by elastic cartilage except for the fatty *lobule* (earlobe). The pattern of whorls and recesses in the auricle direct sound into the auditory canal. The **auditory canal,** or **external acoustic meatus**

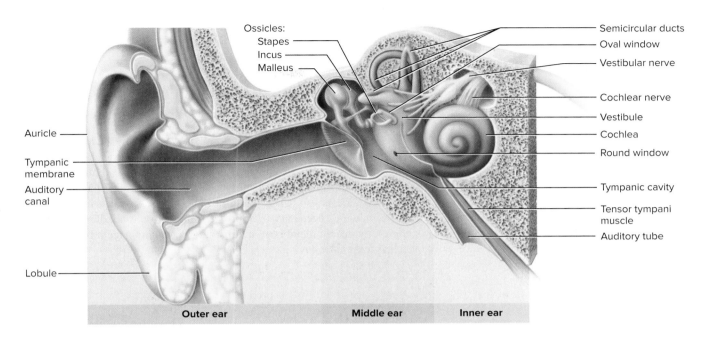

Figure 10.6 General Anatomy of the Ear.

❓ *Cranial nerves are numbered I through XII. Give the name and roman numeral of the thick yellow cranial nerve emerging at the right side of this figure.* **APR**

(me-AY-tus), is a passage through the temporal bone leading to the eardrum. It follows a slightly S-shaped course for about 3 cm. It is lined with skin and possesses ceruminous and sebaceous glands. The secretions of these glands form **cerumen** (earwax), which waterproofs the canal, inhibits bacterial growth, and keeps the eardrum pliable. It also coats the *guard hairs* at the entrance to the canal, making them sticky and more effective at blocking parasites and debris from entering the canal.

The Middle Ear

What we colloquially call the eardrum is more properly known as the **tympanic membrane.** It marks the beginning of the **middle ear.** Medial to the membrane is a tiny air-filled chamber, the **tympanic cavity,** only 2 to 3 mm wide. It contains the body's three smallest bones and its two smallest skeletal muscles. The three bones, called **auditory ossicles,**[5] transfer vibrations from the tympanic membrane to the inner ear. From lateral to medial, they are the **malleus**[6] (attached to the inside of the tympanic membrane), **incus,**[7] and **stapes**[8] (STAY-peez).

The tympanic membrane would not vibrate freely if the air pressure on one side were much greater than pressure on the other side. Therefore, the middle ear is connected to the pharynx by the **auditory (pharyngotympanic** or **eustachian**[9]**) tube.** This tube is normally flattened and closed, but when we yawn or swallow, it opens and allows air to enter or leave the tympanic cavity. This equalizes air pressure on the two sides of the tympanic membrane, but unfortunately it also frequently allows throat infections to spread to the middle ear (see Clinical Application 10.2).

The two muscles of the middle ear are called the **stapedius** (sta-PEE-dee-us) and **tensor tympani** (TEN-sur TIM-pan-eye). They insert on the stapes and malleus, respectively. In response to loud but slowly building noises such as thunder, they contract and inhibit the movement of these bones, protecting sensory cells of the inner ear from destructive overstimulation. They are not adequately effective,

---

[5]*oss* = bone; *icle* = little

[6]*malleus* = hammer, mallet

[7]*incus* = anvil

[8]*stapes* = stirrup

[9]Bartholomeo Eustachio (1520–74), Italian anatomist

Clinical Application 10.2

MIDDLE-EAR INFECTION

Otitis[10] *media* (middle-ear infection) is especially common in children because their auditory tubes are relatively short and horizontal compared to those of adults (fig. 10.7), providing easy passage for microbes to travel from the throat to the middle-ear cavity. Infection can lead to fluid accumulation in the middle ear, which inhibits the vibration of the auditory ossicles, impairs hearing, and causes pain. If untreated, the infection can spread and cause *meningitis,* a potentially deadly infection of the meninges around the brain. Chronic otitis media can also cause fusion of the ossicles, thus causing hearing loss. It is sometimes necessary to drain the fluid from the middle ear by puncturing the tympanic membrane and inserting a tiny drainage tube—a procedure called *tympanostomy.*

Figure 10.7 Auditory Tubes of Children and Adults. (a) The auditory tube in the child is oriented more horizontally compared to that of the sloping tube in the adult. (b) The more nearly vertical tube in the adult effectively drains the middle ear and lessens the risk of infections.

however, at protecting the inner ear (or one's hearing) from constant loud noise such as amplified concert music or from sudden loud noises such as gunshots close to the ear. Such sounds can irreversibly destroy sensory cells of the inner ear and cause permanent hearing impairment.

The Inner Ear

The **inner ear** is a complex region that transforms mechanical energy (vibration) to nerve energy. It is housed in a maze of passages in the temporal bone called the *bony labyrinth* and consists mostly of a complex of fluid-filled chambers and tubes called the **membranous labyrinth** (fig. 10.8). The membranous labyrinth within the bony labyrinth is thus a tube-within-a-tube structure, like a bicycle inner tube within a tire. The membranous labyrinth is filled with a liquid called *endolymph,* similar to intracellular fluid, and surrounded by a liquid called *perilymph,* similar to cerebrospinal fluid.

The stapes of the middle ear has an elongated *base plate* suspended in an opening called the **oval window.** This window marks the beginning of the inner ear. On

---

[10]*ot* = ear; *itis* = inflammation

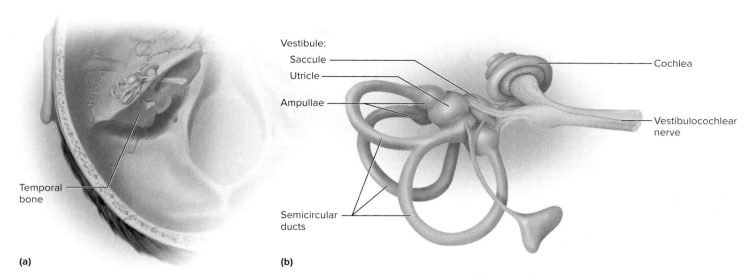

Figure 10.8 The Membranous Labyrinth of the Inner Ear. (a) Position and orientation of inner-ear structures within the temporal bone. (b) Structure of the membranous labyrinth and origins of the vestibulocochlear nerve. APR

the other side of it is a chamber, the **vestibule,** filled with endolymph. Within the vestibule are two pouches called the *saccule* and *utricle,* and arising from it are three loops called *semicircular ducts* and a snail-like coil called the *cochlea* (fig. 10.6). The saccule, utricle, and semicircular ducts are collectively called the **vestibular apparatus,** and are the organs of equilibrium. The cochlea is the organ of hearing.

10.4b Equilibrium

Equilibrium is the sense of body position, movement, and balance. The sense of equilibrium is divided into **static equilibrium,** the sense of orientation of the head in space (whether it is erect or tilted in any direction), and **dynamic equilibrium,** the sense of movement and acceleration. Receptors for both types lie in the saccule and utricle and in the semicircular ducts, as we shall see.

The **saccule**[11] and **utricle**[12] are pouches immediately medial to the middle ear. Each is filled with endolymph and contains a patch of epithelium called a **macula,**[13] composed of sensory **hair cells** and nonsensory supporting cells (fig. 10.9). A hair cell is columnar to pear-shaped, with hairlike microvilli called **stereocilia**[14] on its apical surface. Overlying the macula is a layer of gel called the **otolithic membrane,** containing granules of protein and calcium carbonate called **otoliths.**[15] The otoliths give the membrane added weight and inertia, which enhances its ability to stimulate the hair cells when the body moves.

To understand how this mechanism works, consider the macula that lies almost horizontally on the floor of the utricle. When you tilt your head down to read a book, the heavy otolithic membrane sags downward and bends the stereocilia, stimulating the hair cells (fig. 10.9c). These cells also respond when the body accelerates or decelerates horizontally. For example, if you are sitting in a car that begins to move, the macula moves at the same speed, but the inertia of the otolithic membrane makes it lag behind briefly. This bends the stereocilia backward and stimulates the

---

[11]*saccule* = little sac
[12]*utricle* = little bag
[13]*macula* = spot, patch
[14]*stereo* = solid
[15]*oto* = ear; *lith* = stone

Figure 10.9 The Saccule and Utricle. (a) Locations of the maculae. (b) Structure of a macula. (c) Movement of the otolithic membrane when the head is tilted.

hair cells. If you slow down and stop, the macula slows down at the same rate, but the otolithic membrane briefly keeps moving forward and bends the stereocilia the other way. By comparing input from the utricles of both ears, your brain can evaluate tilts of the head and changes in horizontal motion.

The saccule has an almost vertical macula and its hair cells therefore respond to vertical acceleration and deceleration. If, for example, you are standing in an elevator that begins to move up, the otolithic membrane pulls down on the stereocilia. As the elevator slows to a stop, the macula does too, but the membrane keeps moving upward for a moment. At both the start and stop, the hair cells are stimulated and your brain becomes aware of your vertical movements. These sensations are important in everyday actions such as when you stand up and as your head bobs up and down during walking and running.

The head also rotates, such as when you spin in a chair, walk down a hall and turn a corner, or bend forward to pick something up from the floor. Such movements are detected by the three **semicircular ducts** (fig. 10.10). One of these lies on a plane about 30° from horizontal, and the other two are oriented vertically at right angles to each other. Each duct has a bulb at its base called the **ampulla.** Within the ampulla, there is a mound of sensory epithelium called the *crista ampullaris,* once again composed of hair cells and supporting cells. These are topped by a gelatinous cap called the **cupula,** but this gel has no otoliths. The ducts are filled with endolymph.

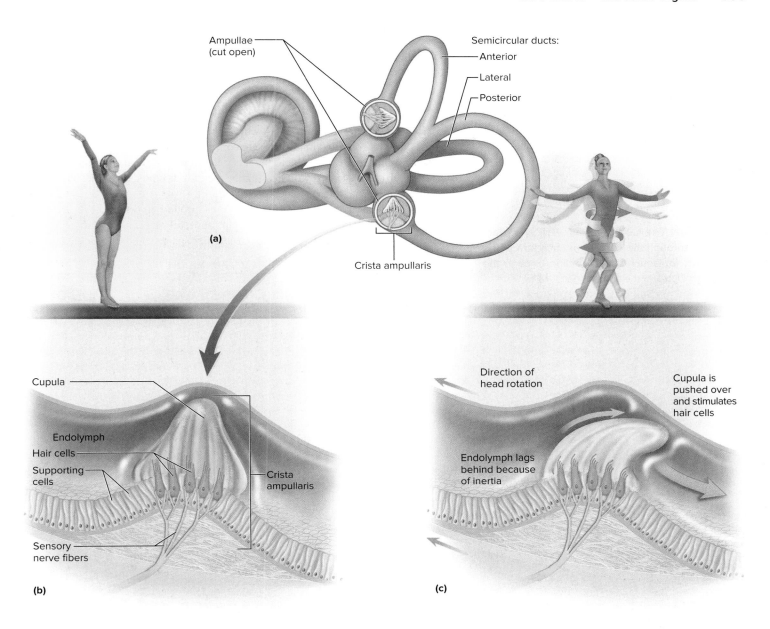

Figure 10.10 The Semicircular Ducts. (a) Structure of the semicircular ducts, with each ampulla opened. (b) Detail of the sensory receptor in the ampulla. (c) Action of the endolymph on the cupula and hair cells when the head rotates.

Suppose you're sitting in a rotating office chair and you spin around to get something from a desk behind you. The horizontal semicircular duct will spin with you, but the fluid within it will lag a bit. This pushes the cupula, bends the stereocilia of the hair cells, and stimulates them. Tilting your head forward or sideways—say, to pick up something on the floor or to lie down in bed—similarly engages the vertical semicircular ducts and stimulates their hair cells. The ducts are arranged to detect rotation in any of the *x, y,* and *z* geometric planes, and movements in a direction between any two of these planes create a mixture of signals from two or more of the ducts in each ear.

When a hair cell is stimulated, it releases a neurotransmitter from synaptic vesicles at its base and stimulates an adjacent nerve fiber. The nerve fibers from the utricle, saccule, and semicircular ducts form the *vestibular nerve,* which joins the *cochlear nerve,* discussed shortly, to form *cranial nerve VIII,* the *vestibulocochlear* nerve described in section 9.4. The nerve fibers lead mostly to the pons of the brainstem. The pons relays signals to several other centers you have studied in the last two chapters: (1) the primary somatosensory cortex, where one becomes conscious of head movements; (2) the spinal cord, where one generates signals to the muscles

336 PART 3 Internal Coordination

that maintain posture and balance; (3) the cerebellum, a center of motor coordination; and (4) midbrain nuclei that activate eye movements to compensate for head movements, keeping one's vision fixed on a given target in spite of head motion.

10.4c The Cochlea and Hearing APR

Hearing is the awareness of sound—waves of molecular motion created by vibrating objects. The outer and middle ear are concerned with collecting airborne sound waves and transferring vibration to the inner ear. In the inner ear, only the **cochlea**[16] (COC-lee-uh) is concerned with hearing. Named for its coiled snail-like shape, this structure converts vibrations to nerve signals.

The cochlea is a fleshy tube that coils around a screwlike bony core. A vertical section through the cochlea cuts the coil at each turn, as shown in **figure 10.11**. The tube is divided into a triangular space called the **cochlear duct,**

---

[16]*cochlea* = snail

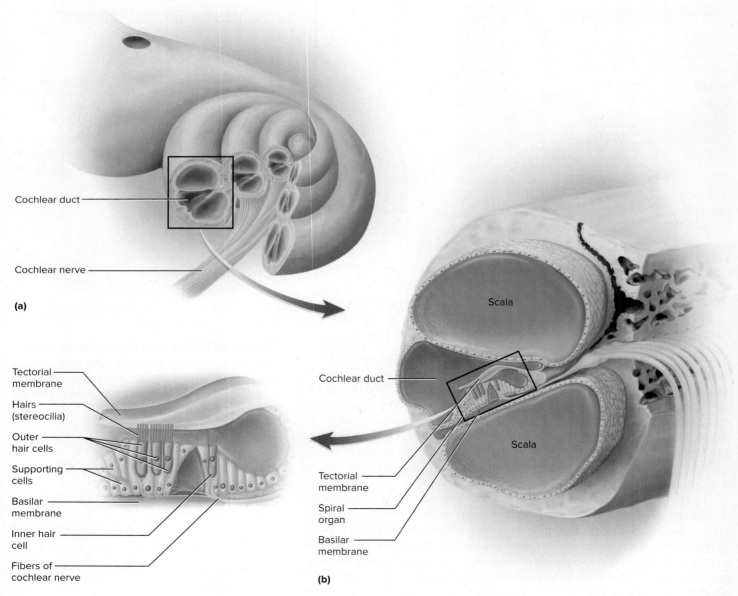

Figure 10.11 The Cochlea. (a) A vertical section, which cuts through the cochlea at each turn of the coiled tube. (b) Detail of one section through the cochlea. (c) Detail of the spiral organ. APR

filled with endolymph, and two larger spaces called *scalae* (SCALE-ee) above and below it, filled with perilymph. A thin membrane separates the upper scala from the cochlear duct and a thicker **basilar membrane** forms the floor of the cochlear duct, separating it from the lower scala. The basilar membrane is a platform for the sensory cells of the cochlea.

The sensory cells of hearing, like those of equilibrium, are hair cells. Together with supporting cells and accessory membranes, they are arranged into a coiled ribbon called the **spiral (acoustic) organ.** Cochlear hair cells are organized into two groups. Everything we hear comes from a single row of about 3,500 **inner hair cells,** each with a row of about 50 or 60 stereocilia on the surface (fig. 10.12). Across from these, farther away from the bony core of the cochlea, are about 20,000 **outer hair cells.** These are arranged in three rows, and each cell bears a V-shaped array of stereocilia. Despite their greater number and arguably more interesting appearance, they do not generate the sounds we hear. They tune the cochlea to enable the brain to better discriminate one pitch of sound from another.

A gelatinous **tectorial**[17] **membrane** lies just above the hair cell stereocilia. Medially, this membrane is anchored to the core of the cochlea. Laterally, it is anchored by the tips of the outer hair cell stereocilia, which are embedded in the gel. The stereocilia of the inner hair cells stop just short of the tectorial membrane and lie free in the endolymph.

The essence of sound and hearing is this: Vibrating objects set air molecules in motion, like a series of colliding billiard balls, until some strike the tympanic membrane and set it in motion. The ossicles of the middle ear transfer this vibration to the fluids of the inner ear. Each time the footplate of the stapes moves in and out at the oval window, it moves the perilymph of the upper scala, which in turn transfers pressure through the endolymph to the basilar membrane of the cochlea (fig. 10.11). The basilar membrane vibrates up and down and the hair cells go along for the ride. With each upward movement, stereocilia of the inner hair cells are pushed against the tectorial membrane above, forcing them to bend over. As we saw in the vestibular apparatus, this bending excites the hair cells and stimulates them to release neurotransmitter from the synaptic vesicles at the base of each cell. The neurotransmitter excites adjacent fibers of the *cochlear nerve* and generates a nerve signal, which we will trace later. Thus, vibrations of the basilar membrane are translated to rhythmic bursts of nerve activity in the cochlear nerve.

Two features of a sound are particularly important to our sense of hearing—its loudness (amplitude) and its frequency (low-pitched or high). The basilar membrane is structured in such a way that it generates a code for each of these sound qualities in the nerve signal.

The louder a sound, the more vigorously the basilar membrane vibrates. This results in stronger signaling of the cochlear nerve fibers by the hair cells, and more action potentials per second in the cochlear nerve fibers to the brain. The brain interprets a high rate of nerve firing as a loud sound and a low firing rate as a quieter sound.

As for the frequency (pitch) of a sound, the basilar membrane tapers from a relatively stiff and narrow proximal end, anchored to the base of the cochlea, to a wide and limber distal end, with no anchorage, at the apex of the cochlea. High-pitched sounds cause the most vibration at the narrow basal end, much as the shortest strings of a piano produce the highest pitch (fig. 10.13). Low-pitched sounds produce the most vibration at the wide distal end, like the longest piano strings producing bass notes. The brain receives signals from a mixture of nerve fibers representing all areas of the cochlea, and decodes the mixture of signals into a perception of pitch. Most things we hear, of course, are not pure tones but complex sounds with a variety

Figure 10.12 Apical Surfaces of the Cochlear Hair Cells.
© Quest/Science Photo Library/Science Source

❓ *Which of the cells in this photo are the source of all sounds that one hears? What do the other cells do, if we cannot hear any of the nerve signals that come from them?*

---

[17]*tect* = roof

Figure 10.13 Frequency Response of the Basilar Membrane of the Cochlea. (a) The cochlea, uncoiled and laid out straight. (b) Sound waves produce a wave of vibration along the basilar membrane. The areas of greatest vibration vary with the frequency of the sound. The amount of vibration is greatly exaggerated in this illustration. (c) The taper of the basilar membrane and its correlation with sound frequencies. High frequencies are best detected by hair cells near the narrow proximal end of the membrane (left), and low frequencies by hair cells near the wide free end (right). (Hz = hertz, or cycles per second, a measure of the frequency of a sound wave)

of pitches, stimulating all parts of the basilar membrane to varying degrees. The brain therefore has to carry out quite a phenomenal task of decoding the complex message. Neuroscientists have achieved a deep understanding of this mechanism, but the details are beyond the scope of this book.

10.4d The Auditory Projection Pathway

Nerve fibers from the basilar membrane feed into the *cochlear nerve,* which then becomes part of the already-discussed *vestibulocochlear* nerve leading to the pons (see fig. 10.8). The pons sends signals back to the cochlea for tuning by the outer hair cells, and back to the middle ear for the reflexes of the stapedius and tensor tympani muscles. It also compares input from the right and left ears to give us a sense of *binaural*[18] *hearing,* the ability to localize sounds in space. The signals for hearing ascend to the inferior colliculi of the midbrain, which further aid in binaural hearing and in processing fluctuations in pitch. Finally, signals ascend farther to be relayed through the thalamus to the **primary auditory cortex** in the temporal lobe of the cerebrum, which is where we become consciously aware of the sound.

---

[18]*bin* = two; *aur* = ears

Before You Go On

Answer these questions from memory to test your understanding of the previous section. Reread the material if you can't answer the questions.

13. What mechanism protects the inner ear from damage by loud sounds? What part of the brain controls this mechanism?
14. Name six places where hair cells can be found in the inner ear. How do hair cells resemble neurons? How do they differ from neurons?
15. Explain how a semicircular duct differs in function from the saccule and utricle.
16. Describe the events beginning with vibration of the tympanic membrane and ending with the release of neurotransmitter from a cochlear hair cell.
17. How can the brain tell a low-pitched sound from a high-pitched sound? A loud sound from a softer sound?

10.5 The Eye and Vision

Expected Learning Outcomes

When you have completed this section, you should be able to

a. describe organs surrounding the eye that protect it and aid in its functions;
b. describe the anatomy of the eye itself;
c. explain how the eye controls light and focuses it on the retina;
d. explain how cells of the retina absorb light and generate nerve signals; and
e. describe the projection pathway that these signals take through the brain.

The eyes are classified as **photoreceptors,** organs specialized to detect light. **Vision,** a step beyond mere photoreception, is the ability to form a recognizable image of an object that emits or reflects light. It begins with focusing light on a membrane called the *retina,* the site of our sensory cells for vision. Here, the light creates a chemical reaction that leads to the generation of a nerve signal. The brain must receive and interpret this nerve signal for the process of vision to be complete.

10.5a Anatomy of the Orbital Region

The eyeball occupies a bony socket called the **orbit.** This general area of the face is the *orbital region.* It contains structures that protect and aid the eye (fig. 10.14):

- The **eyebrows** enhance facial expressiveness and perhaps shield the eye to some degree from glare and forehead perspiration.
- The **eyelids (palpebrae)** block foreign objects from the eye, screen out visual stimuli to facilitate sleep, and blink periodically to moisten the eye with tears and sweep debris from its surface. An eyelid consists mostly of *orbicularis oculi* muscle covered with skin. Along the margin of the eyelid are 20 to 25

tarsal glands that secrete a protective oily film onto the eye surface. The **eyelashes** serve as guard hairs to keep debris from the eye; touching them stimulates hair receptors and triggers the blink reflex.

- The **conjunctiva** (CON-junk-TY-vuh) is a transparent mucous membrane that covers the inner surface of the eyelid and anterior surface of the eyeball except for the region of the cornea. It produces a thin mucous film that prevents drying of the eye. It is richly innervated and very sensitive to pain, and well supplied with minute blood vessels. The appearance of "bloodshot" eyes is due to dilation of these vessels.

- The **lacrimal apparatus** is a group of structures that produce tears and drain them away from the eye. The almond-shaped **lacrimal (tear) gland** is nestled in the superolateral corner of the orbit, between the eyeball and frontal bone. Approximately 12 short ducts carry tears from the gland to the surface of the conjunctiva. Tears cleanse and lubricate the eye surface, supply oxygen and nutrients to the conjunctiva, and contain an antibacterial enzyme to prevent eye infections. After flowing across the eye, tears drain into a pore at the medial corner of each eyelid and from there into a **lacrimal sac** in the medial wall of the orbit. A **nasolacrimal duct** drains the tears into the nasal cavity; watery eyes can therefore lead to a runny nose.

- **Orbital fat** fills the space between the eyeball and bones of the orbit. It cushions the eye, allows it to move freely, and protects the blood vessels, nerves, and muscles of the orbit.

- Six **extrinsic muscles** arise from the walls of the orbit and insert on the surface of the eyeball. The term *extrinsic* distinguishes these from the *intrinsic* muscles inside the eyeball that control the lens and iris. The extrinsic muscles are skeletal muscles that move the eye when we look up, down, or to one side.

Figure 10.14 Accessory Structures of the Orbit. (a) Sagittal section of the orbit. A few of the extrinsic muscles are shown (superior, lateral, and inferior rectus muscles). (b) The lacrimal apparatus for secreting tears and draining them into the nasal cavity. **APR**

❓ *What would be the expected result of an infection that blocked the pores at the medial corner of the eyelids?*

(a)

(b)

10.5b Anatomy of the Eyeball

The eyeball itself is a sphere about 24 mm (1 in.) in diameter. It is composed of three tissue layers that form its wall, optical components that admit and focus light, and neural components that absorb light and generate a nerve signal (fig. 10.15).

1. The outer **fibrous layer** is composed of the tough, protective, white **sclera** ("white of the eye") over most of the eye surface, and the transparent **cornea** over the anterior central region. Only the cornea admits light into the eye.

2. The middle **vascular layer** consists of three regions:
   - The **iris**, an adjustable diaphragm that admits light through its central opening, the **pupil.** The iris contains pigment cells with variable amounts of melanin. If melanin is abundant, the eye color ranges from hazel to brown or black. If melanin is sparse, light reflects from a more posterior layer of the iris and gives it a blue, green, or gray color.
   - The **ciliary body,** a thick ring of smooth muscle tissue that encircles and supports the iris and lens and adjusts lens shape for focusing.
   - The **choroid** (CORE-oyd), a deeply pigmented layer, rich in blood vessels, that underlies and nourishes the retina.

3. The inner **neural layer** consists of the *retina* and the beginning of the *optic nerve*.

The optical components of the eye are transparent elements that admit light rays, bend (refract) them, and focus images on the retina. They include the cornea (already discussed) and the following:

1. The **aqueous humor,** a watery fluid that fills all the space between the cornea and lens. The space from lens to pupil is called the *posterior chamber,* and the space from pupil to cornea is called the *anterior chamber.* The ciliary body secretes aqueous humor into the posterior chamber, the fluid flows

Figure 10.15 The Eye (Horizontal Section). APR

Clinical Application 10.3

GLAUCOMA

Glaucoma[19] is a state of elevated pressure in the eye that occurs when aqueous humor is secreted faster than it is reabsorbed. It results from an obstruction of the scleral venous sinus. Pressure in the anterior and posterior chambers pushes the lens back and puts pressure on the vitreous body. The vitreous body then presses excessively on the retina and choroid and compresses the blood vessels that nourish the retina. Without a good blood supply, retinal cells die and the optic nerve may atrophy, leading to blindness. Often the damage is already irreversible by the time a person notices the first symptoms. Illusory light flashes are an early warning sign of glaucoma. In later stages, the sufferer may experience dimness of vision (origin of the word *glaucoma*), a narrow visual field, and colored halos around artificial lights. The progression of glaucoma can be halted with drugs or surgery, but lost vision cannot be restored. The disease can be detected at an early stage in the course of regular eye examinations. The field of vision is checked, the retina and optic nerve are visually inspected with an ophthalmoscope, and pressure in the eye is checked with an instrument called a *tonometer,* which measures tension on the cornea.

through the pupil into the anterior chamber, and here it is reabsorbed by a circular vein called the *scleral venous sinus* (**fig. 10.16**). Normally the rate of reabsorption balances the rate of secretion, but see Clinical Application 10.3 for an important exception.

2. The **lens** is suspended behind the iris by a ring of fibers, the *suspensory ligament,* that attaches it to the ciliary body. Tension on these fibers somewhat flattens the lens so it is about 9.0 mm in diameter and 3.6 mm thick at the middle. When the lens is removed from the eye and not under tension, it has a more spheroidal shape.
3. The **vitreous body** is a transparent jelly that fills the large space behind the lens. It maintains the spheroidal shape of the eyeball and keeps the retina smoothly pressed against the rear of the eyeball.

The neural components of the eye are the retina and optic nerve. The **retina** is a thin transparent membrane that lines the posterior two-thirds of the eye. It is attached

---

[19]*glauc* = grayness

Figure 10.16 The Flow of Aqueous Humor. Blue arrows indicate the flow of aqueous humor from the ciliary body into the posterior chamber behind the iris, its flow through the pupil into the anterior chamber, and its reabsorption by the scleral venous sinus anterior to the iris.

to the rest of the eyeball at only two points—the **optic disc,** which is a posterior circular area where the **optic nerve** originates, and the **ora serrata,**[20] a scalloped margin that encircles the eye at the anterior edge of the retina, slightly posterior to the lens (fig. 10.15). The rest of the retina is unattached but is held firmly against the wall of the eyeball by the vitreous body.

The retina can be examined with an illuminating and magnifying instrument called an *ophthalmoscope* (off-THAL-mo-scope) (**fig. 10.17**). Directly posterior to the center of the lens, on the visual axis of the eye, is a patch of cells called the **macula lutea**[21] (MACK-you-la LOO-tee-ah), about 3 mm in diameter. In the center of the macula is a pit, the **fovea centralis,** which produces the most finely detailed visual images.

The optic disc lies about 3 mm medial to the fovea. Nerve fibers from all areas of the retina converge here and leave the eye to form the optic nerve. The optic disc is the one area of retina devoid of sensory cells; it forms a *blind spot* in the visual field. To compensate for this, the eye makes tiny, unnoticed scanning movements of the visual field, and the brain "fills in" the blind spot with essentially imaginary information so that we are not distracted by a patch of darkness in the field. Blood vessels also enter and leave at this point. Eye examinations serve for more than evaluating the visual system; they allow for direct, noninvasive examination of these vessels for signs of hypertension, diabetes mellitus, atherosclerosis, and other vascular diseases.

---

[20]*ora* = mouth; *serrata* = scalloped, serrated
[21]*macula* = spot; *lutea* = yellow

(a)

(b)

(c)

Figure 10.17 The Fundus (Rear) of the Eye.
(a) A typical ophthalmoscopic image. (b) Major features of the fundus. Note the blood vessels diverging from the optic disc, where they enter by way of the optic nerve. (c) Use of the ophthalmoscope.
a: © Lisa Klancher c: © Peter Dazeley/Getty Images

❓ *Is figure (a) the subject's right or left eye? How can you tell?*

> **Apply What You Know**
>
> Some have suggested that the blind spot of one eye is compensated for by the other eye, with each eye seeing the area of the visual field that the other eye cannot. What simple experiment could you do that would prove this wrong?

10.5c Forming the Visual Image APR

The visual process begins when light rays enter the eye, focus on the retina, and produce a tiny inverted image. The pupil dilates or constricts to determine how much light enters the eye. Dilation is achieved by the **pupillary dilator,** a system of contractile cells in the iris that radiate from the pupil like the spokes of a wheel. These cells respond to the sympathetic nervous system and dilate the pupil when light intensity falls, when we look from a nearby to a more distant object, or when the body is in a general state of sympathetic arousal. Constriction is achieved by a circle of smooth muscle in the iris called the **pupillary constrictor,** innervated by the parasympathetic nervous system (see fig. 9.18). Pupillary constriction occurs when light intensity rises and when we shift our focus to a relatively close object. The responses to light are called **photopupillary reflexes.**

Image formation depends on **refraction,** the bending of light rays. Light travels at different speeds through media such as air, water, and glass. When light rays strike a denser medium at a 90° angle, they merely slow down; but if they strike at an oblique angle, like the curved off-center areas of the cornea, they also bend. In vision, the greatest amount of refraction, and therefore most focusing, occurs at the air–cornea interface. The lens contributes additional refraction and fine-tunes the focusing process begun by the cornea. The aqueous humor and vitreous body have relatively little effect. An important difference between the cornea and lens is that the curvature of the cornea is fixed, and therefore so is the amount of refraction occurring at the air–cornea interface. The curvature of the lens, however, can be adjusted moment by moment.

When we look at objects more than 6 m away, our eyes are in a relatively relaxed state called **emmetropia** (EM-eh-TRO-pee-ah). Light rays entering the eyes are nearly parallel, the pupil is relatively dilated, and the lens is relatively thin, about 3.6 mm at the center (**fig. 10.18a**). When we look at something closer than 6 m, the light rays from it diverge more strongly as they approach the eyes and thus require more effort to get them to come together again (focus) on the retinas. The eyes must do three things to compensate for divergence of the rays: (1) **convergence**—they must turn medially so that both of them aim at the object (the fixation point); (2) **accommodation**—the lenses thicken to as much as 4.5 mm at the center, increasing their refractive power so they can focus the rays on the retina; and (3) **pupillary constriction**—the pupil narrows to screen out divergent light rays at the periphery of the lens, which could not be effectively focused because they wouldn't pass through very much lens material (**fig. 10.18b**). Inadequate convergence can result from unequal strength of the extrinsic eye muscles and can cause double vision. You can simulate this effect by pressing gently on one eyelid as you view this paragraph. The image will fall on noncorresponding regions of the two retinas and you will see double.

The closest an object can be and still come into focus is called the **near point** of vision. It depends on the flexibility of the lens. The lens stiffens with age, so a 10-year-old child can focus on objects only 9 cm away, but by age 60, the near point averages about 83 cm from the eye. That is, older people have more difficulty focusing on things close to the face—a condition called **presbyopia**[22]—hence the common joke among older people, "My arms have gotten shorter!"

---

[22]*presby* = old age; *opia* = eye condition

Figure 10.18 Focusing and Accommodation in Distance and Near Vision. (a) When we view a distant scene, light rays entering the eye are nearly parallel. The lens is relatively thin and the pupil relatively dilated. (b) When we view a nearby object, light rays are more divergent. The eyes converge, the lens thickens, and the pupil constricts. These reflexes are mediated through the midbrain.

10.5d Generating the Visual Nerve Signal

Sensory transduction, the conversion of light energy into nerve signals, occurs in the retina. To understand the process, we must begin with the retina's cellular layout (**fig. 10.19**). Its outermost layer is a dark *pigment epithelium* that absorbs excess light. Facing this is a layer of receptor cells called **rods** and **cones,** which are packed with **visual pigments** (*rhodopsin* in rods and *photopsin* in cones) to absorb light and begin the process of sensory transduction.

Rods and cones are connected to a thick layer of neurons called **bipolar cells,** and these are connected, directly or indirectly, to a single layer of large round neurons called **ganglion cells.** Ganglion cells are the innermost cells of the retina, adjacent to the vitreous body. Their axons converge at the rear of the eye to form the optic nerve that leaves the eyeball and passes to the brain.

Photoreceptors are named for their shapes (**fig. 10.20**). The slender rods consist of a *cell body* containing the nucleus, an *inner segment* containing mitochondria and other organelles, and a cylindrical *outer segment* that faces the back of the eye. The outer segment contains a stack of membranous discs, like a roll of pennies in a

Figure 10.19 The Retina. (a) Photomicrograph. The sclera and choroid are external to the retina; the pigment epithelium is the most superficial layer of the retina itself. (b) Diagram of the relationships of the retinal cells. APR
a: Dennis Strete/McGraw-Hill Education

paper wrapper. Each disc is packed with molecules of the visual pigment **rhodopsin** (ro-DOP-sin). This is where light absorption occurs, the first step toward generating a visual nerve signal. Rhodopsin consists of a protein called *opsin* and a vitamin A derivative called *retinal* (RET-ih-nal, rhyming with "pal"). Thus, a dietary deficiency of vitamin A can lead to **night blindness,** a difficulty seeing in dim light.

Cones are structurally similar to rods. The outer segment is shorter and more conical, and it has parallel infoldings of the plasma membrane rather than a stack of membranous discs. These infoldings are packed with the cone visual pigment, **photopsin.** Unlike rhodopsin, which has only one chemical form, there are three forms of photopsin that enable us to have color vision. Cones work in a manner similar to rods, but in brighter light. We will examine their properties in the next section.

When a rod is in the dark, unstimulated, it releases a steady stream of inhibitory neurotransmitter from synaptic vesicles in the base of the cell. This inhibits the bipolar cells, the next cells in line. When light enters the eye, rhodopsin absorbs it. Retinal changes shape and separates from the opsin. Purified rhodopsin has a violet color, but when it absorbs light and breaks apart, it becomes colorless. This step in the visual process is therefore called *bleaching*. When this happens, the rod stops releasing neurotransmitter. The bipolar cell, no longer inhibited, stimulates a ganglion cell. The ganglion cell is the only retinal cell type that produces action potentials. It responds to the foregoing changes with a burst of action potentials—a nerve signal that travels out the optic nerve. Cones work in essentially the same way, but they are less sensitive and work only at higher light intensities.

A substantial amount of information processing occurs in the retina before the resulting signal is sent to the brain. Other types of retinal cells, such as *horizontal cells* and *amacrine cells* (fig. 10.19b), are involved in detecting the boundaries of objects, movement, changes in light intensity, and other aspects of the visual scene.

Figure 10.20 Rod and Cone Cells. (a) Rods and cones of a salamander retina (electron micrograph). (b) Structure of human rod and cone cells.
a: © Eye of Science/Science Source

? *Vision requires a steady supply of ATP. Would you expect this to be made in the outer segment, inner segment, or cell body? Explain.*

The Roles of Rods and Cones

A human retina contains about 130 million rods and 6.5 million cones. These two receptor cells are concerned with distinctly different aspects of vision. All rods contain identical rhodopsin molecules. As a result, they all respond to light in the same way and have no basis for distinguishing colors (different wavelengths of light) from each other; essentially, rods produce visual sensations of shades of gray (*monochromatic*, or black-and-white, vision). Furthermore, they are active only in dim light, producing **night (scotopic) vision** and grainy, low-resolution images.

In bright daylight and typical room lighting, rods are "bleached out" and don't respond at all. However, they are very sensitive to low intensities, responding to light even as dim as starlight reflected from white paper. Such dim light stimulates any one rod very weakly, but hundreds of rods converge to stimulate each bipolar cell, and many bipolar cells then collaborate to stimulate each ganglion cell. This extensive convergence of information enables the eye to respond to very dim light, but it also means that each optic nerve fiber going to the brain conveys information from a relatively large patch of retina with hundreds of rods. The resulting image is like a very coarse-grained, overenlarged newspaper photograph or low-resolution image on a computer monitor.

Cones, on the other hand, are somewhat responsive even at moonlight intensity, but function best at daylight and room light intensities; they are responsible for **day (photopic) vision.** Since they act in brighter light than rods do, they do not have to team up to stimulate a bipolar cell; they have far less convergence than the rod pathways. Indeed, in the fovea, there are about 4,000 tiny cones and no rods at all, and each cone has its own bipolar cell and ganglion cell—like a "private line" to the brain. As a result, each optic nerve fiber in this region sends a signal representing only a very tiny area of the retina, and the resulting image is a fine-grained, high-resolution one.

When we look directly at something, we aim each eye so the image falls on the fovea and we see the object as sharply as possible. If you fix your gaze on a few words of this book, for example, you will notice (don't move your eyes!) that the words to the left and right of that are blurry. That's because our peripheral vision is a relatively low-resolution system that serves mainly to alert us to objects or

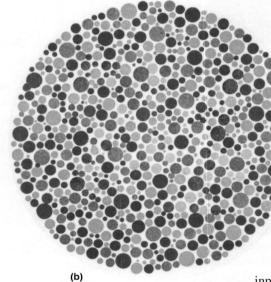

Figure 10.21 Color Vision. (a) Absorption curves of rods and the three types of cones. (b) What number do you see? People with normal trichromatic vision should see 74, but people with red–green color blindness see no discernible number.
b: © Steve Allen/Getty Images

movements in that region, stimulating us to turn our eyes or head to look that way and identify what is there.

Another well-known aspect of cones is that they provide *trichromatic* **color vision.** There are three varieties of photopsin in cones (fig. 10.21): So-called **short-wavelength (S) cones** have a photopsin that best absorbs light in the violet part of the spectrum; **medium-wavelength (M) cones** peak in the green part of the spectrum; and although **long-wavelength (L) cones** peak in the yellow-green wavelengths, they are the only cones whose absorption extends into the red wavelengths, so they bear sole responsibility for our ability to see red. The brain distinguishes colors from each other based on the mixture of signals it gets from these three types of cones.

People with a genetic lack of any one of these three photopsins have **color blindness.** Those with *red–green color blindness,* the most common form, have difficulty distinguishing shades of green, orange, and red from each other (fig. 10.21b). This is a sex-linked genetic trait occurring in about 8% of males but only about 0.5% of females. Such males inherit the gene for red–green color blindness on the X chromosome received from their mothers. Females can also inherit the gene from their mothers, but they also usually inherit a gene for normal color vision on the X chromosome contributed by their fathers, which masks the effect of the maternal gene. The father contributes no X chromosome to his sons, so he cannot contribute a gene to mask the maternal effect on a son's color vision.

10.5e The Visual Projection Pathway

Once nerve signals are generated in the retina, where do they go? The two optic nerves enter the cranial cavity, converge, and form an X called the **optic chiasm**[23] (fig. 10.22). Here, half of the nerve fibers from each eye cross over to the opposite side of the brain. The right cerebral hemisphere thus receives input from the medial (nasal) side of the left eye and the lateral side of the right eye. The left cerebral hemisphere receives the medial fibers from the right eye and lateral fibers from the left. Consequently, the right brain sees things on the left side of the body, the left brain sees things on the right side of the body, and the two overlap in the middle. Recall that the right brain controls most voluntary motor responses on the left side of the body, and vice versa, so each cerebral hemisphere visually monitors the side of the body where it exercises its primary motor control.

Slightly posterior to the optic chiasm, most optic nerve fibers terminate at synapses in the thalamus. New nerve fibers begin here and continue through the white matter as the *optic radiation,* which travels the rest of the way to the **primary visual cortex** in the occipital lobe at the rear of the cerebrum. This is where one becomes aware of the visual stimulus. A stroke that destroys occipital lobe tissue can leave a person blind in one-half of the visual field even if the eyes still function normally. Neurons of the primary visual cortex communicate with neurons of the nearby **visual association area,** just anterior to the primary cortex. The visual association area integrates visual input with memory and enables us to identify and interpret what we are seeing at the moment, as well as to form new short- or long-term visual memories.

---

[23]*chi* = X

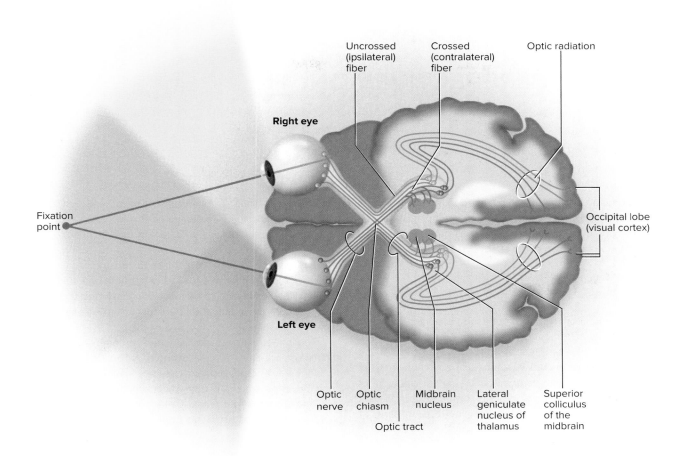

A few optic nerve fibers take a different course. Instead of going to the thalamus, they lead to two midbrain nuclei—one that controls accommodation of the lens and the photopupillary reflex, and one that controls reflexes of the muscles that move the eye (for example, causing us to look toward something we see moving in our peripheral vision).

Figure 10.22 The Visual Projection Pathway to the Brain. Note that half of the nerve fibers from each retina (red and violet) cross over to the opposite side of the brain when passing through the optic chiasm. The right occipital lobe thus monitors the left half of the visual field and vice versa.

❓ *If a cerebral hemorrhage destroyed the optic radiation in just one cerebral hemisphere, where would you expect blindness to be experienced, and why—total blindness in both eyes; total blindness in just one eye; blindness in the medial field of vision of both eyes; blindness in the peripheral field of both eyes; or blindness in the peripheral field of just one eye?*

Before You Go On

Answer these questions from memory to test your understanding of the previous section. Reread the material if you can't answer the questions.

18. Briefly state the function of each of these orbital structures: tarsal glands, lacrimal gland, conjunctiva, and extrinsic muscles.
19. How does the aqueous humor differ from the vitreous body in composition, location, and function?
20. What specific region of the retina produces the most high-resolution images? Why?
21. What component of the eye has the greatest role in refracting light rays to form an image on the retina?
22. Explain why cone cells produce high-resolution images and rod cells do not; why cones provide day vision and rods do not; and why cones provide color vision and rods do not.

PERSPECTIVES ON HEALTH

The most common health problems related to the senses are in vision and hearing. Some of these are commonplace changes that occur almost inevitably with age, whereas others are caused by disease or trauma that can occur at any age.

Hearing Disorders

The most common complaint about hearing is partial deafness. *Deafness* means any hearing loss, whether temporary or permanent. *Conductive deafness* results from conditions that reduce the transfer of vibrations to the inner ear. Such conditions may include a damaged tympanic membrane, otitis media, blockage of the auditory canal with earwax or other matter, and otosclerosis. *Otosclerosis*[24] is fusion of the auditory ossicles to each other or fusion of the stapes footplate to the oval window. Either way, it prevents the ossicles from vibrating freely. Repeated middle-ear infections can cause otosclerosis.

Sensorineural (nerve) deafness results from death of hair cells or any of the nervous elements concerned with hearing. It is a common occupational disease of factory and construction workers, musicians, and other people exposed to frequent and sustained loud sounds. Deafness leads some people to delusions of being talked about, disparaged, or cheated. Beethoven said his deafness nearly drove him to suicide.

Sensorineural deafness is often accompanied by *tinnitus* (ti-NITE-us), a persistent sense of ringing, whistling, clicking, or buzzing in the ear. The physiological basis of tinnitus is not well known, but it may represent an illusion of sensory input from cochlear hair cells that are no longer present, much like the illusion of phantom pain from a limb that has been amputated.

[24]*oto* = ear; *scler* = hardening, stiffening; *osis* = condition, process

Visual Disorders

The most common visual disorders concern inability to focus properly. *Hyperopia* (farsightedness) is an inability to focus on nearby objects because of an abnormally short eyeball. The retina lies in front of the focal point and light rays haven't converged and come into focus by the time they reach it (fig. 10.23b). This is corrected by lenses that cause light rays to converge slightly before they enter the eye, so they reach their focal point sooner on the retina.

Myopia (nearsightedness) is an inability to focus on distant objects because of an abnormally elongated eyeball. Light rays come to a focus before they reach the retina instead of focusing on the retina (fig. 10.23c). The rays have begun to diverge again by the time they reach the retina, so the resulting image is blurry. Myopia is treated with corrective lenses that make the rays diverge slightly before entering the eye, so they converge again at a greater distance behind the cornea—on the retina.

Many people suffer *astigmatism,* a condition in which focusing on objects in one plane, such as the vertical, makes objects in a different plane, such as the horizontal, go out of focus. This is the result of a cornea that is somewhat ovoid, like the back of a spoon, instead of circular. Astigmatism is also treated with corrective lenses.

More serious problems are glaucoma (see Clinical Application 10.3) and cataracts. A *cataract* is a cloudiness of the lens, making the vision dim or milky, as if looking at a scene from behind a waterfall—hence the name.[25] The lenses become more predisposed to cataracts as they thicken with age. Their degeneration is often accelerated and worsened by diabetes mellitus. Heavy smoking and excessive exposure to the ultraviolet rays of sunlight are additional risk factors for cataracts. Cataracts can be treated by replacing the lenses with plastic ones. The artificial lenses improve vision almost immediately, but glasses may still be needed for near vision.

[25]*cataract* = waterfall

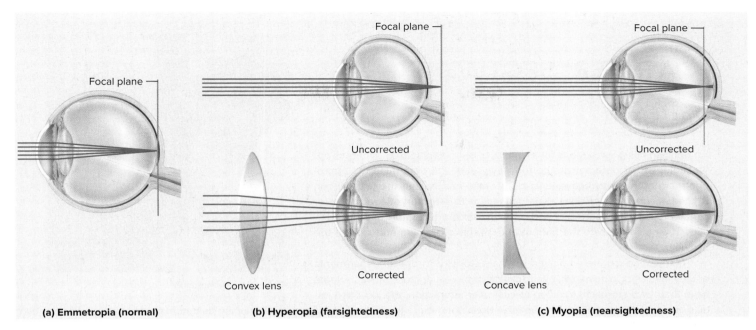

Figure 10.23 Two Common Visual Defects and the effects of Corrective Lenses. (a) The normal emmetropic eye, with light rays converging on the retina. (b) Hyperopia (farsightedness) and the corrective effect of a convex lens. (c) Myopia (nearsightedness) and the corrective effect of a concave lens.

Aging of the Sense Organs

Some sensory functions begin to decline shortly after adolescence, or even earlier. Many schoolchildren and adolescents already experience the beginning of presbyopia, the stiffening of the lenses mentioned earlier, and thus find it harder to focus without corrective lenses. Later in life, people often notice a deficiency of night vision, as more and more light is needed to stimulate the retina. This has several causes: There are fewer receptor cells in the retina, the vitreous body becomes less transparent, the pupil becomes narrower as the pupillary dilators atrophy, and the lenses tend to yellow with age. Adjusting to the dark takes longer because enzymatic reactions in the rod cells are slower. Elderly people also have an elevated risk of glaucoma and cataracts (see Perspectives on Health)—both of which can cause serious visual impairment, even blindness. Having to give up reading and driving can be among the most difficult changes of lifestyle for elderly people.

Auditory sensitivity peaks in adolescence and declines afterward. The tympanic membrane and the joints between the auditory ossicles become stiffer, so vibrations are transferred less effectively to the inner ear, creating a degree of *conductive deafness*. In addition, the numbers of cochlear hair cells and cochlear nerve fibers decline with age, potentially leading to *sensorineural deafness* (see Perspectives on Health). The greatest hearing loss occurs in the frequency range of conversation. The death of receptor cells in the semicircular ducts, utricle, and saccule and of nerve fibers in the vestibular nerve and neurons in the cerebellum results in poor balance and dizziness—a factor in falls that sometimes lead to severely disabling bone fractures.

The senses of taste and smell are blunted as the numbers of taste buds, olfactory cells, and second-order neurons in the olfactory bulbs decline. Elderly people often find food less appealing, so declining sensory function can be a factor in malnutrition.

CAREER SPOTLIGHT

Optician

An optician fills prescriptions for eyeglasses or contact lenses. Whereas opthalmologists or optometrists perform eye tests and determine the appropriate strength of corrective lenses, opticians ensure that the prescription is filled accurately. Opticians assist patients with selection of eyeglasses or contact lenses to ensure that they fit with maximum comfort and functionality. They also educate the patient about proper cleaning and care of the eyewear.

Many opticians are employed by chain stores with optical departments or they may work in hospitals, health clinics, or private doctors' offices. Others have their own stores. In addition to customer interaction, the job often entails business management. The optician does not require the same level of advanced training as optometrists or ophthalmologists. Most opticians obtain a 2-year associate degree from a college with a program in optometric technology. Several states require that opticians be licensed. For further career information, see appendix B.

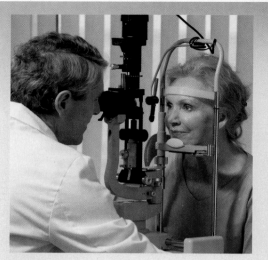
© Huntstock/Getty Images RF

Study Guide

Assess Your Learning Outcomes

To test your knowledge, discuss the following topics with a study partner or in writing, ideally from memory.

10.1 Receptors and Sensations

1. The meaning of *receptor* and distinction between a receptor and a sense organ
2. Why not all sensory signals result in a sensation
3. Four properties of a stimulus—type, location, intensity, and duration—and how the brain is able to interpret each on the basis of the nerve signals it receives
4. Why sense organs are described as transducers
5. The difference between general and special senses, and which senses belong in each category

10.2 The General Senses

1. Five types of simple nerve endings, their structural differences, and the senses for which each is responsible
2. The importance of pain and the harmful consequences of lacking a sense of pain
3. Distinctions between somatic and visceral pain
4. The reasons for referred pain and phantom pain
5. How spinal gating alters a person's sense of pain

10.3 The Chemical Senses—Taste and Smell

1. The locations of taste buds and their relationship to the lingual papillae
2. The cellular organization of a taste bud and the relationship of taste cells to sensory nerve fibers
3. The five primary taste sensations and the types of tastants that produce each sensation
4. The difference between taste and flavor; nongustatory factors that influence flavor
5. The location and cellular structure of the olfactory mucosa
6. The mechanism by which odorants trigger an olfactory nerve signal
7. The cranial nerves that carry gustatory and olfactory stimuli, and the brain regions that receive and integrate these signals
8. The terminal olfactory regions of the brain and the route that olfactory nerve signals travel to get there

10.4 The Ear—Equilibrium and Hearing

1. The three principal regions of the ear and general function of each one
2. Anatomy of the auricle and auditory canal, and functions of cerumen
3. Anatomy of the middle ear, especially its tympanic membrane, auditory tube, auditory ossicles, and muscles; the function of each of these
4. The relationship between the bony labyrinth, membranous labyrinth, perilymph, and endolymph of the inner ear
5. The function and anatomical components of the vestibular apparatus
6. The structure of hair cells and how it relates to the generation of sensory signals
7. The locations and cellular organization of the saccule and utricle, how they generate sensations of motion and body orientation, and how they differ
8. Anatomy of the semicircular ducts, the type of sensation they produce, and how the three ducts in each ear differ from each other in orientation and function
9. The destinations of signals for equilibrium in the brain, and the pathways taken to these destinations
10. Gross anatomy of the cochlea, and the anatomical relationships of its fluid-filled spaces, basilar membrane, and spiral organ
11. Structure of the cochlear hair cells, their relationship to the basilar and tectorial membranes, and the spatial arrangement and functional difference between inner and outer hair cells
12. How sound waves in the auditory canal ultimately result in movement of the basilar membrane and hair cells
13. How hair cell movements result in auditory nerve signals
14. How the brain discriminates between sounds of different loudness and pitch
15. The destinations of signals for hearing in the brain, the pathway taken to these destinations, and types of information processing that occur at different points on this route

10.5 The Eye and Vision

1. Structure and function of components of the orbital region external to the eyeball
2. Anatomy of the lacrimal apparatus and the pathway that tears follow from their origin to their drainage into the nasal cavity
3. The three layers of the eyeball and the components of each one
4. Optical components of the eyeball, their locations, and their functions
5. Neural components of the eyeball and specialized regions of the retina
6. Locations of the pupillary dilator and constrictor, their respective nerve supplies, and the conditions under which they alter the diameter of the pupil

7. Relative contributions of the cornea and lens to the refraction of light rays
8. Responses of the eye to shifting one's focus from a distant object to a nearby one
9. Principal cell types of the retina and their relationships to each other
10. Parts of a rod or cone cell, location of the visual pigment, and how light absorption by the pigment leads to the generation of a nerve signal
11. How and why the rod and cone system differ with respect to light sensitivity, visual resolution, and capacity for color vision
12. Why vision is sharpest at the fovea
13. How the retina generates signals that the brain can interpret as differences in color
14. The destination of visual signals in the brain and the pathways taken to get there

Testing Your Recall

1. All of the following are classified as special senses except
    a. taste.
    b. touch.
    c. vision.
    d. hearing.
    e. equilibrium.

2. A nociceptor is a nerve ending responsible for the sense of
    a. touch.
    b. vibration.
    c. heat or cold.
    d. taste or smell.
    e. pain.

3. The occipital lobe is concerned mainly with the sense of
    a. vision.
    b. proprioception.
    c. pain.
    d. taste.
    e. hearing.

4. A stomachache is classified as
    a. fast pain.
    b. phantom pain.
    c. visceral pain.
    d. somatic pain.
    e. referred pain.

5. Which of these is *not* one of the five primary taste sensations detected by the taste buds?
    a. sweet
    b. spicy
    c. bitter
    d. salty
    e. umami

6. Olfactory cells are found in
    a. the olfactory tract.
    b. the olfactory bulb.
    c. the primary olfactory cortex.
    d. the olfactory mucosa.
    e. the temporal lobe.

7. The malleus, incus, and stapes are components of
    a. the outer ear.
    b. the middle ear.
    c. the vestibular system.
    d. the bony labyrinth.
    e. the inner ear.

8. The malleus, incus, and stapes have their closest anatomical and functional association with
    a. the stapedius and tensor tympani.
    b. the auditory canal.
    c. the auditory tube.
    d. the semicircular ducts.
    e. the cochlear hair cells.

9. A single sensory cell type is associated with all of the following *except*
    a. low-resolution vision.
    b. low-light vision.
    c. black-and-white vision.
    d. photopic vision.
    e. rhodopsin.

10. Nerve fibers of the optic nerve originate from
    a. rod cells.
    b. cone cells.
    c. both rods and cones.
    d. ganglion cells.
    e. bipolar cells.

11. Pressure on a tissue is detected by a simple sense organ called a _____, which has a nerve fiber surrounded by onionlike layers of flattened Schwann cells and fibroblasts.

12. Our sharpest vision is in a retinal pit called the _____ directly behind the central axis of the lens.

13. The optic nerves cross at a point called the _____, where half of the nerve fibers from each eye are routed to each cerebral hemisphere.

14. A person might have dry eyes if the ducts of the _____ glands were overly constricted or obstructed.

15. The sense of vertical movement, as when rising from a chair or riding in an elevator, originates in a patch of sensory cells in a chamber called the _____.

16. _____ is a change in the thickness of the lens to enable the eye to focus on nearby objects.

17. Nerve fibers for the sense of smell originate in the olfactory mucosa and terminate in a part of the brain called the _____.

18. The sensory cells of hearing, called _____, are arranged in four spiral rows in the cochlea.

19. A person can have color blindness if he or she has a defective gene for one of the three types of _____ cells.

20. The position and movements of the body and its limbs are detected by sense organs called _____.

Answers in Appendix A

What's Wrong with These Statements?

Briefly explain why each of the following statements is false, or reword it to make it true.

1. The usual cause of pain is an undesirable overstimulation of receptors whose normal function is touch.

2. The olfactory hairs in the nose and the hairs on cochlear hair cells are actually modified cilia.

3. Differences in the color of light or the pitch of a sound are communicated to the brain by differences in the firing frequencies of the sensory nerve fibers.

4. *Hair receptor* is another name for a hair cell of the inner ear.

5. In vision, most light refraction occurs as light passes through the lens of the eye.

6. The right occipital lobe receives all visual input from the left eye, and the left occipital lobe receives all input from the right eye.

7. Deep bass tones stimulate the same cochlear hair cells as high treble tones, but base and treble notes stimulate them at different frequencies.

8. The auditory (pharyngotympanic) tube lets air into the inner ear to equalize pressure on both sides of the tympanic membrane.

9. After passing across the surface of the eye, tears are reabsorbed by the lacrimal gland and returned to the blood stream.

10. The taste buds, also known as lingual papillae, can be seen on the surface of the tongue with a magnifying glass.

Answers in Appendix A

Testing Your Comprehension

1. You are sitting in a chair attending lecture, your head tilted forward as you take notes. When the lecture ends, you raise your head from your notebook, rise from your chair and leave the room, walk down the hall, and turn a corner headed for your next class. Describe all the actions that would occur in your utricle, saccule, and semicircular ducts to make your brain aware of these movements.

2. Although it is unusual for a female to be color-blind, Julie is one such case. Do you think her father is color-blind? Her mother? Explain.

3. Frank has worked for years as a stage manager for a traveling rock band, and never took precautions to protect his hearing. By the age of 30, he finds he can no longer hear the high notes on an electronic keyboard. What region of his cochlea do you believe is damaged? Name the specific cells in that region that have likely been destroyed.

Design Element Credits: (Camping): monkographic/Shutterstock; (Abseiling): Image Source; (Stethoscope): MedicalRF.com; (Clouds): Pakhnyushchy/shutterstock.com; (Doctors): Purestock/SuperStock; (Mountain climber): McGraw-Hill Education.

Chapter 11

The Endocrine System

The thyroid gland is the body's largest endocrine gland. Thyroid hormone regulates the metabolic rate in adulthood and is important for normal prenatal nervous system development and tissue growth.
Sciepro/Science Photo Library/Getty Images RF

Chapter Outline

- **11.1 Overview of the Endocrine System**
  - 11.1a Hormones and Endocrine Glands
  - 11.1b Comparison of the Nervous and Endocrine Systems
- **11.2 Endocrine Physiology**
  - 11.2a Hormone Structure
  - 11.2b Hormone Receptors and Effects
  - 11.2c Hormone Interactions
  - 11.2d Hormonal Control of Homeostasis
- **11.3 The Hypothalamus and Pituitary Gland**
  - 11.3a Pituitary Anatomy
  - 11.3b Anterior Pituitary Hormones
  - 11.3c Posterior Pituitary Hormones
  - 11.3d Control of the Pituitary Gland
- **11.4 Other Endocrine Glands and Tissues**
  - 11.4a The Pineal Gland
  - 11.4b The Thyroid Gland
  - 11.4c The Parathyroid Glands
  - 11.4d The Thymus
  - 11.4e The Pancreas
  - 11.4f The Adrenal Glands
  - 11.4g The Gonads
  - 11.4h Endocrine Functions of Other Organs and Tissues
- **11.5 Stress Physiology**

Clinical Applications/Perspectives on Health
- Clinical Application 11.1: Hormone Receptors and Therapy
- Clinical Application 11.2: Thyroid Disorders
- Perspectives on Health

End of Chapter
- Aging of the Endocrine System
- Career Spotlight: Diabetes Educator
- Connective Issues
- Study Guide

BASE CAMP

Before ascending to the next level, be sure you're properly equipped with a knowledge of these concepts from earlier chapters.

- Positive and negative feedback and homeostasis (see "Homeostasis and Feedback" in section 1.2d)
- Hydrophilic and hydrophobic qualities of molecules (see "Water" in section 2.2a)
- Chemical structure of steroids and proteins (sections 2.3b and 2.3c)

Anatomy & Physiology Revealed 4.0
Module 8
Endocrine System

The trillions of cells that make up the human body don't function in isolation. They coordinate their activities by communicating through chemical signals, most of which are produced by the nervous and endocrine systems. In previous chapters you learned that the nervous system signals are neurotransmitters. The endocrine signals are hormones. You have undoubtedly heard of estrogen, testosterone, and insulin, examples of hormones that have widespread effects on the body.

Hormones are fundamental to homeostasis. They play key roles in growth and development, metabolism and energy balance, water balance, reproduction, and response to stress. Endocrine disorders such as diabetes mellitus and hypothyroidism are all too common. These diseases are caused by problems with hormone production or cell responses to specific hormones.

In this chapter we will first examine hormone structure and general mechanisms of action, then provide a survey of endocrine glands and tissues, the hormones they produce, and the functions these hormones perform.

11.1 Overview of the Endocrine System

Expected Learning Outcomes

When you have completed this section, you should be able to

a. define *hormone* and *endocrine system;* and
b. describe differences and similarities between the nervous and endocrine systems.

11.1a Hormones and Endocrine Glands

Hormones[1] are chemical messengers secreted by endocrine glands or tissues into the blood. They travel everywhere the blood goes, but they affect only cells that have receptors for them—their **target cells.** Hormones produced in one part of the body often affect very distant organs or cells. For example, those produced by the pituitary gland in the head affect organs and tissues throughout the body.

The **endocrine system** is composed of the glands and cells that secrete hormones (fig. 11.1). The science of **endocrinology** embraces the study of this system and the diagnosis and treatment of its disorders. The most familiar sources of hormones are the endocrine glands such as the pituitary, thyroid, and adrenal glands that were the traditional focus of endocrinologists. They release their secretions (hormones) into the blood—as reflected in the word *endocrine.*[2] This is in contrast to exocrine glands that release their secretions to a local tissue surface by way of ducts (see section 4.5b). In recent decades, scientists have recognized that hormones are secreted by a surprising assortment of cells and organs that aren't traditionally considered to be endocrine glands, such as adipocytes, osteoblasts, and the brain, heart, liver, and small intestine.

---

[1]*hormone* = to excite, set in motion
[2]*endo* = internal; *crin* = to secrete

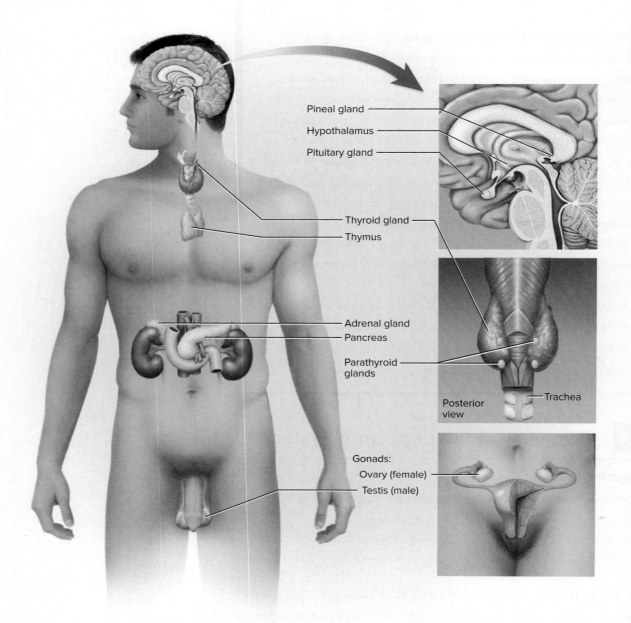

Figure 11.1 Major Organs of the Endocrine System. This system also includes gland cells in many other organs not shown here. **APR**

? *After reading this chapter, name at least three hormone-secreting organs and tissues that are not shown in this illustration.*

11.1b Comparison of the Nervous and Endocrine Systems

One way to better understand the endocrine system is to compare it to the nervous system (**fig. 11.2**). The purpose of both systems is internal communication, but they differ in significant ways (**table 11.1**). For example, the nervous system communicates by means of both electrical and chemical signals, whereas the endocrine system relies solely on chemical messengers. The nervous system responds very quickly to a stimulus and stops responding as soon as stimulation ceases; the endocrine system is slower to respond and stop. The nervous system generally sends its signals to a specific organ, in contrast to the endocrine system whose hormones go everywhere the blood goes.

Despite their differences, the nervous and endocrine systems have much in common. For example, some chemicals such as norepinephrine and dopamine function as both hormones and neurotransmitters. Some hormones are secreted by specialized neurons called *neuroendocrine cells,* which

(a) Nervous system

(b) Endocrine system

Figure 11.2 Communication by the Nervous and Endocrine Systems. (a) A neuron has a long fiber that delivers its neurotransmitter to the immediate vicinity of its target cells. (b) Endocrine cells secrete a hormone into the bloodstream (left). At a point often remote from its origin, the hormone leaves the bloodstream and enters or binds to its target cells (right). **APR**

Table 11.1 Comparison of the Nervous and Endocrine Systems

Nervous System	Endocrine System
Communicates by means of electrical impulses and neurotransmitters	Communicates by means of hormones
Releases neurotransmitters at synapses at specific target cells	Releases hormones into bloodstream for general distribution throughout body
Usually has relatively local, specific effects	Sometimes has very general, widespread effects
Reacts quickly to stimuli, usually within 1–10 ms	Reacts more slowly to stimuli, often taking seconds to days
Stops quickly when stimulus stops	May continue responding long after stimulus stops
Adapts relatively quickly to continual stimulation	Adapts relatively slowly; may respond for days to weeks

functionally belong to both the endocrine and nervous systems. In addition, hormones and neurotransmitters may have similar effects on the same target cells; for example, norepinephrine (a neurotransmitter) and glucagon (a hormone) both target liver cells and cause the release of glucose. Neurons sometimes trigger hormone secretion, and hormones may stimulate or inhibit neurons. Thus, the two systems regulate each other as they coordinate activities of other body systems.

Before You Go On

Answer these questions from memory to test your understanding of the previous section. Reread the material if you can't answer the questions.

1. *Define* endocrine system, hormone, *and* target cell.
2. Compare and contrast how the nervous system and endocrine system coordinate human physiology, and describe some ways in which these two systems overlap and interact.

11.2 Endocrine Physiology

Expected Learning Outcomes

When you have completed this section, you should be able to

a. characterize the three chemical classes of hormones and state examples of each;
b. describe how hormones act on their target cells and receptors;
c. discuss some types of interaction between multiple hormones;
d. explain how the body regulates rates of hormone secretion; and
e. explain how the body rids itself of a hormone whose job is done.

11.2a Hormone Structure

Hormones are classified into three groups on the basis of their chemical structure (fig. 11.3). (1) **Steroid hormones** are lipids synthesized from cholesterol, and share its basic structure of four organic rings with varied side groups. Some familiar steroid hormones are testosterone and estradiol (an estrogen). (2) **Monoamines** are small molecules synthesized from the amino acids tyrosine and tryptophan, such as thyroxine (a thyroid hormone) and epinephrine. (3) **Peptide hormones** are chains ranging from 3 to more than 200 amino acids long; those 50 amino acids or longer are considered proteins, and some are glycoproteins—proteins with carbohydrate attached. Oxytocin is an example of a small peptide hormone, composed of only

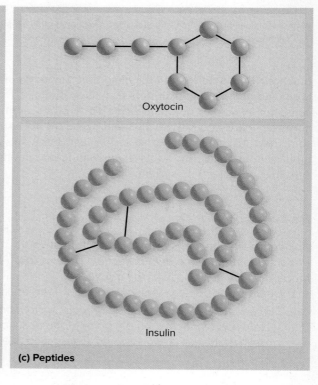

Figure 11.3 The Chemical Classes of Hormones. (a) Two steroid hormones, defined by their four-membered rings derived from cholesterol. (b) Two monoamines, derived from amino acids and defined by their —NH— or —NH$_2$ (amino) groups. (c) A small peptide hormone, oxytocin, and a protein hormone, insulin, defined by their chains of amino acids (the yellow circles).

9 amino acids. Insulin is composed of 51 amino acids, large enough to be regarded as a protein.

11.2b Hormone Receptors and Effects

The function of a hormone is to stimulate a physiological change in a target cell. In order to do this, the hormone must first bind to a receptor—a protein within the target cell or in its plasma membrane. A given hormone fits its receptor much like a key fits only one lock. A "key" such as insulin will fit only an insulin receptor ("lock"), for example; it will not fit an estrogen receptor. Many hormones are in the blood at one time, but any given organ or cell responds only to those for which it has the proper receptors—just as there are many radio broadcast signals in the air at once, but you hear only the one to which you tune your own radio.

The location of the receptor varies with the chemical nature of the hormone. Peptide hormones and most monoamines (such as epinephrine) are hydrophilic and cannot pass into the target cell. They must bind to a receptor at the cell surface (**fig. 11.4a**). The hormone acts as a *first messenger* "knocking at the door" of the target cell. The binding of the hormone to the receptor activates a series of events in the cell, including the production of a *second messenger*. Often, the second messenger is *cyclic adenosine monophosphate (cAMP)*, made by removing two phosphate groups (PP_i) from ATP. Cyclic AMP activates or deactivates enzymes, which causes various metabolic effects in the cell. For example, an activated enzyme can turn metabolic pathways on; it can "wake up" specific genes, leading to the production of new proteins; or it can change the permeability of the plasma membrane, altering the transport of materials into or out of the cell. The binding of glucagon to liver cells, for example, activates a cascade of events that leads to the release of blood sugar (see fig. 11.11).

Steroid hormones, in contrast, are hydrophobic and readily diffuse through the phospholipids of the target-cell membrane. Most of them pass directly into the cell's nucleus and bind to a receptor associated with the DNA (**fig. 11.4b**). This typically causes a gene to be transcribed, which leads to the production of a specific protein. That protein can then alter cell metabolism in various ways. For example, when estrogen enters a cell of the uterus during the first half of a woman's menstrual cycle, it binds to a nuclear receptor and activates a gene for progesterone receptors. The cell produces those receptors, preparing itself to respond to the progesterone that will arrive in the second half of the cycle. The progesterone itself will activate other genes that cause the lining of the uterus to thicken and prepare for the possibility of pregnancy.

Figure 11.4 Hormone Actions on a Target Cell. (a) Peptide hormones cannot enter their target cells; they bind to surface receptors and activate intracellular processes, working through second messengers such as cAMP. (b) Steroid hormones freely enter their target cells and usually bind to receptors in the nucleus associated with specific genes in the DNA. Either process can lead to a great variety of metabolic effects on the target cell. **APR**

? *Would testosterone act by mechanism (a) or mechanism (b)? Explain.*

11.2c Hormone Interactions

We never have just a single hormone in circulation; the blood plasma is always a "cocktail" of many hormones traveling together. Often, two or more of these act on the same target organ and alter each other's effects. There are three types of hormone interactions:

1. **Permissive effects,** in which one hormone enhances the target organ's response to another hormone secreted later. The way that estrogen primes the uterus to respond to the later arrival of progesterone, as just described, is an example of this.
2. **Antagonistic effects,** in which one hormone opposes the action of another. For example, insulin stimulates the liver to absorb glucose and lower its blood concentration, whereas glucagon stimulates the liver to release glucose and raise the blood sugar level.

Clinical Application 11.1

HORMONE RECEPTORS AND THERAPY

In treating endocrine disorders, it is essential to understand the role of hormone receptors. For example, pituitary dwarfism is now rare because it is easily treated in childhood with growth hormone, and genetic engineering has made human growth hormone more available. However, there is another form of dwarfism, *Laron dwarfism,* which results from a defect in the receptor for growth hormone. Administering growth hormone has no benefit to these patients because the body doesn't respond to it. Similarly, *type 2 diabetes mellitus,* the most common form of diabetes, results from an insulin receptor defect or deficiency. No amount of insulin replacement can correct this. Another implication of hormone–receptor physiology for therapy is that estrogen stimulates the growth of some malignant tumors with estrogen receptors. For this reason, estrogen replacement therapy should not be used for women with estrogen-dependent cancer or who are at risk for it.

3. **Synergistic effects,** in which two hormones acting together produce a much stronger response than the sum of their separate effects. In men, for example, testosterone and follicle-stimulating hormone would have very little effect on sperm production if they each worked alone, but when they work together, the testes produce 300,000 sperm per minute.

11.2d Hormonal Control of Homeostasis

The endocrine system precisely regulates numerous physiological variables, most often through negative feedback (see section 1.2c). In a negative feedback loop, the body senses a change in a controlled variable, an endocrine gland responds by altering the rate of hormone secretion, and target cells respond in a way that will ultimately bring the variable back to its original state.

An example of this is the regulation of blood sugar (glucose). When the glucose level rises during digestion of a meal, endocrine cells of the pancreas respond by secreting insulin. Among other effects, insulin targets liver, muscle, and other cells and causes them to absorb glucose from the blood and store it for later use. The feedback mechanism effectively reverses the original change in blood glucose level and maintains homeostasis.

After they are secreted, hormones don't linger in the blood forever. They must be removed once they have served their purpose, or cells will continue to respond and homeostasis will not be maintained. Most hormones are taken up by the liver and kidneys, chemically degraded, and excreted in the bile or urine. Some hormones are removed from the blood very quickly; for example, growth hormone levels decline by 50% within 20 minutes. Others linger longer; thyroid hormone, for example, maintains an effective level in the blood for as long as 2 weeks after its secretion ceases.

Before You Go On

Answer these questions from memory to test your understanding of the previous section. Reread the material if you can't answer the questions.

3. Define the three chemical classes of hormones and give an example of each.
4. Explain what a hormone receptor is and state two places where receptors are located in the target cells.
5. Describe three ways in which hormones can influence each other's actions.
6. Explain how the endocrine system can maintain homeostasis through negative feedback. Give an example.

11.3 The Hypothalamus and Pituitary Gland

Expected Learning Outcomes

When you have completed this section, you should be able to

a. describe the anatomical relationship between the pituitary gland and hypothalamus;
b. explain how the hypothalamus controls pituitary functions; and
c. name the pituitary gland hormones and explain their effects.

This section and the next one survey the major glands of the endocrine system. We begin our exploration with the pituitary and a nearby region of the brain that controls it, the hypothalamus. You may remember that the hypothalamus regulates a wide range of body functions, including water balance, energy balance, growth, and reproduction (see section 9.2a). It fulfills these roles in large part by controlling the production and secretion of hormones from the pituitary gland. The pituitary is an important link in the neural and hormonal regulation of homeostasis.

11.3a Pituitary Anatomy

Even though the **pituitary gland (hypophysis**[3]**)** controls more bodily functions than any other endocrine gland, it is only about as big as a kidney bean. It grows about 50% larger, however, during pregnancy. It is located just below the hypothalamus (see fig. 11.1), sheltered in the *sella turcica*,[4] a saddlelike enclosure in the sphenoid bone (see section 6.3a).

---

[3]*hypo* = below; *physis* = growth
[4]*sella* = saddle; *turcica* = Turkish

The pituitary appears to be a single structure, but functionally and embryologically, it is divided into **anterior** and **posterior lobes.** The anterior pituitary gland develops from a pouch in the roof of the embryonic pharynx (throat) (**fig. 11.5**). Meanwhile, the posterior pituitary gland arises as a downgrowth from the hypothalamus, and retains its connection to the brain throughout life. The two parts of the pituitary gland come to lie side by side before the bone forms around them, and are so closely joined together that they appear to be a single gland.

The hypothalamus is attached to the posterior pituitary gland by a **stalk (infundibulum)** containing bundles of axons (**fig. 11.6a**). These axons originate in two separate groups of neurons (nuclei) in the hypothalamus, represented in blue and green in the figure. These distinct nuclei synthesize two hormones, *oxytocin* and *antidiuretic hormone,* which we will examine shortly, and transport them down the axons for storage in the posterior pituitary. Under appropriate conditions, the hypothalamus sends electrical signals down the axons to stimulate the posterior pituitary to release oxytocin or antidiuretic hormone into the blood.

The hypothalamus connects to the anterior pituitary gland by a network of small blood vessels called the **hypophyseal portal system** (**fig. 11.6b**). A *portal system* is one in which blood flows from one capillary bed to another before returning to the heart. In this specific case, the hypothalamus secretes chemical signals into the capillaries at its end of the portal system; these signals travel a short distance to the anterior pituitary; and here they leave the bloodstream through a second capillary network and tell the anterior pituitary what to do. We will examine these signals shortly in more detail.

Figure 11.5 Embryonic Development of the Pituitary Gland. (a) Sagittal section of the head at 4 weeks, showing the neural and pharyngeal origins of the pituitary gland. (b) Pituitary development at 8 weeks. The anterior pituitary has now separated from the pharynx. (c) Development at 16 weeks. The two lobes are now encased in bone and so closely associated they appear to be a single gland.

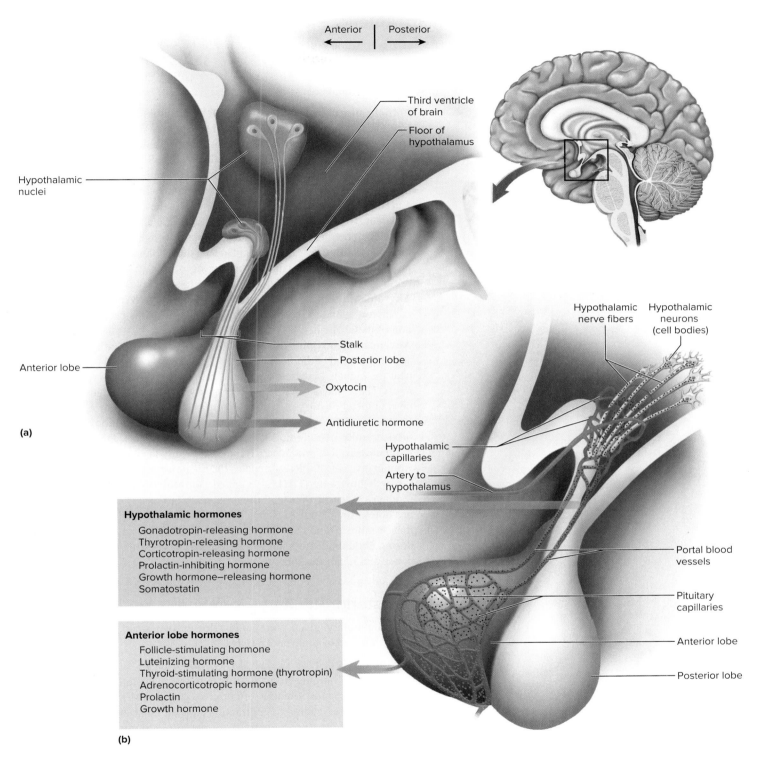

Figure 11.6 Anatomy and Control of the Pituitary Gland. (a) Control of the posterior lobe. The nuclei in blue and green synthesize oxytocin and antidiuretic hormone and transport them down nerve fibers in the stalk for storage in the posterior lobe. Later, nerve signals passing down the stalk trigger the posterior lobe to release these hormones. (b) Control of the anterior lobe. Neurons in the hypothalamus secrete releasing and inhibiting hormones into hypothalamic blood capillaries. These hormones (small black dots) travel down the portal system to capillaries in the anterior lobe. Leaving the bloodstream at that point, they stimulate anterior lobe cells to secrete their hormones, or in other cases inhibit them from doing so. **APR**

? *Which lobe of the pituitary gland is essentially composed of brain tissue?*

11.3b Anterior Pituitary Hormones

The anterior pituitary gland may be small, but the six peptide hormones it produces greatly impact the body, as you will see. The pituitary hormones are summarized in table 11.2. The first two are called **gonadotropins**[5] because they target the gonads (ovaries and testes).

1. **Follicle-stimulating hormone (FSH).** FSH stimulates the development of gametes (sex cells) in women and men. Once females reach reproductive age, FSH stimulates the maturation of bubblelike follicles that contain the eggs. In the process, it stimulates the follicles to secrete estrogen. Therefore, FSH plays an important role in the regulation of monthly ovarian and menstrual cycles. In males, it stimulates the production of sperm.
2. **Luteinizing hormone (LH).** LH also targets the ovaries and testes. In females, it stimulates ovulation, the release of the egg from the follicle. The follicle left behind after ovulation develops into a yellowish structure called the **corpus luteum.**[6] Under the influence of LH, the corpus luteum secretes progesterone, which maintains the lining of the uterus should pregnancy occur. In males, LH stimulates the testes to secrete testosterone.
3. **Thyroid-stimulating hormone (TSH),** also called **thyrotropin.** TSH causes the thyroid gland to secrete thyroid hormone, which is necessary for regulation of metabolism.
4. **Adrenocorticotropic hormone (ACTH),** also called **corticotropin.** ACTH stimulates the adrenal cortex, the outer layer of the adrenal gland near the kidney, to secrete hormones called *glucocorticoids* (primarily *cortisol*). These play a key role in the response to stress and in glucose, fat, and protein metabolism.
5. **Prolactin**[7] **(PRL).** Prolactin is named for its role in lactation. The ability to produce milk is a defining characteristic of mammals. During pregnancy, pituitary cells that produce PRL increase greatly in size and number. After a woman gives birth, PRL stimulates the mammary glands to synthesize milk.
6. **Growth hormone (GH),** also called **somatotropin.** The pituitary gland produces at least a thousand times as much GH as any other hormone. GH promotes growth of tissues throughout the body by stimulating protein synthesis, mitosis, and cell differentiation. The effects are most conspicuous during childhood and adolescence, but GH is important for tissue maintenance and repair throughout life.

11.3c Posterior Pituitary Hormones

The posterior pituitary gland stores and releases two small peptide hormones produced by the hypothalamic nuclei (table 11.2).

1. **Oxytocin**[8] **(OT)** (see fig. 11.3c). Oxytocin has a variety of reproductive functions ranging from childbirth, to breast-feeding, to bonding with offspring. The most obvious effect of oxytocin is stimulation of uterine smooth muscle, leading to contractions during labor and delivery. A synthetic form of oxytocin, Pitocin, is sometimes administered to women to induce labor. During lactation, OT targets cells in the mammary glands that cause milk to flow to the nipple. In both men and women, oxytocin surges during sexual arousal and orgasm. Oxytocin has also been called the "love hormone" because of its importance in forming and maintaining social connections, including bonds between parents and offspring.

---

[5]*trop* = to turn, change
[6]*corpus* = body; *lute* = yellow
[7]*pro* = favoring; *lact* = milk
[8]*oxy* = sharp, quick; *toc* = birth

2. **Antidiuretic hormone (ADH),** also called **vasopressin.** ADH has the same structure as oxytocin except for two of its amino acids. One function of the hypothalamus is to regulate fluid balance in the body. It does this in part by synthesizing ADH, which stimulates the kidneys to retain water and reduce urine output. Alcohol inhibits ADH release, which partially explains the increased volume of urine when one consumes alcoholic beverages.

Table 11.2 Pituitary Gland Hormones

Hormone	Target Organ or Tissue	Principal Effects
Anterior Pituitary		
Follicle-stimulating hormone (FSH)	Ovaries, testes	*Female:* Growth of ovarian follicles and secretion of estrogen *Male:* Sperm production
Luteinizing hormone (LH)	Ovaries, testes	*Female:* Ovulation, maintenance of corpus luteum *Male:* Testosterone secretion
Thyroid-stimulating hormone (TSH)	Thyroid gland	Secretion of thyroid hormone
Adrenocorticotropic hormone (ACTH)	Adrenal cortex	Secretion of glucocorticoids
Prolactin (PRL)	Mammary glands	Milk synthesis
Posterior Pituitary		
Antidiruetic hormone (ADH)	Kidneys	Water retention
Oxytocin (OT)	Uterus, mammary glands	Labor contractions, milk release, social bonding

11.3d Control of the Pituitary Gland

The hypothalamus controls the anterior pituitary gland by secreting several releasing and inhibiting hormones into the portal system. They travel to the anterior pituitary, leave the blood vessels, and either stimulate or suppress the output of the corresponding pituitary hormones. In this way, the hypothalamus regulates the numerous body functions that are affected by the anterior pituitary.

The hypothalamic hormones are listed in the violet box in figure 11.6b. *Releasing hormones* stimulate the anterior pituitary to secrete its hormones (pink box) into the blood. For example, *thyrotropin-releasing hormone (TRH)* stimulates the pituitary to secrete *thyrotropin (thyroid-stimulating hormone, TSH)* into the blood. Hypothalamic inhibiting hormones suppress pituitary output. For example, *prolactin-inhibiting hormone* suppresses prolactin secretion when a woman is not pregnant or nursing.

The hypothalamus controls the posterior pituitary by sending nerve signals down the stalk, commanding it to release ADH when there is a need to reduce urinary water loss, or OT during childbirth and breast-feeding.

Figure 11.7 summarizes the flow of information from the hypothalamus to the anterior pituitary and from the pituitary to its target organs. Superficially, this may seem like a military chain of command from hypothalamus ("the general") to pituitary ("the lieutenant") to the target organs ("the privates"). But unlike a military chain of command, the endocrine system does not work strictly from the top down. Here, the privates also give commands to the lieutenant and even to the general, through communication loops of both negative and positive

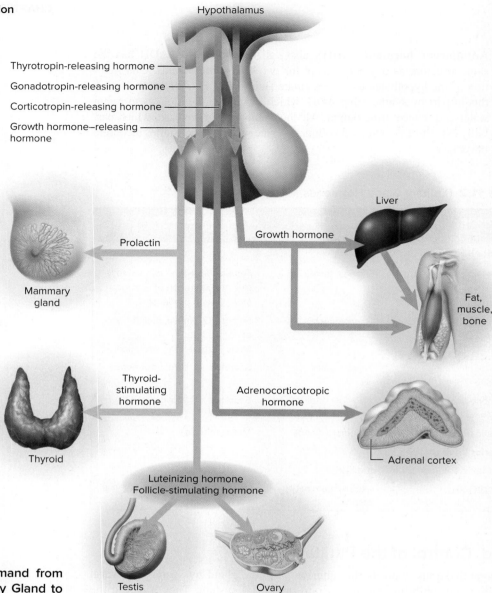

Figure 11.7 The Chain of Command from Hypothalamus to Anterior Pituitary Gland to Target Organs. **APR**

feedback. For example, we have already seen that the hypothalamus secretes TRH to stimulate the pituitary, the pituitary secretes TSH to stimulate the thyroid, and the thyroid responds by secreting thyroid hormone (TH). But TH then feeds back to both the hypothalamus and anterior pituitary to reduce TRH and TSH output, as if saying, "That's enough for now." This keeps TH secretion from becoming excessive. This particular chain of responses is called **negative feedback inhibition.**

By contrast, we saw earlier (see fig. 1.5) that during childbirth, pituitary oxytocin (OT) stimulates labor contractions in the uterus, and stretch receptors in the uterus issue nerve signals back to the hypothalamus to say, "Send more oxytocin!" The hypothalamus commands the posterior pituitary to release more OT. This is a case of **positive feedback control** of the pituitary.

Apply What You Know

Explain the phrase, "The pituitary gland is a link between the neural regulation and hormonal regulation of homeostasis."

Before You Go On

Answer these questions from memory to test your understanding of the previous section. Reread the material if you can't answer the questions.

7. Explain why the anterior and posterior pituitary glands are considered to be separate glands, despite their close physical relationship.
8. Prepare a study aid for yourself by constructing a table with three columns. In the left column, list the six hormones that are synthesized and secreted by the anterior pituitary gland; in the middle column, identify the target organ of each hormone; and in the right column, list the effects of each hormone.
9. How does the hypothalamus control the anterior pituitary gland? How does it control the posterior pituitary gland?
10. List the posterior pituitary gland hormones and explain their functions.
11. Describe the influence of the pituitary target organs over hypothalamic and pituitary function.

11.4 Other Endocrine Glands and Tissues

Expected Learning Outcomes

When you have completed this section, you should be able to

a. describe the structure and locations of the remaining endocrine glands;
b. name the hormones that the glands produce;
c. list the targets and explain the effects of the hormones; and
d. discuss hormones produced by endocrine cells in tissues and organs other than the classic endocrine glands.

Numerous hormones are produced throughout the body. We will begin our discussion with the most superior gland, the pineal gland, and will work our way inferiorly through glands in the neck and trunk. Finally, we will discuss hormones produced by tissues and organs other than those traditionally regarded as endocrine glands.

11.4a The Pineal Gland

The **pineal gland** (PIN-ee-ul) is anatomically associated with the brain. It attaches to the roof of the third ventricle, near the posterior end of the corpus callosum (see figs. 9.2 and 11.1). The pineal gland secretes **melatonin,** a monoamine, primarily during hours of darkness. Melatonin promotes sleep and plays a role in the basic daily cycles of activity known as *circadian rhythms.*

Intriguing recent studies show that increased melatonin levels are associated with decreased incidence of cancer. This finding has implications for our modern lifestyle because exposure to light suppresses the production of melatonin. This may explain the finding that blind women, whose melatonin production is not suppressed by artificial light, have a lower incidence of breast cancer. In contrast, women who work during the night under artificial light (and hence have lower levels of melatonin) have a greater incidence of breast cancer.

The pineal gland is large in children but begins to atrophy at about the age of 7. By adulthood, it is no more than a fibrous, shrunken mass. As it shrinks, granules of calcium phosphate and calcium carbonate called *pineal sand* appear. These tiny grains are visible on X-rays and enable radiologists to determine the position of the pineal gland. This can be clinically useful if the gland has been displaced by a brain tumor or other structural abnormality.

11.4b The Thyroid Gland APR

The **thyroid gland** (fig. 11.8) lies adjacent to the trachea, near the shieldlike *thyroid*[9] *cartilage* of the larynx. It is the largest gland with purely endocrine functions. In adults, it weighs 20 to 25 g, and has a dark reddish brown color due to its unusually great blood flow. It is composed of two lobes that are bulbous at their inferior end and taper superiorly. In most people, a narrow bridge of tissue, the *isthmus,* crosses the front of the trachea and joins the two lobes, and some people have a small third lobe arising medially from the isthmus.

The thyroid gland is composed of sacs called *thyroid follicles*. These are lined with simple cuboidal epithelium composed of **follicular cells,** which synthesize **thyroid hormone (TH).** TH is a monoamine made from two molecules of the amino acid tyrosine. About 90% of the daily output of TH is in a form called **thyroxine,** also known as T_4 or **tetraiodothyronine** (TET-ra-EYE-oh-doe-THY-ro-neen) because it has four iodine atoms (see fig. 11.3b). The other 10% is called T_3 or **triiodothyronine** (try-EYE-oh-doe-THY-ro-neen) because it has only three iodine atoms.

When TH reaches a target cell, T_4 is converted to T_3. T_3 enters the nucleus and binds to receptors associated with the DNA. Its primary effect is to stimulate an increase in metabolic rate, which results in increased consumption of ATP, oxygen, and fuel (and therefore increased appetite) and in heat production. TH secretion rises in cold weather to compensate for the body's heat loss. TH also stimulates growth of the bones, skin, and other tissues; increases alertness; and is crucial to development of the fetal nervous system.

---

[9]*thyr* = shield; *oid* = like, resembling

Figure 11.8 The Thyroid Gland. Gross anatomy, anterior view. Major blood vessels are shown only on the anatomical right. (b) Histology. APR
b: Biophoto Associates/Science Source

❓ *What is the function of the C cells in part (b)?*

Clinical Application 11.2

THYROID DISORDERS

Thyroid disorders result from too little thyroid hormone *(hypothyroidism)* or too much *(hyperthyroidism)*. They occur more frequently in women than in men. The most common cause of hypothyroidism in adults is *Hashimoto disease,* an autoimmune disorder that destroys the thyroid gland. Symptoms include thinning hair and scaly skin, cold intolerance, weight gain, and mental sluggishness. *Graves[10] disease (toxic goiter)* is a form of hyperthyroidism in which dysfunctional antibodies mimic the effect of thyroid-stimulating hormone, causing the thyroid gland to secrete excessive thyroid hormone. The resulting signs and symptoms include rapid weight loss, increased sweating, irritability, heart palpitations, and protruding eyeballs *(exopthalmia).* Graves disease is sometimes treated with radioactive iodine that kills the thyroid cells, or the gland may be surgically removed. Either treatment requires thyroid hormone replacement therapy for the rest of one's life, although this is relatively easy and inexpensive compared to other hormone replacement therapies.

In addition, the thyroid gland has small clusters of **C (clear) cells,** or **parafollicular cells,** nestled between the follicles. They produce a peptide hormone called **calcitonin** in response to elevated blood calcium levels. Calcitonin inhibits the bone-resorbing activity of osteoclasts and stimulates bone deposition by osteoblasts. Calcitonin is significant in children but has less apparent effects in adults. No disease results from an excess or deficiency of calcitonin.

11.4c The Parathyroid Glands APR

Four small, ovoid **parathyroid glands** are usually located on the posterior surface of the thyroid (see fig. 11.1), although their location varies. They secrete **parathyroid hormone (PTH),** which is the dominant calcium-regulating hormone in adults. In response to a blood calcium deficiency, it stimulates osteoclasts and inhibits osteoblasts, thereby dissolving bone and raising the blood calcium level. Note that these effects are opposite from those of calcitonin. PTH also stimulates the kidneys to produce calcitriol (vitamin D_3), which promotes absorption of calcium in the digestive tract. If the parathyroids are accidentally removed in the course of neck surgery, immediate hormone replacement therapy is necessary; a person can otherwise die within a few days of blood calcium deficiency.

11.4d The Thymus

The **thymus** is a bilobed gland in the mediastinum superior to the heart, behind the sternal manubrium. It is the site of maturation for certain white blood cells called T lymphocytes (*T* for *thymus-dependent*), which play a critical role in immunity. The thymus secretes hormones called **thymosin** and **thymopoietin,** which regulate the development and activity of T lymphocytes and stimulate the development of other lymphatic organs. The thymus is discussed more thoroughly in section 14.1c in relation to its immune function.

In the fetus and infant, the thymus is enormous compared to adjacent organs, sometimes extending from near the diaphragm to the neck (**fig. 11.9**). It continues to grow until age 5 or 6, but after the age of 14 it shrinks dramatically. By old age, it is a small fibrous and fatty remnant barely distinguishable from surrounding tissue.

---

[10]Robert James Graves (1769–1853), Irish physician

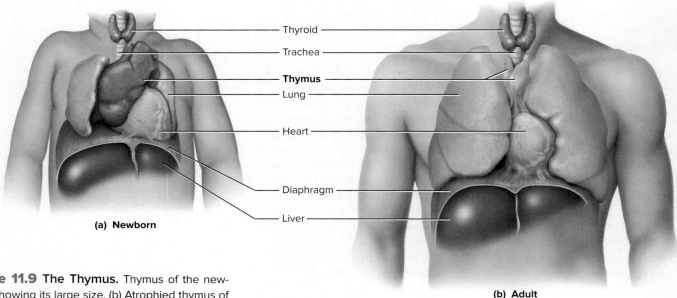

Figure 11.9 The Thymus. Thymus of the newborn, showing its large size. (b) Atrophied thymus of the adult. **APR**

11.4e The Pancreas APR

The **pancreas** is an elongated, spongy organ located below and behind the stomach (fig. 11.10). It is approximately 15 cm long and 2.5 cm wide. One end of it nestles in the curve of the first part of the small intestine, the duodenum. Most pancreatic tissue functions as an exocrine digestive gland, but scattered throughout the exocrine tissue, especially in the tail of the pancreas, are 1 to 2 million little endocrine cell clusters called **pancreatic islets (islets of Langerhans[11])**.

---

[11]Paul Langerhans (1847–88), German anatomist

Figure 11.10 The Pancreas. (a) Gross anatomy of the pancreas and its relationship to the duodenum. (b) Cell types of a pancreatic islet. (c) Light micrograph of a pancreatic islet amid the darker exocrine acini. **APR**
c: ©Ed Reschke

? *What is the function of the exocrine cells in this gland?*

Figure 11.11 Effects of Insulin and Glucagon. (a) Insulin promotes the uptake of blood glucose by cells and its polymerization to make glycogen, an energy-storage carbohydrate. In doing so, it lowers the blood sugar level. (b) Glucagon stimulates some cells, especially in the liver, to break down glycogen and release glucose to the bloodstream, thereby raising blood sugar level. Glycogenesis is the synthesis of glycogen; glycogenolysis is its hydrolysis (breakdown to glucose); and gluconeogenesis is the synthesis of glucose from noncarbohydrates, especially fats and proteins.

Insulin and **glucagon** are the most important pancreatic hormones. They regulate the blood concentration of glucose, a vital fuel for most of the body's tissues. Neurons especially depend on glucose and quickly die if it is not available.

Insulin is produced by **beta (β) cells** of the islets in response to a rise in blood glucose during and after a meal. Insulin promotes the uptake of glucose by liver, muscle, and fat cells, among others, and promotes glycogen synthesis (*glycogenesis*[12]) and fat storage (fig. 11.11a). These processes lower the blood glucose concentration.

Glucagon is produced by **alpha (α) cells** of the islets in response to falling levels of blood glucose. In contrast to insulin, glucagon promotes the mobilization of fuels. Its primary action is on the liver, where it stimulates the breakdown of glycogen (*glycogenolysis*[13]) and the synthesis of glucose from proteins or their amino acids (*gluconeogenesis*[14]) (fig. 11.11b). These processes release glucose into the blood, raising its concentration and making it available to other tissues.

11.4f The Adrenal Glands APR

Each kidney is topped by a hatlike **adrenal gland** (fig. 11.12a). These are relatively large in the fetus and infant but shrink about 50% by the age of 2 years. The adrenal gland has two distinct portions that differ in embryonic origin and the hormones produced—an inner *medulla* surrounded by an outer *cortex*.

---

[12]*glycogen* + *genesis* = creation, formation of
[13]*glycogen* + *lysis* = splitting, breaking apart
[14]*gluco* = glucose, sugar; *neo* = new; *genesis* = creation, formation of

Figure 11.12 The Adrenal Gland. (a) Location and gross anatomy. (b) Histology. APR

Victor Eroschenko

The Adrenal Medulla

The **adrenal medulla** (fig. 11.12a) develops from the same embryonic cells that give rise to the sympathetic nervous system, and its cells are essentially modified sympathetic postganglionic neurons (see "Anatomy of the Sympathetic Division" in section 9.5c). They have no axons or dendrites but secrete their products directly into the blood. They secrete mainly **epinephrine** and **norepinephrine,** two monoamines involved in stress responses. They are discussed later under stress physiology. As you may recall, norepinephrine is also a sympathetic neurotransmitter.

The Adrenal Cortex

The **adrenal cortex** produces steroid hormones *(corticosteroids),* all of which are synthesized from cholesterol. It has three tissue layers that differ in their hormone output (fig. 11.12b).

- The **zona glomerulosa**[15] (glo-MER-you-LO-suh) is the thin, outermost layer of the cortex.
- The **zona fasciculata**[16] (fah-SIC-you-LAH-ta) is the middle and thickest layer of the cortex. Its cells are arranged in parallel cords (fascicles) perpendicular to the gland surface.
- The **zona reticularis**[17] (reh-TIC-you-LAR-iss), is a narrow, innermost layer whose cells form a branching network (reticulum).

---

[15] *glomerul* = little ball; *osa* = full of
[16] *fascicul* = little bundle; *ata* = possessing
[17] *reticul* = little network; *aris* = pertaining to

Aldosterone is produced only in the zona glomerulosa. It stimulates the kidneys to retain sodium in the blood instead of allowing it to be lost in the urine. Water follows the sodium by osmosis. Ultimately, aldosterone slows down urinary water loss and helps to maintain blood volume and pressure. Aldosterone secretion is increased in response to a drop in blood pressure, and is part of the *renin–angiotensin–aldosterone (RAA) system*. Aldosterone is a type of hormone called a *mineralocorticoid*.

Cortisol is produced in the zona fasciculata and zona reticularis. It is secreted in response to adrenocorticotropic hormone (ACTH) from the anterior pituitary gland. Its most noticeable effects are on adipose tissue, muscle, and the liver. It raises the levels of blood glucose and other fuels, providing the body with the energy to respond to stress. Cortisol stimulates muscles to break down their own proteins and release the amino acids into circulation; it stimulates the liver to synthesize glucose from these free amino acids; and it stimulates adipose tissue to break down fats, releasing fatty acids into circulation as a supplemental fuel. These effects are important to the body's ability to adapt to stress and will be discussed later under stress physiology. Cortisol and similar, lesser hormones are called *glucocorticoids*.

Androgens and smaller amounts of **estrogens** are sex hormones produced in both the zona fasciculata and zona reticularis. Androgens stimulate many aspects of male development and reproductive physiology. In men, the testes secrete most of the male hormones, so their adrenal androgens are of relatively minor importance. Androgens are also important in female physiology, however, and the adrenal cortex is their sole source. The principal adrenal androgen is **dehydroepiandrosterone (DHEA)**. In the male fetus, DHEA plays an important role in development of the male reproductive tract. DHEA acts in both sexes to stimulate development of pubic and axillary hair and apocrine sweat glands at puberty, and, in adults, DHEA helps to maintain bone and muscle mass.

> **Apply What You Know**
> The zona fasciculata grows significantly thicker in pregnant women. What do you think could be the benefit of this?

11.4g The Gonads

The **gonads** (ovaries and testes) function as both endocrine and exocrine glands. Their exocrine products are eggs and sperm, and their endocrine products are gonadal hormones, mostly steroids.

The ovaries secrete **estrogen, progesterone,** and **inhibin.** In females of reproductive age, the ovaries undergo monthly cyclical changes that involve hormone production. In the first part of the ovarian cycle, a bubblelike follicle develops, consisting of the egg (oocyte) surrounded by *follicular cells* that pile on top of each other to form layers (fig. 11.13a). Follicular cells produce estrogen. In the middle of the cycle, the egg is released by ovulation. The remains of the follicle become the corpus luteum, a structure that secretes **progesterone.**

Estrogen and progesterone regulate the ovarian and menstrual cycles, prepare the uterus for implantation of the embryo, maintain pregnancy, and contribute to the development of the reproductive system. Inhibin is also secreted by the follicle and corpus luteum. It is an inhibitory signal to the anterior pituitary gland that regulates the secretion of follicle-stimulating hormone (FSH). The ovarian and menstrual cycles and associated hormones are discussed in more detail in chapter 19.

The **testes** secrete **testosterone** and inhibin. Most of the testis is composed of microscopic *seminiferous tubules* that produce sperm. The wall of the tubules is formed of **germ cells** in the process of becoming sperm, and **nurse cells,** which support and protect the germ cells and secrete inhibin. By limiting FSH secretion, inhibin regulates the rate of sperm production. Nestled between the tubules are

Figure 11.13 The Gonads. (a) Histology of an ovarian follicle. (b) Histology of the testis.

a: Ed Reschke/Getty Images b: Ed Reschke

clusters of **interstitial endocrine cells (cells of Leydig**[18]**),** the source of testosterone (**fig. 11.13b**). Testosterone sustains sperm production and the sex drive throughout adolescent and adult life. It stimulates development of the male reproductive system in the fetus, and promotes the build-up of muscle mass in adolescence, among other masculine physical characteristics.

11.4h Endocrine Functions of Other Organs and Tissues

In recent decades, physiologists have discovered several hormones that are secreted by organs and tissues not traditionally regarded as endocrine glands. A list of many of these hormones may be found in **table 11.3**.

- **The heart.** Rising blood pressure stretches the heart wall and stimulates muscle cells in the atria to secrete **natriuretic**[19] **peptides.** These hormones increase urine output and sodium excretion, which leads to a decreased blood volume. The ultimate effect of natriuretic peptides is to lower blood pressure.
- **The liver.** The liver is involved in the production of many hormones: (1) It collaborates with the skin and kidneys to produce **calcitriol (vitamin D$_3$).** (2) It secretes a blood protein called *angiotensinogen,* which is converted to the hormone **angiotensin II,** whose effects are described shortly in relation to the kidneys. (3) It secretes **hepcidin,** a regulator of iron homeostasis. (4) It secretes **insulin-like growth factor I (IGF-I),** a hormone that aids the action of growth hormone.
- **The kidneys.** The kidneys produce most of the body's **erythropoietin (EPO)** (er-RITH-ro-POY-eh-tin), a hormone that stimulates red blood cell production. Like the liver, the kidneys perform certain steps in both calcitriol and angiotensin II synthesis. Angiotensin II has numerous effects. It stimulates the adrenal cortex to secrete aldosterone, thus promoting sodium and water retention, and it constricts blood vessels and increases cardiac output. The overall effect of all of these is to increase blood pressure.
- **The stomach and intestines.** These organs secrete **enteric**[20] **hormones** that coordinate different regions and glands of the digestive system. For example, the stomach secretes **gastrin** when food arrives. Gastrin stimulates the production of hydrochloric acid, which aids protein digestion. Other hormones regulate feelings of hunger and satiety, and are the focus of much interest because of the prevalence of obesity. For example, **ghrelin** (GRELL-in), a "hunger hormone" secreted by the stomach, sends signals to the hypothalamus to stimulate the appetite. In contrast, **peptide YY (PYY),** secreted by the small and large intestines, creates a feeling of satiety and has an appetite-suppressing effect.
- **Adipose tissue.** Fat cells secrete the protein hormone **leptin,** which affects appetite-regulating centers of the hypothalamus. A low level of leptin, signifying a low level of body fat, increases appetite and food intake, whereas a high level of leptin tends to blunt the appetite. Although the discovery of leptin in the 1990s generated a great deal of excitement in the hopes it would offer a treatment for obesity, its therapeutic value has been disappointing. There are a small number of cases of obesity that stem from a leptin deficiency, and these patients benefit from leptin replacement therapy. Obesity usually isn't caused by leptin deficency but rather results from resistance to its actions, so giving the hormone as a treatment has no benefit.
- **Osseous tissue.** Osteoblasts of the bone tissue secrete **osteocalcin,** which not only contributes to the structure of the bone matrix but also stimulates the pancreatic islets and adipose tissue. Its effect is to increase the secretion and action of insulin. It also seems to inhibit weight gain (fat deposition) and the onset of type 2 diabetes mellitus.

---

[18]Franz von Leydig (1821–1908), German histologist
[19]*natri* = sodium; *uretic* = pertaining to urine
[20]*entero* = intestine

Table 11.3 Hormones from Sources Other than the Hypothalamus and Pituitary

Hormones	Source	Target Organs and Tissues	Principal Effects
Melatonin	Pineal gland	Brain	Promotes sleep and plays a role in circadian rhythms
Thymopoietin, thymosin, thymulin	Thymus	Immune cells (T lymphocytes)	Stimulate T lymphocyte maturation and activity
Thyroxine (T_4) and triiodothyronine (T_3)	Thyroid gland	Most tissues	Elevate metabolic rate and heat production; increase respiratory rate, heart rate, and strength of heart beat; stimulate appetite and accelerate breakdown of nutrients; promote growth of skin, hair, nails, and fetal nervous system
Parathyroid hormone (PTH)	Parathyroid glands	Bone, kidneys, small intestine	Raises blood Ca^{2+} level by stimulating bone resorption
Epinephrine, norepinephrine	Adrenal medulla	Most tissues	Increase blood glucose level; promote alertness; raise metabolic rate; raise heart and respiratory rates
Aldosterone	Adrenal cortex	Kidneys	Promotes sodium and water retention; maintains blood pressure and volume
Cortisol	Adrenal cortex	Most tissues	Part of stress response; stimulates fat and protein catabolism and gluconeogenesis; suppresses inflammation
Glucagon	Pancreas	Primarily liver	Stimulates glycogen breakdown; raises blood glucose levels; stimulates fat breakdown and increases fatty acid levels in blood; promotes gluconeogenesis
Insulin	Pancreas	Most tissues	Stimulates glucose absorption and lowers blood glucose levels
Estradiol	Ovaries	Many tissues	Regulates menstrual cycle and pregnancy; prepares mammary glands for lactation; stimulates female reproductive development and adolescent growth
Progesterone	Ovaries	Uterus, mammary glands	Regulates menstrual cycle and pregnancy; prepares mammary glands for lactation
Inhibin	Ovaries and testes	Anterior pituitary gland	Inhibits FSH secretion
Testosterone	Testes	Many tissues	Stimulates sperm production, libido, and musculoskeletal growth; promotes fetal and adolescent reproductive development
Natriuretic peptides	Heart	Kidneys	Lower blood volume and pressure by increasing urine output and sodium excretion
Hepcidin	Liver	Small intestine, liver	Regulates plasma iron level
Insulin-like growth factor I	Liver	Many tissues	Promotes actions of growth hormone
Precursor of calcitriol (vitamin D_3)	Liver	Kidney	Promotes absorption of calcium (see kidneys)
Angiotensinogen	Liver	----	Precursor of angiotensin I (see kidneys)
Angiotensin I	Kidneys	----	Precursor of angiotensin II, a vasoconstrictor
Calcitriol (vitamin D_3)	Kidneys	Small intestine	Promotes absorption of calcium; increases blood calcium level
Erythropoietin	Kidneys	Bone marrow	Promotes red blood cell production; increases oxygen-carrying ability of blood
Gastrin	Stomach and small intestine	Stomach	Stimulates acid secretion
Ghrelin	Stomach and small intestine	Brain	Stimulates hunger
Peptide YY	Stomach and small intestine	Brain	Produces sensation of satiety
Leptin	Adipose tissue	Brain	Decreases appetite
Osteocalcin	Osseous tissue	Pancreas, adipose tissue	Stimulates beta cells to increase insulin secretion; enhances insulin sensitivity and reduces fat deposition

PERSPECTIVES ON HEALTH

Medical problems with the endocrine system center mostly on three dysfunctions: hyposecretion (hormone deficiency), hypersecretion (hormone excess), and hormone inaction, in which a normal amount of hormone is secreted but the body doesn't respond normally to it. Hypo- and hypersecretion disorders have specific names that often refer to the specific endocrine gland, such as *hypo-* and *hyperthyroidism*.

One example of a hyposecretion disorder is *diabetes insipidus,* in which the posterior pituitary puts out little or no antidiuretic hormone. The kidneys, as a result, fail to conserve water; one therefore produces enormous quantities of urine, although the urine is glucose-free (unlike the sugary urine in diabetes mellitus). Hormone hyposecretion is often treated with hormone replacement therapy. An example of a hypersecretion disorder is *pheochromocytoma*[21] (FEE-oh-CRO-mo-sy-TO-muh), a tumor of the adrenal medulla that secretes up to 27 times the normal amounts of epinephrine and norepinephrine. This often causes drenching sweats, crises of severe headaches and hypertension, and sudden death from stroke or cardiac arrhythmia, often before the disease is diagnosed. An example of target-cell insensitivity is *androgen-insensitivity syndrome (AIS),* in which the body lacks testosterone receptors. Even though testosterone is present, it has no effect. In males with AIS, the genitalia show female anatomy, and at puberty, the small amount of estrogen produced by the testes causes males to develop enlarged breasts and other feminine features. Despite their outward appearance, such males are sterile because they do not develop an internal female reproductive tract or ovaries.

Diabetes Mellitus

Diabetes mellitus (DM) is the world's most common metabolic disease. It is the leading cause of adult blindness, renal failure, gangrene, and the necessity for limb amputations.

DM results from the hyposecretion or inaction of insulin. Classic signs and symptoms are "the three polys": *polyuria*[22] (excessive urine output), *polydipsia*[23] (intense thirst), and *polyphagia*[24] (ravenous hunger). These are common patient complaints when first presenting to a physician, and diabetes can be confirmed by further blood and urine tests that reveal *hyperglycemia*[25] (elevated blood glucose), *glycosuria*[26] (glucose in the urine), and *ketonuria* (ketones in the urine).

---

[21]*pheo* = dusky; *chromo* = colored; *cyt* = cells; *oma* = tumor
[22]*poly* = much, excessive; *uri* = urine
[23]*poly* = much, excessive; *dipsia* = drinking
[24]*poly* = much, excessive; *phagia* = eating
[25]*hyper* = excess; *glyc* = sugar, glucose; *emia* = blood condition
[26]*glyco* = glucose, sugar; *uria* = urine condition

There are two forms of diabetes mellitus: type 1 and type 2. *Type 1 diabetes,* formerly called *insulin-dependent* or *juvenile-onset diabetes,* accounts for 5% to 10% of cases in the United States. It is an autoimmune condition that results from destruction of pancreatic beta cells by one's own antibodies. Type 1 diabetes is most often diagnosed around the age of 12 but it can also appear later in life. Treatment requires insulin injections or delivery of insulin by a pump worn on the body, and also includes meal planning, exercise, and self-monitoring of blood glucose levels.

Type 2 diabetes mellitus, formerly called *non-insulin-dependent* or *adult-onset diabetes,* accounts for over 90% of cases. Millions of Americans are prediabetic and are at risk of developing the disease. Type 2 diabetics may have normal or even elevated insulin levels; the problem is *insulin resistance*—a failure of target cells to respond to it. The three major risk factors for type 2 diabetes are heredity, age, and obesity. It tends to run in families, and obesity and physical inactivity increase insulin resistance. Lifestyle changes over recent decades mean that many of us eat too much and do not exercise enough. The incidence of type 2 diabetes has therefore increased, and more and more people are developing it at a young age. The relationship between obesity and diabetes mellitus is complex, but one factor is that adipocytes produce chemical signals that indirectly interfere with glucose transport into most cells. Type 2 diabetes can successfully be managed with a combination of weight loss, careful attention to diet, and exercise, often combined with medications that improve insulin secretion or target-cell sensitivity. Insulin injections may also be employed to supplement insulin-sensitizing drugs.

When cells cannot absorb glucose, they rely on fat and protein for energy. Fat and protein breakdown results in muscular atrophy and weakness. Rapid fat breakdown elevates blood concentrations of free fatty acids and their breakdown products, ketones. As acids, ketones lower the blood pH and produce a condition called *ketoacidosis.* Ketoacidosis depresses the nervous system, produces a deep gasping breathing typical of terminal diabetes, and leads to diabetic coma and death.

In the long term, diabetes mellitus can lead to degenerative neurological and cardiovascular diseases, including *diabetic neuropathy* (nerve damage) and *atherosclerosis* (blockage of blood vessels with lipid deposits). Diabetic neuropathy can make a patient dangerously unaware of minor injuries, which can thus become infected from neglect and result in gangrene and the necessity of amputation. Many diabetics lose their toes, feet, or legs to the disease. Atherosclerosis leads to poor circulation, which contributes to gangrene, kidney failure, and blindness.

Before You Go On

Answer these questions from memory to test your understanding of the previous section. Reread the material if you can't answer the questions.

12. Identify two glands that are larger in children than adults. What hormones do the glands produce and what is their function?
13. What hormone increases the body's heat production in cold weather? What other functions does this hormone have?
14. What hormones are produced by the adrenal cortex and what are their functions?
15. Name one hormone produced by each of the following organs—the heart, kidney, stomach, and ovary—and state the function of each hormone.

11.5 Stress Physiology

Expected Learning Outcomes

When you have completed this section, you should be able to

a. define *stress* in physiological terms; and
b. identify the stress hormones and explain their roles in adapting to stress.

Stress is defined as any situation that upsets homeostasis and threatens one's physical or emotional well-being. Physical causes of stress *(stressors)* include injury, hemorrhage, infection, intense exercise, pain, and malnutrition, among others. Emotional causes include fear, anger, grief, depression, and anxiety. Whatever the cause, we react to stress in a fairly consistent way called the **stress response**, mediated by the endocrine system and sympathetic nervous system. Three stages of stress are recognized, each with its own characteristic hormonal responses.

1. The **alarm reaction** is mediated mainly by norepinephrine from the sympathetic nerves and epinephrine from the adrenal medulla. These chemicals raise the heart rate, blood pressure, and blood glucose level; promote pulmonary ventilation; and prepare the body for action such as fighting or escaping danger. In most cases, we recover from stress during this stage. If a stressful situation persists, however, the increased consumption of liver and muscle glycogen can push one to the next stage.
2. The **stage of resistance** develops as stored glycogen is depleted and the demand for glucose must be met by tapping into alternative fuels. ACTH and cortisol are secreted to break down stored fat and protein into fatty acids, glycerol, and amino acids. The liver utilizes glycerol and amino acids to synthesize more glucose. Fatty acids and glucose are oxidized to make ATP. Sometimes to our misfortune, long-term cortisol release inhibits the immune system. Chronic stress therefore increases one's susceptibility to infections, ulcers, and even cancer.
3. The **stage of exhaustion** sets in as even one's reserve alternative fuels are depleted. When fat stores are gone, the body begins breaking down more and more protein to meet its energy needs. This leads to a progressive wasting away of muscles and weakening of the body. Elevated aldosterone secretion can cause excessive salt and water retention, hypertension, and depletion of potassium from the body. Death can result in this stage from heart failure, kidney failure, or overwhelming infection.

Before You Go On

Answer these questions from memory to test your understanding of the previous section. Reread the material if you can't answer the questions.

16. Identify a variety of physical and emotional causes of stress.
17. Identify the dominant hormones of each stage of stress, and explain how the pathological progression of stress relates to the use or depletion of various body fuels.

Aging of the Endocrine System

The endocrine system shows less degenerative change in old age than any other organ system. Although the reproductive hormones drop sharply in old age and growth hormone and thyroid hormone secretion decline steadily after adolescence, other hormones continue to be secreted at fairly stable levels even into old age. Target-cell sensitivity declines, however, so some hormones have less effect. For example, the pituitary gland is less sensitive to negative feedback inhibition by adrenal glucocorticoids; consequently, ACTH secretion and the response to stress are more prolonged than usual. Diabetes mellitus is more common in old age, largely because target cells have fewer insulin receptors. In part, this is an effect of increased adiposity in old age. The more fat there is at any age, the less sensitive other cells are to insulin. In addition, most people lose muscle mass in old age, and muscle is the body's most significant glucose-buffering tissue.

CAREER SPOTLIGHT

Diabetes Educator

Rolf Bruderer/Getty Images

Diabetes mellitus is the world's most widespread metabolic disease, and people with diabetes often express frustration and dissatisfaction with the care they receive. Patients who can't adequately manage their diabetes suffer needless disability and premature death. A diabetes educator is a professional who assists, educates, and provides emotional support for diabetes patients and their caregivers. Diabetes educators teach techniques of blood glucose monitoring, insulin injection, and the care and use of insulin pumps. They counsel patients on diet, exercise, taking their medication regularly, coping emotionally with the disease, and other skills—all aimed at behavioral changes that will produce better health outcomes. They work in hospitals, physicians' offices, outpatient clinics, pharmacies, home-health-care organizations, and other settings. They are usually able to spend more time with a patient than the physician can, and to provide more emotional support.

    A candidate for diabetes educator must have very good communication skills; a sound knowledge of biological, clinical, and social sciences; and an ability to relate comfortably, patiently, and effectively with people of all kinds. Many diabetes educators are registered nurses who specialize and become diabetes nurse educators (DNEs). Diabetes educators, however, also come from a wide variety of other backgrounds and include dietitians, pharmacists, social workers, exercise physiologists, and mental health professionals. With experience and self-directed online coursework, one may become a Certified Diabetes Educator (CDE) or sit for board exams to become board-certified in advanced diabetes management (BC-ADM). See appendix B for further information on becoming a diabetes educator.

CONNECTIVE ISSUES

Ways in Which the ENDOCRINE SYSTEM Affects Other Organ Systems

All Systems
Growth hormone, insulin, insulin-like growth factors, thyroid hormone, and glucocorticoids affect the development and metabolism of most tissues.

Integumentary System
Sex hormones affect development of body hair and apocrine glands, and estrogen affects subcutaneous fat deposition. The skin carries out the first step of calcitriol (vitamin D_3) production.

Skeletal System
Bone growth and maintenance are affected by growth hormone, parathyroid hormone, calcitonin, calcitriol, estrogen, and testosterone.

Muscular System
Growth hormone and testosterone stimulate muscular growth; insulin regulates glucose uptake by muscle; other hormones affect electrolyte balance, which is critical to muscular function.

Nervous System
Hormones exert negative feedback inhibition on the hypothalamus; several hormones affect nervous system development, mood, and behavior; other hormones affect electrolyte balance, which is critical to neuron function.

Circulatory System
Angiotensin II, aldosterone, ADH, and natriuretic peptides regulate blood volume and pressure; EPO regulates production of red blood cells; epinephrine, thyroid hormone, and other hormones affect heart rate and contraction force.

Lymphatic and Immune Systems
Hormones stimulate lymphoid organs and influence production of immune cells; glucocorticoids suppress immunity and inflammation. Glucocorticoids suppress immunity and inflammation.

Respiratory System
Epinephrine and norepinephrine increase pulmonary airflow.

Urinary System
ADH regulates water excretion; PTH, calcitriol, and aldosterone regulate electrolyte excretion.

Digestive System
Insulin and glucagon regulate nutrient storage and metabolism; enteric hormones regulate gastrointestinal secretion, motility, and appetite.

Reproductive System
Gonadotropins, sex steroids, and other hormones regulate sexual development, sperm and egg production, sex drive, the menstrual cycle, pregnancy, fetal development, and lactation.

Study Guide

Assess Your Learning Outcomes

To test your knowledge, discuss the following topics with a study partner or in writing, ideally from memory.

11.1 Overview of the Endocrine System

1. The meaning of *hormone* and the components of the endocrine system
2. The reason why only some organs respond to a hormone even though it travels throughout the body
3. The distinction between exocrine glands and endocrine glands
4. Functional similarities and differences between the endocrine system and nervous system and examples of how the systems overlap

11.2 Endocrine Physiology

1. The three chemical classes of hormones, with examples of each
2. Why the relationship between a hormone and receptor is described as a "lock-and-key" mechanism
3. How hydrophilic and hydrophobic hormones differ in their actions at a target cell
4. Three ways in which hormones influence each other's action when two or more hormones act on the same target organ
5. How hormone secretion is regulated by negative feedback loops
6. How a hormone's actions are halted once its purpose has been served

11.3 The Hypothalamus and Pituitary Gland

1. The structural relationship and embryonic origins of the two lobes of the pituitary gland
2. How the hypothalamus is functionally connected to each of the lobes
3. Hormones of the anterior pituitary and their target organs and functions
4. Hormones of the posterior pituitary and their target organs and functions
5. How the hypothalamus controls secretion by each of the pituitary lobes
6. How positive and negative feedback from target organs regulate pituitary function

11.4 Other Endocrine Glands and Tissues

1. The location of the pineal gland and the action of melatonin
2. The location and histology of the thyroid gland
3. The chemical nature of thyroid hormone and the differences between T_3 and T_4
4. The effects of thyroid hormone
5. The source and effects of calcitonin
6. The location of the parathyroid glands, their hormone, and its function
7. The location and endocrine function of the thymus
8. Why the pancreas is considered to be both an exocrine and endocrine gland
9. The cells that produce insulin and glucagon and the functions of these hormones
10. The location and two main components of the adrenal glands
11. Hormones of the adrenal medulla and their effects
12. Tissue layers of the adrenal cortex
13. Hormones of the adrenal cortex, their functions, and the tissue layers that produce each one
14. Hormones of the female and male gonads, their cellular sources, and their functions
15. Names and functions of the hormones produced by the heart, liver, kidneys, stomach, intestines, adipose tissue, and bones

11.5 Stress Physiology

1. The physiological definition of *stress*
2. Physical and emotional causes of stress
3. Three stages of the stress response, the major controlling hormones of each, and the events that mark transitions from one stage to the next

Testing Your Recall

1. Which hormone relies on a second messenger to cause a cellular response?
   a. testosterone
   b. glucagon
   c. progesterone
   d. aldosterone
   e. cortisol

2. Which of the following does not influence blood pressure?
   a. aldosterone
   b. natriuretic peptide
   c. epinephrine
   d. cortisol
   e. angiotensin II

3. _____ leads to increased osteoclast activity and elevates blood calcium concentration.
   a. Parathyroid hormone
   b. Calcitonin
   c. Thyroxine
   d. Aldosterone
   e. ACTH

4. Where are receptors for insulin located?
   a. in the pancreatic beta cells
   b. in the blood plasma
   c. on the target-cell membrane
   d. in the target-cell cytoplasm
   e. in the target-cell nucleus

5. Which hormone is *not* released in response to stress?
   a. ACTH
   b. cortisol
   c. leptin
   d. epinephrine
   e. norepinephrine

6. Which hormone would be released if a person was dehydrated?
   a. glucagon
   b. ADH
   c. ACTH
   d. insulin
   e. natriuretic peptide

7. Which gland is derived from an embryonic pouch in the roof of the pharynx?
   a. pineal gland
   b. posterior pituitary gland
   c. thyroid gland
   d. parathyroid gland
   e. anterior pituitary gland

8. Which hormone is secreted by alpha cells?
   a. calcitonin
   b. glucagon
   c. insulin
   d. growth hormone
   e. epinephrine

9. Which hormone is *not* synthesized and secreted by the anterior pituitary gland?
   a. TSH
   b. ACTH
   c. prolactin
   d. FSH
   e. oxytocin

10. What endocrine disorder is characterized by polyuria, polyphagia, and polydipsia?
    a. hypothyroidism
    b. Graves disease
    c. diabetes insipidus
    d. diabetes mellitus
    e. Cushing syndrome

11. In response to stress, _____ stimulates glucose synthesis in liver cells and the release of fatty acids from adipose tissue.

12. Antidiuretic hormone (ADH) is synthesized in the _____ and released by the _____ .

13. The hypothalamus can monitor the body's fat content by the blood concentration of _____ , a hormone secreted by adipocytes.

14. The stomach secretes _____ , a hormone that stimulates appetite.

15. The hypophyseal portal system is a means for the brain to communicate with the _____ .

16. In males, testosterone is mainly secreted by the _____ cells of the testes.

17. Follicle-stimulating hormone (FSH) targets the _____ and _____ .

18. The _____ gland secretes melatonin.

19. The _____ produces hormones that stimulate T lymphocyte maturation.

20. Muscular contractions of the uterus during labor and delivery are stimulated by _____ .

Answers in Appendix A

What's Wrong with These Statements?

Briefly explain why each of the following statements is false, or reword it to make it true.

1. Insulin sensitivity increases when the percentage of body fat rises.

2. Follicle-stimulating hormone and luteinizing hormone target the ovaries but have no effect on the testes.

3. If fatty plaques of atherosclerosis blocked the arteries of the hypophyseal portal system, the function of the ovaries and testes would probably not be affected.

4. The pineal gland and thymus become larger as one ages.

5. The adrenal cortex secretes catecholamines.

6. Glucagon is released immediately after you have eaten a meal.

7. The second stage of the stress response is dominated by the hormone epinephrine.

8. Oxytocin and antidiuretic hormone (ADH) are secreted through a duct leading to the posterior pituitary.

9. Thyroxine reduces metabolic rate and heat production.

10. A peptide hormone can freely enter a target cell, but a steroid hormone cannot.

Answers in Appendix A

Testing Your Comprehension

1. Suppose you were walking down a dark, unfamiliar street when suddenly a large, growling dog leapt out from behind a tree in front of you. List two hormones that would be released in response to your fear, and describe their actions. If you had to run down the street to escape, what hormone would be secreted to support the continued demands for energy?

2. Tom Harley is a motorcycle enthusiast who goes out riding every weekend. One day he rear-ends a car that stops unexpectedly in front of him. Tom falls hard on his face and badly fractures his nose and sphenoid bone. He recovers, but ever since then he has had polyuria—he produces enormous amounts of urine each day and he is continually dehydrated and intensely thirsty. In view of your knowledge of the pituitary gland, explain how his injury could have led to this condition.

3. Ricardo, an 11-year-old boy, is taken to a physician when his parents notice that he is ravenously hungry and thirsty, and also has to go to the bathroom a lot. Tests reveal that he has elevated blood glucose levels and glucose in the urine. What condition does Ricardo have? What hormone is lacking, and how might his symptoms be alleviated?

PART 4 Circulation and Defense

The Circulatory System I

Chapter 12

BLOOD

Chapter Outline

12.1 Introduction
- 12.1a Functions of the Circulatory System
- 12.1b Components and General Properties of Blood
- 12.1c Blood Plasma
- 12.1d How Blood Is Produced

12.2 Erythrocytes
- 12.2a Quantity and Structure
- 12.2b Hemoglobin
- 12.2c The Erythrocyte Life Cycle
- 12.2d Blood Types

12.3 Leukocytes
- 12.3a Structure and Function
- 12.3b Types of Leukocytes
- 12.3c The Leukocyte Life Cycle

12.4 Platelets and Control of Bleeding
- 12.4a Platelet Structure and Function
- 12.4b Platelet Production
- 12.4c Control of Bleeding

Clinical Applications/Perspectives on Health
- Clinical Application 12.1: Sickle-Cell Disease
- Clinical Application 12.2: The Complete Blood Count
- Perspectives on Health

End of Chapter
- Career Spotlight: Phlebotomist
- Connective Issues
- Study Guide

The formed elements of blood. This shows lymphocytes (blue), erythrocytes (red), and platelets (violet) in a mesh of fibrin, a sticky blood-clotting protein.

Dr. Yorgos Nikas/Science Source

BASE CAMP

Before ascending to the next level, be sure you're properly equipped with a knowledge of these concepts from earlier chapters.

- "Osmosis" in section 3.2e
- Red bone marrow (see "Gross Anatomy of Bones" in section 6.1c)

Anatomy & Physiology Revealed 4.0

Module 9 Circulatory System

B**lood has always** had a special mystique. Since ancient times, people have seen blood flow from the body and with it, the life of the individual. Blood is both revered and feared. In various cultures, it is an important part of sacred religious rituals and fertility rites. People also sometimes faint at the sight of blood, and few horror films would be as scary without copious amounts of the red, sticky fluid. Even though blood is one of the most accessible tissues, little was known about its specific functions until the twentieth century. From ancient Egypt to the 1800s, physicians believed they could treat almost any condition—from pneumonia to fevers, inflammation, and headaches—by removing blood. Until the late nineteenth century, surgeons used both lancets and leeches to drain blood with such enthusiasm that leeches in Western Europe were hunted almost to extinction. The benefits of this bloodletting were dubious, and undoubtedly more patients were harmed than were helped.

Over the past several decades, technology has ushered in a new era for **hematology**,[1] the study of blood. Faced with devastating diseases such as AIDS and leukemia, scientists have intensified efforts to understand the blood and, as a result, revolutionary treatments have saved and improved many lives.

12.1 Introduction

Expected Learning Outcomes

When you have completed this section, you should be able to

a. identify the major components of the circulatory system;
b. explain the functions of the circulatory system;
c. describe the components and the physical properties of blood; and
d. summarize the composition of blood plasma.

12.1a Functions of the Circulatory System APR

The **circulatory system** consists of the heart, blood vessels, and blood. The term **cardiovascular system**[2] refers only to the heart and blood vessels, which are discussed in chapter 13.

Single-celled organisms and even the simplest multicellular animals can meet their needs by diffusion of substances from the environment through the plasma membranes of their cells. However, as animals evolved larger bodies, some cells were too far from the surface to have their needs met so simply. Larger animals developed a system of internal passages—blood vessels—to distribute the necessities of life to all cells of the body. A pump, the heart, was necessary to move materials

---

[1] *hem, hemato* = blood; *logy* = study of
[2] *cardio* = heart; *vas* = vessel

along these vessels. And of course there had to be blood itself, the liquid medium of transport for those materials. But the functions of blood are more diverse than that, as you can see below.

Transport:

- The blood carries oxygen from the lungs to all of the body's tissues, while it picks up carbon dioxide from those tissues and carries it to the lungs to be removed from the body.
- It picks up nutrients from the digestive tract and distributes them to all tissues.
- It carries metabolic wastes to the kidneys for removal.
- It carries hormones from endocrine cells to their target cells.
- It transports stem cells to their destinations.

Protection:

- White blood cells destroy harmful microorganisms and cancer cells.
- Antibodies and other blood proteins neutralize toxins and help to destroy pathogens.
- Blood plays several roles in inflammation, a mechanism for limiting the spread of infection and promoting the repair of injured tissue.
- Platelets initiate blood clotting and other processes for minimizing blood loss.

Regulation:

- Blood helps maintain optimal fluid balance and distribution in the body by absorbing or giving off fluid under different conditions.
- Blood proteins stabilize the pH of extracellular fluids by buffering acids and bases.
- Shifts in blood flow help to regulate body temperature by routing blood to the skin for heat loss or routing it deeper to retain heat in the body.

In sum, the blood plays numerous roles in maintaining homeostasis. Considering its vital functions, it is easy to understand why an excessive loss of blood is fatal, and why the circulatory system evolved mechanisms for minimizing such loss.

12.1b Components and General Properties of Blood

Table 12.1 lists several properties of blood. It is a liquid connective tissue composed of plasma and formed elements. **Plasma** is a clear, light-yellow fluid that forms the extracellular matrix of the tissue. **Formed elements** are cells and cell fragments— *red blood cells, white blood cells,* and *platelets* (fig. 12.1). The term *formed elements* denotes that these are membrane-enclosed entities with a specific structure;

Table 12.1 General Properties of Blood

Characteristic	Typical Values for Healthy Adults*
Mean fraction of body weight	8%
Volume in adult body	Female: 4–5 L; male: 5–6 L
Mean temperature	38°C (100.4°F)
pH	7.35–7.45
Hematocrit (packed cell volume)	Female: 37% to 48%; male: 45% to 52%
Hemoglobin	Female: 12–16 g/dL; male: 13–18 g/dL
Mean RBC count	Female: 4.2–5.4 million/μL; male: 4.6–6.2 million/μL
Platelet count	130,000–400,000/L
Total WBC count	5,000–10,000/L

*Values vary slightly depending on the testing methods used.

Figure 12.1 The Formed Elements of Blood.

❓ *What do erythrocytes and platelets lack that the other formed elements have?*

Figure 12.2 Separating Plasma and Formed Elements of Blood. Centrifuging a sample of blood separates the erythrocytes from the leukocytes and platelets (buffy coat) and plasma. The percentage of the volume composed of erythrocytes is called the *hematocrit.*

they can't all be called *cells,* because platelets are merely fragments of certain bone marrow cells.

Most adults have 4 to 6 L of blood. It is typically about 55% plasma and 45% formed elements by volume. These proportions are determined by taking a sample of blood in a tube and spinning it for a few minutes in a centrifuge. The relatively dense formed elements are forced to the bottom of the tube and the plasma rises to the top (**fig. 12.2**). The red blood cells are the densest elements and constitute anywhere from about 35% to 50% of total blood volume depending on one's sex, age, diet, and degree of physical activity. This value, called the *hematocrit,* is commonly used in clinical assessment. The white blood cells and platelets settle in a narrow cream- or buff-colored zone called the *buffy coat* just above the red blood cells. At the top of the tube is the plasma, which accounts for the rest of the volume.

12.1c Blood Plasma

Plasma is a complex mixture of water, proteins, nutrients, nitrogenous wastes, hormones, and gases (**table 12.2**). **Serum** is the fluid that remains when the blood clots and solids are removed.

Protein is the most abundant plasma solute by weight, totaling 6 to 9 grams per deciliter (g/dL). Plasma proteins play a variety of roles including clotting, immune defense, and transport of other solutes such as iron, lipids, and hydrophobic hormones. The three major categories of proteins are albumin, globulins, and fibrinogen. Other plasma proteins are essential for life but are found in far smaller quantities.

Albumin is the smallest and most abundant plasma protein. It transports various solutes and buffers the pH of the plasma. It also makes a major contribution to two important physical properties of blood: viscosity and osmolarity. **Viscosity** refers to the thickness or stickiness of a fluid that results from cohesion of its particles. Whole blood is 4.5 to 5.5 times as viscous as water, mainly because of the red blood cells. Plasma alone, however, is 2.0 times as viscous as water, mainly because of

Table 12.2 Composition of Blood Plasma

Water	92% by weight
Proteins	6–9 g/dL
Nutrients	Glucose, amino acids, lipids (cholesterol, fatty acids, triglycerides, and phospholipids), vitamins, iron, and trace elements
Electrolytes	Salts formed of cations and anions of sodium, potassium, magnesium, calcium, and other elements
Nitrogenous wastes	Urea, uric acid, creatine, creatinine, bilirubin, ammonia
Hormones	Insulin, thyroid hormone, estrogen, and all other hormones
Blood gases	Oxygen, carbon dioxide, and nitrogen

albumin. Viscosity is important in circulation because it partially determines blood flow through the vessels. If the blood is too viscous, it flows sluggishly, whereas if viscosity is reduced, blood flows too easily. Either condition puts a strain on the heart and may lead to cardiovascular problems.

The **osmolarity** of blood is the concentration (molarity) of particles that cannot pass through the walls of the blood vessels (see appendix C); it is determined mainly by the total concentration of protein, sodium, and erythrocytes in the blood. As blood reaches the tissues, fluid moves through the capillary wall, carrying nutrients and oxygen to the surrounding cells. The water must then be reabsorbed to maintain appropriate blood volume and fluid balance. Reabsorption into the bloodstream by osmosis is governed by the relative osmolarity of the blood versus the tissue fluid. If the blood osmolarity is too high, the bloodstream absorbs too much water, and blood volume and pressure are elevated, placing a potentially dangerous strain on the heart and vessels. If osmolarity is low, excess water remains in the tissues, they become edematous (swollen), and blood pressure may drop. Albumin is essential to maintaining optimal osmolarity and, therefore, optimal fluid balance.

The other main categories of protein are globulins and fibrinogen. **Globulins** play important roles in immunity, clotting, and transport. **Fibrinogen** is a soluble precursor of *fibrin,* a sticky protein that forms the framework of a blood clot. The liver supplies nearly all of the plasma proteins except for the antibodies, and secretes as much as 4 g of protein per hour.

> **Apply What You Know**
> How could a disease such as liver cancer or hepatitis result in impaired blood clotting?

12.1d How Blood Is Produced APR

We lose blood continually, not only from bleeding but also as blood cells grow old and die and plasma components are consumed or excreted from the body. Therefore, we must continually replace it.

Each day, an adult typically produces 400 billion platelets, 200 billion RBCs, and 10 billion WBCs. (That is roughly one million RBCs per second!) Production of formed elements is called **hematopoiesis**[3] (he-MAT-o-poy-EE-sis). From infancy onward, the red bone marrow is the site of most blood cell production. **Hematopoietic stem cells (HSCs)** found in red bone marrow have the potential to give rise to all the different kinds of formed elements. Later sections of this chapter explain the formation of specific types.

---

[3]*hemato* = blood; *poiesis* = formation

Blood plasma also requires continual replacement. It is composed mainly of water, which it absorbs from the digestive tract. Nutrients and electrolytes are also absorbed from the digestive tract. The liver continually replenishes plasma proteins.

Before You Go On

Answer these questions from memory to test your understanding of the previous section. Reread the material if you can't answer the questions.

1. Describe the transport, protective, and regulatory functions of the blood.
2. What are the two principal components of blood? List the different kinds of formed elements and their primary functions.
3. What are the three major classes of plasma proteins? Briefly explain their functions.
4. Define *viscosity* and *osmolarity,* and explain how each contributes to normal function of the circulatory system.
5. Why must blood be constantly replaced? What is hematopoiesis, and where does it occur?

12.2 Erythrocytes

Expected Learning Outcomes

When you have completed this section, you should be able to

a. discuss the structure and function of red blood cells;
b. describe the structure and function of hemoglobin;
c. define some clinical measurements of RBC and hemoglobin quantities and give typical values for each;
d. describe the life cycle of erythrocytes; and
e. explain the molecular basis of blood types and their clinical significance.

Erythrocytes[4] **(red blood cells, RBCs)** have two main functions: (1) to pick up oxygen from the lungs and transport it to body tissues, and (2) to pick up carbon dioxide from the tissues and unload it in the lungs. Although severe deficiency of leukocytes or platelets can be fatal within a few days, a severe deficiency of erythrocytes can be fatal within minutes. Major trauma that results in massive blood loss leads rapidly to death because of the loss of oxygen carried by RBCs.

12.2a Quantity and Structure

Erythrocytes are the most abundant formed elements. When blood is viewed with a microscope, the RBCs crowd the field (see fig. 12.11a). Their number is critically important to health because it determines the amount of oxygen the blood can carry. A routine clinical measurement in hematology is the **hematocrit**[5]

---

[4]*erythro* = red; *cyte* = cell
[5]*hemato* = blood; *crit* = to separate

(*packed cell volume, PCV*)—the percentage of blood volume composed of RBCs (fig. 12.2). In women, it normally ranges between 37% and 48% and, in men, between 45% and 52%. The RBC count is normally 4.2 to 5.4 million/microliter (μL) in women and 4.6 to 6.2 million/μL in men. This is often expressed as *cells per cubic millimeter* (mm^3); 1 μL = 1 mm^3. The sex difference occurs because male hormones, androgens, stimulate red blood cell production, whereas estrogens do not. In addition, women of reproductive age may have lower numbers of RBCs due to blood loss during menstruation or because of fetal demands for iron during pregnancy.

Erythrocytes are discoidal, pillowlike cells with a thick rim and a thin sunken center (fig. 12.3). The cells have a diameter of about 7.5 μm. The interior of the cell is curiously devoid of distinguishing features because, unlike most cells, mature RBCs have few or no organelles. The lack of a nucleus and DNA renders them incapable of mitosis and protein synthesis, resulting in an inability for self-repair. The advantage of losing the nucleus during development, however, is that cells acquire their distinctive biconcave shape by caving in at the center where the nucleus used to be. The primary benefit of this shape is a subject of debate. It has been proposed that it increases surface area for diffusion of oxygen. However, in the capillaries where gas exchange occurs, RBCs are actually squeezed into an ovoid or teardrop shape; they are only biconcave in the larger vessels. Alternatively, the shape may enable RBCs to flow smoothly through blood vessels with minimal turbulence, and allow the RBCs to bend and pass through the tiniest blood vessels.

12.2b Hemoglobin

The color of blood is due to **hemoglobin (Hb)**, a red, iron-containing protein. Hemoglobin consists of four polypeptide chains called **globins** (fig. 12.4). Two of these, the *alpha (α) chains*, are 141 amino acids long, and the other two, the *beta (β) chains*, are 146 amino acids long. Each has a nonprotein component bound to it called the **heme** group. The iron atom (Fe) at the center of each heme group is able to bind one O_2 molecule; therefore, each hemoglobin molecule can carry up to four O_2. Hemoglobin also transports about 5% of the CO_2 in the blood, but this is bound to the globin component rather than to the heme. Hemoglobin molecules can transport both gases simultaneously.

RBCs are designed to transport oxygen very efficiently. Lacking organelles, they have more space to pack with hemoglobin. The cytoplasm of an RBC consists mainly of a 33% solution of hemoglobin, with about 280 million molecules per cell. The total **hemoglobin concentration** of whole blood is normally 13 to 18 g/dL in

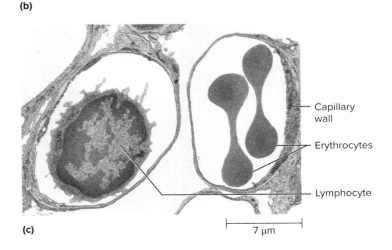

Figure 12.3 The Structure of Erythrocytes. (a) Dimensions and shape of an erythrocyte. (b) Erythrocytes as seen with the scanning electron microscope (SEM). Note their discoidal, concave shape. (c) Erythrocytes in a blood capillary viewed with the transmission electron microscope (TEM). Note the absence of organelles and other internal features in the cells, and how thin the cells are at the center. **APR**

b: studiodav/iStock/Getty Images; c: Thomas Deernick, NCMIR/Science Source

? *Why are erythrocytes caved in at the center?*

Figure 12.4 The Structure of Hemoglobin. The hemoglobin molecule consists of two alpha polypeptides and two beta polypeptides, each bound to a nonprotein heme group that contains an iron atom in the middle.

? *Where does oxygen bind to this molecule?*

men and 12 to 16 g/dL in women. RBCs lack mitochondria, and therefore rely exclusively on anaerobic fermentation to meet their own ATP requirement. If they relied on aerobic respiration, their mitochondria would consume their oxygen cargo instead of delivering it to other cells, like a pizza delivery driver who ate a slice of pizza on the way to your house.

> **Apply What You Know**
> How many molecules of O_2 can be carried in one red blood cell?

Clinical Application 12.1

SICKLE-CELL DISEASE

Sickle-cell disease is a hereditary hemoglobin defect occurring mostly among people of African descent; its symptoms occur in about 1.3% of African Americans, and about 7.7% are asymptomatic carriers with the potential to pass the genetic trait to their children. The disease is caused by a defective gene that codes for the amino acid valine instead of glutamic acid at one position in each beta hemoglobin chain. Manifestation of the disease requires that an individual inherit defective copies of the gene from both parents. This results in an abnormal hemoglobin called *hemoglobin S (HbS)*. HbS turns to gel at low oxygen levels, as when blood passes through the oxygen-hungry skeletal muscles. The RBCs become elongated, stiffened, and pointed (sickle-shaped) (fig. 12.5). These deformed, inflexible cells adhere to each other and clog the tiny blood capillaries, shutting down the blood flow to tissues downstream. This produces severe pain and can lead to heart failure or stroke, among other effects. The spleen removes defective RBCs faster than they can be replaced, thus leading to anemia. Without treatment, a child has little chance of living to age 2, and even with the best treatment, few victims live to age 50. A new technique called gene editing has produced promising results in replacing the damaged DNA in sickle-cell patients, offering hope for future treatments.

Sickle-cell disease originated in areas of Africa where millions of lives have been lost to malaria over thousands of years. Malarial parasites normally invade and reproduce in RBCs but cannot survive in RBCs with HbS. Therefore, the sickle-cell gene confers resistance to malaria, even in individuals who carry only one copy of the gene (are heterozygous for it) and do not have sickle-cell disease. Natural selection has favored the persistence of the gene in certain areas in Africa where the number of lives saved by resistance to malaria has far outweighed the deaths due to sickle-cell disease. In North America, where malaria is rare, the incidence of the gene among those of African descent has declined because there is little evolutionary benefit from possessing it and so many of those who do possess it die young, often without reproducing.

Figure 12.5 Sickle-Cell Disease. One deformed, sickled erythrocyte and three normal erythrocytes.
Eye of Science/Science Source

12.2c The Erythrocyte Life Cycle

An erythrocyte lives for an average of 120 days from the time it is produced in the red bone marrow until it dies. In a state of balance, RBCs are born and die at about equal rates.

The production of red blood cells is called **erythropoiesis** (eh-RITH-ro-poy-EE-sis) (fig. 12.6). It is one aspect of the more general process of hematopoiesis, the production of all formed elements of the blood. Like all formed elements, RBCs trace their origin to a bone marrow stem cell, the hematopoietic stem cell (HSC). HSCs give rise to more specialized cells called **colony-forming units (CFUs),** each type destined to produce one class of formed elements.

The transformation from an HSC to a mature RBC takes 3 to 5 days and involves several steps. When blood oxygen levels fall, the hormone **erythropoietin (EPO)** stimulates the HSCs to transform into cells that multiply and synthesize hemoglobin. Soon, the nucleus and other organelles shrivel and are expelled from these cells, which are now called *reticulocytes*. They leave the bone marrow and enter the circulating blood where they will soon become mature RBCs, ready to do their job of transporting oxygen and carbon dioxide.

As the RBC ages, its membrane proteins deteriorate. Without a nucleus or ribosomes, it cannot synthesize new proteins and its membrane becomes increasingly fragile. Eventually, the cell ruptures as it tries to flex its way through tight capillaries. Many RBCs die in the spleen, which has been called an "erythrocyte graveyard." Its narrow channels severely challenge the ability of old, fragile RBCs to squeeze through the organ and they become trapped, broken up, and destroyed.

When the RBCs rupture, or undergo **hemolysis**[6] (he-MOLL-ih-sis), hemoglobin is released and an empty plasma membrane is left behind. Macrophages digest the membrane fragments and hemoglobin, separating heme from the globin. The globin is hydrolyzed into free amino acids, which become available for protein synthesis or energy, according to the body's needs. The iron released from the heme to the blood is used in the same way as dietary iron. The rest of the heme is eventually converted to a vivid yellow–green pigment called **bilirubin.**[7] The liver removes bilirubin from the blood and secretes it into the bile, which is then stored in the gallbladder. The gallbladder releases the bile to the small intestine, where bacteria convert it to a pigment that colors the feces brown. A high level of bilirubin in the blood causes *jaundice,* a yellowish cast in light-colored skin and the whites of eyes. Jaundice may be a sign of liver disease, bile duct obstruction, or rapid RBC destruction.

---

[6]*hemo* = blood; *lysis* = splitting, breaking
[7]*bili* = bile; *rub* = red; *in* = substance

Figure 12.6 Erythropoiesis. Stages in the development of erythrocytes.

12.2d Blood Types

Transfusions of blood from person to person have saved countless lives. Even in ancient Rome, physicians began to experiment with transfusions, but had mixed results; some lives were saved, but many died of transfusion reactions. It was not until 1900 that a young researcher in Vienna, Karl Landsteiner, discovered blood types A, B, and O. He showed that plasma from one group of people (type A) caused the red cells of another group (type B) to clump, or **agglutinate.** This explained why a transfusion of blood from a person with a different blood type could have disastrous consequences. This concept of compatibility laid the groundwork for successful therapeutic use of transfusions.

Blood types are based on interactions between molecules called *antigens* and *antibodies*. **Antigens** occur on the surfaces of cells and enable the body to distinguish its own cells from foreign matter. When the body detects an antigen of foreign origin, it activates an immune response. One aspect of the response is the production of **antibodies,** proteins produced by certain white blood cells. One method of antibody action is **agglutination,** in which antibody molecules bind to the surface antigens of foreign cells and stick them together.

Human blood groups are determined by genetic variation in antigens on the RBC surface. A person can have the type A antigen, type B, both, or neither, making his or her blood type A, B, AB, or O, respectively (**fig. 12.7** and **table 12.3**). Antibodies that react against A or B antigens appear in the first year of life. A person who has type A antigens on the RBCs has anti-B antibodies in the plasma and a person who has type B antigens has anti-A antibodies in the plasma. A type O person has both anti-A and anti-B, and a type AB person has neither of them. That is, a person has antibodies against any ABO antigen *except* the antigens on one's own RBCs. Consequently, if a type A patient mistakenly received a transfusion with type B blood, the patient's anti-B antibodies would attack the type B erythrocytes. The transfused RBCs would agglutinate, block small blood vessels, burst open *(hemolyze),* and release their hemoglobin. Free hemoglobin can block kidney tubules and cause death from acute renal failure. Another consideration in transfusion compatibility is that antibodies in a mismatched donor's plasma can attack RBCs in the recipient—for example, if type O blood was transfused into a type B patient. In short, a transfusion patient should always receive blood of his or her own type; strictly speaking, there is no such thing as a universal donor or universal recipient, despite the former popularity of these expressions.

(a)

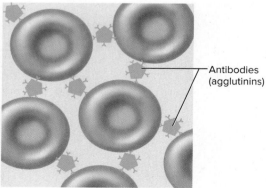

Antibodies (agglutinins)

(b)

Apply What You Know

A person who has type AB blood needs a transfusion. What type or types of blood can he or she receive?

Figure 12.7 Chemical Basis of the ABO Blood Types. (a) The four blood types in the ABO group differ in the terminal sugars of the carbohydrate chains attached to the RBC plasma membranes. Each color represents a different sugar. Note that type AB red cells have both the type A and type B antigens. (b) RBC agglutination in a mismatched transfusion. Five-sided antibodies in the recipient blood plasma bind A or B antigens on the RBCs of the donor blood.

Table 12.3 The ABO Blood Group

Characteristics	Type O	Type A	Type B	Type AB
RBC antigen	Neither A nor B	A	B	Both A and B
Plasma antibody	Anti-A, anti-B	Anti-B	Anti-A	Neither
May safely receive RBCs of	Type O	Type O or A	Type O or B	Type O, A, B, or AB
May safely donate RBCs to	Type O, A, B, or AB	Type A or AB	Type B or AB	Type AB

In 1940, the **Rh blood group** was discovered in rhesus monkeys, for which it is named. *Rh-positive* people have Rh antigens on their RBCs, whereas these are lacking in *Rh-negative* individuals. The Rh blood type is often combined with the ABO type in a single expression such as A+ for type A, Rh-positive, or O– for type O, Rh-negative.

Clinical problems can arise for an Rh+ fetus carried by a woman who is Rh–. During the first pregnancy, there is usually no problem because maternal and fetal blood do not mix until birth, when shearing of the placenta from the uterine wall exposes the mother to the Rh+ fetal blood. She then begins producing antibodies against it that may jeopardize her future pregnancies. If she becomes pregnant with another Rh+ fetus, antibodies she produced when her first baby was born can diffuse through the placenta into the blood of the second fetus and attack its RBCs. The baby may be born with a severe anemia called **hemolytic disease of the newborn,** or **erythroblastosis fetalis.** Pregnant women who are Rh– may be given an Rh immune globulin that binds to the fetal RBC antigens so they cannot stimulate her immune system to produce the antibodies that would cause agglutination. In severe cases of erythroblastosis fetalis, an *exchange transfusion* may be given to completely replace the infant's Rh+ blood with Rh– blood.

Before You Go On

Answer these questions from memory to test your understanding of the previous section. Reread the material if you can't answer the questions.

6. Define hematocrit *and give the normal ranges for women and men.*
7. *Describe the size, shape, and contents of an erythrocyte. How does the lack of mitochondria relate to the function of red blood cells?*
8. *What is the function of hemoglobin? Describe its structure. What is the role of iron? Define* hemoglobin concentration *and provide units of measurement.*
9. *List the stages in the production of an RBC. Why do RBCs die? What happens to their components when they die and disintegrate?*
10. *Describe the antigens that determine the ABO blood types. What are antibodies and antigens and how do they interact to cause a transfusion reaction? What is the clinical significance of the Rh blood group?*

12.3 Leukocytes

Expected Learning Outcomes

When you have completed this section, you should be able to

a. discuss the general function of leukocytes and the specific functions of each individual type;
b. characterize the appearance and relative abundance of each type of leukocyte; and
c. describe the life cycle of leukocytes.

Leukocytes[8] (also called **white blood cells** or **WBCs**) protect us from infectious microorganisms and other pathogens, and we would soon succumb to disease without their defense. When viewed with a microscope, they stand out as vividly stained islands among a sea of the more numerous and pale erythrocytes. They are much more abundant in the body than their low numbers in blood slides would suggest because they spend only a few hours in the bloodstream, then migrate through the walls of capillaries and spend the rest of their lives in connective tissues. The bloodstream is like the subway that the WBCs take to work; in the blood smears, we see only the ones on their commute, not those already at work in the tissues.

12.3a Structure and Function

Leukocytes are not uniform but vary in form and function. In contrast to erythrocytes, they retain their organelles throughout their life; thus, when viewed with a transmission electron microscope, they show a complex internal structure (fig. 12.8). Among their organelles are lysosomes and the usual cast of characters involved in protein synthesis—the nucleus, rough endoplasmic reticulum, ribosomes, and Golgi complex. Leukocytes must synthesize proteins such as lysosomal enzymes to carry out a variety of functions.

12.3b Types of Leukocytes

The five types of leukocytes are distinguished from each other by their relative size and abundance, the size and shape of their nuclei, the presence or absence of cytoplasmic granules, the coarseness and staining properties of those granules, and most importantly their functions.

All WBCs have organelles that appear as granules under the light microscope. Some of these are lysosomes called **nonspecific granules** because they occur in all five WBC types. Others are various kinds of **specific granules** that occur only in three WBC types—neutrophils, eosinophils, and basophils, which are therefore called **granulocytes**. Specific granules stain conspicuously and distinguish each cell type from the others. Granules contain enzymes and other chemicals employed in defense. The two remaining WBC types—monocytes and lymphocytes—are called **agranulocytes** because they lack specific granules.

Bear in mind that the color of WBC nuclei (usually violet) and the colors described here for their granules are not natural colors. They result from the dyes used to stain blood smears, and they can vary depending on what stain was used on any blood smears you may study.

Figure 12.8 The Structure of a Leukocyte. This example is an eosinophil. The lysosomes seen here are the coarse pink granules seen in the cytoplasm of the eosinophil in figure 12.9.
Scott Camazine/Alamy stock photo

---

[8]*leuko* = white; *cyte* = cell

(a) Neutrophils (b) Eosinophil (c) Basophil (d) Lymphocyte (e) Monocyte

Figure 12.9 The Five Kinds of Leukocytes. Note the size of each WBC type compared to the RBCs around it. (a) Neutrophils exhibiting characteristic multilobed nuclei. (b) An eosinophil with a typical bilobed nucleus and coarse pink-staining granules. (c) A basophil, with coarse violet-staining granules obscuring the nucleus. (d) A lymphocyte, almost entirely filled by its nucleus. (e) A monocyte with a typical C-shaped nucleus. **APR**

a–c: Ed Reschke; d–e: Michael Ross/Science Source

Granulocytes

1. **Neutrophils** (NEW-tro-fills) (fig. 12.9a) are the most abundant WBCs, constituting 60% to 70% of the circulating leukocytes. Not only are they easy to find in blood smears, they are also easily recognized by their violet-staining, multilobed nucleus—usually three or four lobes connected by thin strands. The cytoplasm is lightly stippled with tiny, pale granules. Neutrophils are aggressive antibacterial cells. They are quickly attracted to areas of infection or inflammation, where they crawl about phagocytizing and digesting bacteria they encounter. They also secrete a cloud of toxic chemicals (hydrogen peroxide and bleachlike hypochlorite) that kills even greater numbers. The chemicals are lethal to the neutrophils themselves, which typically have a short life span. It has recently been discovered that, in addition to phagocytosis and the chemical killing zone, neutrophils produce webs called *neutrophil extracellular traps* (with the appropriate acronym NET) that consist of proteins and threads of DNA. Pathogen presence triggers the neutrophil to extrude the NET, which traps and kills the invaders. Neutrophils that have died in combat are a major component of pus and are responsible for its milky color. An elevated neutrophil count in the blood is clinically important because it may indicate an infection.

2. **Eosinophils** (EE-oh-SIN-oh-fills) (fig. 12.9b) represent only 2% to 4% of the WBC total, but are also easy to recognize. They usually have a large, bilobed nucleus, like two balloons tied together by a short string. The cytoplasm contains an abundance of large, rosy pink to pinkish-orange specific granules. Eosinophils secrete chemicals that weaken or destroy large parasites such as hookworms or tapeworms, which are too large to be destroyed by any one WBC. They also help to dispose of antigen–antibody complexes formed in immune reactions (see "Humoral Immunity" in section 14.4b). Their numbers rise in parasitic infections, allergies, and diseases of the spleen and central nervous system, and fluctuate between day and night and over the course of the menstrual cycle.

3. **Basophils** (BASE-oh-fills) (fig. 12.9c) are the rarest of WBCs, comprising from less than 0.5% to about 1% of the total. They are recognizable because of their abundant, very coarse granules, usually stained dark violet (sometimes pink). The granules hide the nucleus from view, but the nucleus is large, pale, and typically S- or U-shaped. Basophils secrete **histamine,** a vasodilator that widens blood vessels, speeds the flow of blood to injured tissue, and makes the blood vessels more permeable so that neutrophils and clotting proteins can get into the connective tissues more quickly. They also secrete **heparin,** an anticoagulant that inhibits blood clotting and thus

promotes the mobility of other WBCs in the area. They are important in allergy, inflammation, and attracting neutrophils and other WBCs to sites of infection. An elevated basophil count is commonly seen in diabetes, chickenpox, and various other diseases.

Agranulocytes

4. **Lymphocytes** (LIM-fo-sites) (fig. 12.9d) are second to neutrophils in abundance at 25% to 33% of the WBC count, and are quickly spotted on a blood slide. The lymphocyte count rises in diverse infectious diseases and immune responses. Their dark violet nucleus fills nearly the entire cell. The cytoplasm stains a clear light blue color and forms a narrow and sometimes barely detectable rim around the nucleus. The subclasses of lymphocytes are indistinguishable under the light microscope but have distinct immune functions. Certain types of lymphocytes send signals to WBCs to activate them and coordinate their actions, wheras others destroy cancer cells and infected cells, and yet others are responsible for long-term immunity against pathogens. Their functions are discussed in more detail in chapter 14.
5. **Monocytes** (MON-oh-sites) (fig. 12.9e) are the largest WBCs and comprise about 3% to 8% of the WBC count. The nucleus is large and typically ovoid, kidney-shaped, or horseshoe-shaped. The cytoplasm contains sparse, fine granules. After migrating from the blood into the connective tissues, monocytes transform into even larger cells called **macrophages**[9] (MAC-ro-fay-jez). They consume and destroy microoganisms, foreign debris, and dead or dying body cells, ingesting as much as a whopping 25% of their own volume per hour. They also aid in alerting the immune system to invading pathogens or foreign matter. The monocyte count tends to rise in inflammation and viral infections.

12.3c The Leukocyte Life Cycle

Leukopoiesis (LOO-co-poy-EE-sis), the production of white blood cells, begins with the same hematopoietic stem cells (HSCs) as erythropoiesis. Some HSCs differentiate into distinct types of colony-forming units (CFUs), which then go on to produce any of three cell lines—*myeloblasts*, which produce the three kinds of granulocytes (neutrophils, eosinophils, and basophils); *monoblasts*, which give rise to monocytes; and *lymphoblasts*, which give rise to lymphocytes (fig. 12.10).

Granulocytes and monocytes stay in the red bone marrow until they are needed; the marrow contains 10 to 20 times as many of these cells as the circulating blood does. Lymphocytes begin developing in the bone marrow but do not stay there. Some migrate to the thymus and mature there; then from either the bone marrow or thymus, lymphocytes disperse and colonize the lymph nodes, spleen, tonsils, and mucous membranes. Their life histories and migrations are further discussed in chapter 14.

Circulating leukocytes don't stay in the blood for very long. Granulocytes circulate for 4 to 6 hours and then migrate into the tissues, where they live another 4 or 5 days. Monocytes travel in the blood for 10 to 20 hours, then migrate into the tissues and differentiate into macrophages, which can live for as long as a few years. Lymphocytes are responsible for long-term immunity and therefore may survive for lengths of time ranging from a few weeks to decades. They leave the bloodstream for the tissues and eventually enter the lymphatic system, which empties them back into the blood. They are continually cycled between the blood, lymph, and tissue fluid.

---

[9]*macro* = big; *phage* = eater

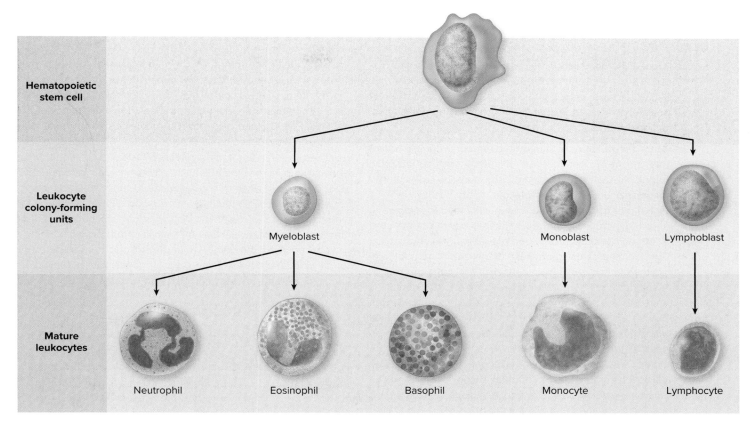

Figure 12.10 **Leukopoiesis.** Stages in the development of leukocytes.

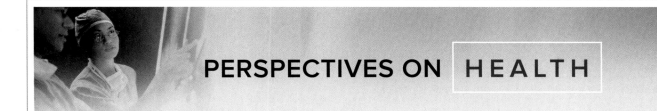

PERSPECTIVES ON HEALTH

Given the vital functions of blood, it is no surprise that disorders can have disastrous consequences for health. A deficiency or excess of a specific cell type underlies many blood diseases. Misshapen or oddly colored cells may also indicate certain underlying conditions.

Erythrocyte Disorders

The two most common RBC disorders are **anemia**[10] (an RBC or hemoglobin deficiency) and **polycythemia**[11] (POL-ee-sih-THEME-ee-uh) (an RBC excess). There are three fundamental causes of anemia:

1. **Reduced erythropoiesis.** When hemoglobin or erythrocytes are produced too slowly to keep up with the rate of RBC death, an individual becomes anemic. One cause of suppressed RBC manufacture or hemoglobin synthesis is a dietary deficiency of minerals or vitamins needed for their production. For example, a lack of iron suppresses erythropoiesis and leads to *iron-deficiency anemia*. Elderly people may experience *pernicious anemia*, an immune disorder that destroys stomach tissue. Normally the stomach produces a protein called *intrinsic factor*, needed for absorbing vitamin B_{12} from the food. Vitamin B_{12} is required for erythropoiesis. The immune attack on the stomach results in a vitamin B_{12} deficiency. Gastric-bypass patients also experience a similar anemia because of the surgical removal of stomach tissue or disconnection of the stomach from the small intestine, where the intrinsic factor is needed. Radiation, viruses, and some poisons can cause anemia by destroying bone marrow. Kidney disease can lead to anemia because of inadequate production of erythropoietin (EPO), a hormone that is the principal stimulus for RBC production.

---

[10]*an* = without; *emia* = blood
[11]*poly* = many; *cyt* = cells; *hemia* = blood condition

(continued on next page)

(continued from previous page)

2. **Rapid RBC destruction.** Excess RBC destruction that exceeds the rate of erythropoiesis results in *hemolytic anemia*. Causes include various poisons, drug reactions, sickle-cell disease, snake venoms, and blood-destroying parasitic infections such as malaria.

3. **Hemorrhage.** Blood loss may lead to *hemorrhagic anemia*. This can be a consequence of trauma such as gunshot, automobile, or battlefield injuries; hemophilia; ruptured aneurysms; or heavy menstruation.

Anemia, whatever its fundamental cause, may lead to **hypoxia,** a deficiency of oxygen in the tissues. In severe cases, the affected person is lethargic and becomes short of breath upon physical exertion. The skin may be pallid because of the lack of hemoglobin.

Polycythemia can result from cancer of the bone marrow or from a multitude of other conditions. Dehydration, for example, is characterized by loss of water from the blood while erythrocytes remain and become highly concentrated. Low oxygen supply (as occurs with smoking, emphysema, sleep apnea, or ascent to high altitude) stimulates erythropoiesis. In the cancerous form, *polycythemia vera*, RBC counts can rise as high as 11 million RBCs/μL and hematocrit as high as 80%. The thick blood "sludges" in the vessels, increases blood pressure, and puts dangerous strain on the cardiovascular system that can lead to heart failure or stroke.

Leukocyte Disorders

A deficiency of WBCs is called **leukopenia**[12] (LOO-co-PEE-nee-uh). A low WBC count may be due to poisoning from lead, mercury, or arsenic, or to radiation exposure. Infectious diseases such as AIDS, measles, mumps, chickenpox, influenza, and poliomyelitis also deplete WBCs. Certain drugs such as glucocorticoids, anticancer drugs, and immunosuppressant drugs given to transplant patients lower the WBC count. A patient with insufficient WBCs can't effectively fight disease and is susceptible to infections and cancer.

An excess of WBCs is **leukocytosis.**[13] A higher-than-normal WBC count can be due to infection or allergy, or even dehydration or emotional disturbances. For clinical purposes, a *differential WBC count*, in which the percentage of each type of WBC is determined, is more useful than a total WBC count. For example, a high neutrophil count is a sign of bacterial infection, whereas a high eosinophil count usually indicates an allergy or a parasitic infection such as worms.

Leukemia is a cancer that originates from mutated hematopoietic cells. Proliferation of the malignant cells results in an extraordinarily high number of circulating leukocytes and their precursors (**fig. 12.11**). Leukemia is classified as myeloid or lymphoid, acute or chronic. **Myeloid leukemia** is characterized by uncontrolled granulocyte production, whereas **lymphoid leukemia** involves uncontrolled lymphocyte or monocyte production. **Acute leukemia** appears suddenly, progresses rapidly, and causes death within a few months if left untreated. **Chronic leukemia** develops more slowly and may be undetected for many months.

As the cancerous cells proliferate uncontrollably, they replace normal bone marrow and a person suffers from a deficiency of normal WBCs, RBCs, and platelets. The numerous cancerous leukocytes that are produced are immature and are incapable of fulfilling their roles as defenders against disease. The lack of functional WBCs renders the patient vulnerable to infection; the diminished quantity of RBCs leads to anemia and fatigue; and the platelet deficiency results in hemorrhaging and impaired blood clotting. Although the immediate cause of death is usually hemorrhage or opportunistic infection, cancerous cells may metastasize from the bone marrow to other organs of the body, where they displace or compete with normal cells. Widespread metastasis is not easily treated by either surgery or radiation therapy and may lead to death of the individual.

Figure 12.11 Normal and Leukemic Blood. (a) A normal blood smear. (b) Blood from a patient with acute monocytic leukemia. Note the abnormally high number of WBCs, especially monocytes.
a: Ed Reschke; b: Leonard Lessin/Science Source

❓ *With all these extra white blood cells, why isn't the body's infection-fighting capability increased in leukemia?*

---

[12]*leuko* = white; *penia* = deficiency
[13]*leuko* = white; *cyt* = cell; *osis* = condition

Clinical Application 12.2

THE COMPLETE BLOOD COUNT

The *complete blood count (CBC)* is a very common clinical test that affords a wealth of information about a patient's health. It includes hematocrit; hemoglobin concentration; total RBC, WBC, and platelet counts; RBC size; and a *differential WBC count,* among other values. The differential WBC count quantifies what percentage of the WBCs consists of each cell type. The relative proportions of one WBC type to another can be valuable indications of various disease states such as infections, inflammation, allergies, cancer, AIDS, and certain metabolic diseases.

Blood counts used to be done by visual examination with a microscope, but now are usually done with an electronic cell counter. Cell counters force a blood sample through a narrow tube with sensors that identify cell types and measure cell sizes and hemoglobin content. They give much faster and more accurate results than visual inspection. However, cell counters still misidentify some cells, and a medical technologist must review the results for suspicious data and visually identify cells that the instrument cannot. If a CBC does not provide enough information or if it suggests certain disorders, additional tests may be done such as coagulation time or a bone marrow biopsy.

Before You Go On

Answer these questions from memory to test your understanding of the previous section. Reread the material if you can't answer the questions.

11. What is the general purpose of WBCs?
12. What is the difference between granulocytes and agranulocytes? Which leukocytes belong to each category?
13. List the five kinds of leukocytes in order of abundance and describe how to identify each one.
14. Explain the functions of each type of leukocyte.
15. What are macrophages? What class of WBCs do they arise from?
16. What does leukopoiesis have in common with erythropoiesis? How does it differ?

12.4 Platelets and Control of Bleeding

Expected Learning Outcomes

When you have completed this section, you should be able to

a. describe the structure and functions of blood platelets;
b. describe platelet production; and
c. describe blood clotting and other mechanisms for controlling bleeding.

Circulatory systems developed very early in animal evolution, and with them mechanisms evolved for stopping leaks, which are potentially fatal. Platelets are the major players in cessation of bleeding and are the focus of this section.

12.4a Platelet Structure and Function

Platelets are small circulating fragments of bone marrow cells called *megakaryocytes*. They are necessary for blood clotting and have other functions listed later in this section. They are among the formed elements of blood, but they are not cells, since they never have a nucleus of their own at any time in their development. They are the second most abundant formed element after erythrocytes. They are so much smaller than RBCs, however (2 to 4 μm in diameter), that they contribute even less than WBCs to total blood volume.

Platelets have a complex internal structure that includes lysosomes, mitochondria, a cytoskeleton, granules filled with platelet secretions, and a system of channels that open onto the platelet surface (the *open canalicular system*) (**fig. 12.12a**). When activated, they form pseudopods and are capable of ameboid movement.

Platelets have a variety of functions:

1. They secrete *vasoconstrictors,* chemicals that cause spasmodic constriction of broken vessels and thus help reduce blood loss.
2. They stick together to form temporary, relatively fragile *platelet plugs* to seal small breaks in injured blood vessels.
3. They secrete *clotting factors* that promote blood clotting.
4. They initiate the formation of a clot-dissolving enzyme that dissolves blood clots that have outlasted their usefulness.
5. They secrete *growth factors* that stimulate mitosis in fibroblasts and smooth muscle and thus help maintain and repair blood vessels.

Figure 12.12 Platelets and Their Production. (a) Structure of blood platelets. (b) Platelet production. The megakaryocyte extends tendrils into the capillaries (sinusoids) of the red bone marrow, where blood flow shears off pieces called proplatelets. These break up into platelets as they travel in the blood stream. Note the size of the megakaryocyte compared to that of the mature RBCs and WBCs around it. **APR**

a: NIBSC/Science Photo Library/Science Source

12.4b Platelet Production

Production of platelets, called **thrombopoiesis,** is one aspect of hematopoiesis. Some hematopoietic stem cells differentiate into **megakaryocytes**[14] (meg-ah-CAR-ee-oh-sites), gigantic cells up to 150 μm in diameter, large enough to be visible to the naked eye. Each has a huge multilobed nucleus and multiple sets of chromosomes (fig. 12.12b).

Platelets arise from long, beaded processes that extend from the surface of the megakaryocyte into the lumen of capillaries in the bone marrow. The force of the blood flow shears the tips of these processes off, releasing them as *proplatelets* that further fragment into individual platelets, especially as they pass through the lungs. About 25% to 40% of the platelets are stored in the spleen and released as needed. The rest circulate freely in the blood and live for 5 or 6 days.

12.4c Control of Bleeding

The circulatory system has evolved mechanisms to seal itself after injury. **Hemostasis,**[15] the cessation of bleeding, occurs in three stages (fig. 12.13):

1. **Vascular spasm,** the prompt constriction of a broken vessel. This narrows the opening and reduces blood loss. It is triggered in part by **serotonin,** a vasoconstrictor secreted by the platelets.
2. A **platelet plug,** a sticky mass of platelets that acts as a stopper to close small breaks in a vessel. In an undamaged blood vessel, the endothelium is smooth and coated with a platelet repellent to prevent adhesion. But when a vessel is broken, platelets adhere to the roughened surfaces and now-exposed collagen fibers of the vessel wall. The platelets then put out long, spiny pseudopods that adhere to other platelets and the vessel wall. The pseudopods contract and pull the wall together, and the platelet mass may reduce or stop minor bleeding.
3. **Coagulation (clotting)** is the last but most effective defense against bleeding. It is a complex process involving more than 30 chemical reactions and a multitude of protein **clotting factors** produced by the liver and platelets. Most of these function as enzymes that activate other enzymes in a cascading chain of events. Ultimately, this series of reactions converts the plasma protein *fibrinogen* to sticky *fibrin*. Fibrin adheres to the wall of the blood

---

[14]*mega* = giant; *karyo* = nucleus; *cyte* = cell
[15]*hemo* = blood; *stasis* = stability

Figure 12.13 Hemostasis. (a) Vascular spasm. Vasoconstriction of a broken vessel reduces bleeding. (b) Platelet plug formation. A platelet plug forms as platelets adhere to exposed collagen fibers of the vessel wall. The platelet plug temporarily seals the break. (c) Coagulation. A blood clot forms as platelets become enmeshed in fibrin threads. This forms a longer-lasting seal and gives the vessel a chance to repair itself.

(a) Vascular spasm

(b) Platelet plug formation

(c) Coagulation

vessel, and as blood cells and platelets arrive, many of them stick to it like insects in a spider web. The resulting mass of fibrin, platelets, and blood cells forms a clot that ideally seals the break in the blood vessel long enough for the vessel to heal.

Once a leak is sealed and the crisis has passed, platelets secrete *platelet-derived growth factor (PDGF)*, a substance that stimulates fibroblasts and smooth muscle to proliferate and replace the damaged tissue of the blood vessel. When tissue repair is completed and the blood clot is no longer necessary, the clot must be disposed of. Platelets then secrete a protein that initiates a series of reactions leading to a fibrin-digesting enzyme called *plasmin*. Plasmin dissolves the old blood clot.

It is important to distinguish coagulation from agglutination. In both reactions, blood cells or platelets aggregate and stick together, so they may seem to be superficially similar processes. However, the mechanisms are entirely different. In coagulation, formed elements are stuck together by fibrin. In the agglutination of blood, RBCs are stuck together by antibody molecules (see "Blood Types" in section 12.2d).

A deficiency of any clotting factor can shut down the coagulation cascade. In **hemophilia,** for example, a hereditary lack of a single protein in the cascade impairs blood clotting and the ability to control bleeding. Most hemophiliacs suffer from a lack of factor VIII; before purified factor VIII became available in the 1960s, more than half of those with hemophilia died before age 5 and only 10% lived to age 21. Although purified factor VIII enabled hemophiliacs to quickly control bleeding and engage in typical activities, it came at a steep price. Many hemophiliacs became infected with hepatitis or HIV carried in the donated plasma. Factor VIII is now obtained from recombinant bacteria rather than from donor blood, so there is no longer any risk of infection to patients.

Unwanted clotting is a more common problem than the inability to form clots. Most strokes and heart attacks are due to **thrombosis**[16]—the abnormal clotting of blood in an unbroken vessel. A **thrombus** (clot) may grow large enough to obstruct a small vessel, or a piece of it may break loose and begin to travel in the bloodstream as an **embolus.**[17] An embolus may lodge in a small artery and block blood flow to tissues downstream. If the vessel supplies a vital organ such as the heart, brain, or lung, *infarction* (tissue death) may result. About 650,000 Americans die annually of *thromboembolism* (traveling blood clots) in the coronary, cerebral, or pulmonary arteries.

Before You Go On

Answer these questions from memory to test your understanding of the previous section. Reread the material if you can't answer the questions.

17. List several functions of blood platelets.
18. How are platelets produced? Describe the size and internal structure of platelets.
19. Explain the three basic mechanisms of hemostasis.
20. What is hemophilia?
21. Define thrombus *and* embolus, *and explain how they may lead to tissue damage.*

---

[16]*thromb* = clot; *osis* = condition
[17]*em* = in, within; *bolus* = ball, mass

CAREER SPOTLIGHT

Phlebotomist

dlewis33/E+/Getty Images

A phlebotomist[18] draws blood for the purposes of clinical testing, blood banking, or transfusion. Although this is often done by nurses and physicians, phlebotomists are employed to reduce the workload on these other professionals, especially in hospitals and blood drives. Some phlebotomists travel to nursing homes, outpatient clinics, or patients' homes to collect samples; some states employ phlebotomists as dialysis technicians. Samples are usually taken by puncturing a vein, but with specialized training, some phlebotomists draw arterial samples, usually from the radial or ulnar arteries at the wrist.

Training and certification requirements vary from state to state. One must have a high-school diploma or general educational development (GED) certificate, then may undertake training ranging from 6 weeks at a trade school or career center to a college associate degree. Training typically includes anatomy, legal aspects of blood collection, patient interaction skills, and blood-collecting techniques. A state may require certification by such agencies as the National Phlebotomy Association or the American Society of Clinical Pathologists. For further information on careers in phlebotomy, see appendix B.

[18]phlebo = vein; tom = cut; ist = one who

CONNECTIVE ISSUES

Ways in Which the BLOOD Affects Other Organ Systems

All Systems
The blood provides oxygen and nutrients to all organs, carries away metabolic wastes, and prevents hyperthermia by transporting heat from the core to the body surface.

Integumentary System
Blood provides the WBCs that combat pathogens that breach the skin surface.

Skeletal System
The blood picks up RBCs, WBCs, and platelets from the red bone marrow; delivers the hormones that regulate the production of these formed elements; and delivers the minerals and regulatory hormones essential for bone metabolism.

Muscular System
The blood delivers oxygen and nutrients, and removes metabolites and heat generated by the muscles.

Nervous System
Cerebrospinal fluid is derived from blood plasma; strokes from cerebral blood clots are a leading cause of death.

Endocrine System
The blood provides precursors for hormone synthesis and distributes all hormones to their target organs.

Lymphatic and Immune Systems
WBCs are key components of the immune system; lymph is derived from and ultimately returned to the blood plasma; and blood-borne lymphocytes and monocytes densely populate the lymphatic organs and tissues.

Respiratory System
The blood transports respiratory gases, and blood pH and the partial pressures of blood gases drive the respiratory rhythm.

Urinary System
Urine production begins as a filtrate of the blood plasma, and water and vital solutes retained by the kidneys are returned to the blood.

Digestive System
The blood transports digested nutrients.

Reproductive System
The blood distributes all sex hormones.

Study Guide

Assess Your Learning Outcomes

To test your knowledge, discuss the following topics with a study partner or in writing, ideally from memory.

12.1 Introduction

1. The components of the circulatory system; the differences between the cardiovascular system and circulatory system
2. The diverse functions of the blood; contributions of blood to homeostasis
3. The two main components of whole blood; the relative amounts of formed elements and plasma in whole blood; and the three main types of formed elements
4. The components of blood plasma and their functions
5. The basis of blood viscosity and osmolarity and their importance in cardiovascular function
6. The formation of blood plasma
7. The definition of *hematopoiesis* and a description of where it occurs
8. The name of the bone marrow cell type that ultimately gives rise to all formed elements

12.2 Erythrocytes

1. Ways of quantifying erythrocytes and hemoglobin levels, and typical values for women and men
2. The structure and function of erythrocytes, and the reason for their unusual structure
3. The structure and function of hemoglobin and how it binds blood gases
4. The process of erythropoiesis; the hormone that stimulates it
5. The life span, death, and disposal of RBCs and hemoglobin
6. The molecular basis of blood types A, B, AB, and O, and definitions of *antigen* and *antibody*
7. The cause and mechanism of a transfusion reaction and why it can lead to death; the meanings of *agglutination* and *hemolysis*
8. The two blood types of the Rh group
9. The cause of hemolytic disease of the newborn; why it is less common in a first pregnancy than in later ones

12.3 Leukocytes

1. The general function of leukocytes (WBCs)
2. The difference between granulocytes and agranulocytes
3. The name, physical characteristics, and functions of each WBC type
4. The cell lines that give rise to WBCs
5. The life history of WBCs after their release from the bone marrow

12.4 Platelets and Control of Bleeding

1. The structure of platelets; why they are not considered to be cells
2. Functions of platelets
3. The site and process of platelet production
4. The general term for cessation of bleeding, and the three stages of the process
5. What happens to a blood clot when vessel repair is completed
6. Differences between coagulation and agglutination
7. The hereditary and molecular bases of hemophilia
8. The differences between thrombosis, thrombus, and embolus; the danger of thromboembolism

Testing Your Recall

1. Which contribute(s) most to viscosity of the blood?
   a. albumin
   b. fibrinogen
   c. globulins
   d. sodium
   e. erythrocytes

2. What would be a normal hematocrit value for an adult male?
   a. 50%
   b. 25%
   c. 4.6 million/μL
   d. between 60% and 70%
   e. not less than 75%

3. Which is *not* a cause of leukocytosis?
   a. dehydration
   b. AIDS
   c. allergy
   d. leukemia
   e. bacterial infection

4. What blood type is a person who has only A antigens on his or her erythrocytes?
   a. type B
   b. type AB
   c. type O
   d. type A
   e. Rh+

5. Which of these is a granulocyte?
   a. a monocyte
   b. an erythrocyte
   c. a lymphocyte
   d. a macrophage
   e. an eosinophil

6. Platelets have all of the following functions *except*
   a. coagulation.
   b. plugging broken blood vessels.
   c. transporting oxygen.
   d. stimulating vasoconstriction.
   e. repairing blood vessels after an injury.

7. Histamine and heparin are secreted by
   a. lymphocytes.
   b. basophils.
   c. monocytes.
   d. eosinophils.
   e. platelets.

8. Which cells fight bacteria by phagocytizing them and bombarding them with hydrogen peroxide?
   a. lymphocytes
   b. macrophages
   c. neutrophils
   d. eosinophils
   e. basophils

9. Which is *not* true of erythrocytes?
   a. They have no nucleus.
   b. They have no mitochondria.
   c. Their cytoplasm consists mostly of a 33% solution of hemoglobin.
   d. They live for about 60 days.
   e. Their main function is to transport O_2 and CO_2.

10. What is the first event that occurs in hemostasis?
    a. vascular spasm
    b. platelet plug formation
    c. coagulation
    d. conversion of fibrinogen to fibrin
    e. secretion of platelet-derived growth factor

11. Production of all the formed elements of blood is called _____.

12. The percentage of blood volume composed of RBCs is called the _____.

13. After monocytes emigrate from the blood into the connective tissues, they become _____.

14. The cessation of bleeding is called _____.

15. _____ is the fluid that remains if all the formed elements and fibrinogen are removed from the blood.

16. The hereditary lack of factor VIII causes a disease called _____.

17. An abnormally low WBC count is called _____.

18. The smallest and most abundant plasma protein, _____, plays a role in maintaining blood osmolarity and viscosity.

19. An excess of RBCs is called _____.

20. The production of red blood cells is called _____.

Answers in Appendix A

What's Wrong with These Statements?

Briefly explain why each of the following statements is false, or reword it to make it true.

1. By volume, the blood usually contains more formed elements than plasma.

2. An increase in the albumin concentration of the blood would tend to reduce blood pressure.

3. Anemia is caused by a low oxygen concentration in the blood.

4. The most important WBCs in combating a bacterial infection are basophils.

5. All mature formed elements have nuclei.

6. Lymphocytes are the most abundant WBCs in the blood.

7. The mechanisms for coagulation and agglutination are essentially the same.

8. Hematopoietic stem cells in bone marrow give rise to all formed elements except lymphocytes.

9. Since RBCs have no nuclei, they do not live as long as granulocytes do.

10. Leukemia is a severe deficiency of WBCs.

Answers in Appendix A

Testing Your Comprehension

1. A patient is found to be seriously dehydrated and to have an elevated RBC count. Does the RBC count necessarily indicate a disorder of erythropoiesis? Why or why not?

2. Patients suffering from renal failure are typically placed on hemodialysis and erythropoietin (EPO) replacement therapy. Explain the reason for giving EPO, and predict what the consequences would be of *not* including this in the treatment regimen.

3. Patients with leukemia are susceptible to infections such as pneumonia and also often suffer from anemia. Explain why these conditions are associated with leukemia.

Design Element Credits: (Camping): monkographic/Shutterstock; (Abseiling): Image Source; (Stethoscope): MedicalRF.com; (Clouds): Pakhnyushchy/Shutterstock.com; (Doctors): Purestock/SuperStock; (Mountain climber): McGraw-Hill Education.

Chapter 13

The Circulatory System II

THE HEART AND BLOOD VESSELS

A colorized image of coronary blood vessels that have been injected with a dye to enhance their visibility in an X-ray. Such an image is called a coronary angiogram.
SPL/Science Source

Chapter Outline

13.1 Overview of the Cardiovascular System
- 13.1a The Pulmonary and Systemic Circuits
- 13.1b Position, Size, and Shape of the Heart
- 13.1c The Pericardium

13.2 Gross Anatomy of the Heart
- 13.2a The Heart Wall
- 13.2b The Chambers
- 13.2c The Valves
- 13.2d Blood Flow Through the Chambers
- 13.2e Coronary Circulation

13.3 Physiology of the Heart
- 13.3a Cardiac Muscle
- 13.3b The Cardiac Conduction System
- 13.3c Electrical Activity and Contraction
- 13.3d Nerve Supply to the Heart
- 13.3e The Electrocardiogram
- 13.3f The Cardiac Cycle
- 13.3g Cardiac Output

13.4 General Anatomy of Blood Vessels
- 13.4a The Vessel Wall
- 13.4b Arteries, Capillaries, and Veins
- 13.4c Variations in Circulatory Routes

13.5 Physiology of Circulation
- 13.5a Blood Pressure
- 13.5b Peripheral Resistance
- 13.5c Regulation of Blood Pressure and Flow
- 13.5d Capillary Fluid Exchange
- 13.5e Venous Return

13.6 Circulatory Routes and Blood Vessels
- 13.6a The Pulmonary Circuit
- 13.6b The Systemic Circuit

Clinical Applications/Perspectives on Health
- Clinical Application 13.1: Cardiac Arrhythmia
- Clinical Application 13.2: Congestive Heart Failure
- Perspectives on Health

End of Chapter
- Aging of the Circulatory System
- Career Spotlight: Electrocardiographic Technician
- Connective Issues
- Study Guide

BASE CAMP

Before ascending to the next level, be sure you're properly equipped with a knowledge of these concepts from earlier chapters.

- Thoracic cavity anatomy (see section 1.3e)
- Desmosomes and gap junctions (see section 3.2d)
- Muscle tissue (see section 4.4b)
- Skeletal muscle excitation and contraction (see section 7.2a)
- Resting membrane potentials and action potentials (see section 8.2a and b)

Module 9
Circulatory System

We are more conscious of our heart than we are of most organs, and more concerned about its possible failure. The Greek scholar Aristotle thought the heart served primarily as the seat of emotion, a view that persisted through the Middle Ages. Even today, the heart is a symbol of emotions, as is reflected in its prominence on Valentine's Day and in common phrases such as "heartfelt" and "heartbroken" to describe various feelings. Recently, scientists have found that there is indeed a physiological emotional connection between the heart and brain; stress, fear, and anxiety can affect cardiac function, and in turn, afferent signals from the heart target emotional centers of the brain. There is even a medical syndrome called takotsubo cardiomyopathy, or "broken-heart syndrome," in which extreme grief or stress causes the heart to weaken and change shape. The metaphorical and biological hearts are perhaps not as far apart as we might think.

The blood vessels, too, have long been the subject of fascination and inquiry, and misunderstandings such as the ancient idea that the arteries carried not blood but a mysterious "vital spirit," and that the veins sent blood to the organs, which then consumed it like food. Not until the seventeenth century was it proven that blood recirculates repeatedly to the heart, an inference soon supported by discovery of the microscopic blood capillaries that connect the arterial and venous systems.

Today, the heart and blood vessels are a central focus of medical research and practice due to the prevalence of cardiovascular disease. It is hard to believe that most of the treatments for heart disease have been developed since the 1950s. Today, treatments that once seemed impossible have become nearly routine, such as coronary bypass surgery, valve replacements, and artificial pacemakers. New frontiers such as artificial hearts and regeneration of cardiac tissue mean that cardiology will continue to be one of the most dramatic and innovative medical fields.

13.1 Overview of the Cardiovascular System

Expected Learning Outcomes

When you have completed this section, you should be able to

a. distinguish between the pulmonary and systemic circulations;
b. describe the general location, size, and shape of the heart; and
c. describe the pericardium, which encloses the heart.

The **cardiovascular system** consists of the heart and the blood vessels, whereas the **circulatory system** includes these and the blood. The heart functions as a muscular pump that keeps blood flowing through the vessels. The vessels deliver blood to

412 PART 4 Circulation and Defense

tissues throughout the body, where it unloads the oxygen and nutrients vital to their survival, and then return it to the heart. **Cardiology** is the study of the heart, clinical evaluation of its functions and disorders, and treatment of cardiac diseases.

13.1a The Pulmonary and Systemic Circuits

The cardiovascular system has two major divisions: a **pulmonary circuit,** which carries blood to the lungs for gas exchange and returns it to the heart, and a **systemic circuit,** which supplies blood to all organs of the body, including the lungs and the wall of the heart itself (fig. 13.1).

The right side of the heart supplies the pulmonary circuit. It receives oxygen-poor blood that has circulated through the body and pumps it into a large artery, the *pulmonary trunk.* From there, the oxygen-poor blood is distributed to the lungs, where it unloads carbon dioxide and picks up a fresh load of oxygen. It then returns this oxygen-rich blood to the left side of the heart by way of *pulmonary veins.*

The left half of the heart supplies the systemic circuit. It pumps blood into the largest artery, the *aorta,* which gives off branches that ultimately deliver oxygen to every organ of the body. After the blood has picked up carbon dioxide from the tissues, it returns to the right heart by way of the two largest veins—the *superior vena cava,* which drains the upper body, and the *inferior vena cava,* which drains everything below the diaphragm.

13.1b Position, Size, and Shape of the Heart

The heart is located in the thoracic cavity in the mediastinum, between the lungs and deep to the sternum (fig. 13.2). Tilted slightly toward the left, about two-thirds of the heart lies to the left of the median plane. The broad superior portion of the heart, the **base,** is the point of attachment for the pulmonary trunk, pulmonary veins, and aorta—the so-called *great vessels.* The inferior end tapers to a blunt point, the **apex,** immediately above the diaphragm.

The normal adult heart weighs about 300 g (10 ounces) and measures about 9 cm (3.5 in.) wide at the base, 13 cm (5 in.) from base to apex, and 6 cm (2.5 in.) from anterior to posterior at its thickest point. Whatever one's body size, the healthy heart is roughly the same size as the fist.

13.1c The Pericardium

The heart is enfolded in a double-walled sac called the **pericardium** (fig. 13.2c). The outer wall, the **pericardial sac (parietal pericardium),** has a tough, superficial *fibrous layer* of dense irregular connective tissue, and a deep *serous layer.* The serous layer turns inward at the base of the heart and forms the **epicardium (visceral pericardium),** which covers the heart surface.

The space between the parietal and visceral pericardium is the **pericardial cavity.** The cavity contains a small amount (5–30 mL) of **pericardial fluid,** exuded by the serous layer of the pericardium. The fluid lubricates the membranes and allows the heart to beat with minimal friction. The pericardium not only reduces friction but also isolates the heart from other thoracic organs and anchors it within the thoracic cavity.

Pericarditis is inflammation of the pericardium, most often caused by viral infection, in which membranes may become roughened and produce a painful *friction rub* with each heartbeat.

Figure 13.1 General Schematic of the Cardiovascular System. APR

? *Are the lungs supplied by the pulmonary circuit, the systemic circuit, or both? Explain.*

CHAPTER 13　The Circulatory System II　413

Figure 13.2 Position and Relationships of the Heart. (a) Cross section of the thorax showing the relationship of the heart to the lungs and thoracic cage. Structures have the same orientation as in the viewer's body. (b) Anterior view with the lungs slightly retracted and the pericardial sac opened. (c) The pericardium and heart wall. A fist in a balloon shows, by analogy, how the double-walled pericardium wraps around the heart. **APR**

? *Does most of the heart lie to the right or left of the median plane?*

Before You Go On

Answer these questions from memory to test your understanding of the previous section. Reread the material if you can't answer the questions.

1. Distinguish between the pulmonary and systemic circuits, and state which part of the heart supplies each one.
2. Make a two-color sketch of the pericardium, using one color for the pericardial sac and another for the epicardium. For the pericardial sac, label both the fibrous and serous layers. Show the relationship between the pericardium, pericardial cavity, and heart wall.

13.2 Gross Anatomy of the Heart

Expected Learning Outcomes

When you have completed this section, you should be able to

a. name and describe the three layers and four chambers of the heart;
b. describe the surface features of the heart;
c. explain the structure and function of the valves;
d. trace the flow of blood through the chambers; and
e. name the vessels that supply blood to the heart wall.

13.2a The Heart Wall

The heart wall consists of three layers—a thick muscular *myocardium* sandwiched between two thin serous membranes, the *epicardium* and *endocardium* (fig. 13.2). The **epicardium,**[1] as we have already seen, covers the surface of the heart. It consists of a simple squamous epithelium overlying a layer of areolar tissue; in most places it is as thin as tissue paper, and translucent so that the underlying muscle shows through. In some areas, it also has a layer of adipose tissue. Some individuals have a thick blanket of fat around the entire heart, but this is not a healthy condition.

The **myocardium,**[2] composed of cardiac muscle, makes up most of the mass of the heart. It does the contractile work of the heart, and its thickness varies greatly according to the workload of individual chambers. The muscle coils around the ventricles in such a way as to create a twisting action when it contracts, "wringing" blood out of the heart like wringing out a wet towel.

---

[1] *epi* = upon; *cardi* = heart
[2] *myo* = muscle; *cardi* = heart

(a) Anterior view

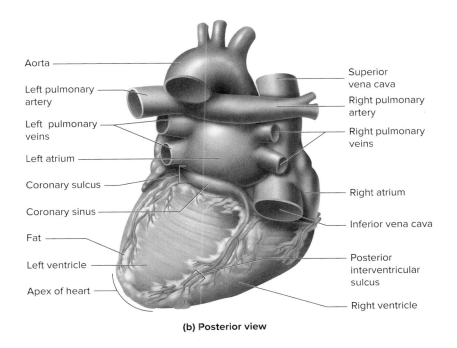

(b) Posterior view

Figure 13.3 External Anatomy of the Heart. The unlabeled coronary blood vessels on the heart surface are identified in figure 13.5. **APR**

The **endocardium**[3] lines the interior of the heart chambers, covers the valve surfaces, and is continuous with the inner lining *(endothelium)* of the blood vessels. Like the epicardium, it is composed of simple squamous epithelium overlying a thin layer of areolar tissue, but it contains no adipose tissue.

The heart also has a connective tissue framework called the **fibrous skeleton,** concentrated especially in the wall between the atria and ventricles and in rings around the valve orifices. It has several roles: (1) It provides attachment for the valves and holds the valve orifices open, yet prevents them from being excessively stretched when blood surges through. (2) It anchors the cardiac muscle cells, giving them something to pull on when they contract. (3) It serves as an electrical insulator between the atria and ventricles, and thus prevents the atria from stimulating the ventricles directly; this is important in the timing and coordination of cardiac excitation and contraction.

13.2b The Chambers

The heart has four chambers that receive and eject blood (**figs. 13.3** and **13.4**). The two smaller, superior chambers are the **right** and **left atria** (AY-tree-uh; singular, *atrium*[4]). They are thin-walled receiving chambers for blood returning to the heart by way of the great veins. Most of the mass of each atrium is on the posterior side of the heart, so only a small portion is visible from the anterior view (fig. 13.3).

---

[3]*endo* = internal; *cardi* = heart
[4]*atrium* = entryway

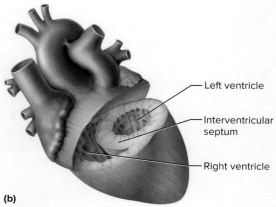

Figure 13.4 Internal Anatomy of the Heart. (a) Anterior view. Arrows indicate the path of blood flow through the heart, with oxygen-poor blood indicated by violet arrows and oxygen-rich blood by orange arrows. (b) Cross section, showing shapes and spatial relationship of ventricles. The right ventricle wraps around the left in a C shape, but both ventricles have the same volume. **APR**

The two inferior chambers, the **right** and **left ventricles,**[5] compose most of the bulk of the heart. They are strong pumps that eject blood into the arteries and keep it flowing around the body. The right ventricle constitutes most of the anterior portion of the heart, whereas the left ventricle forms the apex and inferoposterior portion. The right ventricle wraps partially around the left like a crescent (fig. 13.3b), so in anatomical art and teaching models it often looks smaller than it really is. Both ventricles have the same internal volume, and it is imperative that they pump the same amount of blood.

The boundaries of the four chambers are marked on the surface by three sulci (grooves), which are largely filled with fat and coronary blood vessels. The **coronary**[6] **sulcus** encircles the heart near the base and separates the atria from the ventricles. It is a deep groove that is easily seen when one lifts the edges of the atria (fig. 13.3a). The **anterior** and **posterior interventricular sulci** run obliquely down the front and back of the heart from the base toward the apex. They overlie the wall that separates the right and left ventricles, and therefore mark the boundary between these chambers.

The thickness of the walls varies among the chambers. The atria have thin flaccid walls, corresponding to their light workload—all they do is pump blood to the ventricles immediately below. The walls of the ventricles are much thicker, especially the left, which is two to four times as thick as the right because it pumps blood through the entire body, whereas the right ventricle pumps blood only to the lungs and back. The **interventricular septum** is a thick muscular wall between the two ventricles; the **interatrial septum** lies between the atria.

13.2c The Valves

Four heart valves ensure a one-way flow of blood—one between each atrium and its ventricle and one at the exit from each ventricle to its great artery (fig. 13.4). Each valve consists of two or three flaps of thin tissue called **cusps (leaflets).**

The **atrioventricular (AV) valves** control the opening between each atrium and the ventricle below it. They ensure that blood can't regurgitate back into the atria when the ventricles contract. The **right AV valve** is often called the **tricuspid valve** because it has three cusps. The **left AV valve** is called the **mitral valve** for its resemblance to a miter, the headdress of a church bishop. (It was formerly known also as the *bicuspid valve* from a mistaken belief that it had only two cusps.) Each cusp is anchored to conical **papillary muscles** on the floor of the ventricle by way of stringy **tendinous cords (chordae tendineae)** (KOR-dee ten-DIN-nee-ee). When the ventricles contract, the papillary muscles tense the cords and prevent the AV valves from bulging back into the atria or flipping inside out like windblown umbrellas. Excessive bulging due to slack tendinous cords is called *valvular prolapse*.

The pulmonary and aortic valves (collectively called **semilunar**[7] **valves**) lie between the ventricles and great arteries. The **pulmonary valve** controls the opening from the right ventricle into the pulmonary trunk, and the **aortic valve** controls the opening from the left ventricle into the aorta. Each semilunar valve has three cusps shaped like shirt pockets. When blood is ejected from the ventricles, it opens these valves and presses their cusps against the arterial walls. When the ventricles relax and expand, arterial blood flows backward toward the ventricles and fills the cusps. The inflated pockets meet at the center and quickly seal the opening, so little blood is able to return to the ventricles. Because of the way these cusps are attached to the arterial wall, they cannot prolapse any more than a shirt pocket turns inside out if you jam your hand into it. Therefore, they don't require or possess tendinous cords.

---

[5]*ventr* = belly, lower part; *icle* = little
[6]*coron* = crown; *ary* = pertaining to
[7]*semi* = half; *lunar* = like the moon

13.2d Blood Flow Through the Chambers APR

Until the sixteenth century, anatomists thought that blood flowed directly from the right ventricle to the left through invisible pores in the septum. In actuality, blood in the right and left chambers is kept entirely separate in the adult. Thus, deoxygenated blood returning from the tissues and destined for the lungs doesn't mix with oxygenated blood that has returned from the lungs. Figure 13.4a shows the pathway of blood as it travels from the right atrium through the body and back to the starting point.

Blood that has been through the systemic circuit returns by way of the superior and inferior venae cavae to the right atrium. It flows directly from the right atrium, through the right AV valve, into the right ventricle. When the right ventricle contracts, it ejects this blood through the pulmonary valve into the pulmonary trunk, on its way to the lungs to exchange carbon dioxide for oxygen.

Blood returns from the lungs by way of two pulmonary veins on the left and two on the right; all four of these empty into the left atrium. Blood flows through the left AV valve into the left ventricle. Contraction of the left ventricle ejects this blood through the aortic valve into the ascending aorta, on its way to another trip around the systemic circuit.

13.2e Coronary Circulation

If your heart lasts for 80 years and beats an average of 75 times a minute, it will beat more than 3 billion times and pump more than 400 million liters of blood. It is a remarkably hardworking organ that requires an abundant supply of oxygen and nutrients.

The energetic demands of the heart mean that it receives a disproportionate amount of the body's blood supply. The blood vessels devoted to supplying the heart's own tissues constitute the **coronary circulation** (fig. 13.5). At rest, these vessels supply the heart wall with about 250 mL of blood per minute. This is about 5% of the circulating blood going to meet the metabolic needs of the heart alone, even though the heart is only 0.5% of the body's weight. It receives 10 times its "fair share" to sustain its strenuous workload. Of course, the amount increases dramatically during exercise.

Arterial Supply

Coronary arteries are a focus of attention by health professionals and laypeople alike because their blockage by atherosclerosis can lead to heart attacks (see Perspectives on Health at the end of this section). The pattern of vessel branching is one of the most variable aspects of anatomy; the following description describes the pattern seen in about 70% to 85% of the population.

Immediately after the aorta leaves the left ventricle, it gives off a right and left coronary artery. The openings of the two arteries lie behind the cusps of the aortic valve. When the ventricles relax, a small amount of blood flows backward in the aorta and fills the cusps, allowing flow into the coronary arteries. Unlike most arteries in the body, coronary artery flow peaks during ventricular relaxation rather than ventricular contraction.

- The **left coronary artery (LCA)** curves under the left atrium and travels a short distance before dividing into two branches: the **anterior interventricular** and **circumflex branch** (fig. 13.5a). The anterior interventricular branch is also known clinically as the *left anterior descending (LAD) branch*. It travels down the anterior interventricular sulcus to the apex of the heart, rounds the bend, and continues a short distance up the posterior side. It supplies both ventricles and the anterior two-thirds of their septum. The circumflex branch continues around the left side of the heart in the coronary sulcus.
- The **right coronary artery (RCA)** supplies the right atrium, curving posteriorly to travel along the coronary sulcus. One of its branches, the

Figure 13.5 The Major Coronary Blood Vessels. (a) Anterior view. (b) Posterior view. (c) A polymer cast of the coronary circulation. APR

c: Science Photo Library/Science Source

posterior interventricular branch, runs down the posterior interventricular sulcus and supplies the posterior walls of both ventricles as well as the posterior portion of the interventricular septum. It ends where it joins the anterior interventricular branch of the LCA.

Venous Drainage

Venous drainage refers to the route by which blood leaves an organ. In the heart, after the blood flows through the coronary arteries and passes through capillaries, it drains into several veins that return the blood to the right atrium. Three of the most significant are described below.

- The **coronary sinus** is a large, horizontal, venous tunnel in the coronary sulcus on the posterior side of the heart (fig. 13.5b). It collects blood from numerous veins and then empties into the right atrium.

PERSPECTIVES ON HEALTH

Atherosclerosis and Hardening of the Arteries

As we get older, our arteries become increasingly stiff, a condition called *arteriosclerosis* ("hardening of the arteries"). A contributing factor in arterial stiffness is *atherosclerosis*,[8] the growth of lipid deposits in the arterial walls. Atherosclerosis begins when a vessel is damaged by hypertension, diabetes, or other factors. Macrophages invade the damaged tissue and accumulate cholesterol and fat from the blood. In time, the lesion grows into a fatty *plaque (atheroma)*. Platelets adhere to these plaques and secrete growth factors that stimulate the proliferation of smooth muscle and connective tissue. The plaque grows into a bulging mass of lipid, fiber, and smooth muscle that blocks blood flow through the vessel (**fig. 13.6b**). Blood clots can easily develop on these plaques and may lead to complete obstruction of the vessel, or break free and travel as *emboli* to block smaller vessels downstream.

Degradation and blockage of vessels by atheromas often lead to kidney failure, heart attack, or stroke. Blockage occurs in both arteries and veins, but is more crucial in arteries because this is where it has the potential to shut off blood flow to a vital tissue.

Coronary Artery Disease

Coronary artery disease (CAD) is a degenerative change in the coronary arteries associated with atherosclerosis (fig. 13.6).

[8]*athero* = fatty; *scler* = hard; *osis* = condition

(a) Lumen Artery wall

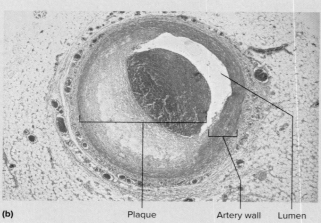
(b) Plaque Artery wall Lumen

(c)

Figure 13.6 Coronary Artery Disease. (a) Cross section of a healthy coronary artery. (b) Cross section of a coronary artery with atherosclerosis. Most of the lumen is obstructed by calcified plaque, and blood flow is severely restricted. (c) Coronary angiogram showing 60% obstruction, at arrow, of the anterior interventricular (left anterior descending, LAD) artery.

a: Ed Reschke/Getty Images; b: Image Source/Getty Images; c: kalewa/Shutterstock

It is the most common cause of heart failure and one of the leading killers in the United States. An early warning symptom of CAD is *angina pectoris,* a transient substernal chest pain that occurs when obstructed coronary arteries are not delivering enough oxygen to some areas of the myocardium. Angina is often a forerunner to myocardial infarction.

One of the treatments for CAD is *balloon angioplasty.* A cardiologist inserts a catheter into the femoral artery at the groin and threads it all the way up to the heart and into the diseased coronary artery. A tiny balloon at the tip of the catheter is then inflated, which crushes the atheroma and plasters it against the vessel wall. This opens the vessel, restores blood flow, and relieves pain, but it is typically a temporary solution and often has to be repeated. It does not reduce the risk of heart attack or death in patients under medical treatment regimens.

A more invasive procedure is *coronary bypass surgery,* in which the chest is opened, the patient is put on a heart–lung bypass machine, and undamaged vessels (usually from the leg or chest) are attached above and below the damaged portion of the coronary artery. This provides a detour around the obstruction so blood can reach downstream tissues.

Prevention of CAD is, of course, preferable to treatment. Regular exercise, a healthy body weight, and a diet rich in fruits and vegetables and low in saturated fat and cholesterol all protect against CAD. Avoiding smoking is also important. For discussion of the so-called "good" and "bad cholesterol" (more correctly known as high- and low-density lipoproteins, or HDLs and LDLs) in relation to cardiovascular health, see "Blood Lipoproteins" in section 18.1e.

While lifestyle plays a role, some risk factors for CAD are unavoidable. One of these is heredity. In a pair of identical twins, if one suffers CAD or dies of a resulting MI, the other is much more likely to do so at a similar age than is a nonidentical sibling with a similar lifestyle. Males and older people are also more at risk of CAD than females and younger people. Drugs that control hypertension and lower cholesterol levels are commonly used to reduce the progression of cardiovascular disease.

- The **posterior interventricular (middle cardiac) vein** leads up the posterior interventricular sulcus and empties into the coronary sinus.
- The **great cardiac vein** travels from the apex up the anterior aspect of the heart, then arcs around the left side and empties into the coronary sinus.

Before You Go On

Answer these questions from memory to test your understanding of the previous section. Reread the material if you can't answer the questions.

3. Which layer of the heart wall is thickest? How do the other two resemble and differ from each other?
4. What are structural and functional differences between the atrioventricular and semilunar valves?
5. Trace the route of the blood through the heart, naming each chamber and valve in order.
6. Trace the routes of blood flow through the coronary arteries and veins that supply the heart wall.

13.3 Physiology of the Heart

Expected Learning Outcomes

When you have completed this section, you should be able to

a. describe the structure of cardiac muscle cells and their attachments to each other;

b. explain why the heart needs such a high capacity for aerobic respiration;

c. list the components of the heart's electrical conduction system, explain how it coordinates the beating of the heart chambers, and compare and contrast action potentials of the pacemaker and myocardium;
d. describe the nerve supply to the heart;
e. diagram the electrocardiogram and label and explain each of its waves;
f. describe the key events that occur through one cycle of cardiac contraction and relaxation;
g. explain what causes the sounds of the heartbeat; and
h. define *cardiac output* and explain what determines it and what chemical agents can modify it.

Despite our justifiable fear of heart failure, the heart is a remarkably hardworking and reliable organ. Two of its most important physiological properties are its rhythmicity and its resistance to fatigue. In this section, we examine the bases for those two properties and the operation of the heart through one cycle of contraction and relaxation.

13.3a Cardiac Muscle

The work of the heart is, of course, done by its muscle cells, called **cardiomyocytes**.[9] Like skeletal muscle fibers, they are striated, but they differ in several other ways. They are short stocky cells rather than slender fibers—typically 50 to 100 μm long and 10 to 20 μm wide (**fig. 13.7**). Most have a single nucleus, often surrounded with glycogen.

The ends of each cardiomyocyte are slightly branched, so the cell somewhat resembles a short log with notches in the ends. Each branch reaches out to connect end to end with another cardiomyocyte. The cells stimulate each other through these connections and therefore act as a unit rather than independently. Their synchronized activity is necessary for the coordinated contractions of the heart chambers.

The ends where two cardiomyocytes meet are marked by **intercalated discs** (in-TUR-kuh-LAY-ted), which appear through the microscope as dark transverse lines, heavier than the striations. The plasma membrane of each cell at this point is folded somewhat like the bottom of an egg carton, increasing the surface area of cell-to-cell contact. The cells are joined at the discs by *desmosomes* and other mechanical junctions that keep them from pulling apart when the myocardium contracts. Electrical *gap junctions* (see "Cell Junctions" in section 3.2d) in the intercalated discs allow ions to flow from one cell directly into the cytoplasm of the next, thus enabling electrical excitation to spread rapidly from cell to cell.

The heart requires a great deal of energy to beat so steadily, year after year without fail. Although it weighs only 300 g, it produces and consumes about 5 kg of ATP per day. Fatty acids and glucose provide most of the energy for ATP production. Cardiomyocytes make ATP almost exclusively by aerobic respiration, and are therefore very resistant to fatigue. To meet this ATP demand, they have exceptionally large mitochondria that make up about 25% of the cell volume, in contrast to skeletal muscle fibers, which have smaller and fewer mitochondria that occupy only 2% of the cell. Cardiomyocytes are vulnerable to oxygen deprivation and need an abundant, reliable supply, which makes the coronary circulation especially critical.

13.3b The Cardiac Conduction System

The heart is described as **autorhythmic** because it beats at its own rhythm without need of stimuli from the nervous system. Instead, the heartbeat is coordinated by

---

[9]*cardio* = heart; *myo* = muscle; *cyte* = cell

Figure 13.7 Cardiac Muscle. (a) Light micrograph. (b) Structure of a cardiomyocyte and its relationship to adjacent cardiomyocytes. All of the colored area is a single cell. Note that it is notched at the ends and typically linked to two or more neighboring cardiomyocytes by the mechanical and electrical junctions of the intercalated discs. (c) Structure of an intercalated disc. **APR**

a: © Ed Reschke

modified cardiomyocytes that have lost the ability to contract, and are specialized to depolarize spontaneously at regular time intervals. These cells are concentrated in two masses called *sinuatrial* and *atrioventricular nodes*. Other noncontractile cardiomyocytes behave much like nerves; together with the nodes, they form the **cardiac conduction system.** Electrical signals travel rapidly through the

Figure 13.8 The Cardiac Conduction System.
Electrical signals travel along the pathways indicated by the arrows. **APR**

❓ *One atrium begins contracting slightly before the other. Which do you think begins first, and why?*

① SA node fires.

② Excitation spreads through atrial myocardium.

③ AV node fires.

④ Excitation spreads down AV bundle.

⑤ Subendocardial branches distribute excitation through ventricular myocardium.

conduction system in the following order, providing the necessary stimulation to excite the far more numerous contractile cardiomyocytes and generate the heartbeat (fig. 13.8):

① The **sinuatrial (SA) node.** This is the heart's pacemaker, a patch of cells in the right atrium near the superior vena cava. These cells depolarize every 0.8 seconds or so, setting off each heartbeat and determining the heart rate—about 70 beats/minute at rest. Heart rate is faster if the pacemaker fires more often.

② **Internodal conduction.** Signals from the SA node spread through both atria, causing them to contract before the ventricles.

③ The **atrioventricular (AV) node.** The AV node is a patch of cells similar to the SA node but located in the interatrial septum just above the tricuspid valve. It acts as the gateway for all electrical signals headed for the ventricles. The signals must pass through this node because the fibrous skeleton blocks them from traveling to the ventricles by any other route. The AV node has fewer gap junctions than the other parts of the conduction system, so it delays the signal for about 100 ms. This allows the ventricles time to fill with blood before the signal travels on and stimulates them to contract.

④ The **atrioventricular (AV) bundle.** This is a cord of modified cardiomyocytes that leaves the AV node and passes electrical signals through the interventricular septum. In the septum, it forks into the **right** and **left bundle branches**, which descend toward the apex of the heart.

⑤ The **subendocardial branches** (*Purkinje fibers*) are a group of large cells that arise from the bundle branches. At the apex of the heart, they turn upward and branch into little vinelike tips throughout the ventricular myocardium. Once the cells in the subendocardial branches have delivered the electrical signal, the contractile cardiomyocytes perpetuate it by passing ions from cell to cell through their gap junctions.

Cardiac signals travel about 1 meter per second (m/s) through the muscle of the atria. When they reach the AV node, they slow down to 0.05 m/s—analogous to a car on a state highway that has to slow down for a small town with one stoplight. This delays the signal enough to allow the ventricles to fill with blood. In the AV bundle and subendocardial branches, the signal speeds up to 4 m/s, like the car speeding up again on the other side of town. As small as the heart is, signals traveling at such a speed can reach all cells of the ventricular myocardium in a small fraction of a second, contributing to their near-simultaneous contraction.

13.3c Electrical Activity and Contraction

We saw in section 8.2a that neurons have a stable voltage across the plasma membrane called a *resting membrane potential,* as they wait for excitation from an external source. Stimulation opens membrane channels, allowing ions to flow through, resulting in an *action potential.* Cells of the cardiac SA node, however, behave differently. They do not rely on an external source of stimulation and they do not have a stable resting potential. Their membranes are leaky, continually allowing sodium ions into the cell. Each time this leakage raises the membrane potential to its threshold voltage, it sets off an action potential; the pacemaker fires and initiates another heartbeat.

The contractile cardiomyocytes, however, do have a stable resting potential, waiting at a voltage of about –90 mV to receive a signal from the cardiac conducting system or from each other. The arrival of a signal opens ion channels in the cell and causes an action potential, which triggers contraction of the cell much like we saw in skeletal muscle. Unlike the action potentials of neurons and skeletal muscle, however, the action potential in cardiac muscle is not simply a quick up-and-down voltage spike. In skeletal muscle, an action potential falls back to the resting level in about 2 milliseconds (ms). In cardiac muscle, however, the voltage remains elevated for about 200 ms (one-fifth of a second). This produces a prolonged contraction of the myocardium, necessary to ensure effective ejection of blood from the ventricles. It also prevents the cardiomyocytes from being stimulated again too soon. Premature restimulation could create a state of tetanus, causing the heart to seize up and cease pumping.

13.3d Nerve Supply to the Heart

Even though the heart has its own pacemaker, it obviously doesn't always beat at the same rate. A typical resting heart rate in adults is about 70 to 75 beats/minute (bpm), whereas vigorous exercise can raise it to as high as 230 bpm. The heart rate is modified by the autonomic nervous system. Sympathetic fibers travel through *cardiac nerves* to the SA and AV notes and the myocardium. They release norepinephrine, which stimulates more rapid firing of the SA node and stronger contractions in the cardiomyocytes. Parasympathetic fibers travel to the heart through the two vagus nerves. In the heart wall itself, they synapse with short postganglionic neurons that lead to the SA and AV nodes; the myocardium receives little or no parasympathetic input. Parasympathetic fibers secrete acetylcholine, which slows the heartbeat by decreasing the speed of depolarization in the pacemaker cells.

13.3e The Electrocardiogram

Next to listening to the heart sounds with a stethoscope, the most common clinical method of evaluating heart function is the **electrocardiogram**[10] **(ECG or EKG).** The depolarization and repolarization of the atrial and ventricular myocardium generate electrical currents that are detectable with electrodes attached to the surface of the skin. The ECG is a composite recording of all action potentials produced by the self-depolarizing nodal cells and the contractile cardiomyocytes that can be amplified and displayed on either a paper chart or a video monitor (**fig 13.9**).

---

[10]*electro* = electricity; *cardio* = heart; *gram* = record of; *EKG* is from the German spelling, Elektrokardiogramm

Figure 13.9 The Electrocardiogram.

Three major events are seen in the ECG: the **P wave, QRS complex,** and **T wave.** The letters don't stand for words, but were used to designate the waves by the German physiologist Willem Einthoven. He developed the technique and made the first successful ECG in 1895, and later was awarded the Nobel Prize in Physiology or Medicine. The P wave reflects depolarization of the atria. The QRS complex represents depolarization of the ventricles. It is the largest wave of the ECG because the ventricles constitute the largest muscle mass in the heart and generate the strongest electrical current. The atria repolarize during the QRS complex but this event is largely masked by depolarization of the ventricles. The T wave represents ventricular repolarization. The waves of the ECG are not action potentials of individual cardiomyocytes but composite voltage changes produced by many cells.

Heart rate can be determined by measuring the time between successive R peaks. We will see in the next section how the waves of the ECG are correlated with the mechanical pumping actions of the heart.

The electrocardiogram is useful because it provides a noninvasive, if indirect, snapshot of heart function and a tool for rapid preliminary diagnosis of a variety of cardiac conditions (see Clinical Application 13.1).

13.3f The Cardiac Cycle

A **cardiac cycle** consists of one complete contraction and relaxation of all four heart chambers, encompassing events from the beginning of one heartbeat to the beginning of the next. Any chamber is said to be in **systole** (SIS-toe-lee) when it contracts and in **diastole** (dy-ASS-toe-lee) when it relaxes; when these words are used without specifying a chamber, however, they usually refer to the ventricles. Excitation of a heart chamber depolarizes its myocardium and leads to systole; repolarization relaxes its muscle and leads to diastole. If the relevant valves are open and allow blood to flow through, systole generally expels blood from a chamber and diastole allows a chamber to refill.

Figure 13.10 illustrates the mechanical events of the cardiac cycle correlated with the electrocardiogram and heart sounds. Closely follow the figure as you study the text.

1. Initially, all four chambers are relaxed, in diastole. The AV valves are open, and without any effort from the heart, blood flows passively through these valves into the ventricles. This alone accounts for more than two-thirds of the ventricular filling.
2. The SA node fires, exciting the atrial myocardium and producing the P wave of the ECG. Atrial systole contributes the last 30% or so of ventricular filling. At the end of atrial systole, each ventricle contains about 130 mL of blood.
3. The AV node fires and electrical excitation spreads throughout the ventricular myocardium, producing the QRS complex in the ECG and setting off ventricular systole. Meanwhile, the atria relax. Pressure in the ventricles rises sharply. Blood surges against the AV valves and forces them shut; the surge against this obstacle creates a vibration heard with the stethoscope as the **first heart sound (S_1)**. At first, the ventricles can't eject any blood because pressure in the aorta and pulmonary trunk is still higher than ventricular pressure and it holds the two semilunar valves closed.
4. When ventricular pressure exceeds pressure in the arteries, the semilunar valves open and blood spurts into the aorta and pulmonary trunk. Ventricular ejection lasts about 200 to 250 ms, the length of time described earlier

Figure 13.10 The Cardiac Cycle. Transitions from phases 1 to 2, 2 to 3, and 4 to 5 correspond to the electrocardiogram waves indicated in red.
APR

? *The ventricles begin contracting at phase 3 but cannot change volume until phase 4. Why?*

when the cardiomyocyte membrane voltage remains elevated. At rest, each ventricle expels about 70 mL of blood; this is called the **stroke volume (SV).** Another 60 mL, almost half of what was in the ventricle, remains behind to mix with fresh blood on the next cycle. In vigorous exercise, however, stroke volume increases and the ventricles may eject as much as 90% of their blood. Diseased hearts may eject much less than 50%.

⑤ The myocardium repolarizes and produces the T wave of the ECG. The ventricles now relax and expand. All four chambers are again in diastole. Very briefly, blood in the aorta and pulmonary trunk surges backward, filling the pocketlike cusps of the two semilunar valves. The three valve cusps come together in the middle of the artery and close the opening. The surge of aortic and pulmonary blood against the closed valves creates a vibration heard as the **second heart sound (S_2).** As the ventricles expand, the AV valves reopen and the ventricles begin to refill with blood pouring down from the atria. The cycle begins anew.

Listening to the heart sounds with a stethoscope is a common part of a physical examination. It is an example of **auscultation** (AWS-cul-TAY-shun), listening to the sounds the body makes. The first and second heart sounds are often described as sounding like a "lubb-dupp." The first heart sound, S_1 ("lubb"), is louder and longer, and the second, S_2 ("dupp"), is softer and shorter.

Heart murmurs are whooshing or swishing sounds that are not part of the "lubb-dupp" sounds. Most heart murmurs are harmless, but some of them indicate valve problems. For example, blood may be regurgitated from the ventricles into the atria during ventricular systole because of structural problems with the AV valves. If the defect is great enough to interfere with the normal flow of blood, valve replacements may be an option. Since the 1960s, there has been a great deal of success in replacing defective valves with either mechanical valves or biological material such as heart valves from pigs. These surgical procedures have prolonged numerous lives.

Clinical Application 13.1

CARDIAC ARRHYTHMIA

Cardiac arrhythmia means any deviation from the normal heart rhythm. It can occur in either the atria or the ventricles. The most common form in the atria is *atrial flutter,* in which the atria show weak chaotic contractions at about 250 to 350 bpm. The ECG shows a "sawtooth" pattern with multiple small atrial peaks between QRS waves. In itself, it is not life-threatening and it can come and go, but it may cause fatigue, pain, dizziness, and breathlessness and it can sometimes lead to more serious conditions such as blood clots in the heart.

The most serious ventricular arrhythmia is *ventricular fibrillation (VF or V-fib),* an uncoordinated contraction of the ventricles often brought on by myocardial infarction (tissue death) or drugs of abuse such as amphetamines. A heart in VF is often said to feel like a squirming "bag of worms." The ECG shows grossly irregular waves of depolarization with no recognizable P, QRS, and T waves. There is no effective pumping of blood from such weak, uncoordinated ventricles, and without treatment, death is imminent. VF can be halted with a strong electrical shock from a *defibrillator,* but this is only a temporary life-prolonging measure. The patient may require a pacemaker, coronary bypass surgery, or other measures to manage the underlying problem.

13.3g Cardiac Output

The purpose of the cardiac cycle is to eject blood into the great arteries. The volume of blood ejected by each ventricle per minute (mL/min.) is called the **cardiac output.** Mathematically, it is the product of stroke volume, SV (mL/beat), and heart rate, HR (beats/min.)—that is, SV × HR = CO. Typical adult resting values are

$$70 \text{ mL/beat (SV)} \times 75 \text{ beats/min. (HR)} = 5{,}250 \text{ mL/min. (CO)}$$

The body's entire volume of blood (usually 4–6 L) passes through the heart every minute. Vigorous exercise can raise CO to as much as 21 L/min., and world-class athletes can pump up to 35 L/min.

Given that SV × HR = CO, there are obviously only two ways to increase cardiac output: Increase the heart rate or the stroke volume (or both). The heart rate increases when the sympathetic nervous system is activated and its neurotransmitter, norepinephrine, and the similar hormone epinephrine are released. Other agents that increase heart rate include thyroid hormone, nicotine, and caffeine. In contrast, the parasympathetic nervous system and its neurotransmitter acetylcholine slow down the heart rate, as do excessive levels of calcium or potassium.

Increasing the stroke volume (and thereby cardiac output) is accomplished by increasing the contraction strength of the cardiomyocytes. Stronger contractions are caused by sympathetic system chemicals epinephrine and norepinephrine, as well as calcium, the hormone glucagon, and the drug digitalis. Conversely, several factors can weaken myocardial contractions and reduce cardiac output, including potassium or calcium excesses, oxygen deficiency, and abnormally low blood pH *(acidosis)*. Note that some agents affect both the heart rate and contraction strength; epinephrine increases both, and potassium excess (hyperkalemia) reduces both.

> **Apply What You Know**
>
> Physical exercise obviously increases cardiac output. Do you think it achieves this through heart rate, contraction strength, or both? Explain.

Before You Go On

Answer these questions from memory to test your understanding of the previous section. Reread the material if you can't answer the questions.

7. Describe the structure of intercalated discs and explain the functional importance of each feature.
8. Why must the signal for each heartbeat slow down at the AV node?
9. Why is it so important that the action potential of a cardiomyocyte remain elevated for an extended time, instead of being a quick up-and-down voltage change like it is in skeletal muscle?
10. Describe the major events in each stage of the cardiac cycle, beginning when all chambers are in diastole.
11. What is represented by the P, QRS, and T waves of the ECG? Draw a simple diagram of an ECG trace and label the waves.
12. What are two ways that cardiac output may be increased? List specific chemicals that speed up or slow down the heart, and that strengthen or weaken its contractions.

13.4 General Anatomy of Blood Vessels

Expected Learning Outcomes

When you have completed this section, you should be able to

a. describe the structure of a blood vessel;
b. compare and contrast arteries, capillaries, and veins; and
c. explain how portal systems and anastomoses differ from the most common route in which blood flows from the heart and back again.

There are three categories of blood vessels—arteries, veins, and capillaries. The three differ in direction of blood flow, the pressure they must withstand, and the corresponding histological structure of their walls.

13.4a The Vessel Wall

The walls of arteries and veins are composed of the following three tissue layers (fig. 13.11):

1. The **tunica interna**[11] (TOO-ni-ca in-TER-nuh) lines the inside of the vessel and is exposed to the blood, which runs in the *lumen* (inner space) of the tube. It consists of a simple squamous epithelium called the **endothelium,** overlying a basement membrane and a sparse layer of loose connective tissue. The endothelium acts as a selectively permeable barrier to materials entering or leaving the bloodstream. It also secretes chemicals that prevent blood cells and platelets from adhering to the vessel wall, and others that dilate or constrict blood vessels and thereby regulate the flow of blood.
2. The **tunica media,** the middle layer, is usually thickest. It consists of smooth muscle, collagen, and elastic tissue. It strengthens the vessels and prevents the blood pressure from rupturing them, and its muscle carries out vasoconstriction and vasodilation.
3. The **tunica externa** is the outermost layer. It consists of loose connective tissue that often blends into surrounding tissues and anchors the blood vessel to them. It is penetrated by small nerves, lymphatic vessels, and small blood vessels *(vasa vasorum)* that serve the tissues of the larger vessel.

In contrast to these, capillaries are very thin-walled vessels with only a tunica interna.

13.4b Arteries, Capillaries, and Veins

All vessels that carry blood away from the heart are classified as **arteries,** whereas all vessels that carry blood back to the heart are classified as **veins. Capillaries** are microscopic vessels that connect the smallest arteries to the smallest veins.

Arteries

Each beat of the heart creates a surge of pressure in the arteries as blood is ejected into them. Arteries withstand these surges because of their resilient structure. They are more muscular than veins, and retain their round shape even when empty. They are divided into three categories by size, but of course there is a gradual transition from one to the next.

---

[11]*tunica* = coat; *interna* = internal

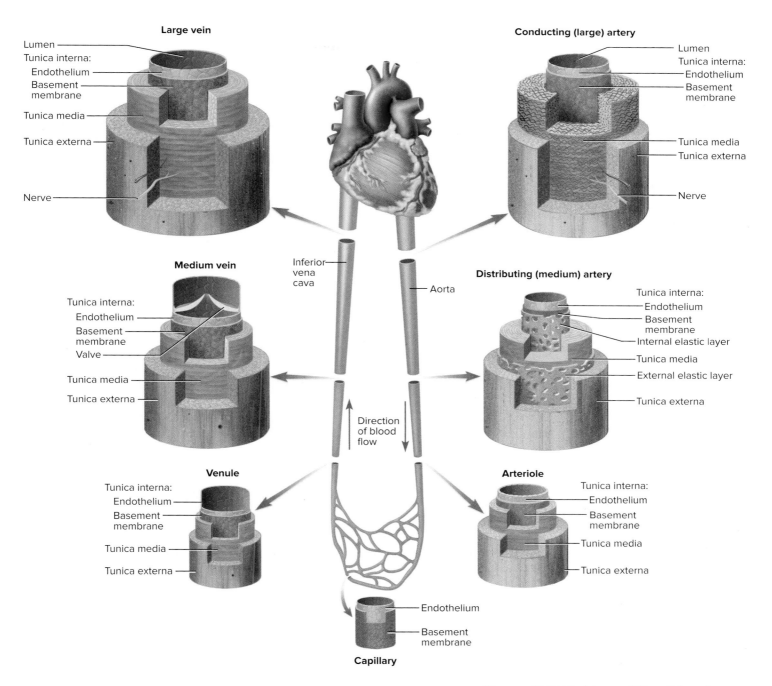

Figure 13.11 Histology of Blood Vessels. Note the changes in wall structure as we progress around the circuit from high-pressure arteries, to blood capillaries (exchange vessels), to low-pressure veins.

? *Why do the arteries need so much more elastic tissue than the veins do?*

1. **Conducting (elastic** or **large) arteries** include the largest arteries, such as the aorta and pulmonary trunk, and may be up to 2.5 cm (1 in.) in diameter. The tunica media is dominated by layers of perforated elastic sheets like slices of Swiss cheese rolled up into a tube; smooth muscle and collagen are present but less visible. The abundance of elastic tissue enables conducting arteries to expand during ventricular systole and recoil during diastole. Their expansion protects smaller downstream arteries by reducing the pressure surge during ejection of blood from the heart, and their recoil prevents blood pressure from dropping too low when the heart relaxes.
2. **Distributing (muscular** or **medium) arteries** are smaller branches that distribute blood to specific organs. You could compare a conducting artery to an interstate highway and distributing arteries to the exit ramps and

432 PART 4 Circulation and Defense

state highways that serve individual towns. These arteries are dominated by smooth muscle in the tunica media, with less abundant elastic tissue. All conducting arteries and most distributing arteries are named and can be traced in a careful dissection or surgery. Examples include the brachial and femoral arteries in the arm and thigh, and the renal artery that supplies blood to the kidney.

3. **Resistance (small) arteries** are usually too variable in location and number to be given individual names. Their tunica media is composed almost entirely of smooth muscle, with little elastic tissue. The smallest of these, called **arterioles,** have only one to three layers of smooth muscle and a very thin tunica externa.

Capillaries

For the blood to serve any purpose, materials such as nutrients, wastes, and hormones must pass between the blood and tissue fluids, through the walls of the vessels. This occurs mostly in the capillaries (**fig. 13.12a**), which are therefore often

Figure 13.12 Blood Capillaries. (a) A capillary bed arising from an arteriole and draining into a venule. Arrows indicate the direction of blood flow. The flow of blood into any given capillary is regulated by a few smooth muscle cells that encircle the capillary mouth and constitute a precapillary sphincter. (b) Cross section of a continuous capillary and an associated cell called a pericyte. (c) Photograph of dye-injected capillary beds.
c: Biophoto Associates/Science Source

called the *exchange vessels* of the cardiovascular system. It is estimated that there are a billion capillaries, enough to ensure that each cell is within four to six cell widths of the nearest capillary.

Capillaries are composed of only an endothelium and basement membrane. Their walls are as thin as 0.2 to 0.4 μm, and they range about 5 to 9 μm in diameter. Since erythrocytes are about 7.5 μm in diameter, they often have to stretch into elongated shapes to squeeze through the smallest capillaries.

Continuous capillaries, the most common type, are tubes of endothelial cells rolled up like burritos and held together by tight junctions (**fig. 13.12b**). Water and small solutes pass through the wall via narrow gaps called **intercellular clefts** between the cells, or by passing through the cells themselves. Continuous capillaries sometimes have cells called *pericytes* associated with them. Among other functions, these can differentiate into new endothelial and smooth muscle cells and contribute to the repair of damaged capillaries and other vessels.

In organs such as the kidneys where rapid filtration is important, or pancreatic islets where it is necessary for the capillaries to take up large molecules (insulin and glucagon), there are special **fenestrated capillaries** with large *filtration pores* in them, somewhat like a sheet of pie dough with a lot of holes punched in it by a small cookie cutter, then rolled into a tube. Intercellular clefts are only about 4 nm wide, but filtration pores range from 20 to 100 nm wide and thus allow much larger molecules to pass through.

> *Apply What You Know*
>
> Considering the functions of the skeletal muscles and the anterior lobe of the pituitary gland, which would you expect to have continuous capillaries and which to have fenestrated capillaries? Explain.

Capillaries are organized into **capillary beds,** webs of 10 to 100 vessels arising from a single arteriole (fig. 13.12). At the beginning of each capillary is a ring of smooth muscle cells called a **precapillary sphincter,** which can dilate to let blood into a capillary or constrict to shut down local blood flow. Blood flow through the capillary beds is regulated to match the metabolic needs of tissues. There isn't enough blood in the body to fill the entire vascular system at once; consequently, about three-quarters of the body's capillaries are shut down at any given time. For example, about 90% of capillaries in skeletal muscle have little or no blood flow during rest. In contrast, during exercise, they receive an abundant flow, while capillary beds elsewhere—for example, in the intestines—shut down to compensate.

Veins

Veins are relatively thin-walled and flaccid, and expand easily to accommodate a greater volume of blood than arteries do. Being farther from the heart, veins have a much lower blood pressure than arteries and their blood flow is steady, rather than pulsating with the heartbeat as it does in arteries. Therefore, veins do not require such muscular or elastic walls to withstand pressure surges.

Small veins merge to form larger and larger ones, like river tributaries, as they approach the heart. In examining the types of veins, we will follow the direction of blood flow, working from smallest to largest vessels.

1. **Venules** receive blood from capillaries. They range up to 1 mm in diameter. The smallest of them have no tunica media and are quite porous; this is where most white blood cells leave the bloodstream to wander among the connective tissues.
2. **Medium veins** range up to 10 mm in diameter and have a thicker tunica media and externa; they are more muscular than venules but less so than

Figure 13.13 Variations in the Systemic Circuit. (a) The most common pathway, in which blood passes through only one capillary bed before returning to the heart. (b) A portal system, in which blood passes through two capillary beds in series before returning to the heart. (c) An arteriovenous anastomosis (shunt), in which blood bypasses the capillaries by flowing directly from an artery to a vein.

(a) Simplest pathway (1 capillary bed)

(b) Portal system (2 capillary beds)

(c) Arteriovenous anastomosis (shunt)

medium arteries. Most veins with individual names are in this category, such as the brachial veins in the arm. Many medium veins in the limbs have **venous valves** that ensure a one-way flow of blood toward the heart, against the pull of gravity. These are discussed later under the subject of venous blood flow.

3. **Large veins** have diameters greater than 10 mm. These include veins that empty into the heart—the two venae cavae and four pulmonary veins.

13.4c Variations in Circulatory Routes

The most common pattern of systemic blood flow is for blood to leave the heart via the arteries, pass through one capillary bed (often quite distant from the heart), then return to the heart via the veins. However, there are important variations on this pattern (**fig. 13.13**). In a **portal system,** the blood flows through two capillary beds in a row before returning to the heart. We have seen one of these already between the hypothalamus and anterior pituitary gland (see "Pituitary Anatomy" in section 11.3a), and we will encounter others in the intestine–liver relationship and in the kidneys. Portal systems are found where a substance is to be picked up by one capillary bed and immediately given off by another.

In contrast, **anastomoses** (ah-NASS-tuh-MO-seez), or **shunts,** are routes in which the blood bypasses capillaries, going directly from an artery to a vein (an *arteriovenous anatomosis*), from one vein to another (a *venous anastomosis*), or from one artery to another (an *arterial anastomosis*). Arteriovenous anastomoses are found at the bases of the fingers, for example, where they can allow blood to detour and avoid the fingers in cold weather in order to reduce heat loss. The resulting lack of warming of the fingers, however, is the reason why frostbite so commonly occurs here. (The same is true of the toes, nose, and ears.) Venous anastomoses are very common in such places as the hand and forearm, where they provide alternative routes of venous return in case one route is blocked. Some of these are often visible through the skin on the back of the hand. One place to find arterial anastomoses is in the coronary blood vessels, allowing blood to bypass an arterial blockage and nourish the heart tissue by an alternative route. They are also common at joints such as the knee and elbow, where they allow alternative routes of blood flow if flexion of a joint temporarily compresses one of the arteries.

Before You Go On

Answer these questions from memory to test your understanding of the previous section. Reread the material if you can't answer the questions.

13. Name the three tunics of a typical blood vessel and explain how they differ from each other.
14. Contrast the tunica media of a conducting artery, arteriole, and medium vein, and explain how the histological differences are related to the functional differences between these vessels.
15. Describe the structural and functional differences between continuous and fenestrated capillaries.
16. Define *portal system* and *anastomosis,* and explain their special functional purposes.

13.5 Physiology of Circulation

Expected Learning Outcomes

When you have completed this section, you should be able to

a. define *blood pressure* and explain how it is commonly measured;
b. explain the relationship between blood pressure, resistance, and flow;
c. describe three factors that determine resistance to blood flow;
d. discuss local, neural, and hormonal control of blood pressure;
e. explain how blood pressure and osmotic pressure interact in capillary fluid exchanges; and
f. describe the mechanisms for returning venous blood to the heart.

We have seen how the heart pumps blood through the blood vessels to deliver it to tissues and organs. Blood **flow** is the amount of blood passing any given point (or entering any given tissue or organ) per unit time. **Perfusion** is the flow relative to a given mass of tissue, such as milliliters of blood per minute per 100 grams of tissue. Thus, a small organ such as the pituitary gland could have greater perfusion but lower flow than a large organ such as a thigh muscle. Blood flow is governed by the fundamental principles that govern the flow of any fluid—pressure and resistance.

13.5a Blood Pressure

Blood pressure (BP) is the force exerted by blood on a vessel wall. Contraction of the heart initiates a wave of pressure that sharply decreases as the blood flows farther and farther away from the ventricle of origin. Blood always flows down a gradient from a point of high pressure to a point of lower pressure. The greater the pressure difference, the greater the flow. Think of a garden hose by analogy. If you slightly open the tap, you get a trickle of water from the end of the hose. Open the tap wider, and the water gushes out. Flow increases because you have increased the pressure difference between the beginning and end of the hose.

Arterial blood pressure is expressed as a ratio of the **systolic pressure** generated by contraction (systole) of the left ventricle, to **diastolic pressure,** the minimum to which the BP falls when the ventricle is in diastole. Both are expressed in millimeters of mercury (mm Hg). A typical healthy adult pressure is 120/80 mm Hg (120 mm Hg systolic, 80 mm Hg diastolic).

Blood pressure measurement is a familiar part of a routine physical examination and is performed with a **sphygmomanometer** (SFIG-mo-meh-NOM-eh-tur). This device consists of an inflatable cuff connected to a rubber bulb for pumping air into it, and a digital sensor, dial gauge, or a calibrated mercury column for measuring air pressure in the cuff. To take a patient's blood pressure, the examiner wraps the cuff snugly around the patient's arm and inflates it with the bulb until it exceeds the systolic blood pressure. Squeezing the arm muscles collapses the brachial artery deep within. Even during systole, the heart cannot force blood through the artery, and there is no blood flow distal to that point. The examiner now listens with a stethoscope at the bend of the elbow (cubital region) while slowly releasing air from the cuff.

At first there is no sound, but as soon as the systolic BP slightly exceeds the cuff pressure, each heartbeat forces the brachial artery open and allows a brief jet of blood to pass through. The vessel collapses again at diastole. This jet of blood and the subsequent surge of blood against the recollapsed artery cause turbulence that the examiner hears as a faint "bump" sound. The cuff pressure at the instant of the first bump is noted as the systolic BP. The examiner continues to release air from the cuff. During this time, a bump is heard every time the artery collapses—that is, once in each heartbeat. But soon the cuff pressure falls to a point that the brachial artery remains open even during diastole, and no further sounds are heard. The point at which the last bump is heard is noted as the diastolic BP.

Blood pressure is a focus of both health providers and patients because persistent high blood pressure is associated with increased risk of cardiovascular disease. Blood pressure often rises in and beyond middle age as the arteries become stiff and less resilient. A persistent resting blood pressure above 130/80 mm Hg is considered to be **hypertension** (high blood pressure). This does not include temporary elevations of BP due to exercise or emotional responses. Weight loss, exercise, and healthy diets can help to lower blood pressure in individuals who have hypertension. Many adults take medication to keep their blood pressure in what is considered to be a healthy range. A chronic low BP is called **hypotension.** It may result from blood loss, dehydration, or inability to regulate blood pressure fluctuation in old age, and is routinely seen in patients approaching the moment of death. There is no particular numerical criterion for hypotension.

Blood pressure is determined by three principal variables: (1) cardiac output, discussed earlier in the chapter; (2) blood volume, which is regulated mainly by the kidneys; and (3) resistance to flow, discussed next.

13.5b Peripheral Resistance

Resistance is a measure of hindrance to blood flow through a vessel caused by friction between the moving fluid and stationary vessel walls. **Peripheral resistance** is opposition to flow in vessels away from the heart, as opposed to resistance encountered in the heart itself. When resistance increases, flow decreases unless the heart pumps harder to compensate for it.

Resistance is determined by three principal variables:

1. **Viscosity,** the "thickness" of the blood. Higher viscosity increases resistance and impedes flow. (Imagine the difference between trying to suck molasses compared to water into a syringe.) The most significant determinants of blood viscosity are the concentrations of erythrocytes and albumin. A deficiency of RBCs or albumin reduces viscosity, and an excess of either increases it.
2. **Vessel length.** The farther a liquid travels through a vessel, the more cumulative resistance it encounters; pressure and flow both decline with distance. This is why, in a reclining person, arteries of the feet have less flow and a weaker pulse than arteries near the heart.
3. **Vessel radius,** the most important variable in flow. Vessel lengths and blood viscosity don't change from one minute to the next, so the only way to control peripheral resistance from moment to moment is by adjusting the radius of the blood vessels. Narrowing of the vessel, **vasoconstriction,** and widening of a vessel, **vasodilation,** are achieved by contraction or relaxation of the muscle in the tunica media.

Radius (r) is a very potent factor in blood flow (F) because flow is proportional to the *fourth power* of radius ($F \propto r^4$); in other words, their relationship is exponential. Doubling the radius of a vessel increases flow by 16 times ($r^4 = 2 \times 2 \times 2 \times 2 = 16$). Arterioles can change their radius as much as threefold, which would result in an 81-fold (3^4) change in blood flow. Going back to our earlier garden hose analogy, you could compare the effect of vasoconstriction on flow to what would happen if you backed your car over the hose.

Apply What You Know

Dehydration can raise the viscosity of the blood even when there is no change in the absolute amount of albumin or number of RBCs in the blood. Explain how.

13.5c Regulation of Blood Pressure and Flow

The perfusion of each organ is not constant but varies depending on its ever-changing demand for blood. During exercise, for example, vasodilation of vessels supplying skeletal muscle and the heart results in a greater percentage of blood being directed to those organs. Perfusion is controlled by local, neural, and hormonal mechanisms.

Local Control

A tissue with a very high metabolic rate needs more blood, oxygen, and nutrients, but such a tissue also produces metabolic by-products such as CO_2, lactic acid, adenosine, and others that relax the smooth muscle of arterioles. The vessels dilate and perfusion increases to meet the demands of the tissue. Other local chemicals released by platelets, endothelial cells, and connective tissue cells—such as histamine and nitric oxide—also dilate vessels and increase perfusion.

Long-term metabolic demand also stimulates new blood vessels to grow into a tissue—a process called **angiogenesis**.[12] A denser capillary network then meets the elevated needs of the tissue. This occurs in such cases as muscular conditioning and in the monthly regrowth of the uterine lining after a woman's menstrual period. Cancerous tumors also induce angiogenesis to feed their growth, and one line of cancer research is to develop drugs that may starve a tumor by blocking angiogenesis.

Neural Control

Blood vessel diameter is also regulated by the autonomic nervous system. Arteries near the heart—especially the aortic arch and internal carotid arteries—have sensory nerve endings called **baroreceptors** that monitor blood pressure (BP) and transmit signals to the medulla oblongata of the brainstem (fig. 13.14). The medulla has a group of neurons called the **vasomotor center** that issues signals to the blood vessels. When blood pressure rises above normal, the vasomotor center sends signals to the blood vessels to dilate, thereby decreasing peripheral resistance; when BP falls below normal, in contrast, the center constricts vessels and thereby raises peripheral resistance and systemic blood pressure. These reactions are called **baroreflexes.** Think of how pressure rises if you squeeze a plastic bottle of water; the same principle applies to regulating BP.

Blood pressure is continuously adjusted in response to a variety of circumstances. When you rise quickly from bed, for example, gravity draws the blood downward and BP in the head and neck falls. The baroreceptors and medulla respond quickly to constrict blood vessels and accelerate the heart, maintaining blood flow to the brain and preventing you from fainting. If the reflex is not quick enough, you may briefly feel dizzy. The vasomotor center also receives input from brain centers concerned with thought and emotion. This is why stress, anger, or excitement raises the blood pressure.

Hormonal Control APR

Some hormones affect blood pressure by constricting or dilating blood vessels.

- **Epinephrine** from the adrenal medulla and **norepinephrine** from sympathetic nerves dilate some vessels, such as arteries of the skeletal muscles, but constrict other vessels, such as those of the digestive tract and skin. During

Figure 13.14 Neural Control of Blood Pressure by Baroreflexes. APR

? *If the blood vessels dilate, does that raise or lower the blood pressure? Explain.*

① Baroreceptors detect degree of stretch in arteries and transmit signals to vasomotor center.

② Vasomotor center integrates input.

③ Outgoing signals are carried by sympathetic nerves to blood vessels.

④ Blood vessels constrict or dilate to adjust blood pressure.

---

[12]*angio* = vessels; *genesis* = production of

exercise, this gives priority to perfusion of the organs where it is needed most, routing blood away from other organs that can temporarily do with less. Epinephrine secreted during exercise or stress can also constrict vessels throughout the body and raise overall blood pressure.

- **Angiotensin II** is a potent vasoconstrictor that raises blood pressure. It is produced through the collaborative action of the liver, kidneys, and lungs (see fig. 16.9). An enzyme required for its synthesis is *angiotensin-converting enzyme (ACE)*. Hypertension is often treated with drugs called ACE inhibitors, which block this enzyme, thus lowering angiotensin II levels and blood pressure.

Certain other hormonal mechanisms can affect blood pressure by regulating urinary loss of water, which in turn affects blood volume.

- **Natriuretic peptides,** secreted by the heart when blood pressure is too high, stimulate the kidneys to excrete more sodium. Water follows by osmosis and is lost from the body, thereby lowering blood volume and pressure. Thus, the more natriuretic peptide secreted, the lower the BP will be.
- **Aldosterone** promotes sodium and water retention and thus raises BP. One way of lowering BP is for the adrenal glands to secrete less aldosterone and allow greater elimination of water from the body.
- **Antidiuretic hormone** also promotes water retention, and here again BP can be lowered by reducing ADH secretion and producing more urine.

The actions of these hormones on the kidneys will be further described in chapter 16.

13.5d Capillary Fluid Exchange

Capillaries are "the business end" of the cardiovascular system. The system would be useless if the blood could not release some materials to the tissues and pick up other materials from them. These exchange processes occur across the thin walls of the capillaries and smallest venules, but capillaries greatly outnumber venules and are the site of most fluid exchange.

Some ways for substances to pass across a vessel wall follow.

1. **Diffusion.** If a substance is more concentrated in the blood than in the surrounding tissue fluid and is capable of crossing the vessel wall, it will leave by diffusion; if it is more concentrated in the tissue fluid, it will tend to enter the blood by diffusion. Substances leaving the systemic blood in this manner include glucose, oxygen, and steroid hormones; substances picked up by the systemic blood in this way include carbon dioxide and other wastes. Oxygen, carbon dioxide, and steroids diffuse through the plasma membranes and cytoplasm of the endothelial cells; hydrophilic substances such as glucose and electrolytes diffuse through the clefts between the cells.
2. **Filtration.** This is a process in which the blood pressure forces fluid through the capillary wall, carrying solutes with it.
3. **Osmosis.** Because of the high concentration of sodium, protein, and erythrocytes in the blood, capillaries have a strong tendency to absorb water from surrounding tissues by osmosis. Solutes dissolved in the water may flow along with it into the blood; this is called *solvent drag*.

Near the arterial end, a typical capillary tends predominantly to give off fluid to the tissues, delivering vital materials to the cells such as oxygen and nutrients. At the venous end, it tends predominantly to absorb fluid, picking up and carrying away such materials as metabolic wastes and excess water (**fig. 13.15**). How can a single capillary do both?

Clinical Application 13.2

CONGESTIVE HEART FAILURE

In certain conditions, the heart fails to pump with sufficient strength to adequately supply blood to tissues and organs. People with heart failure feel fatigued and can't exercise normally because cardiac output fails to keep pace with demands of exercising muscles, for example. Heart failure is often associated with persistent high blood pressure, which causes the heart to work harder to push the blood through the vessels. Over time, the heart wall stretches and the heart becomes noticeably enlarged, but the thin walls contract only weakly. The kidneys respond to heart failure by retaining water and salt, thereby increasing blood volume and blood pressure. The elevated pressure causes excess fluid to filter out of the blood and often to accumulate in the ankles, feet, and lungs, causing edema. The body becomes "congested" with fluid, explaining why the term **congestive heart failure** is used to describe this condition. Heart failure can't be reversed but symptoms can be relieved with the use of diuretics, which act by increasing the volume of urine and thereby reducing fluid in the body. Other medications include drugs that increase the force of cardiac contraction.

Figure 13.15 Capillary Fluid Exchange. (a) At the arterial end of a capillary, blood pressure overrides osmosis and fluid filters from the capillary into the tissues. (b) At the venous end, blood pressure is lower and is now overridden by osmosis, which results in net fluid uptake by the capillary. (c) Excess fluid not reabsorbed by the blood capillary is absorbed from the tissues by a lymphatic capillary. **APR**

When blood enters a capillary from an arteriole, it is under relatively high pressure (typically about 30 mm Hg). This tends to force fluid out through the capillary wall. At the same time, however, the capillary tends to absorb water by osmosis. But since the outward-acting blood pressure overrides the inward-acting osmotic pressure, the effect in most capillaries is for more fluid to leave the capillary at this end than to

Figure 13.16 Venous Valves and the Skeletal Muscle Pump. (a) Muscle contraction squeezes the deep veins and forces blood through the next valve in the direction of the heart. Valves below the point of compression prevent backflow. (b) When the muscles relax, blood flows back downward under the pull of gravity but can flow only as far as the nearest valve.

(a) Contracted skeletal muscles (b) Relaxed skeletal muscles

reenter it. But blood pressure drops along the length of the capillary because of resistance to flow, and by the time blood reaches the venous end of a capillary, it is about 13 mm Hg. Now osmosis overrides the outward blood pressure, so the venous end of a capillary tends to absorb more fluid from the tissues than it gives off.

Typically, though, a capillary reabsorbs only about 85% as much fluid as it filters into the tissue. The excess 15% is reabsorbed by lymphatic vessels, discussed in chapter 14. If the amount of fluid leaving the capillaries is greater than the total amount reabsorbed by the blood capillaries and lymphatics, however, then fluid builds up in the tissue, causing swelling or **edema.**

13.5e Venous Return

After the blood has done its job of exchanging material with the tissues, it has to return to the heart to be repressurized. Flow back to the heart is called **venous return.** Pressure generated by the heartbeat is the most important force that returns blood to the heart, even though pressure is much lower in the veins than it is in arteries. For a person in a sitting or standing position, gravity is also a significant force aiding or opposing venous return. It aids blood in the head and neck—anywhere above the heart—to flow down the superior vena cava into the heart. However, it opposes the venous return of blood from points below the heart such as the abdomen and lower limbs. The pressure gradient generated by the heart isn't sufficient to force all this blood to flow uphill back to the heart, against the pull of gravity. The heart needs help, and it gets this in two main ways.

One is the **skeletal muscle pump** (fig. 13.16). As muscles of the limbs contract and relax, they squeeze the blood vessels that lie among them, promoting a flow of blood. Medium veins in the limbs are equipped with one-way valves. These valves point upward, so as muscles compress a vein, blood can flow upward toward the heart; but when the muscles relax, the valves prevent it from flowing back down toward the hands and feet. Muscular activity therefore causes blood to flow upward, bit by bit, with the valves acting as checkpoints to prevent backflow. However, in many people—especially those who must stand for long periods on their jobs—blood accumulates in the veins and stretches them until the valve cusps cannot meet in the middle. The valves then allow blood to flow back down. This distension of the veins can become permanent, creating **varicose veins** that are often visible through the skin.

The other aid to venous return is the **thoracic pump.** When you inhale, your chest expands and pressure in the thoracic cavity drops below the pressure in your abdominal cavity. The pressure difference squeezes the abdominal portion of the inferior vena cava, pushing the blood upward to the chest like ketchup being squeezed out of a plastic container. Pressure differences between abdominal and thoracic veins help blood flow upward to the heart, which is one reason why exercise and deep breathing are good for circulation.

When the heart fails to pump enough blood to meet the demands of the organs, a person enters a state of **circulatory shock.** One cause of circulatory shock is loss of blood volume by hemorrhaging. Other causes include tumors that compress veins and block the flow of blood, and **venous pooling,** in which blood accumulates in the lower parts of the body instead of returning to the heart. Venous pooling shock may be caused by allergic reactions that result in widespread vasodilation, as in allergies to peanuts or bee stings. Standing still for too long (as in choir or the military) so the skeletal muscle pump is not working can cause blood to pool in the legs, which may lead to fainting (**syncope**) (SIN-co-pee). In many cases the body can recover from mild shock on its own, as when one recovers from a faint, but in more severe cases a person can die from circulatory shock.

Before You Go On

Answer these questions from memory to test your understanding of the previous section. Reread the material if you can't answer the questions.

17. How does a sphygmomanometer measure blood pressure? In a BP such as 120/80 mm Hg, what do the top and bottom numbers represent and what does "mm Hg" mean?
18. List the three main determinants of blood pressure.
19. How do viscosity, vessel length, and vessel radius influence resistance and flow? Why is radius the most important of these?
20. Describe the mechanisms of local, neural, and hormonal control of blood flow. Identify the hormones that affect blood pressure, and explain how they work.
21. Explain how the skeletal muscle pump and thoracic pump contribute to venous return.

13.6 Circulatory Routes and Blood Vessels

Expected Learning Outcomes

When you have completed this section, you should be able to

a. trace the route of blood through the pulmonary circuit;
b. identify the principal systemic arteries and veins of the body; and
c. trace the systemic flow of blood from the heart to any major organ or body region and back to the heart, naming the vessels through which it would travel.

We conclude this chapter with a survey of the major arteries and veins and the routes taken by circulating blood. This discussion is confined to adult circulation; special features of fetal circulation and how it changes after birth are shown in figure 20.8.

13.6a The Pulmonary Circuit

The pulmonary circuit (**fig. 13.17**) serves only to exchange CO_2 for O_2. The lungs receive a separate systemic blood supply, via the *bronchial arteries,* to nourish their tissues. The pulmonary circuit begins with the **pulmonary trunk** arising from the right ventricle. The trunk branches into right and left **pulmonary arteries,** and these give off further branches just before or after entering each lung. Ultimately these arteries lead to a web of capillaries around each pulmonary air sac, or alveolus. Their relationship to the alveolus and the exchange of O_2 and CO_2 are detailed in chapter 15. Leaving the alveolar capillaries, the blood flows into pulmonary venules and larger veins until finally, two **pulmonary veins** from each lung enter the left atrium of the heart (see fig. 13.3). After birth, the pulmonary circuit is the only route in which the arteries carry oxygen-poor blood and the veins carry oxygen-rich blood, the opposite of the systemic system.

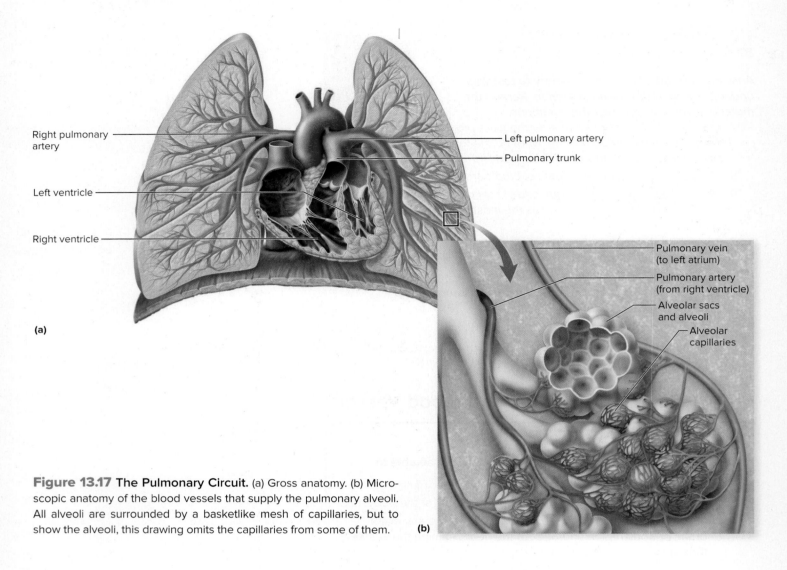

Figure 13.17 The Pulmonary Circuit. (a) Gross anatomy. (b) Microscopic anatomy of the blood vessels that supply the pulmonary alveoli. All alveoli are surrounded by a basketlike mesh of capillaries, but to show the alveoli, this drawing omits the capillaries from some of them.

13.6b The Systemic Circuit

The systemic circuit (figs. 13.18 and 13.19) supplies oxygen and nutrients to all organs and removes their metabolic wastes. Part of it, the *coronary circulation,* was described earlier in this chapter. This section surveys the other major arteries and veins region by region, from head to foot. It must be noted that there is a great deal of variation from person to person in the circulatory system; this description outlines only the most common circulatory pathways.

As we trace the arteries and veins, we will do so in the direction the blood flows. That is, in describing arteries, we begin near the heart and trace them distally, whereas in describing veins, we will begin distally and trace them approaching the heart. In describing the smaller divisions of larger arteries, we speak of them as *branches.* In describing the smaller divisions of larger veins, however, we speak of them as *tributaries,* because like the streams that form the tributaries of a river, they contribute blood to the larger vessel.

The names of blood vessels often give clues to their location or destination. For example, the *brachial artery* runs through the arm region; the *ulnar* and *femoral arteries* are named for adjacent bones; and the *renal vein* drains the kidney. Adjacent veins and arteries often have similar names (*femoral artery* and *femoral vein,* for example). When an artery or vein is unusually short and quickly

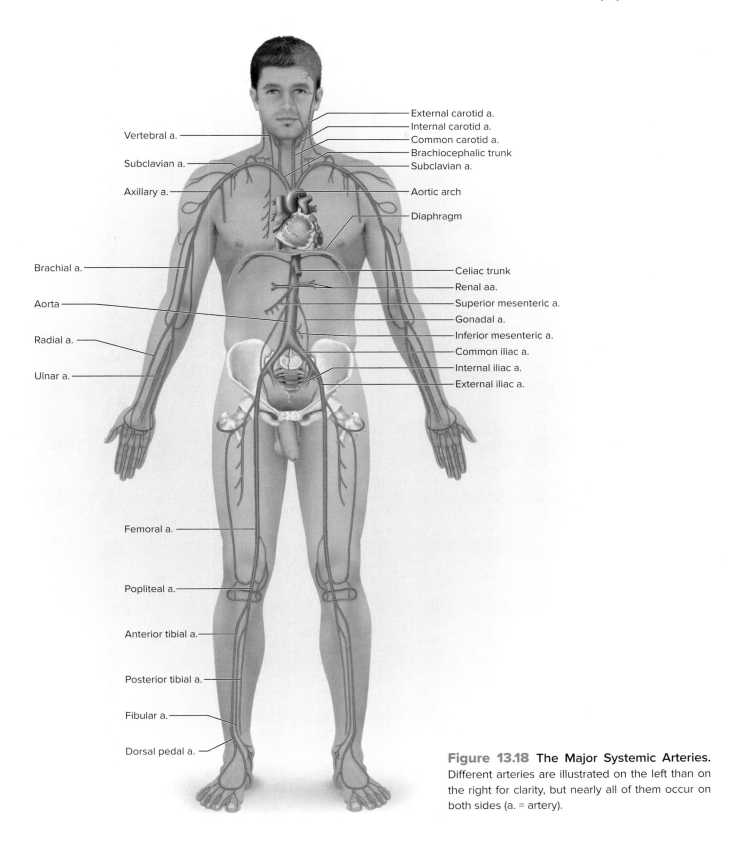

Figure 13.18 The Major Systemic Arteries. Different arteries are illustrated on the left than on the right for clarity, but nearly all of them occur on both sides (a. = artery).

Figure 13.19 The Major Systemic Veins. Different veins are illustrated on the left than on the right for clarity, but nearly all of them occur on both sides (v. = vein, vv. = veins).

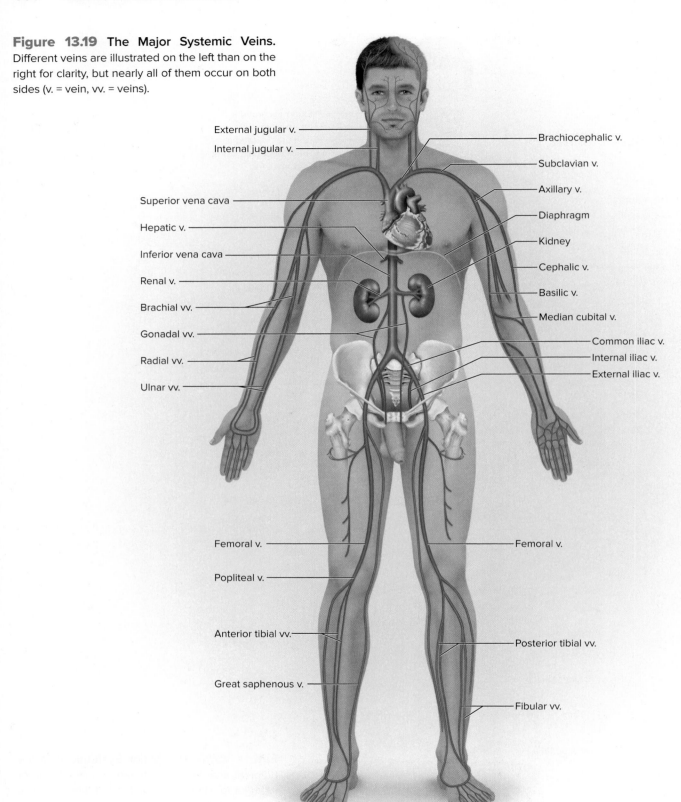

branches, it is called a *trunk*. It is important that you consult the illustrations as you read the text, just as you use a map to visualize the branches of highways and city streets. A verbal description alone doesn't allow effective navigation of the circulatory landscape. On the figures, the abbreviations *a.* and *aa.* mean *artery* and *arteries*, and *v.* and *vv.* mean *vein* and *veins*.

Initial Segments and Branches of the Aorta

The aorta is the body's largest artery. It is divided into three portions—ascending aorta, aortic arch, and descending aorta (fig. 13.20).

- **Ascending aorta.** The ascending aorta begins at the left ventricle and rises for a short distance from the heart. Its only branches are the *coronary arteries* to the heart wall, described earlier.
- **Aortic arch.** The *aortic arch* is the superior ∩-shaped segment, like the handle on a cane. It gives off three major arteries in the following order:
  - **Brachiocephalic[13] trunk** (BRAY-kee-oh-seh-FAL-ic). This short vessel rises from the right side of the aortic arch and soon branches into the *right common carotid* and *right subclavian arteries* described later.
  - **Left common carotid artery** (cah-ROT-id). Unlike the right common carotid, the left one arises directly from the aortic arch.
  - **Left subclavian[14] artery** (sub-CLAY-vee-un). Unlike the right subclavian, the left one arises directly from the aortic arch. This is the third and last artery to arise from the arch before the aorta begins its descent.
- **Descending aorta.** The *descending aorta* passes downward behind the heart, near the vertebral column, through the thoracic and abdominal cavities. It is

---

[13]*brachio* = arm; *cephal* = head
[14]*sub* = below; *clavi* = clavicle, collarbone

Figure 13.20 The Thoracic Aorta and Its Branches. (L. = left, R. = right, a. = artery, aa. = arteries) APR

Figure 13.21 Arteries of the Head and Neck.
(a) Superficial arteries. (b) Base of the brain showing the blood supply to the brainstem, cerebellum, and cerebral arterial circle. (a. = artery) **APR**

❓ *List the arteries, in order, that an erythrocyte must travel to get from the left ventricle to the skin of the left side of the forehead.*

called the *thoracic aorta* above the diaphragm before it passes through an opening called the aortic hiatus. Below the diaphragm it is called the *abdominal aorta*. It ends in the lower abdominal cavity by forking into the lower limbs.

Arteries of the Head and Neck

The head and neck receive blood primarily from two arteries on each side (**fig. 13.21a**).

- **Common carotid arteries.** The *right common carotid artery* splits from the brachiocephalic trunk, whereas the *left common carotid artery* arises directly from the aortic arch. The common carotids pass up the front of the neck, alongside the trachea, where their pulse can be palpated.

- **Vertebral arteries.** These arise from the right and left subclavian arteries and travel up the back of the neck through the transverse foramina of vertebrae C1 through C6. They enter the cranial cavity through the foramen magnum, bound for the brain.

The common carotid arteries have the most extensive distribution of all the head–neck arteries. Near the *laryngeal prominence* ("Adam's apple"), each common carotid splits into external and internal branches.

- **External carotid artery.** This branch ascends the side of the head and provides branches to the thyroid gland, larynx, tongue, teeth, scalp, facial skin and muscles, and chewing muscles.

- **Internal carotid artery.** This branch passes medial to the angle of the mandible and enters the cranial cavity through the carotid canal of the temporal bone. It supplies the eye orbits and about 80% of the cerebrum. Compressing the internal carotids near the mandible can therefore cause loss of consciousness.

The vertebral arteries give rise to branches that supply the cervical vertebrae, spinal cord and its meninges, and deep muscles of the neck. They then enter the foramen magnum and continue as follows (fig. 13.21b):

- **Basilar artery.** Within the cranial cavity, the vertebral arteries converge to form a single median *basilar artery* along the anterior surface of the brainstem. The basilar artery gives off branches to the cerebellum, pons, and inner ear, then divides and flows into the cerebral arterial circle.
- **Cerebral arterial circle (circle of Willis).** The basilar and internal carotid arteries flow into this loop of vessels surrounding the pituitary gland. Most people lack one or more components of this circle; only 20% have a complete arterial circle. The circle gives off *anterior, posterior,* and *middle cerebral arteries* that provide the most significant blood supplies to the cerebrum. Blood supply to the brain is so critical that it is furnished by several arterial anastomoses, so if blood flow along one route is cut off, there are alternative routes through the arterial circle to reach the same brain tissue. Knowledge of the distribution of the arteries arising from the circle is crucial for understanding the effects of blood clots, aneurysms, and strokes on brain function.

Veins of the Head and Neck

- **Dural sinuses.** After blood circulates through the brain tissue, it collects in large thin-walled veins called *dural sinuses*—blood-filled spaces between the layers of the dura mater (see fig. 9.3). The most superficial of these is the *superior sagittal sinus,* which courses along the top of the head just beneath the cranium along the median plane. At the rear of the head, a pair of *transverse sinuses* arches around toward the ears just under the cranial bone (fig. 13.22b). They collect blood from all the other dural sinuses and then exit the cranium through the *jugular foramen* just posterior to each ear.

Blood flows down the neck mainly through three veins on each side, all of which empty into the subclavian vein.

- **Internal jugular[15] veins** (JUG-you-lur). These are continuations of the transverse sinuses after the vessels exit the cranium. They travel deep to the sternocleidomastoid muscles. They also pick up blood from the face, temporal region, and thyroid gland before emptying into the subclavian vein (fig. 13.22a).
- **External jugular veins.** These are smaller veins on the superficial surface of the sternocleidomastoids. They too pass down the neck, draining tributaries from the facial muscles, scalp, and other superficial structures before ending at the subclavian veins.

Arteries of the Thorax

The thoracic aorta issues numerous branches to the vertebrae, spinal cord and its meninges, breasts, lungs, pleurae, pericardium, esophagus, diaphragm, and skin and muscles of the thoracic wall. In the shoulder region, the *subclavian artery* adjacent to the clavicle and the *axillary artery* in the region of the axilla (armpit) give off several branches to the breast, scapula, intercostal muscles, and muscles

---

[15]*jugul* = neck, throat

Figure 13.22 Veins of the Head and Neck. (a) Superficial veins. (b) Base of the brain showing some of the dural sinuses and other drainage from the cerebrum. (Not all of these veins and sinuses are discussed in the text.) (v. = vein)

Figure 13.23 Arteries of the Thorax and Upper Limb. (a. = artery, aa. = arteries) APR

of the pectoral, shoulder, and arm regions. These thoracic arteries are illustrated in figure 13.23, but their names and further description are beyond the scope of this book. Arteries to the breast are larger in females than in males, and enlarge still further during lactation.

Veins of the Thorax

The first three veins below return blood from the shoulder region to the heart, and the azygos system drains the wall and viscera of the thorax (fig. 13.24).

- **Subclavian vein.** This vessel drains the upper limb. It runs with its companion, the subclavian artery, posterior to the clavicle. It receives the external

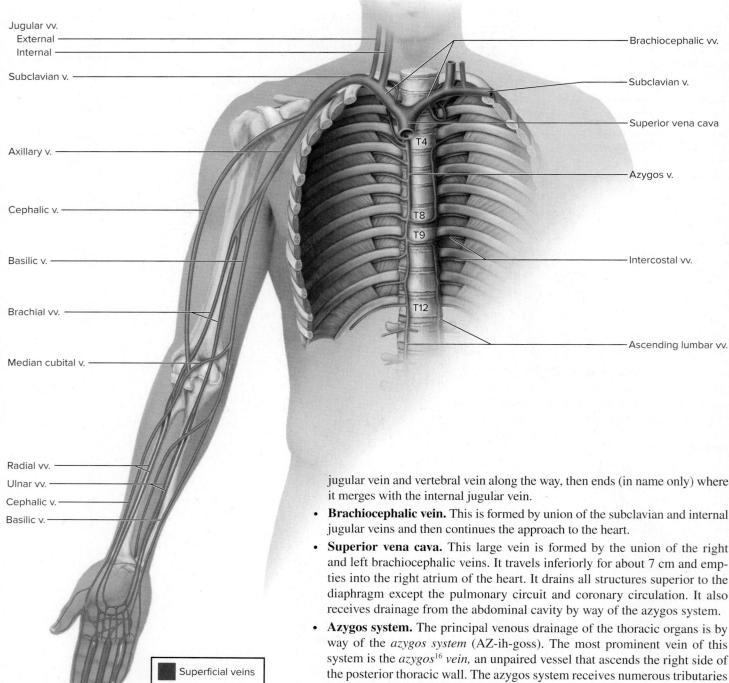

Figure 13.24 Veins of the Thorax and Upper Limb. Note color coding of deep and superficial veins in the upper limb (v. = vein, vv. = veins). APR

jugular vein and vertebral vein along the way, then ends (in name only) where it merges with the internal jugular vein.

- **Brachiocephalic vein.** This is formed by union of the subclavian and internal jugular veins and then continues the approach to the heart.
- **Superior vena cava.** This large vein is formed by the union of the right and left brachiocephalic veins. It travels inferiorly for about 7 cm and empties into the right atrium of the heart. It drains all structures superior to the diaphragm except the pulmonary circuit and coronary circulation. It also receives drainage from the abdominal cavity by way of the azygos system.
- **Azygos system.** The principal venous drainage of the thoracic organs is by way of the *azygos system* (AZ-ih-goss). The most prominent vein of this system is the *azygos*[16] *vein,* an unpaired vessel that ascends the right side of the posterior thoracic wall. The azygos system receives numerous tributaries from many thoracic organs and tissues. A series of *intercostal veins* between the ribs empty in a ladderlike series into the azygos system. This system also receives blood from as far away as the lower limbs and lower abdomen.

Arteries of the Upper Limb APR

In the limbs (appendicular region), the arteries are usually deep and well protected. The upper limb is supplied by a prominent artery that changes name along its course from *subclavian* to *axillary* to *brachial,* then issues branches to the arm, forearm, and hand (fig. 13.23).

---

[16]unpaired; from *a* = without; *zygo* = union, mate

- **Subclavian artery.** The *left subclavian artery* arises from the brachiocephalic trunk and the *right subclavian artery* arises directly from the aortic arch, as described earlier. Each subclavian artery passes posterior to the clavicle, downward over the first rib, and ends in name only at the rib's lateral margin.
- **Axillary artery.** The subclavian changes name to *axillary artery* as soon as it passes the first rib. It continues through the axillary region (armpit) and ends, again in name only, at the neck of the humerus. Beyond this, it is called the brachial artery.
- **Brachial artery** (BRAY-kee-ul). This artery continues from the *axillary artery* down the medial and anterior sides of the humerus and ends just distal to the elbow, supplying the anterior flexor muscles of the arm along the way. At its end, it branches into the next two arteries.
- **Radial artery.** This branch descends the forearm laterally, alongside the radius, nourishing the lateral forearm muscles. The most common place to take a pulse is at the radial artery just proximal to the thumb.
- **Ulnar artery.** This branch descends medially through the forearm, alongside the ulna, nourishing the medial forearm muscles. The radial and ulnar arteries join at the wrist to form arterial arches that issue smaller arteries to the palmar region and fingers.

Veins of the Upper Limb APR

Veins of the appendicular region occur in both deep and superficial groups; you may be able to see several of the superficial ones in your forearms and hands. Veins of the upper limb lead ultimately to the axillary and subclavian veins (fig. 13.24). Beginning distally with the veins that drain the hand, and following the direction of blood flow toward the heart, the most prominent superficial veins are as follows:

A superficial venous network, often visible on the back of the hand, drains the palmar region and fingers and gives rise to the cephalic and basilic veins.

- **Cephalic[17] vein** (sef-AL-ic). This vein begins on the lateral side of the hand, travels up the lateral side of the forearm and arm to the shoulder, and joins the *axillary vein* there. Intravenous fluids are often administered through the distal end of the cephalic vein.
- **Basilic[18] vein** (bah-SIL-ic). This vein begins on the medial side of the hand and travels up the posterior side of the forearm, continuing into the arm. It turns deeper about midway up the arm and joins the *brachial vein* near the axilla. As an aid to remembering which vein is cephalic and which is basilic, visualize your arm held straight away from the torso (abducted) with the thumb up. The cephalic vein runs along the upper side of the arm closer to the head (as suggested by *cephal*, "head"), and the name *basilic* is suggestive of the lower (basal) side of the arm (although the word really means "prominent").
- **Median cubital vein.** This is a short connection between the cephalic and basilic veins that obliquely crosses the cubital fossa. It is often clearly visible through the skin and is a common site for drawing blood.

Deeper venous arches in the hand also receive blood from the fingers and palmar region and give rise to the first two veins below.

- **Radial veins.** A pair of *radial veins* arises from the lateral side of the hand and courses up the forearm alongside the radius. They join near the elbow and form one of the brachial veins.

---

[17]*cephalic* = related to the head
[18]*basilic* = royal, prominent, important

- **Ulnar veins.** A pair of *ulnar veins* runs from the hand up the forearm alongside the ulna. They unite near the elbow to form the other brachial vein.
- **Brachial veins.** The two *brachial veins* continue up the arm, flanking the brachial artery, and converge into a single vein just before the axillary region.
- **Axillary vein.** This vein arises by the union of the brachial and basilic veins. It passes through the axillary region, picking up the cephalic vein along the way. At the lateral margin of the first rib, its name changes to the *subclavian vein*. We have already traced the course of the subclavian vein in discussion of veins of the thorax.

Arteries of the Abdominopelvic Region

The descending aorta passes through an opening in the diaphragm and continues through the abdomen to the level of vertebra L4, where it branches into right and left *common iliac arteries* (fig. 13.25a). Along its descent, it gives rise to the following branches in the order listed. Those indicated in the plural are paired right and left, and those indicated in the singular are solitary median arteries.

- **Celiac trunk.** The short, stubby *celiac*[19] (SEE-lee-ac) trunk supplies the upper abdominal viscera. It goes for only about 1 cm before dividing into three branches: the *common hepatic artery* to the liver, *splenic artery* to the spleen, and *left gastric artery* to the stomach. These arteries and their branches also supply the gallbladder, pancreas, lower esophagus, and greater omentum (fig. 13.25b).
- **Superior mesenteric artery.** This is the most significant intestinal blood supply. It arises from the aorta, not far below the celiac trunk, and fans out in the mesentery to supply nearly all of the small intestine and the proximal half of the large intestine (fig. 13.25c).
- **Renal arteries.** Right and left *renal arteries* supply the kidneys.
- **Gonadal arteries.** More specifically called *ovarian arteries* in the female and *testicular arteries* in the male, these long, slender arteries descend along the posterior body wall to the female pelvic cavity or male scrotum, where they supply the ovaries or testes.
- **Inferior mesenteric artery.** This artery arises medially from the lower aorta and fans out within the mesentery to supply the distal half of the large intestine.
- **Common iliac arteries.** Finally, the aorta forks into these two short arteries at its inferior end.
- **Internal iliac artery.** The common iliac artery descends for about 5 cm along the posterior wall of the pelvic cavity, then divides in two. One division, the *internal iliac artery*, supplies mainly the pelvic wall, gluteal muscles, hip joint, urinary bladder, rectum, the uterus and part of the vagina in women, and the prostate in men. The other division, the *external iliac artery*, supplies the lower limb and is discussed later.

Veins of the Abdominopelvic Region

We will trace the veins in ascending order, following the blood flow back toward the heart.

- **Internal iliac vein.** This vein drains the gluteal muscles, the medial aspect of the thigh, the urinary bladder and rectum, the prostate of the male, and the uterus and vagina of the female.
- **Common iliac vein.** This vein is formed by the union of the internal iliac and an external iliac, described later, which ascends from the lower limb.

---

[19]*celi* = belly, abdomen

- **Inferior vena cava (IVC).** The IVC is the body's largest blood vessel, having a diameter of about 3.5 cm. It is a single median vein formed by the union of the right and left common iliac veins at the level of vertebra L5 (fig. 13.26a). The IVC picks up blood from several veins that drain the lumbar body wall, and the following tributaries, among others. At its superior end, it penetrates the diaphragm and immediately enters the right atrium of the heart from below. It does not receive any thoracic drainage.
- **Gonadal veins.** Like the corresponding arteries, these are called *ovarian veins* in the female and *testicular veins* in the male. They are long, slender veins that originate in the ovaries or testes and climb the posterior abdominal wall. The right gonadal vein empties into the IVC whereas the left gonadal vein empties into the left renal vein.
- **Renal veins.** These veins drain the kidneys into the IVC.

The venous drainage of much of the digestive system forms the **hepatic[20] portal system** (fig. 13.26b). This network of veins receives all of the blood draining from the abdominal digestive tract as well as from the pancreas, gallbladder, and spleen. It is called a *portal system* because it connects capillaries of the intestines and other digestive organs to capillaries in the liver; thus, the blood passes through two capillary beds before it returns to the heart. Intestinal blood is richly laden with nutrients for a few hours following a meal. The hepatic portal system gives the liver first claim to these nutrients before the blood is distributed to the rest of the body. It also allows the blood to be cleansed of bacteria and toxins picked up from the intestines, an important function of the liver. The main components of the hepatic portal system follow. **APR**

- **Inferior mesenteric vein.** This vein arises from a fanlike array of small veins in the mesentery that drain the rectum and distal part of the colon.

---

[20]*hepat* = liver; *ic* = pertaining to

Figure 13.25 Arterial Supply to the Abdominopelvic Region. (a) The abdominal aorta and its branches. (b) Branches of the celiac trunk arising from the upper aorta. (c) Branches of the superior mesenteric artery supplying the intestines. (a. = artery; aa. = arteries) **APR**

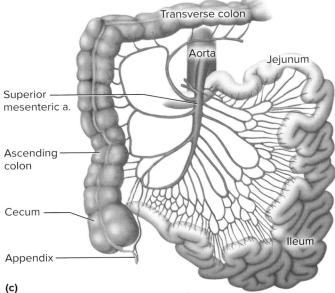

454 PART 4 Circulation and Defense

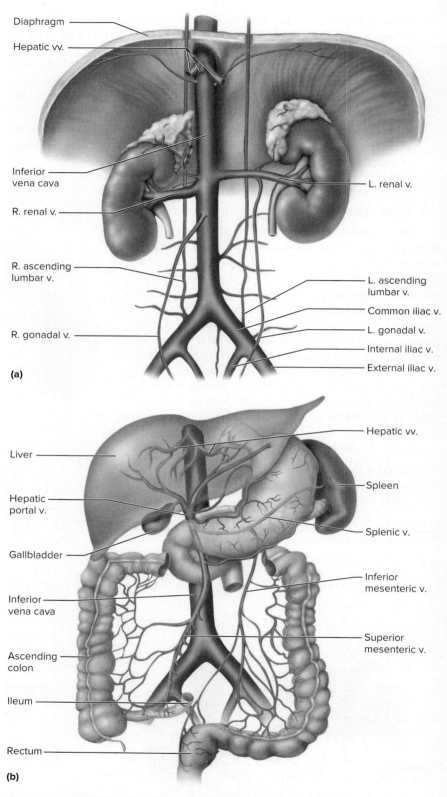

Figure 13.26 Venous Drainage of the Abdominopelvic Region. (a) Anatomy of the inferior vena cava and its tributaries. (b) The hepatic portal system, which drains blood from the major digestive organs and others into the liver. (v. = vein, vv. = veins)

- **Superior mesenteric vein.** This vein begins with a similar fanlike array of veins in the mesentery, but these drain the entire small intestine, the ascending and transverse colon, and stomach.
- **Splenic vein.** This vein drains the spleen and travels across the abdominal cavity toward the liver. Along the way, it picks up veins from the pancreas, then the inferior mesenteric vein, and ends where it meets the superior mesenteric vein.
- **Hepatic portal vein.** This vein is formed by convergence of the splenic and superior mesenteric veins. It enters the inferior surface of the liver and leads ultimately to capillaries called the *hepatic sinusoids*, detailed in chapter 17.
- **Hepatic veins.** Blood from the liver drains into a few short hepatic veins that exit its superior surface and pass only a very short distance before ending high on the inferior vena cava.

Arteries of the Lower Limb APR

As we have already seen, the aorta forks at its lower end into the right and left *common iliac arteries,* and each of these soon divides again into an *internal* and *external iliac artery*. We discussed the internal iliac artery earlier, and we now trace the external iliac artery as it descends into the lower limb (see fig. 13.18).

- **External iliac artery.** The *external iliac artery* passes behind the *inguinal ligament* at the top of the thigh; its continuation is then named the *femoral artery*.
- **Femoral artery.** The *femoral artery* passes superficially through the groin in a region called the *femoral triangle*. Its pulse can be palpated in the upper medial thigh, and this is an important emergency pressure point for stopping arterial bleeding in the lower limb. It gives off several branches that supply the femur and the skin and muscles of the thigh before it descends to the knee.
- **Popliteal artery.** This is a continuation of the femoral artery in the popliteal fossa of the posterior side of the knee. Its branches supply the knee joint, patella, fibula, and some calf muscles. Just below the knee, the popliteal artery branches into the next two arteries.
- **Posterior tibial artery.** This is a continuation of the popliteal artery that passes down the posterior side of the leg, supplying flexor muscles along the way. Inferiorly, it passes behind the medial malleolus of the ankle and into the plantar region of the foot.
- **Anterior tibial artery.** This artery immediately penetrates the interosseous membrane between the upper tibia and fibula and emerges on the anterior side of the membrane. It travels to the ankle, supplying the extensor muscles of the leg along the way.
- **Dorsal pedal artery.** The inferior end of the anterior tibial artery gives rise to the *dorsal pedal artery* on the upper side of the foot. This is a useful site for palpating the pulse in infants and judging the adequacy of lower limb circulation (from strength of the pulse) in adults.

Veins of the Lower Limb APR

As in the upper limb, the lower limb has both superficial and deep veins (see fig. 13.19).

- **Great (long) saphenous**[21] **vein** (sah-FEE-nus). This is a prominent superficial vein and is the longest vein in the body. It arises from the medial side

---

[21]*saphen* = standing

of the foot and travels all the way up the leg and thigh to the inguinal region, where it empties into the *femoral vein* described shortly. It is commonly used as a site for the long-term administration of intravenous fluids; it is a relatively accessible vein in infants and in patients in shock whose veins have collapsed. Portions of this vein are often used as grafts in coronary bypass surgery.

Following are the deep veins of the lower limb:

- **Fibular (peroneal) veins.** Blood draining the foot enters a pair of *fibular veins* on the lateral side of the ankle. These veins run side by side, ascending the back of the leg, and converge like an inverted Y into a single vein.
- **Posterior tibial veins.** Similarly, blood from the foot drains into a pair of *posterior tibial veins* on the medial side of the ankle. These also ascend the leg side by side, embedded deep in the calf muscles. They converge like an inverted Y into a single vein about two-thirds of the way up the tibia.
- **Popliteal vein.** The two inverted Ys just described converge on each other near the knee to form the *popliteal vein*, which passes through the popliteal fossa.
- **Anterior tibial veins.** Two *anterior tibial veins* arise from the superomedial aspect of the foot and travel up the anterior compartment of the leg between the tibia and fibula. They converge just distal to the knee and then flow into the popliteal vein.
- **Femoral vein.** This is a continuation of the popliteal vein into the thigh. It drains blood from the deep thigh muscles and femur.
- **External iliac vein.** This vein is formed by the union of the femoral and great saphenous veins near the inguinal ligament. We already traced the course of the internal iliac vein of the pelvic region. The external and internal iliacs unite to form the *common iliac vein*, also described earlier.

Before You Go On

Answer these questions from memory to test your understanding of the previous section. Reread the material if you can't answer the questions.

22. Contrast the pathways and destinations of the external and internal carotid arteries.
23. Briefly state the organs or parts of organs that are supplied with blood by (a) the cerebral arterial circle, (b) the celiac trunk, (c) the superior mesenteric artery, and (d) the internal iliac artery.
24. If you were dissecting a cadaver, state where would you look to locate (a) the cephalic vein, (b) the dorsal pedal artery, and (c) the femoral artery and vein.
25. If you were dissecting a cadaver, where would you look for the internal and external jugular veins? What muscle would help you distinguish one from the other?
26. Trace one possible path of a red blood cell from the left ventricle to the foot.
27. State two reasons why the great saphenous vein is clinically significant.

Aging of the Circulatory System

Cardiovascular disease (CVD) is a leading cause of death in old age. As cardiac muscle atrophies with age, the heart wall becomes thinner and weaker, resulting in reduced cardiac output. The fibrous components of the heart stiffen, limiting cardiac expansion and weakening its contractions. Degenerative changes in the nodes and conduction pathways increase the risk of cardiac arrhythmias. In addition, coronary atherosclerosis is often found in aging hearts and increases the risk of myocardial infarction (see Perspectives on Health).

Systemic arteries stiffened by arteriosclerosis cannot expand effectively to accommodate the pressure surges of cardiac systole. Consequently, blood pressure may rise steadily with age. Atherosclerosis also narrows the arteries and reduces blood flow to such vital organs as the brain, kidneys, and the heart itself, weakening these organs. Reduced flow to skeletal muscles reduces physical endurance. Atherosclerotic plaques can become focal points of spontaneous blood clotting *(thrombosis)*, especially in the lower limbs, where blood flow is relatively slow and blood clots more easily. About 25% of people over 50 experience venous blockage by blood clots—especially people who don't exercise regularly to stimulate optimal blood flow. Degenerative changes in the veins are most evident in the lower limbs, where the venous valves become weaker and less able to stop the backflow of blood. Chronic stretching of the veins makes varicose veins and hemorrhoids more common in and beyond middle age.

Reduction of cardiovascular function is an inevitable part of aging. However, numerous studies show that exercise, while it doesn't stop the clock from ticking, slows it considerably. Exercise increases flexibility of arteries, lowers resting heart rate, and boosts the ability of the heart to pump blood efficiently. It has the added benefit of lowering blood pressure. In addition, exercise is associated with increased cerebral blood flow and enhanced cognitive function. Staying active, not smoking, and maintaining a normal body-mass index all increase the chance of having good cardiovascular health into old age.

CAREER SPOTLIGHT

Electrocardiographic Technician

© MCT/Getty Images

An *electrocardiographic (ECG or EKG) technician* prepares electrocardiograms (ECGs) for diagnostic, exercise testing, and other purposes. The ECG technician prepares the patient for the test by attaching electrodes to specific sites on the chest and limbs and monitors the equipment while results are recorded. One can become a certified ECG technician through programs at community colleges or vocational colleges. A typical course of training entails 4 months beyond high school and includes anatomy and physiology, medical terminology, interpretation of cardiac rhythms, patient-care techniques, cardiovascular medication, and medical ethics. Many people, however, become ECG technicians through on-the-job training rather than formal programs. Most employers prefer to train people who are already in a health-care profession, such as nurses' aides. With more advanced training, one may become a cardiovascular technologist and assist physicians in diagnosis, cardiac catheterization, echocardiography, and other more specialized skills and for correspondingly better salaries. For further information on a career as an ECG technician or cardiovascular technologist, see appendix B.

CONNECTIVE ISSUES

Ways in Which the CARDIOVASCULAR SYSTEM Affects Other Organ Systems

All Systems
The heart and blood vessels circulate the blood and distribute it throughout the body, delivering hormones and essentials such as nutrients and oxygen, and removing wastes. Capillary filtration and osmosis maintain fluid balance in all organs.

Integumentary System
The routing of blood to and away from the skin is vital to maintaining body temperature.

Skeletal System
The bloodstream picks up RBCs, WBCs, and platelets from the red bone marrow and delivers the hormones that regulate the production of these formed elements. It also provides minerals for bone deposition and delivers hormones that regulate the metabolism of osseous tissue.

Muscular System
Vasodilation in the muscles provides the added oxygen and energy substrates required for exercise and removes the metabolites and heat generated by the muscles.

Nervous System
Endothelial cells of the cerebral blood vessels produce the blood–brain barrier; capillary filtration in the choroid plexuses of the brain produces cerebrospinal fluid; and strokes from cerebral hemorrhage are a leading cause of death.

Endocrine System
The bloodstream transports all hormones, and includes the portal system of blood vessels for hypothalamo–pituitary communication.

Lymphatic and Immune Systems
Capillary filtration produces tissue fluid, which becomes lymph; all lymph ultimately returns to the bloodstream at the subclavian veins; the bloodstream carries the leukocytes and plasma proteins involved in immunity.

Respiratory System
Capillary osmosis and low pulmonary blood pressure prevent the lungs from filling with fluid.

Urinary System
Capillary filtration is the first step in urine production, and capillary reabsorption carries away the water and solutes reabsorbed by the kidneys; kidney function is regulated by several blood-borne hormones.

Digestive System
Capillaries of the intestinal wall pick up and transport digested nutrients; special capillaries (sinusoids) of the liver cleanse the blood of bacteria and other impurities; the circulatory system reabsorbs and recycles bile acids and minerals from the intestines; and blood-borne hormones regulate gastrointestinal activity and appetite.

Reproductive System
The bloodstream transports all sex hormones; vasodilation produces penile erection, enabling intercourse and fertilization; and blood vessels in the scrotum act as a countercurrent heat exchange system that prevents overheating of the testes, which would otherwise halt sperm production.

Study Guide

Assess Your Learning Outcomes

To test your knowledge, discuss the following topics with a study partner or in writing, ideally from memory.

13.1 Overview of the Cardiovascular System

1. Definitions of *cardiovascular system* and *cardiology*
2. The purpose of the pulmonary and systemic circuits and what side of the heart supplies each
3. Description of the location and position of the heart, using precise anatomical terms
4. The relationship between the epicardium, pericardial sac, and pericardial cavity
5. The function of pericardial fluid

13.2 Gross Anatomy of the Heart

1. The three layers of the heart wall and the histology of each
2. Location and functions of the connective tissue of the heart
3. Surface features of the heart and how they relate to its internal structure
4. The reason for variations in wall thickness of different heart chambers
5. Similarities and differences in the structure and function of the atrioventricular valves compared to the semilunar valves
6. The flow of blood through the heart, from the venae cavae to the aorta, naming all chambers, vessels, and valves through which the blood passes
7. Why the heart receives a disproportionate amount of blood relative to its percentage of body mass
8. Names and pathways of the coronary blood vessels

13.3 Physiology of the Heart

1. Structurally distinctive properties of cardiac muscle and the reasons for its special features
2. Components of the cardiac conducting system and their locations
3. The origin and propagation of electrical excitation through the heart
4. The anatomical and functional relationship between the SA node and AV node
5. How and why the electrical activities of the SA node and cardiomyocytes differ from that of skeletal muscle
6. Sympathetic and parasympathetic innervation of the heart
7. The waves of the ECG and the electrical events associated with each
8. Definitions of *systole* and *diastole*
9. The key events in each phase of the cardiac cycle: electrical events, chamber contractions and relaxations, valve opening and closing, sources of the two heart sounds, correlations with the waves of the ECG, and movements of the blood
10. The meaning of *cardiac output* and the two factors that determine it
11. Neural and chemical factors that increase or reduce cardiac output and how they do so

13.4 General Anatomy of Blood Vessels

1. The layers of a blood vessel wall and their structural differences
2. The definitions of *arteries, capillaries,* and *veins*
3. Types or subclasses of arteries, capillaries, and veins; how they differ structurally, and the functional relevance of those differences
4. Functional significance of capillary beds and precapillary sphincters
5. The structure and functional significance of portal systems and anastomoses

13.5 Physiology of Circulation

1. Definitions of *flow* and *blood pressure*
2. The method of measuring blood pressure, and a typical healthy adult value
3. How peripheral resistance influences blood flow; the three variables that govern resistance; and why vessel radius is the most important of these variables
4. How a tissue can regulate its own blood supply independently of nerves and hormones
5. How baroreceptors and the vasomotor center regulate blood flow
6. Hormones that can raise and lower blood pressure, and how they do so
7. Mechanisms of capillary fluid exchange
8. How a capillary can give off fluid at one end and reabsorb it at the other end
9. Mechanisms that allow blood to return to the heart even against the pull of gravity
10. Definition and causes of *circulatory shock*

13.6 Circulatory Routes and Blood Vessels

1. The path of a blood cell from pulmonary trunk to left atrium, naming the blood vessels
2. Segments of the thoracic aorta, the arteries arising from each, and destinations of the blood in those arteries
3. Major arteries of the head and neck, including the routes of blood flow to the brain
4. Veins of the head, neck, and thorax, and the route that blood travels in these veins to the heart
5. Major arteries and veins of the upper limb
6. Major branches of the abdominal aorta and the structures they supply
7. Major abdominal veins and the organs or regions they drain
8. The general function and circulatory routes of the hepatic portal system
9. Major arteries and veins of the lower limb

Testing Your Recall

1. The cardiac conduction system includes all of the following *except*
   a. the SA node.
   b. the AV node.
   c. the tendinous cords.
   d. the bundle branches.
   e. the subendocardial branches.

2. The most prominent vessel draining blood from the anterior ventricular myocardium is
   a. the great saphenous vein.
   b. the anterior interventricular branch.
   c. the middle cardiac vein.
   d. the great cardiac vein.
   e. the not-so-great cardiac vein.

3. The outermost layer of the heart wall is known as
   a. the pericardial sac.
   b. the epicardium.
   c. the parietal pericardium.
   d. the tunica externa.
   e. the serosa.

4. Intestinal blood flows into the liver by way of
   a. the superior mesenteric vein.
   b. the hepatic portal vein.
   c. the abdominal aorta.
   d. the hepatic sinusoids.
   e. the hepatic veins.

5. Ventricular contraction begins at the time of
   a. opening of the AV valves.
   b. the P wave.
   c. the T wave.
   d. the QRS complex.
   e. the second heart sound.

6. Janet is allergic to bee stings. One day, she was stung and her blood pressure dropped so low that she lost consciousness and required emergency treatment. Janet had experienced
   a. myocardial infarction.
   b. cerebral vascular accident.
   c. venous pooling shock.
   d. ventricular fibrillation.
   e. hemorrhagic shock.

7. Which of the following is most responsible for speeding up the heart rate during exercise?
   a. vasoconstriction of peripheral blood vessels
   b. norepinephrine released from sympathetic nerves
   c. acetylcholine released from vagus nerves
   d. atrial natriuretic peptide released from the heart
   e. release of histamine

8. Michael has a tumor that grows to the size of a grapefruit and compresses his external iliac artery. This will result in greatly reduced blood flow to
   a. his lower limb.
   b. his brachial region.
   c. his cerebrum.
   d. his kidneys.
   e. his small and large intestines.

9. What event in the heart corresponds to the first heart sound?
   a. the P wave
   b. the T wave
   c. diastole
   d. closing of the semilunar valves
   e. closing of the AV valves

10. Which type of blood vessels is up to 2.5 cm in diameter and has an abundance of elastic tissue?
    a. continuous capillaries
    b. distributing arteries
    c. venules
    d. large veins
    e. conducting arteries

11. The contraction of any heart chamber is called _____ and its relaxation is called _____.

12. The instrument that measures blood pressure is the _____.

13. A large unpaired artery from the superior abdominal aorta called the _____ supplies the liver, pancreas, stomach, and spleen.

14. The _____ nerves innervate the heart and serve to reduce heart rate.

15. Most of the blood supply to the brain comes from a complex of arteries around the pituitary gland called the _____.

16. The clinical term for high blood pressure is _____.

17. Mechanoreceptors that respond to pressure changes in major arteries near the heart are called _____.

18. Death of cardiac tissue from lack of blood flow is commonly known as a heart attack, but clinically called _____.

19. Repolarization of the ventricles produces the _____ of the electrocardiogram.

20. The _____ is an artery whose branches supply parts of both ventricles and the anterior portion of the interventricular septum.

Answers in Appendix A

What's Wrong with These Statements?

Briefly explain why each of the following statements is false, or reword it to make it true.

1. The blood supply to the myocardium is the coronary circulation; everything but this is called the systemic circuit.

2. The first branch off the aortic arch is the right subclavian artery.

3. Blood pressure can be lowered by increasing the secretion of aldosterone.

4. The pulmonary veins carry blood from the heart to the lungs.

5. Arteries have a series of valves that ensure a one-way flow of blood.

6. If the radius of a blood vessel doubles and all other factors remain the same, blood flow through that vessel will also double.

7. In the hepatic portal system, blood passes through one capillary bed before the blood returns to the heart.

8. An electrocardiogram is a tracing of the action potential of a single cardiomyocyte.

9. There are two ways to increase cardiac output: Increased heart rate and increased blood pressure.

10. The tendinous cords extend from the cusps of the semilunar valves to the papillary muscles in the ventricles.

Answers in Appendix A

Testing Your Comprehension

1. Mr. Jones, 78, dies of a massive myocardial infarction triggered by coronary thrombosis. Upon autopsy, necrotic myocardium is found in the lateral and posterior right ventricle and posterior interventricular septum. Based on the information in this chapter, where in the coronary circulation do you think the thrombosis occurred?

2. Marco and Paul are college roommates. Marco is on the school water polo team, and spends at least an hour a day exercising. Paul does not regularly exercise, and usually plays video games when he is not studying. Marco's resting heart rate is 52 beats per minute, on average, whereas Paul's average heart rate is 75 beats per minute. At rest, Marco's stroke volume (SV) is 100 mL/beat; Paul's is 70 mL/beat. Calculate the resting cardiac output (CO) for Marco and Paul. Explain anatomical and physiological reasons Paul has a higher heart rate.

3. Diuretics are drugs that increase urine output, and are often used to treat hypertension. Explain why they would have the desired effect of lowering blood pressure.

Chapter 14

The Lymphatic System and Immunity

Two natural killer (NK) cells (orange) attacking a cancer cell (red). NK cells are an important defense against cancer, often destroying malignant cells before they can seed the growth of a tumor.
Eye of Science/Science Source

Chapter Outline

14.1 The Lymphatic System
- 14.1a Functions of the Lymphatic System
- 14.1b Lymphatic Vessels and Lymph
- 14.1c Lymphatic Tissues
- 14.1d Lymphatic Organs

14.2 Innate Immunity
- 14.2a Physical Barriers
- 14.2b Protective Cells
- 14.2c Protective Proteins
- 14.2d Protective Processes

14.3 Features of Adaptive Immunity
- 14.3a Antigens
- 14.3b Lymphocytes
- 14.3c Antibodies
- 14.3d Antigen-Presenting Cells

14.4 Cellular and Humoral Immunity
- 14.4a Cellular Immunity
- 14.4b Humoral Immunity
- 14.4c Active and Passive Immunity

Clinical Applications/Perspectives on Health
- Clinical Application 14.1: Lymph Nodes and Metastatic Cancer
- Clinical Application 14.2: Hypersensitivity, Allergies, and Anaphylactic Shock
- Perspectives on Health

End of Chapter
- Aging of the Lymphatic and Immune Systems
- Career Spotlight: Public Health Nurse
- Connective Issues
- Study Guide

BASE CAMP

Before ascending to the next level, be sure you're properly equipped with a knowledge of the following concept from earlier chapters.

- Leukocyte types and their functions, especially lymphocytes (see section 12.3b)
- Capillary fluid exchange (see section 13.5d)

Module 10
Lymphatic System

You may be surprised to know that the human body hosts at least as many bacterial cells as it does human cells. The bacteria and other microorganisms, such as fungi and viruses, that colonize the skin, mucosal surfaces, and especially the gut are collectively called the **microbiome.** We are accustomed to these tiny cohabitants and even depend on them for good health. However, they have also been linked to conditions ranging from obesity, to arthritis, to inflammatory bowel disease. In addition to our regular microscopic inhabitants, we are also constantly exposed to new microbial invaders that enter our body through the food we eat, the water we drink, the air we breathe, and the surfaces we touch. Most of these are harmless, but others have the potential to be pathogenic, or disease-causing. It is a wonder that the body isn't overrun and consumed by invading microbes—which indeed quickly happens when one dies and homeostasis ceases.

What prevents these microbes from taking over? In 1882, Russian zoologist Elie Metchnikoff discovered mobile, phagocytic cells that attacked and seemed to be devouring foreign matter introduced into animals such as starfish larvae. He named the attacking cells *phagocytes* and coined the term *phagocytosis* for the cell process we know so well today. His work set the stage for the science of *immunology,* the study of the *immune system,* and he justifiably received a Nobel Prize in 1908 for this work.

The immune system protects the body from infectious disease. It is not an organ system in itself, but a population of cells that inhabit all of our organs. Immune cells are especially concentrated in a true organ system, the *lymphatic system,* which plays crucial roles not only in immunity but also in body fluid distribution. This chapter explores and integrates the functions of the lymphatic and immune systems.

14.1 The Lymphatic System

Expected Learning Outcomes

When you have completed this section, you should be able to

a. list the functions and basic components of the lymphatic system;
b. discuss how lymph is formed;
c. explain how lymph is returned to the bloodstream; and
d. identify the major lymphatic tissues and organs, and describe their location, structure, and functions.

Our environment is populated by innumerable **pathogens**—bacteria, viruses, fungi, and other microbes that can cause disease. One of our defenses against these is the **lymphatic system** (fig. 14.1), a network of tissues, organs, and vessels that recover tissue fluid, inspect it and cleanse it of pathogens, activate immune responses, and return the fluid to the bloodstream. Components of the system include (1) **lymph,** the fluid that it collects from the tissues and returns to the bloodstream; (2) **lymphatic vessels** (also simply called *lymphatics*), which resemble veins and

Figure 14.1 The Lymphatic System.

transport the lymph; (3) **lymphatic tissue,** composed of loose aggregations of lymphocytes in the connective tissues of various organs such as the digestive and respiratory tracts; and (4) **lymphatic organs,** structures enclosed in a fibrous capsule and containing organized masses of lymphatic tissue. The principal lymphatic organs we will discuss are the tonsils, thymus, spleen, and lymph nodes.

14.1a Functions of the Lymphatic System

The lymphatic system has three functions:

1. **Fluid recovery.** Fluid continually seeps from the blood capillaries into the tissue spaces. The blood capillaries reabsorb about 85% of this fluid and the lymphatic system absorbs the other 15%. If not for the lymphatic system, this 15% would amount to a loss of 2 to 4 L of water and one-quarter to one-half of the protein from the blood plasma per day. One would quickly die from the loss of blood volume and resulting circulatory failure.
2. **Immunity.** Fluid recovered from body tissues is checked by the lymphatic system for toxins, microbes, and other threats. The lymphatic system also guards the openings of the digestive, respiratory, and other tracts. When disease agents are detected, immune cells of the lymphatic system are quickly mobilized to fight them off.
3. **Lipid absorption.** Special lymphatic vessels called *lacteals* in the small intestine absorb dietary lipids. The lipids travel through lymphatic vessels that ultimately empty into the large left subclavian vein. From here, the bloodstream can distribute these lipids throughout the body for storage or immediate use.

14.1b Lymphatic Vessels and Lymph

The lymphatic system recovers excess tissue fluid by means of innumerable microscopic vessels called **lymphatic capillaries** that permeate almost all of the body's tissues. These are similar to blood capillaries, but with important differences. They are closed at one end (**fig. 14.2**) and are essentially sacs of thin, loosely overlapping endothelial cells. The endothelial cells are tethered to surrounding tissue by protein filaments that keep the sac from collapsing. Unlike endothelium of blood capillaries, the cells are not joined by tight junctions but have large gaps between them for passage of proteins, bacteria, and white blood cells.

Figure 14.2 Lymphatic Capillaries.
(a) Relationship of the lymphatic capillaries to a bed of blood capillaries. (b) Uptake of tissue fluid by a lymphatic capillary.

❓ *Why can metastasizing cancer cells get into the lymphatic system more easily than they can enter the bloodstream?*

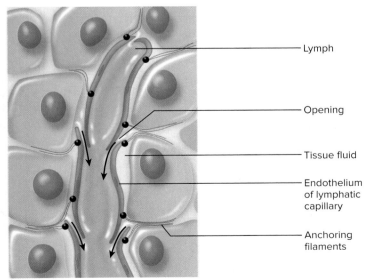

466 PART 4 Circulation and Defense

Figure 14.3 Edema. On the right is a 52-year-old woman with severe edema of the legs and feet; on the left, for comparison, is a 21-year-old woman without edema. Blockage of lymphatic vessels is one of several causes of edema.
MedicImage/Alamy Stock Photo

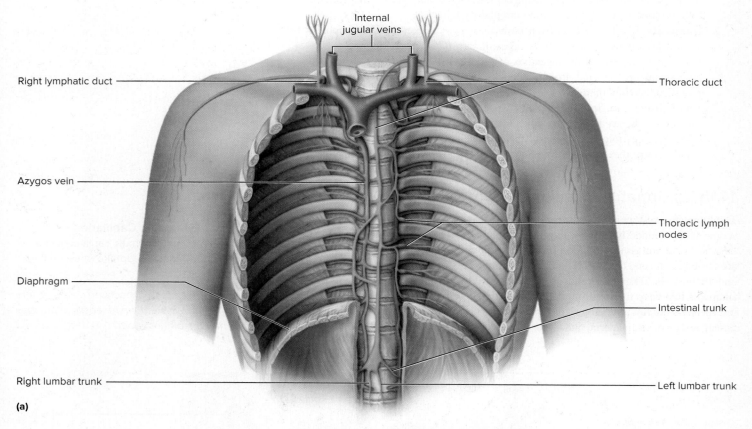

Figure 14.4 Lymphatics of the Thoracic Region. (a) Lymphatics of the thorax and upper abdomen and their relationship to the subclavian veins, where the lymph returns to the bloodstream. (b) Lymphatic drainage of the right mammary and axillary regions. (c) Regions of the body drained by the right lymphatic duct and thoracic duct. **APR**

? Why are the axillary lymph nodes often biopsied in cases of breast cancer?

Overlapping edges of the endothelial cells act as valves that open and close depending on pressure in the surrounding tissue fluid. When tissue fluid pressure is high, the flaps are pushed inward (open) and fluid enters the lymphatic vessel. Inadequate fluid recovery from the tissues can result in **edema,** a swelling and puffiness due to fluid accumulation (fig. 14.3). Once recovered tissue fluid enters the lymphatic vessels, it is called lymph.

As lymphatic capillaries converge, they become larger vessels with different names. Lymphatic capillaries converge to form thin **collecting vessels.** These converge to form **lymphatic trunks,** each of which drains a major portion of the body. For example, the *lumbar trunks* drain the lumbar region as well as the lower limbs. All lymphatic trunks converge to form just two **collecting ducts,** the largest lymphatic vessels (fig. 14.4). The lesser of these, the **right lymphatic duct,** receives lymphatic drainage from the right side of the head and chest and from the right upper limb. It empties into the right subclavian vein. The other one, called the **thoracic duct,** is larger and longer. It receives lymphatic trunks that drain all of the body below the diaphragm, as well as the left side of the head, neck, and chest and the left upper limb. It empties into the left subclavian vein. In summary, the route from the tissue fluid back to the bloodstream is lymphatic capillaries → collecting vessels → six lymphatic trunks → two collecting ducts → subclavian veins (see fig. 14.1). There is a continual recycling of fluid from blood to tissue fluid to lymph and back to the blood.

The larger lymphatic vessels are similar to veins in their structure, with a three-layered wall and valves that ensure a one-way flow of fluid (fig. 14.5). In contrast to blood, the lymph does not flow in response to a pump like the heart. As a result, it flows at even lower pressure and speed than venous blood, propelled by four mechanisms: (1) Lymphatic vessels contract rhythmically when flowing lymph stretches their walls. (2) Contracting skeletal muscles compress the lymphatics and move the lymph, like the *skeletal muscle pump* that moves venous blood. (3) Breathing creates pressure differences between the abdominal and thoracic cavities that cause lymph to flow upward into the thoracic lymphatics, in the same manner as the *thoracic pump* that aids venous blood flow (see section 13.5e for both the skeletal muscle and thoracic pumps). (4) Finally, at the point where the collecting ducts empty into the subclavian veins, the rapidly flowing bloodstream draws lymph into it. Given these mechanisms, it is easy to see why physical exercise significantly increases the rate of lymphatic return.

Lymph is similar to blood plasma but with less protein. Usually clear and colorless, its composition and characteristics vary somewhat from place to place. For example, lymph draining from the small intestine may have a milky appearance because it contains dietary lipids. In addition to proteins, lymph contains lymphocytes, macrophages, bacteria, viruses, cellular debris, and sometimes even traveling cancer cells.

14.1c Lymphatic Tissues

Lymphatic tissues are aggregations of lymphocytes in the connective tissue of mucous membranes and various organs. Lymphatic tissue is particularly abundant in body passages open to the exterior—the respiratory, digestive, urinary, and reproductive tracts—where it is called **mucosa-associated lymphatic tissue (MALT).** It guards the body against environmental pathogens entering the body orifices such as the nose, mouth, anus, vagina, and urethra. In other places, lymphocytes and macrophages congregate in dense masses called **lymphatic nodules,** which come and go as pathogens invade the tissues and the immune system answers the challenge. Abundant lymphatic nodules are a relatively constant feature of lymph nodes, tonsils, and the appendix.

Figure 14.5 Valves in the Lymphatic Vessels.
(a) Photograph of a lymphatic valve. (b) Operation of the valves to ensure a one-way flow of lymph.
a: Dennis Strete/McGraw-Hill Education

? *What would be the consequence if these valves did not exist?*

Figure 14.6 The Tonsils. (a) Locations of the tonsils. (b) Histology of the palatine tonsil. **APR**
b: Biophoto Associates/Science Source

14.1d Lymphatic Organs

In contrast to the lymphatic tissue, lymphatic organs have well-defined anatomical sites and at least partial connective tissue capsules that separate the lymphatic tissue from neighboring tissues. The principal lymphatic organs are the tonsils, thymus, spleen, and lymph nodes.

The Tonsils

The **tonsils** are patches of lymphatic tissue located at the entrance of the pharynx, where they guard against inhaled and ingested pathogens (**fig. 14.6**). Numerous **lingual tonsils** are concentrated in a patch on each side of the root of the tongue; a pair of **palatine tonsils** lies on the lateral walls of the posterior oral cavity; and a single **pharyngeal tonsil (adenoids)** is found medially at the rear of the nasal cavity. Each tonsil has deep pits *(tonsillar crypts)* surrounded by lymphatic nodules. A partial capsule underlies each tonsil and separates it from deeper tissues, whereas the superficial surface of the tonsil is covered by the epithelium of the tongue or pharynx. Inflammation of the tonsils—most often the palatine—is called *tonsillitis,* and is characterized by redness, swelling, and pain. Tonsillitis used to be routinely treated by surgical removal of the tonsils *(tonsillectomy)* but is now usually treated with antibiotics.

The Thymus

The **thymus** (**fig. 14.7**) is located between the heart and the base of the neck. The thymus is completely enclosed in a fibrous capsule. The interior of the thymus is divided into compartments called *lobules,* each of which has a relatively light center called the *medulla,* surrounded by a darker cortex. The lobules contain a network of interconnected *epithelial cells* and a large population of lymphocytes called *T cells* (T for *thymus*). The epithelial cells secrete *thymosin* and other hormones that promote the development and action of T cells. In the cortex, they help to form a *blood–thymus barrier* that isolates developing T cells from blood-borne antigens.

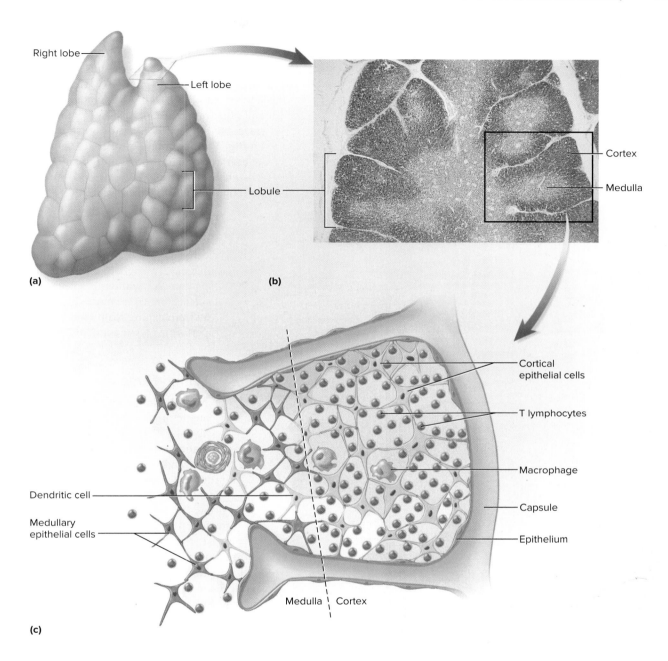

Figure 14.7 The Thymus. (a) Gross anatomy. (b) Histology. (c) Cellular architecture of a lobule.
APR
b: Dennis Strete/McGraw-Hill Education

The importance of thymic signaling molecules is demonstrated by the fact that if the thymus is removed from newborn mammals, they waste away and never develop immunity. The thymus shows profound atrophy as a person ages. Though it is relatively large in children, in old age it may be so shrunken it is difficult to locate. Later in this chapter, we further examine the role of the thymus in maturation of the T cells.

The Spleen

The **spleen** is the body's largest lymphatic organ—up to 12 cm long and usually weighing about 150 g. It is located in the upper left abdomen, tucked snugly between the diaphragm, left kidney, and stomach (fig. 14.8). It has a medial slit, the **hilum,** where the **splenic artery** enters and the **splenic vein** leaves. The spleen's major function is to filter the blood and cleanse it of bacteria and other foreign matter. It is densely stuffed with lymphocytes and macrophages, which monitor the blood for pathogens. The spleen houses a large army of monocytes, which it releases in response to wounds, heart attacks, or microbial infection. The monocytes help to combat pathogens and repair damaged tissues. The spleen is also known as the "erythrocyte graveyard"; old, fragile RBCs rupture as they squeeze through its capillary walls. RBC components such as the membrane fragments are disposed of and free hemoglobin travels to the nearby liver for breakdown and iron recycling. Furthermore, as we'll see later in this chapter, erythrocytes transport foreign antigens to the spleen where the antigens are stripped off and disposed of by the spleen's macrophages.

The spleen is highly vascular and vulnerable to trauma and infection. A ruptured spleen can hemorrhage fatally, but is difficult to repair surgically. Therefore, a once-common procedure in such cases was its removal, *splenectomy*. A person can live without a spleen, but people with splenectomies are more susceptible to infections and premature death, and splenectomy is now performed less commonly than it used to be.

Lymph Nodes

Lymph nodes are the most numerous lymphatic organs, numbering in the hundreds (see fig. 14.1). They lie along the course of the lymphatic vessels and serve two main functions: to filter and cleanse the lymph and to activate an immune response to microbes or other disease agents that may have been picked up from the tissue fluid. Lymph nodes are especially concentrated in the neck, armpit (axillary region), female breast, groin, mediastinum, and mesenteries around the intestines, with smaller clusters in the bend of the elbow and knee. Infections often result in swollen lymph nodes that can be palpated in the neck, armpit, knee, or groin. Lymph nodes also are common sites of metastatic cancer, since loose cancer cells can travel in the lymph, lodge in the nearest node, and seed the growth of a new tumor (see Clinical Application 14.1). A lymph node (**fig. 14.9**) is a small elongated or bean-shaped structure, usually less than 3 cm long. It is often embedded in fat or other connective tissue. The node is enclosed in a fibrous capsule with septa that penetrate and loosely compartmentalize its interior. Inside, an outer, darker, C-shaped *cortex* in the convex regions of the node encircles a lighter, more loosely organized *medulla* that extends to the hilum. The cortex consists mainly of ovoid *lymphatic nodules,* which are richly populated with lymphocytes. The medulla consists of elongated *medullary cords* with lymph-filled sinuses between them. The dense concentration of lymphocytes in the node, especially in the cortex, gives the tissue a granular appearance. Lymphocytes multiply profusely in the nodes, which are a major source of lymphocytes entering the bloodstream.

The curved, convex surface of a lymph node receives several *afferent lymphatic vessels*. Lymph from these vessels percolates slowly through the node and exits through the single *efferent lymphatic vessel* that leaves through the hilum. This creates a bottleneck that slows down lymph flow and allows time for WBCs to cleanse it of foreign matter.

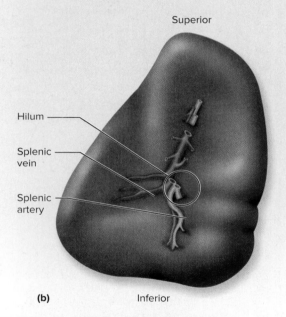

Figure 14.8 The Spleen. (a) Position of the spleen in the left upper quadrant of the abdominal cavity. The stomach has been removed to expose the organs behind it. (b) Gross anatomy of the medial surface. **APR**

a: Dennis Strete/McGraw-Hill Education

Flowing lymph typically passes through one node after another as it finds its way to the subclavian veins, where it reenters the bloodstream. In addition to lymphocytes, the node is populated by phagocytic macrophages. They remove about 99% of the impurities that enter the node (microbes, tissue debris, and other such matter), so by the time the lymph has flowed through a series of nodes and empties into the blood, it is essentially devoid of foreign matter but has numerous lymphocytes that it contributes to the blood.

In addition to simple filtration and phagocytosis, the lymph nodes monitor the collected tissue fluid for foreign *antigens*—macromolecules that activate immune responses. If a macrophage detects and phagocytizes an antigen, it digests it and displays molecular fragments of it bound to proteins on its surface. Lymphocytes detect these displayed fragments and begin an immune response detailed later in this chapter. The macrophages and certain other cells are called **antigen-presenting cells (APCs)** for this behavior. More about their role in immunity will be said later.

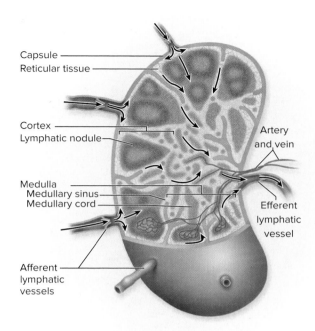

Figure 14.9 Anatomy of a Lymph Node. Partially bisected lymph node showing pathway of lymph flow. **APR**

Clinical Application 14.1

LYMPH NODES AND METASTATIC CANCER

Metastasis is the phenomenon in which cancer cells break free of the original *primary tumor*, travel to other sites of the body, and establish new tumors. Gaps in the walls of lymphatic capillaries allow cancer cells to enter and travel in the lymph. The abnormal cells eventually lodge in a lymph node, multiply there, and destroy it. Cancerous lymph nodes are swollen, relatively firm, and usually painless. Cancer of a lymph node is called *lymphoma*.

Once a tumor is established in one node, cells may emigrate and travel to the next node. If the cancer is detected early enough, it may be eradicated by removing not only the primary tumor, but the lymph nodes downstream from that point. For example, breast cancer is often treated with a combination of lumpectomy (removing the primary tumor and surrounding tissue from the breast) or mastectomy (removing the entire breast) and removal of nearby axillary lymph nodes.

Apply What You Know

Why is it important that afferent and efferent lymphatic vessels ensure that lymph only flows one way in the lymph node?

Before You Go On

Answer these questions from memory to test your understanding of the previous section. Reread the material if you can't answer the questions.

1. List the primary functions and four major components of the lymphatic system.
2. How does fluid get into the lymphatic system? What prevents it from draining back out?
3. What is the difference between a lymphatic tissue and a lymphatic organ? Give an example of each.
4. List the lymphatic organs and describe the location, structure, and function of each.
5. What two functions are performed by the macrophages and lymphocytes found in a lymph node?

14.2 Innate Immunity

Expected Learning Outcomes

When you have completed this section, you should be able to

a. define *immune system, innate immunity,* and *adaptive immunity;*
b. explain the types and importance of physical barriers in defense against pathogens;
c. enumerate the defensive functions of each kind of white blood cell;
d. describe the role of the complement system;
e. explain the process of inflammation; and
f. discuss the mechanism, benefits, and drawbacks of fever.

Infectious diseases have long ravaged human populations, from the bubonic plague in medieval times, to the invasion of smallpox that killed large numbers of indigenous people in the Americas when it was introduced by Europeans, to the shadow of ebola that has recently haunted parts of Africa. The agents of infectious disease are called *pathogens,* and range from viruses and bacteria to fungi, protozoans, and parasitic worms. Our first line of defense against such pathogens is our skin and mucous membranes—physical barriers to invasion. If these barriers are broken, pathogens penetrate the body tissues and other elements of the immune system are called upon to protect us.

The **immune system** consists of physical barriers, immune cells, antimicrobial proteins, and other mechanisms that neutralize and destroy pathogens, ideally before they cause disease. These defenses are divided into *innate immunity* and *adaptive immunity.* **Innate immunity** consists of a variety of defenses that we are born with. Its barriers and protective cells indiscriminately hinder and attack foreign invaders and promote processes such as inflammation and fever that usually inactivate pathogens. Some infections, however, overpower the innate immune response, and **adaptive immunity** is activated. It increases the power to focus on the specific type of pathogen that has invaded and *adapts* to its unique characteristics. It also leaves our body with a memory of the pathogen so we can react more quickly and efficiently if we should encounter it again. These immune defenses keep us healthy most of the time, despite our constant exposure to diverse pathogens. **APR**

14.2a Physical Barriers

The body is especially vulnerable to invasion in places where internal passages open to the exterior environment—the nose, mouth, anus, urethra, and vagina. The mucous membranes that line our respiratory, digestive, urinary, and reproductive tracts, and the skin that covers the body, are physical barriers that microbes cannot easily breach.

Several features make the skin an especially effective barrier. The epidermal cells are joined by tight junctions and are composed mainly of keratin, a tough protein that is difficult to penetrate. Furthermore, the skin surface is inhospitable for microbial growth. It is mostly dry, whereas microbes prefer a moist environment, and it lacks nutrients conducive to their survival. A thin film of acidic sweat and antimicrobial chemicals also retards pathogen growth and survival. Mucous membranes, too, are difficult to penetrate because of tight junctions between the epithelial cells; because the mucus itself is sticky and entraps microbes; and because the mucus contains *lysozyme,* an enzyme that destroys bacteria.

14.2b Protective Cells

Nevertheless, the foregoing barriers are not completely impenetrable. Wounds or insect bites may breach the skin, for example, and the mucous membranes of the internal surfaces may be torn or irritated. Every time we brush our teeth, we create microscopic abrasions of the gums that invite invasion by the many microbes of the oral cavity. If microbes get past such physical barriers, they are attacked by **macrophages** and other *phagocytes* that have a voracious appetite for foreign matter, and they are assaulted by toxic chemicals produced by white blood cells (fig. 14.10).

The five types of leukocytes were described and pictured in section 12.3b. Following is a summary of their most important immune roles.

- **Neutrophils** move in large numbers to sites of infection where they recognize and destroy microorganisms, especially bacteria. They not only voraciously phagocytize bacteria but also secrete a chemical *killing zone* of lethal chemicals, including hydrogen peroxide and hypochlorite, the active ingredient found in chlorine bleach. This cloud of toxic chemicals kills significantly more microbes than phagocytosis does, but also kills the neutrophils themselves. Further, neutrophils produce webs called *neutrophil extracellular traps (NETs)* that consist of proteins and threads of DNA. The nets trap and kill invaders. Neutrophils are turning out to have a multitude of innovative ways to defend us!
- **Eosinophils** weaken or destroy parasites too large for phagocytosis, respond to allergens, and phagocytize and remove antigen–antibody complexes from the tissues.
- **Basophils** secrete histamine and heparin—a vasodilator and anticoagulant, respectively—which enhance the delivery of blood and immune cells to sites of infection or inflammation.
- **Lymphocytes** are of several types with varied roles explained in the ensuing discussion of adaptive immunity, including secretion of antibodies, direct attacks on pathogenic microbes, and destruction of infected and cancerous cells.
- **Monocytes** transform to macrophages in connective tissues, where they function as phagocytes and as messengers to cells that activate adaptive immunity when it is necessary.

Natural killer (NK) cells are lymphocytes that contribute to the innate immune response. They continually patrol the body "on the lookout" for pathogens or diseased host cells. They attack and destroy bacteria, cells infected with viruses, cancer cells, and cells of transplanted tissues and organs. Upon recognition of an enemy cell, the NK cell binds to it and releases proteins called *perforins* (fig. 14.11). This is a "kiss of death"—it creates a hole in the enemy cell membrane that allows a rapid inflow of water and salts. This alone may kill the cell, but in addition, the NK cell secretes protein-degrading enzymes that enter through the hole, destroy the enemy's enzymes, and induce apoptosis (programmed cell

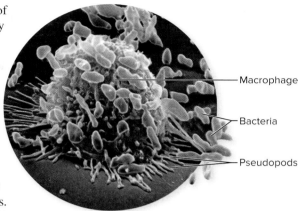

Figure 14.10 Macrophage Phagocytizing Bacteria. Filamentous pseudopods of the macrophage snare the rod-shaped bacteria and draw them to the cell surface, where they are phagocytized.

David M. Phillips/Science Source

Figure 14.11 The Action of a Natural Killer Cell.

death). This is part of our innate immunity; it doesn't rely on prior exposure to a pathogen, and it has no memory of prior exposures. This distinguishes the action of NK cells from those of other lymphocytes discussed later (T and B cells).

14.2c Protective Proteins

Multiple types of proteins provide short-term, nonspecific resistance to pathogens. **Interferons** are a family of proteins secreted by cells in response to infection by viruses. They signal nearby cells to produce other proteins that inhibit viral replication. In addition, they activate natural killer (NK) cells, which destroy infected cells before they can liberate a swarm of newly replicated viruses.

The **complement system** is a group of 30 or more proteins synthesized mainly by the liver. They circulate in the blood in an inactive form; when activated by pathogens, they initiate a complex cascade of chemical events, which ultimately leads to the demise of invading microbes. They were named for the fact that they complement the action of the antibodies that we will study shortly, but they also contribute to innate immunity. Their effects are to stimulate inflammation, to promote phagocytosis of pathogens by neutrophils and macrophages, and to attack and rupture cells that have been invaded.

14.2d Protective Processes

Inflammation

When you suffer a local infection, a splinter in the skin, a tissue irritation such as blisters from tight shoes, or just an itchy mosquito bite, you may notice redness, swelling, heat, and pain in the affected tissue. Those signs and symptoms have traditionally been called the four *cardinal signs* of the defensive process called **inflammation.** Words ending in the suffix *-itis* denote inflammation of specific organs and tissues—for example, *arthritis* is inflammation of the joints, *dermatitis* pertains to the skin, and *gingivitis* is inflammation of the gums. The purposes of inflammation are to (1) rapidly bring phagocytes and immune proteins to the infected area, (2) remove the debris of damaged tissue, and (3) initiate tissue healing and repair.

Almost immediately after invasion by a microorganism, blood vessels dilate and blood flow to the area increases. Tissue damage stimulates basophils of the blood and similar *mast cells* in the connective tissue to release the vasodilator histamine. The cardinal signs of inflammation result from these early responses: (1) Heat and (2) redness result from the increased blood flow; (3) swelling (edema) is due to increased fluid filtration from the capillaries; and (4) pain results from direct injury to the nerves, pressure on the nerves from the edema, and stimulation of pain receptors by lipids called *prostaglandins*.

The increased blood flow rapidly delivers an army of defensive WBCs. The leukocytes wreak destruction upon pathogens that have invaded the body. Neutrophils are the chief enemies of bacteria. **Figure 14.12** summarizes their action. Normally, neutrophils and other leukocytes travel rapidly through capillaries and don't linger to interact with endothelial cells. During infections, however, endothelial cells produce cell-adhesion molecules that snag the WBCs as they pass in the bloodstream, causing them to slowly tumble along the endothelium. **Adhesion** to the vessel wall gives WBCs time to check for signals from nearby injured or infected tissues. In the meantime, chemicals cause the endothelial cells of the blood capillaries and venules to separate a little, widening the intercellular gaps and permitting the WBCs to crawl through in a process called **diapedesis.** The increased permeability also permits fluid, antibodies, and clotting factors and complement proteins to move into the surrounding tissue.

Within an hour of injury or infection, neutrophils begin to accumulate in inflamed tissue and home in on the affected site by **chemotaxis,** moving up a "scent trail" of chemicals released by bacteria and injured tissues. As they encounter

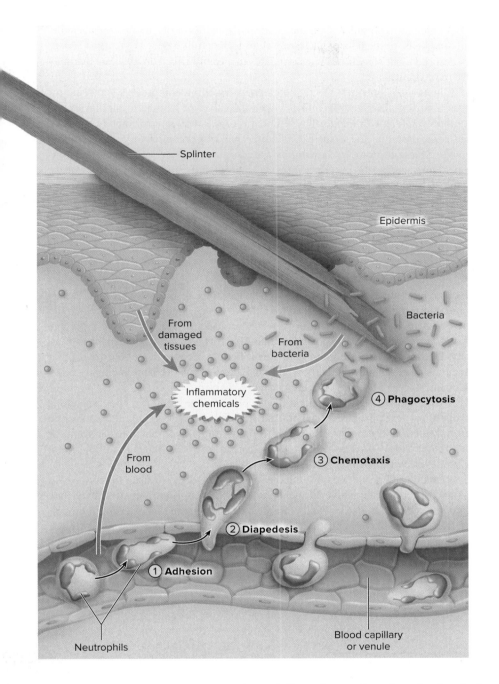

Figure 14.12 Neutrophil Behavior in Inflammation. Tissue injury and infection cause release of inflammatory chemicals from bacteria, the blood, and the damaged tissues themselves. Inflammatory chemicals stimulate neutrophils to adhere to the capillary wall, leave the bloodstream through gaps between the capillary cells (diapedesis), follow the chemical trail by chemotaxis, and phagocytize (engulf) bacteria.

bacteria, neutrophils avidly phagocytize and digest them and produce the chemical killing zone described earlier. They also secrete signaling chemicals that recruit more WBCs, especially monocytes, to the scene.

One priority of inflammation is to prevent pathogens from spreading throughout the body. Pathogens are physically trapped by clots that form in areas adjacent to the injury; a sticky mesh of fibrin walls off the area, so bacteria and other microbes are essentially trapped in a fluid pocket surrounded by a gelatinous, clotted capsule. This makes it difficult for them to escape the attacks by phagocytes, antibodies, and other defenses.

The battle leaves an aftermath of tissue destruction. Monocytes are the major agents of tissue cleanup and repair, arriving within 8 to 12 hours of injury. They emigrate from the bloodstream and turn into macrophages. Macrophages engulf and destroy bacteria and other pathogens, damaged host cells, and dead and dying neutrophils.

As the battle progresses, all the neutrophils and most of the macrophages die. The dead cells, tissue debris, and tissue fluid form a milky pool of **pus.** Pus is usually absorbed, but sometimes it accumulates in a cavity called an **abscess**[1] that may have to be drained.

Edema also contributes to tissue cleanup. Swelling compresses veins and reduces venous drainage, while it forces open the valves of lymphatic capillaries and promotes lymphatic drainage. The lymphatic vessels collect and remove bacteria, dead cells, proteins, and tissue debris better than blood capillaries can.

Vasodilation promotes healing by speeding the delivery of oxygen and amino acids necessary for protein synthesis. Endothelial cells and platelets release chemicals that stimulate tissue regrowth. Meanwhile, the heat of inflamed tissue increases the metabolic rate and the speed of mitosis and tissue repair. Pain also contributes to recovery. It is an alarm signal that calls our attention to the injury and causes us to limit the use of a body part so it has a chance to rest and heal.

Fever

Fever is an abnormal elevation of body temperature resulting from such causes as infection, trauma, and drug reactions. It was long regarded as an undesirable side effect of illness, and efforts were made (and still are) to reduce it for the sake of comfort. It is now recognized, however, as a defensive mechanism that, in moderation, does more good than harm. People recover from colds and other infectious diseases more quickly when they allow a fever to run its course rather than using fever-reducing medications such as aspirin.

When neutrophils and macrophages phagocytize bacteria, they secrete a **pyrogen,**[2] or fever-producing agent, that stimulates the hypothalamus to secrete a *prostaglandin.* The prostaglandin, in turn, raises the hypothalamic set point for body temperature—say to 39°C (102°F) instead of 37°F. Aspirin and ibuprofen reduce fever by inhibiting prostaglandin synthesis.

When the set point rises, the cutaneous blood vessels constrict to reduce heat loss and a person may shiver to generate heat. During the onset of a fever, one experiences chills, feels cold and clammy to the touch, and has a rising temperature (fig. 14.13). In the next stage, the body temperature hovers around the higher set point for as long as the pathogen is present. The elevated temperature inhibits

---

[1]*ab* = away; *scess (from cedere)* = to go
[2]*pyro* = fire, heat; *gen* = producing

Figure 14.13 The Course of a Fever.

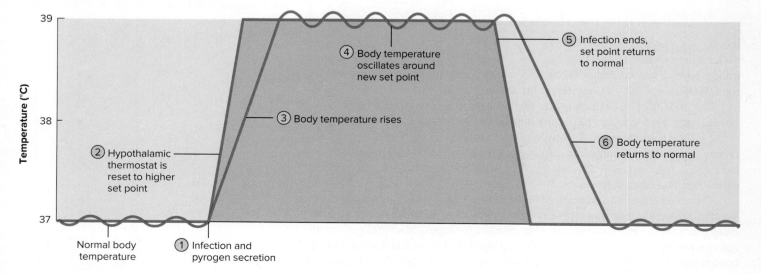

reproduction by bacteria and viruses, promotes interferon activity, and increases metabolic rate and speeds tissue repair. When the infection is defeated, pyrogen secretion ceases and the hypothalamic thermostat is set back to normal. This activates heat-losing mechanisms, especially cutaneous vasodilation and sweating. The skin is warm and flushed during this stage. The result is falling temperature until it returns to normal.

Even though most fevers are beneficial, excessively high temperature can be dangerous. It speeds up enzymatic pathways, thus disrupting metabolic coordination and leading to cellular dysfunction. Fevers above 40.5°C (105°F) can make a person delirious. Convulsions and coma ensue at higher temperatures, and death or irreversible brain damage commonly results from fevers that range from 44° to 46°C (111° to 115°F).

Before You Go On

Answer these questions from memory to test your understanding of the previous section. Reread the material if you can't answer the questions.

6. List several features that make the skin an effective barrier against pathogens.
7. Explain the roles of macrophages, natural killer (NK) cells, and neutrophils in innate immune defense.
8. How do interferons and complement proteins protect against disease?
9. Explain the causes of the four cardinal signs of inflammation.
10. How does inflammation promote the destruction of pathogens?
11. List three benefits of fever.

14.3 Features of Adaptive Immunity

Expected Learning Outcomes

When you have completed this section, you should be able to

a. compare and contrast adaptive and innate immunity;
b. contrast cellular immunity and humoral immunity;
c. explain what antigens are and discuss their role in immunity;
d. describe the role played by lymphocytes in the immune response;
e. compare and contrast the developmental processes of T and B lymphocytes;
f. describe the structure and function of antibodies; and
g. explain the role of antigen-presenting cells in immunity.

Most infections are stopped by the physical and chemical barriers associated with innate immunity, but sometimes the innate immune response is not enough to resist or overcome the infection, especially in people who are malnourished, poorly housed, deprived of sleep, or stressed in other ways. When innate defenses fail, the more targeted and forceful defenses of the adaptive immune response are employed. The rest of this chapter is about **adaptive immunity,** which relies on the body's ability to recognize foreign invaders individually, respond to them selectively, and remember them so it can defeat them more quickly in future exposures. Two characteristics distinguish adaptive immunity from innate immunity:

1. **Specificity.** Whereas innate immunity is not targeted to a particular pathogen, adaptive immunity is so sharply focused on a specific invader that a particular response is of little use against infection by a different pathogen. The adaptive response that is tailored to the polio virus will not be effective against the measles virus, for example. There are billions of pathogens that our immune system is able to respond to on an individual basis, a remarkable feat.
2. **Memory.** When reexposed to the same pathogen, the body reacts so quickly that there is no noticeable illness. The reaction time for inflammation and other innate defenses, by contrast, is just as long for later exposures as for the initial one.

There are two forms of adaptive immunity: cellular and humoral. **Cellular (cell-mediated) immunity** employs lymphocytes that directly attack and destroy foreign cells or diseased host cells. It is a means to rid the body of pathogens that reside inside our cells, where they are inaccessible to attack by antibodies—for example, intracellular bacteria, viruses, yeasts, and protozoans. Cellular immunity also acts against parasitic worms, cancer cells, and cells of transplanted organs and tissues.

Humoral (antibody-mediated) immunity employs proteins called antibodies, which don't directly destroy a pathogen but tag it for destruction by other mechanisms. The term *humoral* comes from the fact that many of the antibodies are dissolved in body fluids, once called "humors." Humoral immunity is effective against extracellular microbes and noncellular entities such as toxins, venoms, and allergens.

Both forms of adaptive immunity rely on the body's ability to recognize specific foreign invaders and mount a targeted response to them. How this occurs is the focus of the following sections. Before we address those mechanisms (the "plot"), let us look at the "cast of characters" that play various roles in the drama of immunity.

14.3a Antigens

An **antigen**[3] is any molecule that triggers an immune response. Some antigens are components of plasma membranes and bacterial cell walls; others are venoms or toxins. Most antigens are complex molecules unique to each organism—usually proteins, polysaccharides, glycoproteins, or glycolipids. Their uniqueness enables the body to distinguish its own ("self") molecules from those of any other individual or organism ("nonself"). The immune system "learns" to distinguish self- from nonself-antigens prior to birth; thereafter, it normally attacks only nonself-antigens.

Small molecules such as simple sugars and amino acids are not generally antigenic. They are too universal to all individuals (and species) to help the immune system distinguish self from nonself; if the immune system did attack them, it would be attacking substances on which our lives depend. However, certain small molecules called **haptens**[4] *become* antigenic by binding to proteins of one's own body, creating an antigenic complex. After the first exposure, the hapten alone can activate an allergic response. Haptens include penicillin and some other drugs, and various chemicals in cosmetics, detergents, industrial chemicals, animal dander, and poison ivy (Clinical Application 14.2).

---

[3]acronym from *antibody generating*
[4]*hapt* = grasp, seize

Clinical Application 14.2

HYPERSENSITIVITY, ALLERGIES, AND ANAPHYLACTIC SHOCK

Hypersensitivity is an excessive, harmful immune reaction to antigens. It includes reactions to tissues and organs transplanted from another person, abnormal reactions to one's own tissues *(autoimmunity)*, and *allergies*,[5] which are reactions to environmental antigens. Such antigens, called *allergens*, occur in dust; mold; animal dander; toxins from poison oak and other plants; and foods such as nuts, eggs, shellfish, and gluten, a protein in wheat flour and some other grain products. Drugs such as penicillin act as allergens in some people.

Hypersensitivity may be characterized by a very rapid, acute response, or may exhibit a slower onset (1–3 hours after exposure) and last longer (10–15 hours). The most common allergies are acute in nature and begin seconds after exposure. They usually subside after 30 minutes but can be severe or even fatal. Allergens bind to the membranes of basophils and stimulate secretion of histamine and other inflammatory chemicals that trigger glandular secretion, vasodilation, increased capillary permeability, smooth muscle spasms, and other effects. The clinical signs include local edema, mucus hypersecretion, watery eyes, runny nose, hives (red itchy skin), and sometimes breathing problems, cramps, diarrhea, and vomiting. Examples of acute sensitivity are asthma (see Clinical Application 15.1) and food allergies.

Anaphylaxis[6] (AN-uh-fih-LAC-sis) is an immediate and sometimes severe acute reaction. Local anaphylaxis can be relieved with antihistamines. *Anaphylactic shock*, however, is a widespread hypersensitivity characterized by dyspnea (labored breathing), widespread vasodilation, and a sudden drop in blood pressure that can lead to circulatory shock. Extreme cases can cause death due to suffocation or circulatory shock. Antihistamines are inadequate to counter anaphylactic shock, but epinephrine relieves symptoms by dilating bronchioles, increasing cardiac output, and restoring blood pressure.

14.3b Lymphocytes

The two broad categories of lymphocytes involved in adaptive immunity are *T* and *B lymphocytes*. T lymphocytes are named for the thymus, where they spend time as part of their maturation process. B lymphocytes are named for an organ *(bursa)* in chickens where they were discovered, but the *B* can serve as a reminder that they remain in the bone marrow throughout their maturation process. Once mature, both types populate the lymph nodes, mucous membranes, and other lymphatic tissues and organs throughout the body, and remain on constant surveillance for pathogens. They have the ability to recognize and respond to an almost limitless variety of foreign agents. In addition, T cells specialize in recognizing and destroying body cells that have gone awry and become cancerous.

T Lymphocytes

T lymphocytes (T cells) have a life history involving three stages and anatomical stations in the body. We can loosely think of these as their "birth," their "training" or maturation, and finally their "deployment" to locations where they will carry out their immune functions.

T cells are "born" in the red bone marrow as descendants of the hematopoietic stem cells described in section 12.1. Immature T cells travel from the bone marrow to

---

[5]*allo* = altered; *erg* = action, reaction
[6]*ana* = against; *phylax* = protection

the thymus, where they mature into fully functional T cells (fig. 14.14). Their "training" in the thymus is a complex multistep process. The essence of it is that young T cells develop antigen receptors on the cell surface and are then tested to see if they will respond to foreign antigens and yet not react against one's self-antigens. About 98% of them fail this test (and you thought your final exams were hard!)—either they fail to respond to foreign antigens or they overreact to self-antigens, signifying that they could be harmful to one's own tissues. These self-reactive cells are destroyed or rendered inactive, leaving only 2% with the ability to respond selectively to foreign antigens. Now considered to be **immunocompetent** but **self-tolerant,** they multiply and build up a huge population of T cells. They now "graduate from school" and disperse to the lymph nodes and other organs, ready to do battle.

T cells mature mostly during fetal life and early childhood. Although the thymus gradually atrophies as a person ages, it continues to produce *thymosin,* a hormone that promotes the proliferation of T cells throughout the body.

B Lymphocytes

B lymphocytes (B cells) mature in the bone marrow, but they undergo a similar process of testing and selection, then likewise disperse and colonize lymphatic tissues and organs throughout the body (fig. 14.14). B cells are abundant in lymph nodes, mucous membranes, and the spleen. They differ from T cells in their mode of attack against foreign invaders, as we will soon see in more detail.

14.3c Antibodies

Pathogens may be found either in infected cells or outside cells in extracellular fluid. In humoral immunity, activated B cells differentiate into *plasma cells.* These cells secrete proteins called **antibodies,** which bind to pathogens and ultimately lead to their destruction. Antibodies are also known as **immunoglobulins (Ig)** or **gamma globulins.** The basic structural unit of an antibody is a Y-shaped protein with four

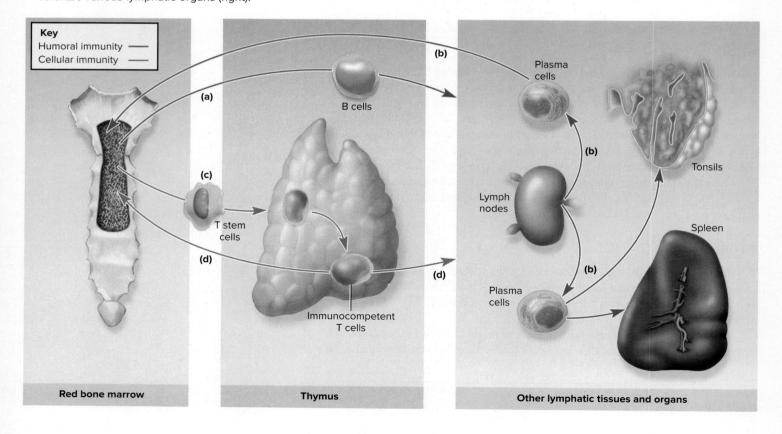

Figure 14.14 The Life History and Migrations of B and T Cells. T cells are represented by the red pathways and B cells by the violet. (a) B cells achieve immunocompetence in the red bone marrow (left), and many emigrate to lymphatic tissues and organs, including the lymph nodes, tonsils, and spleen (right). (b) Plasma cells develop in the lymph nodes and other sites, then migrate back to the bone marrow and to other lymphatic organs, where they spend a few days secreting antibodies. (c) T stem cells migrate from the bone marrow and attain immunocompetence in the thymus. (d) Immunocompetent T cells emigrate from the thymus and recolonize the bone marrow or colonize various lymphatic organs (right).

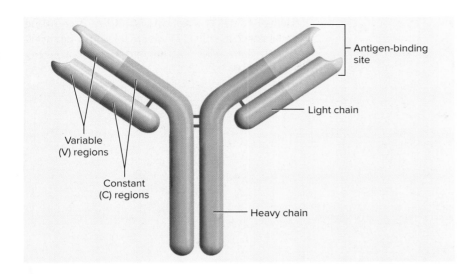

Figure 14.15 Antibody Structure. A molecule of IgG.

Table 14.1 Antibody Classes

Class	Location and Function
IgA	Found in blood plasma, mucus, tears, milk, saliva, and intestinal secretions. Provides passive immunity to the newborn.
IgD	A membrane protein of B cells; thought to function in activation of B cells by antigens.
IgE	A membrane protein of basophils and mast cells. Stimulates them to release histamine and other chemical mediators of inflammation and allergy; important in hypersensitivity reactions; attracts eosinophils to sites of parasitic infection.
IgG	Constitutes about 80% of circulating antibodies in blood plasma. IgG and IgM are responsible for most adaptive immune responses against invaders.
IgM	Constitutes about 10% of circulating antibodies in plasma. When found in the membrane of B cells, it acts as part of the antigen receptor. It is also secreted by the cell in the first part of the immune response.

polypeptide chains—two *heavy chains* about 400 amino acids long and the two *light chains* about half that long (fig. 14.15). The arms have a terminal *variable (V) region* that determines what antigens an antibody will bind to, and the rest of the antibody is a *constant (C) region* that determines the functional class to which an antibody belongs. There are five classes of antibodies named **IgA, IgD, IgE, IgG, and IgM,** which differ in their functions (table 14.1). The human immune system is believed to be capable of producing at least 10 billion and perhaps up to 1 trillion different antibodies. Any one individual has a much smaller subset of these, but nevertheless, such diversity helps to explain why we can deal with the staggering diversity of antigens that exist in our environment.

14.3d Antigen-Presenting Cells

T lymphocytes cannot recognize unprocessed antigens; they must first be introduced to them by *antigen-presenting cells (APCs)* that process and display antigen fragments. Antigen-presenting cells include *macrophages* of various kinds, *dendritic cells* of the skin and mucous membranes, and even B lymphocytes, which can play an antigen-presenting role in addition to their other functions.

When an APC encounters an antigen, it internalizes it by phagocytosis, digests it with lysosomal enzymes, and displays the relevant (antigenic) fragments attached to proteins on its surface (fig. 14.16). This is like an invitation to the immune system to launch an attack on the antigenic invader. We shall see shortly how the immune cells respond in different forms of adaptive immunity.

Figure 14.16 The Action of Antigen-Presenting Cells (APCs). Stages in the processing and presentation of an antigen by an APC such as a macrophage. **APR**

With so many cell types involved in immunity, it is no surprise that chemical messengers are required to coordinate their activities. Lymphocytes and APCs communicate with a great variety of chemical signals called **interleukins.**[7]

Before You Go On

Answer these questions from memory to test your understanding of the previous section. Reread the material if you can't answer the questions.

12. List two ways in which adaptive immunity differs from innate immunity.
13. What are the two forms of adaptive immunity?
14. Define *antigen* and explain what role antigens have in immunity.
15. What role does the thymus play in the life history of a T cell?
16. State at least two differences between B cells and T cells.
17. Describe the structure, location, and basic function of antibodies.
18. Describe how antigen-presenting cells process an antigen.

---

[7]*inter* = between; *leuk* = leukocytes

14.4 Cellular and Humoral Immunity

Expected Learning Outcomes

When you have completed this section, you should be able to

a. list the types of lymphocytes involved in cellular immunity and describe the roles they play;
b. describe the process of antigen presentation and T cell activation;
c. describe how T cells destroy infected cells;
d. compare and contrast humoral immunity and cellular immunity;
e. explain how B cells recognize and respond to an antigen;
f. describe how antibodies lead to destruction of pathogens;
g. explain the role of memory cells in immunity; and
h. define active and passive immunity.

The two forms of adaptive immunity, cellular and humoral, differ in the types of lymphocytes involved and their mode of attack. Thus, they are not redundant but complementary methods of defeating pathogens and remembering them so those pathogens are less able to make one ill in the future.

14.4a Cellular Immunity

Cellular (cell-mediated) immunity is a form of specific defense in which T lymphocytes directly attack and destroy cells that have become infected or turned cancerous. The sacrifice of infected cells prevents the spread of infection to healthy cells. The main classes of T cells involved in cellular immunity follow.

1. **Cytotoxic T (T_C) cells** are "effector" cells that selectively carry out the attack on infected or cancerous cells.
2. **Helper T (T_H) cells** promote the action of T_C cells, but also play key roles in humoral immunity and nonspecific defense.
3. **Memory T (T_M) cells** are descended from the cytotoxic T cells and are responsible for "remembering" the pathogen that infected the destroyed cells.

Both cellular and humoral immunity occur in three stages: recognition, attack, and memory (or "the three Rs of immunity"—recognize, react, and remember). In cellular immunity, the key events of each stage are as follows (fig. 14.17):

① Recognition. Cytotoxic and helper T cells patrol the lymph nodes and other tissues, inspecting the body's cells as though looking for trouble. If a cell has turned malignant, it will exhibit abnormal *host (self) antigens* on its surface. When a cytotoxic T (T_C) cell encounters the abnormal antigen, it is activated and multiplies rapidly and produces a clone of itself—a large number of identical T_C cells programmed against that antigen. This process is called **clonal selection.** There is now a multitude of T cells that have been programmed to attack malignant cells displaying the same antigen as was on the cell that was initially encountered.

The process is slightly different in cells that are infected by viruses or other outside interlopers. When antigen-presenting cells (APCs) such as neutrophils or macrophages bump into an infected cell, they display *foreign antigens* on their surface. Helper T (T_H) cells respond to the APCs by secreting chemical signals (interleukins) with multiple effects: They attract "reinforcements" to the battlefield (neutrophils, macrophages, and natural killer cells), and they stimulate clonal selection and maturation of T and B cells as illustrated. (We will discuss the B cells later and focus on the T cells for now.)

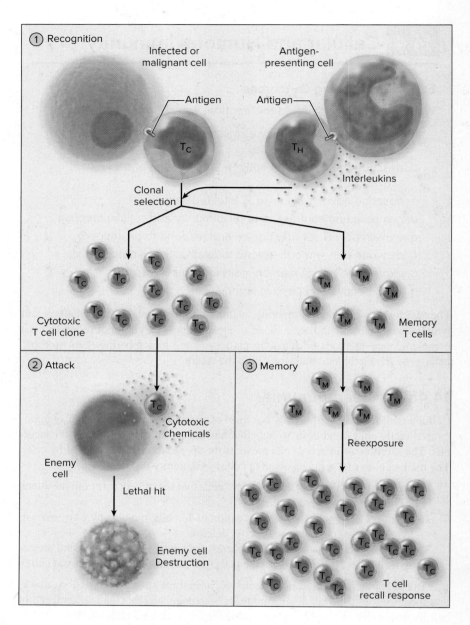

Figure 14.17 The Three Stages of Cellular Immunity.

Figure 14.18 T Cell Attack. Four cytotoxic T cells have bound to this cancer cell and delivered a lethal hit of toxic chemicals that will destroy it. **APR**
Steve Gschmeissner/Science Source

② **Attack.** An army of T cells is now prepared to go on the attack against the enemy. T_C cells are the only lymphocytes that attack and kill enemy cells directly. A T_C cell "docks" on the enemy cell surface and delivers a **lethal hit** of chemicals that will destroy it (**fig. 14.18**), then goes off in search of other enemy cells, leaving these chemicals behind to do their work. The toxic chemicals include (a) *perforins* and *granzymes,* which destroy a cell in the manner we saw earlier for natural killer (NK) cells (see fig. 14.11); (b) *interferons,* which inhibit the replication of viruses and activate NK cells, among other effects; and (c) *tumor necrosis factor,* which aids in macrophage activation and kills cancer cells, sometimes seeming to miraculously "melt away" malignant tumors. As helper T cells recruit more and more T_C cells, the immune response exerts an overwhelming force against malignant cells or those infected with foreign pathogens and destroys them.

③ **Memory.** Adaptive immunity is defined in part by the ability to remember this encounter so the body can respond more quickly next time. The clonal selection described at step ① produces not only T_C cells, but also memory T (T_M) cells. These long-lived cells are far more numerous than the original population of T cells that was able to respond to that specific pathogen, and they can be activated more quickly than was possible in the first-time response. Upon reexposure to the same pathogen later in life, T_M cells mount a rapid **T cell recall response,** where T_C cells destroy the invading pathogen so quickly we don't even notice any illness. That is, we are immune to that disease.

Note that helper T cells are necessary to stimulate an attack against foreign antigens. Indeed, T_H cells play a central coordinating role in all of the body's main lines of defense. When they encounter an antigen-presenting cell bearing a foreign antigen, they secrete various signaling chemicals that promote innate, humoral, and cellular immunity (fig. 14.19). The human immunodeficiency virus (HIV) preferentially attacks T_H cells, thus striking at the crossroad of all forms of immunity and explaining why AIDS is so devastating.

14.4b Humoral Immunity

Humoral immunity employs B lymphocytes. It is a more indirect means of fighting pathogens than cellular immunity. Instead of directly attacking cells, humoral immunity uses antibodies that bind to antigens and tag them for destruction by other means. But like cellular immunity, humoral immunity works in three stages—recognition, attack, and memory.

Figure 14.19 The Central Role of Helper T Cells in Immunity. **APR**

Recognition

Figure 14.20 depicts the following stages of antigen recognition leading up to the attack. These events occur mainly in our lymph nodes, where there is a dense population of B cells ready to respond to antigens delivered to them.

① **Antigen binding.** A B cell has thousands of surface receptors for one particular type of antigen. B cell activation begins when that antigen binds to several of these receptors.
② **Antigen presentation.** The B cell internalizes and processes the antigen and displays fragments of it on its plasma membrane—thus B cells, like macrophages, act as antigen-presenting cells. A helper T cell binds to the antigen–B cell complex and secretes an interleukin that activates the B cell.
③ **Clonal selection.** The B cell responds to the interleukin by dividing into a clone of identical B cells programmed against the antigen.
④ **Differentiation.** Some members of the clone become memory B cells while others enlarge and differentiate into **plasma cells,** which were so named in the nineteenth century because they have a large volume of cytoplasm (then called plasma) compared to other lymphocytes. Their volume is due to a large amount of rough endoplasmic reticulum (fig. 14.21), which you will remember as a cell's apparatus for protein synthesis. Their product is antibodies, which carry out the attack stage that we examine next. About 10% of the B cells in a given clone stay in lymph nodes, but the rest leave, take up residence elsewhere, and produce antibodies until they die.

Apply What You Know

Consider the function of rough endoplasmic reticulum and the chemical nature of antibodies. Explain how the difference between a B cell and plasma cell relates to plasma cell function.

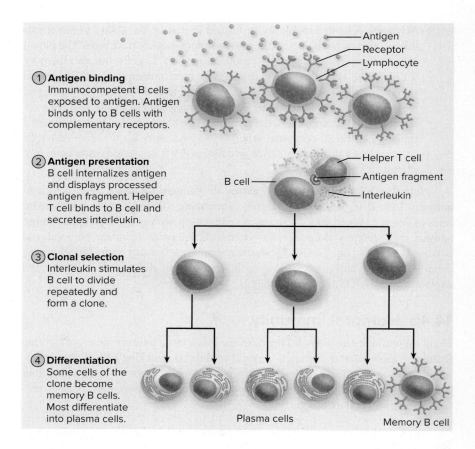

Figure 14.20 The Recognition Stage of Humoral Immunity

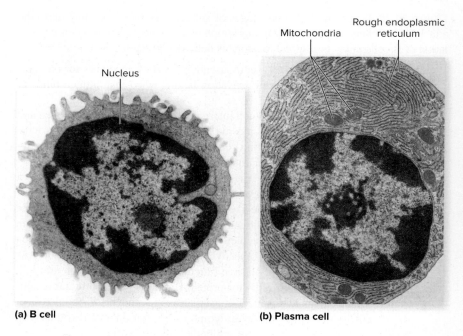

Figure 14.21 B Cell and Plasma Cell. (a) B cells have little cytoplasm and scanty organelles. (b) A plasma cell, which differentiates from a B cell, has an abundance of rough endoplasmic reticulum.
a–b: Don W. Fawcett/Science Source

Attack

The plasma cell is a remarkably efficient protein factory, pumping out 2,000 antibody molecules per second. This comes at a cost to self-maintenance, however, and a plasma cell dies after only 4 to 5 days of such intense activity. The antibodies remain and lay the groundwork for the attack phase. Although antibodies do not, themselves, destroy enemy antigens, they can use any of the following means to render them harmless and mark them for destruction.

- **Neutralization.** Only certain regions of an antigen molecule are pathogenic—for example, the parts of a toxin molecule or virus that enable them to bind to human cells. Antibodies can neutralize an antigen by binding to these active regions and masking their effects.
- **Activation of the complement system.** Activation of the complement system is the most important way antibodies provide defense against foreign cells such as bacteria. When an antibody of the IgG or IgM class binds to an antigen, complement proteins bind to receptors on the antibody. This initiates the cascade of events that leads to destruction of the invader by complement mechanisms described earlier (see section 14.2c).
- **Agglutination** is the clumping of enemy cells by antibodies. It occurs in mismatched blood transfusions and even more importantly as a defense against bacteria. An antibody molecule has binding sites for antigens at its Y-shaped tips; it can bind to antigen molecules on two or more enemy cells at once and stick them together (fig. 14.22a). This immobilizes alien cells and prevents them from spreading through the tissues. Further, neutrophils and macrophages can phagocytize agglutinated clusters of bacteria more efficiently than phagocytizing bacteria one at a time.
- **Precipitation** is a similar process in which antigen molecules (not whole cells) are clumped by adhesion to antibodies (fig. 14.22b). This creates large *antigen–antibody complexes* that can be removed and degraded by eosinophils, or they can bind to red blood cells and travel with them to the liver and spleen, where they are stripped off the RBC and disposed of. This process of ridding the body of agglutinated antigens is called *immune clearance*.

Figure 14.22 Agglutination and Precipitation by Antibodies. (a) Agglutination of foreign erythrocytes by IgM. (b) Precipitation of free antigens by an antibody such as IgG, forming an antigen–antibody complex.

Memory

When a person is exposed to a particular antigen for the first time, the immune reaction is called the **primary response** (fig. 14.23). The appearance of protective antibodies is delayed for 3 to 6 days while B cells multiply and differentiate into plasma cells. As the plasma cells begin secreting antibody, the IgM level *(titer)* in the blood rises, followed in a few days by IgG. The IgM and IgG titers fall to low levels within a month or so.

The primary response, however, leaves one with an immune memory of the antigen. We saw that during clonal selection, some members of the clone become **memory B cells** rather than plasma cells. Memory B cells, found mainly in the lymph nodes, mount a very quick **secondary response** if reexposed to the same antigen (fig. 14.23). Plasma cells form within hours, and the antibody titer rises much higher than before and peaks within a few days. The response is so rapid that the antigen has little chance to exert a noticeable effect on the body, and no illness results. The IgG level remains elevated for weeks to years, conferring lasting protection. Memory doesn't last as long in humoral immunity, however, as it does in cellular immunity.

14.4c Active and Passive Immunity

Up to this point, we have examined immunity only as a process that occurs in response to unintentional but natural exposure to environmental antigens. However, there are other varieties of immunity bestowed by intentional antigen exposure or by acquiring antibodies produced by another individual.

Any case in which the body produces its own antibodies or T cells is called **active immunity.** What we have studied so far is considered

Figure 14.23 The Primary and Secondary Responses in Humoral Immunity. The left panel shows the primary response, seen on first exposure to a given antigen. The right panel shows the secondary response, seen in an immune person on reexposure. The individual is exposed to the antigen on day 0 in both cases. Note the differences in the speed of response, the height of the antibody titer, and the rate of decline in the antibody titer.

natural active immunity, because it results from natural exposure to antigens. In contrast, *artificial active immunity* is a process in which we deliberately subject ourselves to antigens in order to stimulate antibody or T cell production and produce an immune memory that protects us against future disease—in other words, **vaccination.**

Because of vaccination, certain diseases that were once dreaded harbingers of death are now little more than ghostly memories. A vaccine contains dead or disabled pathogens that are injected or given orally. This stimulates the production of antibodies but does not make the person ill. Vaccination gives the immune system the opportunity to gain experience with the antigen so that it "remembers" it. If one ever encounters it again, a rapid secondary response occurs and prevents it from causing disease. Spectacular success has been achieved with certain vaccines; for example, the World Health Organization launched a campaign in the 1960s to eradicate smallpox worldwide, and no cases have been known to occur naturally since 1977. Smallpox is the only disease thus far to be eradicated, but there is some hope that the world may be rid of polio in the not-too-distant future. A global initiative to eradicate polio was begun in 1988, when there were some 350,000 cases in 125 countries. Thanks to vigorous vaccination efforts, by 2019 that number had dropped to fewer than 100 cases. New cases occur today only in Pakistan and Afghanistan, but the persistence of polio in those countries and a few others is a source of concern.

At one time, optimists hoped that vaccines would wipe out infectious disease. We now know that infectious diseases are so complex that we will never completely conquer them—we are locked in an arms race that has been carried out for millions of years. For example, despite great effort, scientists have yet to develop effective vaccines against global diseases such as malaria, tuberculosis, or AIDS. The global disruption caused by the recent coronavirus pandemic is a stark reminder that despite technological and scientific advances, infectious disease continues to be part of the human story.

We can also acquire disease resistance by **passive immunity,** bestowed by either natural or artificial means. *Natural passive immunity* is the process in which a fetus or infant acquires antibodies from the mother through the placenta before birth or via the breast milk after. It provides important early protection of the infant until it can build its own immunity by exposure to environmental antigens. *Artificial passive immunity* can be acquired at any age by injecting blood serum from another person or from animals (such as horses) that have made antibodies against a specific antigen. Such *immune serum* is used for emergency treatment of snakebites, botulism, tetanus, rabies, and ebola, among others. This century-old medical practice has attracted renewed interest as a possible treatment for patients who are severely ill with COVID-19, the disease caused by the coronavirus (SARS-CoV-2).

Before You Go On

Answer these questions from memory to test your understanding of the previous section. Reread the material if you can't answer the questions.

19. Name four types of lymphocytes involved in cellular immunity. Which of these is also essential to humoral immunity?
20. What are the three phases of cellular and humoral immune responses?
21. Explain how helper and cytotoxic T cells are activated.
22. Describe how cytotoxic T cells destroy target cells.
23. Compare and contrast a B cell and a plasma cell.
24. Discuss three ways antibodies act against antigens.
25. Why does the secondary immune response prevent a pathogen from causing disease, yet the primary response does not?
26. Explain how vaccination works.

PERSPECTIVES ON HEALTH

AIDS: A Global Epidemic

AIDS (acquired immunodeficiency syndrome) is a modern plague that affects every part of the world, but incredibly, was unknown before the early 1980s. Globally, nearly 38 million are infected with the virus that causes AIDS. Each year over 700 thousand people die of the disease. Sub-Saharan Africa is the most heavily affected region, with as many as 25% of adults infected in some areas. The disease has had profound demographic, social, and economic impacts in many parts of the world.

AIDS is caused by *HIV (human immunodeficiency virus)*, which is spread by contact with body fluids such as blood, semen, vaginal secretions, and breast milk. It is most frequently passed from person to person during sexual activity, but may also be spread through infected needles, or from mothers to their infants in breast milk. HIV primarily infects helper T cells. Figure 14.24a shows components of the virus. Its surface molecules bind to corresponding receptors on the plasma membrane of the helper T (T_H) cells. Once inside the cell, the virus may remain latent for anywhere from 2 to 15 years. This is why most people have no clinical symptoms for a period of time after infection. Once the virus becomes active in T_H cells, it replicates and the cells eventually die. As the cells die, millions of new infectious viral particles spew forth into the bloodstream and infect more T cells (fig. 14.24b). A healthy T_H count is 600 to 1,200 cells/μL; a criterion for AIDS is a count less than 200/μL. The scarcity of T_H cells disables the specific immune response and the patient succumbs to opportunistic infections such as pneumonia or tuberculosis or dies from cancer—diseases that are normally kept under control by adaptive immunity.

There is currently no cure for AIDS. The disease can be prevented by avoiding contact with body fluids such as blood or semen that might be infected. Antiretroviral medications that disable the HIV virus and prevent its replication have dramatically improved the survival of HIV-positive persons. However, the cost of these drugs means that they are out of reach for many, especially in less-developed countries where the disease has spread at an alarming rate (mostly through heterosexual activity). The AIDS epidemic has triggered unprecedented efforts to develop a vaccine or cure, but the challenges to doing so are daunting. The virus is one of the fastest-evolving microbes that exists. It reproduces incredibly rapidly—a single virus can produce billions of copies in a day. In addition, it accumulates numerous mutations as it copies its genetic material, and therefore is constantly reinventing itself. It thus presents a moving target that frustrates attempts to develop a vaccine with an effective antigen.

HIV and AIDS dramatically illustrate the importance of understanding the fundamental mechanisms of the immune system. They also highlight the vulnerability of the human species to disease, even in this era of sophisticated medical intervention.

Figure 14.24 The Human Immunodeficiency Virus (HIV). (a) Structure of the virus. (b) Viruses emerging from a dying helper T cell. Each virus can now invade a new helper T cell and produce a similar number of descendants.

b: NIBSC/SPL/Science Source

Aging of the Lymphatic and Immune Systems

The amounts of lymphatic tissue and red bone marrow decline with age. Consequently, there are fewer hematopoietic stem cells, disease-fighting leukocytes, and antigen-presenting cells (APCs). Also, the lymphocytes produced in these tissues often fail to mature or become immunocompetent. Both humoral and cellular immunity depend on APCs and helper T cells, and therefore both types of immune response are blunted. As a result, an older person is less protected against cancer and infectious diseases. It becomes especially important in old age to be vaccinated against influenza and other acute seasonal infections.

CAREER SPOTLIGHT

Public Health Nurse

Public health nurses are registered nurses (RNs) who usually work in government, for private nonprofit agencies, or in clinics. They focus on the overall health of communities, working to educate individuals, families, and other groups about health-care issues, disease prevention, nutrition, and child care. They collaborate with community leaders, teachers, parents, and physicians in ensuring the health of the community. The tasks of a public health nurse focus on prevention and education and include vaccinating children against childhood diseases, administering flu shots, and screening populations for hypertension. Qualifications vary from state to state—some require a bachelor's degree in nursing whereas in other states an associate degree from a community college suffices. See appendix B for further information on careers in public health nursing.

Christopher Futcher/Getty Images

CONNECTIVE ISSUES

Ways in Which the LYMPHATIC AND IMMUNE SYSTEMS Affect Other Organ Systems

All Systems
Lymphatics drain excess tissue fluid and remove cellular debris and pathogens from tissues. The immune system defends other systems from infection and cancer.

Integumentary System
Lymphatic drainage prevents cutaneous edema; dendritic cells of the immune system densely populate the skin and guard against invasion.

Skeletal System
Interleukins affect the rate of hematopoiesis in the red bone marrow.

Muscular System
Lymphatics prevent muscular edema.

Nervous System
Microglia of the immune system protect tissues of the central nervous system from agents of disease.

Endocrine System
Lymph transports some hormones.

Circulatory System
Lymphatics return fluid to the bloodstream and help to sustain blood volume; the spleen and other lymphatic organs prevent accumulation of debris and foreign matter in blood; the spleen disposes of old RBCs.

Respiratory System
Alveolar macrophages remove debris from lungs; lymphatics keep the pulmonary alveoli from filling with fluid.

Urinary System
Lymphatics absorb fluid and proteins in the kidneys, helping the kidneys to concentrate the urine and conserve water.

Digestive System
Lymph absorbs dietary lipids from the small intestine and transports them to the liver; hepatic macrophages play a defensive role in the liver.

Reproductive System
Ovaries and testes require immune barriers to protect eggs and sperm against immune attack.

Study Guide

Assess Your Learning Outcomes

To test your knowledge, discuss the following topics with a study partner or in writing, ideally from memory.

14.1 The Lymphatic System

1. Definitions of *lymphatic system, immune system,* and *immunology*
2. The types of pathogens
3. Major components and functions of the lymphatic system
4. Types and structure of lymphatic vessels
5. Origin and composition of lymph and the route of its flow
6. Lymphatic tissues, their locations and histology, and how they differ from lymphatic organs
7. The types, locations, structure, and function of tonsils
8. The location, anatomy, and functions of the thymus
9. The location, anatomy, and functions of the spleen
10. Anatomy, histology, and functions of lymph nodes, and areas of lymph node concentration in the body
11. The relationship of lymph nodes to lymphatic vessels and how this relates to their lymph-cleansing function

14.2 Innate Immunity

1. Definitions of *innate* and *adaptive immunity* and differences between the two
2. The role of skin and mucous membranes as a first line of defense
3. How each leukocyte type functions in immune defense
4. The role of natural killer (NK) cells in recognizing and destroying enemy cells
5. The role of interferons in defense
6. The definition of *complement system* and ways in which it leads to the destruction of pathogens
7. The definition of *inflammation* and its four cardinal signs
8. Definitions of *adhesion, diapedesis,* and *chemotaxis* and the role of neutrophils in inflammation
9. The mechanisms by which inflammation combats a pathogen and promotes healing
10. Benefits and dangers of fever
11. The roles of pyrogens, prostaglandins, and the hypothalamus in fever

14.3 Features of Adaptive Immunity

1. Two characteristics of adaptive immunity that distinguish it from innate immunity
2. The two types of adaptive immunity, and their similarities and differences
3. Definition of *antigen* and why antigens are fundamental to specific immunity
4. How haptens differ from other antigens, how they become antigenic, and where one might encounter haptens in everyday life
5. Similarities and differences between T and B lymphocytes
6. The life history of a T cell from its "birth" until it is ready to do battle
7. Similarities and differences between the life history of a B cell and that of a T cell
8. The basic structure and varied classes of antibodies, and how antibody structure relates to the ability to recognize and bind a specific antigen
9. Some types of antigen-presenting cells and their role in the immune response
10. Definition of *interleukins* and their role in immunity

14.4 Cellular and Humoral Immunity

1. The four types of T cells associated with cellular immunity
2. Differences between T_H cells and T_C cells in the types of antigens they recognize
3. How antigen-presenting cells help T cells recognize enemy antigens
4. The roles of clonal selection and interleukins in cellular and humoral immunity
5. The nature of a T_C cell's lethal hit on an enemy cell
6. The mechanism of memory and the T cell recall response in cellular immunity
7. The process of antigen recognition and B cell activation in humoral immunity
8. The relationship between a B cell and a plasma cell, and their structural and functional differences
9. Why we say only that antibodies indirectly promote the destruction of antigens rather than destroying them directly
10. Three ways in which antibodies promote antigen destruction
11. The mechanism of memory and the secondary response in humoral immunity; similarities and differences between memory in the two forms of adaptive immunity
12. The difference between active and passive immunity and the reason artificial immunity is associated with vaccination

Testing Your Recall

1. Classic signs of inflammation include all of the following *except*
   a. edema.
   b. *fever.*
   c. redness.
   d. pain.
   e. heat.

2. Which of the following cells are involved in innate but not adaptive immunity?
   a. helper T cells
   b. cytotoxic T cells
   c. *natural killer cells*
   d. B cells
   e. plasma cells

3. Histamine produced by _____ acts to _____ blood vessels.
   a. neutrophils; constrict
   b. *basophils; dilate*
   c. eosinophils; dilate
   d. NK cells; constrict
   e. cytotoxic T cells; rupture

4. Neutrophils use all of the following to combat bacteria except
   a. phagocytosis.
   b. hydrogen peroxide.
   c. *antibodies.*
   d. extracellular traps.
   e. hypochlorite.

5. All of the following are important in innate immunity *except*
   a. histamine.
   b. pyrogens.
   c. inflammation.
   d. *clonal selection.*
   e. interferons.

6. Where do T cells attain immunocompetence?
   a. in the red bone marrow
   b. in the lymph nodes
   c. *in the thymus*
   d. in the spleen
   e. in the mucosa-associated lymphatic tissue

7. The only lymphatic organ with both afferent and efferent lymphatic vessels is
   a. the spleen.
   b. *a lymph node.*
   c. a palatine tonsil.
   d. the pharyngeal tonsil.
   e. the thymus.

8. Antibodies cannot get inside a host cell to destroy a viral infection. Therefore, we use _____ to combat viral infections.
   a. interleukins
   b. complement
   c. humoral immunity
   d. haptens
   e. *cytotoxic T cells*

9. Which statement is *not* true of antibodies?
   a. *They directly kill infected cells by using a lethal hit.*
   b. They are proteins.
   c. They have constant regions that define the antibody class.
   d. They are produced by plasma cells.
   e. They activate the complement system.

10. All lymph ultimately reenters the bloodstream at what point?
    a. the right atrium
    b. the common carotid artery
    c. the internal iliac veins
    d. *the subclavian veins*
    e. the inferior vena cava

11. Any organism capable of causing disease is called a/an _pathogen_.

12. The lymphatic vessels of the small intestine that absorb dietary lipids are called _lacteals_.

13. Cells that inhibit T-cell responses once a pathogen has been eradicated are called _____.

14. _____ are small proteins, synthesized in response to viral infections, that inhibit viral replication and activate natural killer (NK) cells.

15. B cells become _plasma_ cells before they begin to secrete antibodies.

16. Cells that process antigens and display fragments of them to activate the adaptive immune response are called _antigen-presenting cells_.

17. Tonsillitis most often involves inflammation of the _palatine_ tonsils.

18. The _thymus_ is located in the mediastinum, and is large in childhood but atrophies after about age 15.

19. _neutrophils_ are phagocytic cells that wander in connective tissues and kill bacteria by releasing a lethal cloud of chemicals, including hydrogen peroxide and hypochlorite.

20. Basophils release a chemical called _histamine_ that causes vasodilation.

Answers in Appendix A

What's Wrong with These Statements?

Briefly explain why each of the following statements is false, or reword it to make it true.

1. Cytotoxic T cells play roles in ~~both innate and~~ adaptive immunity. [only]
2. Perforins have a role only in adaptive immunity.
3. Lymphatic capillaries are ~~less~~ [more] permeable than blood capillaries.
4. T lymphocytes are involved exclusively in cellular [and humoral] immunity.
5. Once a person has been vaccinated against a disease, if they encounter the antigen later in life, a rapid ~~primary~~ [secondary] response will occur.
6. Lymph nodes are populated by ~~B cells but not T cells~~ [Both].
7. HIV infects primarily ~~B cells~~ [Helper T cells].
8. Most plasma cells form in the ~~red bone marrow~~ [lymph nodes].
9. The complement proteins of the blood plasma come primarily from ~~plasma cells~~ [liver].
10. Lymphatic nodules are a prominent feature of the tonsils, lymph nodes, ~~thymus,~~ and ~~spleen~~.

Answers in Appendix A

Testing Your Comprehension

1. Severe combined immunodeficiency disease (SCID) is a genetic disorder that results in scarcity or absence of both T and B cells from birth. Explain why children with SCID must live in protective enclosures, such as sterile plastic chambers and suits. How are SCID and AIDS similar? How do they differ?

2. In treating a woman for malignancy in the right breast, a surgeon removes some of her axillary lymph nodes. Following the surgery, the patient experiences edema of her right arm. Explain why.

3. A burn research center uses mice for studies of skin grafting. To prevent rejection of grafted tissues, the mice are thymectomized at birth (their thymus is removed). Even though B cells do not develop in the thymus, the mice show no humoral response and are very susceptible to infection. Explain why the removal of the thymus would improve the success of skin grafts but adversely affect humoral immunity.

PART 5 Intake and Output

The Respiratory System

Chapter 15

Chapter Outline

15.1 Functions and Anatomy of the Respiratory System
- 15.1a Respiratory Functions
- 15.1b Respiratory Anatomy

15.2 Pulmonary Ventilation
- 15.2a The Mechanics of Breathing
- 15.2b Measurements of Ventilation
- 15.2c Neural Control of Breathing

15.3 Gas Exchange and Transport
- 15.3a Composition of Air
- 15.3b Alveolar Gas Exchange
- 15.3c Gas Transport
- 15.3d Systemic Gas Exchange
- 15.3e Blood Gases and the Respiratory Rhythm

Clinical Applications/Perspectives on Health
- Clinical Application 15.1: Asthma
- Clinical Application 15.2: High-Altitude Physiology
- Perspectives on Health

End of Chapter
- Aging of the Respiratory System
- Career Spotlight: Respiratory Therapist
- Connective Issues
- Study Guide

Branching of the bronchial trees into the lungs, seen from below in this artistic conception. Each lung receives one main bronchus, which ultimately branches into 8 million microscopic alveolar sacs and 150 million alveoli, where the bloodstream picks up oxygen.
Anatomical Travelogue/Science Source

BASE CAMP

Before ascending to the next level, be sure you're properly equipped with a knowledge of these concepts from earlier chapters.

- Thoracic cavity anatomy (see section 1.3e)
- Types of epithelial tissue (see section 4.2)
- Serous and mucous membranes (see section 4.5b)
- Bones of the nasal region (see section 6.3a)
- Brainstem anatomy (see section 9.2a)
- Pulmonary blood circulation (see section 13.6a)

 Anatomy & Physiology Revealed® 4.0 | Module 11 Respiratory System

Breath represents life. The first breath of a baby and the last gasp of a dying person are two of the most dramatic moments of human experience. It is evident that one can't live very long without breathing. But why do we breathe? It comes down to the fact that our cells, tissues, and organs depend on the energy carrier ATP to sustain their activities. By far the greatest amount of ATP synthesis occurs in mitochondria, requires oxygen, and produces carbon dioxide. Breathing supplies oxygen and eliminates carbon dioxide. The respiratory system consists essentially of passageways that deliver air to the lungs, where oxygen diffuses into the blood and carbon dioxide diffuses out. The respiratory system collaborates closely with the circulatory system to provide oxygen to the tissues and remove carbon dioxide from them.

15.1 Functions and Anatomy of the Respiratory System

Expected Learning Outcomes

When you have completed this section, you should be able to

a. explain the functions of the respiratory system;
b. name and describe the structures of the respiratory system; and
c. trace the passage of air from the nose to the alveoli.

15.1a Respiratory Functions

The obvious function of the respiratory system is to obtain oxygen (O_2) from the atmosphere and to expel the carbon dioxide (CO_2) generated by our metabolism. However, the respiratory system has a broader range of functions than commonly supposed:

- **Gas exchange.** Exchange of oxygen and carbon dioxide between the atmosphere and blood occurs in the lungs.
- **Communication.** The babbling of babies, complex speech, and vocalizations such as laughing and crying result from air passing through respiratory passageways.
- **Sense of smell.** Chemoreceptors in the nose allow us to perceive odors.
- **Acid–base balance.** CO_2 reacts with water to produce acid. Therefore, if our blood pH rises too high, we can lower it by slowing our breathing to let a little more CO_2 accumulate. Conversely, if the pH falls too low, it can be corrected (raised) by exhaling CO_2 faster than the body produces it.
- **Expulsion of abdominal contents.** A breath-holding Valsalva maneuver helps to expel abdominal contents during urination, defecation, and childbirth.

15.1b Respiratory Anatomy

The **respiratory system** consists of air passages from the nose to the lungs and gas-exchange surfaces in the lungs themselves (**fig. 15.1**). The air we breathe flows through a series of ever smaller tubes until it finally reaches millions of tiny sacs

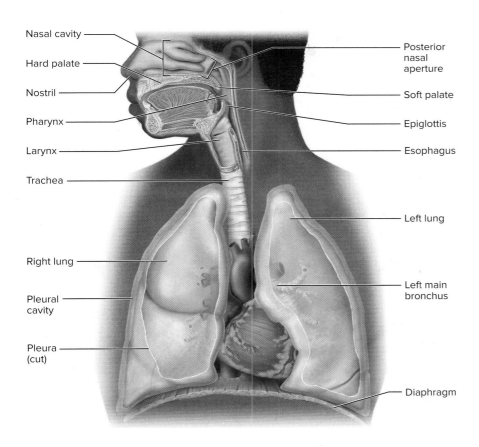

Figure 15.1 The Respiratory System. APR

deep in the lungs; there, the thin walls allow oxygen and carbon dioxide to be quickly exchanged between the air and blood. The respiratory system is often divided into the **upper respiratory tract,** which includes structures in the head and neck (from nose through larynx), and the **lower respiratory tract,** which includes structures in the thorax (from trachea through lungs).

The Nose

The lower half of the nose is shaped mostly by hyaline cartilage and the upper half by bone. You can easily find the boundary between bone and cartilage by palpating your own nose. During normal breathing, air is drawn through the **nares** (NAIR-eze) or **nostrils,** the anterior boundary of the nose (fig. 15.2). The nostrils have stiff **guard hairs,** or **vibrissae** (vy-BRISS-ee), that prevent insects and large airborne particles from entering. The nose extends posteriorly to the **posterior nasal apertures,** or **choanae**[1] (co-AH-nee), that lead into the throat.

The **nasal cavity** is divided into right and left halves by a wall, the **nasal septum,** composed of bone and hyaline cartilage. The roof and walls of the cavity are formed by the ethmoid and sphenoid bones. The floor is formed by the **palate,** which separates the nasal cavity from the oral cavity and allows you to breathe while chewing food. The palate is divided into a bony *hard palate* anteriorly and a fleshy *soft palate* posteriorly. The lateral walls of the nasal cavity have inward folds called **nasal conchae** that reach almost all the way to the septum, thus occupying most of the cavity. They cause the airflow to be quite turbulent; as the inhaled air contacts the mucous membrane covering the conchae, it is warmed, humidified, and cleansed of most of its debris (such as particles of dust and pollen).

---

[1]*choana* = funnel

498 PART 5 Intake and Output

Figure 15.2 Anatomy of the Upper Respiratory Tract. (a) Median section of the cadaver head. (b) Internal anatomy. (c) The nasal septum and regions of the pharynx. APR

a: Rebecca Gray/McGraw-Hill Education

? *Draw or describe a line across part (b) of this figure to indicate the boundary between the upper and lower respiratory tract.*

Most of the nasal cavity is lined with **respiratory epithelium** of the ciliated, pseudostratified columnar type (see fig. 4.7). Nasal mucus is secreted by *goblet cells* of the respiratory epithelium, glands deeper in the mucosa, and the *paranasal sinuses,* which drain into the nose. Inhaled dust, pollen, bacteria, and other foreign matter cling to the sticky mucus, which is swept by cilia into the pharynx. The mucus and its load of debris are then swallowed and most foreign material is kept from the lungs. A small portion of the epithelium in the roof of the nasal cavity is **olfactory epithelium,** responsible for the sense of smell. It is also pseudostratified columnar, but its cilia are immobile and serve to bind odor molecules rather than move mucus.

The Pharynx

The **pharynx**[2] (FAIR-inks), or throat, is a muscular funnel that facilitates swallowing and speech. It extends from the posterior nasal apertures to the larynx and esophagus and consists of three regions: the *nasopharynx, oropharynx,* and *laryngopharynx* (fig. 15.2c). The nasopharynx is a space superior to the soft palate that serves as a passageway for air. It houses the pharyngeal tonsil, a patch of lymphoid tissue that is well positioned to respond to airborne pathogens. The **auditory (eustachian) tube** from the middle ear opens into the nasopharynx. This explains why an infection in the nose and throat may cause a middle-ear infection *(otitis media)* (see Clinical Application 10.2).

The oropharynx lies between the soft palate and root of the tongue. It contains the palatine and lingual tonsils and serves as a passageway for both food and air. The laryngopharynx (la-RIN-go-FAIR-inks) lies mostly behind the larynx, ending where the esophagus begins. The nasopharynx is lined with pseudostratified columnar epithelium, whereas the oropharynx and laryngopharynx are lined with the more abrasion-resistant stratified squamous epithelium to accommodate the passage of food.

The Larynx

The **larynx** (LAIR-inks), or "voice box," is a chamber composed mainly of cartilage and muscle (fig. 15.3). Its main function is to prevent food and drink from entering the airway, but it has evolved the additional role of sound production.

[2]*pharynx* = throat

Figure 15.3 Anatomy of the Larynx. Most muscles are removed in order to show the cartilages.

Figure 15.4 Endoscopic View of the Larynx. Looking into the glottis with a laryngoscope.
Phototake

The larynx is framed by nine cartilages. The largest of these is the **thyroid[3] cartilage,** named for its shieldlike shape; it has an anterior peak *(laryngeal prominence)* commonly called the "Adam's apple." It is especially visible in men because testosterone promotes its growth during puberty. It forms most of the "box" of the voice box and provides anterior anchorage for the vocal cords. A large *cricoid cartilage* and its ligaments attach the inferior end of the larynx to the trachea. Other smaller cartilages provide support to soft tissues, operate the vocal cords, and shape the epiglottis above.

The **epiglottis**[4] is a spoon-shaped flap that seals the laryngeal opening during swallowing to keep food and liquid out of the airway. When you swallow, the root of the tongue pushes the epiglottis downward while muscles attached to the thyroid cartilage pull the larynx up to meet it. The epiglottis caps the opening like a lid and diverts food and drink to the esophagus, behind the larynx. If food or liquid gets past the epiglottis, it triggers a *cough reflex* to expel it and, ideally, prevent it from getting as far as the trachea or lungs. The epiglottis and smaller cartilages of the larynx serve as important landmarks for the placement of a breathing tube by emergency personnel, a procedure called *endotracheal intubation.*

Within the larynx, the **vocal cords** stretch like a V from the midpoint of the thyroid cartilage in front to two small, mobile cartilages posteriorly (**fig. 15.4**). The vocal cords and the opening between them are collectively called the *glottis*. Airflow through the glottis vibrates the vocal cords and produces sound. Under the control of several small muscles, the cartilages at the posterior ends of the cords pivot, altering the tension on the cords and the angle between them. This alters the pitch of the voice. The larynx alone produces only crude sounds that some have compared to a hunter's duck call; the pharynx, oral cavity, cheeks, tongue, and lips shape these sounds into intelligible words. Longer and less taut vocal cords produce a lower-pitched sound than shorter or tauter cords; for this reason, men generally have lower-pitched voices than women, and adults lower than children.

The Trachea

The **trachea,**[5] or "windpipe," is about 12 cm (4.5 in.) long and 2.5 cm (1 in.) in diameter (**fig. 15.5**). At its inferior end, it forks into right and left *bronchi*. It is lined with a ciliated pseudostratified columnar epithelium rich in mucus-producing goblet cells. Cilia sweep the mucus and foreign particles upward toward the throat, away from the lungs. This **mucociliary escalator** effectively removes potentially harmful debris, including microorganisms that could otherwise cause pulmonary infections.

The trachea is supported by 16 to 20 C-shaped rings of hyaline cartilage, some of which can be felt at the front of the neck. Similar to the wire spiral in a vacuum cleaner hose, the cartilage rings reinforce the trachea and keep it from collapsing when we inhale. The open part of the C faces posteriorly and allows the esophagus to expand and push into the muscular posterior wall of the trachea when food is swallowed.

> **Apply What You Know**
> Chemicals in cigarette smoke paralyze cilia of the respiratory epithelium. How does this affect the mucociliary escalator? What are the possible consequences for a person who smokes?

---

[3]*thyr* = shield; *oid* = resembling
[4]*epi* = above, upon
[5]*trache* = rough

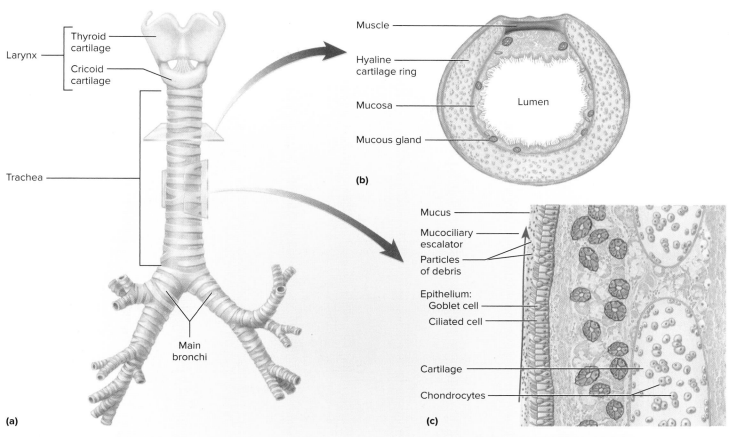

Figure 15.5 Anatomy of the Lower Respiratory Tract.
(a) Anterior view. (b) Cross section of the trachea showing the C-shaped cartilage rings. (c) Longitudinal section of the trachea showing the action of the mucociliary escalator. (d) Tracheal epithelium showing ciliated cells and goblet cells. **APR**

Prof. P.M. Motta/Univ. "La Sapienza," Rome/Science Source

❓ Why do inhaled objects more often go into the right main bronchus than into the left?

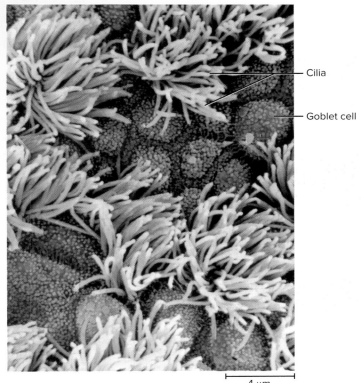

The Lungs, Pleurae, and Bronchial Tree

The **lungs** are soft, spongy organs that occupy most of the thoracic cavity. They are separated from each other by a partition called the **mediastinum,** which contains the heart, esophagus, and major blood vessels. The lungs are roughly cone-shaped with a broad *base* inferiorly; a superior peak, the *apex;* a broad, curved *costal surface* that lies against the rib cage; and a *mediastinal surface* facing the heart (fig. 15.6). Each lung has a slit *(hilum)* on the mediastinal surface through which it receives bronchi, blood vessels, nerves, and lymphatics. The two lungs are not symmetrical. The right lung is divided into three lobes, whereas the left has two. The left lung has an indentation called the *cardiac impression* to accommodate the left-leaning apex of the heart.

A serous membrane, the **pleura,**[6] lines the thoracic wall and adheres to the lungs. It consists of two layers—the **visceral pleura** that forms the outer surface of the lung, and the **parietal pleura** that lines the inner surface of the rib cage. The space between

---

[6]*pleur* = side of body, rib

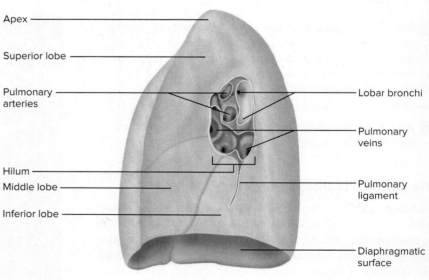

Figure 15.6 Gross Anatomy of the Lungs.
APR

them, called the **pleural cavity** (fig. 15.7), contains only a thin film of **pleural fluid**. This is a lubricant that reduces friction as the lungs expand and contract. Infection of the pleurae produces a condition called *pleurisy*, in which the membranes may become roughened and rub together, making each breath a painful experience. Chest wounds can allow air to enter the pleural cavity—a condition called *pneumothorax*. The membranes then separate and the lung collapses. This is treated by drawing the air out through a *chest tube* inserted between the ribs.

The **bronchial tree** is a series of branching passages that carry air from the trachea to the gas-exchange surfaces in the lungs. It begins where the trachea splits into two **main (primary) bronchi** (BRONK-eye) that lead to the lungs (fig. 15.6a). The right main bronchus is slightly wider and more vertical than the left one.

Like the trachea, the main bronchi are supported by rings of hyaline cartilage and have a layer of smooth muscle in the wall. The main bronchi branch into **lobar (secondary) bronchi** that supply each lobe of the lungs. These branch into **segmental (tertiary) bronchi,** which lead eventually to **bronchioles** (BRON-kee-oles), tubes less than 1 mm in diameter and with no cartilage at all (fig. 15.8). Smooth muscle is the main supportive tissue of the bronchioles, and unhindered by cartilage, they have more capacity to dilate and constrict than the larger bronchi do. The changes in diameter, **bronchodilation** and **bronchoconstriction,** regulate airflow.

The bronchioles branch into smaller divisions called *terminal bronchioles.* From the nasal cavity through terminal bronchioles, the airway is called the **conducting zone** because it functions only for airflow. Its walls are too thick for efficient gas diffusion to and from the blood. Beyond this, the passages have such thin walls that they do exchange gases with the blood and therefore they constitute the **respiratory zone.**

Figure 15.7 Lungs in Relation to the Thoracic Cavity. This cross section of the cadaver is oriented the same way as the reader's body. The pleural cavity is especially evident where the left lung has shrunken away from the thoracic wall, but in a living person the lung fully fills this space, the parietal and visceral pleurae are pressed together, and the pleural cavity is only a potential space between the membranes, as on the right side of this photograph.

Rebecca Gray/Don Kincaid/McGraw-Hill Education

Figure 15.8 Histology of the Lung. (a) Light micrograph. (b) Scanning electron micrograph. Note the spongy texture of the lung.

a: MICROSCAPE/Science Source; b: Biophoto Associates/Science Source

> ### Clinical Application 15.1
>
> **ASTHMA**
>
> Asthma sufferers are familiar with the symptoms of an attack—shortness of breath, wheezing, and coughing. It affects about 10% of children in the United States, making it the most common cause of chronic illness among children. Exposure to a trigger such as cat dander or pollen causes swollen and inflamed airways and release of chemicals that cause violent and sudden spasms of the bronchi. In the short term, asthma can be medically managed with inhaled drugs called *β-adrenergic agonists* that mimic the effect of epinephrine (adrenaline) and dilate the bronchioles. Such inhalers treat acute symptoms but not the underlying causes. In more severe cases, patients receive inhaled corticosteroids on a daily basis to suppress inflammation. Recently, researchers have tried to answer the question of why asthma is so prevalent in modern populations, in contrast to its relative scarcity even as late as the nineteenth century. A hypothesis that has received some support is that enhanced hygiene in modern societies has led to reduced exposure to bacteria and parasitic worms in childhood. The immune system may not develop as robustly in individuals who lack such exposure to pathogens. In addition, women in industrialized countries have the choice of whether to breast-feed or bottle feed. Individuals who are not breast-fed may lack stimulants to the developing immune system that are present in breast milk. This evolutionary point of view provides us with insight into why the disease has become one of the most prevalent conditions in the modern world.

The passages of the respiratory zone, in order, are *respiratory bronchioles,* some of which are still ciliated; then *alveolar ducts,* which have a simple squamous epithelium and no cilia; and finally, *alveolar sacs,* which are clusters of tiny air sacs called *alveoli.*

In summary, the path of airflow is as follows. The first several passages belong to the conducting zone: nasal cavity → pharynx → trachea → main bronchus → lobar bronchus → segmental bronchus → bronchiole → terminal bronchiole. Then begins the respiratory zone: respiratory bronchiole → alveolar duct → alveolus.

The Alveoli

An **alveolus** (AL-vee-OH-lus) is a pouch about 0.2 to 0.5 mm in diameter (**fig. 15.9**). Each lung contains approximately 150 million alveoli, which provide a large surface area for gas exchange. Alveoli are lined mostly with *squamous alveolar cells* whose thinness allows for rapid gas diffusion between air and the blood. In addition, *great alveolar cells* are found in the walls of each alveolus. They secrete *pulmonary surfactant,* a substance that prevents the walls of the alveoli from sticking together. Without surfactant, the surfaces of the alveoli would cling together like wet paper, making it difficult for them to inflate during inhalation. The importance of surfactant is apparent when it is lacking, as it often is in premature infants whose immature lungs don't produce surfactant on their own. This used to be a significant cause of death among those born prematurely, but treatment with artificial surfactant has substantially improved their chances of survival. A third cell type in the alveolus is the *alveolar macrophages,* which wander the alveoli and phagocytize dust particles or pathogens that escape the mucociliary escalator. They are the most numerous of all cells in the lung, and as many as 100 million perish each day as they ride up the mucociliary escalator to be swallowed and digested, thus ridding the lungs of debris.

Each alveolus is covered with a web of blood capillaries. Gases are exchanged across a thin **respiratory membrane** that consists of the squamous alveolar cell, the squamous endothelial cell that lines the capillary, and their shared thin basement membrane. The total thickness of the respiratory membrane is only 0.5 μm—a very minimal barrier to gas exchange. By comparison, a red blood cell has a diameter 15 times as great, about 7.5 μm.

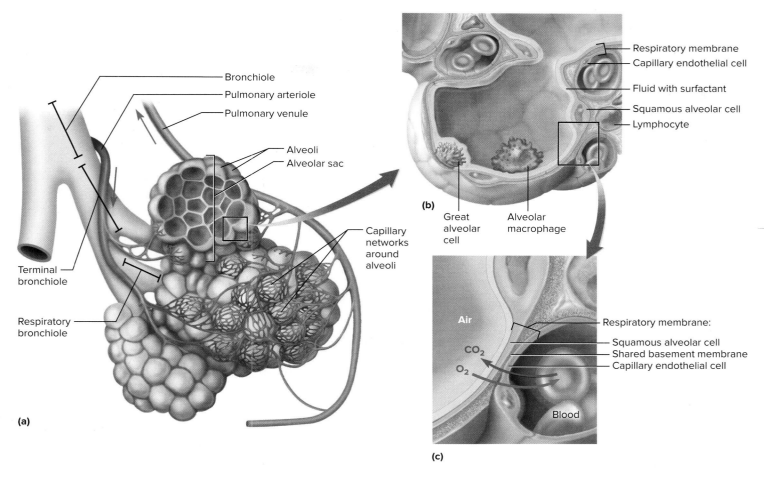

Figure 15.9 Pulmonary Alveoli. (a) Clusters of alveoli and their blood supply. (b) Structure of an alveolus. (c) Structure of the respiratory membrane.

Blood Supply to the Lungs

The lungs receive two blood supplies—pulmonary and systemic. The pulmonary arteries lead to the capillaries surrounding the alveoli. Their only purpose is gas exchange. The pulmonary veins lead immediately back to the heart (see "The Pulmonary Circuit" in section 13.6a). The systemic supply comes from about five *bronchial arteries* arising from the thoracic aorta (see fig. 13.22). They serve to nourish the lung tissues and pleurae, leading to capillaries in the bronchi and other lung tissues but not to the alveoli. *Bronchial veins* leaving the lungs drain indirectly, via the *azygos system* (see fig. 13.23), into the superior vena cava.

Before You Go On

Answer these questions from memory to test your understanding of the previous section. Reread the material if you can't answer the questions.

1. A dust particle is inhaled and gets into an alveolus without being trapped along the way. Describe the path it takes, naming all air passages from the external nares to the alveolus. What would happen to it after arrival in the alveolus?
2. Explain how the structure of the larynx prevents choking and enables the production of sound.
3. Describe the relationship between the visceral pleura, parietal pleura, lungs, and thoracic wall. Why can we say the lungs are "surrounded by" the pleural cavity but not "in" the pleural cavity?
4. List the cell types found in the alveoli and explain their functions.

15.2 Pulmonary Ventilation

Expected Learning Outcomes

When you have completed this section, you should be able to

a. explain how differences in pressure between the atmosphere and lungs account for the flow of air into and out of the lungs;
b. name the muscles of respiration and describe their roles in inspiration and expiration;
c. define the clinical measurements of pulmonary ventilation; and
d. describe the brainstem centers that control breathing and the inputs they receive from other parts of the nervous system.

15.2a The Mechanics of Breathing

Breathing, or **pulmonary ventilation,** consists of **inspiration** (inhaling) and **expiration** (exhaling). One complete inspiration and expiration is called a **respiratory cycle.**

Airflow into and out of the lungs is driven by differences in pressure between the atmosphere and the lungs. Like blood and other fluids, air flows from a point of high pressure to a point of lower pressure. If **atmospheric (barometric) pressure** is greater than **intrapulmonary pressure** (pressure inside the lungs), air flows into the lungs; if intrapulmonary pressure exceeds atmospheric pressure, air flows out. A person breathes by raising and lowering the intrapulmonary pressure so that it cyclically falls below and rises above the atmospheric pressure.

Pressure changes inside the lungs that drive ventilation result from contraction and relaxation of the diaphragm and other muscles that alter pressure in the thoracic cavity. For any container of a given volume, if the volume increases, the pressure falls, and vice versa. Consider a syringe, for example. If you pull back the plunger, you increase the volume within the syringe, causing the internal pressure to drop. When it falls below the surrounding air pressure, air flows into the syringe. Conversely, if you push the plunger in, you reduce the volume and increase the pressure within the syringe. When its internal pressure rises above the surrounding air pressure, air flows out. The respiratory muscles of the thoracic cavity have the same effect on volume and pressure within the thorax, and on airflow into and out of the lungs.

When dealing with thoracic pressures, we must use a new unit of measure, different from what you're accustomed to. In chapter 13, we used millimeters of mercury (mm Hg) as a measure of blood pressure. This is a measure of how high up a vacuum tube a force such as blood pressure or the weight of the atmosphere can push a column of mercury. Mercury is a very heavy liquid, so we use it because pressures can be measured with a relatively short column of mercury, as in the sphygmomanometers of a doctor's office. But the pressures in respiratory airflow are so small that they couldn't move a mercury column much at all; mercury-based instruments aren't sensitive enough. Respiratory physiologists therefore traditionally used water columns, which are more sensitive, and the pressures are measured in centimeters of water (cm H_2O). One mm Hg is about 13.6 mm H_2O (~1.4 cm H_2O). Small pressure changes will move a column of water more than a column of mercury; one can see them and measure them more accurately.

The changes in pressure during inspiration and expiration are shown in figure 15.10. At the beginning (step ① in the figure), there is no movement of the thoracic cage, no difference between the air pressure within the lungs and the ambient atmospheric pressure, and no airflow. When the ribs swing up and out during inspiration (step ② in the figure), the parietal pleura follows. The visceral pleura (lung surface) clings to the parietal pleura and is pulled outward with it; the entire lung expands along with the thoracic cage. As in the syringe analogy,

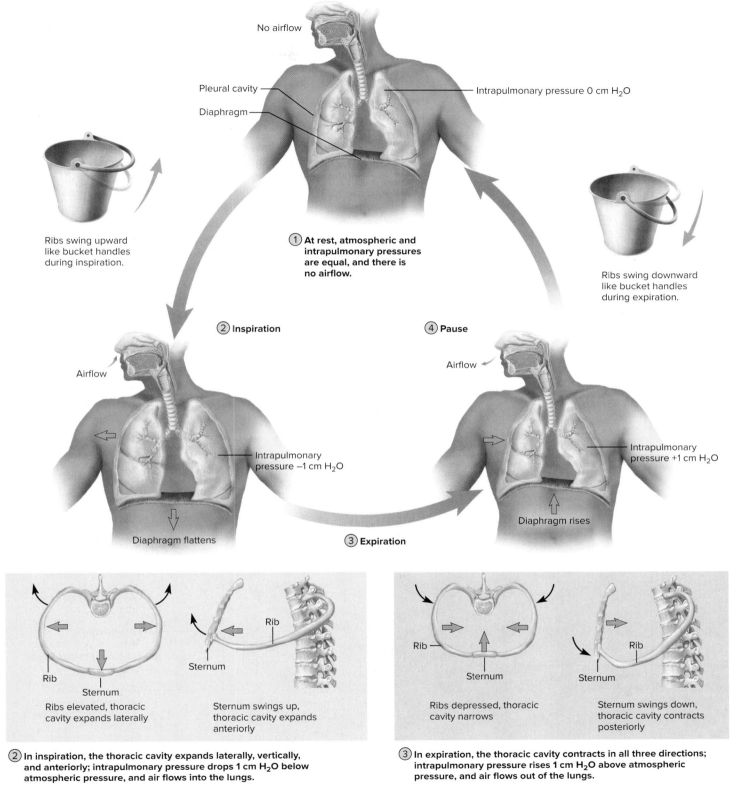

Figure 15.10 The Respiratory Cycle. All pressures given here are relative to atmospheric pressure, which is considered to be zero as a point of reference. Note that pressures governing respiratory airflow are measured in cm H_2O (centimeters of water), not mm Hg. Like bucket handles, each rib is attached at both ends (to the spine and sternum) and swings up and down during inspiration and expiration. **APR**

the increase in volume leads to a decrease in pressure. So now there is a pressure gradient from the atmospheric pressure to the negative pressure in the alveoli, and air flows into the lungs.

During relaxed breathing, expiration is a passive process achieved mainly by the elastic recoil of the thoracic cage. Many structures of the thorax are elastic: the costal cartilages, the ligaments that attach the ribs to the spine, and elastic tissue in the bronchi and bronchioles. As the diaphragm and other muscles of inspiration relax, the structures spring back to resting size and the thoracic cage decreases in volume and compresses the lungs. This raises the intrapulmonary pressure above atmospheric pressure and air flows out of the lungs (step ③ in the figure).

The principal muscle (prime mover) of pulmonary ventilation is the **diaphragm,** the muscular dome that separates the thoracic and abdominal cavities (fig. 15.11). It alone accounts for about two-thirds of the pulmonary airflow. When relaxed, it bulges upward and presses against the base of the lungs, and the lungs are at their minimum volume. When the diaphragm contracts, it flattens, dropping about 1.5 cm in relaxed inspiration and as much as 7 cm in deep breathing. This enlarges the thoracic cavity (including the lungs), reduces intrapulmonary pressure, and produces an inflow of air. Other muscles assist the diaphragm. The **external intercostal muscles** pull the ribs upward and the sternum forward, increasing the diameter of the chest. In quiet breathing, the dimensions of the thoracic cage increase by only 2 to 5 mm in each direction, but combined with the movement of the diaphragm, this increases

Figure 15.11 The Respiratory Muscles. Boldface indicates the principal respiratory muscles; the others are accessory. Blue arrows indicate muscular actions of inspiration and green arrows indicate expiration. The arrows show the direction of muscle action. Muscles listed on the left are active during inspiration and those on the right are active during forced expiration. Note that the diaphragm is active in both phases.

thoracic volume by approximately 500 mL. Thus, during a typical inspiration, about 500 mL of air flows into the respiratory tract.

Quiet expiration is mostly passive, and doesn't require muscular action. During forced expiration—for example, when singing or shouting, coughing or sneezing, exercising, or playing a wind instrument—the *rectus abdominis* pulls down on the sternum and lower ribs, while part of the **internal intercostal muscles** pulls the other ribs downward. These actions reduce the chest diameter and help to expel air more rapidly and thoroughly. Singers and public speakers employ "abdominal breathing," in which the *transverse abdominal* and *abdominal oblique muscles* compress the abdominal cavity and push some of the viscera such as the stomach and liver upward against the diaphragm. This increases the pressure in the thoracic cavity and helps to expel air. Breathing is also aided by several other muscles of the neck, abdomen, lower back, and even the pelvic floor.

Not only does abdominal pressure affect thoracic pressure, the opposite is also true. Depression of the diaphragm raises abdominal pressure and helps to expel the contents of certain abdominal organs, thus aiding in childbirth, urination, defecation, and vomiting. During such actions, we often consciously or unconsciously employ the **Valsalva[7] maneuver.** This consists of taking a deep breath, holding it by closing the glottis, and then contracting the abdominal muscles to raise abdominal pressure and push the organ contents out.

15.2b Measurements of Ventilation

Measurements of pulmonary ventilation are used to assess the severity of respiratory disease or to monitor a patient's improvement or disease progression. To measure pulmonary function, a subject breathes into a device called a **spirometer,**[8] which recaptures the expired breath and records the rate of breathing and volumes of air that enter and leave the lungs. The measurement of pulmonary function is called **spirometry.** Figure 15.12 shows values for a healthy adult. These values are proportional to body size; consequently, they are generally higher for men than for women.

One value obtained with a spirometer is **tidal volume,** the amount of air inhaled or exhaled in one relaxed breath. The average tidal volume is about 500 mL. Other volumes include the **inspiratory reserve volume,** the amount of air that can be drawn into the lungs beyond the normal quiet inspiration (typically about 3,000 mL), and **expiratory reserve volume,** the amount of air that can be expelled beyond the normal quiet expiration (typically about 1,200 mL). **Vital capacity** is the maximum amount of air that can be exhaled after the deepest possible inspiration. It is the sum of tidal volume, inspiratory reserve volume, and expiratory reserve volume. It represents the maximum ability to ventilate lungs in one breath, and is an especially important measure of pulmonary health. Even when one exhales as much as possible, an amount of air called the **residual volume** remains in the lungs and mixes with fresh air on the next inspiration. The residual volume prevents collapse of the alveoli and allows gas exchange to continue even between inspirations.

Spirometry is especially valuable for the diagnosis and assessment of *restrictive* and *obstructive* lung disorders. **Restrictive disorders** are conditions associated with poor pulmonary *compliance*—the ability of the lungs to expand when pressure changes. Because of differences in age or pulmonary health, one person's

Figure 15.12 Spirometry. The wavy line indicates inspiration when it rises and expiration when it falls.
BSIP/Science Source

❓ *Why does accurate spirometry require the subject to wear a nose clip, as the person in the photo is doing?*

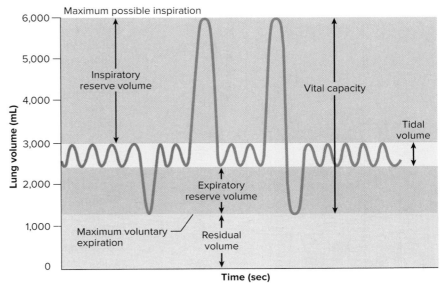

---

[7]Antonio Maria Valsalva (1666–1723), Italian anatomist
[8]*spiro* = breath; *meter* = measuring device

lungs may expand more readily than another person's in response to an identical pressure change in the thorax. Compliance is compromised when the lungs are stiffened by a buildup of scar tissue in diseases such as tuberculosis and black lung disease. Low compliance results in reduced vital capacity. **Obstructive disorders** interfere with airflow by narrowing or blocking airways, making it harder to inhale or exhale. Examples include asthma, emphysema, and chronic bronchitis (see Clinical Application 15.1 in section 15.1b and Perspectives on Health at the end of this section). Obstructive disorders are assessed by having a person exhale as rapidly as possible into a spirometer. A healthy adult should be able to expel 75% to 85% of the vital capacity in 1.0 second. A value lower than 75% indicates the patient probably has an obstructive disorder.

> **Apply What You Know**
>
> Suppose a person has a tidal volume of 500 mL, an inspiratory reserve volume of 2,800 mL, and an expiratory reserve of 1,000 mL. Calculate vital capacity. If this person can exhale 2,500 mL in 1.0 second with maximum effort, what percentage of the vital capacity is this? What does the percentage indicate about this person's health?

15.2c Neural Control of Breathing

Breathing, like the heartbeat, is automatic; we breathe without thinking about it, which is a good thing or we would fear going to sleep. However, in contrast to the heart, the lungs don't have a built-in pacemaker. Instead, they rely on messages from the nervous system to coordinate the complex skeletal muscle actions required to sustain ventilation. The unconscious cycle of breathing is controlled by respiratory centers in the brain, specifically in the medulla oblongata and pons (fig. 15.13).

The principal pacemaker for breathing is the **ventral respiratory group (VRG)** of neurons in the medulla oblongata. It contains *inspiratory neurons* that, at rest, typically fire for about 2 seconds and issue nerve signals to the respiratory muscles via the spinal cord, paired *phrenic nerves* to the diaphragm, and paired *intercostal nerves* to the intercostal muscles. The VRG also contains *expiratory neurons,* which then begin to fire and inhibit the inspiratory neurons for about 3 seconds. The typical resting breathing rhythm of about 12 breaths/min. results from these neuron groups in the VRG "taking turns" in this manner.

Of course, we don't always breathe at that relaxed rate. The medulla and pons receive input from several sources that influence one's respiratory rate and depth. *Chemoreceptors* monitor the pH of the blood and cerebrospinal fluid as well as blood CO_2 and O_2; *stretch receptors* in the pleurae and bronchial tree monitor expansion of the lungs; *irritant receptors* in the bronchial tree inhibit respiration in response to smoke, noxious fumes, and other agents; centers in the hypothalamus and limbic system can alter one's respiration in response to emotions; and cerebral centers enable us to voluntarily modify our respiration, as in speaking, singing, laughing, swimming, or crying.

Figure 15.13 Respiratory Control Centers in the Central Nervous System. The ventral respiratory group (VRG) is the pacemaker that generates the rhythm of breathing. Its output goes to spinal integrating centers that give rise to the intercostal and phrenic nerves, which innervate the intercostal muscles and diaphragm. A brainstem integrating center processes information from multiple sources and then issues output to the VRG to adjust the respiratory rhythm.

PERSPECTIVES ON HEALTH

Smoking and Respiratory Health

Smoking cigarettes increases the risk of premature death due to emphysema, cancer, and cardiovascular disease. The history of respiratory diseases is linked to the use of cigarettes, which were not manufactured until the end of the nineteenth century. At the beginning of the twentieth century, lung cancer was uncommon, but by the 1990s, it had become the leading cause of cancer deaths in the United States. The incidence and mortality from lung cancer are decreasing in the United States due to decades of public education and tobacco control policies, but are increasing elsewhere in the world. Still, approximately 14% of adults in the United States smoke.

Chronic Obstructive Pulmonary Disease

Chronic obstructive pulmonary disease (COPD) is a collective term for chronic bronchitis and emphysema. It is among the leading causes of disability and death in the United States, and is almost always caused by cigarette smoking.

Chronic bronchitis begins with cellular abnormalities brought on by persistent exposure to cigarette smoke. Many ciliated cells of the respiratory tract transform into nonciliated squamous cells. Meanwhile, cilia on cells that retain them are paralyzed by toxins in the smoke. The lack of competent cilia impairs the mucociliary escalator, which fails to transport debris and microorganisms away from the lungs to the throat. Attempts to clear the airway result in a persistent cough, the "smoker's hack." Inhaled particulates in smoke bypass the damaged cilia and travel as far as the alveoli. Alveolar macrophages ingest these, especially tar, and fill with black material, leading to the black color of smokers' lungs. The immune function of the macrophages is also impaired. Stagnation of the respiratory mucus and immune suppression of the macrophages invite recurring infections and inflammation, leading to shortness of breath *(dyspnea)* and oxygen deficiency in the blood and tissues *(hypoxia)*.

In emphysema, the alveoli become hyperinflated and many of them burst. The lungs can become permeated with grape-size and larger spaces that take the place of functional alveoli. As the disease progresses, symptoms include dyspnea, extreme fatigue, inability to tolerate even slight physical exertion such as walking across a room, and mental sluggishness due to inadequate oxygenation of the brain.

Chronic hypoxia from COPD stimulates the kidneys to secrete erythropoietin, which stimulates the production of more erythrocytes (see section 12.2). A sharp rise in RBC count increases blood viscosity and pressure, puts a strain on the heart, and thus leads to cardiac disease as a further consequence of smoking.

Lung Cancer

Cigarette smoke contains over 60 carcinogens, substances known to cause cancer. Accumulation of mutations results in the formation of lung tumors (fig. 15.14), over 90% of which develop in the mucous membranes of the large bronchi. In the most common form of lung cancer, *squamous-cell carcinoma,* the bronchial epithelium changes from the ciliated pseudostratified type to the stratified squamous type. Proliferating cells invade the tissues under the epithelium, producing bleeding lesions. Dense masses of fibrous tissue (collagen) replace functional lung tissue. In the most dangerous form of lung cancer, *small-cell (oat-cell) carcinoma,* malignant cells quickly spread beyond the lungs and establish secondary (metastatic) tumors in the mediastinum and other organs.

By the time lung cancer is diagnosed, it usually has already metastasized to other organs such as the heart, bones, liver, lymph nodes, and brain. The chance of recovery is poor relative to other cancers, with only 18% of patients surviving for 5 years after diagnosis.

The recent legalization of marijuana for recreational use in many states has led to increased use of the drug. Smoking marijuana has been linked to changes in the respiratory epithelium and chronic bronchitis. Studies to date, however, have been limited, and more data are needed to assess the effects of marijuana smoke on lung function and overall public health.

Recently, the use of e-cigarettes (vaping) has skyrocketed, especially among young people. Several serious cases of vaping-associated lung disease have occured across the United States. Vaping presents new public health challenges, and there are few short-term and no long-term data about its health effects.

(a) Healthy lung, mediastinal surface

(b) Smoker's lung with carcinoma

Figure 15.14 Effect of Smoking. The healthy lung (a) sharply contrasts with the blackened, cancerous smoker's lung (b).

a: Dennis Strete/McGraw-Hill Education; b: Biophoto Associates/Science Source

Before You Go On

Answer these questions from memory to test your understanding of the previous section. Reread the material if you can't answer the questions.

5. Explain pressure changes that occur in the thorax during inspiration and expiration. Why does contraction of the diaphragm cause inspiration but contraction of the abdominal muscles causes expiration?
6. What is the prime mover of respiration? What other muscles act as synergists during inspiration and forced expiration?
7. Define *tidal volume, inspiratory reserve volume, and expiratory reserve volume.* How is vital capacity calculated? Explain why vital capacity is useful for assessing pulmonary function.
8. Name the principal brainstem center that sets the respiratory rhythm, and explain how it does so.
9. Identify several sources of input that the brainstem center receives and uses to modify the respiratory rhythm under special circumstances.

15.3 Gas Exchange and Transport

Expected Learning Outcomes

When you have completed this section, you should be able to

a. define *partial pressure* and explain how it affects diffusion of gases across the respiratory membrane;
b. describe gas exchange in the lungs and systemic capillaries;
c. describe how O_2 and CO_2 are transported in the blood; and
d. explain the effect of blood gases and pH on the respiratory rhythm.

15.3a Composition of Air APR

Ultimately, respiration is about gases, especially oxygen and carbon dioxide. The respiratory structures and mechanisms discussed so far only set the stage for delivering oxygen to the tissues and returning carbon dioxide to the lungs where it can be expelled. To grasp how these processes occur, it is necessary to begin with the composition of the air we inhale.

The air we breathe consists of 78.6% nitrogen, 20.9% oxygen, 0.04% carbon dioxide, traces of several other gases, and variable amounts of water vapor (0.5% on a typical cool clear day). Atmospheric pressure is the sum of the pressures exerted separately by each of the gases. The separate contribution of each gas is called its **partial pressure,** and is symbolized with a P followed by the formula of the gas, such as P_{O_2}. For example, oxygen accounts for 20.9% of the volume of air. At the average sea-level atmospheric pressure of 760 mm Hg, its partial pressure is therefore:

$$P_{O_2} = 0.209 \times 760 \text{ mm Hg} = 159 \text{ mm Hg}$$

When there is a difference between the partial pressure of a gas at one point and its partial pressure at another point, we say there is a **partial-pressure gradient.** Movement of a gas from the point of high partial pressure to the point of low partial pressure is movement *down the gradient.*

15.3b Alveolar Gas Exchange APR

Alveolar gas exchange (external respiration) refers to the *loading* of O_2 and *unloading* of CO_2 in the pulmonary alveoli. It depends on the tendency of gases to diffuse down their partial-pressure gradients. As we saw earlier, the P_{O_2} at sea level is 159 mm Hg. As the air is breathed into the alveoli, it becomes humidified and its partial pressure of H_2O rises, making the proportion of oxygen lower. By the time the air reaches the alveoli, it has a P_{O_2} of about 104 mm Hg, compared to 40 mm Hg in the blood of the surrounding capillaries (fig. 15.15). Oxygen

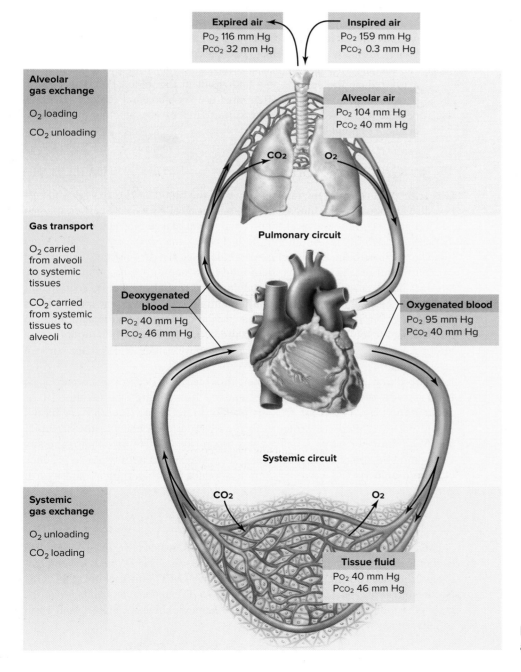

Figure 15.15 Change in P_{O_2} and P_{CO_2} along the Circulatory Route.

therefore diffuses down its pressure gradient across the respiratory membrane from the alveolar air into the blood. In contrast, carbon dioxide has a P_{CO_2} of about 46 mm Hg in the capillary blood and 40 mm Hg in the inhaled air, so it diffuses from the capillary into the alveolus.

Alveolar gas exchange is subject to several factors that affect diffusion through the respiratory membrane:

1. **Partial-pressure gradient.** The greater the difference in partial pressure between one point and another, the faster the gas will diffuse. This is why a patient's blood oxygen level can be improved by giving oxygen-enriched air.
2. **Membrane thickness.** Normally the respiratory membrane is very thin, but in conditions such as pneumonia, it can be thickened by fluid accumulation (pulmonary edema). Gases then have to travel farther from the alveolar air space to the blood, and as a result, oxygen can't get to the red blood cells quickly enough to fully load their hemoglobin, and the blood can become oxygen-deficient (a state called *hypoxemia*).
3. **Membrane surface area.** Each lung has about 70 m^2 of respiratory membrane, normally a more-than-adequate surface area for gas exchange. Degenerative pulmonary diseases such as emphysema, however, dramatically reduce the surface area and result in low blood P_{O_2}. This is why persons with emphysema often must use supplemental oxygen—to compensate for the reduced alveolar surface area by accentuating the pressure gradient in point 1 of this list.

15.3c Gas Transport

Gas transport refers to the manner in which the blood carries oxygen and carbon dioxide between the lungs and systemic tissues. As we saw earlier, oxygen has a higher P_{O_2} in the alveoli than in the blood arriving at an alveolus, so it diffuses into the blood. About 98.5% of it enters the erythrocytes (RBCs) and binds to the iron in the hemoglobin molecules (see fig. 12.4). Hemoglobin with oxygen bound to it is called **oxyhemoglobin.** Since the hemoglobin molecule has four heme groups, and each heme can bind one O_2 molecule, each hemoglobin can carry up to four O_2. The remaining 1.5% of the transported O_2 is dissolved gas in the blood plasma. The poison carbon monoxide also binds to the iron, and much more tightly than oxygen does. It prevents hemoglobin from binding oxygen, and this is why breathing carbon monoxide in the air can be fatal.

The blood also transports carbon dioxide. The P_{CO_2} in systemic tissue fluid is typically about 46 mm Hg, and in the arriving arterial blood it is about 40 mm Hg. Therefore, CO_2 diffuses from the tissue fluid into the blood. About 5% of it binds to proteins such as hemoglobin and plasma albumin, and 5% is transported as dissolved gas much like the CO_2 in champagne and carbonated beverages. Ninety percent, however, reacts with water to produce **carbonic acid** (H_2CO_3), which then dissociates into bicarbonate and hydrogen ions:

$$CO_2 + H_2O \rightarrow H_2CO_3 \rightarrow HCO_3^- + H^+$$

This reaction occurs primarily within the RBCs, which have an enzyme called **carbonic anhydrase** that greatly accelerates the process. The H^+ binds to hemoglobin and thus does not greatly affect the blood pH; the HCO_3^- diffuses out of the RBC into the blood plasma. Carbon dioxide binds to a different site on hemoglobin (the globin) than the oxygen does (the heme), so hemoglobin can transport both gases simultaneously.

Clinical Application 15.2

HIGH-ALTITUDE PHYSIOLOGY

One of the most thrilling adventures one can have is to climb mountains, but climbing at high altitudes poses physiological challenges. In the past several decades, many people have attempted to ascend Mount Everest, which at 8,848 m (29,029 ft) is the highest point on the earth's surface. The atmospheric pressure at the peak is approximately 255 mm Hg, or about one-third of the pressure at sea level. Oxygen is still approximately 21% of the mix of gases in the air; thus, the P_{O_2} is only 53 mm Hg. At very high altitudes, the P_{O_2} gradient between the alveolus and the blood capillaries is not great enough to drive the diffusion of O_2 to the blood.

At high altitude, the body responds in the short term by increasing ventilation and heart rate, resulting in enhanced oxygen delivery to the tissues. Climbers attempt to offset the lack of oxygen by a process called *acclimatization*; they ascend to increasingly higher elevations over a period of several days or even weeks. The slow ascent leads to physiological changes that help one cope with the "thin air." For example, within a few days at high altitude, red blood cell production increases due to increased levels of erythropoietin (EPO) secreted by the kidneys. As a result, the hematocrit, normally about 40% to 45%, can rise to over 55%. In addition, the mitochondria become more efficient at using O_2 to produce ATP. Despite acclimatization, ascent to very high altitudes (over 20,000 ft) often leads to symptoms such as fatigue, headache, nausea, and a decline in cognitive ability.

15.3d Systemic Gas Exchange

Systemic gas exchange (internal respiration) means the unloading of O_2 and loading of CO_2 at the systemic capillaries. Blood arriving here typically has a P_{O_2} of about 95 mm Hg and the surrounding tissue fluid has about 40 mm Hg. Oxygen therefore *dissociates* from the oxyhemoglobin molecule and diffuses out of the blood into the tissue fluid. Typically, the blood gives up about 22% of its oxygen in one pass through a systemic capillary bed, so it still has 78% of its load as it returns to the heart and lungs.

The amount of oxygen unloaded, however, is not constant in all tissues or at all times. Active tissues need more oxygen, and fortunately, hemoglobin adjusts its unloading to variations in local demand. In highly active tissue such as exercising muscle, it can unload up to 80% of its oxygen. There are multiple mechanisms by which hemoglobin adjusts O_2 unloading to the metabolic state of a local tissue. A few of these (not a complete list) follow.

1. Active tissues consume O_2 rapidly, lowering the P_{O_2} of the tissue fluid. If there is a greater difference between the P_{O_2} of the blood and tissue fluid, then O_2 diffuses more rapidly into the tissue. This is a case of the partial-pressure gradient issue explained earlier.
2. Active tissues produce more CO_2 than less active ones. CO_2 generates acid through the carbonic acid reaction shown above, lowering the pH of the tissue fluid. Hemoglobin responds to the lower pH by releasing more O_2—a phenomenon called the *Bohr effect*.
3. Active tissues generate more heat than less active ones. Hemoglobin releases O_2 more easily at higher temperatures, thus unloading more O_2 to the warmest, most active tissues.

15.3e Blood Gases and the Respiratory Rhythm

We breathe more heavily at some times (during exercise, for example) than at others. Pulmonary ventilation varies as it adjusts to the body's rate of CO_2 production and demand for O_2. The effect is to maintain a stable blood pH (7.40 ± 0.05) and systemic arterial P_{O_2} and P_{CO_2} of 95 and 40 mm Hg, respectively. Regulation of these values depends on input to brainstem respiratory centers from the previously described chemoreceptors that monitor the composition of blood and cerebrospinal fluid. The most potent stimulus for breathing is pH, followed by CO_2. Perhaps surprisingly, the least significant is O_2.

A blood pH below 7.35 is known as **acidosis.** The corrective homeostatic response to acidosis is an increase in the rate and depth of breathing, "blowing off" CO_2 faster than the body produces it. As CO_2 is eliminated from the body, the carbonic acid reaction shifts to the left:

$$CO_2 + H_2O \leftarrow H_2CO_3 \leftarrow HCO_3^- + H^+$$

Thus, the H^+ on the right is consumed, raising the pH and ideally returning it to normal.

A pH greater than 7.45 is **alkalosis.** The response to this is a reduced respiratory rate and depth, allowing some CO_2 to accumulate and lower the pH to normal. You can see that P_{CO_2} and pH are very closely linked in regulating respiration. Oxygen normally has little effect because hemoglobin is almost saturated with O_2 anyway when it leaves the lungs, and little can be gained from heavier breathing. Oxygen does become more influential, however, in conditions involving high altitude (aviation and mountain climbing) and in diseases such as emphysema and pneumonia.

In summary, the main chemical stimulus to pulmonary ventilation is a change in pH that stimulates chemoreceptors of the brain and arteries. A pH change derives from changes in P_{CO_2} of the arterial blood. Therefore, changes in levels of carbon dioxide in the blood indirectly influence pulmonary ventilation. Ventilation is adjusted to maintain arterial pH at about 7.40. Normally, arterial P_{O_2} has relatively little effect on respiration.

Before You Go On

Answer these questions from memory to test your understanding of the previous section. Reread the material if you can't answer the questions.

10. Identify three factors that affect the efficiency of alveolar gas exchange.
11. How is most oxygen transported in the blood?
12. What are the three ways in which blood transports CO_2? Which one is most significant?
13. Give three reasons why highly active tissues extract more oxygen from the blood than less active tissues.
14. What is the most potent chemical stimulus to respiration?
15. Explain how changes in pulmonary ventilation can correct pH imbalances.

Aging of the Respiratory System

Respiratory failure is a leading cause of death in elderly people, but a gradual decline of respiratory function begins as early as the 30s. This decline is one of several factors in the gradual loss of physical endurance. The costal cartilages and joints of the thoracic cage become less flexible, and the lungs have less elastic tissue and fewer alveoli. Vital capacity, forced expiratory volume, and other values fall. Elderly people are also less capable of clearing the lungs of irritants and pathogens and are therefore increasingly vulnerable to respiratory infections. Pneumonia causes more deaths than any other infectious disease and is often contracted in hospitals and nursing homes. The chronic obstructive pulmonary diseases (COPDs)—emphysema and chronic bronchitis—are more common in old age since they represent the cumulative effects of a lifetime of degenerative change. COPDs also contribute to cardiovascular disease, hypoxemia (low blood Po_2), and hypoxic degeneration in all organ systems.

CAREER SPOTLIGHT

Respiratory Therapist

The respiratory therapist (RT) is an essential part of medical teams that deal with respiratory health. An RT specializes in assessment and treatment of respiratory conditions such as asthma or emphysema, and conditions that stem from other causes but affect breathing, such as strokes, heart attack, and spinal cord trauma. An RT is skilled in the use of equipment such as mechanical ventilators, which assist with breathing, and spirometers, devices that measure respiratory volumes. He or she may work in a variety of settings including hospital intensive care units, emergency rooms, or neonatal units; on emergency flight teams; or in home health care. To become an RT, students usually must complete either an associate degree or certificate program at a community college, or a 4-year bachelor's degree. One must pass national exams to become either a certified respiratory therapist or registered respiratory therapist. See appendix B for further information on careers in respiratory therapy.

Rick Brady/McGraw-Hill Education

CONNECTIVE ISSUES

Ways in Which the RESPIRATORY SYSTEM Affects Other Organ Systems

All Systems
The respiratory system provides oxygen to all tissues and removes their carbon dioxide.

Integumentary System
Respiratory dysfunctions can cause such skin discolorations as cyanosis (blueness) or erythema (redness).

Skeletal System
Any respiratory disorder that causes hypoxemia stimulates accelerated erythropoiesis in the red bone marrow.

Muscular System
Acid–base imbalances caused by respiratory dysfunction can affect neuromuscular function. Acidosis, for example, makes muscular reflexes more sluggish.

Nervous System
Respiratory control is necessary in carrying out vocalizations initiated by the nervous system; acid–base balances resulting from respiratory dysfunction can make neurons insufficiently or excessively excitable.

Endocrine System
Lungs contain angiotensin-converting enzyme (ACE), which converts angiotensin I to the hormone angiotensin II; hypoxemia stimulates the secretion of the hormone erythropoietin.

Circulatory System
By participation in angiotensin II production, lungs contribute to hormonal control of blood flow and blood pressure; variations in respiratory rate stabilize blood pH; thoracic pump aids in venous return of blood to the heart; obstruction of pulmonary circulation can lead to right-sided heart failure.

Lymphatic and Immune Systems
Thoracic pump aids lymph flow.

Urinary System
The Valsalva maneuver aids in emptying the urinary bladder; respiratory and urinary systems collaborate in regulating acid–base balance.

Digestive System
The Valsalva maneuver aids in vomiting and defecation.

Reproductive System
The Valsalva maneuver aids in childbirth.

Study Guide

Assess Your Learning Outcomes

To test your knowledge, discuss the following topics with a study partner or in writing, ideally from memory.

15.1 Functions and Anatomy of the Respiratory System

1. Five functions of the respiratory system
2. Function of the vibrissae (guard hairs)
3. Anterior and posterior boundaries of the nose
4. Structure and functional significance of the nasal conchae
5. Histology of the respiratory epithelium and the functional significance of its ciliated cells and goblet cells
6. Three divisions of the pharynx
7. Major structural features of the larynx and the function of the epiglottis
8. How the vocal cords produce sound; how their sound is transformed into intelligible words; and why men, women, and children differ in the pitch of the voice
9. Structure of the trachea and the functional significance of its cartilage, smooth muscle, and epithelial type
10. The nature and functional importance of the mucociliary escalator
11. Gross anatomy of the lungs and their spatial relationships with other thoracic structures
12. Structure and function of the pleurae
13. Anatomy and histology of the bronchial tree from the main bronchi through the bronchioles
14. Distinctions between the upper and lower respiratory tract and the conducting and respiratory zones of the tract
15. Components of the respiratory zone of the airway and the relationship of their special histology to their function
16. Cellular organization of the alveoli, the function of each cell type, and the structure of the respiratory membrane
17. The source and function of pulmonary surfactant
18. The two sources of blood to the lungs and their functional difference

15.2 Pulmonary Ventilation

1. Definitions of *inspiration*, *expiration*, and *respiratory cycle*
2. How respiratory airflow is governed by atmospheric and intrapulmonary pressures
3. The principal muscles of inspiration and how their actions result in an inflow of air
4. Role of the pleurae in inspiration
5. The normal mechanism of relaxed expiration, and why it does not require a muscular effort
6. Muscles that assist with forced expiration
7. How the Valsalva maneuver is performed and what bodily functions it aids
8. Definitions of the lung volumes and capacities measured with a spirometer, and why this is done
9. Definitions and examples of restrictive and obstructive lung disorders
10. Locations of the principal pacemaker of breathing, how its neurons control the respiratory cycle, and the inputs it receives to modify respiratory rate and depth

15.3 Gas Exchange and Transport

1. The composition of the air we breathe
2. The meaning of *partial pressure* and how the partial pressures of gases such as O_2 and CO_2 are calculated
3. The meaning of *alveolar gas exchange* and how it is governed by gradients of partial pressure
4. Three factors that can alter the efficiency of alveolar gas exchange
5. How O_2 and CO_2 are transported in the blood; which of the methods depend on hemoglobin and how they relate to hemoglobin structure
6. The chemical equation for the formation of carbonic acid from water and CO_2; how it relates to CO_2 transport and blood pH
7. The meaning of *systemic gas exchange* and how it relates to partial-pressure gradients
8. How hemoglobin is able to release more oxygen to some tissues than to others in accordance with the metabolic state of different tissues
9. How respiration relates to the pH of the blood
10. Definitions of *acidosis* and *alkalosis* and how the respiratory system responds to each
11. Why P_{O_2} normally has little effect on pulmonary ventilation

Testing Your Recall

1. Which structure seals the opening to the larynx and diverts food and drink to the esophagus?
   a. glottis
   b. epiglottis ✓
   c. vocal cord
   d. thyroid cartilage
   e. cricoid cartilage

2. The structures responsible for warming and humidifying air as it travels through the nasal cavity are
   a. choanae.
   b. conchae. ✓
   c. nares.
   d. vibrissae.
   e. goblet cells.

3. In speech, crude sounds are converted to recognizable words primarily by
   a. the trachea.
   b. the thyroid cartilage.
   c. the vocal cords.
   d. the epiglottis.
   e. the lips and tongue. ✓

4. A restrictive lung disorder is most likely to result in
   a. slower airflow.
   b. a reduced vital capacity. ✓
   c. alkalosis.
   d. pulmonary scar tissue.
   e. a COPD.

5. The source of pulmonary surfactant is
   a. the visceral pleura.
   b. tracheal glands.
   c. alveolar capillaries.
   d. squamous alveolar cells.
   e. great alveolar cells. ✓

6. Which of the following are fewest in number but largest in diameter?
   a. alveoli
   b. terminal bronchioles
   c. alveolar ducts
   d. bronchi ✓
   e. respiratory bronchioles

7. Which of the following is/are *not* lined with pseudostratified columnar epithelium?
   a. nasal cavity
   b. trachea
   c. nasopharynx
   d. alveolar ducts ✓
   e. main bronchi

8. What is the average P_{O_2} at sea level?
   a. 159 mm Hg ✓
   b. 100 mm Hg
   c. 0.03 mm Hg
   d. 40 mm Hg
   e. 46 mm Hg

9. Which of these values would normally be the highest?
   a. tidal volume
   b. inspiratory reserve volume
   c. expiratory reserve volume
   d. vital capacity ✓
   e. It would differ from one person to another.

10. During inspiration, most of the airflow results from the action of
    a. the external intercostal muscles.
    b. the internal intercostal muscles.
    c. the diaphragm. ✓
    d. the rectus abdominis muscle.
    e. the abdominal oblique muscles.

11. The process of measuring pulmonary function is called __spirometry__.

12. The __ventral respiratory group__ is a respiratory pacemaker in the medulla oblongata.

13. The largest cartilage of the larynx is the __thyroid__ cartilage.

14. The last line of defense against inhaled particles is phagocytic cells called __alveolar macrophages__.

15. A serous membrane that clings to the inside walls of the rib cage and the superior surface of the diaphragm is the __parietal pleura__.

16. In the blood, CO_2 combines with H_2O to form __carbonic acid__.

17. Nerve endings that monitor the pH and chemical composition of the blood are called __chemoreceptors__.

18. The mechanism that moves debris-laden mucus up the bronchi and trachea to the pharynx, where it can be swallowed, is called the __mucociliary escalator__.

19. Each lung receives a main bronchus, blood vessels, and nerves through a slit called the __hilum__.

20. __Asthma__ is a condition caused by bronchospasm that results in shortness of breath, wheezing, and coughing.

Answers in Appendix A

What's Wrong with These Statements?

Briefly explain why each of the following statements is false, or reword it to make it true.

1. An average tidal volume is 1,000 mL.
2. Restrictive lung diseases are characterized by narrowed or blocked airways.
3. Most of the CO_2 transported by the blood is in the form of dissolved gas.
4. The palatine tonsils are found in the nasopharynx.
5. Each lung contains approximately 1 million alveoli.
6. Pulmonary surfactant is produced by squamous alveolar cells.
7. The most important factor that influences breathing rate and depth is P_{O_2}.
8. Reduced respiratory rate and depth of breathing may be associated with acidosis.
9. The P_{O_2} in air freshly drawn into the alveoli is 40 mm Hg.
10. When intrapulmonary pressure exceeds atmospheric pressure, air flows into the lungs.

Answers in Appendix A

Testing Your Comprehension

1. In emphysema, there is a substantial loss of pulmonary alveoli and gas-exchange surface area, resulting in a low blood P_{O_2}. Explain why emphysema would also result in a high blood P_{CO_2} and acidosis.
2. Why is it more sensible for the hilum of each lung to be on its medial surface than it would be for it to be located at its apex, base, or costal surface?
3. The rhythmic beating of the heart is driven by a pacemaker located in the heart itself, yet the rhythmic expansion and contraction of the lungs are controlled by a pacemaker in the brainstem. Why do you think this is necessary, rather than the lungs having their own pacemaker?

The nasal conchae have inward folds on the lateral walls of the cavity that reach almost all the way to the septum. They cause the airflow to be quite turbulent as the inhaled air contacts the mucus membrane covering the conchae, it is warmed, humidified, and cleansed of most of its debris.

Chapter 16

The Urinary System

Each kidney has about 1 million of these microscopic, blood-filtering structures called glomeruli. An arteriole (yellow) feeds into the glomerulus (pink), a ball of blood capillaries that filter the blood and begin the process of urine formation.
© Susumu Nishinaga/Science Source

Chapter Outline

16.1 Functions of the Urinary System
- 16.1a Functions of the Kidneys
- 16.1b Metabolic Wastes and Excretion

16.2 Anatomy of the Kidney
- 16.2a Gross Anatomy
- 16.2b Renal Circulation
- 16.2c The Nephron

16.3 Glomerular Filtration
- 16.3a Filtration Pressure
- 16.3b Glomerular Filtration Rate
- 16.3c Regulation of Glomerular Filtration

16.4 Tubular Reabsorption and Secretion
- 16.4a The Proximal Convoluted Tubule
- 16.4b The Nephron Loop
- 16.4c The Distal Convoluted Tubule
- 16.4d Tubular Secretion

16.5 Water Conservation
- 16.5a The Collecting Duct
- 16.5b Control of Water Loss

16.6 Urine Storage and Elimination
- 16.6a The Ureters
- 16.6b The Urinary Bladder
- 16.6c The Urethra
- 16.6d Voiding Urine
- 16.6e Urine Volume and Properties

16.7 Fluid, Electrolyte, and Acid–Base Balance
- 16.7a Fluid Balance
- 16.7b Electrolyte Balance
- 16.7c Acid–Base Balance

Clinical Applications/Perspectives on Health
- Clinical Application 16.1: Kidney Stones
- Clinical Application 16.2: Urinary Tract Infection
- Perspectives on Health

End of Chapter
- Aging of the Urinary System
- Career Spotlight: Dialysis Technician
- Connective Issues
- Study Guide

BASE CAMP

Before ascending to the next level, be sure you're properly equipped with a knowledge of these concepts from earlier chapters.

- Peritoneum and retroperitoneal organs (see section 1.3e)
- Acids, bases, and pH (see section 2.2b)
- Osmosis and carrier-mediated transport (see section 3.2e)
- Blood pressure, resistance, and flow (see section 13.5a)
- Capillary fluid exchange (see section 13.5d)

Anatomy & Physiology Revealed 4.0
Module 13
Urinary System

An inevitable effect of metabolism is the production of poisonous wastes, so all living organisms, from bacteria to plants to humans, must have a way of eliminating them. In mammals such as ourselves, removing metabolic wastes and toxins from the blood and excreting them from the body is a task of the urinary system. The kidneys do much more than this, however, as you will see. They have close functional relationships with the endocrine, circulatory, and respiratory systems that we have covered in recent chapters. They are involved in regulating fluid, electrolyte, and acid–base balances, among other aspects of homeostasis. This explains why patients with kidney failure have a number of problems that may include anemia, hypertension, and electrolyte imbalances.

Anatomically, the urinary system is closely associated with the reproductive system. The two systems have a shared embryonic development, and in many animals, the eggs and sperm are emitted through the urinary tract. In human males, the urethra continues to serve as a passage for both urine and sperm. Thus, the urinary and reproductive systems are often collectively called the *urogenital* or *genitourinary system,* and *urologists* treat both urinary and male reproductive disorders.

16.1 Functions of the Urinary System

Expected Learning Outcomes

When you have completed this section, you should be able to

a. name and locate the organs of the urinary system;
b. list several functions of the kidneys in addition to urine formation;
c. define *excretion* and identify the systems that excrete wastes; and
d. name the major nitrogenous wastes and identify their sources.

The **urinary system** is one of the anatomically simpler systems of the body, having only six principal organs: two **kidneys,** two **ureters,** the **urinary bladder,** and the **urethra** (fig. 16.1). Most of this chapter concerns the kidneys, the organs of urine production. The others—the organs of urine elimination—are treated in section 16.6.

16.1a Functions of the Kidneys APR

The kidneys play more roles than are commonly realized:

- They filter the blood plasma and excrete toxic metabolic wastes.
- They regulate blood volume, pressure, and osmolarity by regulating water output.

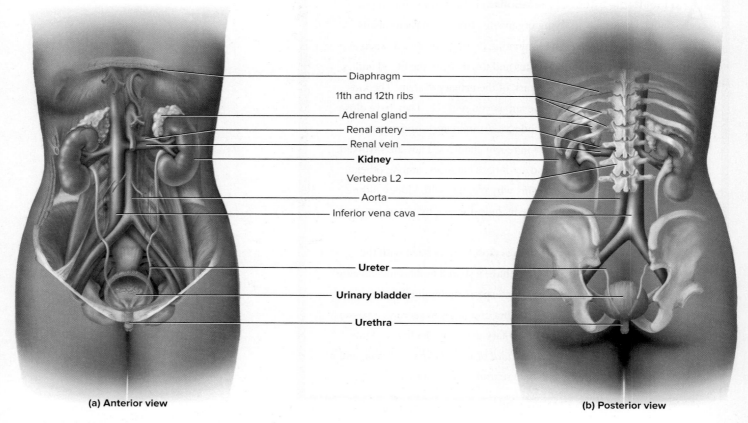

Figure 16.1 The Urinary System. (a) Anterior view, showing the left kidney typically a little higher than the right. (b) Posterior view, showing partial protection of the kidneys by the lower ribs. **APR**

- They regulate the electrolyte and acid–base balance of body fluids.
- They secrete the hormone *erythropoietin,* which stimulates the production of red blood cells and thus supports the oxygen-carrying capacity of the blood.
- They carry out the final step in synthesizing the hormone *calcitriol (vitamin D)* and thereby contribute to calcium homeostasis and bone metabolism.
- They detoxify some drugs.

16.1b Metabolic Wastes and Excretion

Excretion is the process of separating wastes from body fluids and eliminating them. A **waste** is any substance that is useless or present in excess of the body's needs. A **metabolic waste,** more specifically, is a waste substance produced by the body—for example, as a by-product of ATP production or protein catabolism. The urinary system excretes a variety of metabolic wastes, toxins, and drugs in the urine. Other systems also excrete wastes; for example, the respiratory system excretes carbon dioxide, and the digestive and integumentary systems excrete water and salts.

Among the most toxic of our metabolic wastes are small nitrogen-containing compounds called **nitrogenous wastes.** About 50% of the nitrogenous waste is **urea,** a by-product of protein metabolism. Protein breakdown produces ammonia, which is exceedingly toxic. The liver quickly converts ammonia to urea, which is somewhat less toxic. The urea is carried in blood plasma to the kidneys where it is filtered into the urine and excreted.

Other nitrogenous wastes in the urine include **uric acid** and **creatinine** (cree-AT-ih-neen), produced by the catabolism of nucleic acids and creatine phosphate, respectively. Although less toxic than ammonia and less abundant than urea, these too are far from harmless.

Before You Go On

Answer these questions from memory to test your understanding of the previous section. Reread the material if you can't answer the questions.

1. State four functions of the kidneys other than forming urine.
2. List four nitrogenous wastes and their metabolic sources.
3. Name some wastes eliminated by systems other than the urinary system.

16.2 Anatomy of the Kidney

Expected Learning Outcomes

When you have completed this section, you should be able to

a. describe the location and general appearance of the kidney and its relationship to neighboring organs;
b. identify the major external and internal features of the kidney;
c. trace the flow of blood through the kidney;
d. describe the components of a nephron; and
e. trace the flow of fluid through the renal tubules.

The kidneys lie against the posterior abdominal wall at the level of vertebrae T12 to L3 (fig. 16.1). About half of the left kidney is within the rib cage, as rib 12 crosses it. The right kidney is slightly lower because the space above it is occupied by the large right lobe of the liver. The kidneys are retroperitoneal, along with the ureters, urinary bladder, renal artery and vein, and adrenal glands, which lie against the posterior body wall outside the peritoneal cavity (see fig. 1.13).

16.2a Gross Anatomy APR

Each kidney weighs about 150 g and measures about 11 cm long, 6 cm wide, and 3 cm thick—roughly the size of a bar of bath soap. The lateral surface is convex, and the medial surface is concave and has a slit, the **hilum,** where it receives the renal blood vessels, nerves, lymphatics, and ureter (fig. 16.2).

Connective tissue layers anchor the kidney and protect it from trauma. It is surrounded by a thick layer of fat called the *perirenal fat capsule,* which cushions it and helps to hold it in place. The outermost layer of the kidney itself is the *fibrous (renal) capsule,* which encloses the glandular tissue like plastic wrap.

The glandular, urine-forming tissue of the kidney has a C shape enclosing a medial space called the **renal sinus.** The outer rind of the kidney, the **renal cortex,** is about 1 cm thick in most places. Its inner tissue, the **renal medulla,** is divided into 6 to 10 cones called **renal pyramids,** separated from each other by extensions of the cortex called **renal columns.** The broad base of each pyramid faces the cortex, and its blunt apex, the **renal papilla,** faces the sinus. One pyramid and the overlying cortex constitute one *lobe* of the kidney.

The papilla of each pyramid nestles in a cup called a **minor calyx**[1] (CAY-lix), which collects its urine. Two or three minor calyces (CAY-lih-seez) converge to

---

[1]*calyx* = cup

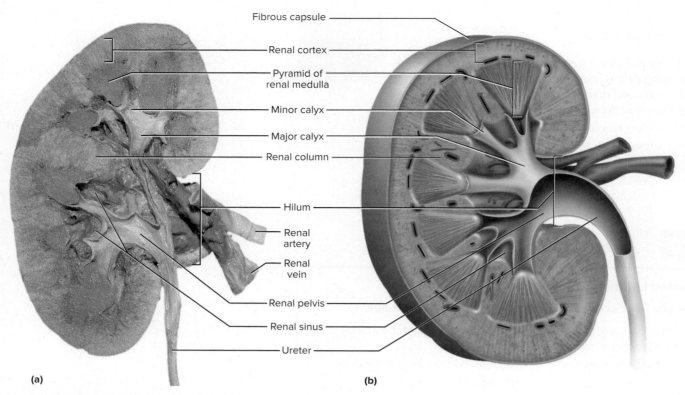

Figure 16.2 Gross Anatomy of the Kidney.
(a) Photograph of frontal section. (b) Major anatomical features. APR

a: © Rebecca Gray/McGraw-Hill Education

form a **major calyx,** and two or three major calyces converge in the sinus to form the **renal pelvis.**[2] The funnel-like renal pelvis converges on the ureter, a tube that leads to the urinary bladder.

16.2b Renal Circulation

Although the kidneys account for only 0.4% of the body weight, they receive about 21% of the cardiac output. This is a reflection of how important they are in eliminating toxic waste and regulating blood volume and composition.

Each kidney is supplied by a **renal artery** arising from the aorta. Branches of the renal artery lead to *arcuate arteries* that arch over the base of each renal pyramid, and from here, slender *cortical radiate arteries* penetrate into the cortex and ascend toward the kidney surface (fig. 16.3). Like the trunk of a pine tree, a cortical radiate artery gives off a series of lateral branches called **afferent arterioles.**

Each afferent arteriole supplies the capillaries of one functional unit of the kidney, called a *nephron* (fig. 16.4). Blood leaves the nephron by way of an **efferent arteriole,** which usually leads to a web of **peritubular capillaries** that surround the winding tubule of the nephron. From here, the blood flows through a series of veins that parallel the arteries and have similar names, ultimately leading to the **renal vein.** This vein exits the kidney at the hilum and leads to the inferior vena cava.

The foregoing description pertains only to the blood supply of the renal cortex. The renal medulla receives only 1% to 2% of the total blood flow, yet this is crucial to renal function. It is supplied by relatively straight, small blood vessels called the **vasa recta,**[3] which arise from the efferent arterioles of the deepest nephrons and then drain into the *arcuate veins.*

---

[2]*pelvis* = basin
[3]*vasa* = vessels; *recta* = straight

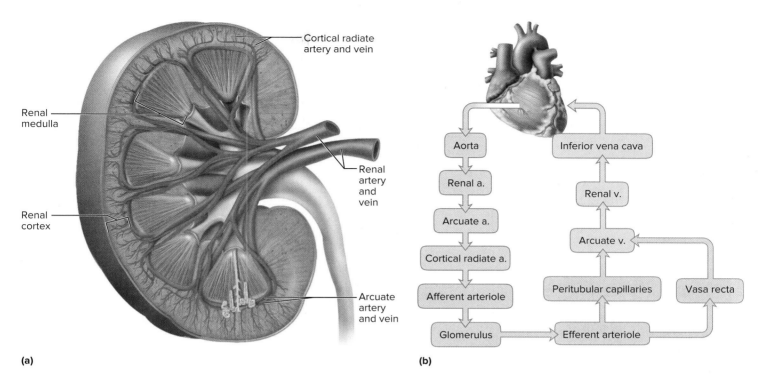

Figure 16.3 Renal Circulation. (a) The larger blood vessels of the kidney. (b) Flowchart of renal circulation (omitting some minor vessels).

16.2c The Nephron

Each kidney has about 1.2 million urine-forming functional units called **nephrons**[4] (NEF-rons). This is where the blood plasma is filtered and processed to form urine. To understand how just one of these works is to understand nearly everything about how the whole kidney works. Each nephron is composed of two principal parts: a *renal corpuscle,* which filters the blood plasma, and a long coiled *renal tubule,* which converts the filtrate to urine (fig. 16.4).

The Renal Corpuscle

The **renal corpuscle** is composed of a ball of capillaries called a **glomerulus**[5] (glo-MERR-you-lus) and a two-layered **glomerular capsule** (formerly called the *Bowman capsule*) that encloses it like a cup (fig. 16.5). The outer layer of the capsule is a simple squamous epithelium, and the inner layer consists of elaborate cells called **podocytes**[6] wrapped around the capillaries of the glomerulus (see fig. 16.8). The two layers are separated by a filtrate-collecting **capsular space.**

The afferent arteriole delivers blood to the glomerulus and the efferent arteriole carries it away. The two arterioles lie snugly side by side, penetrating the capsule at a single opening on one side. The afferent arteriole is significantly larger than the efferent arteriole. Thus, the glomerulus has a large inlet and a small outlet—resulting in unusually high blood pressure in the glomerulus, which is crucial to the kidney's task of blood filtration. The filtrate enters the capsular space and then passes to the renal tubule, described next.

---

[4]*nephro* = kidney
[5]*glomer* = ball; *ulus* = little
[6]*podo* = foot; *cyte* = cell

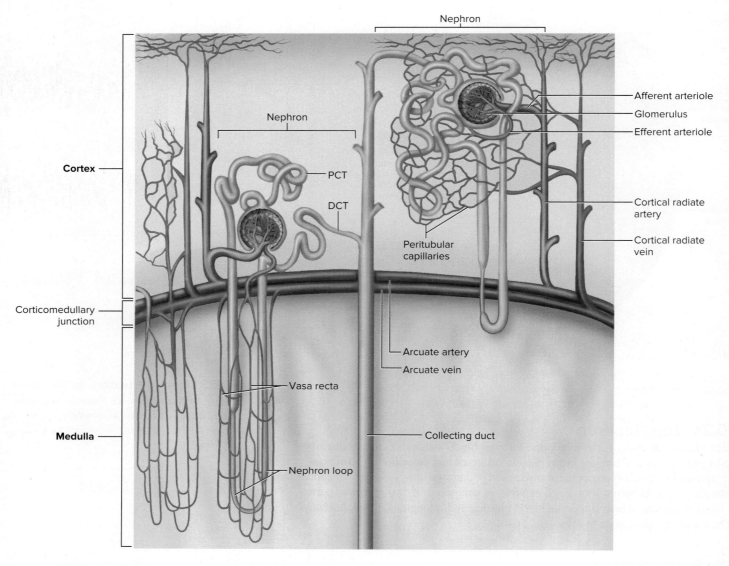

Figure 16.4 Blood Circulation in the Nephron. For clarity, vasa recta are shown only on the left and peritubular capillaries only on the right. In nephrons close to the medulla (left), the efferent arteriole gives rise to the vasa recta. In nephrons farther out on the cortex (right), the nephron loop barely dips into the renal medulla and the efferent arteriole gives rise to peritubular capillaries (DCT = distal convoluted tubule; PCT = proximal convoluted tubule).

The Renal Tubule

The **renal tubule** leads away from the glomerular capsule on the side opposite from the two arterioles. It ends at the tip of a renal pyramid. It is about 3 cm long and divided into four regions: the *proximal convoluted tubule, nephron loop, distal convoluted tubule,* and *collecting duct* (**fig. 16.6**). The first three of these are parts of one nephron; the collecting duct receives fluid from many nephrons. Each region of the tubule has unique physiological properties and roles in the production of urine:

1. The **proximal convoluted tubule (PCT)** arises from the glomerular capsule. It is long and compactly coiled. It is lined with simple cuboidal epithelium with dense microvilli that facilitate absorption of materials.
2. The **nephron loop** (formerly called the *loop of Henle*) is a long U-shaped portion of the renal tubule that begins where the PCT straightens. The *descending limb* of the loop dips into the medulla and, at its deep end, turns 180° and forms the *ascending limb*. The ascending limb travels back to the cortex, parallel and close to the descending limb. (In nephrons that are relatively far from the medulla, the loops are much shorter or even nonexistent, and many do not penetrate deeply enough to reach the medulla.) Some

Figure 16.5 The Renal Corpuscle. (a) Anatomy of the corpuscle. (b) A vascular cast of the glomerulus and nearby arteries. Note the difference in diameters between the afferent and efferent arterioles.

b: © Steve Gschmeissner/Science Source

❓ *Which is larger, the afferent or efferent arteriole? How does the difference affect glomerular blood pressure?*

portions of the loop have a simple cuboidal epithelium and therefore look thicker than others; thus, they are called the *thick segments*. The rest of the loop has a simple squamous epithelium and is called the *thin segment*. The thin segment is very permeable to water, whereas the rest is less so. The thick segments are heavily engaged in the active transport of ions, so they require a lot of ATP and the cells here are packed with mitochondria, accounting for their thickness. The functional significance of the thick and thin segments will be discussed later.

3. The **distal convoluted tubule (DCT)** begins shortly after the ascending limb reenters the cortex. It is shorter and less coiled than the proximal convoluted tubule. The DCT is the end of the nephron.
4. The **collecting duct (CD)** receives fluid from the DCTs of several nephrons as it passes through the cortex before descending into the medulla. Numerous collecting ducts converge toward the apex of a renal pyramid and end in pores at the conical tip of the papilla. Urine drains from these pores into the minor calyx that encloses the papilla.

The flow of fluid from the point where it is filtered from the blood to the point where urine leaves the body is glomerular capsule → proximal convoluted tubule → nephron loop → distal convoluted tubule → collecting duct → minor calyx → major calyx → renal pelvis → ureter → urinary bladder → urethra.

Before You Go On

Answer these questions from memory to test your understanding of the previous section. Reread the material if you can't answer the questions.

4. Define the two parts of a renal corpuscle, and explain what distinguishes a renal corpuscle from a nephron.
5. Trace the path taken by one red blood cell from the renal artery, through the renal cortex, to the renal vein.
6. Trace a drop of urine on its route from the glomerular capsule to the point where it leaves the body.

Figure 16.6 Microscopic Anatomy of the Renal Tubule. (a) Location of the nephrons in one wedge-shaped lobe of the kidney. (b) Structure of a nephron. For clarity, the nephron is stretched out to separate the convoluted tubules and the nephron loop is greatly shortened. (c) The true proportions of the nephron loops relative to the convoluted tubules. Three nephrons are shown. Their proximal and distal convoluted tubules are commingled in a single tangled mass in each nephron. Note the extreme lengths of the nephron loops. The deepening yellow color indicates increasing concentration as the fluid passes through the nephron.

16.3 Glomerular Filtration

Expected Learning Outcomes

When you have completed this section, you should be able to

a. describe the process by which the kidney filters the blood plasma; and
b. describe how the sympathetic nervous system, hormones, and the kidney itself regulate filtration.

The kidney converts blood plasma to urine in four stages: glomerular filtration, tubular reabsorption, tubular secretion, and water conservation (fig. 16.7), which will be explored in this order. As it is transformed to urine, the fluid is referred to by different names that reflect its changing composition: (1) The fluid in the capsular space, called **glomerular filtrate,** is similar to blood plasma except that it has almost no protein. (2) The fluid from the proximal convoluted tubule through the distal convoluted tubule is called **tubular fluid.** It differs from the glomerular filtrate because of substances removed and added by the tubule cells. (3) The fluid is called **urine** once it enters the collecting duct, since it undergoes little alteration beyond that point except for a change in concentration.

Figure 16.7 Basic Stages of Urine Formation.
APR

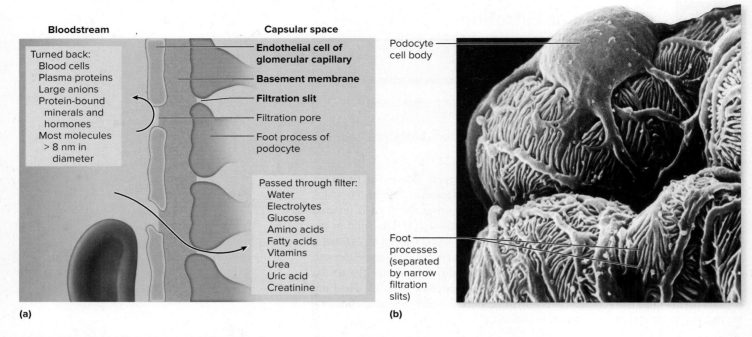

Figure 16.8 The Glomerular Filtration Membrane. (a) Components of the filtration membrane. (b) Blood capillaries in the glomerulus closely wrapped in spidery podocytes.
b: © Don W. Fawcett/Science Source

Glomerular filtration, the first step of urine formation, is a process in which water and some solutes in the blood plasma are filtered from the capillaries of the glomerulus into the capsular space of the nephron. The fluid passes through three barriers that constitute the **filtration membrane (fig. 16.8)**:

1. **The endothelium of the capillary.** The glomerulus is composed of *fenestrated capillaries* perforated by large filtration pores (see "Capillaries" in section 13.4b). Like fenestrated capillaries elsewhere, these are highly permeable, but their pores are small enough to exclude blood cells from passing to the filtrate.
2. **The basement membrane.** This is a glycoprotein gel that holds back most protein and other particles too large to pass through its fine mesh. Passing large molecules through it would be like trying to force sand through a kitchen sponge; a little plasma protein gets through, but not much.
3. **Filtration slits.** The podocytes of the glomerular capsule are shaped somewhat like octopi, with bulbous cell bodies and several thick arms. Each arm has numerous little extensions called *foot processes (pedicels[7])* that wrap around the capillaries and interdigitate with each other, like wrapping your hands around a pipe and lacing your fingers together.

Small molecules pass freely through the filtration membrane into the capsular space. These include water, electrolytes, glucose, fatty acids, amino acids, nitrogenous wastes, and vitamins. Larger particles are turned back by the filtration membrane and left behind in the blood.

16.3a Filtration Pressure

Glomerular filtration is driven by the same basic pressures that drive capillary fluid exchange in other places (see "Capillary Fluid Exchange" in section 13.5d). However, blood pressure is much higher in the glomerulus than in most other capillaries—about 60 mm Hg compared with 10 to 15 mm Hg elsewhere. This results from the fact that the afferent arteriole is substantially larger than the efferent arteriole, giving the glomerulus a large inlet and small outlet. Two forces oppose

---

[7]*pedi* = foot; *cel* = little

the blood pressure in the glomerulus: (1) fluid pressure in the capsular space, and (2) osmotic pressure of the blood. Normally, the glomerular blood pressure remains high enough to override these, so these capillaries engage solely in filtration. They reabsorb little or no fluid.

The high blood pressure in the glomeruli makes the kidneys especially vulnerable to hypertension, which can have devastating effects on renal function. Hypertension ruptures glomerular capillaries and leads to scarring of the kidneys *(nephrosclerosis)*. It promotes atherosclerosis of the renal blood vessels, just as it does elsewhere in the body, and thus diminishes renal blood supply. Over time, hypertension often leads to renal failure.

16.3b Glomerular Filtration Rate

Glomerular filtration rate (GFR) is the amount of filtrate formed per minute by the two kidneys combined. In adult females, the GFR is about 105 mL/min.; in males, about 125 mL/min. These rates are equivalent to 150 L/day in females and 180 L/day in males—impressive numbers considering that this is about 60 times the amount of blood plasma in the body. Obviously, only a small portion of this is eliminated as urine. An average adult reabsorbs 99% of the filtrate and excretes 1 to 2 L of urine per day.

16.3c Regulation of Glomerular Filtration

GFR must be precisely controlled. If it is too high, fluid flows through the renal tubules too rapidly for them to reabsorb the usual amount of water and solutes. Urine output rises and creates a threat of dehydration and electrolyte depletion. If GFR is too low, fluid flows too slowly and the tubules reabsorb wastes that should be eliminated in the urine. GFR is adjusted through neural, hormonal, and local mechanisms that control glomerular blood pressure.

1. **Neural control.** The sympathetic nervous system richly innervates the kidneys. During exercise or in conditions such as circulatory shock, it constricts the afferent arterioles, reduces GFR, and reduces urine output. This conserves water and redirects blood from the kidneys to the heart, brain, and skeletal muscles where it is more urgently needed. Urine volume may be reduced during participation in a sporting event for this reason.
2. **Hormonal control.** Several hormones also regulate GFR. When blood pressure (BP) drops, the kidneys secrete the enzyme **renin** (REE-nin). Renin converts *angiotensinogen,* a blood plasma protein, into *angiotensin I.* An enzyme found in the lungs, *angiotensin-converting enzyme (ACE),* further converts this to **angiotensin II,** a hormone that helps to restore fluid volume and blood pressure (fig. 16.9). Angiotensin II makes one thirsty and encourages water intake; it stimulates the adrenal cortex to secrete **aldosterone,** which promotes retention of sodium and water in the body; and it stimulates the posterior pituitary to secrete **antidiuretic hormone (ADH),** which promotes water retention. Aldosterone and ADH reduce urinary water loss from the body, while thirst, of course, builds fluid volume and raises BP.

    Angiotensin II is also a potent vasoconstrictor. Widespread vasoconstriction raises one's overall blood pressure (BP), while constriction of the efferent arteriole (the glomerular outlet) sustains glomerular blood pressure and filtration. By stabilizing GFR, angiotensin II ensures that the kidneys continue to filter waste from the body even when systemic blood pressure is abnormally low.
3. **Local control (renal autoregulation).** Aside from the neural and hormonal "remote control" mechanisms, the nephrons also monitor their own performance and independently regulate their GFRs. They maintain a relatively stable GFR in spite of changes in arterial blood pressure. A rise in BP stretches the afferent arteriole and makes it reflexively constrict, so

Figure 16.9 The Renin–Angiotensin–Aldosterone Mechanism. This chain of events is activated by a drop in blood pressure and acts to raise it again.

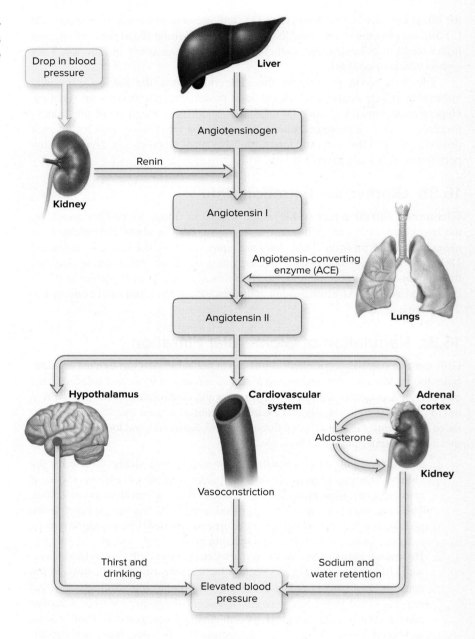

there is little increase in blood flow into the glomerulus or in GFR. If BP falls, in contrast, the afferent arteriole relaxes and allows more blood into the glomerulus, so GFR doesn't fall very much. These local mechanisms ensure that urine output changes very little even when one's overall blood pressure changes.

In addition, as tubular fluid flows through the final portion of the nephron loop, it passes a group of sensory cells that monitor its flow rate and osmolarity. A structure called the **juxtaglomerular**[8] **apparatus** (JUX-tuh-glo-MER-you-lur) is found where the nephron loop comes into contact with the afferent and efferent arterioles (**fig. 16.10**). Two special cell types occur here: (1) the *macula densa,*[9] a patch of closely spaced epithelial cells at the end of the nephron loop on the side of the tubule facing the arterioles; and (2) *granular cells,* which are enlarged smooth muscle cells

---
[8]*juxta* = next to; *glomer* = ball; *ul* = little; *ar* = pertaining to
[9]*macula* = spot, patch; *densa* = dense

Figure 16.10 The Juxtaglomerular Apparatus.

located mostly in the afferent arteriole, directly across from the macula densa. When stimulated by the macula, they dilate or constrict the arterioles. They also secrete the renin that initiates the hormonal mechanisms described earlier.

In spite of these controls, extreme drops in blood pressure can cause the GFR and urine output to fall precipitously or even cease. This is seen, for example, in severe circulatory shock from hemorrhage.

To summarize the events thus far: Glomerular filtration occurs because the high blood pressure of the glomerular capillaries overrides reabsorption. The filtration membrane allows water and most plasma solutes into the capsular space while retaining formed elements and most protein in the bloodstream. Glomerular filtration is maintained at a fairly steady rate in spite of normal variations in systemic blood pressure. This stability is achieved by nervous, hormonal, and local control mechanisms.

Before You Go On

Answer these questions from memory to test your understanding of the previous section. Reread the material if you can't answer the questions.

7. Name the four stages in which blood plasma is converted to urine.
8. Explain why, in view of the structure of the glomerular filtration membrane, the glomerular filtrate has a different composition from the blood.
9. Assume a person is exercising, sweating profusely, and experiences a significant drop in body fluid and blood volume. Describe the homeostatic mechanisms that would help the kidneys maintain a normal GFR.

16.4 Tubular Reabsorption and Secretion

Expected Learning Outcomes

When you have completed this section, you should be able to

a. describe how the renal tubules reabsorb water and useful solutes from the glomerular filtrate and return them to the blood; and
b. describe how the tubules secrete solutes from the blood into the tubular fluid.

The next two steps of converting glomerular filtrate to urine are *tubular reabsorption* and *tubular secretion,* the removal of some chemicals from the tubular fluid and addition of other chemicals to it. These two processes occur simultaneously, so we discuss them together in this section. We will trace the course of the tubular fluid through the nephron, from the proximal convoluted tubule through the distal convoluted tubule, and see how it is modified at each point along the way. Refer to figure 16.7 to put these processes into perspective.

16.4a The Proximal Convoluted Tubule

Tubular reabsorption is the process of reclaiming water and solutes from the tubular fluid and returning them to the blood. The proximal convoluted tubule (PCT) reabsorbs about 65% of the tubular fluid, while it also removes some substances from the blood and secretes them into the tubule for disposal in the urine. The importance of the PCT is reflected in its relatively great length and a shaggy fringe *(brush border)* of tall microvilli that greatly increase its absorptive surface area. It requires a great deal of energy to transport materials across the epithelial cells. About 6% of the calories you burn each day, and of your ATP budget, goes just to this process in the PCTs of your kidneys.

The PCT reabsorbs a greater variety of chemicals than any other part of the nephron, including the following substances and many others (**fig. 16.11**):

- **Electrolytes.** Sodium reabsorption is the key to almost everything else, because it creates an osmotic and electrical gradient that drives the reabsorption of water and the other solutes. Transport proteins in the plasma membranes of tubular cells are responsible for sodium uptake from the tubular fluid; on the other side of the cell, Na^+–K^+ pumps move Na^+ to the extracellular fluid, and from there it passes to the blood. Chloride ions (Cl^-) follow Na^+ by electrical attraction. Potassium, magnesium, calcium, and phosphate ions are reabsorbed by membrane transport proteins and by leaking through the spaces between epithelial cells.
- **Glucose.** Glucose is cotransported with Na^+. Normally, all glucose in the tubular fluid is reabsorbed and there is none in the urine.
- **Nitrogenous wastes.** Some urea in the tubular fluid is reabsorbed along with water in the PCT, but about half of it passes in the urine. Thus, the kidneys don't completely clear the blood of this waste, but keep its concentration down to a safe level.
- **Water.** The kidneys reduce about 180 L of glomerular filtrate to 1 or 2 L of urine each day, so obviously water reabsorption is a significant function. About two-thirds of the water is reabsorbed by osmosis in the PCT.

After water and solutes cross the tubule epithelium, they are reabsorbed into the blood by the peritubular capillaries.

There is a limit to the amount of solute that the renal tubule can reabsorb because there are limited numbers of transport proteins in the plasma membranes. If all the transporters are occupied as solute molecules go by, no more can be

Figure 16.11 Reabsorption and Secretion of Solutes and Water in the Renal Tubule.

reabsorbed and the excess passes through the tubule and into the urine. For example, all of the glucose that enters the renal tubule is normally reabsorbed; the number of transporters is adequate to handle the number of glucose molecules passing down the tubule in a given time. But at any blood glucose level above 220 mg/dL or so, glucose filters into the tubules faster than the tubules can reabsorb it, and the excess passes in the urine—a condition called **glycosuria**[10] (GLY-co-soo-ree-uh). In untreated diabetes mellitus, the plasma glucose concentration may exceed 400 mg/dL, so glycosuria is one of the classic signs of this disease.

16.4b The Nephron Loop

The primary function of the nephron loop is to generate an osmotic gradient that enables the collecting duct to concentrate the urine and conserve water, as explained later. In addition, however, the loop reabsorbs water and salts. The thin segment of the loop is very permeable to water but not to solutes. As tubular fluid flows down the descending limb, about 15% of the water is reabsorbed and returned to the bloodstream. Cells in the thick segment of the ascending limb are actively engaged in salt reabsorption, but are impermeable to water. Therefore, the ascending limb selectively reabsorbs salts from the fluid flowing through it.

16.4c The Distal Convoluted Tubule

Fluid arriving in the distal convoluted tubule (DCT) still contains about 20% of the water and 7% of the salts from the glomerular filtrate. If this were all passed as urine, it would amount to 36 L/day and we could hardly ever leave

---

[10]*glycos* = sugar; *uria* = urine condition

the bathroom! Obviously, a great deal of fluid reabsorption still has to occur. The DCT reabsorbs variable amounts of water and salts, but much less than the PCT. In keeping with its lesser (yet not unimportant) role in reabsorption, it is much shorter than the PCT and doesn't have such a prominent brush border of microvilli. The DCT is regulated by several hormones—particularly aldosterone and natriuretic peptides.

Aldosterone, the "salt-retaining hormone," is secreted by the adrenal cortex when the blood pressure or Na^+ concentration falls or its K^+ concentration rises. It stimulates the ascending limb of the nephron loop, the DCT, and the collecting duct to reabsorb more Na^+ and secrete more K^+. Its net effect is that the body retains Na^+ and water, urine volume is reduced, and the urine has an elevated K^+ concentration. The retention of salt and water helps to maintain blood volume and pressure.

Natriuretic peptides are secreted by the heart in response to high blood pressure. Their effect is opposite to aldosterone—to *increase* excretion of Na^+ and water in the urine, and therefore reduce blood volume and pressure.

16.4d Tubular Secretion

Tubular secretion is a process in which the renal tubule extracts chemicals from blood of the peritubular capillaries and secretes them into the tubular fluid (see fig. 16.7). In the proximal convoluted tubule and nephron loop, it serves three purposes: (1) acid–base balance, through secretion of varying proportions of hydrogen ions (H^+) to bicarbonate ions (HCO_3^-); (2) waste removal, through extracting urea, uric acid, bile acids, and ammonia from the blood; and (3) clearance of drugs and contaminants from the body, such as morphine, penicillin, and aspirin (see fig. 16.11). One reason so many drugs must be taken three or four times a day is to keep pace with this rate of clearance and maintain a therapeutically effective drug concentration in the blood.

In summary, the PCT reabsorbs about 65% of the glomerular filtrate and returns it to the blood of the peritubular capillaries. Much of this reabsorption occurs by osmosis linked to the transport of sodium ions. The nephron loop reabsorbs another 25% of the filtrate. The DCT reabsorbs more sodium, chloride, and water, but its rates of reabsorption are subject to control by hormones, especially aldosterone and natriuretic peptides. Drugs, wastes, and other solutes from the blood are secreted into the tubular fluid. The DCT essentially completes the process of determining the chemical composition of the urine. The principal function left to the collecting duct is to conserve water.

Before You Go On

Answer these questions from memory to test your understanding of the previous section. Reread the material if you can't answer the questions.

10. List five substances that are reabsorbed in the proximal convoluted tubule. Which of these does the most to drive reabsorption of the others? Explain how it does that.
11. Explain why glucose appears in the urine of a person with diabetes.
12. How do the thin and thick segments of the nephron loop differ in their absorptive functions?
13. Contrast the effects of aldosterone and natriuretic peptides on the distal convoluted tubule and on blood volume and pressure.

16.5 Water Conservation

Expected Learning Outcomes

When you have completed this section, you should be able to

a. explain how the collecting duct and antidiuretic hormone regulate the volume and concentration of urine;
b. explain how the kidney maintains an osmotic gradient in the renal medulla that enables the collecting duct to function; and
c. describe the hormonal mechanism for adjusting the body's rate of water loss to its state of hydration or dehydration.

The kidneys serve not just to eliminate metabolic waste from the body but also to prevent excessive water loss to support the body's fluid balance. As the kidney returns water to the tissue fluid and blood, the fluid remaining in the renal tubule, and ultimately passed as urine, becomes more and more concentrated. In this section, we examine the kidney's mechanism for conserving water and concentrating the urine.

16.5a The Collecting Duct

The collecting duct (CD) begins in the renal cortex, where it receives tubular fluid from numerous nephrons. As it passes through the medulla, it usually reabsorbs water and concentrates the urine. When urine enters the upper end of the CD, it is isotonic with blood plasma—300 milliosmoles per liter (mOsm/L) (see appendix C for explanation of this unit of measurement). By the time it leaves the lower end, it can be up to four times as concentrated—that is, highly hypertonic to the blood plasma.

Two facts enable the collecting duct to produce such hypertonic urine: (1) The osmolarity of the tissue fluid surrounding it is four times as high in the lower medulla as it is in the cortex. (2) The medullary portion of the CD is more permeable to water than to solutes. Therefore, as urine passes down the CD through the increasingly salty medulla, water leaves the tubule by osmosis, most NaCl and other wastes remain behind, and the urine becomes more and more concentrated (fig. 16.12).

It may seem surprising that the tissue fluid is so hypertonic in the deep medulla compared to the cortex and upper medulla. We would expect the salt to diffuse toward the cortex until it was evenly distributed through the kidney. However, there is a mechanism that overrides this—the nephron loop. The loop continually recaptures salt and returns it to the deep medullary tissue through a complex mechanism called a *countercurrent multiplier*. It is called a *multiplier* because it multiplies the salinity deep in the medulla and a *countercurrent* mechanism because it is based on fluid flowing in opposite directions in the descending and ascending limbs of the loop. The different permeabilities of the thin segment (to water only) and thick segment (to salts only) are also important to this mechanism. Without the action of this countercurrent multiplier, the collecting duct would be unable to absorb water and produce such concentrated urine.

16.5b Control of Water Loss

Just how concentrated the urine becomes depends on the body's state of hydration. If you are dehydrated, you produce scanty and very concentrated urine. Your high blood osmolarity stimulates the release of **antidiuretic hormone (ADH)** from the posterior pituitary gland. ADH induces the cells of the collecting duct to synthesize *aquaporins* (water-channel proteins) and install them in the plasma

Figure 16.12 Water Reabsorption by the Collecting Duct. Note that the osmolarity of the tissue fluid increases fourfold from 300 mOsm/L in the cortex to 1,200 mOsm/L deep in the medulla. Urine concentration increases proportionately as water leaves the duct through its aquaporins.

membrane, so more water can pass through the epithelium. More water is reabsorbed and is carried away by the surrounding blood capillaries *(vasa recta)*. Urine output is consequently reduced, and its osmolarity can be as high as 1,200 mOsm/L—four times the osmolarity of the blood plasma.

By contrast, if you drink a large volume of water, you soon produce a large volume of hypotonic urine. Hydration inhibits ADH secretion, and the tubule cells remove aquaporins from the plasma membrane. The CD reabsorbs NaCl but is less permeable to water. More water passes through as abundant, dilute urine, with an osmolarity as low as 50 mOsm/L. This response is called *water diuresis*[11] (DY-you-REE-sis).

To summarize what we have studied in this section, the collecting duct can adjust water reabsorption to produce urine as hypotonic (dilute) as 50 mOsm/L or as hypertonic (concentrated) as 1,200 mOsm/L, depending on the body's need for water conservation or removal. In a state of hydration, ADH is not secreted and the cortical part of the CD reabsorbs salt without reabsorbing water; the water remains to be excreted in the dilute urine. In a state of dehydration, ADH is secreted, the medullary part of the CD reabsorbs water, and the urine is more concentrated. The CD is able to do this because it passes through a salinity gradient in the medulla from 300 mOsm/L near the cortex to 1,200 mOsm/L near the papilla. This gradient is produced by a countercurrent multiplier of the nephron loop, which concentrates NaCl in the lower medulla. The vasa recta are arranged in such a way that they can supply blood to the medulla without subtracting from its salinity gradient.

Before You Go On

Answer these questions from memory to test your understanding of the previous section. Reread the material if you can't answer the questions.

14. Explain why the osmotic gradient of the renal medulla is necessary for the ability of the collecting duct to concentrate the urine.
15. Discuss the role of the nephron loop in maintaining that osmotic gradient.
16. Predict how a lack of ADH would affect the volume of urine, and explain why.

---

[11]*diuresis* = passing urine

PERSPECTIVES ON HEALTH

Renal Insufficiency and Dialysis

Renal insufficiency is a state in which the kidneys cannot maintain homeostasis due to extensive destruction of their nephrons. Some causes of nephron destruction include the following:

- Hypertension
- Atherosclerosis of the renal arteries
- Chronic or repetitive kidney infections
- Trauma from such causes as blows to the lower back or continual vibration from machinery
- Poisoning by heavy metals and solvents
- Blockage of renal tubules with proteins small enough to be filtered by the glomerulus—for example, myoglobin released by skeletal muscle damage and hemoglobin released by a transfusion reaction
- Glomerulonephritis, an autoimmune disease of the glomerular capillaries

Nephrons can regenerate and restore kidney function after short-term injuries. Even when some of the nephrons are irreversibly destroyed, others hypertrophy and compensate for their lost function. Indeed, a person can survive on as little as one-third of one kidney. When 75% of the nephrons are lost, however, urine output may be as low as 30 mL/h compared with the normal rate of 50 to 60 mL/h. This is insufficient to rid the body of toxic substances, including nitrogenous wastes. The level of nitrogenous waste in the blood is typically expressed as *blood urea nitrogen (BUN)*. An elevated BUN may indicate renal insufficiency, and can progress to a dangerous state called *uremia*. Convulsions, coma, and death can follow within a few days. Renal insufficiency also tends to cause anemia because the diseased kidneys produce too little erythropoietin (EPO), the hormone that stimulates red blood cell formation.

Hemodialysis is a procedure for artificially clearing wastes from the blood when the kidneys are not adequately doing so. Blood is pumped from the radial artery to a *dialysis machine* (artificial kidney) and returned to the patient by way of a vein. In the dialysis machine, the blood flows through tubes of selectively permeable membrane surrounded by dialysis fluid. Urea, potassium, and other solutes that are more concentrated in the blood than in the dialysis fluid diffuse through the membrane into the fluid, which is discarded. Glucose, electrolytes, and drugs can be administered by adding them to the dialysis fluid so they will diffuse through the membrane into the blood. People with renal insufficiency accumulate substantial amounts of body water between treatments, and dialysis serves also to remove this excess water. Patients are typically given EPO to compensate for its lack of production by the kidneys.

16.6 Urine Storage and Elimination

Expected Learning Outcomes

When you have completed this section, you should be able to

a. describe the functional anatomy of the ureters, urinary bladder, and male and female urethra;
b. explain how the nervous system and urethral sphincters control the voiding of urine; and
c. describe some physical and chemical properties of urine.

Urine is produced continually, but fortunately it doesn't drain continually from the body. Urination is episodic—occurring when we allow it. This is made possible by an apparatus for storing urine and by neural controls for its timely release.

16.6a The Ureters

The renal pelvis funnels urine into the ureter, a muscular tube that extends to the urinary bladder. The ureter is about 25 cm long and reaches a maximum diameter of about 1.7 cm near the bladder. The ureters pass posterior to the bladder and enter it from below, pierce its muscular wall, and open onto its floor. A small flap of mucosa acts as a valve at the opening of each ureter into the bladder, preventing urine from backing up into the ureter when the bladder contracts.

The ureter consists mostly of a thick tube of smooth muscle, surrounded externally by a sheath of loose fibrous connective tissue that binds it to the body wall, and lined internally by a mucosa with a transitional epithelium, an epithelial type unique to the urinary tract (see fig. 4.12). When urine enters the ureter and stretches it, the muscular wall contracts and initiates a peristaltic wave that milks the urine from the renal pelvis down to the bladder. These contractions occur every few seconds to every few minutes, proportional to the rate at which urine enters from above. The lumen of the ureter is very narrow and is easily obstructed or injured by kidney stones (Clinical Application 16.1).

16.6b The Urinary Bladder

The urinary bladder (fig. 16.13) is a muscular sac on the floor of the pelvic cavity posterior to the pubic symphysis. Its wall contains three layers of smooth muscle that are collectively called the **detrusor**[12] (deh-TROO-zur). The bladder is lined internally with an especially thick transitional epithelium that provides protection from the caustic urine that is stored there. The openings of the two ureters and the urethra mark a smooth triangular area called the *trigone*[13] (TRY-goan) on the bladder floor. This is a common site of bladder infection (Clinical Application 16.2).

Clinical Application 16.1

KIDNEY STONES

A *renal calculus*[14] (kidney stone) is a hard granule composed usually of calcium phosphate or calcium oxalate. Renal calculi form in the renal pelvis and are usually small enough to pass unnoticed in the urine flow. Some, however, grow as large as several centimeters and block the renal pelvis or ureter, which can lead to the destruction of nephrons as pressure builds in the kidney. A large, jagged calculus passing down the ureter stimulates strong contractions that can be excruciatingly painful. It can also damage the ureter and cause *hematuria* (blood in the urine). Causes of renal calculi include hypercalcemia, dehydration, pH imbalances, frequent urinary tract infections, or an enlarged prostate gland causing urine retention. Calculi are sometimes treated with stone-dissolving drugs, but often they require surgical removal. A nonsurgical technique called *shock wave lithotripsy*[15] uses ultrasound to pulverize the calculi into fine granules easily passed in the urine.

The bladder is highly distensible. As it fills, it expands superiorly, the wrinkled mucosa flattens, and the epithelium thins from five or six cell layers to only two or three. A moderately full bladder contains about 500 mL of urine and extends about 12.5 cm from top to bottom. The maximum capacity is 700 to 800 mL.

---

[12]*de* = down; *trus* = push
[13]*tri* = three; *gon* = angle
[14]*calc* = calcium, stone; *ul* = little
[15]*litho* = stone; *tripsy* = crushing

(a) Female

(b) Male

16.6c The Urethra

The urethra conveys urine from the urinary bladder to the **external urethral orifice,** the point of exit from the body. In the female, the urethra is a tube 3 to 4 cm long bound to the anterior wall of the vagina by fibrous connective tissue. Its orifice lies between the vaginal orifice and clitoris. The male urethra is about 18 cm long and passes through the penis. It has three regions: the *prostatic urethra* immediately inferior to the bladder, where it passes through the prostate and receives not only urine but also semen; the *membranous urethra,* a short, thin-walled segment that passes through the pelvic floor; and the *spongy urethra,* the portion contained in the penis, passing through a tissue called the *corpus spongiosum* (see fig. 19.1).

In males, the detrusor is thickened near the urethra to form an involuntary (smooth muscle) **internal urethral sphincter.** This sphincter contracts during ejaculation to prevent semen from being forced into the urinary bladder. Females do not have an internal urethral sphincter, but both sexes have a voluntary (skeletal muscle) **external urethral sphincter** that encircles the urethra where it passes through the

Figure 16.13 The Urinary Bladder and Urethra. Frontal sections. (a) Female. (b) Male. **APR**

? *Why do women lack the internal urethral sphincter present in men? What male-specific function does this sphincter serve?*

Clinical Application 16.2

URINARY TRACT INFECTION

Infection of the urinary bladder is called *cystitis.*[16] It is especially common in females because bacteria such as *Escherichia coli* can travel easily from the perineum up the short urethra. Because of this risk, young girls should be taught never to wipe the anus in a forward direction. Cystitis is frequently triggered in women by sexual intercourse ("honeymoon cystitis"). If cystitis is untreated, bacteria can spread up the ureters and cause *pyelitis,*[17] infection of the renal pelvis. If it reaches the renal cortex and nephrons, it is called *pyelonephritis.* Kidney infections can also result from invasion by blood-borne bacteria. Urine stagnation due to renal calculi or prostate enlargement increases the risk of infection.

---

[16]*cyst* = bladder; *itis* = inflammation
[17]*pyel* = pelvis; *itis* = inflammation

pelvic floor (the membranous urethra in the male). This sphincter provides voluntary control over the voiding of urine.

16.6d Voiding Urine APR

When the bladder is filling, the sympathetic nervous system normally prevents premature leakage of urine. It issues nerve fibers from the spinal cord to the detrusor to *relax* it, preventing bladder contraction. The external urethral sphincter, innervated by somatic nerve fibers, remains contracted and prevents escape of urine.

The act of urinating, called **micturition**[18] (MIC-too-RISH-un), is initiated by an autonomic spinal *micturition reflex.* Filling of the bladder excites stretch receptors in the bladder wall. They issue signals to the spinal cord, which returns signals to the bladder via parasympathetic nerves. Parasympathetic stimulation contracts the detrusor. In very young children and in people with spinal cord injuries that disconnect the brain from the lower spinal cord, this alone results in emptying of the bladder.

Normally, however, one also has voluntary control over urination. Stretch signals from the bladder ascend the spinal cord to a nucleus in the pons called the **micturition center.** This nucleus integrates information about bladder tension with information from other brain centers such as the cerebral cortex. Thus, urination can be inhibited by knowledge that the circumstances are inappropriate. If it is desirable and appropriate to urinate, signals descend from the pons to inhibit the spinal somatic neurons that keep the external urethral sphincter constricted. The sphincter then relaxes, and micturition proceeds. If the bladder is not full enough to trigger the micturition reflex, but one wishes to "go" anyway because of a long drive or lecture coming up, the Valsalva maneuver (see section 15.2) can be used to compress the bladder and excite the stretch receptors early, thereby getting the reflex started.

16.6e Urine Volume and Properties

An average adult produces 1 to 2 L of urine per day. An output in excess of 2 L/day is called **diuresis** or **polyuria**[19] (POL-ee-YOU-ree-uh). Fluid intake and some drugs can temporarily increase output to as much as 20 L/day. Diseases such as diabetes may be characterized by chronic polyuria. Low urine output, called **oliguria,**[20] can result from kidney disease, dehydration, circulatory shock, and prostate enlargement, among other causes. If urine output drops to less than 400 mL/day, unsafe concentrations of wastes accumulate in the blood plasma. **Uremia** is a condition of dangerously high nitrogenous waste levels in the blood, usually indicating kidney failure.

Because it is one of the most accessible body fluids, urine has been used for centuries to assess health. It is still common to provide a urine sample during a medical checkup. The basic composition and properties of urine are as follows:

- **Appearance.** Urine varies from almost colorless to deep amber, depending on the body's state of hydration. The yellow color of urine is due to a pigment produced by the breakdown of hemoglobin from expired erythrocytes. Unusual colors can be imparted to it by certain foods, vitamins, drugs, and metabolic diseases. **Pyuria**[21] (pus in the urine) can indicate a kidney infection. **Hematuria** (blood in the urine) may be due to a urinary tract infection, various kidney diseases, trauma, or kidney stones.
- **Odor.** Fresh urine has a distinctive but not necessarily repellent odor. As it stands, however, bacteria multiply, degrade urea to ammonia, and produce the pungent odor typical of stale wet diapers. Asparagus, other foods, and some metabolic diseases can impart distinctive aromas to the urine. Diabetes mellitus gives it a sweet, "fruity" odor of acetone.

---

[18]*mictur* = to urinate
[19]*poly* = many, much; *ur* = urine; *ia* = condition
[20]*olig* = few, a little; *ur* = urine; *ia* = condition
[21]*py* = pus; *ur* = urine; *ia* = condition

- **Specific gravity.** This is a ratio of the density (g/mL) of a substance to the density of distilled water. Distilled water has a specific gravity of 1.000 by definition, and urine ranges from 1.001 when it is very dilute to 1.028 when it is very concentrated.
- **Osmolarity.** Urine can have an osmolarity as low as 50 mOsm/L in a very hydrated person or as high as 1,200 mOsm/L in a dehydrated person. Compared with the osmolarity of blood (300 mOsm/L), then, urine can be either hypotonic or hypertonic.
- **pH.** The pH of urine ranges from 4.5 to 8.2 but is usually about 6.0 (mildly acidic). This is one reason the urinary tract needs to be protected by transitional epithelium, a tissue found only in this area.
- **Chemical composition.** Urine averages 95% water and 5% solutes by volume; table 16.1 lists several of the major solutes. It is abnormal to find glucose, free hemoglobin, albumin, or bile pigments in the urine; their presence can serve as important indicators of disease.

Apply What You Know

Predict the effect of dehydration on the specific gravity of urine.

Table 16.1 Properties and Composition of Urine

Physical Properties		
Specific gravity	1.001–1.028	
Osmolarity	50–1,200 mOsm/L	
pH	4.5–8.2 (average 6.0)	
Solutes	**Concentration (mg/dL)***	**Output (g/day)****
Urea	1,800	21
Creatinine	150	1.8
Ammonia	60	0.68
Chloride	533	6.4
Sodium	333	4.0
Potassium	166	2.0
Phosphate	83	1
Calcium	17	0.2
Magnesium	13	0.16

*Typical values for a young adult male
**Assuming a urine output of 1.2 L/day

Before You Go On

Answer these questions from memory to test your understanding of the previous section. Reread the material if you can't answer the questions.

17. Where does the ureter begin and end? What prevents contraction of the bladder from forcing urine back up to the kidneys?
18. Explain involuntary and voluntary control of urination.
19. Define *polyuria, hematuria,* and *uremia,* and identify one possible cause of each.

16.7 Fluid, Electrolyte, and Acid–Base Balance

Expected Learning Outcomes

When you have completed this section, you should be able to

a. name the major fluid compartments;
b. list the body's sources of water and routes of water loss;
c. describe mechanisms of regulating water intake and output;
d. list the functions of sodium and potassium;
e. explain how electrolyte balance is regulated; and
f. describe three ways the body regulates pH.

In the course of producing urine and eliminating metabolic wastes, the kidneys play a vital role in the homeostatic regulation of the body's fluid, electrolyte, and acid–base balance. Imbalances in these variables can easily have life-threatening consequences. They are closely regulated by the collaborative effort of the urinary, respiratory, and digestive systems, coordinated by nervous and hormonal mechanisms.

16.7a Fluid Balance

Our internal landscape is watery. About 45% to 60% of the adult body is made up of water, depending on proportions of fat and lean body mass. Adipose tissue contains very little body water, whereas lean mass, which is mostly skeletal muscle, contains significant amounts of water. Women on average have more adipose tissue than men, and therefore usually have less water relative to total body mass. Active people have relatively more muscle mass and therefore have more body water relative to weight.

Fluids are distributed between two main areas, or **fluid compartments.** About two-thirds of the body water is in the **intracellular compartment,** within the cells (*intracellular fluid*). This compartment is separated from the **extracellular compartment** (*extracellular fluid*) by plasma membranes. The extracellular compartment includes all the fluids outside cells—mainly *tissue fluid* between the cells, *plasma* in the blood vessels, and *lymph* in lymphatic vessels.

> **Apply What You Know**
>
> Review synovial fluid of the joints (see section 6.5a), cerebrospinal fluid of the brain (see section 9.1d), and aqueous humor of the eye (see section 10.5b) if necessary, and explain whether (and why) you would classify those as belonging to the intracellular or extracellular compartment.

Fluid balance means a state in which our average daily water gains and losses are equivalent (about 2,500 mL/day for the average adult) and body water is properly distributed among the fluid compartments (not abnormally pooled somewhere, as in edema). A little of our body water comes from metabolic processes such as aerobic respiration, but for the most part, fluid balance is a matter of intake versus output—the amount we ingest in food and drink versus the amount we lose in various ways (fig. 16.14). Most of our water loss is through urine, fecal moisture, and the breath. We also lose water through the skin in two ways—sweating and *cutaneous transpiration*. Sweat is produced by cutaneous

Figure 16.14 Typical Water Gains and Losses in a State of Fluid Balance.

? *How would the right side of this figure change if the person had an abnormally high rate of ADH secretion?*

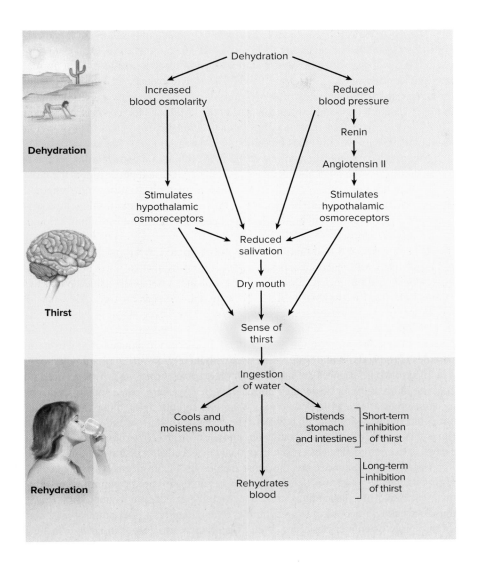

Figure 16.15 Dehydration, Thirst, and Rehydration.

glands, usually in response to heat. We're less conscious of cutaneous transpiration, the diffusion of water through the epidermis to the surface, where it evaporates.

Fluid intake is governed mainly by thirst, which is controlled by mechanisms shown in figure 16.15. Dehydration raises blood osmolarity and reduces blood volume. The rise in osmolarity stimulates neurons called *osmoreceptors* in the hypothalamus, which leads to the sensation of thirst and a desire to drink. Normally, we drink long before there is any significant deficit in blood volume or significant rise in osmolarity. In a more significant state of dehydration, however, the hormonal mechanisms of angiotensin II and antidiuretic hormone, detailed earlier, come into play to accentuate the thirst and reduce urinary loss of water. Contrary to popular myth, it is not absolutely necessary to drink 8 glasses (64 ounces) of water per day. Hydration requirements vary depending on how much a person exercises, for example, and, usually, feeling thirsty motivates us to take in enough fluids.

Fluid output, as we have seen, is governed mainly by varying urine volume and osmolarity, and these are regulated mainly by antidiuretic hormone.

16.7b Electrolyte Balance

Electrolytes are salts that dissociate into free ions in water. They are essential for a tremendous variety of cellular functions, such as nerve and muscle action potentials, bone building, enzyme activation, pH balance, osmotic effects on fluid balance, and more.

The major cations from electrolytes are sodium (Na^+), potassium (K^+), and calcium (Ca^{2+}), and the major anions are chloride (Cl^-), bicarbonate (HCO_3^-), and phosphates (P_i). In a state of **electrolyte balance,** each of these is within its own normal range of concentration in the body fluids. Electrolyte balance is achieved mainly through equivalent daily rates of absorption by the intestines and excretion by the kidneys. Calcium regulation was explained under "Mineral Homeostasis" in section 6.2c. Anion homeostasis tends to come about automatically as an effect of cation regulation, so here we will focus on sodium and potassium.

Electrolyte concentrations differ in the intracellular and extracellular compartments. Sodium is the most abundant extracellular cation, and potassium is the most abundant in the intracellular fluid. These two ions play pivotal roles in resting and action potentials, and sodium is the most significant solute in determining the distribution of water among fluid compartments, because water follows sodium through the process of osmosis.

Both sodium and potassium balance are regulated by aldosterone. Aldosterone promotes Na^+ reabsorption and K^+ excretion by the kidneys. A rise in aldosterone secretion thus results in less Na^+ but more K^+ in the urine. When the aldosterone level is especially high, the urine can be sodium-free. Aldosterone has little effect on plasma sodium *concentration,* however, because when Na^+ is reabsorbed, a proportionate amount of water is reabsorbed along with it; more fluid is retained, but its concentration is unchanged. In contrast to aldosterone, natriuretic peptides cause sodium and water to be excreted, thereby lowering blood volume.

Electrolyte imbalances are named with the prefixes *hyper-* and *hypo-* to indicate concentrations above or below their normal ranges. *Hyponatremia,* a sodium deficiency, sometimes affects athletes who lose water and sodium through copious sweating but compensate by drinking plain water. The resulting dilution of sodium in the body fluids can result in weakness, confusion, and even death. *Hypernatremia,* a sodium excess, is less common but can result from badly managed administration of I.V. fluids.

Potassium imbalances are the most dangerous of all electrolyte imbalances. *Hypokalemia,* a potassium deficiency, can result from chronic vomiting or diarrhea, excessive use of laxatives, or aldosterone hypersecretion. Symptoms include muscle weakness and cardiac arrhythmia. *Hyperkalemia,* a potassium excess, can set in rapidly when a crush injury (such as getting an arm caught in farm or factory machinery) ruptures large numbers of cells and releases intracellular K^+ into the extracellular fluids, or it can have a slower onset in cases of aldosterone hyposecretion or renal failure. Hyperkalemia is associated with muscle cramps, diarrhea, vomiting, and atrial fibrillation. Cardiac arrest can result from potassium excess or deficiency, but the risk of this is especially great when there is an abrupt potassium overload. Clinical administration of I.V. potassium must be done with extreme care for such reasons.

16.7c Acid–Base Balance

Our life-sustaining metabolic pathways depend on enzymes, and enzymes, being proteins, are very pH-sensitive. Conditions that are even slightly too acidic or basic can cause life-threatening disruption of enzyme function. Consequently, the pH of our blood and tissue fluid is maintained within a very narrow margin—7.35 to 7.45. **Acid–base balance,** or **pH balance,** means the regulation of blood pH within this range.

A pH above this range is called **alkalosis,** and a pH below this range is called **acidosis.** Alkalosis causes nerve and muscle cells to be hyperexcitable and can lead to consequences as severe as convulsions and respiratory paralysis. Acidosis depresses the nervous system and can lead to coma and death. Acidosis is a common cause of death among diabetics who do not manage their condition well. Our homeostasis is constantly challenged by acids of metabolic origin—lactate (lactic acid) from anaerobic fermentation, carbonic acid from carbon dioxide, and fatty acids from fat catabolism. Acidosis is therefore more common than alkalosis.

The body regulates pH in three ways: chemical buffers, the respiratory system, and the urinary system.

Chemical Buffers

A chemical buffer is a mixture that binds excess H^+ and removes it from solution when there is an excess (low pH), or releases H^+ into solution when there is a deficiency (high pH). Buffer solutions therefore tend to have a stable pH even if challenged by the addition of more acid or base. For example, bicarbonate ions (HCO_3^-) and the amino groups (—NH_2) of proteins can bind and neutralize H^+, thereby raising pH, as follows:

$$H^+ + HCO_3^- \rightarrow H_2CO_3$$
$$H^+ + -NH_2 \rightarrow -NH_3^+$$

Conversely, if pH rises too high, carbonic acid (H_2CO_3) and the carboxyl groups (—COOH) of proteins can liberate hydrogen ions and lower it:

$$H_2CO_3 \rightarrow HCO_3^- + H^+$$
$$-COOH \rightarrow -COO^- + H^+$$

Bicarbonate, proteins, and phosphates are the most significant chemical buffers in the body. Chemical buffers can restore and stabilize normal pH in a fraction of a second, but they are relatively limited in the amount of excess acid or base they can neutralize.

The Respiratory Buffer

The respiratory system takes longer than chemical buffers to act—a few minutes—but it can neutralize more acid or base than the chemical buffers alone. It depends on the relationship between carbon dioxide and pH, and the ability to vary the breathing rhythm so we retain more CO_2 in the body or expel it faster.

Carbon dioxide reacts with water to release acid and lower the pH:

$$CO_2 + H_2O \rightarrow H_2CO_3 \rightarrow HCO_3^- + H^+$$

A CO_2 excess in the body therefore tends to cause acidosis, a common occurrence in lung diseases such as emphysema where the body cannot expel CO_2 as fast as it is made. On the other hand, if there is a CO_2 deficiency in the body fluids, these reactions happen in reverse to raise the CO_2 level but lower the H^+ concentration, thus raising pH:

$$HCO_3^- + H^+ \rightarrow H_2CO_3 \rightarrow CO_2 + H_2O$$

We saw in section 15.2c that the brainstem respiratory centers are very sensitive to pH changes. Acidosis stimulates one to breathe faster to expel CO_2 faster than it is produced, thus raising the pH to normal. Alkalosis stimulates one to breathe more slowly, thus accumulating CO_2 and lowering the pH to normal.

The Renal Buffer

Of all buffering mechanisms, the kidneys are slowest to act and require several hours or even days to correct a pH imbalance, but they can rid the body of more acid than the chemical or respiratory buffer system can. The mechanisms are complex and beyond the scope of this book, but the essence of it is this: When there is an excess of acid (H^+) in the blood, it reacts with blood bicarbonate ions to form carbonic acid, which breaks down to H_2O and CO_2, like the last of the reactions shown in the previous section. These move from the blood into the cells of the renal tubule. These epithelial cells have an enzyme that reverses this reaction and generates HCO_3^- and H^+ again. The HCO_3^- is returned to the blood and the H^+ is secreted into the tubular fluid—a case of the *tubular secretion* phenomenon described earlier in this chapter. Thus, H^+ is removed from the blood and appears in the tubular fluid, although indirectly through the carbonic acid reactions. In the tubular fluid, some of this H^+ reacts with bicarbonate ions to again produce H_2O and CO_2, and some remains as free H^+, giving the urine its characteristic mildly acidic pH, about 6.0.

Ammonia (NH_3) in the tubular fluid also can react with H^+ and chloride ions to produce ammonium chloride, NH_4Cl. This mechanism can come to the aid of acid–base balance when the body is under a particularly heavy acid challenge, and the urine then shows up to 10 times as much ammonium chloride as usual. A high urine ammonium chloride level can be a sign of an acid-generating disease such as diabetes mellitus.

Before You Go On

Answer these questions from memory to test your understanding of the previous section. Reread the material if you can't answer the questions.

20. What are the major fluids of the extracellular compartment?
21. What is the most important mechanism for regulating fluid intake? What hormone is most important in regulating fluid output?
22. What two organ systems are most important in regulating electrolyte balance?
23. What are the terms for sodium and potassium excess and deficiency? What are causes and consequences of sodium and potassium imbalances?
24. What hormones are most important in regulating sodium balance? Describe their actions.
25. What are the terms for excessively high and low pH? State some causes and effects of each.
26. Describe how chemical buffers, the respiratory system, and the kidneys resist or compensate for shifts in body fluid pH.

Aging of the Urinary System

Nephrons are abundant early in life, but the number declines markedly with age. By age 85, there can be as much as a 40% reduction in the number of functional nephrons and the size of the kidneys. In addition, the renal blood supply and GFR are significantly reduced. The result is that although the aged kidneys function adequately under normal conditions, they cope less efficiently with illness and in clearance of medications. In men, an enlarged prostate may compress the urethra and make it more difficult to void urine. In women, strain on the pelvic floor muscles during childbirth makes incontinence more likely later in life.

CAREER SPOTLIGHT

Dialysis Technician

Dialysis technicians support the care of patients with kidney failure who require dialysis. The job entails setting up and operating hemodialysis machines. In addition, the technician prepares the patient for dialysis and monitors both the patient and the machine during the process. The technician must learn the scientific principles of dialysis, understand the equipment involved, and respond to physical and emotional needs of the person undergoing treatment. Dialysis technicians work closely with a team of nurses and physicians who care for the patient. They work at hemodialysis facilities in hospitals and outpatient facilities, or may assist patients with in-home dialysis. Training is provided on the job or through vocational schools or community colleges. Dialysis technicians must pass a state exam to become certified. See appendix B for further information on dialysis technician careers.

© Life in View/Science Source

CONNECTIVE ISSUES

Ways in Which the URINARY SYSTEM Affects Other Organ Systems

All Systems
The urinary system serves all other systems by eliminating metabolic wastes and maintaining fluid, electrolyte, and acid–base balance.

Integumentary System
Renal control of fluid balance is essential for sweat secretion; the epidermis is normally a barrier to fluid loss—thus, the skin and kidneys collaborate to maintain fluid balance.

Skeletal System
Renal control of calcium and phosphate balance and role in calcitriol synthesis are essential for bone deposition.

Muscular System
Renal control of Na^+, K^+, and Ca^{2+} balance is important for muscle contraction.

Nervous System
The nervous system is very sensitive to fluid, electrolyte, and acid–base imbalances that may result from renal dysfunction.

Endocrine System
Renin secretion by the kidneys leads to angiotensin synthesis and aldosterone secretion; the kidneys produce erythropoietin.

Circulatory System
The kidneys control blood pressure more than any other organ; erythropoietin from the kidneys regulates hematocrit; the kidneys regulate plasma composition.

Lymphatic and Immune Systems
The acidity of urine destroys pathogens and thus defends against infection.

Respiratory System
The urinary and respiratory systems collaborate to regulate acid–base balance.

Digestive System
The kidneys excrete toxins absorbed by the digestive tract; the kidneys excrete hormones and metabolites after the liver deactivates them; calcitriol synthesized by the kidneys regulates Ca^{2+} absorption by the small intestine.

Reproductive System
The urethra serves as a common passageway for urine and semen in males; the urinary system of pregnant woman eliminates metabolic wastes of the fetus.

Study Guide

Assess Your Learning Outcomes

To test your knowledge, discuss the following topics with a study partner or in writing, ideally from memory.

16.1 Functions of the Urinary System

1. The principal organs of the urinary system
2. Functions of the kidneys
3. Definitions of *excretion, waste,* and *metabolic waste*
4. The types and sources of nitrogenous wastes

16.2 Anatomy of the Kidney

1. The anatomical position of the kidneys; their size, shape, and connective tissue coverings
2. Internal organization of the glandular and collecting tissues of the kidney
3. The route of blood flow through the kidney
4. Structure of the nephron
5. The route of fluid flow through the nephron as it becomes urine

16.3 Glomerular Filtration

1. The four stages in urine formation, and the changing names of the fluid at various stages
2. The meaning of *glomerular filtration* and the site of its occurrence
3. Components of the filtration membrane and how they affect the composition of the filtrate
4. Factors that determine filtration pressure and glomerular filtration rate (GFR)
5. The mechanism of sympathetic nervous control of glomerular filtration
6. Hormones that influence glomerular filtration, and their mechanisms of action
7. Mechanisms of local control by which a nephron regulates its own glomerular filtration
8. Structure of the juxtaglomerular apparatus and how it regulates GFR

16.4 Tubular Reabsorption and Secretion

1. Definitions of *tubular reabsorption* and *tubular secretion*
2. Structural adaptations of the proximal convoluted tubule (PCT) for its role in absorption
3. Substances reabsorbed by the PCT, the mechanisms of reabsorption, and how sodium reabsorption drives the others
4. Where the reabsorbed materials go after they leave the renal tubule
5. Limitations on solute reabsorption and why excess levels of blood glucose result in glycosuria
6. Structure of the nephron loop and how it relates to tubular reabsorption
7. Substances that are reabsorbed by the distal convoluted tubule (DCT)
8. Hormones that influence DCT function, and their modes of action
9. The role of tubular secretion in the composition of the urine

16.5 Water Conservation

1. The basic function of the collecting duct
2. The significance of tissue fluid osmolarity and selective permeability of the collecting duct in urine formation and water conservation
3. How the nephron loop creates the osmotic gradient of the renal medulla necessary for water conservation
4. Hormonal control of collecting duct function and urine volume and concentration

16.6 Urine Storage and Elimination

1. Anatomy and peristaltic function of the ureters
2. Anatomy of the urinary bladder, its spatial relationship with the ureters and urethra, and the changes it undergoes as it fills and stretches
3. Structure of the female and male urethra
4. The two urethral sphincters, their locations, and differences in their muscular composition and nervous control
5. Components and actions of the spinal micturition reflex
6. The mechanism of voluntary control over micturition
7. Typical daily output of urine, factors that influence it, and consequences of insufficient output
8. Physical and chemical properties of normal urine

16.7 Fluid, Electrolyte, and Acid–Base Balance

1. Typical body water content and reasons for its variability
2. The body's fluid compartments and how they differ
3. The meaning of fluid balance
4. Major sources of water and means of water loss
5. The principal factors that regulate water intake and output
6. The meanings of electrolytes and electrolyte balance
7. The most physiologically important ions that arise by electrolyte dissociation in water
8. Hormonal control of sodium and potassium balance
9. The names and pathological consequences of sodium and potassium deficiencies and excesses
10. The meaning of acid–base balance and why it is crucial to one's metabolic pathways
11. The normal range of blood pH and terms for deviations above and below this range
12. Three major chemical buffer systems in the body and examples of how they stabilize pH
13. How respiration can adjust body pH
14. How the kidneys excrete acid and adjust body pH
15. Relative importance, speed, and effectiveness of the chemical, respiratory, and renal buffer systems

Testing Your Recall

1. Micturition occurs when the _____ contracts.
   a. detrusor
   b. internal urethral sphincter
   c. external urethral sphincter
   d. muscular wall of the ureter
   e. muscular wall of the urethra

2. The compact ball of capillaries in a nephron is called
   a. the nephron loop.
   b. the peritubular plexus.
   c. the renal corpuscle.
   d. the glomerulus.
   e. the vasa recta.

3. Which of these is the most abundant nitrogenous waste in the blood?
   a. uric acid
   b. urea
   c. ammonia
   d. creatinine
   e. albumin

4. Which of the following is excluded by the filtration membrane from passing into the filtrate?
   a. glucose
   b. formed elements of blood
   c. sodium
   d. water
   e. nitrogenous wastes

5. What percentage of glomerular filtrate is reabsorbed by the proximal convoluted tubule?
   a. 6%
   b. 25%
   c. 45%
   d. 65%
   e. The amount varies depending on a person's state of hydration.

6. Which of the following does *not* result from dehydration?
   a. increased production of angiotensin II
   b. stimulation of thirst
   c. increased production of ADH
   d. increased blood osmolarity
   e. increased urine volume

7. Each nephron receives its blood supply directly from
   a. a renal artery.
   b. an interlobar artery.
   c. a segmental artery.
   d. an efferent arteriole.
   e. an afferent arteriole.

8. Determine the correct order of flow through the renal tubule, if we let
   CD = collecting duct,
   DCT = distal convoluted tubule,
   GC = glomerular capsule,
   PCT = proximal convoluted tubule, and
   NL = nephron loop.
   a. PCT–GC–NL–DCT–CD
   b. CD–DCT–PCT–GC–NL
   c. PCT–DCT–NL–CD–GC
   d. GC–PCT–NL–DCT–CD
   e. GC–CD–PCT–NL–DCT

9. Which of the following is/are secreted into the renal tubule?
   a. water
   b. glucose
   c. bicarbonate ions
   d. sodium chloride
   e. albumin

10. How would increased ADH secretion change the urine?
    a. reduced volume
    b. increased volume
    c. reduced osmolarity
    d. reduced urea concentration
    e. reduced potassium concentration

11. In lung diseases such as emphysema where exhalation is impeded, an excess of CO_2 may cause acidosis.

12. glomerular filtration rate is the amount of filtrate formed per minute by the two kidneys combined.

13. The two ureters carry urine from the kidneys to the urinary bladder.

14. The macula densa is a group of epithelial cells of the nephron loop that monitors the flow or composition of the tubular fluid.

15. To enter the capsular space, filtrate must pass between foot processes of the podocytes, cells that form part of the filtration membrane in the glomerular capsule.

16. The nephron is the functional unit of the kidney where urine is produced.

17. The descending limb of the nephron loop is permeable to water, whereas the ascending limb of the loop is permeable to salt.

18. The internal urethral sphincter is under involuntary control and relaxes during the male micturition reflex, but contracts during ejaculation.

19. Aldosterone stimulates the ascending limb of the nephron loop, the DCT, and the collecting duct to reabsorb more sodium and secrete more potassium.

20. The fibrous capsule tightly wraps the kidney and protects it from trauma and infection.

Answers in Appendix A

What's Wrong with These Statements?

Briefly explain why each of the following statements is false, or reword it to make it true.

1. Aldosterone has no effect on the distal convoluted tubule.

2. Loss of water through the skin occurs only through sweating.

3. The kidney has more collecting ducts than nephrons.

4. Sodium is reabsorbed only in the proximal convoluted tubule.

5. Natriuretic peptides increase blood volume by promoting sodium reabsorption in the kidney.

6. If all other conditions remain the same, dilation of the afferent arteriole reduces the glomerular filtration rate.

7. Angiotensin II lowers blood pressure.

8. The minimum osmolarity of urine is 300 mOsm/L, equal to the osmolarity of the blood.

9. The sympathetic nervous system sends signals that contract the detrusor and relax the internal urethral sphincter.

10. A normal GFR is less than 100 mL/min.

Answers in Appendix A

Testing Your Comprehension

1. Animals that live in freshwater, such as muskrats and beavers, have very short nephron loops. In view of the function of the nephron loop in humans, explain why it is less necessary in such aquatic mammals. As a result of this difference, would you expect their urine to be hypotonic or hypertonic to ours? Why?

2. Hypertension (high blood pressure) is often treated with medications called ACE inhibitors, which interfere with the angiotensin-converting enzyme that is necessary to convert angiotensin I to angiotensin II. Explain why ACE inhibitors are prescribed to people who suffer from chronic hypertension.

3. Cholera is a bacterial infection common in impoverished countries and areas stricken by warfare and natural disasters. It causes profuse diarrhea and often death, especially in children. In view of section 16.7 of this chapter, why do you think these children could die in a few days from unrelenting diarrhea? Be as detailed as you can about the physiological mechanisms.

The Digestive System

Chapter **17**

Chapter Outline

17.1 Overview of the Digestive System
- 17.1a Digestive System Functions
- 17.1b General Anatomy

17.2 The Mouth Through Esophagus
- 17.2a The Mouth
- 17.2b Saliva and Salivary Glands
- 17.2c The Pharynx
- 17.2d The Esophagus
- 17.2e Swallowing

17.3 The Stomach
- 17.3a Anatomy
- 17.3b Gastric Secretions
- 17.3c Gastric Motility
- 17.3d Regulation of Gastric Function
- 17.3e Protection of the Stomach

17.4 The Liver, Gallbladder, and Pancreas
- 17.4a The Liver
- 17.4b The Gallbladder and Bile
- 17.4c The Pancreas

17.5 The Small Intestine
- 17.5a Anatomy
- 17.5b Intestinal Secretion and Motility

17.6 Chemical Digestion and Absorption
- 17.6a Carbohydrates
- 17.6b Proteins
- 17.6c Fats
- 17.6d Vitamins and Minerals
- 17.6e Water

17.7 The Large Intestine
- 17.7a Anatomy
- 17.7b Intestinal Bacteria
- 17.7c Absorption and Motility
- 17.7d Defecation

Clinical Applications/Perspectives on Health
- Clinical Application 17.1: Peptic Ulcer
- Clinical Application 17.2: Lactose Intolerance
- Perspectives on Health

End of Chapter
- Aging of the Digestive System
- Career Spotlight: Dental Hygienist
- Connective Issues
- Study Guide

The small intestine absorbs nutrients across epithelial tissue such as that shown in this microscope image. Columnar cells are interspersed with bright blue mucus-producing goblet cells.
© Steve Gschmeissner/Science Source

BASE CAMP

Before ascending to the next level, be sure you're properly equipped with a knowledge of these concepts from earlier chapters.

- Body cavities and membranes (see section 1.3e)
- Peritoneum and intra- and retroperitoneal organs (see section 1.3e)
- Carbohydrate, fat, and protein structures (see section 2.3a, b, and c)
- pH and enzyme function (see sections 2.2b and 2.3c)
- Epithelial tissues (see section 4.2)

Module 12
Digestive System

Eating is a fundamental human behavior, required for survival and central to social interactions and cultural identity. The food we eat is not yet in a form that can be used either for energy or as structural building blocks for tissues. The digestive system is essentially a disassembly line that breaks food into small components such as amino acids and monosaccharides that can be transported into cells. Consider what happens when we eat a steak, for example. The proteins that make up beef muscle differ very little from that of your own muscles. Even if they were identical, however, proteins such as myosin in the steak could not be directly absorbed, transported in the blood, and incorporated into your muscles. The ingested proteins must be dismantled into amino acids before they are useful to cells. Once digested, the amino acids in the beef proteins might be used to build your own muscle myosin, but could equally wind up in your insulin, collagen, or any other protein. After the food has been digested, the nutrients are absorbed in the digestive tract and then distributed to the tissues. The study of the digestive tract and the diagnosis and treatment of its disorders is called **gastroenterology.**[1]

17.1 Overview of the Digestive System

Expected Learning Outcomes

When you have completed this section, you should be able to

- a. list the functions and major physiological processes of the digestive system;
- b. distinguish between mechanical and chemical digestion;
- c. list the regions of the digestive tract and the accessory organs of the digestive system;
- d. identify the histological layers of the digestive tract;
- e. describe the special nervous network of the digestive tract; and
- f. describe the relationship between the digestive tract, peritoneum, and mesenteries.

---

[1]*gastro* = stomach; *entero* = intestine; *logy* = study of

17.1a Digestive System Functions

The **digestive system** processes food, extracts nutrients from it, and eliminates the residue. It does this in five stages:

1. **ingestion,** the selective intake of food;
2. **digestion,** the breakdown of food into a form usable by the body;
3. **absorption,** the uptake of nutrient molecules into the epithelial cells of the digestive tract and then into the blood or lymph;
4. **compaction,** absorption of water and consolidation of the indigestible residue into feces; and
5. **defecation,** the elimination of feces.

The digestion phase has two facets, mechanical and chemical. **Mechanical digestion** is the physical breakdown of food into smaller particles, achieved by the cutting and grinding action of the teeth and the churning movements of the stomach. Mechanical digestion exposes more food surface to the action of digestive enzymes.

Chemical digestion employs enzymes to break dietary macromolecules into their basic building blocks: polysaccharides into monosaccharides, proteins into amino acids, fats into monoglycerides and fatty acids, and nucleic acids into nucleotides. All chemical digestion is by the process of enzymatic hydrolysis, breaking covalent bonds by adding water to them (see section 2.3). The digestive enzymes that carry this out are produced by the salivary glands, stomach, pancreas, and small intestine.

Some nutrients are already present in usable form in the ingested food and are absorbed without being broken down by digestion: vitamins, minerals, cholesterol, and water.

17.1b General Anatomy

The two anatomical subdivisions of the digestive system are the digestive tract and the accessory organs (fig. 17.1). The **digestive tract** (alimentary[2] canal) is a tube that extends from mouth to anus, and measures about 5 m (16 feet) long in a living person (but longer in the cadaver). It includes the mouth, pharynx, esophagus, stomach, small intestine, and large intestine. Part of the digestive tract, the stomach and intestines, constitutes the *gastrointestinal (GI) tract.* The **accessory organs** are the teeth, tongue, salivary glands, liver, gallbladder, and pancreas. We will first explore the structure of the system as a whole and then we will further explore the structure and functions of each region.

The digestive tract is essentially a tube open to the environment at both ends. With slight variations from one region to another, most of it follows the same basic structural plan—a wall composed of four main tissue layers, from inner to outer surface (fig. 17.2):

1. The **mucosa (mucous membrane)** lines the lumen. It consists of an inner epithelium, a loose connective tissue layer called the *lamina propria,* and a thin layer of smooth muscle. Simple columnar epithelium lines the digestive tract from the stomach through the large intestine; stratified squamous epithelium is found from the oral cavity through the esophagus and in the lower anal canal, where

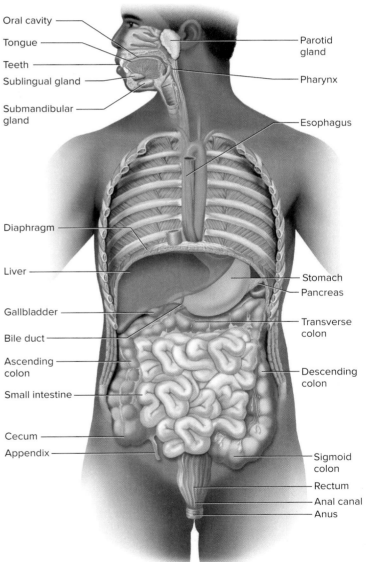

Figure 17.1 The Digestive System. **APR**

---

[2]*aliment* = food

Figure 17.2 Tissue Layers of the Digestive Tract. Cross section of the esophagus just below the diaphragm. The layers shown here are typical of most of the digestive tract.

the tract is subject to more abrasion. The mucosa often exhibits an abundance of lymphocytes and lymphatic nodules—the *mucosa-associated lymphatic tissue (MALT)* (see section 14.1c).

2. The **submucosa** is a layer of loose connective tissue containing blood vessels, lymphatic vessels, and nerves.
3. The **muscularis externa** consists of smooth muscle, usually arranged in two layers: a deep layer whose muscle fibers encircle the tract and a more superficial layer whose fibers run longitudinally. The muscularis externa is responsible for *motility,* muscle contractions that mix and propel contents through the digestive tract.
4. A thin **serosa (serous membrane)** covers the outer surface of most of the digestive tract. It consists of areolar tissue topped by a simple squamous epithelium. The serosa begins in the lower 3 to 4 cm of the esophagus and ends just before the rectum. The pharynx, most of the esophagus, and the rectum are covered instead by a fibrous connective tissue layer called the **adventitia** that blends into the connective tissues of adjacent organs.

A network of neurons called the **enteric[3] nervous system (ENS)** branches throughout the submucosa and muscularis externa of the esophagus, stomach, and intestines. The ENS coordinates muscle activity responsible for mechanical mixing of food in the stomach and propulsion of material through the intestines, and regulates numerous secretions in different gut sections that maintain the optimal pH and biochemical environment for enzyme function. This system is thought to have more neurons than the spinal cord! Neurons in the gut produce some of the

---

[3]*enter* = intestine

same neurotransmitters as are found in the brain, such as dopamine and serotonin. Given the complexity of digestive system function, it is not surprising that there is extensive neural coordination; more surprising is comparatively how little we know about it. While the ENS can function independently of the CNS, there is two-way communication between the two, linking emotional and cognitive brain centers with peripheral intestinal functions. This so called gut–brain axis is influenced by the gut microbiome, the collection of bacteria in the intestines. The role of the system in both health and disease has become more evident, and its dysfunction has been implicated in conditions ranging from obesity, to irritable bowel syndrome, to Parkinson disease.

The stomach and intestines undergo such strenuous contractions as they process food that they need freedom to move in the abdominal cavity. Thus, they are not tightly bound to the abdominal wall, but rather are loosely suspended from it by connective tissue sheets called **mesenteries** (fig. 17.3; see also figs. 1.13 and 1.14). The *parietal peritoneum,* a serous membrane that lines the abdominal cavity, turns inward along the posterior midline and forms a two-layered mesentery that extends to the digestive tract. At the stomach and small intestine, the two layers separate and pass around opposite sides of the organ, forming the serosa. The mesenteries hold the abdominal viscera in their proper relationship to each other and prevent the small intestine from becoming twisted and tangled by changes in body position and by the intestine's own contractions. They also provide passage for blood vessels and nerves that supply the digestive tract, and contain many lymph nodes and lymphatic vessels and often a substantial amount of fat. Much of the adipose tissue in obese individuals is in the mesenteries.

A mesentery called the *lesser omentum* extends from the right superior margin of the stomach to the liver. A larger mesentery, the *greater omentum,* hangs from the inferior margin of the stomach and loosely covers the small intestine like an apron. The omenta have a loosely organized, lacy appearance due to irregularities in the thickness and amount of fat.

Some digestive organs such as the stomach and liver are *intraperitoneal*—enclosed by serosa on all sides. Others such as the pancreas and duodenum are *retroperitoneal*—lying against the posterior body wall and covered by peritoneum only on the anterior surface (see fig. 1.13).

Figure 17.3 Mesenteries of the Digestive Tract. (a) Shown with apronlike greater omentum overhanging the small intestine in its natural position. (b) Greater omentum and small intestine retracted to expose other mesenteries of the small and large intestines. APR

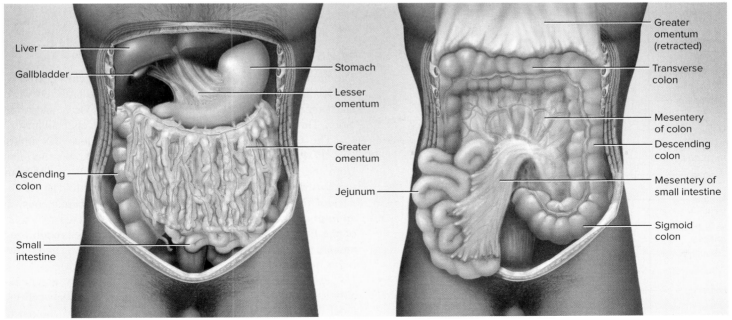

Before You Go On

Answer these questions from memory to test your understanding of the previous section. Reread the material if you can't answer the questions.

1. What is the difference between mechanical and chemical digestion?
2. List the tissue layers of the digestive tract from inside to outside. Describe the characteristics and functions of each.
3. Name the independent network of neurons in the digestive tract and identify its functions.
4. Distinguish between the meanings of peritoneum and mesenteries, and explain the function of the mesenteries.
5. Define intraperitoneal and retroperitoneal. Name an organ in each category.

17.2 The Mouth Through Esophagus

Expected Learning Outcomes

When you have completed this section, you should be able to

a. describe the gross anatomy of the digestive tract from the mouth through esophagus;
b. describe the general structure of a tooth, and identify the types of teeth and their respective positions;
c. list the functions of saliva and describe its composition; and
d. describe the process of swallowing.

Figure 17.4 The Oral Cavity. **APR**

17.2a The Mouth

The mouth is also known as the **oral,** or **buccal, cavity** (BUCK-ul). Its functions include ingestion (food intake), taste, chewing, mechanical and chemical digestion, swallowing, speech, and respiration. The mouth is lined with stratified squamous epithelium.

The cheeks, lips, palate, and tongue enclose the mouth (**fig. 17.4**). The **vestibule** is the space that separates the teeth from the cheeks and lips—the space where you insert your toothbrush to brush the outer surfaces of the teeth. The fleshiness of the lips and cheeks is due mainly to subcutaneous fat and the orbicularis oris and buccinator muscles (see "Muscles of the Face" in section 7.4b). The lips and cheeks retain food and are essential for articulate speech and for sucking and blowing actions, including suckling by infants.

The **tongue,** although muscular and bulky, is a remarkably agile and sensitive organ. It manipulates food between the teeth while it avoids being bitten, and it is sensitive enough to feel a stray hair in a bite of food. Its surface is covered with nonkeratinized stratified squamous epithelium and exhibits bumps and projections called *lingual papillae,* the site of the taste buds (see fig. 10.4). Its anterior two-thirds, called the *body,* occupies the oral cavity; the posterior one-third, the *root,* occupies the first part of the pharynx. The body is attached to the floor of the mouth by a median fold called the *lingual frenulum.*[4]

The **palate** separates the oral cavity from the nasal cavity, making it possible to breathe while chewing food. Its anterior portion, the *hard (bony) palate,* consists of horizontal plates of the maxillae and palatine bones covered with a thin mucous membrane. The posterior portion, the *soft palate,* has no bone but is more spongy, containing glandular tissue and skeletal muscle. A conical median projection, the *uvula,*[5] is clearly visible at the rear of the mouth.

The teeth are collectively called the **dentition.** They serve to *masticate* the food, breaking it into smaller pieces, making it easier to swallow and speeding up chemical digestion by exposing more surface area to enzymatic action. Adults normally have 16 teeth in the mandible and 16 in the maxilla. From the midline to the rear of each jaw, there are two incisors, a canine, two premolars, and up to three molars (**fig. 17.5**). The chisel-like **incisors** bite off a piece of food, and the slightly pointed **canines** puncture and shred it. The

---

[4]*frenulum* = little bridle
[5]*uvula* = little grape

Figure 17.5 The Dentition and Structure of a Tooth and Its Alveolus. (a) The deciduous (baby) teeth and ages at which they emerge from the gums (erupt). (b) The adult teeth and ages of eruption. Each figure shows only the upper teeth. The ages at eruption are composite ages for the corresponding upper and lower teeth. Generally, the lower (mandibular) teeth erupt somewhat earlier than their upper (maxillary) counterparts. (c) Structure of a tooth and periodontal tissues. This particular example is a molar. **APR**

? *Which teeth are absent from a 3-year-old child?*

(a) Deciduous (baby) teeth

(b) Permanent teeth

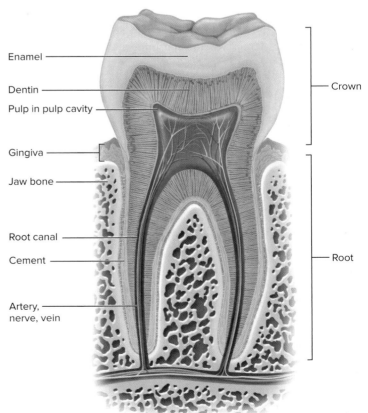

(c)

premolars and **molars** are "grinders" with broad, lumpy surfaces adapted for crushing and grinding food.

Regions of a tooth are defined by their relationship to the gum, or **gingiva** (JIN-jih-vuh): The **crown** is the portion above the gum, and the **root** is the portion inserted into the tooth socket below the gum. Internally, the **pulp cavity** in the crown and the narrow **root canal** in the root are occupied by **pulp**—a mass of loose connective tissue, blood and lymphatic vessels, and nerves.

Most of a tooth consists of hard yellowish tissue called **dentin,** covered with **enamel** in the crown and neck and **cement** in the root. Enamel is a nonliving, hardened secretion, whereas dentin and cement are living connective tissues with cells or cell processes embedded in a calcified matrix. Damaged dentin and cement can regenerate, but damaged enamel cannot—it must be artificially repaired.

17.2b Saliva and Salivary Glands

Saliva is a watery solution that moistens the mouth, cleanses the teeth, inhibits bacterial growth, and dissolves molecules so they can stimulate the taste buds. It contains mucus that lubricates food and binds particles together into a slippery, easily swallowed mass. *Salivary amylase* begins to digest starch in our food as we chew, and *lingual lipase* begins fat digestion once we swallow and stomach acid activates the enzyme.

We produce about 1.0 to 1.5 liters of saliva per day. Some of this, secreted throughout the day whether we are eating or not, comes from numerous small salivary glands scattered throughout the mucosa of the tongue, lips, cheeks, and palate. The presence of food in the mouth stimulates secretion of more abundant saliva from three pairs of major salivary glands external to the oral cavity (fig. 17.6):

1. The **parotid**[6] **glands** are located just beneath the skin anterior to the earlobes. Their ducts open into the mouth opposite each second upper molar tooth. *Mumps* is an inflammation and swelling of the parotid gland caused by a virus.
2. The **submandibular glands** are located halfway along the medial side of the body of the mandible. Their ducts empty into the mouth at a papilla beneath the tongue, near the lower central incisors.
3. The **sublingual glands** are located in the floor of the mouth. They have multiple ducts that empty into the mouth beneath the tongue.

17.2c The Pharynx

The pharynx is a muscular funnel that connects the oral cavity to the esophagus and the nasal cavity to the larynx; thus, it is a point where the digestive and respiratory tracts intersect. It has a deep layer of longitudinally oriented skeletal muscle and a superficial layer of circular skeletal muscle. The circular muscle is divided into superior, middle, and inferior **pharyngeal constrictors,** which force food downward during swallowing. Section 15.1b and figure 15.2 describe the three regions of the pharynx (*nasopharyx, oropharynx,* and *laryngopharynx*).

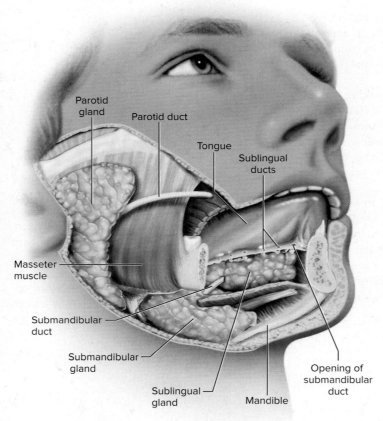

Figure 17.6 The Salivary Glands. Part of the mandible has been removed to expose the sublingual gland medial to it. **APR**

---

[6]*par* = next to; *ot* = ear

17.2d The Esophagus

The **esophagus** is a straight muscular tube 25 to 30 cm long (see fig. 17.1). It begins posterior to the trachea near the inferior border of the larynx. It passes through the mediastinum, penetrates the diaphragm at an opening called the *esophageal hiatus*, then continues another 3 to 4 cm to meet the stomach at the **cardiac orifice** (named for its proximity to the heart). A constriction called the **lower esophageal sphincter** prevents stomach contents from regurgitating into the esophagus, thus protecting the esophageal mucosa from the erosive effect of stomach acid. "Heartburn" has nothing to do with the heart, but is the burning sensation produced by acid reflux into the esophagus.

The wall of the esophagus is organized into the tissue layers described earlier, with some regional specializations (see fig. 17.2). The mucosa has a nonkeratinized stratified squamous epithelium. The submucosa contains glands that secrete lubricating mucus. The muscularis externa is voluntary skeletal muscle in the upper esophagus, but transitions to involuntary smooth muscle in the lower esophagus. When the esophagus is empty, the contraction of its muscular wall folds the mucosa and submucosa into longitudinal ridges, giving the lumen a starlike shape in cross section. Most of the esophagus is covered with a connective tissue adventitia; the short segment below the diaphragm is covered by a serosa.

① **Oral phase.** The tongue forms a food bolus and pushes it into the laryngopharynx.

17.2e Swallowing

Swallowing, or **deglutition** (DEE-glu-TISH-un), is a complex action involving over 22 muscles in the mouth, pharynx, and esophagus. It is coordinated by the **swallowing center** in the medulla oblongata, and occurs in three phases (**fig. 17.7**).

① The **oral phase** is under voluntary control. The tongue collects food, presses it against the palate to form a soft cohesive mass called a **bolus,** and pushes it posteriorly. Food accumulates in the oropharynx in front of the blade of the epiglottis. When the bolus reaches a critical size, the epiglottis tips posteriorly and the bolus slides around it through a space on each side, into the laryngopharynx.

② The **pharyngeal phase** is involuntary and is initiated when food contacts touch receptors of the laryngopharynx. The soft palate and root of the tongue block food from entering the nasal cavity or reentering the mouth. The vocal cords adduct (come together) in the middle, and the larynx is pulled up against the epiglottis, preventing food from entering the larynx and trachea. These actions also widen the upper end of the esophagus to receive the bolus. The pharyngeal constrictors contract in order from superior to inferior to drive the bolus downward into the esophagus. Breathing is automatically suspended for a moment during this phase.

③ The **esophageal phase** is a wave of involuntary contraction called **peristalsis,** controlled jointly by the brainstem swallowing center and the enteric nervous system in the esophageal wall. The esophagus constricts above the bolus and relaxes below it, propelling the bolus along. When you are standing or sitting upright, most food and liquid drop through the esophagus by gravity faster than the peristaltic wave can catch up to it. Peristalsis, however, propels larger food pieces and ensures that you can swallow regardless of the body's position—while lying down or even standing on your head! A second wave of peristalsis follows if the first wave is insufficient to clear food from the esophagus. Liquid normally reaches the stomach in 1 to 2 seconds and a food bolus in 4 to 8 seconds. As a bolus reaches the lower end of the esophagus, the lower esophageal sphincter relaxes to let it pass into the stomach.

② **Pharyngeal phase.** The palate, tongue, vocal cords, and epiglottis block the oral and nasal cavities and airway while pharyngeal constrictors push the bolus into the esophagus.

③ **Esophageal phase.** Peristalsis drives the bolus downward, and relaxation of the lower esophageal sphincter admits it into the stomach.

Figure 17.7 Swallowing. See numbered steps in the text for further explanation. **APR**

? *What actions prevent the pharynx from forcing food back into the mouth or nose?*

Before You Go On

Answer these questions from memory to test your understanding of the previous section. Reread the material if you can't answer the questions.

6. Identify the diverse functions of the mouth and which parts perform each function.
7. Describe the spatial relationship of dentin, enamel, and cement in a tooth. Name the four kinds of teeth and state how many of each type a person has.
8. Name the major salivary glands, state their locations, and describe the composition of the saliva.
9. Where is the cardiac orifice? Name the sphincter found here and discuss its function.
10. Explain the three stages of swallowing. Which is voluntary and which are involuntary?

17.3 The Stomach

Expected Learning Outcomes

When you have completed this section, you should be able to

a. describe the gross and microscopic anatomy of the stomach;
b. identify the cell types of the gastric mucosa and their functions;
c. identify the secretions of the stomach and their functions;
d. describe how the nervous system and hormones regulate the stomach; and
e. explain why the stomach doesn't digest itself.

The stomach is a muscular sac in the upper left abdominal cavity immediately inferior to the diaphragm. It functions primarily as a food-storage organ, and holds 1.0 to 1.5 L after a typical meal. When extremely full, it may hold up to 4 L and extend nearly as far as the pelvis.

Well into the nineteenth century, authorities regarded the stomach as essentially a grinding chamber, fermentation vat, or cooking pot. We now know that it mechanically breaks down and liquefies the food and begins the chemical digestion of proteins and fat. This produces a pasty mixture of semidigested food called **chyme**[7] (pronounced "kime"). Most chemical digestion occurs after the chyme passes on to the small intestine.

17.3a Anatomy

The stomach is J-shaped and divided into four regions (fig. 17.8). (1) The *cardiac part* is a small area within about 3 cm of the cardiac orifice. (2) The *fundus* is the superior dome that nestles against the diaphragm. (3) The *body* makes up

---

[7]*chyme* = juice

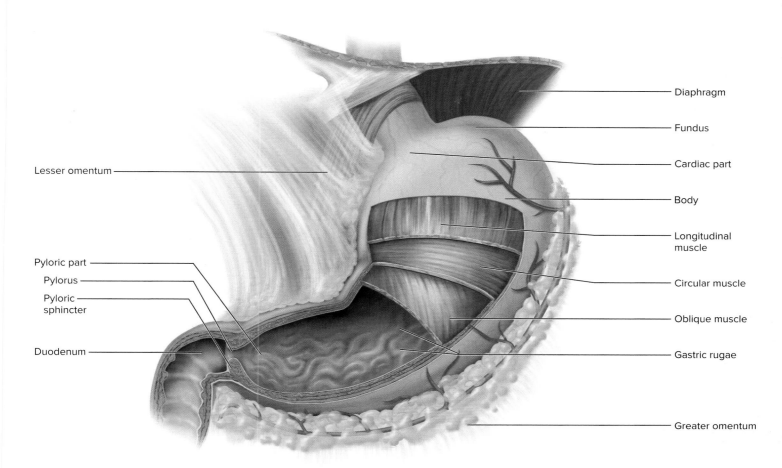

Figure 17.8 Gross Anatomy of the Stomach
APR

❓ *How does the muscularis externa of the stomach differ from that of the esophagus? Why?*

the greatest part of the stomach distal to the cardiac orifice. (4) The *pyloric part* is a slightly narrower pouch at the inferior end that terminates at the **pylorus,**[8] a narrow passage into the duodenum. The pylorus is surrounded by a thick ring of smooth muscle, the **pyloric sphincter,** which regulates the movement of chyme into the duodenum.

The stomach wall has the typical layers of the digestive tract with some variations. The mucosa is covered with simple columnar epithelium (fig. 17.9). The mucosa and submucosa are flat and smooth when the stomach is full, but as it empties, these layers form conspicuous longitudinal wrinkles called *gastric rugae*[9] (ROO-jee). The lamina propria is almost entirely occupied by glands, to be described shortly. The muscularis externa has three layers rather than two: an outer longitudinal, middle circular, and inner oblique layer, correlated with the strong churning actions of the stomach.

Apply What You Know

Contrast the epithelium of the esophagus with that of the stomach. Why is each epithelial type best suited to the function of its respective organ?

---

[8]*pylorus* = gatekeeper
[9]*rugae* = folds, creases

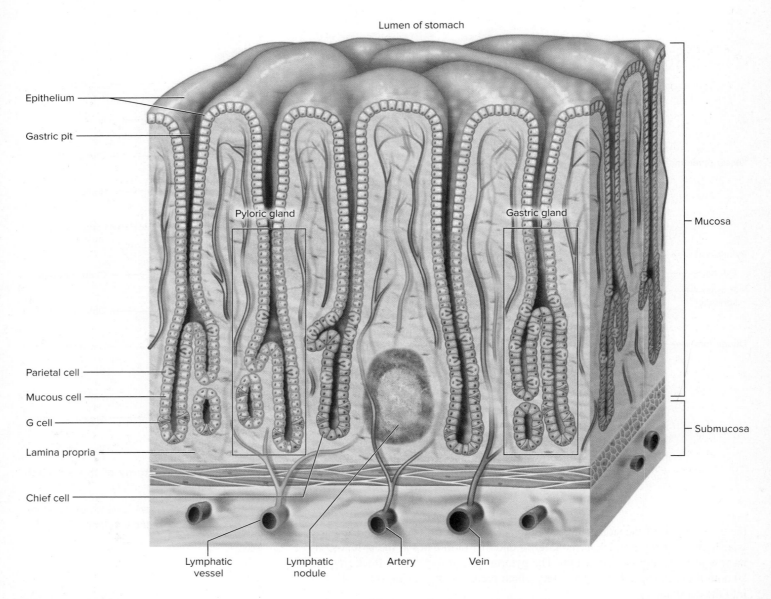

Figure 17.9 Microscopic Anatomy of the Gastric Mucosa. Both gastric and pyloric glands are shown, although pyloric glands are found in the lower end of the stomach and gastric glands in the upper to middle stomach (fundus and body). **APR**

The gastric mucosa is pocked with depressions called *gastric pits* lined with the same columnar epithelium as the surface. Two or three simple tubular glands open into the bottom of each gastric pit. Near the esophagus and pylorus, the stomach has *cardiac* and *pyloric glands,* respectively; their main secretion is mucus. But in most of the stomach, we find **gastric glands.** These three glands have the following cell types:

- **Mucous cells** secrete mucus.
- **Regenerative cells,** found in the base of the pit and neck of the gland, divide rapidly and produce a continual supply of new cells to replace cells that die.
- **Parietal cells,** found mostly in the upper half of the gland, secrete *hydrochloric acid, intrinsic factor,* and an appetite-regulating hormone, *ghrelin* (see section 18.5a).
- **Chief cells,** so named because they are the most numerous, secrete the enzymes *pepsinogen* and *gastric lipase.*
- **Enteroendocrine cells,** concentrated especially in the lower end of a gland, secrete hormones and local chemical messengers that regulate digestion. These include, for example, the *G cells* in figure 17.9, which secrete the regulatory hormone *gastrin.*

17.3b Gastric Secretions

The gastric glands produce 2 to 3 L of *gastric juice* per day, composed mainly of water and the following solutes:

- **Hydrochloric acid (HCl)** is produced by the parietal cells. The gastric juice can have a pH as low as 0.8, more acidic than the battery acid of a car and low enough to cause a serious chemical burn on the skin. Stomach acid has several functions: (1) It activates the enzymes *pepsin* and *lingual lipase;* (2) it converts dietary iron to a form that the body can absorb and use; (3) it breaks up connective tissues and plant cells, helping to liquefy food and form chyme; and (4) it contributes to innate immunity by destroying most ingested pathogens.
- **Pepsin** is an enzyme that digests proteins to shorter peptide chains, which then pass to the small intestine where their digestion is completed. Chief cells secrete an inactive precursor called *pepsinogen,* and hydrochloric acid converts it to pepsin.
- **Gastric lipase,** also secreted by chief cells, digests a small percentage of the dietary fat. Most fat, however, is digested in the small intestine.
- **Intrinsic factor** is a glycoprotein secreted by the parietal cells. It is essential for the absorption of vitamin B_{12} by the small intestine. Without vitamin B_{12}, hemoglobin cannot be synthesized and anemia develops. The secretion of intrinsic factor is the only indispensable function of the stomach. Digestion can continue following removal of the stomach *(gastrectomy),* but a person must then take vitamin B_{12} by injection, or vitamin B_{12} and intrinsic factor orally.
- **Hormones and local chemical messengers** are produced by enteroendocrine cells. These include *gastrin, serotonin, histamine, somatostatin,* and several others. Some are peptides identical to those produced in the central nervous system, and have thus been called **gut–brain peptides.** Their roles in digestion and appetite regulation are discussed in later sections of this chapter and in section 18.5a. Table 17.1 summarizes the principal gastric secretions.

Table 17.1 Major Secretions of the Gastric Glands

Secretory Cells	Secretion	Function
Mucous cells	Mucus	Protects mucosa from HCl and enzymes
Parietal cells	Hydrochloric acid	Activates pepsin; helps liquefy food; converts dietary iron to usable form; destroys ingested pathogens
	Intrinsic factor	Enables small intestine to absorb vitamin B_{12}
	Ghrelin	Stimulates the sense of hunger
Chief cells	Pepsinogen	Converted to pepsin, which digests protein
	Gastric lipase	Digests fat
Enteroendocrine cells	Gastrin	Stimulates gastric glands to secrete HCl and enzymes; stimulates intestinal motility; relaxes ileocecal valve
	Serotonin	Stimulates gastric motility
	Histamine	Stimulates HCl secretion
	Somatostatin	Inhibits gastric secretion and motility; delays emptying of stomach; inhibits secretion by pancreas; inhibits gallbladder contraction and bile secretion; reduces blood circulation and nutrient absorption in small intestine

17.3c Gastric Motility

As you begin to swallow, the swallowing center of the medulla oblongata signals the stomach to relax, thus preparing it to receive food. The arriving food also causes smooth muscle to relax and the stomach stretches to accommodate more food.

Peristalsis

Soon, the stomach begins rhythmic peristaltic contractions that churn the food, mix it with gastric juice, and promote its physical breakup and chemical digestion. When these waves reach the pyloric sphincter, they squeeze it shut. The stomach thus squirts only about 3 mL of chyme into the duodenum at a time. Receiving such small amounts of chyme enables the duodenum to neutralize the stomach acid and digest nutrients little by little. A typical meal is emptied from the stomach in about 4 hours, but it takes less time if the meal is more liquid and as long as 6 hours if the meal is high in fat.

Vomiting

Vomiting is the forceful ejection of stomach and intestinal contents from the mouth. It involves multiple muscular actions integrated by the *emetic center* of the medulla oblongata. Vomiting is commonly induced by overstretching of the stomach, chemical irritants such as alcohol and bacterial toxins, visceral trauma, intense pain, or psychological and sensory stimuli that activate the emetic center (repugnant sights, smells, and thoughts). Vomiting is usually preceded by nausea, and is often accompanied by tachycardia, profuse salivation, and sweating.

Chronic vomiting can cause dangerous fluid, electrolyte, and acid–base imbalances. In cases of frequent vomiting, as in the eating disorder *bulimia,* the tooth enamel becomes severely eroded by the hydrochloric acid in the chyme. Aspiration (inhalation) of this acid is very destructive to the respiratory tract. Many have died from aspiration of vomit when they were unconscious or semiconscious. This is the reason that surgical anesthesia, which may induce nausea, must be preceded by fasting until the stomach and duodenum are empty.

17.3d Regulation of Gastric Function

Gastric secretion and motility are divided into three stages called the *cephalic, gastric,* and *intestinal phases,* based on whether the stomach is being controlled by the brain, by itself, or by the small intestine, respectively. These phases overlap and all three can occur simultaneously (fig. 17.10).

(1) The **cephalic phase** is the stage in which the stomach responds to the mere sight, smell, taste, or thought of food. These sensory and mental inputs converge on the hypothalamus, which relays signals to the medulla oblongata. Vagus nerve fibers from the medulla stimulate the enteric nervous system, which, in turn, stimulates gastric activity.

(2) The **gastric phase** is a period in which swallowed food activates gastric activity. About two-thirds of gastric secretion occurs during this phase. Ingested food stimulates gastric activity both by stretching the stomach and by raising the pH of its contents. Three chemical signals activate HCl secretion by the parietal cells: **acetylcholine** from the vagus nerves

Figure 17.10 Neural and Hormonal Control of the Stomach. See numbered steps in the text for further explanation.

and **histamine** and **gastrin** from the enteroendocrine cells of the gastric glands. Gastrin and acetylcholine also signal the chief cells to secrete pepsinogen.

③ The **intestinal phase** occurs as chyme enters the duodenum and triggers local nervous reflexes and the secretion of regulatory hormones. Acid and semidigested fats in the duodenum trigger inhibitory signals, sent to the stomach by way of the enteric nervous system. In addition, signals from the duodenum to the medulla oblongata inhibit vagal stimulation of the stomach and activate sympathetic neurons, which inhibit gastric activity. Furthermore, duodenal enteroendocrine cells release the hormones **secretin** and **cholecystokinin (CCK),**[10] which suppress gastric activity. Collectively, all these neural and hormonal signals from the duodenum are like a message to the stomach to slow down and give the duodenum time to process the chyme it already contains before sending it more.

17.3e Protection of the Stomach

One might think that the stomach would be its own worst enemy; it is, after all, made of meat. Some people enjoy haggis and tripe, dishes made from animal stomachs, and have no difficulty digesting those. Why, then, doesn't the human stomach digest itself? The stomach is protected in three ways from the harsh acidic and enzymatic environment it creates:

1. The mucosa is coated with thick, highly alkaline mucus that neutralizes HCl and resists the enzymes.

---

[10]*chole* = bile; *cysto* = bladder (gallbladder); *kin* = action

(a) Normal

(b) Peptic ulcer

(c) *Helicobacter pylori*

Figure 17.11 Peptic Ulcer. (a) A healthy mucosa in the lower esophagus where it opens into the stomach. The small white spots are reflections of light from the endoscope. (b) A bleeding peptic ulcer. A peptic ulcer typically has an oval shape and yellow-white color. Here the yellowish floor of the ulcer is partially obscured by black blood clots, and fresh blood is visible around the margin of the ulcer. (c) *Helicobacter pylori,* the bacterium responsible for most peptic ulcers.

a: © CNRI/Science Source; b: © CNRI/SPL/Science Source; c: © Eye of Science/Science Source

Clinical Application 17.1

PEPTIC ULCER

Inflammation of the stomach, called *gastritis,* can lead to a *peptic ulcer* as pepsin and hydrochloric acid erode the stomach wall (fig. 17.11). Peptic ulcers occur even more commonly in the duodenum and occasionally in the esophagus. If untreated, they can perforate the organ and cause fatal hemorrhaging or peritonitis. Most such fatalities occur in people over age 65.

Most ulcers are caused by a bacterium, *Helicobacter pylori,* which invades the mucosa of the stomach and duodenum and opens the way to chemical damage to the tissue. Other risk factors include smoking and the use of aspirin and other nonsteroidal anti-inflammatory drugs (NSAIDs). NSAIDs suppress the synthesis of prostaglandins, which normally stimulate the secretion of protective mucus and acid-neutralizing bicarbonate. Aspirin itself is an acid that directly irritates the gastric mucosa. Ulcers are often successfully treated with antibiotics against *Helicobacter,* combined with proton pump inhibitors that suppress the secretion of acid.

For a long time, ulcers were thought to be linked to psychological stress, but in the 1980s, when *Helicobacter pylori* was discovered to be associated with peptic ulcers, emphasis on psychological factors fell out of favor. Recent studies, however, indicate that psychological stress does play a role in peptic ulcers. This, along with evidence that the brain–gut connection influences gastrointestinal diseases, has led many researchers and clinicians to reconsider the role of psychological stress in the development of peptic ulcers.

2. The epithelial cells are joined by tight junctions that prevent gastric juice from seeping between them and digesting the underlying tissue.
3. In spite of these other protections, the stomach's epithelial cells live only 3 to 6 days and are then sloughed off into the chyme and digested with the food. They are replaced just as rapidly, however, by cell division in the gastric pits.

The breakdown of these protective mechanisms can result in inflammation and peptic ulcer (Clinical Application 17.1).

Before You Go On

Answer these questions from memory to test your understanding of the previous section. Reread the material if you can't answer the questions.

11. Where is the pyloric part of the stomach? Discuss the functional importance of the pyloric sphincter.
12. List five kinds of cells in gastric glands and state their functions.
13. Name the chemical messengers that regulate gastric function and state the effect of each.
14. Compare and contrast the three phases of gastric regulation. State whether gastric secretion and motility increase or decrease in each phase. Summarize the roles of sympathetic, parasympathetic, and enteric nervous stimulation over the course of these three phases.
15. Describe three ways the stomach is protected from self-digestion.

PERSPECTIVES ON HEALTH

Acute Gastrointestinal Illness

In recent years, headlines about contaminated apple juice, hamburger, and spinach, as well as outbreaks of illness on cruise ships, have focused public attention on contagious gastrointestinal illnesses. Acute gastrointestinal illness is caused by a variety of bacteria and viruses, and is marked by diarrhea, nausea, vomiting, and abdominal pain. *Diarrhea* occurs when the large intestine is irritated by pathogens and feces pass through too quickly for adequate water reabsorption. In most cases, the person recovers after a few unpleasant days; in severe cases, however, fluid loss can lead to dehydration, electrolyte imbalance, shock, and even death. When one is suffering from gastrointestinal illness, it is important to remain hydrated, preferably with beverages that contain electrolytes.

Pathogens that cause gastrointestinal disease are transmitted via contaminated water or food. Before widespread efforts in the nineteenth century to improve sanitation and provide clean drinking water, gastrointestinal infections such as cholera were devastatingly common. Gastrointestinal infections continue to cause illness and death throughout the world, especially in developing countries. Globally, diarrheal disease is the second leading cause of death in children less than 5 years of age.

Three types of bacteria are responsible for many cases of gastrointestinal disease in the United States—*Salmonella, Campylobacter,* and *Shigella. Salmonella* is usually transmitted by eating foods contaminated with animal feces, such as poultry or eggs. Thorough cooking kills the bacteria. *Campylobacter* is often contracted by eating or coming into contact with raw or undercooked poultry or other meat, as when one cuts a chicken on a cutting board and then prepares other food such as a salad on the same board without washing it thoroughly. Some people acquire *Campylobacter* through contact with feces of an infected pet dog or cat. *Shigella* is transmitted from the stool or soiled fingers of one person to the mouth of another. This may occur, for example, when food handlers neglect to wash their hands with soap after they use the bathroom. *Shigella* can also be spread by flies that land in infected feces and then transmit the pathogen when they land on food. *Shigella* outbreaks sometimes occur in day-care centers when caretakers don't carefully clean their hands or surfaces where diapers are changed.

Globally, rotavirus is the most common cause of diarrheal disease in infants and young children. Severe cases require hospitalization, where fluids are replaced intravenously. The virus spreads rapidly through fecal contamination, hand-to-mouth contact, or physical contact with infected persons or toys. Rotavirus is life-threatening in infants, who can suffer massive loss of fluid. There is no specific drug treatment for rotavirus infection; viruses do not respond to antibiotics. Before the introduction of a rotavirus vaccine in 2006, almost all children in the United States were likely to be infected with rotavirus before their fifth birthday. Rotavirus is still the most common cause of diarrheal disease among infants and children, especially in low-income countries, but access to the vaccine has significantly reduced the number of childhood deaths.

17.4 The Liver, Gallbladder, and Pancreas

Expected Learning Outcomes

When you have completed this section, you should be able to

a. describe the gross and microscopic anatomy of the liver, gallbladder, and pancreas;

b. discuss the products and functions of each; and

c. explain how hormones regulate secretions of the liver and pancreas.

The small intestine receives not only chyme from the stomach but also secretions from the liver and pancreas, which enter the digestive tract near the junction of the stomach and small intestine. These secretions are so important to the digestive processes of the small intestine that it is necessary to understand them before we continue with intestinal physiology.

17.4a The Liver APR

The **liver** (fig. 17.12) is a reddish brown gland located immediately inferior to the diaphragm, filling most of the right hypochondriac and epigastric regions. It is the body's largest gland, weighing about 1.4 kg (3 pounds). It has a tremendous variety of functions, but only one of them, the secretion of bile, contributes to digestion. Others are discussed in "Liver Functions in Metabolism" in section 18.3c, which provides a more thorough physiological basis for understanding the nondigestive roles of the liver.

Gross Anatomy

The liver has four lobes—the right, left, quadrate, and caudate lobes. From an anterior view, we see only a large *right lobe* and smaller *left lobe*. They are separated by the *falciform*[11] *ligament,* a sheet of mesentery that suspends the liver from the diaphragm and anterior abdominal wall. The *round ligament,* also visible anteriorly, is a fibrous remnant of the umbilical vein, which carries blood from the umbilical cord to the liver of a fetus.

From the inferior view, we see a squarish *quadrate lobe* next to the gallbladder and a tail-like *caudate*[12] *lobe* posterior to that. The gallbladder is a sac adhering to a depression between the right and quadrate lobes. Between the quadrate and caudate lobes is an opening through which the hepatic portal vein and hepatic artery enter and the bile passages leave. The posterior aspect of the liver has a deep groove that accommodates the inferior vena cava. The liver is covered by a serosa except for its superior surface where it adheres to the diaphragm.

Microscopic Anatomy

The interior of the liver is filled with innumerable tiny cylinders called **hepatic lobules,** about 2 mm long by 1 mm in diameter. A lobule consists of a **central vein** passing down its core, surrounded by radiating sheets of cuboidal cells called **hepatocytes** (fig. 17.12d). Imagine spreading a book wide open until its front and back covers touch. The pages of the book would fan out around the spine somewhat like the plates of hepatocytes fan out from the central vein of a liver lobule.

Each plate of hepatocytes is an epithelium one or two cells thick. The spaces between the plates are blood-filled channels called **hepatic sinusoids.** These are bounded by a leaky fenestrated endothelium that freely allows blood plasma into the space around the hepatocytes, where it contacts their brush border of microvilli. Blood filtering through the sinusoids comes directly from the stomach and intestines via the hepatic portal vein. After a meal, the hepatocytes absorb glucose, amino acids, iron, vitamins, and other nutrients from it for metabolism or storage. They also remove and degrade hormones, toxins, bile pigments, and drugs. At the same time, they secrete albumin, lipoproteins, clotting factors, angiotensinogen, and other products into the blood. Between meals, they break down stored glycogen and release glucose into the circulation. The sinusoids also contain phagocytic cells called **hepatic macrophages,** which remove bacteria and debris from the blood.

---

[11]*falci* = sickle; *form* = shape
[12]*caud* = tail

CHAPTER 17 The Digestive System 573

Figure 17.12 The Liver. (a) Location; note that most of the liver lies within the rib cage. (b) Anterior view. (c) Inferior view. (d) Histology. Note the relationship of the hepatic lobules to the blood vessels and bile passages. **APR**

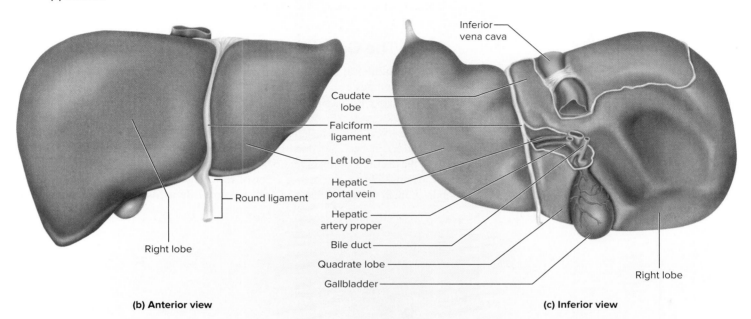

After filtering through the sinusoids, blood flows into the central vein and from there to vessels that lead to a few hepatic veins. These leave the liver at its superior surface and drain immediately into the inferior vena cava.

The liver secretes about 500 to 1,000 mL of bile per day into narrow channels between sheets of hepatocytes. Bile exits the liver by way of the *common hepatic duct.* A short distance farther on, this is joined by the *cystic duct* coming from the gallbladder (fig. 17.13). Their union forms the **bile duct,** which descends through the lesser omentum toward the duodenum. Near the duodenum, the bile duct joins the duct of the pancreas and forms an expanded chamber called the **hepatopancreatic ampulla.** The ampulla terminates at a fold of tissue on the duodenal wall, the *major duodenal papilla.* Bile and pancreatic juices flow into the duodenum through an opening in the papilla.

17.4b The Gallbladder and Bile

The **gallbladder** is a pear-shaped sac on the underside of the liver that stores and concentrates bile. It is about 10 cm long and internally lined by a highly folded mucosa with a simple columnar epithelium. **Bile** is a fluid containing minerals, cholesterol, phospholipids, bile pigments, and bile acids. When released from the liver, bile is a light yellow-green color, but when concentrated by the gallbladder, it becomes deep green. Its principal pigment is **bilirubin,** derived from the decomposition of hemoglobin. Further metabolic derivatives of bilirubin are responsible for the brown color of feces and yellow color of urine; without bile, human feces have a chalky gray color. **Bile acids (bile salts)** are steroids

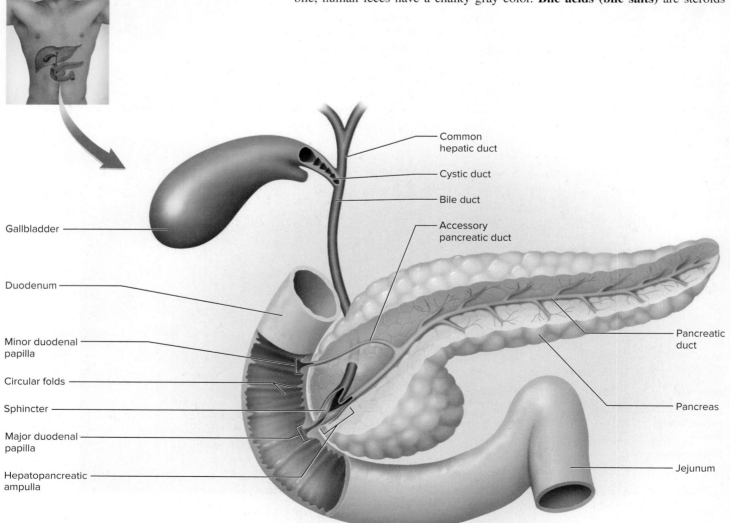

Figure 17.13 Gross Anatomy of the Gallbladder, Pancreas, and Bile Passages. The liver is omitted to show more clearly the gallbladder, which adheres to its inferior surface, and the ducts that emerge from the liver tissue. APR

synthesized from cholesterol. They aid in fat digestion and absorption. All other components of the bile are wastes destined for excretion in the feces. When these waste products become excessively concentrated, they may form gallstones.

17.4c The Pancreas

The **pancreas** (fig. 17.13) is a spongy, flattened, retroperitoneal gland pressed between the posterior body wall and lower margin of the stomach. It measures 12 to 15 cm long and about 2.5 cm thick. The pancreas is both an endocrine and exocrine gland. Its endocrine part is the *pancreatic islets,* the source of insulin and glucagon, described in section 11.4e. About 99% of the pancreas is exocrine tissue with a digestive function.

The exocrine pancreas secretes about 1,500 mL of **pancreatic juice** per day. This fluid flows to the duodenum by way of the **pancreatic duct,** which runs lengthwise through the gland and joins the bile duct at the hepatopancreatic ampulla. Pancreatic juice is an alkaline solution of mainly sodium bicarbonate and enzymes. Two of the enzymes, **trypsin** and **chymotrypsin** (KY-mo-TRIP-sin), digest dietary protein; **pancreatic amylase** digests starch; **pancreatic lipase** digests fat; and **ribonuclease** and **deoxyribonuclease** digest RNA and DNA, respectively. We will later examine the mode of action of these enzymes on the nutrients.

Pancreatic juice and bile are not secreted continually, but only when needed. The following signal molecules activate their secretion:

1. **Acetylcholine (ACh)** is secreted by the vagus nerves and enteric plexus even before food is swallowed. It stimulates pancreatic secretion, but the fluid is retained in the pancreatic duct until chyme enters the duodenum.
2. **Cholecystokinin (CCK)** is secreted by the mucosa of the duodenum in response to fat in the chyme. It stimulates enzyme secretion by the pancreas, contraction of the gallbladder, and relaxation of the sphincter of the ampulla so the pancreatic secretions and bile can enter the duodenum and mix with the fatty chyme.
3. **Secretin** is produced by the duodenum in response to the acidity of the chyme. It stimulates both the liver and pancreas to secrete sodium bicarbonate, which buffers the stomach acid and protects the duodenum from its erosive effect. The rise in duodenal pH due to this bicarbonate also activates digestive enzymes of the pancreatic juice and intestinal mucosa.

Before You Go On

Answer these questions from memory to test your understanding of the previous section. Reread the material if you can't answer the questions.

16. What does the liver contribute to digestion?
17. Where is bile secreted? Where is it stored? Where does it perform its action?
18. List three enzymes secreted by the pancreas and state the function of each.
19. What stimulates cholecystokinin (CCK) secretion, and how does CCK affect other parts of the digestive system?

17.5 The Small Intestine

Expected Learning Outcomes

When you have completed this section, you should be able to

a. describe the gross and microscopic anatomy of the small intestine, including the differences in its three regions;
b. state how and for what functional reason the mucosa of the small intestine differs from that of the stomach;
c. give two reasons why the small intestine needs such a large surface area, and four structural features that meet that need; and
d. describe the types of movement that occur in the small intestine.

Nearly all chemical digestion and nutrient absorption occur in the small intestine. It efficiently carries out these tasks partly because it is the longest part of the digestive tract—about 5 m long (ranging 3 to 7 m) in a living person, and up to 8 m long in the cadaver, where there is no muscle tone. The term *small* intestine refers not to its length but to its diameter—about 2.5 cm (1 in.).

17.5a Anatomy

The small intestine is a coiled mass filling most of the abdominal cavity inferior to the stomach and liver. It is divided into three regions (**fig. 17.14**): the duodenum, jejunum, and ileum.

The **duodenum**[13] (dew-ODD-eh-num, DEW-oh-DEE-num) constitutes the first 25 cm (10 in.). It begins at the pylorus, arcs around the head of the pancreas, and curves to the left. Its name refers to its length, about equal to the width of 12 fingers. Along with the pancreas, most of the duodenum is retroperitoneal. The duodenum receives stomach contents, pancreatic juice, and bile. The stomach acid is neutralized both by the alkaline pancreatic juice and bicarbonate-rich mucus secreted by *duodenal glands* in the submucosa. The thick mucus serves to protect the mucosa from the corrosive stomach acid. Pepsin from the stomach is inactivated by the elevated duodenal pH, and pancreatic enzymes take over the job of chemical digestion. Fats are physically broken up (emulsified) by the bile acids.

The **jejunum**[14] (jeh-JOO-num) is the first 40% of the small intestine beyond the duodenum—roughly 1.0 to 1.8 m in a living person. Its name refers to the fact that early anatomists typically found it to be empty. The jejunum is located largely in the upper left quadrant of the abdomen. Its wall is relatively thick and muscular, and it has an especially rich blood supply and redder color than the rest of the small intestine. Most digestion and nutrient absorption occur here.

The **ileum**[15] forms the last 60% of the small intestine (about 1.6 to 2.7 m). It lies mostly in the lower right quadrant of the abdomen. Compared with the jejunum, its wall is thinner, less muscular, less vascular, and has a paler pink color. It has prominent lymphatic nodules in clusters called **aggregated lymphoid nodules** (formerly called *Peyer patches*), which are readily visible to the naked eye and become progressively larger approaching the large intestine.

---

[13] *duoden* = 12
[14] *jejun* = empty, dry
[15] from *eilos* = twisted

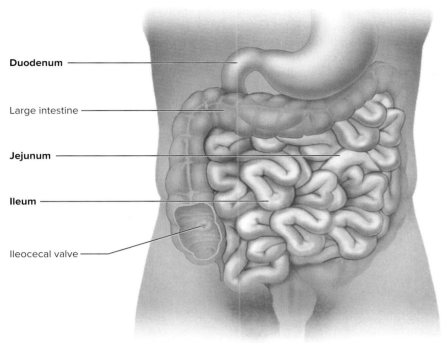

Figure 17.14 Gross Anatomy of the Small Intestine. Note the spatial relationships of its three main regions: duodenum, the first 25 cm past the stomach; jejunum, mostly in the upper left abdominal quadrant; and ileum, mostly in the lower right abdominal quadrant, ending at the ileocecal valve.

The end of the small intestine is the **ileocecal junction** (ILL-ee-oh-SEE-cul), where the ileum joins the *cecum* of the large intestine. The muscularis of the ileum is thickened at this point to form a sphincter, the **ileocecal valve,** which protrudes into the cecum and regulates the passage of food residue into the large intestine. Both the jejunum and ileum are intraperitoneal and thus covered with a serosa, which is continuous with the complex, folded mesentery that suspends the small intestine from the posterior abdominal wall.

Effective digestion and absorption require that the small intestine have a large internal surface area. This is provided by its relatively great length and by three kinds of internal folds or projections: *circular folds, villi,* and *microvilli.* If the mucosa were smooth, like the inside of a hose, it would have a surface area of about 0.3 to 0.5 m^2, but with these surface elaborations, its actual surface area is about 200 m^2—clearly a great advantage for nutrient absorption.

Circular folds, the largest elaborations of the intestinal wall, are transverse ridges up to 10 mm high that project into the lumen (fig. 17.13). They occur from the duodenum to the middle of the ileum, where they cause the chyme to flow on a spiral path along the intestine. This slows its progress, causes more contact with the mucosa, and promotes more thorough mixing and nutrient absorption.

Villi (VIL-eye; singular, *villus*) are small finger- to tongue-shaped projections about 0.5 to 1.0 mm high (fig. 17.15). They give the mucosa a fuzzy texture like a terrycloth towel. Villi are covered with simple columnar epithelium composed of two kinds of cells: columnar **absorptive cells (enterocytes),** which absorb digested nutrients, and **goblet cells,** which secrete a coat of protective mucus over the intestinal surface. The core of a villus is filled with loose connective tissue. Embedded in this are an arteriole, a capillary network, a venule, and a lymphatic capillary called a **lacteal** (LAC-tee-ul). Most absorbed nutrients pass into the blood capillaries, but most lipids are picked up by the lacteal. The lipids impart a milky appearance for which the lacteal is named.[16] The lacteal connects with larger lymphatic vessels that ultimately carry the lipids to the bloodstream (see "Lymphatic Vessels and Lymph" in section 14.1b).

---

[16]*lact* = milk

Microvilli form a fuzzy **brush border** on the surface of each absorptive cell. Being only about 1 μm high, microvilli can be resolved only with the electron microscope. They greatly increase the absorptive surface area of the small intestine and possess **brush border enzymes**—components of the plasma membrane that carry out some of the final stages of chemical digestion.

On the floor of the small intestine, between the bases of the villi, there are numerous pores that open into tubular glands called **intestinal crypts.** In the upper half, they consist of absorptive cells and goblet cells like those of the villi. The lower half is dominated by dividing epithelial cells. In its life span of 3 to 6 days, an epithelial cell migrates up the crypt to the tip of the villus, where it is sloughed off and digested.

> ### Apply What You Know
> Both the small intestine and the proximal convoluted tubule of the kidney (see section 16.4a) are tubular structures specialized for absorption. Identify at least two features that they have in common to serve this purpose.

Figure 17.15 Intestinal Villi. (a) Villi. Each villus is about 1 mm high. (b) Structure of a villus. APR

a: © Meckes/Ottawa/Science Source

(a)

(b)

(a) Segmentation

(b) Peristalsis

Figure 17.16 Contractions of the Small Intestine. (a) Segmentation, in which circular constrictions of the intestine cut into the contents, churning and mixing them. (b) Peristalsis, in which successive waves of contraction overlap each other. Each wave travels partway down the intestine and milks the contents toward the colon.

17.5b Intestinal Secretion and Motility

The small intestine secretes 1 to 2 L of alkaline *intestinal juice* per day, which helps to mix nutrients with digestive enzymes and protect the mucosa from stomach acid.

It exhibits two kinds of contractions—*segmentation* and *peristalsis*. These serve to (1) mix chyme with intestinal juice, bile, and pancreatic juice, allowing them to neutralize acid and digest nutrients more effectively; (2) churn chyme and bring it into contact with the mucosa for digestion and absorption; and (3) move residue toward the large intestine.

Segmentation (fig. 17.16a) is a movement in which ringlike constrictions appear at several places at once along the intestine and then relax as new constrictions form elsewhere. This is the most common movement of the small intestine. Its effect is to knead or churn the contents so the chyme is thoroughly mixed with the digestive enzymes and digestion is more thorough, and to ensure that all of the chyme contacts the absorptive surface so the nutrients can be absorbed from it. When most nutrients have been absorbed and little remains but undigested residue, segmentation declines and peristalsis begins (fig. 17.16b). These waves of contraction, somewhat like you would use to squeeze toothpaste from a tube, move chyme toward the colon over a period of about 2 hours.

Before You Go On

Answer these questions from memory to test your understanding of the previous section. Reread the material if you can't answer the questions.

20. What four aspects of its structure give the small intestine such a large surface area?
21. Why are mucous glands more important in the duodenum than in other regions of the small intestine?
22. Sketch or describe a villus and label its epithelium, brush border, lamina propria, blood capillaries, and lacteal. How does the function of the blood capillaries differ from that of the lacteal?
23. Distinguish between segmentation and peristalsis of the small intestine. How do these differ in function?

17.6 Chemical Digestion and Absorption

Expected Learning Outcomes

When you have completed this section, you should be able to

a. describe how each major class of nutrients is chemically digested and name the enzymes involved; and
b. describe how each type of nutrient is absorbed by the small intestine.

As food material passes from the mouth through the small intestine, nutrients are chemically degraded and absorbed. In this section, we focus on how this occurs, especially for the major classes of nutrients—carbohydrates, proteins, and lipids. These are summarized and compared in figure 17.17.

17.6a Carbohydrates

Most digestible dietary carbohydrate is starch. Cellulose is indigestible and is not considered here, although its importance as dietary fiber is discussed in the next chapter.

Starch (fig. 17.17a) is digested to the disaccharide maltose, and finally to glucose, which is absorbed by the small intestine. The process begins in the mouth, where salivary amylase begins to break down the starch into maltose and other small carbohydrate chains *(oligosaccharides)*. Salivary amylase functions best at pH 6.8 to 7.0, typical of the oral cavity. It is quickly denatured upon contact with stomach acid, although can continue digesting starch for up to 2 hours in the stomach if it is in the middle of a starchy food mass such as a pancake breakfast, where it can escape contact with the acid for a while. The stomach itself, however, produces no carbohydrate-digesting enzymes. About 50% of dietary starch is digested to maltose before it reaches the duodenum. Other dietary sugars include sucrose (cane or table sugar) and lactose (milk sugar), but these are not acted upon until they reach the small intestine.

Starch digestion resumes in the small intestine when the chyme mixes with *pancreatic amylase.* This breaks the remaining starch and oligosaccharides down to maltose. Carbohydrate digestion is completed when the maltose and other disaccharides contact the brush border of the epithelial cells. Here, they are acted upon by three brush border enzymes—*maltase,* which splits the maltose into glucose molecules; *sucrase,* which splits sucrose into glucose and fructose; and *lactase,* which digests lactose to glucose and galactose. In most populations, however, lactase activity ceases or declines to a low level after age 4 and lactose becomes indigestible (see Clinical Application 17.2).

The monosaccharide products of carbohydrate digestion—glucose, fructose, and galactose—are immediately absorbed by carrier-mediated transport across the absorptive cell surface. Within the cell, most of the fructose is converted to glucose. Glucose, galactose, and any remaining fructose then pass out the base of the epithelial cells by facilitated diffusion, pass into the blood capillaries of the villi, and are carried off by the hepatic portal system to the liver for metabolism or storage.

17.6b Proteins

Protein (fig. 17.17b) is digested by enzymes called **proteases.** There are no proteases in the saliva; protein digestion begins in the stomach. Here, *pepsin* digests 10% to 15% of it into shorter peptides and a small amount of free amino acids. Pepsin has

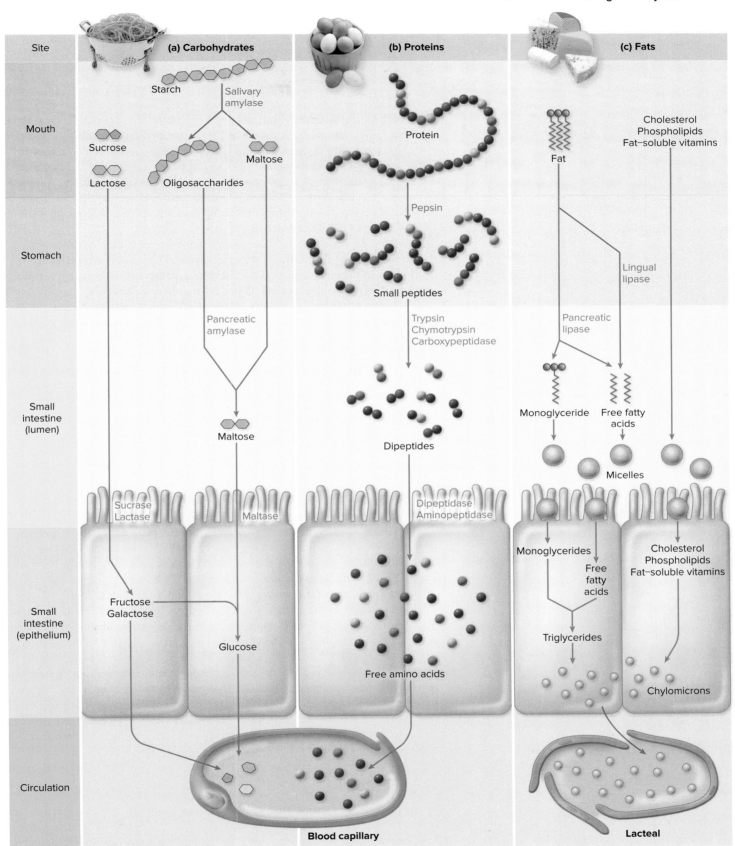

Figure 17.17 Digestion and Absorption of the Three Major Nutrient Classes. Enzyme names are indicated in red.
a: © Burke/Triolo Productions/Getty Images; b: © Pixtal/agefotostock; c: © Pixtal/AGE Fotostock

❓ As they leave the small intestine, sugars and amino acids arrive at the liver before any other organ, but lipids do not. Why not? Based on sections 13.6b and 14.1b, trace the route that a dietary fat would have to take to reach the liver.

Clinical Application 17.2

LACTOSE INTOLERANCE

Homo sapiens is a strange species indeed—the only mammal that continues to drink milk after infancy and, moreover, drinks the milk of other species! To digest milk sugar, one needs the enzyme lactase, and understandably, lactase production ceases in other mammals after they are weaned. It also ceases in the majority of humans worldwide. Only in people of western and northern Europe, a few pastoral tribes of Africa, and their descendants in the Americas and elsewhere, does lactase production continue into adulthood. These populations have a long history of milking domestic animals that has led to the persistence of lactase production.

People without lactase have *lactose intolerance*. If they consume milk, lactose passes undigested into the large intestine. This causes osmotic retention of water in the colon and leads to diarrhea. Lactose fermentation by colonic bacteria produces gas, resulting in painful cramps and flatulence.

Lactose intolerance occurs in about 15% of American whites; 90% of American Blacks, who are predominantly descended from nonpastoral African tribes; 70% or more of Mediterraneans; and nearly all people of Asian descent, including Native Americans descended from early migrants into North, Central, and South America. People with lactose intolerance can consume products such as yogurt and cheese, in which bacteria have broken down the lactose, and they can digest milk with the aid of lactase drops or tablets.

---

an optimal pH of 1.5 to 3.5; thus, it is inactivated when it passes into the duodenum and mixes with the alkaline pancreatic juice (pH 8).

In the small intestine, three proteases from the pancreas continue the job—*trypsin, chymotrypsin,* and *carboxypeptidase.* Finally, the remaining dipeptides and other short chains are taken apart, one amino acid at a time, by two brush border enzymes—*dipeptidase* and *aminopeptidase.*

Amino acid absorption is similar to that of monosaccharides. There are several transport proteins in the epithelial cell membranes for different classes of amino acids. At the basal surfaces of the cells, amino acids exit by facilitated diffusion, enter the blood capillaries of the villus, and are carried away in the hepatic portal circulation.

Amino acids absorbed by the small intestine come not only from dietary protein but also from sloughed epithelial cells digested by the same enzymes, and from the enzymes digesting each other. Thus, the body recycles its own proteins rather than losing the valuable amino acids from these cells and enzymes in the feces. The amino acids from the last two sources total about 30 g/day, compared with about 44 to 60 g/day from the diet.

17.6c Fats

Fat (triglyceride) digestion (fig. 17.17c) is complicated by the fact that fats are hydrophobic and can't enter the intestinal blood capillaries in significant quantities. Fat digestion, therefore, occurs in a more roundabout way than carbohydrate and protein digestion.

Saliva contains an enzyme called *lingual lipase,* which is activated by stomach acid and digests some fat there. Most fat, however, leaves the stomach undigested. Before being passed along to the small intestine, it is churned and broken up into small **emulsification droplets** by the stomach. In the small intestine, the droplets are coated with certain components of the bile—lecithin and bile acids—to prevent them from recombining into larger globules. This *emulsification* process is much like scrubbing a greasy frying pan with dish soap. It exposes far more of the fat's surface to enzyme action.

Lingual lipase no longer works at the pH of the small intestine, but *pancreatic lipase* takes over. It splits two of the fatty acids from each triglyceride molecule, leaving free fatty acids and a *monoglyceride*. These are taken up along with dietary cholesterol, phospholipids, and fat-soluble vitamins into small droplets called **micelles**[17] (my-SELLS), a component of the bile produced by the liver. When micelles contact the brush border, the lipids leave them and are transported into the absorptive epithelial cells. Monoglycerides and free fatty acids are smaller than triglycerides and easier for the enterocytes to absorb, but once inside the enterocytes, they are reassembled into triglycerides. The enterocyte Golgi complex then packages all of these lipids into another type of droplet called **chylomicrons**[18] (KY-lo-MY-crons), coated with phospholipids and protein. These leave the base of the cells and enter the lacteal within the villus; they are too large to penetrate into the blood capillaries. From the lacteals, the white, fatty lymph *(chyle)* flows through larger and larger lymphatic vessels, eventually entering the bloodstream at the left subclavian vein (see "Lymphatic Vessels and Lymph" in section 14.1b). The further fate of dietary fat is described in "Lipids" in section 18.1e.

17.6d Vitamins and Minerals

Fat-soluble vitamins A, D, E, and K are absorbed with other lipids as just described. If they are ingested without fat-containing food, they are not absorbed at all but are passed in the feces and wasted. Water-soluble vitamins (the B complex and vitamin C) are absorbed by simple diffusion, with the exception of vitamin B_{12}. This is an unusually large molecule that can only be absorbed if it binds to intrinsic factor from the stomach. Minerals (electrolytes) are absorbed along the entire length of the small intestine.

Iron and calcium are unusual in that they are absorbed in proportion to the body's need, whereas other minerals are absorbed at fairly constant rates regardless of need. Iron absorption is regulated by the liver hormone *hepcidin,* which limits absorption and protects us from a toxic iron overload. Calcium is absorbed mostly by diffusion through the gaps between epithelial cells in the jejunum and ileum, but in the duodenum its absorption depends on membrane transport proteins and is regulated by parathyroid hormone and calcitriol (vitamin D).

17.6e Water

The digestive system is one of several systems involved in water balance. The digestive tract receives about 9 L of water per day. Most of this, about 6.7 L, consists of the digestive system's own secretions—saliva, gastric juice, bile, pancreatic juice, and intestinal juice. The rest (2.3 L) is from food and drink. The small intestine absorbs about 8 L of this, the large intestine absorbs about 0.8 L, and the remaining 0.2 L is voided in the daily fecal output. Water is absorbed by osmosis, following the absorption of salts and organic nutrients that create an osmotic gradient from the intestinal lumen to the ECF.

---

[17]*mic* = grain, crumb; *elle* = little
[18]*chylo* = juice; *micr* = small

Before You Go On

Answer these questions from memory to test your understanding of the previous section. Reread the material if you can't answer the questions.

24. Name the enzymes that digest starch, fat, and protein. Name the end products of their digestion.
25. Describe how digested sugars and amino acids get from the intestinal lumen into the blood.
26. Describe how lipids get from the intestinal lumen into the lymph and ultimately the blood.
27. What is the difference between fat-soluble and water-soluble vitamins in terms of absorption?
28. What are all the sources of water in the digestive tract? How much of it is lost by defecation, and what happens to the rest of it?

17.7 The Large Intestine

Expected Learning Outcomes

When you have completed this section, you should be able to

a. describe the gross anatomy and mucosa of the large intestine;
b. state the physiological significance of intestinal bacteria;
c. discuss the types of contractions that occur in the colon; and
d. explain the neurological control of defecation.

The large intestine (**fig. 17.18**) receives about 500 mL of indigestible food residue per day, reduces it to about 150 mL of feces by absorbing water and salts, and eliminates the feces by defecation.

17.7a Anatomy

The **large intestine** measures about 1.5 m (5 feet) long and 6.5 cm (2.5 in.) in diameter. It begins with the **cecum,**[19] a pouch in the lower right abdominal quadrant inferior to the ileocecal valve. Attached to the lower end of the cecum is the **appendix,** a slender, closed tube 2 to 7 cm long. The appendix is densely populated with lymphocytes and is a significant source of immune cells.

The **colon** is that portion of the large intestine between the ileocecal valve and the rectum (not including the cecum, rectum, or anal canal). It is divided into the ascending, transverse, descending, and sigmoid regions. The **ascending colon** begins at the ileocecal valve and passes up the right side of the abdominal cavity. It makes a 90° turn near the right lobe of the liver and becomes the **transverse colon.** This passes horizontally across the upper abdominal cavity and turns 90° downward near the spleen. Here it becomes the **descending colon,** which passes down the left side of the abdominal cavity. Ascending, transverse, and descending colons thus form a squarish, three-sided frame around the small intestine.

---

[19]cec = blind

Figure 17.18 The Large Intestine. APR

❓ *Which anal sphincter is controlled by the autonomic nervous system? Which is controlled by the somatic nervous system? Explain the basis for your answers.*

The muscularis externa of the colon has longitudinal smooth muscle cells concentrated especially in three thickened, ribbonlike strips, each called a **taenia coli**[20] (TEE-nee-ah CO-lye). The muscle tone of the taeniae coli (plural) contracts the colon lengthwise and causes its wall to bulge, forming pouches called **haustra**[21] (HAW-stra; singular, *haustrum*).

The pelvic cavity is narrower than the abdominal cavity, so the colon turns medially along the hip bone, then downward, forming a roughly S-shaped portion called the **sigmoid**[22] **colon.** In the pelvic cavity, the large intestine continues as the **rectum,**[23] about 15 cm long.

The final 3 cm of the large intestine is the **anal canal,** which passes through the pelvic floor and terminates at the anus. Prominent *hemorrhoidal veins* form superficial plexuses in the walls of the anus. Unlike veins in the limbs, they lack valves and are particularly subject to distension and venous pooling. Hemorrhoids are permanently distended veins that protrude into the anal canal or form bulges external to the anus. They often develop in obesity and pregnancy, where pressure in the abdominopelvic cavity impedes venous return and causes blood pressure to back up in the hemorrhoidal veins.

---

[20]*taenia* = ribbon; *coli* = of the colon
[21]*haustr* = to draw
[22]*sigm* = sigma or S; *oid* = resembling
[23]*rect* = straight

Most of the large intestine is lined with a simple columnar epithelium containing an abundance of mucus-secreting goblet cells. The lower anal canal is lined with abrasion-resistant stratified squamous epithelium. Cancer of the colon and rectum (colorectal cancer) is the second most common cause of cancer deaths in the United States. It usually originates with the mucus-secreting goblet cells and, if untreated, may metastasize to distant organs. Screening for colorectal cancer often involves endoscopic examination of the colon, a procedure called a colonoscopy, that is typically recommended for those of age 50 or older.

The anus is regulated by two sphincters: an **internal anal sphincter** composed of smooth muscle of the muscularis externa, and an **external anal sphincter** composed of skeletal muscle of the pelvic floor.

17.7b Intestinal Bacteria

The healthy adult human gut harbors more than 1,000 species of bacteria, collectively called the **microbiome.** The greatest density of gut microorganisms is found in the large intestine. We provide them with room and board, while they provide us with nutrients from our food that we are not equipped to extract on our own. For example, some bacteria in the colon digest cellulose, pectin, and other plant polysaccharides for which we have no digestive enzymes, and we absorb the resulting sugars. We therefore get more nutrition from our food because of these bacteria than we would get without them. One person may even get more calories than another from the same amount of food because of differences in their bacterial populations. Some bacteria also synthesize B vitamins and vitamin K. Vitamin K absorbed by the colon is important because the diet alone usually does not provide enough to ensure adequate blood clotting.

The normal gut microbes also deter dangerous pathogens that can't compete successfully with the normal residents for nutrients and space. When a person takes antibiotics to combat a bacterial infection, normal gut bacteria are killed along with the initial invader, potentially allowing other pathogenic strains to establish themselves. This can lead to further disease and sometimes even death. Studies have linked intestinal microbial communities with a surprising array of conditions ranging from obesity to gastrointestinal disorders such as inflammatory bowel disease.

One of the less desirable and sometimes embarrassing products of these bacteria is intestinal gas **(flatus).** This is a mixture of bacterial gases and nitrogen from swallowed air. Painful cramping can result when undigested nutrients pass into the colon and furnish an abnormal substrate for bacterial action—for example, in lactose intolerance. Flatus is composed of nitrogen (N_2), carbon dioxide (CO_2), hydrogen (H_2), methane (CH_4), hydrogen sulfide (H_2S), and two amines—indole and skatole. Indole, skatole, and H_2S produce the odor of flatus and feces, whereas the others are odorless. The hydrogen is combustible and has sometimes ignited during the use of electrical cauterization in surgery.

17.7c Absorption and Motility

The large intestine takes 36 to 48 hours to reduce the residue of a meal to feces, with the residue spending the longest time (about 24 h) in the transverse colon. It reabsorbs water and electrolytes (especially NaCl). The feces consist of about 75% water and 25% solids. The solids are about 30% bacteria, 30% undigested dietary fiber, 10% to 20% fat, and smaller percentages of protein, epithelial cells, mucus, and salts. The fat is not from the diet but from broken-down epithelial cells and bacteria.

Colonic motility is predominantly a type of segmentation called **haustral contractions,** which occur about every 30 minutes. Distension of a haustrum with feces stimulates it to contract. This churns and mixes the residue, promotes water and salt absorption, and passes the residue distally to another haustrum. Stronger contractions called **mass movements** occur one to three times a day, last about 15 minutes, and move residue for several centimeters at a time. Mass movements occur especially in the transverse to sigmoid colon, and often within an hour after breakfast, moving the feces that accumulated and stretched the colon overnight.

17.7d Defecation

Like urination, the elimination of feces is not a continual event but is episodic and is controlled by neurological reflexes and sphincters; thus, it ordinarily occurs only in appropriate circumstances and times. Filling of the stomach with a new meal triggers a *gastrocolic reflex* through the enteric nervous system, inducing mass movements that propel feces from the upper colon toward the rectum. Stretching of the rectum then sets off both enteric and spinal reflexes called **defecation reflexes,** which account for the urge to defecate.

One of the defecation reflexes, called the *intrinsic defecation reflex,* is controlled through the enteric plexus of the colonic wall. Stretch signals from the rectum are conducted through the plexus (1) to the descending and sigmoid colon, further activating a peristaltic wave that drives feces downward, and (2) to the internal anal sphincter, causing it to relax. In an infant, this alone results in defecation. The other reflex, called the *parasympathetic defecation reflex,* is spinal. Stretch signals are conducted to the spinal cord and return to the rectum through a visceral reflex arc via parasympathetic nerve fibers in the pelvic nerves. These signals likewise intensify peristalsis and relax the internal anal sphincter.

Obviously and fortunately, once we have acquired bowel control in childhood, we are not at the mercy of these two involuntary reflexes. The external anal sphincter is skeletal muscle and therefore under voluntary control. Feces usually are voided only if we voluntarily relax that sphincter in addition to the foregoing involuntary reflexes. The external anal and urinary sphincters are controlled together by inhibitory signals from the brainstem. This is why we find it hard to defecate without also urinating. Some spinal cord injuries, however, disconnect the brainstem from the lower spinal cord and abolish this voluntary control. The resulting inability to voluntarily retain the feces is called *fecal incontinence.*

Defecation is also aided by the voluntary Valsalva maneuver, in which a breath hold and contraction of the abdominal muscles increase abdominal pressure, compress the rectum, and squeeze the feces from it.

Apply What You Know

One way to "potty train" a child or housebreak a puppy is to put the child on the toilet or the puppy outdoors within half an hour after a meal. Explain why this strategy would promote the desired behavioral modification, in view of what you now know about the neurological control of the colon.

Before You Go On

Answer these questions from memory to test your understanding of the previous section. Reread the material if you can't answer the questions.

29. List the parts of the large intestine in order from the ileocecal valve to the external anal sphincter.
30. Compare and contrast the mucosa and muscularis externa of the small and large intestines.
31. Define microbiome, *and discuss advantages and disadvantages of the abundant bacterial load in the large intestine.*
32. Compare and contrast the defecation and urination (micturition) reflexes.

Aging of the Digestive System

Restaurants commonly offer a "seniors' menu" and reduced prices to customers over 60 or 65 years old, knowing that they tend to eat less. There are multiple reasons for this reduced appetite. Older people have lower metabolic rates and tend to be less active than younger people and, hence, need fewer calories. The stomach atrophies with age, and it takes less to fill it up. For many, food has less esthetic appeal in old age because of diminished senses of smell, taste, and even vision. In addition, older people secrete less saliva, making food less flavorful and swallowing more difficult. Reduced salivation also makes the teeth more prone to caries (cavities). Dentures are an unpleasant fact of life for many people over 65 who have lost their teeth to caries and periodontitis. Atrophy of the epithelium of the oral cavity and esophagus makes these surfaces more subject to abrasion and may further detract from the ease of chewing and swallowing.

The reduced mobility of old age makes shopping and meal preparation more troublesome, and with food losing its sensory appeal, some decide that it simply isn't worth the trouble. However, one's protein, vitamin, and mineral requirements remain essentially unchanged, so vitamin and mineral supplements may be needed to compensate for reduced food intake and poorer intestinal absorption. Malnutrition is common among older people and is an important factor in anemia, reduced immunity, and generally declining health in old age.

As the gastric mucosa atrophies, it secretes less acid and intrinsic factor. Acid deficiency reduces the absorption of calcium, iron, zinc, and folic acid. Heartburn becomes more common as the weakening lower esophageal sphincter fails to prevent the reflux of stomach contents into the esophagus. The most common digestive complaint of older people is constipation, which results from reduced muscle tone and weaker peristalsis of the colon. This seems to stem from a combination of factors: atrophy of the muscularis externa, reduced sensitivity to neurotransmitters that promote motility, less fiber and water in the diet, and less exercise. The liver, gallbladder, and pancreas show only slightly reduced function, but the drop in liver function reduces the rate of drug deactivation and can contribute to overmedication of the elderly.

CAREER SPOTLIGHT

Dental Hygienist

© MIXA/Getty Images

Dental hygienists are licensed health-care professionals who promote oral health and wellness. Their job includes assessment of the health of the oral cavity and surrounding structures of the head and neck to inspect for abnormalities or diseases such as oral cancer. They review health and dental history. They clean teeth by removing calculus and plaque. An important aspect of their care is teaching patients oral-hygiene techniques. They also apply cavity-preventive agents such as sealants and fluoride. In addition, they produce dental radiographs (X-rays) for the dentist to interpret.

The majority of dental hygienists are employed by dentists and work in general dental offices. However, they may also serve in hospitals, nursing homes, public-health clinics, and schools. Dental hygienists must complete an accredited degree program and then pass both written and clinical examinations in order to be licensed by the state in which they practice. See appendix B for further career information in dental hygiene.

CONNECTIVE ISSUES

Ways in Which the DIGESTIVE SYSTEM Affects Other Organ Systems

All Systems
The digestive system provides nutrients for all tissues.

Integumentary System
Vitamin C absorbed in the small intestine is necessary for collagen production and skin integrity.

Skeletal System
The small intestine adjusts calcium absorption in proportion to the needs of the skeletal system; dental decay or periodontitis can lead to bone infections and deterioration of mandibular and maxillary bone.

Muscular System
Between meals glucose is drawn from the liver's glycogen stores, supplying energy for skeletal muscle movement.

Nervous System
Gut–brain peptides produced in the stomach and small intestine stimulate brain centers that control appetite; chronic vomiting or diarrhea can cause electrolyte and pH imbalances that interfere with nervous system function.

Endocrine System
The stomach and small intestines produce some hormones and thus play an endocrine role; feeding stimulates the secretion of growth hormone and insulin; the liver degrades hormones and influences the rate of hormone clearance from the blood.

Circulatory System
The digestive tract absorbs fluid needed to maintain blood volume; the liver degrades heme from dead erythrocytes and secretes clotting factors, blood albumin, and other plasma proteins; the liver regulates blood glucose level; the intestinal epithelium stores iron and releases it as needed for hemoglobin synthesis; dental decay or periodontitis can lead to heart disease (infectious endocarditis).

Lymphatic and Immune Systems
The GI mucosa is a site of lymphocyte production; maternal IgA absorbed by the infant intestine confers passive immunity; acid, lysozyme, mucus, and other digestive secretions provide innate immunity against pathogens.

Respiratory System
Contraction of abdominal muscles aids in expiration by pushing digestive organs upward against the underside of the diaphragm.

Urinary System
The intestines complement the kidneys in water and electrolyte reabsorption; the liver and kidneys collaborate in calcitriol synthesis; the liver synthesizes urea, and the kidneys excrete it.

Reproductive System
The digestive system provides nutrients for fetal growth and lactation.

Study Guide

Assess Your Learning Outcomes

To test your knowledge, discuss the following topics with a study partner or in writing, ideally from memory.

17.1 Overview of the Digestive System

1. Definition of *gastroenterology*
2. Five stages through which the digestive system processes food
3. Distinction between mechanical and chemical digestion and general purposes served by each
4. The two anatomical subdivisions of the digestive system
5. Layers of the digestive tract
6. The enteric nervous system and its function
7. Anatomical relationships among the parietal peritoneum, mesentery, greater omentum, and lesser omentum
8. Definitions of *intraperitoneal* and *retroperitoneal*

17.2 The Mouth Through Esophagus

1. Seven functions of the mouth
2. Anatomy and histology of the lips, cheeks, tongue, and palate
3. Tissues of the teeth and the types and numbers of teeth from midline to rear of the jaw
4. The three major pairs of major salivary glands; components of saliva and their functions
5. Anatomy of the pharynx and the digestive role of its sphincters
6. Anatomy of the esophagus, histology of its wall, and importance of its lower sphincter
7. The phases of swallowing

17.3 The Stomach

1. Definition of *chyme*
2. Role of the stomach in both mechanical and chemical digestion
3. Shape of the stomach, its four regions, and its pyloric sphincter
4. Histology of the stomach wall
5. Cells found in the gastric glands and their functions
6. Functions of hydrochloric acid
7. Two food macromolecules that are chemically digested in the stomach; the enzymes involved
8. Secretions of parietal and chief cells and the purposes they serve
9. The importance of intrinsic factor
10. Contractions of the stomach and the ejection of chyme into the duodenum
11. Vomiting, including its stimuli and risks
12. Events of the cephalic, gastric, and intestinal phases of gastric activity, and their neural and hormonal regulation
13. Three ways the stomach protects itself from self-digestion

17.4 The Liver, Gallbladder, and Pancreas

1. How liver function contributes to digestion
2. Gross anatomy of the liver
3. The structure of a hepatic lobule
4. The name and functions of the cuboidal epithelial cells of the liver
5. Structure of the gallbladder and bile passages
6. The composition and digestive role of bile
7. The location and gross anatomy of the pancreas
8. Function of the bicarbonate in pancreatic juice
9. Pancreatic enzymes and their individual functions
10. Hormonal regulation of pancreatic function

17.5 The Small Intestine

1. Length and diameter of the small intestine
2. The three regions of the small intestine, how they differ histologically, and what marks the beginning and end of each
3. Names of the internal folds and projections that increase the surface area of the small intestine
4. The structure of a villus, including its types of surface epithelial cells and the contents of its core
5. Differences between intestinal segmentation and peristalsis, and the function of each

17.6 Chemical Digestion and Absorption

1. Chemical digestion of starch in the mouth; the role and sources of amylase; how pH affects amylase
2. The actions of maltase, sucrase, and lactase and site where they perform chemical digestion
3. The absorption and fate of dietary monosaccharides
4. Sources of amino acids absorbed by the small intestine
5. The definition of *protease*, the names of proteases involved in digestion, and the anatomical sites where different proteases work
6. How amino acids are absorbed by the small intestine

7. The role of bile and emulsification in fat digestion
8. The role of lipase and chylomicrons in fat digestion
9. The absorption of fat-soluble vitamins, water-soluble vitamins, and minerals
10. Varying modes of vitamin absorption in the small intestine
11. The hormonal regulation of calcium and iron absorption
12. The sources, absorption, and elimination of water in the digestive tract

17.7 The Large Intestine

1. Segments of the large intestine; the distinction between the large intestine and colon
2. Comparison of the wall of the small and large intestines and functional explanation of the anatomical differences
3. Reasons for the histological difference between the colon and anal canal
4. Definition of *microbiome* and beneficial actions of intestinal bacteria; the composition of intestinal gas
5. Absorptive functions of the large intestine
6. Modes of contraction and neural control of the colon
7. The mechanism of defecation

Testing Your Recall

1. Which of the following enzymes acts in the stomach?
   a. chymotrypsin
   b. pepsin
   c. lactase
   d. dipeptidase
   e. sucrase

2. Which of the following digests proteins?
   a. trypsin
   b. lipase
   c. maltase
   d. sucrase
   e. amylase

3. The lacteals absorb
   a. chylomicrons.
   b. micelles.
   c. emulsification droplets.
   d. low-density lipoproteins.
   e. monosaccharides.

4. The folds visible on the inside of an empty stomach are
   a. circular folds.
   b. gastric rugae.
   c. haustra.
   d. microvilli.
   e. villi.

5. Which answer correctly lists the order of tissue layers of the small intestine, beginning with the innermost layer?
   a. mucosa, submucosa, serosa, muscularis externa
   b. submucosa, mucosa, muscularis externa, serosa
   c. serosa, muscularis externa, submucosa, mucosa
   d. mucosa, submucosa, muscularis externa, serosa
   e. mucosa, muscularis externa, submucosa, serosa

6. The gallbladder
   a. stores and concentrates bile.
   b. synthesizes cholesterol.
   c. synthesizes bile.
   d. secretes lipase.
   e. breaks down bilirubin.

7. Which structure is *not* lined with simple columnar epithelium?
   a. ascending colon
   b. duodenum
   c. stomach
   d. rectum
   e. anal canal

8. All of the following contribute to the absorptive surface of the small intestine except
   a. its length.
   b. the brush border.
   c. haustra.
   d. circular folds.
   e. villi.

9. Which of the following secrete hydrochloric acid?
   a. hepatocytes
   b. parietal cells
   c. chief cells
   d. pancreatic cells
   e. enteroendocrine cells

10. Which of these is the voluntary phase of swallowing?
    a. oral phase
    b. cephalic phase
    c. gastric phase
    d. secretory phase
    e. pharyngeal phase

11. Most of the bulk of a tooth is composed of a yellowish tissue called __dentin__.

12. The three regions of the small intestine from proximal to distal are the __duodenum__, __jejunum__, and __ileum__.

13. The __falciform ligament__ suspends the liver from the abdominal wall and separates its right and left lobes.

14. Pepsinogen is produced by __chief__ cells.

15. Haustra are pouchlike outfoldings of the __colon__.

16. The digestive tract has its own independent nervous plexus called the __enteric__ nervous system.

17. __amylase__, an enzyme that breaks down starch, is present in both pancreatic juice and __saliva__.

18. The three pairs of major salivary glands are the _parotid_, _submandibular_, and _sublingual_.

19. What structure has hard and soft regions and makes it possible to continue breathing while chewing food? _Palate_

20. _Peristalsis_ is a wave of muscular contraction produced by the muscularis externa that passes along material in the gastrointestinal lumen.

Answers in Appendix A

What's Wrong with These Statements?

Briefly explain why each of the following statements is false, or reword it to make it true.

1. Micelles are absorbed into lacteals.

2. Taeniae coli are found in the small intestine.

3. The ileocecal valve regulates the passage of chyme between the stomach and the duodenum.

4. Vitamin D is necessary for the intestinal absorption of iron.

5. The only monosaccharide absorbed in the small intestine is glucose.

6. Proteases are involved in lipid digestion.

7. The first phase of swallowing is the pharyngeal phase.

8. Most of the water in the digestive tract is reabsorbed in the small intestine.

9. Intrinsic factor stimulates the release of bile into the duodenum.

10. The jejunum is part of the large intestine.

Answers in Appendix A

Testing Your Comprehension

1. Which do you think would have the most severe effect on digestion: surgical removal of the stomach, gallbladder, or pancreas? Explain.

2. Explain why most lipids must be absorbed by the lacteals rather than by the blood capillaries of a villus.

3. Follow a peanut butter sandwich through the digestive tract, beginning with the mouth. What major nutrient classes are in the sandwich? Where do mechanical and chemical digestion occur? Discuss at least three specific enzymes in your answer.

Nutrition and Metabolism

Chapter 18

Chapter Outline

18.1 Nutrition
- 18.1a Calories
- 18.1b Nutrients
- 18.1c Carbohydrates
- 18.1d Fiber
- 18.1e Lipids
- 18.1f Proteins
- 18.1g Minerals and Vitamins

18.2 Carbohydrate Metabolism
- 18.2a Glucose Catabolism
- 18.2b Glycolysis
- 18.2c Anaerobic Fermentation
- 18.2d Aerobic Respiration
- 18.2e Overview of ATP Production
- 18.2f Glycogen Metabolism

18.3 Lipid and Protein Metabolism
- 18.3a Lipids
- 18.3b Proteins
- 18.3c Liver Functions in Metabolism

18.4 Metabolic States and Metabolic Rate
- 18.4a The Absorptive State
- 18.4b The Postabsorptive State
- 18.4c Metabolic Rate

18.5 Appetite Regulation

18.6 Body Heat and Thermoregulation
- 18.6a Body Temperature
- 18.6b Heat Production and Loss
- 18.6c Thermoregulation
- 18.6d Disturbances of Thermoregulation

Clinical Applications/Perspectives on Health
- Clinical Application 18.1: Hepatitis
- Clinical Application 18.2: Hypothermia and Frostbite
- Perspectives on Health

End of Chapter
- Career Spotlight: Dietitian
- Study Guide

Mitochondria are the cell's "powerhouses," the organelles that make most of the ATP needed to sustain life. If all the body's mitochondria suddenly ceased to function, life itself would cease within one minute.

© Keith R. Porter/Science Source

BASE CAMP

Before ascending to the next level, be sure you're properly equipped with a knowledge of these concepts from earlier chapters.

- Composition and structure of polysaccharides, disaccharides, and monosaccharides (see section 2.3a)
- Composition and structure of triglycerides and of saturated versus unsaturated fats (see section 2.3b)
- Composition and structure of amino acids and proteins (see section 2.3c)
- Structure and function of ATP (see section 2.3e)
- Metabolism, catabolism, and anabolism (see section 2.4c)
- Actions of insulin and glucagon (see section 11.4e)

Anatomy & Physiology Revealed 4.0

Module 12
Digestive System

Food is our source of energy for cellular functions and the raw material for building tissues and organs. Understanding food and nutrition is far from simple, however. We are surrounded by conflicting advice and opinions about what constitutes optimal nutrition. Through most of human history, people struggled to obtain enough food; many modern humans face the opposite problem—access to too much food, leading to obesity and associated health problems such as diabetes and cardiovascular disease. In chapter 17, we saw how the digestive system breaks nutrients down into usable forms and absorbs them into the blood and lymph. We now consider these nutrients in more depth, focusing on metabolism, the chemical changes that capture energy from food and make it available for cellular functions. In addition, we will explore regulation of appetite, body weight, and energy balance, and the related topics of body heat and thermoregulation.

18.1 Nutrition

Expected Learning Outcomes

When you have completed this section, you should be able to

a. define *calorie* and identify the three primary sources of dietary calories;
b. define *nutrient*, identify the six major categories of nutrients, and distinguish between macronutrients and micronutrients;
c. identify the major forms and sources of dietary carbohydrates;
d. name some forms of dietary fiber, explain why fiber is important in the diet yet is not a nutrient, and distinguish water-soluble from water-insoluble fiber;
e. discuss the forms of dietary lipid and explain why lipids are a more efficient means of energy storage than carbohydrates are;
f. describe the forms of blood lipoproteins and explain their respective functions;
g. discuss the body's protein and amino acid requirements; and
h. describe the minerals and vitamins required of the diet.

Good nutrition is part of a healthy lifestyle. Nutrients acquired from food support growth and maintenance of tissues and fuel cellular activities. In this section, we will discuss the energy content of food and specific nutrients.

18.1a Calories

We eat to acquire energy—the capacity to do work. Energy is measured in calories, a subject familiar to anyone concerned about the control of body weight. One calorie is the amount of heat that will raise the temperature of 1 g of water 1°C. The dietary

Calories (with a capital *C*) listed on a food package are actually **kilocalories (kcal)**; one kilocalorie is 1,000 calories. For most adults, a healthy diet ranges from 2,000 to 2,500 kcal/day.

Nearly all dietary calories come from carbohydrates, proteins, and fats. Carbohydrates and proteins yield about 4 kcal/g when they are completely oxidized, and fats yield about 9 kcal/g.

When a chemical is described as **fuel** in this chapter, we mean it is oxidized solely or primarily to extract energy from it. The extracted energy is usually used to make adenosine triphosphate (ATP), the energy carrier necessary for most cellular work.

18.1b Nutrients

A **nutrient** is any ingested chemical that is incorporated into the body's tissues and used for fuel, growth, repair, or maintenance of the body. Nutrients fall into six major classes: water, carbohydrates, lipids, proteins, minerals, and vitamins. Water, carbohydrates, lipids, and proteins are considered **macronutrients** because they must be consumed in relatively large quantities (table 18.1). Minerals and vitamins (see table 18.2) are called **micronutrients** because only small quantities are required.

The **recommended daily allowance (RDA)** of a nutrient is a liberal but safe estimate of the daily intake that would meet the nutritional needs of most healthy people. Consuming less than the RDA of a nutrient doesn't necessarily mean you will be malnourished, but the probability of malnutrition increases in proportion to the amount of the deficit and how long it lasts.

Many nutrients can be synthesized by the body when they are unavailable from the diet. The body is incapable, however, of synthesizing minerals, most vitamins, eight of the amino acids, and one to three of the fatty acids. These are called **essential nutrients** because it is essential that they be included in the diet.

18.1c Carbohydrates

A well-nourished adult has about 440 g of carbohydrate, most of it in three places: 325 g of muscle glycogen, 90 to 100 g of liver glycogen, and 15 to 20 g of blood glucose. Carbohydrates—sugars and starch—serve primarily as fuel for cellular

Table 18.1 Macronutrient Requirements and Functions

Nutrient	Daily Requirement	Representative Functions
Water	2.7–3.7 L	Solvent; coolant; dilutes and eliminates metabolic wastes; supports blood volume and pressure; reactant or product in many metabolic reactions
Carbohydrates	130 g	Fuel; component of glycoproteins, glycolipids, nucleic acids, and ATP; some structural functions
Lipids	20% to 35% of total calories	Fuel; plasma membrane structure; myelin sheaths of nerve fibers; hormones; insulation; protective padding around organs
Proteins	46–56 g	Muscle contraction; structure of cellular membranes and extracellular material; major component of connective tissues; enzymes; some hormones; antibodies; transport of plasma lipids; oxygen binding and transport; blood clotting; blood viscosity and osmolarity; emergency fuel

work and are required in greater amounts than any other nutrient except water. The brain alone consumes about 120 g of glucose per day. Carbohydrates also serve as a structural component of other molecules (table 18.1).

Dietary carbohydrates are derived primarily from plants in the form of polysaccharides (complex carbohydrates), disaccharides, and monosaccharides. The only significant polysaccharide nutrient is starch, obtained principally from grains and root vegetables such as potatoes. Cellulose is also an important dietary polysaccharide, but it cannot be considered a nutrient because it is not digested and does not enter our tissues (see the following section on fiber). Dietary disaccharides include sucrose, refined from sugarcane and sugar beets, and lactose, found in milk. A third disaccharide, maltose, comes mostly from the digestion of starch.

Digestion of starch and disaccharides ultimately results in the production of three monosaccharides: glucose, fructose, and galactose. Glucose is the only monosaccharide present in the blood in significant quantity and is therefore known as *blood sugar*. Its concentration is normally maintained at 70 to 110 mg/dL in peripheral venous blood. This level is precisely regulated by insulin and glucagon. Most body cells rely on a combination of carbohydrates and fats to meet energy needs, but some cells such as neurons and erythrocytes depend almost exclusively on blood glucose. Even a brief period of **hypoglycemia** (glucose deficiency) causes nervous system disturbances felt as weakness or dizziness.

> **Apply What You Know**
> Glucose concentration is about 15 to 30 mg/dL higher in arterial blood than in most venous blood. Explain why.

Ideally, most carbohydrate intake should be in the form of starch. This is partly because foods that provide starch also usually provide other nutrients. Simple sugars, by contrast, not only provide empty calories but also promote tooth decay. In a typical American diet, only 50% of the carbohydrates come from starch and the other 50% from sucrose and corn syrup. Carbohydrate consumption in the United States has become excessive over the past century due to increased use of sugar in processed foods. A century ago, Americans consumed an average of 1.8 kg (4 pounds) of sugar per person per year. Now, with sucrose and high-fructose corn syrup so widely used in foods and beverages, the average American ingests over 28 kg (more than 60 pounds) of added cane sugar and 21 kg (46 pounds) of corn syrup per year. A single nondiet soft drink contains 38 to 43 g (about 8 teaspoons) of sugar per 355 mL (12-ounce) serving.

18.1d Fiber

Dietary fiber refers to all fibrous materials of plant and animal origin that resist digestion. Most is plant matter—the carbohydrates cellulose and pectin and such noncarbohydrates as gums and lignin. The recommended daily allowance is about 30 g of fiber, but average intake varies greatly from country to country—from 40 to 150 g/day in India and Africa to only 12 g/day in the United States.

Fiber is not considered a nutrient because it is never absorbed into the human tissues, but it is an essential component of the diet. **Water-soluble fiber** includes pectin and certain other carbohydrates found in oats, beans, peas, carrots, brown rice, and fruits. It reduces blood cholesterol levels.

Water-insoluble fiber includes cellulose and lignin. It apparently has no effect on cholesterol levels, but it absorbs water and swells, thereby softening the stool and increasing its bulk by 40% to 100%. This stretches the colon, stimulates peristalsis, and quickens the passage of feces. In doing so, water-insoluble fiber reduces the risk of constipation and diverticulitis. Recent studies indicate that dietary fiber, especially from whole grains, plays a role in prevention of colon cancer.

18.1e Lipids

Fats and other lipids serve as energy-storage molecules and building blocks for cell membranes and steroid hormones. Fat (triglyceride, see section 2.3b) is the main component of adipose tissue, whose various structural and metabolic functions are discussed in section 4.3. In addition to fat, the body contains significant amounts of other lipids such as phospholipids and cholesterol. These are major structural components of plasma membranes and myelin. Cholesterol is also a precursor of steroid hormones such as estrogen, testosterone, and cortisol, and of the bile acids that aid in fat digestion.

Fat and Energy Storage

Fat constitutes about 15% of the body mass of a healthy man in his early 20s, and about 25% of the body mass of a woman. Obesity, an excess of body fat, is discussed in Perspectives on Health at the end of this section. Fat is an efficient way to store energy. It contains over twice as much energy as carbohydrates (9 kcal/g of fat compared with 4 kcal/g of carbohydrate). Typical fat reserves in a healthy young man contain enough energy for 119 hours of running, whereas his carbohydrate stores would suffice for only 1.6 hours. Furthermore, fat is hydrophobic, contains almost no water, and is therefore a more compact energy-storage substance than carbohydrates. Carbohydrates are hydrophilic, absorb water, and thus expand and occupy more space in the tissues.

Dietary Fat and Cholesterol

Dietary fat may be saturated or unsaturated. Saturated fats are predominantly of animal origin. They occur in meat, egg yolks, and dairy products, but also in some plant products such as coconut and palm oils. Unsaturated fats come mostly from nuts, seeds, fish, and vegetable oils. Partially hydrogenated oils (so-called *trans* fats) are similar to naturally occurring unsaturated fats, but are manufactured by adding hydrogen to oils to make them solid. They were widely used in processed foods and home cooking beginning in 1911, but the U.S. Food and Drug Administration banned them in 2015 because of a demonstrated link to heart disease (see Clinical Application 2.2).

A certain amount of dietary fat is necessary for health. Whereas the body can synthesize most fatty acids, **essential fatty acids** are those we cannot manufacture and must obtain from food. While some fats promote health, fat should account for no more than 35% of your daily caloric intake, and no more than 10% of your fat intake should be saturated fat. Americans typically obtain 40% to 50% of calories from fat.

Cholesterol is necessary for cell membrane stability and steroid hormone production, yet excessive levels of cholesterol in the blood serum increase the risk of cardiovascular disease. It is desirable to maintain a serum cholesterol concentration of 200 mg/dL or less. This is determined, however, more by heredity and physiology than by diet. We synthesize 85% of our cholesterol internally, and when we reduce dietary cholesterol intake, the body simply synthesizes more to make up for it. Restricting dietary cholesterol alone lowers the serum cholesterol level by no more than 5%. Dietary cholesterol is no longer considered to be a nutritional concern for most people. More important is the fact that certain saturated fatty acids (SFAs) raise serum cholesterol level. Some food advertising is deceptive on this point. It may truthfully advertise a food as being cholesterol-free, but neglect to mention that it contains SFAs that can raise the consumer's cholesterol level anyway. A moderate reduction of saturated fatty acid intake can lower blood cholesterol by 15% to 20%, which is considerably more effective than reducing dietary cholesterol alone. The richest dietary source of cholesterol is egg yolks, but it is also prevalent in dairy products, shellfish, organ meats such as liver and brains, and other mammalian meat.

Blood Lipoproteins

Lipids must be transported to all cells of the body, yet they are hydrophobic and don't dissolve in the aqueous blood plasma. To overcome this, they are transported in the form of complexes called **lipoproteins**—tiny droplets with a core of cholesterol and triglycerides and a coating of proteins and phospholipids.

Lipoproteins are classified into categories by their density: **high-density lipoproteins (HDLs), low-density lipoproteins (LDLs),** and **very-low-density lipoproteins (VLDLs).** The higher the ratio of lipid to protein, the lower the density (easy to remember if you just bear in mind that oil and fat float in water because of their relatively low density). VLDLs, produced in the liver, transport triglycerides to adipose tissue for storage; when the triglycerides are removed, the VLDLs become LDLs, which contain mostly cholesterol (**fig. 18.1a**). Cells that need cholesterol, such as those that synthesize steroid hormones, absorb LDLs from the blood by receptor-mediated endocytosis.

High-density lipoproteins start as protein shells produced in the liver. They travel in the blood and pick up excess cholesterol and phospholipids from the tissues. The next time the HDL circulates through the liver, cholesterol is removed and eliminated in the bile (**fig. 18.1b**). Therefore, HDLs are a vehicle for removing excess cholesterol from the body. A high concentration of HDL protects against atherosclerosis, whereas a high LDL concentration is a warning sign because it signifies a high rate of cholesterol deposition in the arteries. Elevation of HDLs is promoted by a diet low in calories and saturated fats and by regular aerobic exercise. LDL levels rise in response to saturated fat consumption, cigarette smoking, and stress. In the popular press, HDLs and LDLs are often nicknamed "good cholesterol" and "bad cholesterol," respectively, but as we have seen, they are not cholesterol alone but complexes of proteins and diverse lipids.

18.1f Proteins

Proteins play an enormous diversity of roles in cellular processes and cell and tissue structure. They serve as signaling molecules (hormones and growth factors); protective antibodies; and membrane receptors, pumps, and ion channels. They function in oxygen transport (hemoglobin), muscle contraction (myosin and actin), and as enzymes that catalyze nearly every major reaction of cellular metabolism. Fibrous proteins such as collagen, elastin, and keratin make up much of the structure of skin, hair, nails, bones, cartilages, tendons, and ligaments. No other class of biomolecules has such a broad variety of functions.

Protein constitutes about 12% to 15% of the body's mass; about 65% of your protein is in your skeletal muscles. For persons of average weight, the RDA of protein is 46 to 56 g, depending on age and sex. Multiplying your weight in pounds by 0.37 or your weight in kilograms by 0.8 gives an estimate of your RDA of protein in grams. A higher intake is recommended, however, under conditions of tissue building or repair, such as stress, infection, injury, intense exercise, and pregnancy. Infants and children require more protein

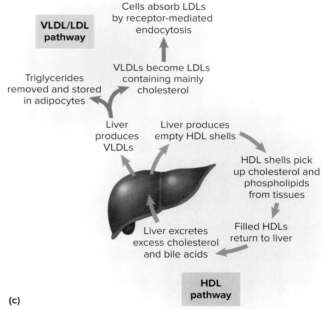

than adults relative to body weight because proteins are necessary for building tissues during growth.

Total protein intake is not the only measure of dietary adequacy. The nutritional value of a protein depends on whether it supplies the right amino acids in the proportions needed to make human proteins. Adults can synthesize 12 of the 20 amino acids from other organic compounds when they are not available from the diet, but there are 8 **essential amino acids** that we cannot synthesize. These must be obtained by including the right proteins in the diet.

Cells don't store surplus amino acids for later use. When a protein is to be synthesized, all of the amino acids necessary must be present at once; if even one is missing, the protein can't be made. High-quality **complete proteins** are those that provide all of the essential amino acids in the necessary proportions for human tissue growth and maintenance. Lower-quality **incomplete proteins** lack one or more essential amino acids. For example, grains are low in lysine, and legumes such as beans are low in methionine.

The animal proteins in meat, eggs, and dairy products closely match human proteins in amino acid composition. Thus, animal products provide high-quality complete protein, whereas plant proteins are incomplete. We typically use 70% to 90% of the animal protein we eat but only 40% to 70% of the plant protein. It therefore takes a larger serving of plant protein than animal protein to meet our needs—for example, we need 400 g (about 14 ounces) of rice and beans to provide as much usable protein as 115 g (about 4 ounces) of hamburger. However, this doesn't mean that your dietary protein must come from meat; indeed, about two-thirds of the world's population receives adequate protein from diets containing very little animal protein. We can combine plant foods so that one provides what another lacks—beans and rice, for example, are a complementary combination of legume and grain. Beans provide the isoleucine and lysine lacking in grains, whereas rice provides the tryptophan and cysteine lacking in beans. In contrast, soybeans contain all of the essential amino acids and provide complete protein.

Figure 18.1 Lipoproteins. (a) Structure of a lipoprotein. (b) Lipid and protein composition of three types of serum lipoproteins. (c) Pathways of lipoprotein processing.

? *Why is a high HDL:LDL ratio healthier than a high LDL:HDL ratio?*

Reducing meat intake while increasing plant intake has advantages. Among other considerations, plant foods provide more vitamins, minerals, and fiber, less saturated fat, and no cholesterol. Also important in our increasingly crowded world is the fact that producing meat requires far more land than producing food crops and destroys more natural habitat.

18.1g Minerals and Vitamins

Minerals and vitamins are required in small quantities but have very potent effects on physiology. Minerals are inorganic elements that plants extract from soil or water and introduce into the food web. Vitamins are small organic compounds that are not used as fuel, but serve predominantly as coenzymes or cofactors that enable enzymes to function. Deficiencies or excessive amounts lead to disease states and are potentially lethal. Table 18.2 summarizes requirements and dietary sources of some important minerals and vitamins.

Minerals constitute about 4% of the body mass, with three-quarters of this being the calcium and phosphorus in the bones and teeth. Phosphorus is also a key structural component of phospholipids and ATP. Calcium, iron, magnesium, and manganese function as cofactors for enzymes. Iron is essential to the oxygen-carrying capacity of hemoglobin. Many mineral salts function as electrolytes. For example, sodium chloride (table salt) is required for nerve and muscle function and maintenance of blood volume. The recommended sodium intake is 1.1 g/day, but a typical U.S. diet contains about 3 to 7 g/day. This is due not just to the use of table salt but more significantly to the large amounts of salt in processed foods, much of it "disguised" in soy sauce, MSG (monosodium glutamate), baking soda, and baking powder. Broadly speaking, the best sources of minerals are vegetables, legumes, milk, eggs, fish, shellfish, and some other meats. Cereal grains are a relatively poor source, but processed cereals may be mineral-fortified.

Vitamins were originally named with letters in the order of their discovery, but they also have chemically descriptive names such as ascorbic acid (vitamin C) and riboflavin (vitamin B_2). Most vitamins must be obtained from the diet (table 18.2), but the body synthesizes some of them from precursors called *provitamins*—for example, vitamin D from cholesterol and vitamin A from carotene, abundantly present in carrots, squash, and other yellow vegetables and fruits. Vitamin K and folic acid are produced by the bacteria of the large intestine, which supplement the amounts obtained from the tabulated dietary sources.

Vitamins are classified as water-soluble or fat-soluble. **Water-soluble vitamins** are absorbed with water from the small intestine, dissolve freely in the body fluids, and are quickly excreted by the kidneys. They cannot be stored in the body and therefore seldom accumulate to excess. The water-soluble vitamins are ascorbic acid (vitamin C) and the B vitamins. Ascorbic acid promotes hemoglobin synthesis, collagen synthesis, and sound connective tissue structure; it is an antioxidant that scavenges free radicals and possibly reduces the risk of cancer. The B vitamins function as **coenzymes**—small organic molecules that assist enzymes by transferring electrons from one metabolic reaction to another, making it possible for enzymes to catalyze these reactions. Some of their functions are discussed later in this chapter as we consider carbohydrate metabolism.

Fat-soluble vitamins are absorbed with dietary lipids. They are more varied in function than water-soluble vitamins. Vitamin A is a component of the visual pigments and promotes epithelial tissue maintenance. Vitamin D promotes calcium absorption and bone mineralization. Vitamin K is essential to blood clotting. Vitamins A and E are antioxidants.

It is common knowledge that various diseases result from vitamin deficiencies, but it is less well known that vitamin excess also causes disease. A *deficiency* of vitamin A, for example, can result in night blindness, dry skin and hair, a dry

Table 18.2 Mineral and Vitamin Requirements and Sources

Mineral	RDA (mg)	Some Dietary Sources*
Major Minerals		
Calcium	1,000	Milk, fish, shellfish, greens, tofu, orange juice
Phosphorus	700	Red meat, poultry, fish, eggs, milk, legumes, whole grains, nuts
Sodium	1,500	Table salt, processed foods; usually present in excess
Chloride	2,300	Table salt, some vegetables; usually present in excess
Magnesium	310–420	Milk, greens, whole grains, nuts, legumes, chocolate
Potassium	4,700	Red meat, poultry, fish, cereals, spinach, squash, bananas, apricots
Trace Minerals		
Zinc	8–11	Red meat, seafood, cereals, wheat germ, legumes, nuts, yeast
Iron	8–15	Red meat, liver, shellfish, eggs, dried fruits, legumes, nuts, molasses
Fluoride	3–4	Fluoridated water and toothpaste, tea, seafood, seaweed
Iodine	0.15	Marine fish, fish oils, shellfish, iodized salt

Vitamin	RDA (mg)	Some Dietary Sources
Water-Soluble Vitamins		
Ascorbic acid (C)	75–90	Citrus fruits, strawberries, tomatoes, greens, cabbage, cauliflower, broccoli, brussels sprouts
B complex		
Thiamin (B_1)	1.1–1.2	Red meat, liver, other organ meats, eggs, greens, asparagus, legumes, whole grains, seeds, yeast
Riboflavin (B_2)	1.1–1.3	Widely distributed, and deficiencies are rare; all types of meat, milk, eggs, greens, whole grains, apricots, legumes, mushrooms, yeast
Pyridoxine (B_6)	1.3	Red meat, poultry, bananas, potatoes, chickpeas, pistachios
Cobalamin (B_{12})	0.002	Fish, shellfish, liver, poultry, dairy products
Niacin (nicotinic acid, B_3)	14–16	Readily synthesized from tryptophan, which is present in any diet with adequate protein; red meat, liver, other organ meats, poultry, fish, apricots, legumes, whole grains, mushrooms
Panthothenic acid (B_5)	5	Widely distributed, and deficiencies are rare; red meat, liver and other organ meats, eggs, green and yellow vegetables, legumes, whole grains, mushrooms, yeast
Folic acid (folacin, B_9)	0.4	Eggs, liver, greens, citrus fruits, legumes, whole grains, seeds
Biotin (B_7)	0.03	Red meat, liver, other organ meats, eggs, cheese, cabbage, cauliflower, bananas, legumes, nuts
Fat-Soluble Vitamins		
Retinol (A)	0.7–0.9	Fish oils, eggs, cheese, milk, greens, other green and yellow vegetables and fruits, margarine
Calcitriol (D)	0.005	Formed by exposure of skin to sunlight; fish, fish oils, fortified milk
α-tocopherol (E)	15	Fish oils, greens, seeds, wheat germ, vegetable oils, margarine, nuts
Phylloquinone (K)	0.09–0.12	Most of the RDA is met by synthesis by intestinal bacteria; liver, greens, cabbage, cauliflower

*"Red meat" refers to mammalian muscle such as beef and pork. Liver is specified separately and refers to beef, pork, and chicken livers, which are similar for most nutrients. "Other organ meats" refers to brain, pancreas, heart, kidney, and so on.

conjunctiva and cloudy cornea, and increased incidence of urinary, digestive, and respiratory infections. This is the world's most common vitamin deficiency. An *excess* of vitamin A, however, may cause anorexia, nausea and vomiting, headache, pain and fragility of the bones, hair loss, an enlarged liver and spleen, and birth defects.

Some people take *megavitamins*—doses 10 to 1,000 times the RDA—thinking that they will improve their health or athletic performance. Vitamins aren't burned as fuel and do not supply energy to the body; small amounts fully meet the body's metabolic needs. There is no evidence that vitamin supplements improve performance except when used to correct a dietary deficiency. Megadoses of fat-soluble vitamins can be especially toxic.

PERSPECTIVES ON HEALTH

Obesity

Obesity is clinically defined as a weight more than 20% above the recommended norm for one's age, sex, and height. In the United States, over 70% of adults are overweight, and over 40% are obese. About 20% of youth aged 2 to 19 years are obese. *Body mass index (BMI)* is a useful tool for assessing obesity. $BMI = W/H^2$, where W is weight in kilograms and H is height in meters. (If using weight in pounds and height in inches, $BMI = 703 \, W/H^2$.) A BMI of 20 to 25 is considered to be optimal for most people. A BMI over 27 is defined as overweight, and above 30 is considered obese.

The obesity epidemic is a serious health concern. Excess fat increases a person's risk of atherosclerosis, hypertension, and joint degeneration; cancer of the breast, uterus, and liver in women; and cancer of the colon, rectum, and prostate gland in men. Excess thoracic fat impairs breathing and results in increased blood P_{CO_2}, sleepiness, and reduced vitality. Obesity is also a significant impediment to successful surgery.

The rise in obesity has paralleled an increased incidence of type 2 diabetes mellitus. High levels of abdominal fat are associated with insulin insensitivity and lead to high blood sugar and all of the resulting complications. Until recently, type 2 diabetes almost exclusively developed in people who were age 50 and older, but the incidence of type 2 diabetes is rising in young people as more children become obese.

Heredity plays a significant role in obesity. However, a predisposition to obesity is often greatly worsened by overfeeding in infancy and childhood. Consumption of excess calories in childhood causes adipocytes to increase in size and number. In adulthood, adipocytes don't multiply except in some cases of extreme weight gain; their number remains constant. Weight gains result from changes in cell size (cellular hypertrophy).

As so many dieters learn, it is very difficult to substantially reduce one's adult weight. Most diets are unsuccessful over the long run as dieters lose and regain weight over and over. From an evolutionary standpoint, this is not surprising. The body's appetite- and weight-regulating mechanisms are geared toward stimulation of appetite and eating rather than restraint. In our evolutionary past, scarcity of food was a more common problem than food surplus, and ability to store fat was a key survival adaptation. Recent changes in human diet and activity have been too rapid for the gene pool to respond, with the result that billions of people around the world are overfat. Former survival mechanisms have now become mechanisms of pathology.

Understandably, pharmaceutical companies are keenly interested in developing effective weight-control drugs. There could be an enormous profit, for example, in drugs that affect appetite-regulating hormones—for example, by inhibiting the appetite-stimulating hormone *ghrelin* or enhancing or mimicking the appetite-suppressing hormone *leptin* (see section 18.5a). Such efforts have so far met with little success. An integrative approach to obesity requires better understanding of appetite and body weight regulation, as well as emphasis on increased exercise and consumption of healthy foods.

Before You Go On

Answer these questions from memory to test your understanding of the previous section. Reread the material if you can't answer the questions.

1. What are the four classes of macronutrients?
2. Why is fiber important in the diet, and why isn't it considered a nutrient?
3. Why is fat more efficient than carbohydrates for energy storage?
4. Contrast the functions of VLDLs, LDLs, and HDLs. Explain why a high level of blood HDLs is desirable, but a high level of LDLs is undesirable.
5. Why do some proteins have more nutritional value than others?
6. Define the terms *mineral* and *vitamin*. Give several important examples of each.

18.2 Carbohydrate Metabolism

Expected Learning Outcomes

When you have completed this section, you should be able to

a. define *metabolism, catabolism,* and *anabolism;*
b. describe the basic purposes and products of glycolysis, anaerobic fermentation, and aerobic respiration;
c. be able to trace the energy content of glucose through all these stages of metabolism, stating where the energy is contained by the end of each stage; and
d. describe the body's modes of glycogen synthesis and breakdown, and its synthesis of glucose from other noncarbohydrates.

Metabolism is the sum of all chemical changes in the body, and consists of two classes of reactions: energy-requiring synthesis reactions called **anabolism,** and energy-releasing breakdown reactions called **catabolism.** In this section, we focus mainly on the energy-releasing catabolism of glucose and the use of that energy to make ATP.

18.2a Glucose Catabolism

Most dietary carbohydrate is burned as fuel within a few hours of absorption. Glucose is the major product of carbohydrate digestion, and provides us with quick energy when it is totally oxidized. The overall reaction for this is

$$C_6H_{12}O_6 + 6\ O_2 \rightarrow 6\ CO_2 + 6\ H_2O$$

The function of this reaction is not to produce carbon dioxide and water but to transfer energy from glucose to ATP.

If the preceding reaction were carried out in a single step, it would generate a short, intense burst of heat—like the burning of paper, which has the same chemical equation. Not only would this be useless to the body's metabolism, but it would kill the cells. In the body, however, the process is carried out in a series of small steps, each controlled by a separate enzyme. Energy is released in small manageable amounts, and as much of it as possible is transferred to ATP. The rest is released as heat.

Major pathways of glucose catabolism (fig. 18.2) follow:

1. **Glycolysis,** which splits a glucose molecule into two molecules of pyruvate, takes place in the cytoplasm.
2. **Anaerobic fermentation** processes pyruvate in a way that allows glycolysis to continue functioning in a state of oxygen deficiency.
3. **Aerobic respiration** occurs in the presence of oxygen and oxidizes pyruvate to carbon dioxide and water. The steps associated with aerobic respiration take place in the mitochondria.

Coenzymes are vitally important to these reactions. Enzymes remove electrons (as hydrogen atoms) from the intermediate compounds of these pathways and transfer them to coenzymes, which then donate the electrons to other compounds later in one of the reaction pathways. Coenzymes are the temporary carriers of the energy extracted from glucose metabolites. The enzymes of glucose catabolism cannot function without their coenzymes.

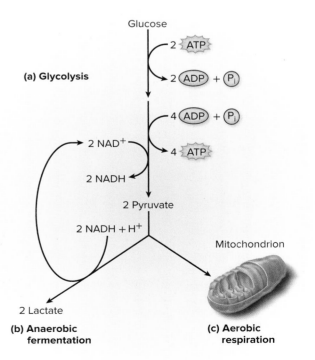

Figure 18.2 Pathways of Glucose Catabolism. (a) Glycolysis. (b) Anaerobic fermentation. (c) Aerobic respiration.

Two coenzymes of special importance are **NAD⁺** (nicotinamide adenine dinucleotide) and **FAD** (flavin adenine dinucleotide). Both are derived from B vitamins: NAD⁺ from niacin and FAD from riboflavin.

18.2b Glycolysis

Glycolysis[1] consists of a series of enzymatic steps that convert glucose, a six-carbon sugar (C_6), into 2 three-carbon (C_3) molecules of **pyruvate (pyruvic acid)** (fig. 18.2a). Glycolysis requires an "initial investment" of 2 ATP, but pays it back "with interest"—generating 4 ATP, for a net yield of 2 ATP. It also requires 2 molecules of the coenzyme NAD⁺. As glycolysis extracts energy from glucose, it transfers it to NAD⁺, creating a higher-energy form, NADH. At the end of glycolysis, some of the energy that was in the glucose has escaped as heat, but energy still useful to the body remains in the 2 ATP, 2 NADH, and mostly, the 2 pyruvate.

18.2c Anaerobic Fermentation

The fate of pyruvate depends on whether or not oxygen is available. In an exercising muscle, the demand for ATP may exceed the supply of oxygen. The only ATP the cells can make under these circumstances is the 2 ATP produced by glycolysis. Cells without mitochondria, such as erythrocytes, are also restricted to making ATP by this method.

In the absence of oxygen, a cell resorts to a one-step reaction called *anaerobic fermentation*. (This is often inaccurately called *anaerobic respiration*, but strictly speaking, that is a process found only in certain bacteria.) In this pathway, pyruvate is reduced to **lactate (lactic acid)** (fig. 18.2b) and generates NAD⁺. NAD⁺ is necessary to keep glycolysis running; this is the key purpose of anaerobic fermentation, not lactate production.

Most cells can survive brief periods of oxygen deficiency by generating ATP indirectly in this way. However, oxygen is required to produce enough ATP to sustain the body for more than a few minutes. Skeletal muscle is relatively tolerant of anaerobic fermentation, and cardiac muscle less so. The brain employs almost no anaerobic fermentation.

18.2d Aerobic Respiration

In the presence of oxygen, pyruvate enters the mitochondria and the pathway of *aerobic respiration* (fig. 18.2c). Most of our ATP is produced through oxidation of pyruvate in two principal steps:

- **matrix reactions,** so named because their controlling enzymes are in the fluid of the mitochondrial matrix, and
- **membrane reactions,** so named because their controlling enzymes are bound to the membranes of the mitochondrial cristae.

The Matrix Reactions

Most of the matrix reactions (fig. 18.3) constitute a series of enzymatic steps called the **citric acid (Krebs**[2]**) cycle.** For every glucose molecule that enters glycolysis, all of the matrix reactions occur twice (once for each pyruvate). As the products of pyruvate pass through the cycle, carbon atoms are stripped away as CO_2 and ultimately exhaled as metabolic waste. The energy extracted from pyruvate is transferred to ATP, more NADH, and the reduced (high-energy) form of the other coenzyme, $FADH_2$. At the end of the cycle, there is nothing left of the organic matter of the

---

[1] *glyco* = sugar; *lysis* = splitting
[2] Sir Hans Krebs (1900–1981), German biochemist

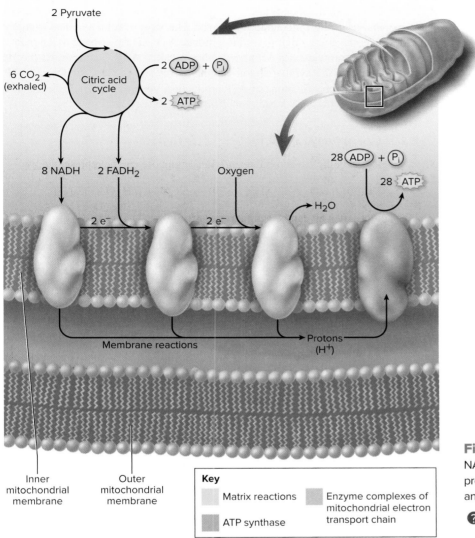

Figure 18.3 Aerobic Respiration. Of the 8 NADH generated by the matrix reactions, 2 are produced where pyruvate enters the citric acid cycle and 6 are generated by the citric acid cycle itself.

❓ *What two vitamins are needed in the diet to make this process work? Why?*

glucose; its carbon atoms have all been carried away. What remains of value at the end of the matrix reactions is the energy content of another 2 ATP, 8 NADH, and 2 FADH$_2$ (from two pyruvate molecules). These reduced coenzymes provide energy for the relatively massive production of ATP in the membrane reactions that follow.

The Membrane Reactions

The membrane reactions (fig. 18.3) further oxidize NADH and FADH$_2$ and transfer their energy to ATP. This is accomplished by passing their electrons along a series of *enzyme complexes* called the **mitochondrial electron-transport chain,** bound to the inner mitochondrial membrane. The members of the chain are arranged in a precise order that enables each one to receive a pair of electrons and pass them to the next—like a row of people passing along a hot potato. By the time the "potato" reaches the last member in the chain, it is relatively "cool"—its energy has been used in the process of making ATP. The final electron acceptor in the chain is oxygen. Each oxygen atom (O) receives two electrons (2 e⁻) from the transport chain and two protons (2 H$^+$) from the matrix reactions, generating water (H$_2$O) as an end product. The energy liberated from electron transfer is used to drive the enzyme complexes that remove H$^+$ from the matrix and pump it into the space between the inner and outer membranes. This creates a very high H$^+$ concentration and a positive charge between the membranes, compared with a low H$^+$ concentration and a negative charge in the mitochondrial matrix. The steep electrochemical gradient provides potential energy that is used to produce ATP. H$^+$ flows back to the matrix through channel proteins in the inner membrane called **ATP synthase.** As the H$^+$ flows through, the ATP synthase harnesses the energy and uses it to synthesize ATP. This is a relatively prolific mode of ATP production, generating up to 28 more ATP in contrast to the 2 from glycolysis and 2 from the matrix reactions.

18.2e Overview of ATP Production

In summary, the complete aerobic respiration of glucose can be represented by

$$C_6H_{12}O_6 + 6\ O_2 + 32\ ADP + 32\ P_i \rightarrow 6\ CO_2 + 6\ H_2O + 32\ ATP$$

and the sources of ATP can be summarized as follows:

$$\begin{array}{rl} 2 & \text{ATP from glycolysis (net yield),} \\ 2 & \text{ATP from matrix reactions,} \\ +\ 28 & \text{ATP from membrane reactions} \\ \hline 32 & \text{ATP per glucose} \end{array}$$

It is clear that oxygen is required for the most efficient ATP production, and that most of the ATP in our bodies is generated by the membrane reactions of the mitochondria.

> **Apply What You Know**
>
> Your friend who is also taking this course begins taking extra vitamins because she points to this discussion and says, "Look, they provide energy for making ATP. They'll give me extra energy; pep me up." But you point to the discussion of vitamins in the previous section and say, "Well look here; it says vitamins don't provide energy to the body." Now you're both confused. Is your book contradicting itself? Should you start taking the extra vitamins, or is your friend wasting money on a misconception? Explain.

18.2f Glycogen Metabolism

ATP is quickly used after it is synthesized—it is an *energy-transfer* molecule, not an *energy-storage* molecule. Therefore, if the body has an ample amount of ATP and there is still more glucose in the blood, it converts the glucose to other compounds better suited for energy storage—namely, glycogen and fat. Here, we consider the synthesis and use of **glycogen,** a large polymer of glucose stored in the liver, muscles, and a few other organs (see fig. 11.11). Fat is discussed later.

Glycogenesis, the synthesis of glycogen, is stimulated by insulin. It consists of polymerization reactions that link thousands of glucose molecules into branched chains of glycogen for storage. The average adult body contains about 400 to 450 g of glycogen: nearly one-quarter of it in the liver; three-quarters of it in the skeletal muscles; and small amounts in cardiac muscle, the uterine lining, and other tissues.

Glycogenolysis, the hydrolysis of glycogen, is stimulated by glucagon and epinephrine between meals. It releases glucose when new glucose is not being absorbed from the intestine. The liver releases glucose into the blood, maintaining blood sugar levels between meals. Muscle and other glycogen-storing tissues aren't as good at sharing as the liver—they use the free glucose only for their own needs.

Table 18.3 Terminology Related to Glucose and Glycogen Metabolism

Anabolic (Synthesis) Reactions	
Glycogenesis	The synthesis of glycogen by polymerizing glucose
Gluconeogenesis	The synthesis of glucose from noncarbohydrates such as glycerol and amino acids (from fat and protein)
Catabolic (Breakdown) Reactions	
Glycogenolysis	The hydrolysis of glycogen to release glucose
Glycolysis	The splitting of glucose into two molecules of pyruvate in preparation for anaerobic fermentation or aerobic respiration

Gluconeogenesis[3] occurs when even the stored glycogen isn't enough to meet our energy needs. This is the conversion of noncarbohydrates such as glycerol and amino acids to glucose. Since the glycerol comes from the breakdown of fat, gluconeogenesis is our means of weight loss when we control calorie intake to "burn off" fat. Using amino acids to make glucose, however, is less healthy; it means we're breaking down protein and may be losing muscle mass. This can be a sign of disease (as in cancer and uncontrolled diabetes), excessive dieting, or starvation.

The distinctions between the foregoing similar terms are summarized in table 18.3.

Before You Go On

Answer these questions from memory to test your understanding of the previous section. Reread the material if you can't answer the questions.

7. In the laboratory, glucose can be oxidized in a single step to CO_2 and H_2O. Why do living cells do it in so many little steps?
8. What are the end products of glycolysis? By the end of glycolysis, where is the energy that used to be in the glucose?
9. What is the main advantage of aerobic respiration over anaerobic fermentation?
10. What is the total yield of ATP for all steps from glucose through the end of aerobic respiration? State what quantities of ATP are produced at each major stage.
11. What conditions promote glycogenesis, glycogenolysis, and gluconeogenesis?

18.3 Lipid and Protein Metabolism

Expected Learning Outcomes

When you have completed this section, you should be able to

a. describe the process that uses the energy of triglycerides for synthesizing ATP, including the fates of the glycerol and fatty acid components;
b. describe the process of protein catabolism and why it generates nitrogenous wastes; and
c. identify several nondigestive functions of the liver.

---

[3]*gluco* = sugar, glucose; *neo* = new; *genesis* = production of

In the foregoing discussion, glycolysis and the mitochondrial reactions were treated from the standpoint of carbohydrate oxidation. When blood glucose concentration is low, most cells can break down other nutrient molecules to produce ATP. Here, we examine metabolic pathways associated with the oxidation of lipids and proteins as fuel.

18.3a Lipids

The total amount of triglycerides stored in the body's adipocytes usually remains quite constant, but there is continual turnover as lipids are released, transported in the blood, and either oxidized for energy or redeposited in other adipocytes. As we all know, a diet high in sugars causes us to gain weight, mostly in the form of fat. When sugars and amino acids exceed the body's immediate needs, they can be used to synthesize triglyceride precursors, glycerol and fatty acids. These can then be condensed to form a triglyceride, which is stored in the adipose tissue. Triglyceride synthesis is called **lipogenesis**.

In contrast, if glucose supplies are low, fat is used for fuel. Breaking down fat begins with **lipolysis** (lih-POLL-ih-sis)—splitting triglycerides into glycerol and fatty acids. Lipolysis is stimulated by epinephrine, norepinephrine, leptin, glucocorticoids, thyroid hormone, and growth hormone. The glycerol and fatty acids are further oxidized by separate pathways (fig. 18.4). Glycerol enters the pathway

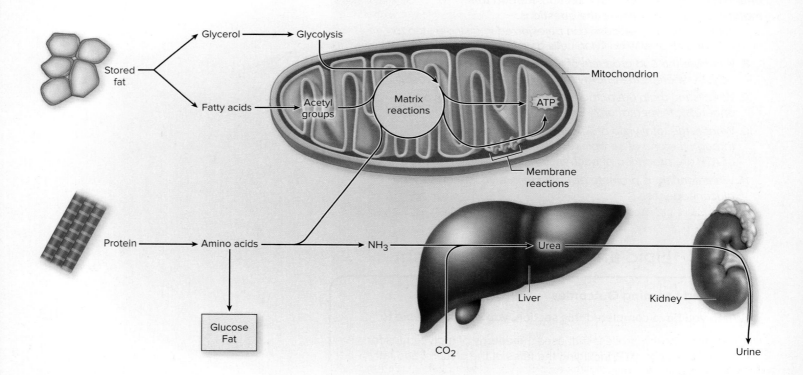

Figure 18.4 Fat and Protein Catabolism. Pathways for burning fat and protein as fuel to generate ATP.

of glycolysis. It generates only half as much ATP as glucose, however, because it is a C_3 compound (compared with glucose, C_6); thus, it leads to the production of only half as much pyruvate.

The fatty acids are oxidized in the mitochondria into 2-carbon fragments called *acetyl groups* and fed into the matrix reactions. A typical fatty acid of 16 carbon atoms can yield 129 molecules of ATP—obviously a much richer source of energy than a glucose molecule.

Excessively rapid fat breakdown, as in diabetes mellitus, can produce an overload of small organic acids called *ketone bodies*. Their accumulation can cause a pH imbalance called *ketoacidosis*. In untreated diabetes, cells cannot take up glucose and therefore oxidize fats. This leads to a telltale sweet odor on the breath from exhaled acetone (one of the ketone bodies); a diagnostic presence of ketone bodies in the urine; and, in extreme cases, a ketoacidosis-induced coma and death.

> **Apply What You Know**
> Explain why extreme low-carbohydrate diets can elevate blood ketone bodies and even cause ketoacidosis.

18.3b Proteins

Amino acids are primarily used for protein synthesis—the process of building the thousands of different proteins necessary for cellular function and structure. Each protein requires a specific combination of amino acids, assembled in proper order according to instructions contained in the DNA. Protein manufacture, described in section 3.3e, is stimulated by various hormones such as growth hormone and thyroid hormone. We need a constant supply of amino acids because proteins are continually degraded and rebuilt.

Amino acids can also be converted to glucose or fat or used directly as fuel. Although they are not the preferred fuel of the body, amino acids may be converted to pyruvate or one of the acids of the citric acid cycle (fig. 18.3). Before they enter the citric acid cycle, amino acids are **deaminated**—the amino group ($-NH_2$) is removed and becomes ammonia (NH_3), which is extremely toxic to cells. The liver quickly combines ammonia with carbon dioxide to produce a less toxic waste, urea, which is then excreted in the urine. Urea is the body's main nitrogenous waste. When a diseased liver cannot produce urea, ammonia accumulates in the blood and death from *hepatic coma* may ensue within a few days.

18.3c Liver Functions in Metabolism

The liver plays a wide variety of roles in the processes discussed in this chapter—especially carbohydrate, lipid, and protein metabolism. Although it is connected to the digestive tract and generally regarded as a digestive gland, the overwhelming majority of its functions are nondigestive (**table 18.4**). All of these are performed by the cuboidal hepatocytes described in section 17.4a. Such functional diversity is remarkable in light of the uniform structure of these cells. Because of the numerous critical functions performed by the liver, degenerative liver diseases such as hepatitis, cirrhosis, and liver cancer are especially life-threatening (Clinical Application 18.1).

Table 18.4 Functions of the Liver

Carbohydrate Metabolism

Converts dietary fructose and galactose to glucose. Stabilizes blood glucose concentration by storing excess glucose as glycogen (glycogenesis); releasing glucose from glycogen when needed (glycogenolysis); and synthesizing glucose from fats and amino acids (gluconeogenesis) when glucose demand exceeds glycogen reserves.

Lipid Metabolism

Carries out most of the body's fat synthesis and synthesizes cholesterol and phospholipids; produces VLDLs to transport lipids to adipose tissue and other tissues for storage or use; and stores fat in its own cells. Carries out most oxidation of fatty acids; produces ketone bodies. Produces HDL shells, which pick up excess cholesterol from other tissues and return it to the liver; excretes the excess cholesterol in bile.

Protein and Amino Acid Metabolism

Carries out most deamination of amino acids. Converts ammonia to urea. Synthesizes nonessential amino acids.

Synthesis of Plasma Proteins

Synthesizes nearly all the proteins of blood plasma, including albumin, globulins, fibrinogen, and several clotting factors.

Vitamin and Mineral Metabolism

Converts vitamin D_3 to calcidiol, a step in the synthesis of calcitriol; stores a 3- to 4-month supply of vitamin D. Stores a 10-month supply of vitamin A and enough vitamin B_{12} to last for several years. Stores and releases iron as needed. Excretes excess calcium by way of the bile.

Digestion

Synthesizes bile acids and lecithin, which emulsify fat and promote its digestion.

Disposal of Drugs, Toxins, and Hormones

Detoxifies alcohol, antibiotics, and many other drugs. Degrades hormones, removing them from circulation.

Clinical Application 18.1

HEPATITIS

Hepatitis, inflammation of the liver, is usually caused by one of three strains of the hepatitis virus. They differ in mode of transmission, severity of the resulting illness, affected age groups, and the best strategies for prevention. Hepatitis A is common and mild. Over 45% of people in urban areas of the United States have had it. It spreads rapidly in such settings as day-care centers, and it can be acquired by eating uncooked seafood such as oysters. Hepatitis B and C are far more serious. Both are transmitted sexually and through blood and other body fluids; the incidence of hepatitis C surpasses AIDS as a sexually transmitted disease. They may be passed among injecting drug users who share needles and may be transmitted by using nonsterile instruments during body piercings or tattoos. Initial signs and symptoms of hepatitis include fatigue, nausea, vomiting, and weight loss. The liver becomes enlarged and tender. Jaundice, or yellowing of the skin, tends to follow as hepatocytes are destroyed, bile passages are blocked, and bile pigments accumulate in the blood. Hepatitis A causes up to 6 months of illness, but most people recover and then have permanent immunity to it. Hepatitis B and C, however, often lead to chronic hepatitis, which can progress to cirrhosis or liver cancer. More liver transplants are necessitated by hepatitis C than by any other cause.

Before You Go On

Answer these questions from memory to test your understanding of the previous section. Reread the material if you can't answer the questions.

12. What two products result from the hydrolysis of triglycerides? How is each of them used to make ATP, and how does the ATP yield from this compare to the yield from glucose?
13. Why can an abnormally rapid rate of fat oxidation result in a pH imbalance in the body?
14. What must be done to an amino acid before it can enter the citric acid cycle? What happens to the by-product of this preparatory reaction?
15. List at least six nondigestive functions of the liver.

18.4 Metabolic States and Metabolic Rate

Expected Learning Outcomes

When you have completed this section, you should be able to

a. distinguish the absorptive (fed) state from the postabsorptive (fasting) state;
b. explain what happens to carbohydrates, fats, proteins, and amino acids in each of these states;
c. describe the hormonal and nervous regulation of each state;
d. define *metabolic rate* and *basal metabolic rate;* and
e. describe some factors that alter the metabolic rate.

Your metabolism changes from hour to hour depending on how long it has been since your last meal. For about 4 hours during and after a meal, you are in the **absorptive (fed) state.** This is a time in which nutrients are being absorbed and may be stored or used immediately to meet energy and other needs. The **postabsorptive (fasting) state** prevails in the late morning, late afternoon, and overnight. During this time, the stomach and small intestine are empty and the body's energy needs are met from stored fuels.

18.4a The Absorptive State

In the absorptive (fed) state, blood glucose is readily available for ATP synthesis. It serves as the primary fuel and spares the body from having to draw on stored fuels. The status of major nutrient classes during this phase is as follows:

- **Carbohydrates.** Absorbed sugars are transported by the hepatic portal system to the liver. Most glucose passes through the liver and becomes available to cells everywhere in the body. Glucose in excess of immediate need, however, is absorbed by the liver and may be converted to glycogen or fat. Most fat synthesized in the liver is released into the circulation.
- **Fats.** Unlike most nutrients, dietary fats enter into lymphatic capillaries (lacteals) rather than the blood capillaries (see fig. 17.15). Fats are not initially processed by the liver, but instead flow with the lymph into a large vein above the heart. The fats are distributed throughout the body in systemic blood

vessels and are taken up by tissues, especially adipose and muscular tissue, for which they are a major energy source.
- **Amino acids.** Amino acids, like sugars, circulate first to the liver. Most pass through and become available to other cells for protein synthesis. Some, however, are removed by the liver and are (1) used for protein synthesis, (2) deaminated and used as fuel for ATP synthesis, or (3) deaminated and used for fatty acid synthesis.

Regulation of the Absorptive State

The absorptive state is regulated largely by insulin, which is secreted by beta cells of the pancreas in response to elevated blood glucose and amino acid levels. Its release is also stimulated by the intestinal hormones gastrin, secretin, and cholecystokinin. Insulin, well known for lowering blood sugar, has more functions than one might suppose.

- Within minutes, it increases the cellular uptake of glucose by as much as 20-fold. As cells absorb glucose, the blood glucose concentration falls.
- It stimulates glucose oxidation, glycogenesis, and fat synthesis.
- It stimulates the active transport of amino acids into cells and promotes protein synthesis.

18.4b The Postabsorptive State

The essence of the postabsorptive (fasting) state is to prevent blood glucose concentration from falling below a range of about 90 to 100 mg/dL. This is especially critical to the brain, which normally doesn't use alternative energy substrates. The postabsorptive status of major nutrients is as follows:

- **Carbohydrates.** Glucose is drawn from the body's glycogen reserves (glycogenolysis) or synthesized from other compounds (gluconeogenesis). The liver usually stores enough glycogen after a meal to support 4 hours of postabsorptive metabolism before significant gluconeogenesis occurs.
- **Fats.** Adipocytes and hepatocytes hydrolyze fats and convert the glycerol to glucose. Free fatty acids (FFAs) can be used as a source of energy by most body cells.
- **Proteins.** If glycogen and fat reserves are depleted, the body begins to use proteins as fuel.

Regulation of the Postabsorptive State

Postabsorptive metabolism is more complex than the absorptive state. It is regulated mainly by the sympathetic nervous system and glucagon. As blood glucose levels drop, insulin secretion declines and pancreatic alpha cells secrete glucagon. Glucagon promotes glycogenolysis and gluconeogenesis, which raise the blood glucose level, and it promotes fat breakdown and a rise in FFA levels. Thus, it makes both glucose and lipids available for fuel.

The sympathetic nervous system also promotes glycogenolysis and fat breakdown, especially under conditions of injury, fear, anger, and other forms of stress. Adipose tissue is richly innervated by the sympathetic nervous system. Adipocytes, hepatocytes, and muscle cells also respond to epinephrine from the adrenal medulla. In circumstances with a threat of injury, the sympathetic system mobilizes stored energy reserves and makes them available to meet the demands of tissue repair. Stress also stimulates the release of cortisol, which promotes fat and protein catabolism and gluconeogenesis (see section 11.5).

18.4c Metabolic Rate

Metabolic rate refers to the amount of energy liberated in the body per unit of time, expressed in such terms as kcal/h or kcal/day. Metabolic rate depends on physical activity, mental state, absorptive or postabsorptive status, thyroid hormone and other hormones, and other factors. Your **basal metabolic rate (BMR)** is a baseline when you're awake but relaxed, in a postabsorptive state 12 to 14 hours since the last meal. It is not the minimum metabolic rate needed to sustain life, however. When you're asleep, your metabolic rate is slightly lower than the BMR. **Total metabolic rate** is the sum of BMR and energy expenditure for voluntary activities, especially muscular contractions.

The BMR of an average adult is about 2,000 kcal/day for a male and slightly less for a female. Roughly speaking, therefore, we must therefore consume at least 2,000 kcal/day to fuel our essential metabolic tasks—active transport, muscle tone, brain activity, cardiac and respiratory rhythms, renal function, and other essential processes. Even a relatively sedentary lifestyle requires another 500 kcal/day to support a low level of physical activity, and someone who does hard physical labor may require as much as 5,000 kcal/day.

Aside from physical activity, some factors that raise the total metabolic rate (TMR) and caloric requirements include pregnancy, anxiety (which stimulates epinephrine release and muscle tension), fever (TMR rises about 14% for each 1°C of body temperature), and thyroid hormone. TMR is relatively high in children and declines with age. Therefore, as we reach middle age we often find ourselves gaining weight with no apparent change in food intake.

Some factors that lower TMR include apathy, depression, and prolonged starvation. In weight-loss diets, loss is often rapid at first and then goes more slowly. This is partly because the initial loss is largely water weight and partly because the TMR drops over time, fewer dietary calories are "burned off," and there is more fat synthesis even with the same caloric intake. As we reduce food intake, the body reduces its metabolic rate to conserve body mass—thus making weight loss all the more difficult.

Before You Go On

Answer these questions from memory to test your understanding of the previous section. Reread the material if you can't answer the questions.

16. Define absorptive state *and* postabsorptive state. *In which state is the body storing excess fuel? In which state is it drawing from these stored fuel reserves?*
17. *What hormone primarily regulates the absorptive state, and what are the major effects of this hormone?*
18. *What is the main priority of postabsorptive metabolism? What are the principal nervous and hormonal means of regulating it?*
19. Define basal metabolic rate *and explain how it differs from total metabolic rate. What factors raise the total metabolic rate?*
20. *List a variety of factors and conditions that raise your total metabolic rate above your basal metabolic rate.*

18.5 Appetite Regulation

Expected Learning Outcomes

When you have completed this section, you should be able to

a. explain the concept of energy balance;
b. name the hormones that regulate short- and long-term appetite and describe the effects of each; and
c. name the appetite-stimulating and -inhibiting signals secreted in the hypothalamus, and describe their effects.

The subjects of nutrition and metabolism quickly bring to mind the subject of body weight and the popular desire to control it. Weight is determined by the body's **energy balance**—if energy intake and output are equal, body weight is stable. We gain weight if energy intake exceeds output and lose weight if output exceeds intake.

Despite variation in daily food intake and energy expenditure, energy balance is closely regulated, and most people maintain a fairly constant body weight. However, humans, unlike other animals, often ignore feelings of satiety to eat more than is comfortable. In modern times, access to abundant, appealing, high-calorie food is coupled with a relatively inactive lifestyle. Consequently, many people have excess amounts of body fat and obesity has become common. Evidence from twin and adoption studies and family pedigrees indicates that about 30% to 50% of the variation in human weight is due to heredity; the rest is due to factors such as levels of physical activity and eating habits.

The struggle for weight control often seems to be a struggle against the appetite. Since the early 1990s, physiologists have discovered several hormones called **gut–brain peptides** that act as two-way chemical signals between the gastrointestinal tract and brain and play roles in controlling short- and long-term appetite and body weight. The following three peptides work over the short term—minutes to hours—first making one feel hungry and motivated to eat, then causing a feeling of satiety that signals an end to the meal.

- **Ghrelin**[4] is secreted by parietal cells in the stomach. It has been called the "hunger hormone" because it is a powerful appetite stimulator. Shortly after eating, ghrelin concentration in the blood drops.
- **Peptide YY (PYY)** is secreted by cells in the ileum and colon that sense that food has arrived even as it enters the stomach. The primary effect of PYY is to signal satiety and terminate eating. Thus, ghrelin is one of the signals that begins a meal, and PYY is one of the signals that ends it. PYY remains elevated well after a meal. It acts as a kind of brake that prevents the stomach from emptying too quickly; thus, it prolongs the sense of satiety.
- **Cholecystokinin (CCK)** is secreted by cells in the duodenum and jejunum. It stimulates the brain and produces an appetite-suppressing effect. Thus, it joins PYY as a signal to stop eating. (Earlier, we saw

---

[4]named partly from *ghre* = growth, and partly as an acronym derived from *growth hormone releasing*

that CCK also regulates activity of the gallbladder and pancreas; see section 17.4.)

Other peptides regulate appetite, metabolic rate, and body weight over the longer term, thus governing one's average rate of caloric intake and energy expenditure over periods of weeks to years. The following two members of this group work as "adiposity signals," informing the brain about how much fat we have and activating mechanisms for adding or reducing fat. Adipose tissue is increasingly seen as an important source of multiple hormones that influence the body's energy balance.

- **Leptin**[5] is secreted mainly by adipocytes throughout the body. Its level is proportional to fat stores, so this is the brain's primary way of knowing how much body fat we have. Leptin *inhibits* fat deposition and stimulates lypolysis (fat breakdown). Animals with a leptin deficiency or a defect in leptin receptors overeat and become extremely obese. With few exceptions, however, obese humans are not leptin-deficient and are not aided by leptin injections. More commonly, human obesity is linked to unresponsiveness to leptin—a receptor defect rather than a hormone deficiency.
- **Insulin** is secreted by pancreatic beta cells in response to eating. As we have seen, insulin lowers glucose levels in the blood by causing it to be transported into cells. In addition, like leptin, it signals the status of the body's fat stores, but in contrast to leptin, insulin *promotes* fat deposition. When average daily insulin secretion rises, food intake and body weight increase.

An important brain center for appetite regulation is the hypothalamus. All five of the aforementioned peptides have receptors in the hypothalamus, which in turn, secretes signals according to information it receives (fig. 18.5). One of these is **neuropeptide Y** (NPY), a potent appetite stimulant. Ghrelin stimulates NPY secretion, whereas insulin, PYY, and leptin inhibit it. The hypothalamus also secretes **melanocortin,** which inhibits eating. Leptin stimulates melanocortin secretion and inhibits the secretion of appetite stimulants called *endocannabinoids,* named for their resemblance to the tetrahydrocannabinol (THC) of marijuana.

Hunger is stimulated not only by chemical signals but also by movements of the stomach and intestinal walls. Mild **hunger contractions** begin soon after the stomach is emptied and increase in intensity over a period of hours. They can become quite a painful and powerful incentive to eat, yet they don't affect the amount of food consumed. Food intake is terminated by more than hormonal signals. Merely chewing and swallowing food briefly satisfy the appetite. Inflating the stomach with a balloon inhibits hunger even in an animal that hasn't actually swallowed any food. Satiety produced by these mechanisms, however, is short-lived. Lasting satiety depends on the absorption of nutrients into the blood.

Appetite is a question of not merely *how much* but also *what kind* of food is consumed. Humans are adapted to eat a wide variety of foods including plants and meat. Some foods provide nutrients that others do not. Different neurotransmitters seem to govern the appetite for different classes of nutrients. For example, *norepinephrine* stimulates the appetite for carbohydrates, *galanin* for fatty foods, and *endorphins* for protein.

---

[5]*lept* = thin

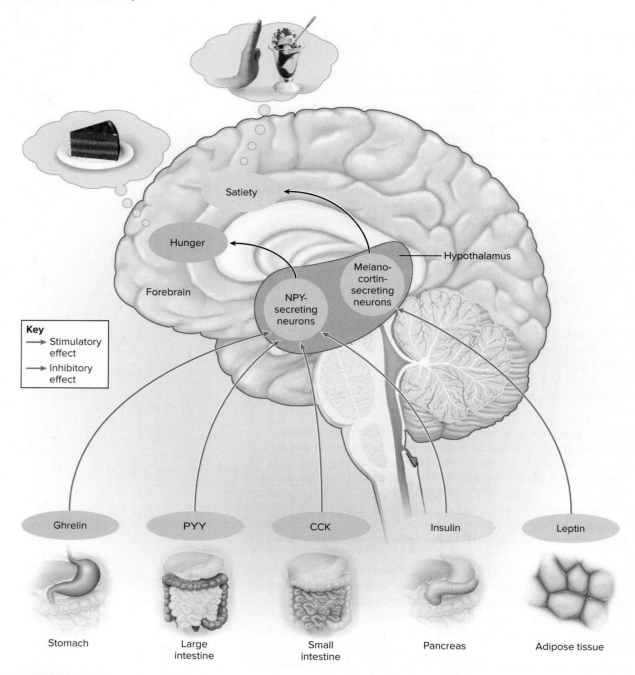

Figure 18.5 Gut–Brain Peptides in Appetite Regulation. Peptides produced by the digestive tract, pancreas, and adipose tissue stimulate or inhibit the brain and regulate appetite.

Before You Go On

Answer these questions from memory to test your understanding of the previous section. Reread the material if you can't answer the questions.

21. How does the effect of ghrelin differ from the effects of peptide YY and cholecystokinin? Where is each of these hormones secreted?
22. Contrast the effects of leptin and insulin on long-term energy balance.
23. Where are neuropeptide Y and melanocortin secreted? What are their effects?
24. Identify some neurotransmitters that stimulate the appetite for specific categories of nutrients.

18.6 Body Heat and Thermoregulation

Expected Learning Outcomes

When you have completed this section, you should be able to

a. state the normal ranges of human core and shell temperature;
b. identify the principal sources of body heat;
c. define and contrast three modes of heat loss and their relative contributions to total heat loss;
d. describe how the hypothalamus monitors body temperature and how the heat-losing and heat-promoting centers work; and
e. name and describe the pathological states of excessively high and low body temperatures.

The enzymes that control our metabolism depend on an optimal, stable temperature. In order to maintain this, heat loss from the body must be matched by heat generation. **Thermoregulation,** the balance between heat production and loss, is therefore a critically important aspect of homeostasis.

18.6a Body Temperature

Body temperature fluctuates about 1°C (1.8°F) in a 24-hour cycle. It tends to be lowest in the early morning and highest in the late afternoon. Temperature also varies from one place in the body to another, and from person to person. Thus, it is hard to specify any body temperature as "normal."

The most important measure is the **core temperature**—the temperature of organs in the cranial, thoracic, and abdominal cavities. It can be closely estimated from rectal temperature, which is relatively easy to measure. It is typically 37.2° to 37.6°C (99.0°–99.7°F) but may be as high as 39°C (102.2°F) during vigorous exercise such as running for prolonged periods.

Shell temperature is the cooler temperature closer to the surface, especially skin and oral temperature. Adult oral temperature is typically 36.6° to 37.0°C (97.9°–98.6°F) but may be as high as 40°C (104°F) during hard exercise. Shell temperature fluctuates as a result of processes that serve to maintain a stable core temperature.

18.6b Heat Production and Loss

Most body heat comes from metabolic reactions related to nutrient oxidation and ATP use. At rest, most heat is generated by the brain, heart, liver, and endocrine glands. Skeletal muscles contribute about 20% to 30% of the total resting heat. However, during vigorous exercise, they produce 30 to 40 times as much heat as the rest of the body.

The body loses heat in four ways: radiation, conduction, convection, and evaporation.

1. **Radiation** is the loss of thermal energy through infrared (IR) rays emitted by all moving molecules. We absorb heat from every object around us, but it is especially noticeable coming from the sun or a heat lamp. Since we are usually warmer than the objects around us, however, we usually lose more heat this way than we gain.
2. **Conduction** is the transfer of kinetic energy from molecule to molecule as they collide with one another. Heat from your body warms the

clothes you wear, the chair you sit in, and the air around you through conduction.

3. **Convection** is the movement of warm air away from the body and its replacement by a current of cooler air (or water, if one is immersed in a cool lake or bath). It accelerates the process of conductive heat loss. Convection can occur spontaneously as the lower density of warm air or water makes it rise, with cooler air or water coming from below to replace it, but it can also be accelerated by a breeze or a fan forcing air movement.

4. **Evaporation** underlies the cooling effect of perspiration. It requires an appreciable input of heat to overcome the tendency of water molecules to cling together by hydrogen bonding, so when water molecules do break free and evaporate, they carry a substantial amount of heat with them. In extreme conditions, the body can lose 2 L or more of sweat and, with it, up to 600 kcal of heat per hour. Air movement (convection) also speeds evaporative heat loss, as we feel most dramatically when we're wet with sweat and a cooling breeze picks up or we stand in front of a fan.

The relative amounts of heat lost by the different methods depend on prevailing conditions. A nude body in still air at 21°C (70°F) loses about 60% of its heat by radiation, 18% by conduction, and 22% by evaporation. If air temperature is higher than skin temperature, evaporation becomes the only means of heat loss because radiation and conduction add more heat to the body than they remove. Hot, humid weather hinders even evaporative cooling because there is less of a humidity gradient from skin to air.

18.6c Thermoregulation

Thermoregulation is achieved through several negative feedback loops. The hypothalamus monitors the temperature of the blood and receives signals from nerve endings *(thermoreceptors)* in the skin. In turn, its heat-sensing neurons send signals to hypothalamic nuclei called the *heat-losing* and *heat-promoting centers*.

When the body is overheated, as from exercise or a hot environment, the **heat-losing center** stimulates the dilation of cutaneous blood vessels. This increases blood flow through the skin and thus promotes heat loss to the surrounding air (**fig. 18.6a**). If this fails to restore normal temperature, the heat-losing center triggers sweating.

In a cold environment, the **heat-promoting center** of the hypothalamus activates mechanisms to conserve body heat or generate more. By way of the sympathetic nervous system, it causes cutaneous vasoconstriction. This retains warm blood deeper in the body and less heat is lost through the skin (**fig. 18.6b**). If this fails to restore or maintain normal core temperature, the body resorts to shivering. On a cold day, you may notice that your muscles become tense and you begin to shiver. Muscle contraction causes heat to be released when ATP is produced, and shivering can increase the body's heat production as much as fourfold.

A more long-term mechanism for generating heat is to raise metabolic rate. After several weeks of cold weather, for example, metabolic rate increases as much as 30%. Activation of the sympathetic nervous system raises the metabolic rate, with the result that more nutrients are burned as fuel. We consume more calories to "stoke the furnace" and, consequently, have greater appetites in the winter than in the summer. Infants generate heat by breaking down *brown fat,* a tissue in which lipolysis is not linked to ATP synthesis (see

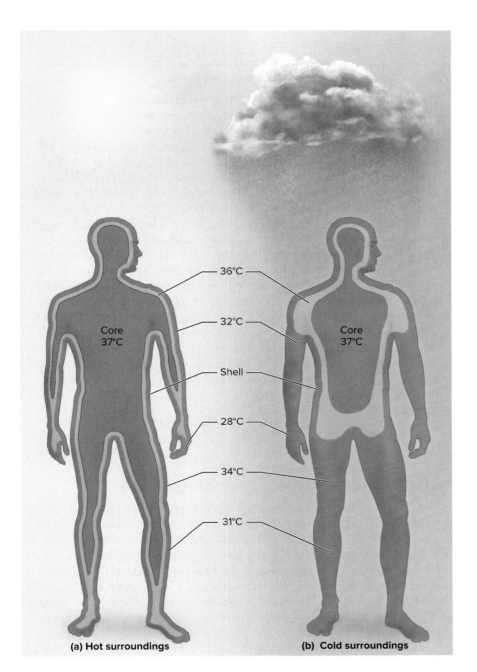

Figure 18.6 Distribution of Body Heat in Hot and Cold Surroundings. (a) In a hot environment, cutaneous vasodilation carries heat close to the body surface and the shell is relatively thin. (b) In a cold environment, cutaneous vasoconstriction retains heat deeper in the body and the insulating shell is significantly thicker. It is easy to see from this why frostbite so often affects extremities such as fingers and toes.

"Adipose Tissue" in section 4.3), so all the energy released from the fat is in the form of heat.

In addition to these physiological mechanisms, humans and other animals practice **behavioral thermoregulation.** For example, just getting out of the sun greatly cuts down heat gain by radiation; shedding heavy clothing helps to cool the body; or putting on a sweater or adding another blanket to the bed conserves heat.

In summary, you can see that thermoregulation is a function of multiple organs: the brain, autonomic nerves, skin, blood vessels, and skeletal muscles.

Clinical Application 18.2

HYPOTHERMIA AND FROSTBITE

Hypothermia occurs when the body can't generate heat as quickly as it is lost to the environment. All of us have experienced a drop in body temperature in response to cold weather; a decrease in body temperature of 1° to 2°C (1.8° to 3.6°F) leads to shivering, goose bumps, and numb hands. Exposure to extreme cold, however, can lead to hypothermia, when body temperature drops to a level where the hypothalamus can no longer produce an adequate response, and cellular metabolic processes slow to the point of disrupting normal body function. Mountain climbers exposed to low temperatures and windy conditions must be careful to notice the symptoms of hypothermia in themselves and their companions—poor muscle coordination, difficulty speaking, blue fingertips and lips, and impaired cognitive ability. Drinking warm liquids, getting to shelter, and warming with blankets or by sharing body heat are essential to prevent a further slide toward nervous system depression and possible death. Another concern is frostbite, which results from exposure of body surfaces to temperatures low enough to freeze tissues. It occurs most often in regions of high surface area and exposure—fingers and toes, earlobes, and the tip of the nose. If intracellular water freezes, the ice crystals expand and rupture cells, and tissues die. In severe cases, gangrene follows thawing and the affected areas must be amputated.

18.6d Disturbances of Thermoregulation

Either excessively low body temperature, called **hypothermia,** or excessively high body temperature, called **hyperthermia,** can disrupt metabolism to the point of death. Hypothermia can result from exposure to cold weather or immersion in icy water (see Clinical Application 18.2). If the core temperature falls below 33°C (91°F), the metabolic rate drops so low that heat production can't keep pace with heat loss, and the temperature falls even more. Death from cardiac fibrillation may occur below 32°C (90°F), but some people survive body temperatures as low as 29°C (84°F) in a state of suspended animation. A body temperature below 24°C (75°F) is usually fatal.

Hyperthermia is characterized by heat cramps, heat exhaustion, and heatstroke. **Heat cramps** are painful muscle spasms that result from excessive electrolyte loss in the sweat. They occur especially when a person begins to relax after strenuous exertion and heavy sweating. **Heat exhaustion** results from more severe water and electrolyte loss and is characterized by low blood pressure, dizziness, vomiting, and sometimes fainting. **Heatstroke (sunstroke)** occurs when the core body temperature is over 40°C (104°F); the skin is hot and dry; and the subject exhibits nervous system dysfunctions such as delirium, convulsions, or coma. It is also accompanied by tachycardia, hyperventilation, inflammation, and multiorgan dysfunction; it is often fatal.

Section 14.2d describes the mechanism of fever and its importance in combating infection. Fever is a normal protective mechanism that should be allowed to run its course if temperature isn't extraordinarily high. A body temperature above 42° to 43°C (108°–110°F), however, is dangerous. At a core temperature of 44° to 45°C (111°–113°F), metabolic dysfunction and neurological damage can be fatal.

Before You Go On

Answer these questions from memory to test your understanding of the previous section. Reread the material if you can't answer the questions.

25. Define *core* and *shell* temperature. Why can't we define shell temperature in terms of a fixed anatomical boundary or distance from the body surface?
26. What is the primary source of body heat? What are some lesser sources? How does the proportion of heat produced by skeletal muscles change during exercise compared to rest?
27. Distinguish between radiation, conduction, convection, and evaporation as modes of heat loss. Consider a person outdoors on a day when the air temperature is 38°C (101°F); what would be the person's sole mode of body heat loss?
28. Explain how the hypothalamus acts to return body temperature to normal if it senses that it is too warm or too cool.
29. Describe the effects of hyperthermia and hypothermia.
30. Distinguish between heat cramps, heat exhaustion, and heatstroke.

CAREER SPOTLIGHT

Dietitian

A registered dietitian (RD) is a food and nutrition expert. To become an RD requires earning a bachelor's degree, completing an internship approved by the American Dietetic Association, and passing a national examination. Registered dietitians focus on prevention and treatment of disease using a nutritional approach. They work in a variety of settings, often as members of a medical team in hospitals, clinics, or other care facilities such as nursing homes. They offer nutritional counseling in schools, child-care centers, and correctional facilities. They may also provide nutritional support to both professional and recreational athletes in sports facilities. Dietitians serve in the food and nutrition industry and are also active in nutrition research.

A dietetic technician, registered (DTR), is a nutrition practitioner who has completed an associate degree at an accredited institution, an approved internship, and a national DTR examination. The majority of DTRs work with RDs as nutrition counselors in hospitals, clinics, and other health-care facilities; public health settings such as schools and child-care centers; and weight-management centers. The obesity epidemic and studies that link chronic diseases to diet have focused our attention on nutrition, and have highlighted the key role that these professions play in disease prevention and management. See appendix B for further information on careers in dietetics.

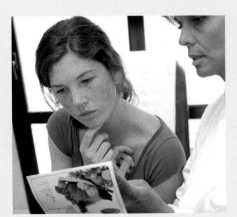
© Burger/Phanie/Science Source

Study Guide

Assess Your Learning Outcomes

To test your knowledge, discuss the following topics with a study partner or in writing, ideally from memory.

18.1 Nutrition

1. Definitions of *calorie* (with a lowercase *c*) and *kilocalorie* (kcal)
2. The definition of *nutrient*, and why some substances are important in the diet yet cannot be considered nutrients
3. The energy content of carbohydrates and proteins compared with that of fat
4. The meanings of *macronutrients* and *micronutrients*, with examples of each
5. Why certain fatty acids and amino acids are called "essential"
6. The chief dietary mono-, di-, and polysaccharides and some sources of each
7. Examples of dietary fiber and why fiber is important
8. Reasons why fat is superior to carbohydrates for long-term energy storage
9. Differences between saturated and unsaturated fats and some dietary sources of each
10. Some foods high in cholesterol, and why cholesterol is important for normal body function
11. The relative effect of dietary cholesterol versus other factors that influence one's blood cholesterol level
12. How blood lipoproteins enable lipids to be transported in spite of their hydrophobic nature
13. The three kinds of blood lipoproteins and the differences between their composition and function
14. The true meanings of so-called *good cholesterol* and *bad cholesterol*, and why these blood lipoproteins have acquired those colloquial expressions
15. Several functions of dietary proteins
16. How proteins of plant and animal origin differ in nutritional value, and how one can obtain all necessary amino acids from a diet free of animal protein
17. The definition of *essential amino acid* and how essential amino acids are obtained
18. The best dietary sources of minerals; important minerals in the body and their general functions
19. The general functions of vitamins and how they differ from other small organic compounds
20. The difference between water-soluble and fat-soluble vitamins; examples of each
21. The hazards of excessive vitamin intake

18.2 Carbohydrate Metabolism

1. The definition of *metabolism* and its two subdivisions
2. The equation for the complete oxidation of glucose
3. Why the human body can't simply break down glucose in one reaction step
4. The three major pathways of glucose catabolism
5. The two coenzymes involved in glucose oxidation and the general role that they have in common
6. The end products of glycolysis
7. The purpose of anaerobic fermentation, and its main reactant and end product
8. Where aerobic respiration occurs in the cell; differences between the matrix and membrane reactions; and the relative ATP yield of each
9. Distinctions between glycolysis, glycogenesis, glycogenolysis, and gluconeogenesis, including the purpose, input, and output (products) of each

18.3 Lipid and Protein Metabolism

1. The breakdown products of triglyceride hydrolysis and how each of these can be used to make ATP
2. The relative ATP yield of fat versus glucose oxidation
3. The cause and pathological effects of ketoacidosis
4. The process of catabolizing proteins for fuel and how it gives rise to ammonia and urea
5. Five (or more) nondigestive functions of the liver

18.4 Metabolic States and Metabolic Rate

1. The distinction between the absorptive and postabsorptive metabolic states
2. How carbohydrates, fats, and proteins (or amino acids) are processed during each of those states
3. The hormonal and nervous mechanisms for regulating these two states
4. Definitions of *basal* and *total metabolic rate*, and some typical values of each
5. Factors that raise and lower metabolic rate

18.5 Appetite Regulation

1. The concept of energy balance
2. The roles of ghrelin, peptide YY, and cholecystokinin as short-term regulators of appetite
3. The two long-term adiposity signals to the brain and their effects on body fat stores
4. The role of the hypothalamus in appetite regulation; two peptides produced by the hypothalamus that stimulate or inhibit appetite

18.6 Body Heat and Thermoregulation

1. The difference between core temperature and shell temperature, and where one would place a thermometer to measure each
2. The source of most body heat when one is at rest
3. The four ways in which the body loses heat
4. How the hypothalamus monitors and regulates body temperature
5. How the body core and shell change in hot and cold conditions
6. The terms for abnormally high and low body temperatures and causes and effects of each

Testing Your Recall

1. _____ are not used as fuel and are required in relatively small quantities.
   a. Micronutrients
   b. Macronutrients
   c. Essential nutrients
   d. Proteins
   e. Lipids

2. The only significant digestible polysaccharide in the diet is
   a. glycogen.
   b. cellulose.
   c. starch.
   d. maltose.
   e. fiber.

3. Which of the following store(s) the greatest amount of energy for the smallest amount of space in the body?
   a. glucose
   b. triglycerides
   c. glycogen
   d. proteins
   e. vitamins

4. A/an _____ is a nutrient that cannot be synthesized by the body and therefore must come from the diet.
   a. glucose molecule
   b. saturated fat
   c. high-density lipoprotein
   d. extrinsic fatty acid
   e. essential fatty acid

5. Which of the following is most likely to make you hungry?
   a. leptin
   b. ghrelin
   c. cholecystokinin
   d. peptide YY
   e. melanocortin

6. Which of the following normally releases glucose from the liver between meals?
   a. insulin
   b. glycogenesis
   c. glycolysis
   d. glycogenolysis
   e. gluconeogenesis

7. What function is served by oxygen in the process of aerobic respiration?
   a. It provides the energy necessary to get glycolysis started.
   b. It oxidizes glucose to pyruvate.
   c. It converts pyruvate to lactate.
   d. It transfers high-energy electrons between mitochondrial enzymes.
   e. It serves as the final electron acceptor in the electron-transport chain.

8. Glycolysis has a net yield of _____ molecules of ATP for each molecule of glucose.
   a. 2
   b. 4
   c. 28
   d. 32
   e. 38

9. Which of these occurs in the mitochondrial matrix?
   a. glycolysis
   b. glycogenesis
   c. the electron-transport chain
   d. the citric acid cycle
   e. anaerobic fermentation

10. Absorbing heat from the sun is an example of
    a. conduction.
    b. convection.
    c. hyperthermia.
    d. radiation.
    e. evaporation.

11. A/an _____ protein lacks one or more essential amino acids.

12. Lipoproteins called _____ are sometimes nicknamed "good cholesterol," whereas others called _____ are characterized as "bad cholesterol" because they are associated with atherosclerosis.

13. Synthesis of glucose from amino acids or triglycerides is called _____.

14. The major nitrogenous waste resulting from protein catabolism is _____.

15. The organ that synthesizes the nitrogenous waste in question 14 is the _____.

16. The hormone _____ is the main regulator of the absorptive state.

17. The temperature of organs in the body cavities is called _____.

18. The appetite hormones leptin and ghrelin act on the _____, an important brain center for appetite regulation.

19. In the process of anaerobic fermentation, pyruvate is reduced to _____.

20. A biochemist or physiologist would generally use the term _____ and the abbreviation _____ for what a dietitian would call one Calorie or 1,000 calories.

Answers in Appendix A

What's Wrong with These Statements?

Briefly explain why each of the following statements is false, or reword it to make it true.

1. Ghrelin and leptin are two hormones that stimulate the appetite.

2. Water is not considered a nutrient because we don't digest it and it contains no calories.

3. A gram of sugar contains more than twice as much energy as a gram of fat.

4. Most of the body's cholesterol comes from the diet.

5. Most of the liver's functions are digestive.

6. Even under anaerobic conditions, the mitochondria can produce a little ATP.

7. Reactions occurring in the mitochondrial matrix produce more ATP than glycolysis and the mitochondrial membrane reactions combined.

8. Gluconeogenesis occurs especially in the absorptive state during and shortly after a meal.

9. Dangerously high levels of blood ketone bodies can result from excessively low-fat diets.

10. At a comfortable air temperature, the body loses more heat by conduction than by any other means.

Answers in Appendix A

Testing Your Comprehension

1. Cyanide blocks the transfer of electrons from the final cytochrome to oxygen in the electron-transport chain. In light of this, explain why it causes sudden death. Also explain whether cyanide poisoning could be treated by giving a patient supplemental oxygen, and justify your answer.

2. People with type 1 diabetes mellitus inject insulin to help lower blood sugar. Explain why they may sometimes become hypoglycemic (experience low blood sugar). What hormone would be secreted to counteract the effects of hypoglycemia?

3. Your friend is trying to reduce his cholesterol and is discouraged. He says to you, "I only eat foods that say they are cholesterol-free on the labels! How could I still have high cholesterol?" Explain to your friend why he still may have high cholesterol even if he eats foods that are advertised as cholesterol-free.

PART 6 Human Life Cycle

The Reproductive System

Chapter 19

Chapter Outline

19.1 Essentials of Sexual Reproduction
19.2 The Male Reproductive System
    19.2a Anatomy
    19.2b From Puberty to Andropause
19.3 The Female Reproductive System
    19.3a The Ovaries and Reproductive Tract
    19.3b External Genitalia
    19.3c Breasts
    19.3d From Puberty to Menopause
19.4 The Production and Union of Sex Cells
    19.4a Meiosis and Gametogenesis
    19.4b Spermatogenesis, Sperm, and Semen
    19.4c Oogenesis and the Ovarian–Menstrual Cycle
    19.4d Uniting the Gametes
19.5 Pregnancy, Childbirth, and Lactation
    19.5a Pregnancy
    19.5b Childbirth
    19.5c Postpartum Changes
    19.5d Lactation

Clinical Applications/Perspectives on Health
- Clinical Application 19.1: Prostate Diseases
- Clinical Application 19.2: Cervical Cancer
- Perspectives on Health

End of Chapter
- Aging of the Reproductive System
- Career Spotlight: Midwife
- Connective Issues
- Study Guide

Only one sperm will enter this egg; the egg has mechanisms for blocking all other sperms from entering.
© David M. Phillips/Science Source

BASE CAMP

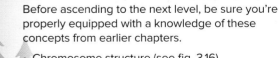

Before ascending to the next level, be sure you're properly equipped with a knowledge of these concepts from earlier chapters.

- Chromosome structure (see fig. 3.16)
- Mitosis (see section 3.4b)
- Gonadotropins of the pituitary (see section 11.3b)

Module 14
Reproductive System

From all we have learned of the structure and function of the human body, it seems a wonder that it works at all! The fact is, however, that even with modern medicine we cannot keep it working forever. The body suffers various degenerative changes as we age, and eventually our time is up and we must say good-bye. Yet our genes live on in new containers—our offspring. In this chapter, we examine the male and female roles in producing a new life; and in chapter 20, we examine the human life history from prenatal development to old age.

19.1 Essentials of Sexual Reproduction

Expected Learning Outcomes

When you have completed this section, you should be able to

a. explain why sexual reproduction in humans requires two different types of gametes;
b. enumerate the functions of the male and female reproductive systems; and
c. distinguish between the gonads of the two sexes, and between the internal and external genitalia.

The essence of sexual reproduction is that it is biparental—the offspring receive genes from two parents and therefore are not genetically identical to either one. To achieve this, the parents must produce **gametes**[1] (sex cells) that meet and combine their genes in a **zygote**[2] (fertilized egg). The gametes require two properties to succeed at this: motility so they can establish contact, and enough cytoplasm to divide into a multicellular embryo. A single cell cannot perform both roles well, because to carry ample cytoplasm means to be relatively heavy, which is inconsistent with motility. Therefore, these tasks are divided between two kinds of gametes. The small motile one—little more than DNA with a propeller—is the **sperm,** and the large one is the **egg (ovum).** The **reproductive systems** of the two sexes jointly produce and unite these gametes. In mammals, the female is the parent who harbors the fetus, gives birth, and nourishes the infant.

The organs that produce the gametes—the *testes* of the male and *ovaries* of the female—are collectively called the **gonads**[3] or *primary sex organs.* In addition to producing gametes, they produce sex hormones with widespread effects throughout the body, such as bone growth and maintenance, muscle growth, hair growth, and fat deposition. The other organs necessary for reproduction are called *secondary sex organs,* such as the penis, uterus, mammary glands, and others to be considered in this chapter.

---

[1]*gam* = marriage, union
[2]*zygo* = yoke, union
[3]*gon* = seed

According to location, reproductive organs are also classified as **external** and **internal genitalia.** The external genitalia are located in the perineum between the thighs. Most of them are externally visible or located superficially beneath the skin. The internal genitalia are located mainly in the pelvic cavity, except for the male testes and some associated ducts in the scrotum.

Before You Go On

Answer these questions from memory to test your understanding of the previous section. Reread the material if you can't answer the questions.

1. Define *gonad in a way that includes the word* gamete(s).
2. Distinguish between a gamete and a zygote.
3. Distinguish between internal and external genitalia.

19.2 The Male Reproductive System

Expected Learning Outcomes

When you have completed this section, you should be able to

a. describe the anatomy of the male reproductive tract;
b. trace the pathway taken by a sperm cell from its formation to ejaculation, naming all the passages that it travels;
c. describe the structure, locations, and functions of the male accessory glands, scrotum, and penis; and
d. discuss male sexual development from puberty through andropause.

19.2a Anatomy APR

This section surveys the male reproductive system (**fig. 19.1**), which is concerned with the production and delivery of sperm. We will proceed roughly in the order of sperm formation, transport, and emission, beginning with the testes, continuing through the duct system and accessory organs, and ending with the penis.

The Testes

The **testes (testicles)** produce both sex hormones and sperm. They are oval and slightly flattened, about 4 cm long and 2.5 cm wide (**fig. 19.2**). The testis has a white fibrous capsule with extensions *(septa)* that divide the interior tissue into about 300 wedge-shaped *lobules*. Each lobule contains one to three **seminiferous**[4] **tubules** (SEM-ih-NIF-er-us)—slender ducts up to 70 cm long each, in which the sperm are produced. Between the tubules are clusters of **interstitial endocrine cells,** the source of testosterone.

---

[4]*semin* = seed, sperm; *fer* = to carry

628 PART 6 Human Life Cycle

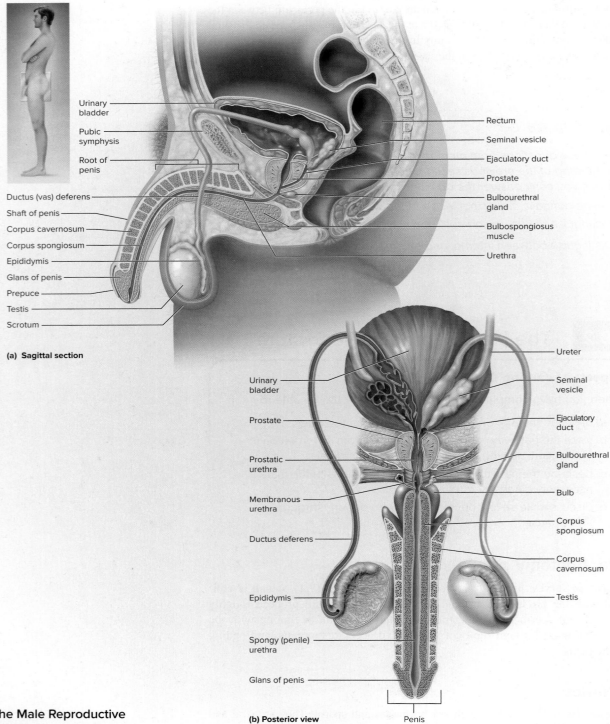

Figure 19.1 The Male Reproductive System. (a) Sagittal section. (b) Posterior view. APR

A seminiferous tubule has a narrow lumen lined by a thick epithelium (fig. 19.2c, d). The epithelium consists of several layers of **germ cells** in the process of becoming sperm, and a much smaller number of tall **nurse cells (supporting cells, sustentocytes**[5]**),** which support and protect the germ cells and promote

---

[5]*sustento* = support

Figure 19.2 The Testis and Associated Structures. (a) The scrotum is opened and folded downward to reveal the testis and associated organs. (b) Anatomy of the testis, epididymis, and spermatic cord. (c) Electron micrograph of a seminiferous tubule. (d) Light micrograph. **APR**
a: © Dennis Strete/McGraw-Hill Education; c: © Steve Gschmeissner/Science Source; d: © Ed Reschke

their development. The germ cells depend on the nurse cells for physical support, nutrients, waste removal, growth factors, protection from the immune system, and other needs. The nurse cells also secrete a hormone, *inhibin,* that regulates the rate of sperm production, as we will see later.

The Epididymis, Ductus Deferens, and Urethra

Mature sperm drift slowly down the seminiferous tubules and leave the testis by way of about 12 short ducts on its posterior side. These lead to an organ called the **epididymis**[6] (EP-ih-DID-ih-miss) that adheres to the posterior side of the testis (fig. 19.2a, b). The epididymis, a site of sperm maturation and storage, contains a single coiled duct embedded in connective tissue. The duct is about 6 m (20 ft) long, but it is so slender and highly coiled that it's packed into an epididymis only 7.5 cm long. It reabsorbs about 90% of the fluid secreted by the testis. Sperm mature as they travel down the epididymis over a period of about 20 days. They are stored here and in the adjacent portion of the ductus deferens, which follows. Stored sperm remain fertile for 40 to 60 days, but if they become too old without being ejaculated, they disintegrate and the epididymis reabsorbs them.

The duct of the epididymis straightens out at the inferior end, turns 180°, and becomes the **ductus deferens.** This is a muscular tube about 45 cm long and 2.5 mm in diameter. It has a very narrow lumen and a thick wall of smooth muscle important in ejaculation. It is also called the *vas deferens,* which is the basis of the term *vasectomy,* the male method of contraception that entails cutting out a short segment of the duct at the rear of the scrotum to interrupt the flow of sperm. The ductus deferens travels upward through a passage in the groin, the **inguinal canal,** and enters the pelvic cavity. There, it turns medially and approaches the urinary bladder. Posterior to the bladder, the ductus deferens ends by uniting with the duct of the seminal vesicle, a gland considered later. The two ducts unite into a short (2 cm) passage called the **ejaculatory duct,** which passes through the prostate gland and empties into the urethra (fig. 19.1).

The male urethra is shared by the reproductive and urinary systems, although it cannot pass urine and semen simultaneously. It is about 20 cm long and consists of three regions: the *prostatic urethra* embedded in the prostate, the short *membranous urethra* passing through the pelvic floor, and the long *spongy (penile) urethra* passing through the penis (fig. 19.1).

The Scrotum

The testes are contained in the **scrotum,**[7] a pendulous pouch of skin, muscle, and fibrous connective tissue (**fig. 19.3**). The scrotum and penis are the male external genitalia and occupy the **perineum** (PERR-ih-NEE-um), a diamond-shaped area between the thighs bordered by the pubic symphysis, ischial tuberosities, and coccyx. The left testis is usually suspended lower than the right so the two are not compressed against each other. The scrotum is divided into right and left compartments by a median septum.

The scrotum also contains the epididymis and spermatic cord associated with each testis. The **spermatic cord** is a fibrous bundle that contains the ductus deferens, blood and lymphatic vessels, and nerves. It passes upward behind and superior to the testis, then anterior to the pubis and into the inguinal canal.

Sperm production requires a temperature of about 35°C—or 2°C cooler than the temperature in the pelvic cavity—and the scrotum has three mechanisms for controlling testicular temperature. In response to cold, a muscle of the spermatic cord called the **cremaster**[8] contracts and draws the testes closer to the body, while a layer of smooth muscle beneath the skin within the **dartos fascia** contracts and tautens the scrotum. When warmer, the cremaster and dartos both relax and the testes are suspended farther away from the pelvic floor. In addition, the spermatic

---

[6]*epi* = upon; *didym* = twins, testes
[7]*scrotum* = bag
[8]*cremaster* = suspender

Figure 19.3 The Scrotum and Spermatic Cord. The cremaster, pampiniform plexus, and dartos fascia help regulate the temperature of the testis. **APR**

cord has a network of veins, the **pampiniform plexus,** that carries heat away from the arterial blood on its way to the testis, preventing the warm blood from overheating the testis.

The testes begin their fetal development high in the abdominal cavity near the kidneys. As the fetus grows, the testes descend and eventually pass through the inguinal canal into the scrotum. About 3% of boys are born with the testes still in the pelvic cavity, a condition called **cryptorchidism (undescended testes).** If uncorrected, this condition will lead to sterility because sperm cannot develop at the pelvic cavity temperature. The condition can be corrected with a testosterone injection or a minor surgery to bring the testes into the scrotum.

The Seminal Vesicles

The **seminal vesicles (seminal glands)** are a pair of glands on the posterior side of the urinary bladder (see fig. 19.1). Each has a connective tissue capsule, smooth muscle, and a single, highly convoluted and branched duct that exits the gland and joins the ductus deferens. The vesicles produce a yellowish secretion that forms about 60% of the semen. Its composition and functions are discussed later.

The Prostate

The **prostate**[9] (PROSS-tate) surrounds the urethra and ejaculatory duct immediately inferior to the urinary bladder (see Clinical Application 19.1). It measures about

---

[9]*pro* = before; *stat* = to stand; avoid the misspelling *prostrate*

Clinical Application 19.1

PROSTATE DISEASES

The prostate remains at a stable size from age 20 to 45 or so, then begins to grow slowly. By age 70, over 90% of men show some degree of noncancerous enlargement called *benign prostatic hyperplasia (BPH)*. The main concern in BPH is that it compresses the urethra, makes it harder to empty the bladder, requires more frequent urination since the bladder refills more quickly, and may promote bladder and kidney infections.

Prostate cancer is the second most common cancer in men (after lung cancer). Approximately 11.6% of men will be diagnosed with prostate cancer at some point in their life. Tumors tend to form near the periphery of the gland, where they don't obstruct urine flow and easily go unnoticed until they cause pain. Prostate cancer can metastasize to nearby lymph nodes and then to the lungs and other organs. Once it metastasizes, it has a 50% to 90% mortality rate, but is highly survivable if detected and treated before then. It can be detected by palpation through the rectal wall, called *digital rectal examination (DRE)*, or by blood tests for certain prostatic enzymes. One of these enzymes is known as *prostate-specific antigen,* so the blood assay is called a PSA test. Prostate cancer grows so slowly that if it is discovered late in life, the risks of surgery may be greater than the risk of metastasis; a patient and doctor may opt just to keep an eye on it and forego treatment if it presents no imminent threat.

2 × 4 × 3 cm and opens into the prostatic urethra by numerous pores in the urethral wall. The thin, milky secretion of the prostate constitutes about 30% of the semen. Its functions, too, are considered later.

The Bulbourethral Glands

The **bulbourethral glands** are brownish, spherical glands about 1 cm in diameter, with a short duct to the uretha at the root of the penis (see fig. 19.1). During sexual arousal, they produce a clear slippery fluid that provides some of the lubrication for intercourse, but more importantly, neutralizes the acidity of residual urine in the urethra (thus protecting the sperm).

The Penis

The **penis**[10] serves to deposit semen in the vagina. The externally visible portion of it, composed of the **shaft** and **glans**[11] (**fig. 19.4**), is half its total length. In the flaccid (nonerect) state, the external portion measures about 8 to 10 cm (3–4 in.) long and 3 cm in diameter; the typical dimensions of an erect penis are 13 to 18 cm (5–7 in.) long and 4 cm in diameter. The glans is the expanded head at the distal end of the penis with the external urethral orifice at its tip.

The skin is loosely attached to the shaft, allowing for movement and expansion during erection. It continues over the glans as the **prepuce (foreskin).** The inner surface of the prepuce and facing surface of the glans

---

[10]*penis* = tail
[11]*glans* = acorn

Figure 19.4 Anatomy of the Penis. (a) Superficial dissection of shaft, lateral view. (b) Cross section at midshaft. **APR**

are covered with a thin mucous membrane similar to the inner surface of the eyelid. In infancy and childhood, these two membranes are fused to each other; attempts to retract the foreskin of an infant or child can be injurious. The membranes separate gradually in later childhood and adolescence. The adult prepuce remains anchored to the proximal margin of the glans by a ventral fold of tissue called the *frenulum*. The prepuce is the most densely innervated and sensitive region of the penis, with an abundance of tactile corpuscles and nerves. Circumcision thus removes the most sensitive part of the penis. The exposed mucous membrane of the glans then transforms to a thin, drier, and less sensitive epidermis.

Internally, the penis consists mainly of three cylindrical bodies called **erectile tissues** (fig. 19.4b), which fill with blood during sexual arousal and account for its enlargement and erection. A single erectile body, the **corpus spongiosum,** passes along the ventral side of the penis and encloses the urethra. It expands at the distal end to fill the entire glans. The upper side of the penis has a pair of erectile tissues called **corpora cavernosa** (singular, *corpus cavernosum*), which extend only as far as the margin of the glans. They are separated by a median septum and sheathed in a tight fibrous tunic, which plays a role in erection explained later.

All three cylinders of erectile tissue are spongy in appearance and contain tiny blood sinuses called **lacunae (cavernous spaces).** The partitions between lacunae, called **trabeculae,** are composed of connective tissue and smooth muscle. In the flaccid penis, muscle tone collapses the lacunae, which then form tiny slits in the tissue.

At the body surface, the penis turns 90° posteriorly and continues inward as the **root.** The corpus spongiosum ends internally as a dilated **bulb,** which is ensheathed in the *bulbospongiosus muscle.* The corpora cavernosa diverge like the arms of a Y and attach the penis to the pubic arch of the pelvic girdle.

19.2b From Puberty to Andropause

Unlike any other organ system, the reproductive system remains dormant for several years after birth. Around age 10 to 12 in most boys and 8 to 10 in girls, a surge of pituitary gonadotropins awakens the reproductive system and begins preparing it for adult reproductive function. **Adolescence**[12] is the period from the onset of gonadotropin secretion and reproductive development until a person attains full adult height, usually in the late teens to early twenties. **Puberty**[13] is the first few years of adolescence, until the first menstrual period in girls and the first ejaculation of viable sperm in boys. In North America, this is typically attained around age 12 in girls and 14 in boys.

As the brain matures, the hypothalamus begins to secrete **gonadotropin-releasing hormone (GnRH).** GnRH stimulates the anterior pituitary gland to secrete two *gonadotropins*—**follicle-stimulating hormone (FSH)** and **luteinizing hormone (LH).** Although named for their functions in females, LH also stimulates the interstitial endocrine cells of the testis to secrete testosterone, and FSH renders the seminiferous tubules more sensitive to testosterone. Without FSH, the tubules don't respond to testosterone at all. Testosterone produces many conspicuous effects on the adolescent male:

- Growth of the sex organs, beginning with the testes and scrotum, then the penis and internal ducts and glands. This is accompanied by sperm production.
- A burst of generalized body growth, especially lengthening of the limb bones and increased muscle mass under the influence of both testosterone and growth hormone. This is accompanied by enlargement of the larynx, deepening of the voice, and an elevated metabolic rate and appetite.
- Appearance of pubic hair, axillary hair, and later, the facial hair, along with the apocrine sweat glands in these areas.
- Awakening of the sex drive, or **libido,** accompanied by frequent erection and spontaneous ejaculation, especially during sleep (*nocturnal emission* or "wet dreams").

Throughout adulthood, testosterone sustains the male reproductive tract, sperm production, and libido. The nurse cells also secrete a hormone called **inhibin.** By inhibiting FSH but not LH secretion, this hormone can slow down sperm production without reducing the body's other responses to testosterone.

Testosterone secretion peaks at about 7 mg/day at age 20. As men age, testosterone levels decline, but the amount of decrease varies considerably between individuals. Since testicular hormones normally inhibit pituitary gonadotropin secretion, the declining level of testosterone is accompanied by a rising level of FSH and LH after age 50. These hormonal changes may produce effects that some call **andropause** or **male climacteric.** Andropause has little or no noticeable effect in most men, but in some cases there are mood changes, hot flashes, or illusions of suffocation—symptoms similar to those of menopause

---

[12]*adolesc* = to grow up
[13]*pub* = body hair, or *puber* = grown up

in women. Men normally produce sperm to the end of life, although in reduced numbers.

We will soon explore the process of sperm production and physiology of intercourse for the male, but first we will survey the female reproductive system. We can then take a more integrative view of what sperm and egg production, as well as the responses of intercourse, have in common for both sexes.

Before You Go On

Answer these questions from memory to test your understanding of the previous section. Reread the material if you can't answer the questions.

4. What are seminiferous tubules? What types of cells compose them? What is their relationship to the interstitial endocrine cells?

5. Describe the pathway taken by the sperm from the time they leave the testis until they are ejaculated. What point in this pathway is the main storage place for sperm awaiting ejaculation?

6. Describe three ways in which the body regulates the temperature of the testes, and explain why this is important.

7. Name the three accessory glands of the male reproductive system, describe their locations, and state their functions.

8. What are the two kinds of erectile tissues in the penis, and where are they located?

9. What pituitary hormones regulate the secretion and action of testosterone? Describe several effects of testosterone.

19.3 The Female Reproductive System

Expected Learning Outcomes

When you have completed this section, you should be able to

a. describe the anatomy and histology of the ovaries;

b. describe the gross anatomy and histology of the female reproductive tract;

c. describe the structure and function of the glands and other accessory organs of the female reproductive system; and

d. discuss female sexual development from puberty through menopause.

The female reproductive system is more complex than the male's because it serves more purposes. Whereas the male needs only to produce and deliver gametes, the female must do these as well as provide nutrition and safe harbor for embryonic and fetal development, then give birth and nourish the infant. Furthermore, female reproductive physiology is more conspicuously cyclic, and female hormones are secreted in a more complex sequence compared with the relatively steady, simultaneous secretion of regulatory hormones in the male. **APR**

19.3a The Ovaries and Reproductive Tract

The female *reproductive tract* consists of a pair of uterine (fallopian) tubes, the uterus, and the vagina. The ovaries, strictly speaking, are not part of the reproductive tract, but are most conveniently considered here as a starting point for our description.

The Ovaries

The **ovaries**[14] (**fig. 19.5**) are the female gonads; they produce egg cells (ova) and sex hormones. Each ovary lies in a shallow depression of the posterior pelvic wall and is held in place by several connective tissue ligaments (see fig. 19.7). The ovary is almond-shaped, about 3 cm long, 1.5 cm wide, and 1 cm thick. It is enclosed in a white fibrous capsule, and internally it is loosely divided into a central core *(medulla)* of fibrous connective tissue and blood vessels, and an outer zone *(cortex)* where the eggs develop.

Unlike the testis, the ovary has no ducts. Instead, each egg develops in its own bubblelike **follicle** and is released by *ovulation*, the bursting of the follicle. The cortex is the site of many thousands of ovarian follicles. A child's ovaries are smooth-surfaced, but during the reproductive years, they become corrugated with bulges created by growing follicles. Figure 19.5 shows several types of follicles that coincide with different stages of egg maturation, as discussed later. The wall of each follicle, composed in mature follicles of inner, multilayered *granulosa cells* and an outer husk called a *theca*, is the source of the female sex hormones, particularly *estrogens* and *progesterone*. We will later examine these follicles, their cyclic development, and their endocrine functions in more detail.

[14]*ov* = egg; *ary* = place for

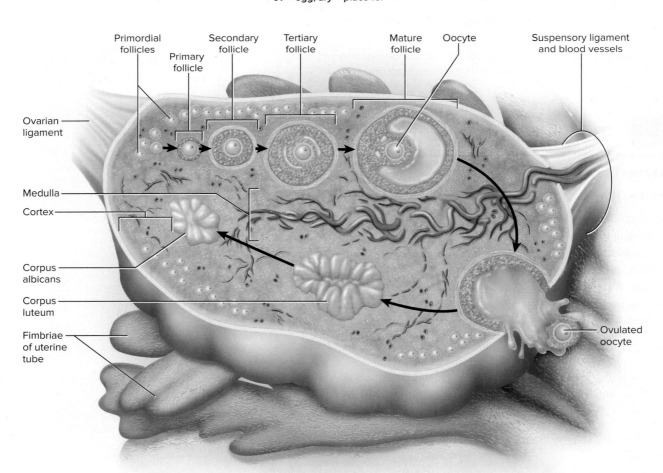

Figure 19.5 Structure of the Ovary. Arrows indicate the developmental sequence of the ovarian follicles; the follicles do not migrate around the ovary. Not all the follicle types illustrated here are present simultaneously. **APR**

Figure 19.6 The Female Pelvic Cavity and Reproductive Organs (Sagittal Section). APR

The Uterine Tubes

The **uterine (fallopian[15]) tube** is a muscular canal about 10 cm long that extends from each ovary to the uterus (fig. 19.6). At the distal (ovarian) end, it flares into a trumpetlike **infundibulum**[16] with feathery projections called **fimbriae**[17] (FIM-bree-ee) that catch an ovulated egg. The mucosa of the uterine tube is folded into longitudinal ridges and has an epithelium of ciliated cells and a smaller number of secretory cells (fig. 19.7a). The cilia beat toward the uterus and, with the help of muscular contractions of the tube, convey the egg or developing embryo in that direction.

The Uterus

The **uterus**[18] (figs. 19.6 and 19.7) is a muscular chamber that opens into the roof of the vagina and usually tilts forward over the urinary bladder. It harbors the fetus, provides nutrition, and expels the fetus at the end of its development. It is somewhat pear-shaped, with a broad superior curvature called the **fundus,** a midportion called the **body,** and a cylindrical inferior end called the **cervix.** The nonpregnant uterus measures about 7 cm from cervix to fundus, 4 cm wide at its broadest point on the fundus, and 2.5 cm thick, but it expands enormously during pregnancy to accommodate the growing fetus.

---

[15]Gabriele Fallopio (1523–62), Italian anatomist and physician
[16]*infundibulum* = funnel
[17]*fimbria* = fringe
[18]*uterus* = womb

Figure 19.7 The Female Reproductive Tract.
(a) Electron micrograph of the mucosa of the uterine tube, showing ciliated cells (yellow) and mucous cells (red). (b) Posterior view of the reproductive tract and associated organs. (c) Relationship of the ovary to ligaments of the uterus and uterine tube. APR
a: © SPL/Science Source

The lumen is roughly triangular, with its upper corners opening into the uterine tubes and the lower end communicating with the vagina by way of a narrow passage through the cervix called the **cervical canal** (fig. 19.7b). The canal contains glands that secrete mucus, thought to inhibit the spread of microorganisms from the vagina into the uterus.

The uterine wall consists of a thin external serosa called the *perimetrium,* a thick middle layer called the *myometrium,* and an inner mucosa called the *endometrium.* The **myometrium**[19] consists mainly of smooth muscle and is responsible for the labor contractions that help to expel the fetus.

The **endometrium**[20] has a simple columnar epithelium and deep glandular pits with coiled *spiral arteries* between them (see fig. 19.16). The superficial half to two-thirds of the endometrium, called the **functional layer,** is shed in each menstrual period. The deeper layer, called the **basal layer,** stays behind and regenerates a new functional layer in the next cycle. When pregnancy occurs, the endometrium is the site of attachment of the embryo and forms part of the *placenta* from which the fetus is nourished.

---

[19]*myo* = muscle; *metr* = uterus
[20]*endo* = inside; *metr* = uterus

Clinical Application 19.2

CERVICAL CANCER

Cervical cancer occurs most often among women from ages 30 to 50, especially those who smoke, began sexual activity at an early age, or have histories of frequent sexually transmitted diseases or cervical inflammation. It is often caused by the human papillomavirus (HPV) (see Perspectives on Health in the next section). Cervical cancer usually begins in the lower cervix, develops slowly, and remains a local, easily removed lesion for several years. If the cancerous cells spread to the subepithelial connective tissue, however, the cancer is said to be *invasive* and is much more dangerous.

The best protection against cervical cancer is early detection by means of a *Pap*[21] smear (fig. 19.8). Loose cells are scraped from the cervix and microscopically examined for signs of *dysplasia* (abnormal development) or *carcinoma* (cancer). Different grades of abnormality call for measures ranging from a repeat smear in a few months to a biopsy. Confirmed cases of cervical cancer can be treated with radiation therapy, electrosurgical excision, or hysterectomy.[22] From age 21 to 65, women are advised to have a Pap smear every 3 years. Pap smears are not recommended before age 21 or after 65.

The Vagina

The **vagina**[23] (see figs. 19.6 and 19.7) is a tube 8 to 10 cm long that allows for the discharge of menstrual fluid, receipt of the penis and semen, and birth of a baby. The vaginal wall is thin but very distensible. It has no glands, but it is lubricated by serous fluid seeping through its wall and mucus draining from the cervix. The vaginal epithelium is nonkeratinized stratified squamous—the form of tissue best adapted for lubrication and abrasion resistance. Its cells are rich in glycogen. Bacteria ferment this to lactic acid, which produces a low vaginal pH (about 3.5–4.0) that inhibits the growth of pathogens.

---

[21]George N. Papanicolaou (1883–1962), Greek-American physician and cytologis
[22]*hyster* = uterus; *ectomy* = cutting out
[23]*vagina* = sheath

Figure 19.8 Pap Smears. (a) These are smears of normal epithelial cells from the cervix. (b) In the malignant (cancerous) cells, note the loss of cell volume and the greatly enlarged nuclei.
a: © SPL/Science Source; b: © Parviz M. Pour/Science Source

(a) Normal cells

(b) Malignant cells

Figure 19.9 The Female Perineum. (a) Surface anatomy. (b) Subcutaneous structures. APR

At the vaginal orifice, the mucosa folds inward and forms a membrane, the **hymen,** which stretches across the opening. The hymen is often torn during childhood or adolescence by ordinary physical activity, the use of tampons, or medical examination with a speculum, or it may tear at the first act of sexual intercourse.

19.3b External Genitalia

The external genitalia of the female are collectively called the **vulva**[24]; they include the mons pubis, labia majora and minora, clitoris, vaginal orifice, and accessory glands and erectile tissues (fig. 19.9).

The **mons**[25] **pubis** is an anterior mound of adipose tissue and skin overlying the pubic symphysis, bearing most of the pubic hair. The **labia majora**[26] (singular, *labium majus*) are a pair of thick folds of skin and adipose tissue forming the lateral borders of the vulva; they bear pubic hair only on their lateral surfaces. Medial to these are the hairless, much thinner **labia minora**[27] (singular, *labium minus*). The area enclosed by them, called the **vestibule,** contains the urinary and vaginal orifices. At the anterior margin of the vestibule, the labia minora meet and form a hoodlike **prepuce** over the clitoris.

The **clitoris**[28] (CLIT-er-is, cli-TOR-is) consists of a pair of corpora cavernosa, similar to the dorsal half of the penis; it has no corpus spongiosum and it does not enclose the urethra. Only its head, the **glans,** is external, protruding slightly from the prepuce; a leglike *crus* anchors it to the pelvic girdle on each side. The function of the clitoris is entirely sensory, serving as the primary center of sexual stimulation.

Just deep to the labia majora, a pair of subcutaneous erectile tissues called the **vestibular bulbs** brackets the vagina like parentheses. They become congested with blood during sexual excitement and cause the vagina to tighten somewhat around the penis, enhancing sexual stimulation for both partners. On each side of the vagina is a pea-size **greater vestibular gland** (also called the Bartholin gland) with a short duct opening into the vestibule or lower vagina. These glands moisten the vulva, and during sexual excitement they provide most of the lubrication for intercourse. The vestibule is also lubricated by a number of **lesser vestibular glands.** A pair of mucous **paraurethral glands** (formerly called the Skene glands) opens into the vestibule near the external urethral orifice. They eject fluid, sometimes abundantly, during orgasm. They arise from the same embryonic structure as the male's prostate, and their fluid is similar to the prostatic secretion.

19.3c Breasts

The **breast** (fig. 19.10) is a mound of tissue overlying the pectoralis major in both the male and female. The female breast markedly enlarges at puberty and remains so for life, but usually contains very little mammary gland. The **mammary gland** develops within the breast during pregnancy, remains active in the lactating breast, and atrophies when a woman ceases to nurse.

The breast has two principal regions: the conical to pendulous **body,** with the nipple at its apex, and an extension toward the armpit called the **axillary tail.** Lymphatics of the axillary tail are especially important as a route of breast cancer metastasis, making this region particularly important to include in breast self-examination.

---

[24]*vulva* = covering
[25]*mons* = mound, eminence; *pub* = body hair
[26]*labia* = lip; *major* = larger, greater
[27]*labia* = lips; *minor* = smaller, lesser
[28]origin uncertain; possibly from *kleis* = door key, or *klei* + *ein* = to close

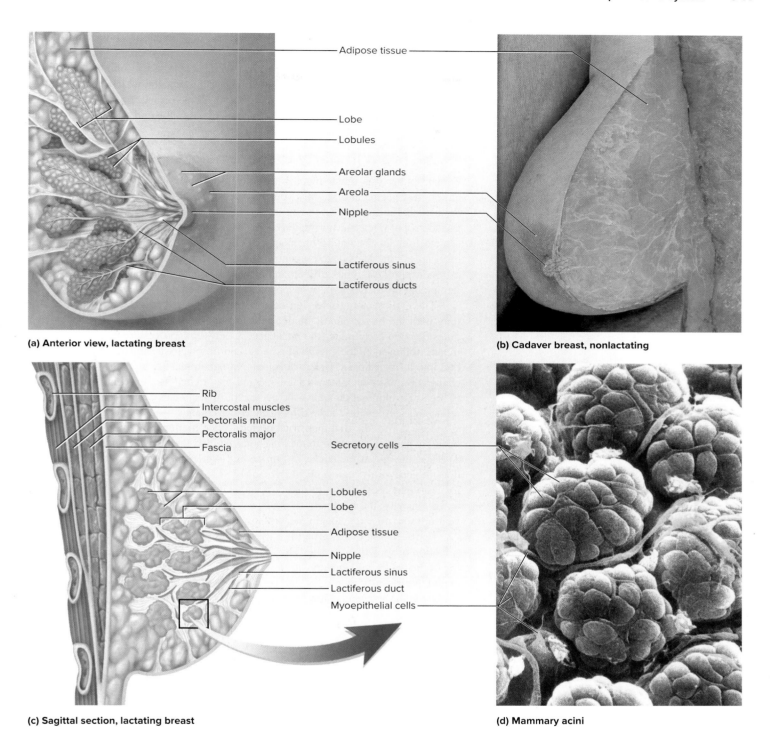

Figure 19.10 The Breast. Parts (a), (c), and (d) depict the breast in a lactating state. Some of the features in (a) and (c) are absent from the nonlactating breast in part (b). The cluster of lobules boxed in (c) would contain numerous microscopic acini like the ones in (d). APR

b: © The University of Toledo/McGraw-Hill Education;
d: © Dr. Donald Fawcett/Science Source

The nipple is surrounded by a circular colored zone, the **areola.** Dermal blood capillaries and nerves come closer to the surface here than in the surrounding skin and make the areola more sensitive and deeper in color. The areola has sparse hairs and **areolar glands,** visible as small bumps on the surface. The areolar glands and sebaceous glands of the region oil the skin and help to prevent chapping in a nursing mother.

Internally, the nonlactating breast is mostly adipose and collagenous tissue (fig. 19.10b). Breast size is determined by the adipose tissue and is unrelated to the amount of milk the mammary gland can produce. Even in the nonlactating breast, the region of the areola exhibits a system of short **lactiferous**[29] **ducts** leading to the nipple. During pregnancy, these branch and develop into 15 to 20 lobes of glandular tissue arranged radially around the nipple. Each lobe has numerous secretory acini (fig. 19.10d) drained by one lactiferous duct, which dilates to form a *lactiferous sinus* just before opening onto the skin surface. Lactation and functional details of the mammary acinus are discussed at the end of this chapter.

19.3d From Puberty to Menopause

Female puberty is under much the same hormonal control as that of the male, although it begins at a slightly younger age—typically age 9 or 10 in the United States and Europe, but as young as age 3 in 1% to 3% of girls. Rising levels of gonadotropin-releasing hormone (GnRH) stimulate the anterior pituitary gland to secrete follicle-stimulating hormone (FSH) and luteinizing hormone (LH). FSH, especially, stimulates development of the ovarian follicles, which, in turn, secrete estrogens, progesterone, inhibin, and a small amount of androgen. **Estrogens**[30] are feminizing hormones with widespread effects on the body.

The earliest noticeable sign of female puberty is breast development, beginning with the *breast bud,* a small conical elevation of the areola. The lobules and ducts develop early under the influence of estrogens, progesterone, and other hormones, followed by the growth of adipose and fibrous tissue and development of the bud into a fuller breast. Breast development is complete around age 20, but minor changes occur in each menstrual cycle and major changes in pregnancy. The onset of breast development is soon followed by the appearance of pubic and axillary hair, sebaceous (oil) glands, and axillary (underarm) glands.

Next comes the first menstrual period, called **menarche**[31] (men-AR-kee). This occurs when a girl attains about 17% body fat, and therefore depends on nutrition. The average age at menarche in Europe and America is now about 12. Menarche doesn't necessarily signify fertility. A girl's first few cycles are typically *anovulatory* (no egg is ovulated), probably because it takes an egg cell about 290 days to reach maturity and readiness to ovulate. Most girls don't ovulate regularly until a year or more after they begin menstruating.

Estrogen stimulates growth of the ovaries and secondary sex organs, and the secretion of growth hormone. This brings about a rapid increase in height, widening of the pelvis, and fat deposition, especially in the breasts and hips. Inhibin is secreted in girls, as it is in boys, and regulates the development of eggs and follicles.

Around the age of 45 to 55, women, like men, go through a midlife change in hormone secretion called the **climacteric.** In women, it is accompanied by **menopause,** the cessation of menstruation. A female is born with hundreds of thousands of eggs in her ovaries, each in its own follicle. The older she gets, the fewer follicles remain. Climacteric begins not at any specific age, but when she has only about 1,000 follicles left. Even the remaining follicles secrete less estrogen and progesterone. The drop in ovarian hormones brings about various degrees of atrophy in the uterus, vagina, breasts, skin, and bones. Blood vessels constrict and dilate in response to shifting hormone balances, and the sudden dilation of cutaneous arteries may cause **hot flashes**—a spreading sense of heat from the abdomen to the thorax, neck, and face, commonly accompanied by sweating and sometimes headaches resulting from the sudden dilation of arteries in the head. It is difficult to precisely establish the time of menopause because the menstrual periods can stop for several months and then begin again. Menopause is generally considered to have occurred when there has been no menstruation for a year or more.

---

[29]*lact* = milk; *fer* = to carry
[30]*estro* = desire, frenzy; *gen* = to produce
[31]*men* = monthly; *arche* = beginning

Before You Go On

Answer these questions from memory to test your understanding of the previous section. Reread the material if you can't answer the questions.

10. What structure in the ovary serves the same purpose as seminiferous tubules in the testis? In what ways does it differ from a seminiferous tubule?

11. What structure transports an egg from the ovary to the uterus? What feature within this structure moves the egg along this route?

12. Name the upper, middle, and lower parts of the uterus and the three layers of the uterine wall.

13. Identify all structures enclosed within the labia majora, including the subcutaneous glands, and describe their relative positions.

14. What is the difference between the breast and mammary gland? What traces of mammary gland can be found in the nonlactating breast?

15. What hormone from the brain stimulates the onset of puberty? How does that hormone lead to the production of estrogen?

19.4 The Production and Union of Sex Cells

Expected Learning Outcomes

When you have completed this section, you should be able to

a. explain the relevance of meiosis to sexual reproduction, state the stages of meiosis, and describe how it differs from mitosis;

b. describe the stages in the production of sperm and eggs, how these stages relate to meiosis, and the major differences between sperm and egg production;

c. give a functional description of the major components of semen and sperm cells;

d. relate the process of egg production to the cyclic changes in the ovary and uterus;

e. describe the production of eggs and how it is correlated with cyclic changes in the ovaries and uterus; and

f. describe the physiological processes that occur in the male and female during sexual intercourse.

The essence of reproduction is the production and fusion of gametes—eggs and sperm. We've examined male and female structures associated with reproduction. Now let's see how sex cells are produced and how they get together.

19.4a Meiosis and Gametogenesis

Sexual reproduction greatly increases genetic diversity, thereby enhancing species survival, but it poses a problem: Human cells normally have 46 chromosomes, and two cells—one from each parent—unite to produce a child. What's to prevent the child from having 92 chromosomes per cell, and the generation after that having 184, and so on? Sexual reproduction has to avoid such doubling of the chromosome number in each new generation. This is where **meiosis**[32] comes in.

To understand meiosis, it is important to know that the chromosomes of most cells occur in 23 pairs—designated *chromosomes 1* through *22* and one pair of *sex chromosomes,* designated *X* and *Y* in the male and two *X*s in the female. In all 23 pairs, one chromosome was contributed by the individual's mother and the other by the father. Within each pair except the male's XY, the two chromosomes look identical, carry genes for the same traits, and are called **homologous chromosomes.** Any cell that has 23 chromosome pairs (46 chromosomes in all) is called a **diploid** cell, symbolized $2n$. In contrast, the germ cells (sperm, eggs, and some of the developmental stages leading to them) have 23 unpaired chromosomes, and are called **haploid** (n). One function of meiosis is to reduce diploid stem cells to haploid germ cells; this is why meiosis is also called *reduction division.*

Meiosis has many similarities to mitosis (see section 3.4), but several important differences:

1. Mitosis serves a range of functions including fetal development, childhood growth, and tissue maintenance and repair, whereas meiosis is used only for the production of gametes.
2. Mitosis maintains a constant number of chromosomes from parent cell to daughter cells, whereas meiosis reduces the chromosome number from diploid to haploid.
3. Mitosis doesn't change the genetic makeup of the chromosomes. In an early stage of meiosis, however, the chromosomes of each pair join and exchange portions of their DNA. This creates new combinations of genes—some from one's mother and some from one's father on each chromosome—so the chromosomes we pass to our offspring are not the same ones that we inherited from our parents.
4. In mitosis, each parent cell produces only two daughter cells. In meiosis, it produces four. In the male, four equal-size sperm develop from each original germ cell. In the female, though, three of the four daughter cells are tiny cells that soon die, and only one large, mature egg is produced.

Meiosis consists of two cell divisions in succession (**fig. 19.11**). The first division produces two haploid daughter cells, and the second division divides each of these into two more, so that the ultimate outcome is four haploid cells. In the first division *(meiosis I),* after two homologous chromosomes exchange segments of their DNA, they migrate to separate daughter cells. The daughter cells have only 23 chromosomes, but each chromosome retains its double-stranded form, composed of two chromatids joined at the centromere (see fig. 3.19, section 3.4). In the second division *(meiosis II),* each centromere divides and each daughter cell now receives 23 single-stranded chromosomes.

19.4b Spermatogenesis, Sperm, and Semen

Both sperm and egg production rely on mitosis *and* meiosis. Sperm production is called **spermatogenesis.** In the average young man, it occurs at the remarkable rate of about 300,000 sperm per minute, or 400 million per day. Yet these sperm constitute only about 10% of the fluid that is ejaculated, called the *semen.* The germ

---

[32]*meio* = less, fewer

Figure 19.11 Meiosis. For simplicity, the cell is shown with only two pairs of homologous chromosomes. Human cells begin meiosis with 23 pairs. APR

? *Although we pass the same genes to our offspring as we inherit from our parents, we do not pass on the same chromosomes. What event in this figure accounts for the latter fact?*

cells that undergo spermatogenesis are nestled in the nurse cells of the seminiferous tubules until the process is complete and they are released into the tubule lumen.

Spermatogenesis

At birth, the testes contain stem cells called *spermatogonia,* which lie dormant for years around the perimeter of the seminiferous tubules. At puberty, testosterone reactivates them and initiates spermatogenesis. The essential steps of spermatogenesis are as follows (**fig. 19.12**), numbered to match the figure.

① **Spermatogonia** divide by mitosis and sustain a man's lifetime supply of sperm. One daughter cell from each division remains near the tubule wall, maintaining a lifelong population of stem cells. The other daughter cell migrates slightly away from the tubule wall on its way to dividing and differentiating into sperm.

② The latter cell enlarges and becomes a **primary spermatocyte.** Since this cell is about to undergo meiosis and become genetically different from other cells of the body, it must be protected from the immune system. Ahead of the primary spermatocyte, the tight junction between two nurse cells is dismantled, while a new tight junction forms behind the spermatocyte. The spermatocyte moves toward the lumen, like an astronaut passing through a double-door airlock, and is now protected by the barrier closing behind it.

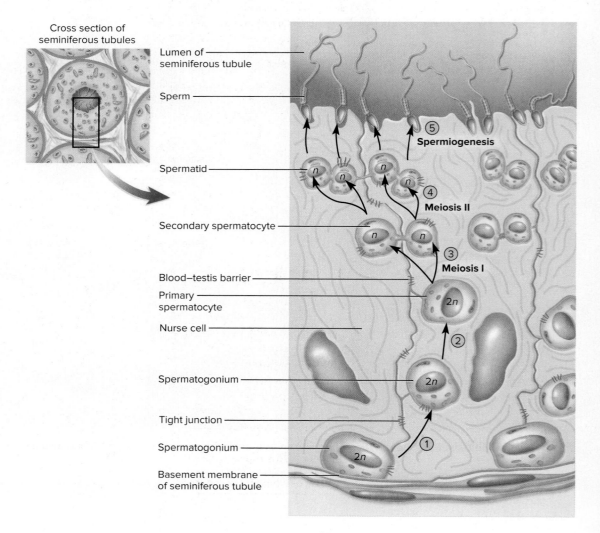

Figure 19.12 Spermatogenesis. The process proceeds from the bottom of the figure (periphery of the seminiferous tubule) to the top (lumen of the tubule). The daughter cells from secondary spermatocytes through spermatids remain connected by slender cytoplasmic processes until spermiogenesis is complete and individual spermatozoa are released. See text for explanation of the numbered steps ($2n$ = diploid cells, n = haploid cells). **APR**

3. Now safely isolated from blood-borne antibodies, the primary spermatocyte undergoes meiosis I, dividing into two equal-size, haploid and genetically unique **secondary spermatocytes.**
4. Each secondary spermatocyte undergoes meiosis II, dividing into two **spermatids**—a total of four for each spermatogonium.
5. A spermatid divides no further, but transforms into a single sperm cell. This transformation, called **spermiogenesis,** consists essentially of sprouting a tail (flagellum) and shedding excess cytoplasm, making the sperm a lightweight, mobile, self-propelled cell. It will not move under its own power, however, until ejaculation.

When fully formed, the sperm depart from their supportive nurse cells and are washed down the seminiferous tubule by a slow flow of fluid. After division of the spermatogonium, it takes about 70 days for the one daughter cell to become mature sperm.

Sperm

The **spermatozoon (sperm cell)** has two parts: a pear-shaped head and a long flagellum, or tail (fig. 19.13). The **head** is filled mostly by two structures: a nucleus and acrosome. The most important of these is the nucleus, which fills most of the head and contains a haploid set of condensed chromosomes. The **acrosome**[33] is a lysosome in the form of a thin cap covering the apical half of the nucleus. It contains enzymes that a successful sperm later uses to penetrate the egg.

The **tail,** about 50 μm long, is divided into three regions called the midpiece, principal piece, and endpiece. The **midpiece** is the thickest part. It contains large mitochondria coiled around the core of the flagellum. They produce the ATP needed for sperm movement. The **principal piece** constitutes most of the tail and consists of the axoneme surrounded by a sheath of supportive fibers, which stiffen the tail and enhance its propulsive power. The principal piece provides most propulsion of the sperm. It tapers to a short **endpiece** composed of axoneme only. You could think of the sperm midpiece as the "motor," the principal piece as the "propeller," and the head or its nucleus as the "cargo."

Semen

The fluid expelled in orgasm is called **semen**[34] **(seminal fluid).** A typical ejaculation is 2 to 5 mL of semen, composed mainly of seminal vesicle fluid (60%), prostatic fluid (30%), and sperm (10%), with a trace of other secretions. Most sperm emerge in the first one or two jets of semen. The semen usually has a **sperm count** of 50 to 120 million sperm/mL. A sperm count lower than 20 to 25 million sperm/mL is usually associated with **infertility (sterility),** the inability to fertilize an egg.

The prostate contributes a thin, milky white fluid containing calcium and other ions. It contains a protein-hydrolyzing enzyme called *serine protease* (also known as prostate-specific antigen, PSA; see Clinical Application 19.1). The seminal vesicles contribute a viscous yellowish fluid. This is the last component of the semen to emerge, and it flushes remaining sperm from the urethra. It contains fructose and other carbohydrates that provide nourishment for the sperm, as well as prostaglandins that enhance sperm motility and stimulate peristaltic contractions of the uterus. (Prostaglandins were discovered in and named for the prostate, but are more abundant in the seminal vesicle secretion.)

A well-known property of semen is its stickiness. This is due to a fibrinlike protein, *semenogelin,* whose production is initiated by a clotting

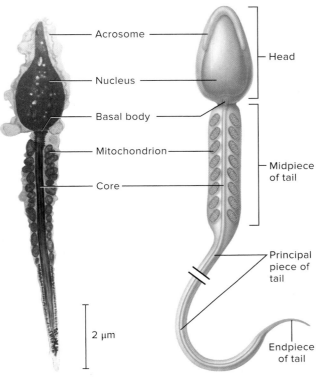

Figure 19.13 The Mature Spermatozoon. (a) Head and part of the tail of a spermatozoon (electron micrograph). (b) Sperm structure.
a: © BSIP SA/Alamy Stock Photo

---
[33]*acro* = tip, peak; *some* = body
[34]*semen* = seed

enzyme produced by the prostate gland. Semenogelin entangles the sperm, sticks to the walls of the vagina and cervix, and ensures that the semen doesn't simply drain back out of the vagina. It may also promote the uptake of sperm-laden clots of semen into the uterus.

About 20 to 30 minutes after ejaculation, the prostatic enzyme serine protease breaks down semenogelin and liquifies the semen. The sperm, which lay still until then, now become very active, thrashing with their tails and crawling up the mucosa of the vagina and uterus. Their motility depends on ATP, which is made in the midpiece mitochondria by oxidizing the fructose in the semen.

19.4c Oogenesis and the Ovarian–Menstrual Cycle

In the absence of pregnancy, women have two interrelated monthly cycles controlled by shifting patterns of hormone secretion: the **ovarian cycle,** consisting of events in the ovaries, and the **menstrual cycle,** consisting of parallel changes in the uterus. The ovarian cycle is concerned with *oogenesis,* the production of eggs, and *folliculogenesis,* the parallel developments in the follicles that enclose them.

Oogenesis

Oogenesis[35] (OH-oh-JEN-eh-sis), like spermatogenesis, produces a haploid gamete by means of meiosis. A major difference, however, is that whereas males produce sperm continually, oogenesis is conspicuously cyclic and usually produces only one mature egg per month. It is accompanied by cyclic changes in hormone secretion and in the histological structure of the ovaries and uterus; the uterine changes result in the monthly menstrual flow. Oogenesis occurs in the following stages (**fig. 19.14**):

1. Stem cells called **oogonia** (OH-oh-GO-nee-uh) multiply by mitosis in the ovaries of a female fetus. Most of them degenerate and die before she is born.
2. Shortly before birth, the surviving oogonia transform into **primary oocytes** and proceed as far as early meiosis I, but then go into developmental arrest until puberty. No oogonia remain after birth. The term *egg,* or *ovum,* applies loosely to any stage from the primary oocyte to the time of fertilization. At birth, there are approximately 2 million primary oocytes, most of which degenerate and die during childhood, until only about 200,000 remain at the onset of puberty. This is a woman's lifetime supply of eggs, but it is ample. Even if a female ovulated every 28 days from age 14 to 50, she would ovulate at most 480 times. All of the other eggs degenerate without ovulating between puberty and menopause.
3. Oogenesis resumes in adolescence, when FSH stimulates monthly cohorts of about two dozen oocytes to complete meiosis I. Each primary oocyte divides into two haploid daughter cells of unequal size and different destinies. It is important to produce an egg with as much cytoplasm as possible, because if fertilized, it must divide repeatedly and produce numerous daughter cells. Splitting each oocyte into four equal but small parts would run counter to this purpose. Therefore, meiosis I produces a large daughter cell called the **secondary oocyte** and a much smaller one called the **first polar body.** The polar body disintegrates; it is merely a means of discarding the extra set of chromosomes. The secondary oocyte proceeds as far as metaphase II, then arrests until after ovulation. If it is not fertilized, it dies and never finishes meiosis.
4. If fertilized, it completes meiosis II and casts off a **second polar body,** which disposes of one chromatid from each chromosome. The chromosomes of the large remaining egg unite with those of the sperm, resulting in a diploid zygote. Further development of the zygote is discussed in chapter 20.

---

[35]*oo* = egg; *genesis* = production

Figure 19.14 Oogenesis (Left) and Corresponding Development of the Follicle (Right). APR

1–2: © Ed Reschke; 3: © Alvin Telser/McGraw-Hill Education; 4: © Ed Reschke/Getty Images; 5: © Petit Format/Science Source; 6: © Alvin Telser/McGraw-Hill Education

? *How many eggs are produced per primary oocyte? How does this compare to the number of sperm per primary spermatocyte? What is the functional importance of this difference?*

Folliculogenesis

As an egg undergoes oogenesis, the follicle around it undergoes as many as five stages of growth, collectively called **folliculogenesis.**

1. **Primordial follicles.** A primordial follicle consists of a primary oocyte surrounded by a single layer of squamous *follicular cells* (fig. 19.14). Primordial follicles begin to appear in the fetus as early as the twelfth week. They persist into adulthood, with most of them waiting at least 13 years and some as long as 50 years before they develop any further.
2. **Primary follicles.** The follicular cells become cuboidal but still form just one layer around the now larger primary oocyte.
3. **Secondary follicles.** The follicular cells divide and pile on top of each other in layers; the follicular cells are now called **granulosa cells.** They secrete a layer of clear gel around the oocyte called the **zona pellucida,** while the connective tissue around the follicle condenses to form a fibrous husk, or **theca.**
4. **Tertiary follicles.** Next, the follicle cells secrete a fluid that accumulates in little pools in the follicle wall. As they enlarge, the pools merge and become a single fluid-filled cavity, the **antrum.** On one side of the antrum, a mound of granulosa cells covers the oocyte and secures it to the follicle wall. The innermost layer of cells around the egg creates a barrier that ensures that nothing from the bloodstream can get to the oocyte except by going through (not between) these cells. The granulosa cells selectively allow nutrients and hormones to pass through to the egg, while screening out antibodies and other potentially harmful chemicals. The theca develops a rich supply of blood vessels and collaborates with the follicle cells to secrete sex hormones, especially estrogen. The theca absorbs cholesterol from the blood and converts it to androgens (androstenedione and a lesser amount of testosterone). These diffuse inward to the granulosa cells, and they convert the androgens to estrogens, especially estradiol. After ovulation, they will also transform into the ovary's progesterone-secreting cells, as we will see shortly. Thus, you can see that the ovarian follicles are a woman's main source of her dominant sex hormones.
5. **Mature (graafian[36]) follicles.** Normally only one follicle in each monthly cohort becomes a mature follicle, destined to ovulate while the rest degenerate.

The Ovarian Cycle

We can now relate oogenesis and folliculogenesis to the timetable of events in the monthly *sexual cycle* (**fig. 19.15**). The cycle varies from 20 to 45 days in length, differing from person to person and from month to month in the same person, but it averages 28 days, so we will use this as a basis for discussion. As you study this cycle, bear in mind that hormones of the hypothalamus regulate the pituitary gland; pituitary hormones regulate the ovaries; and the ovaries, in turn, secrete hormones that regulate the uterus. That is, the basic hierarchy of control can be represented: hypothalamus → pituitary → ovaries → uterus. However, the ovaries also exert feedback control over the hypothalamus and pituitary.

We begin with a brief preview of the cycle. It starts with a 2-week *follicular phase.* The first 3 to 5 days are when **menstruation** occurs, in which blood and endometrial tissue are discharged vaginally. While this is happening, a cohort of ovarian follicles grows until one of them ovulates around day 14. Endometrial tissue is regenerated and thickens between the end of menstruation and ovulation. After ovulation, the remainder of that follicle becomes a body called the *corpus luteum.* Over the next 2 weeks, called the *luteal phase,* the corpus luteum stimulates endometrial secretion, making the endometrium thicken still more. If pregnancy does not occur, the endometrium breaks down again in the last 2 days. As loose tissue and blood accumulate, menstruation begins and the cycle starts over. Now the details.

---

[36]Reijnier de Graaf (1641–73), Dutch physiologist and histologist

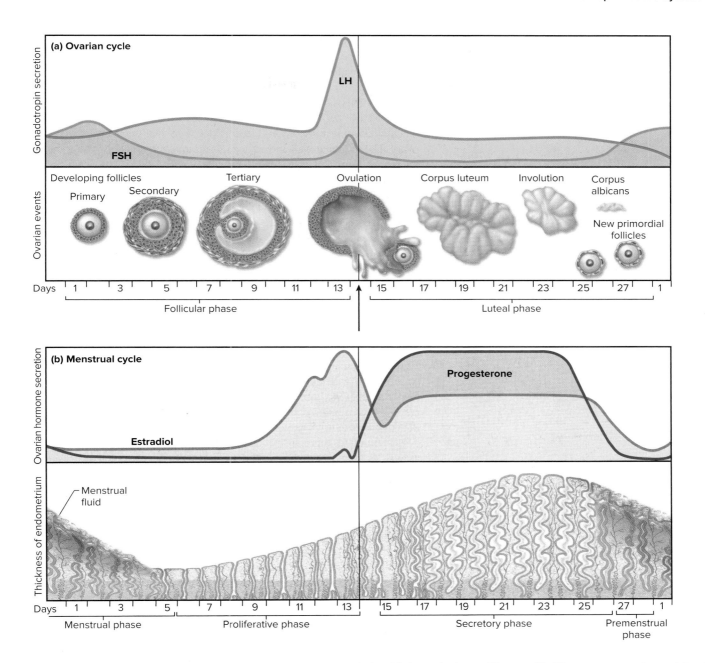

Figure 19.15 The Female Sexual Cycle.
(a) The ovarian cycle (events in the ovary). (b) The menstrual cycle (events in the uterus). **APR**

1. **Follicular phase.** This phase, from day 1 typically through day 13, is marked by the growth of a cohort of follicles stimulated by **follicle-stimulating hormone (FSH).** As they grow, these follicles secrete **estradiol** (ESS-tra-DI-ol) (one of the *estrogens*). By unknown means, one follicle has already been selected during the previous month's cycle to become the **dominant follicle**—the one destined to ovulate in this cycle. The dominant follicle is the most sensitive of all of them to FSH, luteinizing hormone (LH), and estradiol, so it races ahead of the others toward the day of ovulation. At the same time, estradiol and **inhibin** from the ovaries indirectly suppress pituitary FSH secretion. Therefore, FSH levels decline but the dominant follicle heightens its own sensitivity to the FSH that remains, so all of the less sensitive follicles degenerate while the dominant follicle continues to grow. It reaches a diameter of about 20 mm and bulges from the ovarian surface like a great blister, ready to ovulate. It is now considered a mature follicle. The follicular phase is the most variable part of the cycle in length, which makes it difficult to reliably predict the date of ovulation for the purposes of family planning or pregnancy avoidance.

2. **Ovulation.** Ovulation, typically around day 14, is the rupture of the mature follicle and the release of its egg and some of the surrounding cells. On the day before ovulation, there is a sharp spike in **luteinizing hormone (LH)**, (fig. 19.15a). This induces several momentous events. The primary oocyte completes meiosis I, giving off the first polar body and becoming a haploid secondary oocyte. Follicular fluid builds rapidly and the follicle swells to as much as 25 mm in diameter—about twice the usual thickness of the whole ovary. The follicular wall and adjacent ovarian tissue weaken. With mounting internal pressure and a weakening wall, the mature follicle approaches rupture. Meanwhile, the uterine tube prepares to catch the oocyte. It swells with edema; its fimbriae envelop and caress the ovary in synchrony with the woman's heartbeat; and its cilia create a gentle current in the nearby peritoneal fluid.

   Ovulation itself takes only 2 or 3 minutes. A nipplelike protrusion appears over the follicle; it seeps fluid for 1 or 2 minutes; and then the follicle bursts. The remaining fluid oozes out, carrying the oocyte and its companion cells (see photo in figure 19.14). These are normally swept up by the ciliary current and taken into the uterine tube.

   Follicle maturation and ovulation normally occur in only one ovary per cycle, and the ovaries usually alternate from month to month. Some women can feel the moment of ovulation as a slight twinge of abdominal pain.

3. **Luteal phase.** Days 15 to 28, from just after ovulation to the onset of menstruation, are called the **luteal (postovulatory) phase.** When the follicle ruptures, it collapses. Follicular and theca cells multiply and fill the antrum, and a dense bed of blood capillaries grows amid them. The follicle is now called the **corpus luteum,**[37] named for a yellow lipid that accumulates in its cells.

   LH stimulates this transformation from ruptured follicle to corpus luteum, as well as its continued growth. The corpus luteum secretes estrogen and especially **progesterone.** As we'll see in the next section, progesterone prepares the uterus for the possibility of pregnancy (and is named for this fact[38]). At the same time, hormones from the corpus luteum exert negative feedback inhibition on the pituitary and suppress its output of FSH and LH. (This is the basis for hormonal birth control pills, which usually contain a combination of estrogen and progesterone and suppress ovulation.)

   In the absence of pregnancy, the corpus luteum begins shrinking around day 22 (8 days after ovulation) and eventually becomes an inactive bit of scar tissue. The shrinkage is called *involution*[39] and the resulting scar is the *corpus albicans.*[40] With the waning of ovarian steroid secretion, the pituitary is no longer inhibited and FSH levels begin to rise again, ripening a new cohort of follicles.

The Menstrual Cycle

The menstrual cycle consists of uterine changes in parallel with the ovarian events—a buildup of endometrium through most of the cycle, followed by its breakdown and vaginal discharge. The menstrual cycle is divided into a *proliferative phase, secretory phase, premenstrual phase,* and *menstrual phase* (fig. 19.16). The menstrual phase averages 5 days long, and the first day of noticeable vaginal discharge is defined as day 1 of the sexual cycle. But menstruation is best understood after you become acquainted with the buildup of endometrial tissue that precedes it. Thus, we begin our survey of the cycle with the proliferative phase.

---

[37]*corpus* = body; *lute* = yellow
[38]*pro* = favoring, promoting; *gest* = pregnancy; *sterone* = steroid hormone
[39]*in* = inward; *volution* = turning
[40]*corpus* = body; *albicans* = white

Figure 19.16 Endometrial Changes Through the Menstrual Cycle. (a) Late proliferative phase. The endometrium is 2 to 3 mm thick and has relatively straight, narrow endometrial glands. Spiral arteries penetrate upward between the endometrial glands. (b) Secretory phase. The endometrium has thickened to 5 to 6 mm by accumulating glycogen and mucus. The endometrial glands are much wider and more distinctly coiled, showing a zigzag or "sawtooth" appearance in histological sections. (c) Menstrual phase. Dying tissue has begun to fall away from the uterine wall, with bleeding from broken blood vessels and pooling of blood in the tissue and uterine lumen.

1. **Proliferative phase.** In this phase, the uterus rebuilds the layer of tissue lost in the last menstruation. At day 5 (or the end of menstruation), the endometrium is about 0.5 mm thick and consists only of the permanent *basal layer* of tissue that is never shed. As a new cohort of follicles develops in the ovaries, they secrete more and more estrogen. Estrogen stimulates mitosis in the basal layer as well as a prolific regrowth of blood vessels, thus regenerating the *functional layer* that was lost at menstruation (fig. 19.16a). By day 14, the endometrium is 2 to 3 mm thick. Estrogen also stimulates endometrial cells to produce progesterone receptors, priming them for the progesterone-dominated phase to follow.
2. **Secretory phase.** In this phase, the endometrium thickens still more, but because of secretion and fluid accumulation rather than mitosis. This phase extends from day 15 (after ovulation) to day 26 of a typical cycle. After ovulation, the corpus luteum secretes mainly progesterone, which stimulates the endometrial glands to secrete glycogen. The glands grow wider, longer, and more coiled, and the endometrium swells with tissue fluid (fig. 19.16b). By the end of this phase, the endometrium is 5 to 6 mm thick—a soft nutritious bed available for embryonic development in the event of pregnancy.
3. **Premenstrual phase.** If there is no pregnancy, the last 2 days of the cycle are marked by endometrial degeneration. This comes about because the dying corpus luteum secretes less and less progesterone. The loss of progesterone triggers spasmodic contractions of the spiral arteries of the endometrium. The interrupted blood flow leads to tissue necrosis and sometimes menstrual cramps. As the endometrial glands and blood vessels degenerate, pools of blood accumulate in the tissue. Pieces of dying endometrium fall away from the uterine wall, mix with blood and serous fluid in the lumen, and form the **menstrual fluid** (fig. 19.16c). Eventually, all of the functional layer is lost in the menstrual fluid and only the basal layer remains.
4. **Menstrual phase (menses).** This phase begins when enough fluid has accumulated in the uterus to be discharged from the vagina. The first day of discharge marks day 1 of the menstrual cycle. The average woman expels about 40 mL of blood and 35 mL of serous fluid over a 5-day period.

To summarize the integration of the ovarian and uterine cycles: The ovaries go through a follicular phase characterized by growing follicles, then ovulation, and then a postovulatory (mostly luteal) phase dominated by the corpus luteum. The uterus, in the meantime, goes through a menstrual phase in which it discharges its functional layer; then a proliferative phase in which it replaces that tissue by mitosis; then a secretory phase in which the endometrium thickens by the accumulation of secretions; and finally, a premenstrual phase in which the functional layer breaks down again. The first half of the cycle is governed largely by follicle-stimulating hormone (FSH) from the pituitary and estradiol from the ovaries. Ovulation is triggered by luteinizing hormone (LH) from the pituitary, and the second half of the cycle is governed mainly by LH from the pituitary and progesterone from the ovaries.

The timetable of ovulation and menstruation, especially the menstrual period starting so conspicuously every 28 days or so, easily creates a misconception that it takes 28 days for an egg and follicle to mature and ovulate. The reality is that it can take anywhere from roughly 12 to 50 *years* for each one to develop—from development of the primary oocytes before a girl is born to ovulation of the last secondary oocyte in menopause. Most of this time, the follicles lie dormant, with their eggs in early meiosis I, waiting for a "wake-up call" from the pituitary in the form of follicle-stimulating hormone. Starting in puberty, about a dozen follicles each month respond to this and resume their development. But even then, it takes about 290 days for one member of this cohort, the dominant follicle, to mature and ovulate. So any egg released from its follicle began its meiosis before the girl was born, "went to sleep" for many years, then "woke up" and resumed its development 290 days before ovulation. The reason for the 28-day periodicity of ovulation is that these cohorts mature in overlapping waves, with one cohort each month finally producing its mature follicle and its candidate for fertilization, the ovulated secondary oocyte.

19.4d Uniting the Gametes APR

We have seen how the sperm and eggs are produced; the purpose of reproduction is to get them together. The physiology of *sexual intercourse (coitus)* was a taboo subject for many decades until, in the 1950s, William Masters and Virginia Johnson daringly launched the first physiological studies of sexual response in the laboratory. Thanks to their pioneering work and those who have followed, sexual response is now studied on a level as matter-of-fact as the physiology of digestion or respiration.

Anatomical Foundations

To understand sexual function, we must give closer attention to the blood circulation and nerve supply to the genitalia. Each internal iliac artery of the male gives off a **penile artery** that enters the root of the penis and divides in two. One branch, the **dorsal artery,** travels dorsally along the penis not far beneath the skin (see fig. 19.4), supplying blood to the skin, fascia, and corpus spongiosum. The other branch, the **deep artery,** travels through the core of the corpus cavernosum and gives off smaller arteries that penetrate the trabeculae and empty into the lacunae. When the deep artery dilates, the lacunae fill with blood and the penis becomes erect. Blood drains from the penis through a median *deep dorsal vein,* which runs between the two dorsal arteries near the penile surface. The blood supply to the female's clitoris is essentially the same except that there is no corpus spongiosum.

The penis and clitoris are richly innervated by sensory and motor nerve fibers. The glans of each has an abundance of tactile, pressure, and temperature receptors. Signals travel out of the penis or clitoris by way of a pair of prominent *dorsal nerves* that lead to the sacral spinal cord. Sensory fibers of the penile shaft and scrotum of the male and the labia and vestibule of the female are also highly important to sexual stimulation.

Both autonomic and somatic motor fibers carry impulses from the spinal cord to the genitalia. Sympathetic fibers from spinal cord levels T12 to L2 innervate the arteries of the penis and clitoris, the accessory glands, and the male spermatic ducts. They dilate the arteries and can induce erection even when the sacral region of the spinal cord is damaged. The arteries also receive parasympathetic fibers from spinal cord segments S2 to S4 of the spinal cord. These nerves are involved in an autonomic reflex arc that causes erection in response to direct stimulation of the genitalia and perineal region.

Sexual Response

At the onset of sexual arousal, the genitalia become swollen with blood, muscle tension rises throughout the body, and the heart rate, blood pressure, and respiratory rate rise. The bulbourethral glands of the male and greater vestibular glands of the female secrete their fluids during this phase, with the female secretion, especially, providing lubrication for intercourse. The vagina contains no glands but becomes moistened by seepage of serous fluid from its wall.

The most obvious and necessary manifestation of male sexual arousal is **erection** of the penis. This is an autonomic reflex mediated mainly by parasympathetic nerve fibers from the spinal cord. These nerve fibers dilate the deep arteries of the corpora cavernosa, causing rapid inflow of blood. As lacunae nearest the deep arteries fill with blood, they compress lacunae closer to the periphery of the erectile tissue. This is where blood leaves the erectile tissues, so compression blocks the outflow of blood. Each corpus cavernosum is wrapped in a tight fibrous tunic or sleeve. This tunic cannot expand very much laterally, so it contributes to the tension and firmness of the corpus cavernosum and forces the penis to expand more in length than in width. Once entry into the vagina is achieved, tactile and pressure sensations produced by vaginal massaging of the penis further accentuate the erection reflex. In the female, the clitoris becomes similarly engorged and erect, swelling to two or three times its usual size. But since the clitoris cannot swing upward away from the body like the penis, it tends to withdraw beneath the prepuce.

The male corpus spongiosum has neither a central artery nor a fibrous tunic. It swells and becomes more visible as a cordlike ridge along the ventral surface of the penis, but it doesn't become nearly as engorged and hardened as the corpora cavernosa.

> ### Apply What You Know
> Why is it important that the corpus spongiosum not become as engorged as the corpora cavernosa? What might be the consequence if it did so?

Orgasm[41] (**climax**) is a short but intense reaction that lasts from 3 to 15 seconds; it is usually marked by the discharge of semen in males and by vaginal and uterine contractions in females. From the standpoint of producing offspring, the most significant aspect of male orgasm is the **ejaculation**[42] of semen into the vagina. It occurs in two stages called *emission* and *expulsion*. In **emission,** the sympathetic nervous system induces peristalsis in the ductus deferens, which propels sperm from the epididymis to the prostatic urethra. The prostate expels its own fluid into the urethra, followed quickly by the seminal vesicles. The contractions and fluid flow in this phase create an urgent sensation that ejaculation is inevitable.

---

[41]*orgasm* = swelling
[42]*e* = *ex* = out; *jacul* = to throw

Semen in the urethra activates both somatic and sympathetic reflexes that result in its **expulsion.** Sensory signals travel to an integrating center in the upper lumbar region of the spinal cord. Sympathetic signals return to the prostate and seminal vesicles, causing them to expel more fluid into the urethra. The internal urethral sphincter constricts so urine cannot enter the urethra and semen cannot enter the bladder. Somatic motor signals from the spinal cord stimulate the **bulbospongiosus muscle,** which ensheaths the root of the penis (see fig. 19.1). In five or six spasmodic contractions, this muscle compresses the urethra and forcibly expels the semen. Most sperm emerge in the first milliliter of semen, mixed primarily with prostatic fluid. The seminal vesicle secretion follows and flushes most remaining sperm from the urethra.

In female orgasm, the lower end of the vagina gives three to five strong contractions about 0.8 seconds apart, while the cervix plunges spasmodically into the vagina and into the pool of semen, should this be present. The uterus exhibits peristaltic waves of contraction, which might help to draw semen from the vagina.

Following orgasm, sympathetic stimulation constricts the arteries and reduces the flow of blood into the penis and clitoris. It also causes contraction of the smooth muscle in the trabeculae of the erectile tissues, which squeezes blood from the lacunae. The penis may remain semierect long enough to continue intercourse, but gradually it becomes soft and flaccid again. In men, this is usually followed by a *refractory period* of anywhere from 10 minutes to a few hours, in which it is usually impossible to attain another erection and orgasm. Women do not have a refractory period and may quickly experience additional orgasms.

Sperm can make it well into the uterine tubes in half an hour or less after ejaculation. For reasons explained in section 20.1, however, it is about 10 hours before they become capable of penetrating an egg and initiating a pregnancy.

Before You Go On

Answer these questions from memory to test your understanding of the previous section. Reread the material if you can't answer the questions.

16. Why is meiosis necessary in sexual reproduction? At what stages in meiosis do we see a change in the number of chromosomes per cell and the number of chromatids per chromosome?
17. List the stages of spermatogenesis; identify which stages involve mitosis, meiosis I, and meiosis II; and state the chromosome number at each stage. Then do the same for oogenesis.
18. Identify the main parts of a mature spermatozoon and the function of each part.
19. Spermatogenesis produces four functional gametes per spermatogonium, and oogenesis produces only one gamete per oogonium. Why?
20. Name the five kinds of ovarian follicles and identify the essential features that distinguish one from another.
21. Describe the changes occurring in the ovaries and uterus in the 2 weeks before ovulation and in the 2 weeks after.
22. What is a corpus luteum? Where does it come from and what purpose does it serve?
23. Describe some similarities and differences in the physiology of intercourse and orgasm between the male and female.

PERSPECTIVES ON HEALTH

Sexually Transmitted Diseases

Sexually transmitted diseases (STDs) have been well known at least since the writings of Hippocrates. Here, we discuss three bacterial STDs—gonorrhea, chlamydia, and syphilis—and three viral STDs—genital herpes, genital warts, and hepatitis. AIDS is discussed in chapter 14 in Perspectives on Health.

All of these STDs have an *incubation period* in which the pathogen multiplies without symptoms, and a *communicable period* in which one can transmit the disease to others even in the absence of symptoms. Aside from recent advances with human papillomavirus (HPV), there are few vaccines for STDs, because the bacteria and viruses live inside host cells and go undetected by the immune system. STDs can cause fetal deformity, stillbirth, and neonatal death (see section 20.3).

Gonorrhea (GON-oh-REE-uh) is caused by the bacterium *Neisseria gonorrhoeae*. Galen, thinking the pus discharged from the penis was semen, named the disease *gonorrhea* ("flow of seed"). Symptoms include abdominal discomfort, genital pain and discharge, painful urination, and abnormal uterine bleeding. Most infected women are asymptomatic, but it can cause scarring of the uterine tubes, resulting in infertility. Gonorrhea is treated with antibiotics.

Nongonococcal urethritis (NGU) is any urethral inflammation caused by agents other than the gonorrhea bacterium. NGU often produces pain or discomfort on urination. The most common bacterial NGU is infection by *Chlamydia trachomatis*. Most chlamydia infections are asymptomatic, but they may cause urethral discharge and pain in the testes or pelvic region. Gonorrhea and chlamydia frequently occur together.

Pelvic inflammatory disease (PID) is bacterial infection of the female pelvic organs, usually with *Chlamydia* or *Neisseria*. It often results in sterility and may require surgical removal of infected uterine tubes or other organs. PID is responsible for many cases of ectopic pregnancy (see Clinical Application 20.1).

Syphilis is caused by a corkscrew-shaped bacterium named *Treponema pallidum*. After an incubation period of 2 to 6 weeks, a small, hard lesion called a *chancre* (SHAN-kur) appears at the site of infection—in males, usually on the penis and therefore noticeable, but in females, sometimes out of sight in the vagina. It disappears in 4 to 6 weeks, ending the first stage of syphilis and often creating an illusion of recovery. A second stage ensues, however, with a widespread pink rash, other skin eruptions, fever, joint pain, and hair loss. This subsides in 3 to 12 weeks, but symptoms can come and go for up to 5 years. A person is contagious even when symptoms are not present. The disease may progress to a third stage, *tertiary syphilis (neurosyphilis)*, with cardiovascular damage and brain lesions that can cause paralysis and dementia. Syphilis is treatable with antibiotics.

Genital herpes is one of the most common STDs in the United States, with 20 to 40 million infected people at a given time. It is usually caused by the *herpes simplex virus type 2 (HSV-2)*. After an incubation period of 4 to 10 days, the virus causes blisters on the penis of the male; on the labia, vagina, or cervix of the female; and sometimes on the thighs and buttocks of either sex. Over 2 to 10 days, these blisters rupture, seep fluid, and begin to form scabs. The initial infection may be painless or it may cause intense pain, urethritis, and watery discharge from the penis or vagina. The lesions heal in 2 to 3 weeks and leave no scars.

During this time, however, HSV-2 colonizes sensory nerves and ganglia. Here the virus can lie dormant for years, later migrating along the nerves and causing epithelial lesions anywhere on the body. The movement from place to place is the basis of the name *herpes*.[43] Most infected persons have five to seven recurrences, ranging from several years apart to several times a year. An infected person is contagious to a sexual partner when the lesions are present and sometimes even when they are not. HSV-2 may increase the risk of cervical cancer and AIDS.

Human papillomavirus (HPV) is the most common sexually transmitted infection in the United States. An HPV infection may resolve on its own without causing any health problems; yet nearly all cases of cervical cancer are caused by HPV, and it is also linked to cancers of the oropharynx, penis, vagina, and anus. In addition, HPV can cause *genital warts (condylomas)*. In the male, lesions usually appear on the penis, perineum, or anus; in the female, they are usually on the cervix, vaginal wall, perineum, or anus. Lesions are sometimes small and almost invisible. Genital warts are sometimes treated with cryosurgery (freezing and excision), laser surgery, or interferon. An HPV vaccine became available in 2006, and infection rates in young people have declined as a result of vaccinations.

Hepatitis B and *C,* introduced in Clinical Application 18.1, are inflammatory liver diseases caused by the hepatitis B and C viruses (HBV, HCV). Although they can be transmitted by means other than sex, they are becoming increasingly common as STDs. Hepatitis C threatens to become a major epidemic of the twenty-first century. It already far surpasses the prevalence of AIDS and is the leading reason for liver transplants in the United States.

---

[43]*herp* = to creep

19.5 Pregnancy, Childbirth, and Lactation

Expected Learning Outcomes

When you have completed this section, you should be able to

a. itemize the major hormones of pregnancy and describe their effects;
b. describe the effects of pregnancy on a woman's body;
c. explain what happens in each stage of childbirth;
d. discuss the hormonal control of lactation; and
e. discuss the composition of colostrum and breast milk.

This section treats pregnancy from the maternal standpoint—adjustments of the woman's body to pregnancy and the mechanism of childbirth—followed by the physiology of lactation. Development of the fetus is described in chapter 20.

19.5a Pregnancy

Pregnancy (gestation) lasts an average of 266 days from **conception** (fertilization of the egg) to childbirth; however, the gestational calendar is usually measured from the first day of the woman's last menstrual period (LMP), since a woman is more likely to know that date than when she conceived. Thus, the birth is predicted to occur 280 days (40 weeks) from the LMP. The duration of pregnancy, called its *term,* is commonly described in 3-month intervals called **trimesters.**

Pregnancy places considerable stress on a woman's body and requires adjustments in nearly all the organ systems. Some of these effects are as follows:

Integumentary system. The skin grows to accommodate expansion of the abdomen and breasts and the added fat deposition in the hips and thighs. Stretching of the dermis often tears the connective tissue and causes *stretch marks (striae).* Increased melanin synthesis commonly darkens the areola of the breast, making the nipple a more easily seen target for a nursing infant.

Circulatory system. A woman gains 1 to 2 liters of blood over the course of a pregnancy, with about 625 mL/min. going to supply the placenta. Another effect of pregnancy is that the growing uterus puts pressure on the large pelvic blood vessels, thus interfering with venous return from the lower body. This can result in hemorrhoids, varicose veins, and edema of the feet.

Respiratory system. Breathing also becomes more labored as the growing uterus puts pressure on the diaphragm from below. In the last month, however, the fetus drops lower in the abdominopelvic cavity, taking some pressure off the diaphragm and allowing the woman to breathe more easily.

Urinary system. A pregnant woman's urine output is slightly elevated, enabling her to dispose of both her own and the fetus's metabolic wastes. As the pregnant uterus compresses the bladder and reduces its capacity, urination becomes more frequent.

Digestive system. For many women, one of the first signs of pregnancy is **morning sickness**—nausea, especially after rising from bed, in the first few months of gestation. This sometimes progresses to vomiting, occasionally severe enough to require hospitalization to stabilize a woman's fluid and electrolyte balance.

Metabolism and nutritional needs. Pregnancy raises a woman's metabolic rate and stimulates appetite. Even in the last trimester, however, a woman needs only an extra 300 kcal/day. A healthy average weight gain is 11 kg (24 pounds) but some women gain as much as 34 kg (75 pounds). About 3 kg (7 pounds) of the healthy weight gain is the fetus; the rest is the placenta, fetal membranes, amniotic fluid, uterus, breasts, maternal body fluids, and fat. Pregnancy warrants taking supplemental iron, vitamin D, and vitamin K. Extra folic acid can prevent the risk of neurological disorders in the fetus such as spina bifida, but it has to be taken even before pregnancy. By the time a woman is aware she is pregnant, it is too late for extra folic acid to be effective.

19.5b Childbirth

The uterus weighs about 50 g when a woman is not pregnant and about 900 g (2 pounds) by full term. Most of this increase is due to muscle growth in preparation for childbirth. Ancient authorities thought that the fetus kicked against the uterus and pushed itself out head first. Expulsion of the fetus, however, is achieved only by the contractions of the mother's uterine and abdominal muscles. That doesn't mean the fetus is just a passive passenger. Evidence suggests that the fetus generates chemical signals that help initiate labor.

During gestation, the uterus exhibits relatively weak **Braxton Hicks**[44] **contractions.** These become stronger in late pregnancy and often send women rushing to the hospital with "false labor." At full term, however, these contractions transform suddenly into the more powerful **labor contractions.** True labor contractions mark the onset of **parturition** (PAR-too-RISH-un), the process of giving birth.

The increasing contractility of the uterus may stem from the rising ratio of estrogen to progesterone toward the end of pregnancy. Also, as the pregnancy nears full term, the posterior lobe of the pituitary releases increasing amounts of **oxytocin (OT),** while progesterone stimulates the uterine muscle to develop more OT receptors. Oxytocin directly stimulates the uterine contractions. Stretching of the uterus by the growing fetus also plays a role in initiating labor contractions. This is probably why twins are born an average of 19 days earlier than solitary infants.

Labor contractions typically begin about 30 minutes apart, then become more frequent and intense, eventually occurring every 1 to 3 minutes. It is important that they be intermittent rather than one long, continual contraction. Each contraction compresses blood vessels and reduces blood flow to the placenta, so the uterus must periodically relax to restore flow and oxygen delivery to the fetus. Contractions are strongest in the fundus and body of the uterus and weaker near the cervix, thus pushing the fetus downward.

According to the *positive feedback theory of labor,* contractions are induced by stretching of the cervix. This triggers a reflex contraction of the uterine body that pushes the fetus downward and stretches the cervix still more. Thus, there is a self-amplifying cycle of stretch and contraction. In addition, stretching triggers a reflex through the spinal cord, hypothalamus, and posterior pituitary. The pituitary releases oxytocin, which is carried in the blood and stimulates the uterine muscle. This, too, is a positive feedback cycle: cervical stretching → oxytocin secretion → uterine contraction → cervical stretching (see fig. 1.5).

As labor progresses, a woman feels a growing urge to "bear down." A somatic reflex arc extends from the uterus to the spinal cord and back to the skeletal muscles of the abdomen. Contraction of these muscles—partly reflexive and partly voluntary—aids in expelling the fetus, especially when combined with breath holding (the Valsalva maneuver) for increasing intra-abdominal pressure.

Labor occurs in three stages (**fig. 19.17**), each lasting longer in a woman giving birth for the first time (a *primipara*) than in a woman who has given birth before (*multipara*).

---

[44] John Braxton Hicks (1823–97), British gynecologist

1. **Dilation stage.** This stage is marked by the **dilation** (widening) of the cervical canal and **effacement** (thinning) of the cervix. The cervix dilates to about 10 cm (the diameter of the baby's head), and usually the fetal membranes rupture and discharge the *amniotic fluid* (the "breaking of the waters"). This stage typically lasts 8 to 24 hours in a primipara but as little as a few minutes in a multipara.
2. **Expulsion (second) stage.** This stage typically lasts 30 to 60 minutes in a primipara and as little as 1 minute in a multipara. It begins when the baby's head enters the vagina and lasts until the baby is entirely expelled. Delivery of the head is the most difficult part, with the rest of the body following much more easily.
3. **Placental (third) stage.** The uterus continues to contract after expulsion of the baby. The placenta, however, is a nonmuscular organ that cannot contract, so it buckles away from the uterine wall. The placenta and fetal membranes *(afterbirth)* are then expelled by uterine contractions. The membranes must be carefully inspected to be sure everything has been expelled. If any of these structures remain in the uterus, they can cause postpartum hemorrhaging. About 350 mL of blood is typically lost in the placental stage, but uterine contractions compress the blood vessels and prevent more extensive bleeding.

19.5c Postpartum Changes

Over the first 6 weeks **postpartum** (after birth), the mother's anatomy and physiology stabilize and her reproductive organs return nearly to their condition prior to pregnancy. The shrinkage of the uterus during this period is called **involution** (the same word as for the shrinkage of the corpus luteum described earlier). It is achieved through the autolysis (self-digestion) of uterine cells by their own enzymes. For about 10 days, this produces a vaginal discharge called **lochia**[45] (LO-kee-ah), which is bloody at first and then turns clear and serous. Breast-feeding hastens involution because (1) it suppresses estrogen secretion, which would otherwise cause the uterus to remain more flaccid; and (2) it stimulates oxytocin secretion, which causes the myometrium to contract and firm up the uterus sooner.

19.5d Lactation

Lactation is the synthesis and ejection of milk from the mammary glands. It lasts for as little as a week in women who don't breast-feed their infants, but it can continue for many years as long as the breast is stimulated by a nursing child or mechanical device (breast pump). Numerous studies conducted before the widespread marketing of artificial infant formulas suggest that worldwide, women traditionally nursed their infants until a median age of about 2.8 years.

Development of the Mammary Glands in Pregnancy

In nonpregnant, non-nursing women, the mammary gland is represented by only a few milk ducts that penetrate from the nipple a short way into the breast tissue. In pregnancy, the high estrogen level causes the ducts to grow and branch extensively. Growth hormone, prolactin, and other hormones also contribute to this development. Once the ducts are complete, progesterone stimulates the budding and development of acini at the ends of the ducts. The acini are organized into grapelike clusters (lobules) within each lobe of the breast (see fig. 19.10).

---

[45]*lochos* = childbirth

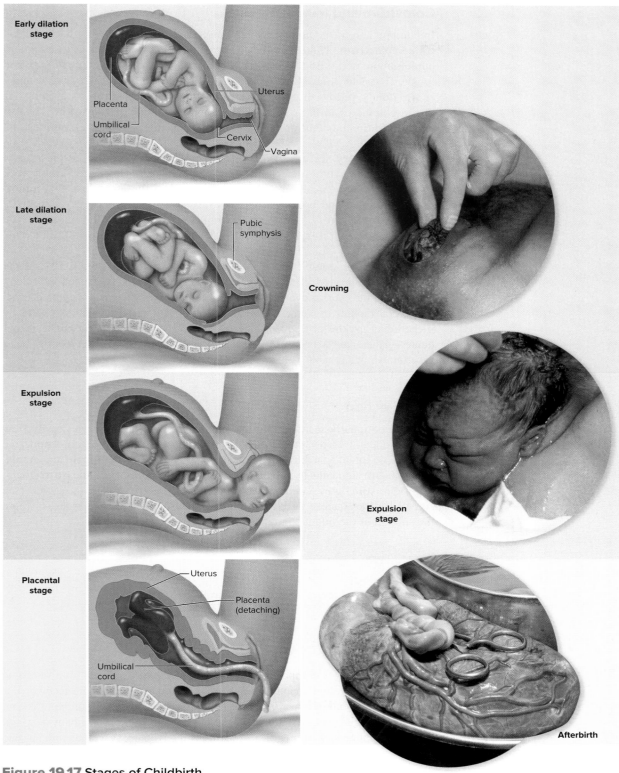

Figure 19.17 Stages of Childbirth.
Top: Petit Format/Science Source
bottom: © Medicshots/Alamy Stock Photo

Colostrum and Milk Synthesis

In late pregnancy, the mammary acini and ducts are distended with a secretion called **colostrum.** This is similar to breast milk in protein and lactose content but has about one-third less fat. It is the infant's only natural source of nutrition for the first 1 to 3 days postpartum. Colostrum has a thin watery consistency and a cloudy yellowish color. The daily volume of colostrum is at most 1% of the volume of milk secreted later. A major benefit of colostrum is that it contains antibodies that protect the infant from infection and gastrointestinal inflammation.

The pituitary secretes prolactin beginning 5 weeks into the pregnancy, but no milk is synthesized until after birth. Placental steroids inhibit milk synthesis, but when the placenta is discharged in the afterbirth, this inhibition is removed and prolactin begins to exert its effect. Milk is synthesized in increasing quantity over the following week. Each time the infant nurses, there is a spike in prolactin level. This stimulates the synthesis of milk for the next feeding.

If the mother doesn't nurse, she stops lactating in about a week. Even if she does nurse, milk production declines after 7 to 9 months. In women who don't breast-feed, the ovarian cycle resumes in a few weeks, whereas it may be delayed for several months in women who do. Lactation suppresses ovulation, especially if the infant suckles at frequent intervals, including throughout the night. In our hunter–gatherer ancestors, this ensured a relatively long space between births, but it is certainly not a reliable form of contraception for modern women.

Milk Ejection

Milk is continually secreted into the mammary acini, but it doesn't easily flow into the ducts. Its flow, called **milk ejection,** is controlled by a neuroendocrine reflex. The infant's suckling stimulates nerve endings of the nipple and areola, which in turn signal the hypothalamus and posterior pituitary to release oxytocin. Oxytocin stimulates contractile *myoepithelial cells* around the acini (see fig. 19.10d) to squeeze milk down into the ducts and lactiferous sinuses, where the suction of the infant's mouth can draw it out.

> ### Apply What You Know
> When a woman is nursing her baby at one breast, would you expect only that breast, or both breasts, to eject milk? Explain why.

Breast Milk

Breast milk changes composition over the first 2 weeks, varies from one time of day to another, and changes even during the course of a single feeding. For example, at the end of a feeding there is less lactose and protein in the milk, but six times as much fat, as there is at the beginning. Therefore, a baby may not receive complete nourishment if its feeding is interrupted. A woman nursing one baby eventually produces about 1.5 L of milk per day; women with twins produce more. Lactation places a great demand on the mother for calcium, phosphate, protein, fat, and other nutrients, and requires appropriate dietary adjustments so that mineralization of the infant's skeleton doesn't come at the expense of the mother's bone tissue.

It used to be common to feed infants with cow's milk, but this is a poor substitute for colostrum and breast milk. Table 19.1 compares their composition. Calves and babies have different patterns of growth and different nutritional requirements.

Table 19.1 A Comparison of Colostrum, Human Milk, and Cow's Milk

Constituents	Human Colostrum	Human Milk	Cow's Milk
Total protein (g/L)	22.9	10.6	30.9
Lactalbumin (g/L)	–	3.7	25.0
Casein (g/L)	–	3.6	2.3
Immunoglobulins (g/L)	19.4	0.09	0.8
Fat (g/L)	29.5	45.4	38.0
Lactose (g/L)	57	71	47
Calcium (mg/L)	481	344	1,370
Phosphorus (mg/L)	157	141	910

Cow's milk is more difficult for an infant to digest because of its excessive protein content; it causes more diaper rash; and it fails to provide the immunity and other benefits of breast-feeding.

Before You Go On

Answer these questions from memory to test your understanding of the previous section. Reread the material if you can't answer the questions.

24. What are the roles of oxytocin, prostaglandins, and uterine stretching in childbirth?
25. In childbirth, what are dilation, effacement, and crowning? In what stage of childbirth does each one occur?
26. How does colostrum differ from breast milk? What purposes does it serve?
27. Contrast the roles of prolactin and oxytocin in breast-feeding.

Aging of the Reproductive System

Male reproductive function changes relatively gradually after middle age, with declining testosterone secretion, sperm count, and libido. Men remain fertile (able to father children) well into old age, but by age 65, sperm count is about one-third of what it was in a man's 20s. Erectile dysfunction (ED)—the inability to maintain an erection long enough to complete intercourse—becomes more common for a variety of reasons, including atherosclerosis, medications, and psychological factors. About 20% of men in their 60s and 50% of men in their 80s experience some degree of ED. Nearly all men with ED, however, remain able to ejaculate.

In women, the changes in reproductive function are more abrupt and pronounced. Over the course of menopause, gametogenesis ceases, active follicles are depleted, and the ovaries stop secreting sex hormones. This can result in vaginal dryness, genital atrophy, and reduced libido. With the loss of ovarian steroids, postmenopausal women have elevated risks of atherosclerosis and osteoporosis. Women who become pregnant in their 30s and 40s or beyond have an increasing risk of having infants with chromosomal birth defects such as Down syndrome, for a reason explained in the next chapter (see "Birth Defects" in section 20.3c).

CAREER SPOTLIGHT

Midwife

A midwife is a person who provides prenatal counseling and care to expectant mothers, attends the birth, and provides postpartum care to the mother and infant. Midwives encourage natural birth with minimal technological intervention. They work with relatively healthy women and low-risk pregnancies, but are trained to recognize complications and to refer these to obstetricians for advanced care. Midwives often attend births in the home or at a birthing center, but in some hospitals today, the majority of babies are delivered by midwives. The term *midwife* derives from Old English "with *(mid)* woman *(wif),*" referring to the pregnant woman and not the practitioner; both women and men work as midwives.

© Ruth Jenkinson/MIDIRS Science Source

The best career opportunities and salaries go to certified nurse midwives (CNMs), who must earn a nursing degree and then pursue about 2 years of specialized training in midwifery, often leading to a master's degree. CNMs must pass a national board exam, and can then be licensed in any of the 50 states. Most CNM programs are given at universities, and most CNMs practice in hospitals.

There also are *direct-entry midwives* who do not hold nursing degrees but have more limited employment opportunities. These range from lay midwives (LMs), educated through self-study and apprenticeship, to certified professional midwives (CPMs), who complete formal programs of study. Prerequisites for such programs vary by state but may include general biology, microbiology, human anatomy and physiology, mathematics, English, and social science. Many states do not recognize or license direct-entry midwives. Where licensed, these midwives usually practice outside of hospitals and sometimes establish their own birthing centers.

See appendix B for further information on careers in midwifery.

CONNECTIVE ISSUES

Ways in Which the REPRODUCTIVE SYSTEM Affects Other Organ Systems

Integumentary System
At puberty, androgens stimulate development of body hair and apocrine glands and increased secretion by sebaceous glands; estrogens stimulate fat deposition and breast development in females; pregnancy may cause stretch marks and pigmentation changes in skin.

Skeletal System
Androgens and estrogens stimulate bone deposition and adolescent skeletal growth and maintain adult bone density.

Muscular System
Androgens and estrogens stimulate muscle growth.

Nervous System
Sex steroids stimulate the brain and libido; hormones from the gonads and placenta exert negative feedback control of the hypothalamus.

Endocrine System
The gonads and placenta are part of the endocrine system and secrete androgens, estrogens, progesterone, and pregnancy hormones.

Circulatory System
Androgens stimulate erythropoiesis; estrogens may inhibit atherosclerosis in females; pregnancy increases blood volume and cardiac output and sometimes causes varicose veins.

Lymphatic and Immune Systems
The blood–testis barrier in testes and zona pellucida in ovaries isolate sperm and eggs from the immune system. Sexually transmitted diseases present challenges to the immune system.

Respiratory System
Pulmonary ventilation increases in response to sexual arousal and in pregnancy.

Urinary System
Sexual arousal constricts the internal urethral sphincter; prostatic hyperplasia may impede urine flow; pregnancy crowds the urinary bladder.

Digestive System
The fetus crowds digestive organs, contributing to heartburn and constipation. Pregnancy is often associated with nausea, and both pregnancy and lactation increase nutritional requirements.

Study Guide

Assess Your Learning Outcomes

To test your knowledge, discuss the following topics with a study partner or in writing, ideally from memory.

19.1 Essentials of Sexual Reproduction

1. The meanings of *gamete* and *zygote,* and why sexual reproduction employs two different kinds of gametes
2. The meanings of *gonads, primary* and *secondary sex organs,* and *external* and *internal genitalia*

19.2 The Male Reproductive System

1. Gross anatomy of the testis and the internal organization of its tissues
2. Structure of the seminiferous tubules and the functions of its germ cells and nurse cells
3. Anatomy and functions of the epididymis and ductus deferens
4. Structure of the scrotum and the components of the spermatic cord
5. The mechanisms by which the scrotum and spermatic cord regulate the temperature of the testes, and the reason why this temperature regulation is so important
6. The locations and structure of the seminal vesicles and prostate, and their contributions to the semen
7. The location and function of the bulbourethral glands
8. Gross anatomy of the penis and the microscopic anatomy of its three erectile tissues
9. Hormonal events that bring on puberty, and the bodily changes that these hormones produce
10. The hormonal events of andropause and the effects of these changes experienced by some men

19.3 The Female Reproductive System

1. Location, gross anatomy, and internal structure of the ovaries
2. The regions and internal histology of the uterine tubes
3. The regions and shape of the uterus
4. The three layers of the uterine wall and two subdivisions of the endometrium
5. The gross anatomy and histology of the vagina
6. Anatomy of the vulva, including the locations and structures of the mons pubis, labia majora and minora, clitoris, prepuce, and vaginal orifice
7. The locations and functions of the vestibular bulbs, greater and lesser vestibular glands, and paraurethral glands
8. External regions and internal structure of the nonlactating breast
9. Hormonal events that bring on puberty, and the bodily changes that these hormones produce
10. The hormonal events of female climacteric and menopause and the effects of these changes experienced by some women

19.4 The Production and Union of Sex Cells

1. The homologous pairs of human chromosomes, the sex chromosomes, and the distinction between diploid and haploid cells
2. The reason that meiosis is integral to sexual reproduction, how it converts diploid cells to haploid cells, and how it contributes to genetic diversity
3. Similarities and differences between mitosis and meiosis
4. The stages of meiosis and principal events that characterize each stage
5. The stages of spermatogenesis, the points at which mitosis and meiosis are involved, and the point at which the chromosome number changes from diploid to haploid
6. Structure of the mature spermatozoon and the functions of its parts
7. Constituents of semen, their sources, and their functions
8. The stages of oogenesis, and the similarities and differences between oogenesis and spermatogenesis
9. The stages of folliculogenesis, the features that distinguish ovarian follicles of each type, the correlation between folliculogenesis and oogenesis, and the structure of a mature follicle
10. The phases and events of the ovarian cycle, how they relate to changes in the follicles, and the hormones that dominate each phase
11. The phases of the menstrual cycle, the endometrial changes that occur in each, and how these relate to events in the ovaries and their hormonal output
12. Circulatory anatomy, erectile tissues, and innervation of the penis and clitoris and how these relate to erectile function
13. The mechanism of erection, physiology of orgasm and ejaculation, and processes that follow orgasm in both sexes

19.5 Pregnancy, Childbirth, and Lactation

1. The timetable of pregnancy
2. Adaptations to pregnancy seen in a woman's integumentary, circulatory, respiratory, urinary, and digestive systems, and in her metabolism and nutritional needs
3. Factors responsible for the increase in uterine contractility toward the end of pregnancy, for the onset of true labor, and for its progression
4. The stages of labor and the defining events of each stage
5. Changes in the maternal body over the first 6 weeks after childbirth
6. Development of the mammary glands during pregnancy and the hormones that regulate this
7. The nature of colostrum, how it differs from milk, and what purposes it serves
8. The influence of breast-feeding on ovarian cycling and fertility
9. The roles of prolactin and oxytocin in lactation, and the mechanism by which oxytocin stimulates milk ejection
10. The composition of breast milk and reasons why it is superior to cow's milk for infant nutrition

Testing Your Recall

1. Thinning of the cervical tissue in childbirth is called
   a. climacteric.
   b. effacement.
   c. lochia.
   d. involution.
   e. dilation.

2. Prior to ejaculation, sperm are stored mainly in
   a. the seminiferous tubules.
   b. the seminal vesicles.
   c. the ejaculatory ducts.
   d. the epididymis.
   e. the prostate.

3. The hormone that triggers the onset of puberty in both sexes is
   a. gonadotropin-releasing hormone.
   b. human chorionic gonadotropin.
   c. follicle-stimulating hormone.
   d. testosterone.
   e. estrogen.

4. The 46 chromosomes of a diploid set separate into two haploid sets at
   a. prophase I.
   b. prophase II.
   c. anaphase I.
   d. anaphase II.
   e. telophase II.

5. Sperm motility requires ATP, which is produced in the sperm's
   a. acrosome.
   b. nucleus.
   c. midpiece.
   d. polar body.
   e. zona pellucida.

6. Menstruation is brought on by a drop in the level of
   a. progesterone.
   b. estrogen.
   c. oxytocin.
   d. FSH.
   e. LH.

7. The ovaries secrete all of the following *except*
   a. estrogen.
   b. progesterone.
   c. androgen.
   d. follicle-stimulating hormone.
   e. inhibin.

8. The greatest fraction of the volume of the semen comes from
   a. the testes.
   b. the epididymis.
   c. the prostate.
   d. the corpus cavernosum.
   e. the seminal vesicles.

9. The tissue lost in menstruation is
   a. the functional layer of the endometrium.
   b. the basal layer of the endometrium.
   c. the myometrium.
   d. the infundibulum.
   e. the corpus luteum.

10. Testosterone is secreted by
    a. the hypothalamus.
    b. the anterior pituitary.
    c. the posterior pituitary.
    d. interstitial endocrine cells of the testis.
    e. germ cells of the testis.

11. A fertilized egg is called a/an _____.

12. Meiosis I produces two secondary spermatocytes in a male, but produces a secondary oocyte and _____ in the female.

13. A tertiary follicle is defined by the appearance of a cavity called the _____.

14. Days 15 to 28 of the ovarian cycle are called the _____ phase.

15. The male ejaculatory duct is formed by the convergence of a ductus deferens and the duct from a _____.

16. The hormone that stimulates both labor contractions and milk ejection is _____.

17. Sperm develop in microscopic ducts called _____ in the testes.

18. In the first 1 to 3 days postpartum, the mammary glands secrete _____ rather than milk.

19. Ovulation is triggered by a spike in the secretion of _____ around day 13 of a typical ovarian cycle.

20. In males and females, the prepuce covers part of the penis or clitoris, respectively, called the _____.

Answers in Appendix A

What's Wrong with These Statements?

Briefly explain why each of the following statements is false, or reword it to make it true.

1. Testosterone is secreted by the nurse cells of the testes.

2. A sperm carries its DNA in the thick midpiece region at the base of the tail.

3. More than half of the volume of the semen comes from the prostate.

4. Like the penis, the clitoris has three erectile tissues—a corpus spongiosum and a pair of corpora cavernosa.

5. Androgens play a significant role in puberty only in males.

6. The vagina is lubricated by an abundance of mucous glands in its wall.

7. Usually, there are no lactiferous ducts in the breast; these begin to develop only in pregnancy.

8. Climacteric is a midlife change after which men and women are no longer fertile.

9. Spermiogenesis is a stage in which the male sex cells complete meiosis II and become mature sperm cells.

10. Orgasm is also known as *climacteric* in both males and females.

Answers in Appendix A

Testing Your Comprehension

1. A breast-feeding mother leaves her baby at home and goes grocery shopping. She hears someone else's baby crying in another aisle, and notices her blouse becoming wet with a little exuded milk. Explain the physiological connection between hearing that sound and the ejection of milk.

2. Explain why sperm and eggs cannot be produced by mitosis alone. Why is meiosis also necessary?

3. What cells in the ovary perform essentially the same function as the interstitial endocrine cells of the testis? Explain your answer.

Chapter 20
Human Development and Aging

Chapter Outline

20.1 Fertilization and Preembryonic Development
- 20.1a Sperm Migration and Capacitation
- 20.1b Fertilization and Meiosis II
- 20.1c Major Stages of Prenatal Development
- 20.1d The Preembryonic Stage

20.2 The Embryonic and Fetal Stages
- 20.2a Prenatal Nutrition
- 20.2b Embryonic Membranes
- 20.2c Embryonic Folding and Organ Development
- 20.2d Fetal Development

20.3 The Neonate
- 20.3a Adapting to Life Outside the Womb
- 20.3b Premature Infants
- 20.3c Birth Defects

20.4 Aging, Senescence, and Death
- 20.4a Exercise and Rates of Senescence
- 20.4b Theories of Senescence
- 20.4c Life Expectancy and Death

Clinical Applications/Perspectives on Health
- Clinical Application 20.1: Ectopic Pregnancy
- Clinical Application 20.2: The Thalidomide Tragedy
- Perspectives on Health

End of Chapter
- Career Spotlight: Genetic Counselor
- Study Guide

A new life in the making (MRI scan).
© Simon Fraser/Science Source

BASE CAMP
Before ascending to the next level, be sure you're properly equipped with a knowledge of these concepts from earlier chapters.

- Anatomy and histology of the uterus (see "The Uterus" in section 19.3a)
- Structure of the spermatozoon (see "Sperm" in section 19.4b)
- Meiosis (section 19.4a)

Module 14
Reproductive System

Perhaps the most dramatic, miraculous aspect of human life is the transformation of a one-celled fertilized egg into an independent, fully developed individual. From the beginning of recorded thought, people have pondered how a baby forms in the mother's body and how two parents can produce another human being who, although unique, possesses characteristics of each. In his quest to understand prenatal development, Aristotle speculated that the hereditary traits of a child resulted from the mixing of the male's semen with the female's menstrual blood. In the seventeenth century, some scientists thought that the head of the sperm had a miniature human curled up in it, while others thought that the miniature person existed in the egg and the sperm were parasites in the semen. The modern science of **embryology**—the study of prenatal development—was not born until the nineteenth century when advances in microscope technology allowed for better study of the early stages. In the last part of the twentieth century, and continuing today, studies of regulatory genes that control prenatal development and evolution have tremendously deepened our insight into human development.

Embryology can be viewed as one aspect of **developmental biology,** a broader science that embraces changes in form and function from fertilized egg through old age. Human development from conception to death is the scope of this chapter.

20.1 Fertilization and Preembryonic Development

Expected Learning Outcomes

When you have completed this section, you should be able to

a. describe the processes of sperm migration and fertilization;
b. explain how an egg prevents fertilization by more than one sperm;
c. outline the timetable of prenatal development from the two perspectives of clinical trimesters and developmental stages;
d. describe the major events that transform a fertilized egg into an embryo; and
e. describe the implantation of the preembryo in the uterine wall.

Authorities attach different meanings to the word *embryo.* Some use it to denote stages beginning with the fertilized egg or at least with the two-celled stage produced by its first division. Others first apply the word to an individual 16 days old, when it consists of three tissue layers called the *ectoderm, mesoderm,* and *endoderm.*

The events leading up to that stage are called *embryogenesis*, and the first 16 days after fertilization are called the *preembryonic stage*. This is the sense in which we will use these terms in this book. We begin with the process of **fertilization (conception),** in which an egg and sperm unite and their chromosomes are combined.

20.1a Sperm Migration and Capacitation

A human egg survives only 12 to 24 hours after ovulation if it isn't fertilized, yet it takes about 72 hours to reach the uterus. Usually only one egg is ovulated, so a sperm must find the correct uterine tube and travel well up it to meet the egg before it dies. The vast majority of sperm never make it that far—they are destroyed by vaginal acidity, fail to get through the cervical canal, succumb to attack by leukocytes in the uterus, or travel up the wrong tube. The average ejaculation discharges about 200 million sperm into the vagina, yet only about 200 of these—1 in a million—make it to the vicinity of an egg.

Sperm can reach the distal region of the uterine tube in half an hour or less, but they can't fertilize an egg for about 10 hours. First they must undergo a process called **capacitation** that makes it possible to penetrate an egg. This happens as fluids of the female reproductive tract dilute inhibitory factors in the semen and weaken the membrane covering the head of the sperm.

Sperm survive for up to 6 days after ejaculation, so there is little chance of pregnancy from intercourse occurring more than a week before ovulation. Fertilization also is unlikely if intercourse takes place more than 14 hours after ovulation, because the egg would no longer be viable by the time the sperm became capacitated. For those wishing to conceive a child, the optimal window of opportunity is therefore from a few days before ovulation to 14 hours after. Those wishing to avoid pregnancy, however, should allow a wider margin of safety for variations in sperm and egg longevity, capacitation time, and time of ovulation.

20.1b Fertilization and Meiosis II

When the sperm approach the egg, proteins on the egg's extracellular coat, the *zona pellucida,* recognize and bind them. This initiates the *acrosomal reaction* in which enzymes released from the acrosome allow the sperm to tunnel through the zona pellucida. The first sperm to arrive at the zona pellucida is not the one to fertilize the egg. It requires many sperm to clear a path through the barriers around the egg (fig. 20.1). Ultimately, one sperm reaches the egg surface and fuses to its plasma membrane, initiating reactions in the egg that prevent more sperm from passing through.

The sperm nucleus enters the egg and adds its haploid *(n)* set of chromosomes to the haploid set of egg chromosomes, producing a diploid (2*n*) set. Fertilization by two or more sperm, called **polyspermy,** would produce a triploid (3*n*) or larger set of chromosomes and the egg would die of a "gene overdose." It is important for the egg to prevent this, and it has two mechanisms for doing so. First, the binding of a sperm to the egg triggers a rapid inflow of Na^+ that depolarizes the egg membrane; this inhibits the attachment of any more sperm. Second, the egg undergoes a *cortical reaction,* in which secretory vesicles just beneath the egg membrane undergo exocytosis. They release a secretion that swells with water, pushes any remaining sperm away from the egg, and creates an impenetrable **fertilization membrane** between the egg and zona pellucida.

A secondary oocyte begins meiosis II before ovulation and completes it only if fertilized. Through the formation of a second polar body, the fertilized egg discards one chromatid from each chromosome. The sperm and egg nuclei then swell

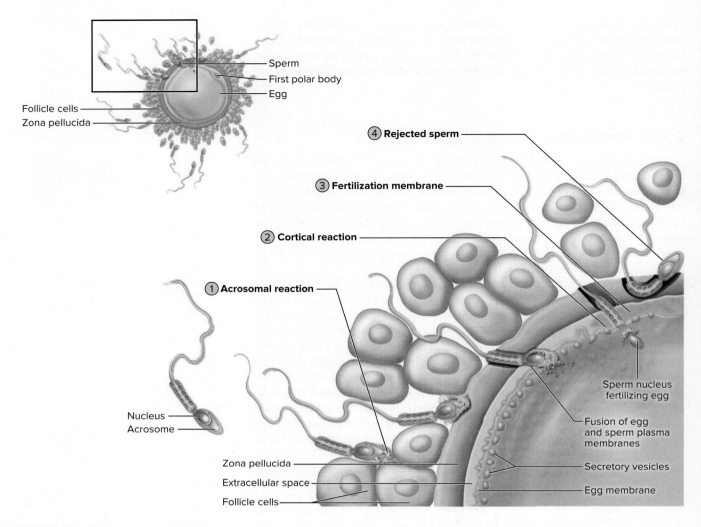

Figure 20.1 Fertilization and Prevention of Polyspermy. **APR**

and rupture, spilling their chromosomes into a single diploid set. The fertilized egg, now called a **zygote,** is ready for its first mitotic division.

We will use the term *conceptus* for everything that arises from this zygote—not only the future fetus but also the placenta, umbilical cord, and membranes associated with it.

20.1c Major Stages of Prenatal Development

Clinically, the course of a pregnancy is divided into 3-month intervals called **trimesters:**

1. The **first trimester** extends from fertilization through 12 weeks. This is a precarious stage in which more than half of all embryos die. The conceptus is most vulnerable to stress, drugs, and nutritional deficiencies during this time.
2. The **second trimester** (weeks 13 through 24) is a period in which the organs complete most of their development. It becomes possible with sonography to see details of anatomy in the fetus. By the end of this trimester, the fetus looks distinctly human and, with intensive care, infants born at the end of the second trimester have a chance of survival.

Table 20.1 The Stages of Prenatal Development

Stage	Age*	Major Developments and Defining Characteristics
Preembryonic Stage		
Zygote	0–30 hours	A single diploid cell formed by the union of egg and sperm
Cleavage	30–72 hours	Mitotic division of the zygote into smaller, identical cells called blastomeres
Morula	3–4 days	A spheroidal stage consisting of 16 or more blastomeres
Blastocyst	4–16 days	A fluid-filled, spheroidal stage with an outer mass of trophoblast cells and inner mass of embryoblast cells; becomes implanted in the endometrium; inner cell mass forms an embryonic disc and differentiates into the three primary germ layers
Embryonic Stage	16 days–8 weeks	A stage in which the primary germ layers differentiate into organs and organ systems; ends when all organ systems are present
Fetal Stage	8–38 weeks	A stage in which organs grow and mature to the point of being capable of supporting life independently of the mother

*From the time of fertilization

3. In the **third trimester** (week 25 to birth), the fetus grows rapidly and the organs develop enough to support life outside the womb. At 35 weeks from fertilization, the fetus typically weighs about 2.5 kg (5.5 pounds). It is considered mature at this weight, and usually survives if born early. Most twins are born at about 35 weeks and solitary infants around 38 weeks.

From a more biological than clinical standpoint, human development is divided into the *preembryonic, embryonic,* and *fetal stages.* The timetable and landmark events that distinguish these are outlined in table 20.1 and described in the following section.

20.1d The Preembryonic Stage

The **preembryonic stage** comprises the first 16 days of development, leading to the existence of an embryo. It involves three major processes: cleavage, implantation, and embryogenesis.

Cleavage

Cleavage refers to mitotic divisions that occur in the first 3 days while the conceptus migrates down the uterine tube (fig. 20.2). The first cleavage occurs about 30 hours after fertilization and produces the first two daughter cells, or **blastomeres.**[1] These divide simultaneously at shorter and shorter time intervals, doubling the number of

---

[1] *blasto* = bud, precursor; *mer* = segment, part

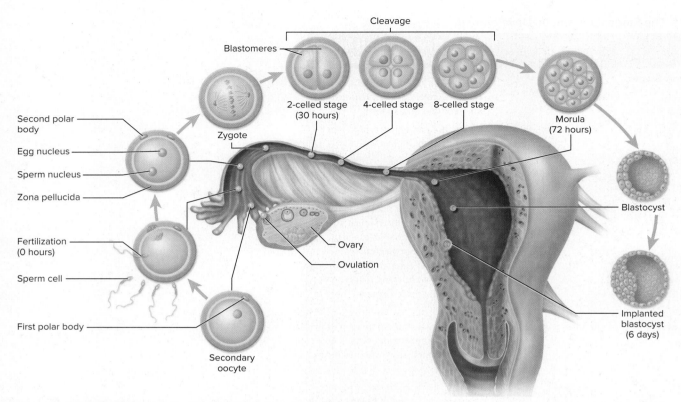

Figure 20.2 Migration of the Conceptus. The egg is fertilized in the distal end of the uterine tube, and the preembryo begins cleavage as it migrates to the uterus.

Why can't the egg be fertilized in the uterus?

blastomeres each time. By the time the conceptus arrives in the uterus, about three to four days after ovulation, it consists of 16 or more cells and is called a **morula** because of its resemblance to a mulberry (*morum* in Latin).[2] The morula is no larger than the zygote; cleavage merely produces smaller and smaller blastomeres. This increases the ratio of cell surface area to volume, which favors rapid nutrient uptake and waste removal, and it produces a larger number of cells from which to form different embryonic tissues.

The morula lies free in the uterine cavity for 4 to 5 days and divides into 100 cells or so. Meanwhile, the zona pellucida disintegrates and releases the conceptus, which is now at a stage called the **blastocyst**—a hollow sphere with an outer layer of squamous cells called the **trophoblast**[3] and an inner cell mass called the **embryoblast** (fig. 20.3a). The trophoblast is destined to form part of the placenta and play an important role in nourishment of the embryo, whereas the embryoblast is destined to become the embryo itself.

In some cases, the embryoblast splits into two cell masses, resulting in identical *monozygotic twins,* which usually share the same placenta. Nonidentical *dizygotic twins* and most other multiple births occur when two or more eggs are ovulated in the same month and fertilized by separate sperm. They are no more or less genetically similar than other siblings. They attach separately to the uterine wall and each forms its own placenta.

> **Apply What You Know**
> Your friend John says that he and his sister Beth are identical twins. Even though they do look strikingly similar, you are skeptical of his claim. Other than tactful silence, how might you respond?

---

[2]*mor* = mulberry; *ula* = little
[3]*tropho* = food, nourishment; *blast* = producing

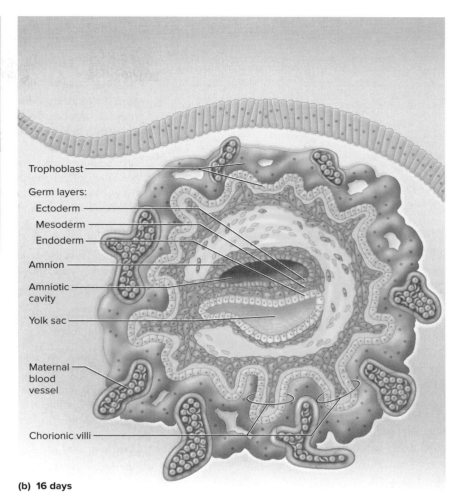

Figure 20.3 Implantation. (a) Early implantation about 8 days after ovulation. The trophoblast has begun growing rootlets, which penetrate the endometrium. (b) By 16 days, the conceptus is completely covered by endometrial tissue. The embryo is now flanked by a yolk sac and amnion and is composed of three primary germ layers.

Implantation

About 6 days after ovulation, the blastocyst adheres to the endometrium and becomes embedded in it—a process called **implantation.** The trophoblast cells on the attached side proliferate and grow into the uterus like little roots, digesting endometrial cells along the way. The endometrium reacts to this injury by growing over the blastocyst and eventually covering it (fig. 20.3). Implantation takes about a week and is completed about the time the next menstrual period would have occurred if the woman had not become pregnant.

Another role of the trophoblast is to secrete the hormone **human chorionic gonadotropin (HCG)** (CORE-ee-ON-ic go-NAD-o-TROPE-in). HCG stimulates the corpus luteum to secrete estrogen and progesterone, and progesterone prevents menstruation. Pregnancy tests are based on detecting HCG in a woman's urine. The level of HCG in the mother's blood rises until the end of the second month. During this time, the trophoblast develops into a membrane called the *chorion,* which takes over the role of the corpus luteum and makes HCG unnecessary. The ovaries then become inactive for the rest of the pregnancy, but estrogen and progesterone levels rise dramatically as they are secreted by the ever-growing chorion.

Clinical Application 20.1

ECTOPIC PREGNANCY

In about 1 out of 300 pregnancies, the blastocyst implants somewhere other than the uterus, producing an *ectopic*[4] *pregnancy.* Most cases are *tubal pregnancies,* implantation in the uterine tube. This usually occurs because the conceptus encounters an obstruction resulting from earlier pelvic inflammatory disease, tubal surgery, previous ectopic pregnancies, or other causes. The uterine tube cannot expand enough to accommodate the growing conceptus for long; unless detected and treated early, the tube usually ruptures within 12 weeks, with potentially fatal hemorrhaging. Occasionally, a conceptus implants in the abdominopelvic cavity, producing an *abdominal pregnancy.* It can grow anywhere it finds an adequate blood supply—for example, on the outside of the uterus, colon, or bladder. About 1 pregnancy in 7,000 is abdominal. This is a serious threat and usually requires aborting the pregnancy to save the woman's life, but about 9% of abdominal pregnancies go full term and end in delivery by cesarean section.

Embryogenesis

During implantation, the embryoblast undergoes **embryogenesis**—arrangement of the blastomeres into the three **primary germ layers:** *ectoderm, mesoderm,* and *endoderm.* At the beginning of this phase, the embryoblast separates slightly from the trophoblast, creating a narrow space between them called the **amniotic cavity** lined by a membrane, the *amnion.* The embryoblast flattens into an **embryonic disc** composed of two cell layers—a deep layer facing the amniotic cavity and a superficial layer facing away. The superficial layer produces a fluid-filled *yolk sac.* Now the embryonic disc is flanked by two spaces: the amniotic cavity on one side and the yolk sac on the other.

Meanwhile, the disc elongates and a thickened ridge forms along the midline, with a **primitive groove** down its middle. These events make the embryo bilaterally symmetric and define its future right and left sides, dorsal and ventral surfaces, and head and tail ends.

The next step is **gastrulation**—multiplying cells migrate medially toward the primitive groove and down into it (**fig. 20.4**). They become a layer called **endoderm,** which will become the inner lining of the digestive tract among other things. A day later, migrating cells form a third layer between the first two, called **mesoderm.** Once this is formed, the upper surface is called **ectoderm.** The ectoderm and endoderm are epithelia composed of tightly joined cells, but the mesoderm is a more loosely organized tissue. Mesoderm gives rise to skeletal muscle as well as a loose, gelatinous, fetal connective tissue called **mesenchyme.** Mesenchyme in turn gives rise to such tissues as smooth and cardiac muscle, bone, blood, and adipose tissue. Precise control of these embryonic cell migrations is crucial, because any misplacement of cells at this stage results in gross deformities of the embryo and a likelihood of prenatal death.

Once the three primary germ layers are formed, embryogenesis is complete and the individual is considered an **embryo.** It is about 2 mm long and 16 days old at this point.

---

[4]*ec* = out of; *top* = place

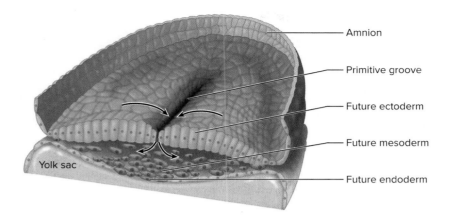

Figure 20.4 Formation of the Primary Germ Layers (Gastrulation). The embryonic disc at 16 days, corresponding to the center of figure 20.3b. Surface cells migrate into the primitive groove, first forming a layer of endoderm beneath, then filling the space between these first two layers with mesoderm. Upon completion of this process, the uppermost layer is considered ectoderm.

? *Name any three adult organs that develop from the red cells in the middle.*

Before You Go On

Answer these questions from memory to test your understanding of the previous section. Reread the material if you can't answer the questions.

1. How soon can a sperm reach an egg after ejaculation? How soon can it fertilize an egg? What accounts for the difference?
2. Describe two ways a fertilized egg prevents the entry of excess sperm.
3. In the blastocyst, what are the cells called that eventually give rise to the embryo? What are the cells that carry out implantation?
4. In which clinical trimester does a conceptus reach the embryonic stage? The fetal stage?
5. What defines a true embryo and at what age is this stage reached?

20.2 The Embryonic and Fetal Stages

Expected Learning Outcomes

When you have completed this section, you should be able to

a. identify the major tissues derived from the primary germ layers;
b. explain how the conceptus is nourished before the placenta takes over this function;
c. describe the formation of the placenta and list its multiple functions;
d. describe the four embryonic membranes and their functions;
e. describe the conversion of a flat embryonic disc into a cylindrical form enclosing a primitive gut; and
f. describe how the fetal circulatory system differs from that of a neonate.

Sixteen days after conception, the germ layers are present and the **embryonic stage** of development begins. Over the next 6 weeks, a placenta forms on the uterine

Table 20.2 Derivatives of the Three Primary Germ Layers

Germ Layer	Major Derivatives
Ectoderm	Epidermis; hair follicles and arrector pili muscles; cutaneous glands; nervous system; adrenal medulla; pineal and pituitary glands; lens, cornea, and intrinsic muscles of the eye; internal and external ear; salivary glands; epithelia of the nasal cavity, oral cavity, and anal canal
Mesoderm	Bones; bone marrow; cartilage; skeletal, cardiac, and most smooth muscle; cartilage; adrenal cortex; middle ear; dermis; blood; blood and lymphatic vessels; bone marrow; lymphoid tissue; epithelium of kidneys, ureters, gonads, and genital ducts; mesothelium of the abdominal and thoracic cavities
Endoderm	Most of the mucosal epithelium of the digestive and respiratory tracts; mucosal epithelium of urinary bladder and parts of urethra; epithelial components of accessory reproductive and digestive glands (except salivary glands); thyroid and parathyroid glands; thymus

wall and becomes the embryo's primary means of nutrition, while the germ layers differentiate into organs and organ systems—a process called **organogenesis** (table 20.2). Although these organs are still far from functional, it is their presence at 8 weeks that marks the transition from the embryonic stage to the fetal stage. Here we will see how the embryo becomes a fetus, how the membranes collectively known as the "afterbirth" develop around the fetus, and how the conceptus is nourished throughout its gestation.

20.2a Prenatal Nutrition

Over the course of gestation, the conceptus is nourished in three different, overlapping ways. As it travels down the uterine tube and lies free in the uterine cavity before implantation, it is nourished by a glycogen-rich secretion of the uterine tubes and endometrial glands called *uterine milk*.

As it implants, the conceptus shifts to **trophoblastic nutrition,** in which it consumes so-called *decidual*[5] *cells* of the endometrium. Progesterone from the corpus luteum stimulates these cells to proliferate and accumulate a store of glycogen, protein, and lipid. As the conceptus burrows into the endometrium, the trophoblast digests them and relays the nutrients to the embryoblast. This is the only mode of nutrition for the first week after implantation. It remains the dominant source of nutrients through the end of week 8; the period from implantation through week 8 is therefore called the **trophoblastic phase** of the pregnancy. Trophoblastic nutrition wanes as placental nutrition takes over, and ceases entirely by the end of week 12.

In **placental nutrition,** nutrients diffuse from the mother's blood through the placental wall into the fetal blood. The **placenta**[6] is a disc-shaped organ attached to the uterus on one side and, on the other side, attached by way of an **umbilical cord** to the fetus (fig. 20.5). The uterine surface of the placenta bears shaggy **chorionic villi** (CORE-ee-ON-ic VILL-eye) embedded in the endometrium. When fully developed, the placenta is about 20 cm in diameter, 3 cm thick, and weighs about one-sixth as much as the newborn infant.

---

[5]*decid* = falling off
[6]*placenta* = flat cake

Figure 20.5 The Placenta and Embryonic Membranes. (a) A 12-week fetus. The placenta is completely formed. (b) A portion of the mature placenta and umbilical cord, showing the relationship between fetal and maternal circulation. (c) The fetal side of the placenta, showing blood vessels, the umbilical cord, and some of the amniotic sac attached to the lower left margin. (d) The maternal (uterine) side, where chorionic villi give the placenta a rougher texture.
c and d: © Dr. Kurt Benirschke

The placenta begins to develop about 11 days after conception, becomes the dominant mode of nutrition around the beginning of week 9, and is the sole mode of nutrition from the end of week 12 until birth. The period from week 9 until birth is called the **placental phase** of the pregnancy.

The development of the placenta begins during implantation, as extensions of the trophoblast penetrate more and more deeply into the endometrium, like the roots of a tree penetrating into the nourishing soil of the uterus. These roots are the early chorionic villi. As they invade uterine blood vessels, they become surrounded by pools of maternal blood. These pools eventually merge to form a single blood-filled sinus. Maternal blood stimulates rapid growth of the villi, which become branched and treelike. Mesoderm grows into the villi and gives rise to the blood vessels that connect to the embryo by way of the umbilical cord.

The umbilical cord contains two **umbilical arteries** and one **umbilical vein.** Pumped by the fetal heart, blood flows to the placenta by way of the umbilical arteries and then returns to the fetus by way of the umbilical vein. The chorionic villi are thus *filled with* fetal blood and *surrounded by* maternal blood (fig. 20.5b); the two bloodstreams do not mix unless there is damage to the placental barrier. The barrier, however, is only 3.5 μm thick—half the diameter of one red blood cell. As development progresses, this barrier becomes more and more permeable to chemicals. Oxygen and nutrients pass from the maternal blood, through the barrier, into the fetal blood, while fetal wastes pass the other way to be eliminated by the mother. Unfortunately, the placenta is also permeable to nicotine, alcohol, and most other drugs that may be present in the maternal bloodstream and harmful to the embryo.

Table 20.3 describes the many functions of the placenta.

> *Apply What You Know*
>
> What basic law of diffusion determines that at the placenta, fetal wastes pass from the fetal blood into the maternal blood, and oxygen passes from the maternal blood into the fetal blood, rather than in the opposite directions? (See "Transport Through Cellular Membranes" in section 3.2e.)

20.2b Embryonic Membranes

The placenta is only one of several accessory organs that develop alongside the embryo; there are also four embryonic membranes called the *amnion, yolk sac, allantois,* and *chorion* (**figs. 20.5a** and **20.6**). The **amnion** is a transparent

Table 20.3 Functions of the Placenta

Nutritional roles	Permits nutrients such as glucose, amino acids, fatty acids, minerals, and vitamins to diffuse from the maternal blood to the fetal blood; stores nutrients such as carbohydrates, protein, iron, and calcium in early pregnancy and releases them to the fetus later, when fetal demand is greater than the mother can absorb from the diet
Excretory roles	Permits nitrogenous wastes such as ammonia, urea, uric acid, and creatinine to diffuse from the fetal blood to the maternal blood
Respiratory roles	Permits O_2 to diffuse from mother to fetus and CO_2 from fetus to mother
Endocrine roles	Secretes estrogens, progesterone, human chorionic gonadotropin, and other hormones; allows hormones synthesized by the conceptus to pass into the mother's blood and maternal hormones to pass into the fetal blood
Immune roles	Transfers maternal antibodies (especially IgG) into fetal blood to confer passive immunity on fetus

Figure 20.6 Embryonic Folding. The right-hand figures are cross sections cut about midway along the figures on the left. From 22 to 28 days, the two ends of the embryo curl toward each other (left-hand figures) until the embryo assumes a C shape, and the flanks of the embryo fold laterally (right-hand figures), converting the flat embryonic disc into a more cylindrical body and eventually enclosing a body cavity.

sac that encloses the embryo and is penetrated only by the umbilical cord. It is filled with **amniotic fluid** (see fig. 20.7d), which protects the embryo from trauma, infection, and temperature fluctuations; allows the freedom of movement important to muscle development; enables the embryo to develop symmetrically; prevents body parts from adhering to each other, such as an arm to the trunk; and stimulates lung development as the fetus "breathes" the fluid. The fetus also repeatedly swallows amniotic fluid and urinates it back into the amniotic sac. At term, the amnion contains 700 to 1,000 mL of fluid, released vaginally when the amnion ruptures during labor (the "breaking of the waters").

The **yolk sac,** named for its role in egg-laying vertebrates, is a small bag suspended from the underside of the embryo. It encloses the egg yolk in reptiles and birds, but in humans and other mammals, it contributes to the formation of

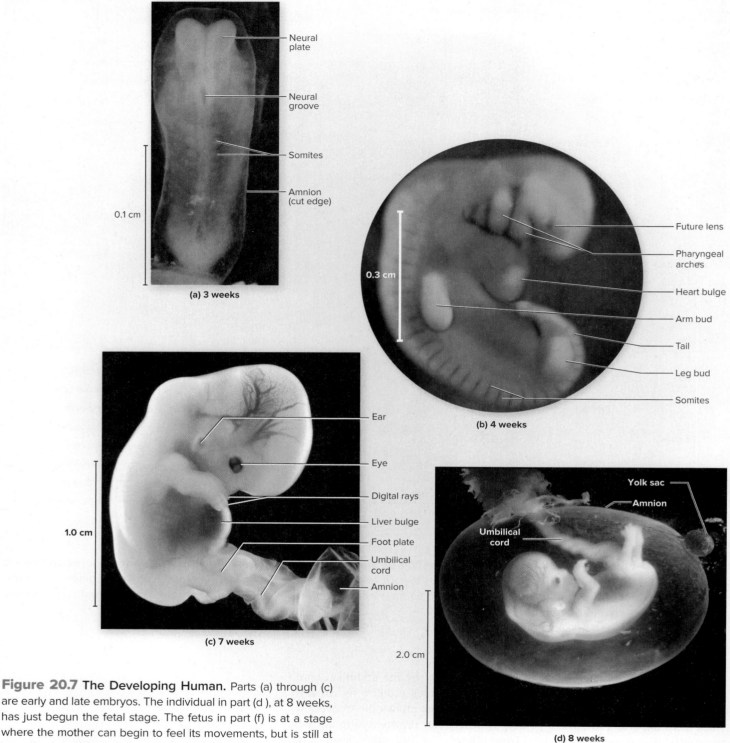

Figure 20.7 The Developing Human. Parts (a) through (c) are early and late embryos. The individual in part (d), at 8 weeks, has just begun the fetal stage. The fetus in part (f) is at a stage where the mother can begin to feel its movements, but is still at least 8 weeks from being marginally viable if born prematurely, and is 18 weeks short of full term.

a: © Human Developmental Anatomy Center, National Museum of Health and Medicine, Silver Spring, MD, USA; b: © Anatomical Travelogue/Science Source; c: Neil Harding/The Image Bank/Getty Images d: © Dr G. Moscoso/Science Source; e: © John Watney/Science Source; f: © Biophoto Associates/Science Source

(continued on next page)

(e) 12 weeks

(f) 20 weeks

Figure 20.7 *(continued)*

the digestive tract and produces the first blood cells and the stem cells that will become future eggs or sperm.

The **allantois** (ah-LON-toe-iss) begins as an outgrowth of the tail end of the gut. It forms the foundation for the umbilical cord and becomes part of the urinary bladder.

The **chorion** is the outermost membrane, enclosing all the others and the embryo. Initially, it has shaggy chorionic villi around its entire surface (see fig. 20.3b), but as the pregnancy advances, the villi of the placental region grow and branch while the rest of them degenerate. The chorion thus becomes smooth surfaced away from the placental region (fig. 20.5a). The chorion secretes pregnancy-regulating hormones and, through its role in the placenta, provides an avenue of fetal nutrition.

20.2c Embryonic Folding and Organ Development

One of the major transformations in the embryonic stage is conversion of the flat embryonic disc into a somewhat cylindrical form. This occurs during week 4 as the embryo rapidly grows and folds around the yolk sac (see fig. 20.6). As the head and tail ends curve around the ends of the yolk sac, the embryo becomes C-shaped, with the head and tail almost touching. At the same time, the lateral margins of the disc fold around the sides of the yolk sac to form the ventral surface of the embryo. This lateral folding encloses a longitudinal channel, the *primitive gut,* which later becomes the digestive tract.

As a result of embryonic folding, the entire surface becomes covered with ectoderm, which later produces the epidermis of the skin. In the meantime, the mesoderm splits into two layers. One of them adheres to the ectoderm and the other to the endoderm, thus opening a body cavity between them. This cavity is soon subdivided by a diaphragm, giving rise to the thoracic cavity and peritoneal cavity. By the end of week 5, the thoracic cavity further divides into pleural and pericardial cavities.

Two more significant events in organogenesis are the appearance of a **neural tube,** which later becomes the brain and spinal cord, and segmentation of the mesoderm into blocks of tissue called **somites** (fig. 20.7a, b), which later give rise to the vertebral column, skeletal muscles, and dermis of the skin.

We cannot delve at greater length into development of all the organ systems here, but this description is at least enough to see how some of them begin to form. By the end of 8 weeks, all of the organ systems are present, the individual is about 3 cm long, and it is now considered a **fetus** (fig. 20.7d). Some bones have just begun to calcify and the skeletal muscles exhibit weak spontaneous contractions. The heart, beating since the fourth week, now circulates blood. The head is nearly half the total body length.

20.2d Fetal Development

The fetus is the final stage of prenatal development, from the start of the ninth week until birth. The organs that formed during the embryonic stage now undergo growth and cellular differentiation, acquiring the functional capability to support life outside the mother.

The circulatory system shows the most conspicuous changes from the prenatal state to the neonatal (newborn) state (**fig. 20.8**). The unique aspects of fetal circulation are the umbilical–placental circuit and the presence of three circulatory shortcuts called *shunts.* The internal iliac arteries give rise to the two umbilical arteries, which pass into the umbilical cord. The arterial blood discharges its wastes in the placenta, loads oxygen and nutrients coming from the mother, and returns to the fetus by way of a single umbilical vein, which leads toward the liver. The umbilical vein is depicted in red because of its well-oxygenated blood. A little venous blood filters through the liver to nourish it, but most of it bypasses the liver by way of a shunt called the **ductus venosus,** which leads to the inferior vena cava.

In the inferior vena cava, placental blood mixes with venous blood from the fetus's body and flows to the right atrium of the heart. After birth, the right ventricle pumps all of its blood into the lungs, but there is little need for this in the fetus because the lungs are not yet functional. Therefore, most fetal blood bypasses the pulmonary circuit. Some goes directly from the right to left atrium through a passage in the interatrial septum called the **foramen ovale.** Some also goes into the right ventricle and is pumped into the pulmonary trunk, but most of this is shunted directly into the aorta by way of a short passage called the **ductus arteriosus.** The lungs receive only a trickle of blood, sufficient to meet their metabolic needs during development. Blood leaving the left ventricle enters the general systemic circulation. This circulatory pattern changes dramatically at birth, when the neonate is cut off from the placenta and the lungs expand with air. Those changes will be described later.

Fetal growth is charted by weight and body length. Body length is customarily measured from the crown of the head to the curve of the buttocks in a sitting position *(crown-to-rump length, CRL),* thus excluding the lower limbs. Full-term fetuses have an average CRL of about 36 cm (14 in.) and average weight of about 3.0 to 3.4 kg (6.6–7.5 pounds). The fetus gains about 50% of its birth weight in the last 10 weeks. Additional aspects of embryonic and fetal development are listed in **table 20.4** and depicted in figure 20.7.

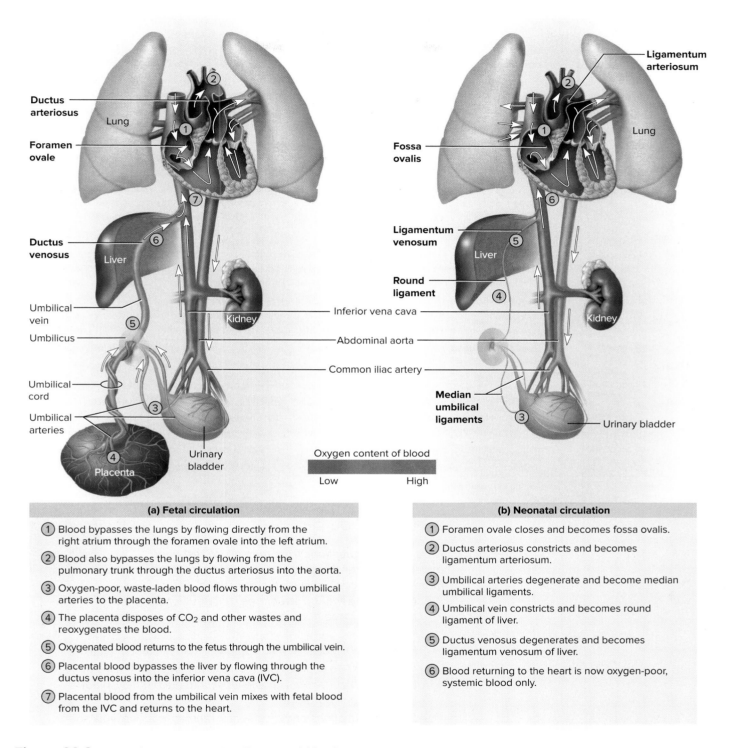

Figure 20.8 Blood Circulation in the Fetus and Newborn. Boldfaced terms in (a) indicate the three shunts in the fetal circulation, which allow most blood to bypass the liver and lungs. Boldfaced terms in (b) indicate the postpartum vestiges of fetal structures.

Table 20.4 Major Events of Prenatal Development

End of Week	Crown-to-Rump Length (CRL) and Weight	Developmental Events
First Trimester		
4	0.6 cm; <1 g	Vertebral column and central nervous system begin to form; limbs represented by small limb buds; heart begins beating around day 22; no visible eyes, nose, or ears
8	3 cm; 1 g	Eyes form, eyelids fused shut; nose flat, nostrils evident but plugged with mucus; head nearly as large as the rest of the body; brain waves detectable; bone calcification begins; limb buds form paddlelike hands and feet with ridges called *digital rays*, which then separate into distinct fingers and toes; blood cells and major blood vessels form; genitals present but sexes not yet distinguishable
12	9 cm; 45 g	Eyes well developed, facing laterally; eyelids still fused; nose develops bridge; external ears present; limbs well formed, digits exhibit nails; fetus swallows amniotic fluid and produces urine; fetus moves, but too weakly for mother to feel it; liver is prominent and produces bile; palate is fusing; sexes can be distinguished
Second Trimester		
16	14 cm; 200 g	Eyes face anteriorly, external ears stand out from head, face looks more human; body larger in proportion to head; skin is bright pink, scalp has hair; joints forming; lips exhibit sucking movements; kidneys well formed; digestive glands forming and *meconium*[7] (fetal feces) accumulating in intestine; heartbeat can be heard with a stethoscope
20	19 cm; 460 g	Body covered with fine hair called *lanugo*[8] and cheeselike sebaceous secretion called *vernix caseosa*,[9] which protects it from amniotic fluid; skin bright pink; brown fat forms and will be used for postpartum heat production; fetus is now bent forward into "fetal position" because of crowding; *quickening* occurs—mother can feel fetal movements
24	23 cm; 820 g	Eyes partially open; skin wrinkled, pink, and translucent; lungs begin producing surfactant; rapid weight gain
Third Trimester		
28	27 cm; 1,300 g	Eyes fully open; skin wrinkled and red; full head of hair present; eyelashes formed; fetus turns into upside-down *vertex position;* testes begin to descend into scrotum; marginally viable if born at 28 weeks
32	30 cm; 2,100 g	Subcutaneous fat deposition gives fetus a more plump, babyish appearance, with lighter, less wrinkled skin; testes descending; twins usually born at this stage
36	34 cm; 2,900 g	More subcutaneous fat deposited, body plump; lanugo is shed; nails extend to fingertips; limbs flexed; firm hand grip
38	36 cm; 3,400 g	Prominent chest, protruding breasts; testes in inguinal canal or scrotum; fingernails extend beyond fingertips

Before You Go On

Answer these questions from memory to test your understanding of the previous section. Reread the material if you can't answer the questions.

6. Name three mature structures that develop from the ectoderm, three from the mesoderm, and three from the endoderm.
7. How is trophoblastic nutrition different from placental nutrition?
8. Identify the two sources of blood to the placenta. Where do these two bloodstreams come closest to each other? What keeps them separated?
9. State the functions of the placenta, amnion, chorion, yolk sac, and allantois.
10. What developmental characteristic distinguishes a fetus from an embryo? At what gestational age is this attained?
11. Identify the three circulatory shunts of the fetus. Why does the blood take these "shortcuts" before birth?

---

[7]*mecon* = poppy juice, opium; refers to an appearance similar to black tar opium
[8]*lan* = down, wool
[9]*vernix* = varnish; *case* = cheese; *osa* = full of

PERSPECTIVES ON HEALTH

Methods of Contraception

Contraception means any procedure or device intended to prevent pregnancy (the presence of an implanted conceptus in the uterus). This essay summarizes the most popular methods and some issues involved in choosing among them.

Behavioral Methods

Abstinence (refraining from intercourse) is, obviously, a completely reliable method if used consistently. The *fertility awareness-based method* (sometimes called the *rhythm method*) relies on avoiding intercourse near the time of expected ovulation. Among typical users, it has a 25% failure rate, partly due to lack of restraint and partly because it is difficult to predict the exact date of ovulation. Intercourse must be avoided for at least 7 days before ovulation so there will be no surviving sperm in the reproductive tract when the egg is ovulated, and for at least 2 days after ovulation so there will be no fertile egg present when sperm are introduced.

Withdrawal (coitus interruptus) requires the male to withdraw the penis before ejaculation. This often fails because of lack of willpower, because some sperm are present in the preejaculatory fluid, and because sperm ejaculated anywhere in the vulva can potentially get into the reproductive tract.

Barrier and Spermicidal Methods

Barrier methods are designed to prevent sperm from getting into or beyond the vagina. They are most effective when used with chemical spermicides, available as nonprescription foams, creams, and jellies. Second only to birth-control pills in popularity is the male *condom,* a sheath usually made of latex, worn over the penis. Female condoms that cover the vulva and line the vagina are also available. Condoms are the only contraceptive methods that also protect against disease transmission. Condoms have the advantages of being inexpensive and requiring no medical examination or prescription.

The *diaphragm* is a latex dome worn over the cervix to block sperm migration. It requires a physical examination and prescription to ensure proper fit, but is otherwise comparable to the condom in convenience and reliability, provided it is used with a spermicide. Without a spermicide, it is not very effective. Unlike the male and female condoms, the diaphragm and other methods that follow offer no protection from sexually transmitted diseases.

The *sponge* is a concave foam disc inserted before intercourse to cover the cervix. It contains spermicide and acts by absorbing semen and killing the sperm. It requires no prescription or fitting. The sponge provides protection for up to 12 hours, and must be left in place for 6 hours after intercourse.

Hormonal Methods

Most hormonal methods of contraception are aimed at preventing ovulation. They mimic the negative feedback effect of ovarian hormones on the pituitary gland, inhibiting FSH and LH secretion so follicles do not mature. For most women, they are highly effective and present minimal complications.

The oldest and still the most widely used hormonal method in the United States is the *combined oral contraceptive (birth-control pill).* It is composed of estrogen and progestin, a synthetic progesterone. It must be taken daily, at the same time of day, for 21 days each cycle. The 7-day withdrawal allows for menstruation. Side effects include an elevated risk of heart attack or stroke in smokers and in women with a history of diabetes, hypertension, or clotting disorders.

Other hormonal methods avoid the need to remember a daily pill. One option is a skin patch that releases estrogen and progestin transdermally. It is changed at 7-day intervals (three patches per month and 1 week without). The NuvaRing is a soft flexible vaginal ring that releases estrogen and progestin for absorption through the vaginal mucosa. It must be worn continually for 3 weeks and removed for the fourth week of each cycle. Medroxyprogesterone (trade name Depo-Provera) is a progestin administered by injection every 3 months. It provides highly reliable, long-term contraception, although in some women it causes headaches, nausea, or weight gain.

Some drugs can be taken orally after intercourse to prevent implantation of a conceptus. These are called emergency contraceptive pills (ECPs), or "morning-after pills." An ECP is a high dose of estrogen and progestin or a progestin alone. It can be taken within 72 hours after intercourse and induces menstruation within 2 weeks. ECPs inhibit ovulation, inhibit sperm or egg transport in the uterine tube, and prevent implantation. They do not work if a blastocyst is already implanted.

Intrauterine Devices

Intrauterine devices (IUDs) are springy, often T-shaped devices inserted through the cervical canal into the uterus. Some IUDs act by releasing a synthetic progesterone, but most have a copper wire wrapping or copper sleeve. IUDs irritate the uterine lining and interfere with blastocyst implantation, and copper IUDs also inhibit sperm motility. An IUD can be left in place for 5 to 12 years.

(continued on next page)

Surgical Methods

People who are confident that they do not want more children (or any) often elect to be surgically sterilized. This entails the cutting and tying or clamping of the genital ducts, thus blocking the passage of sperm or eggs. Surgical sterilization has the advantage of convenience, since it requires no further attention. Its initial cost is higher, however, and for people who later change their minds, surgical reversal is much more expensive than the original procedure and is often unsuccessful. *Vasectomy* is the severing of the male's ductus (vas) deferens, done through a small incision in the back of the scrotum. In *tubal ligation,* the uterine tubes are tied or cut. This can be done through small abdominal incisions to admit a cutting instrument and laparoscope (viewing device).

Issues in Choosing a Contraceptive

We have not considered all the currently available methods of contraception or all the issues important to the choice of a contraceptive. No one method can be recommended as best for all people. Many issues enter into one's choice, including personal preference, pattern of sexual activity, medical history, religious views, convenience, initial and ongoing costs, and disease prevention. For most people, however, the two primary issues are safety and reliability. Further information about options and proper use of contraceptives should be sought from a health department, college health service, physician, reproductive health clinic, or comparable resources.

20.3 The Neonate

Expected Learning Outcomes

When you have completed this section, you should be able to

a. describe how the respiratory and circulatory systems change and affect each other at birth;
b. explain why the neonate faces unique challenges in immunity, thermoregulation, and fluid balance;
c. describe the most critical physiological problems of a premature infant; and
d. discuss and classify several common causes of birth defects.

Development is by no means complete at birth. For example, the liver and kidneys still are not fully functional, most joints are not yet ossified, and myelination of the nervous system is not completed until adolescence. Indeed, humans are born in a very immature state compared with other mammals—a fact necessitated by the narrow outlet of the female pelvis.

20.3a Adapting to Life Outside the Womb

The first 4 weeks of postpartum life constitute the **neonatal period.** This period is a time in which the neonate must adapt to life outside the mother's body. Here we consider some of the major adjustments of the neonate to this new way of life.

Respiration

The most obvious adaptation to external life is that the infant must breathe on its own. It is an old misconception that a neonate must be spanked to stimulate it to breathe. During birth, CO_2 accumulates in the baby's blood and strongly stimulates the respiratory chemoreceptors. Unless the infant is depressed by oversedation of the mother, it normally begins breathing spontaneously. It requires a great effort, however, to take the first few breaths and inflate the collapsed alveoli. For the first 2 weeks, a baby takes about 45 breaths per minute, but subsequently stabilizes at about 12 breaths per minute.

Circulation

When the lungs first expand with air, resistance and blood pressure in the pulmonary circulation drop rapidly and pressure in the right side of the heart falls below that in the left. Blood flows briefly from the left atrium to the right through the foramen ovale, pushing two flaps of tissue into place to close this shunt. Usually these flaps fuse and permanently seal the foramen during the first year, leaving a depression, the *fossa ovalis,* in the interatrial septum. Pressure changes in the pulmonary trunk and aorta also cause the ductus arteriosus to collapse. It closes permanently around 3 months of age, leaving a cord, the *ligamentum arteriosum,* between the two vessels. After the umbilical cord is clamped and cut, the umbilical arteries and vein collapse and become fibrotic (fig. 20.8).

Immunity

Cellular immunity begins early in fetal development, but the immune responses of the neonate are still weak. Fortunately, an infant is born with a near-adult level of antibodies in the IgG class acquired from the mother through the placenta. Maternal IgG breaks down rapidly after birth, but its level remains high enough for 6 months to protect the infant from measles, diphtheria, polio, and most other infectious diseases. By 6 months, the infant's own IgG reaches about half the typical adult level. The lowest total (maternal + infant) level of IgG exists around 5 to 6 months of age, and respiratory infections are especially common at that age. A breast-fed neonate also acquires protection from the IgA present in the colostrum.

Thermoregulation

Thermoregulation is another critical aspect of neonatal physiology. An infant has a large ratio of surface area to volume, so it loses heat easily. One of its defenses against hypothermia is brown fat, a special adipose tissue deposited from weeks 7 to 20 of fetal development. The mitochondria of brown fat release all the energy of fatty acids as heat rather than using it to make ATP; thus, this is a heat-generating tissue. Nevertheless, body temperature is more variable in infants and children than in adults.

Fluid Balance

The kidneys are not fully developed at birth and cannot concentrate the urine as much as a mature kidney can. Consequently, infants have a relatively high rate of water loss and require more fluid intake, relative to body weight, than adults do.

20.3b Premature Infants

Infants born before the start of week 37 are generally considered **premature.** They have multiple difficulties with respiration, thermoregulation, excretion, digestion, and liver function. Neonates born at that age average about 2.6 kg (5.8 lb). Most neonates weighing 1.5 to 2.5 kg are viable, but only with neonatal intensive care. Neonates weighing under 500 g rarely survive.

Infants born before 7 months commonly suffer **infant respiratory distress syndrome (IRDS),** the most common cause of death in premature babies. It occurs because the lungs have not produced enough pulmonary surfactant (see "The Alveoli" in section 15.1b). Consequently, when the infant exhales, the alveoli do not remain open but collapse like sheets of wet paper clinging together. A great deal of effort is needed to reinflate them. The infant becomes exhausted by the high energy demand of breathing. IRDS can be treated by ventilating the lungs with oxygen-enriched air at a positive pressure to keep the lungs inflated between breaths, and by administering surfactant as an inhalant.

A premature infant has an incompletely developed hypothalamus and therefore can't effectively thermoregulate. Body temperature must be controlled by placing the infant in a warmer.

It is difficult for premature infants to ingest milk because of their small stomach volume and undeveloped sucking and swallowing reflexes. Some must be fed by nasogastric or nasoduodenal tubes.

The liver is also poorly developed, and bearing in mind its very diverse functions (see table 18.4), you can probably understand why this would have several serious consequences. The premature liver doesn't synthesize sufficient albumin, so the baby suffers hypoproteinemia. This upsets the osmotic balance between capillary filtration and reabsorption, leading to edema. The infant bleeds easily because of a deficiency of the clotting factors synthesized by the liver. Jaundice is common in neonates, especially premature babies, because the liver cannot dispose of bile pigments efficiently.

20.3c Birth Defects

A birth defect, or **congenital anomaly**,[10] is the abnormal structure or position of an organ at birth, resulting from a defect in prenatal development. The study of birth defects is called **teratology**.[11] Birth defects are the single most common cause of infant mortality in North America. Some congenital defects aren't detected until months to years after birth. Thus, by the age of 2 years, 6% of children are diagnosed with congenital anomalies, and by age 5 the incidence is 8%. The following sections discuss some known causes of congenital anomalies, but in many cases, the cause is unknown.

Mutations and Genetic Anomalies

Genetic errors are the most common cause of birth defects, accounting for one-third of all cases and 85% of those with identifiable causes. Some defects stem from **mutations,** or changes in DNA structure—for example, achondroplastic dwarfism, microcephaly (abnormal smallness of the head), stillbirth, and some childhood cancer. Mutations can occur through errors in DNA replication during the cell cycle or under the influence of environmental agents called **mutagens,** including viruses, radiation, and some chemicals.

Some of the most common genetic disorders result, however, from the failure of homologous chromosomes to separate during meiosis. Homologous chromosomes pair up during prophase I and normally separate from each other at anaphase I (see fig. 19.11). This separation, called *disjunction,* produces daughter cells with 23 chromosomes each. In **nondisjunction,** a pair of chromosomes fails to separate. Both chromosomes of that pair then go to the same daughter cell, which receives 24 chromosomes while the other daughter cell receives 22. **Aneuploidy**[12] (AN-you-PLOY-dee), the resulting presence of an extra chromosome or lack of one, accounts for about 50% of spontaneous abortions. Aneuploidy can be detected prior to birth by *amniocentesis*—the examination of cells in a sample of amniotic fluid—or by *chorionic villus sampling (CVS),* the biopsy of cells from the chorion.

---

[10]*con* = with; *gen* = born; *a* = without; *nomaly* = evenness, symmetry
[11]*terato* = monster; *logy* = study of
[12]*an* = not, without; *eu* = true, normal; *ploid, from diplo* = double, paired

One example of aneuploidy involving the sex chromosomes is **Klinefelter[13] syndrome.** This occurs when the two X chromosomes of the female fail to separate in oogenesis, and the resulting XX egg is fertilized by a Y-bearing sperm. The resulting XXY zygote has 47 chromosomes instead of 46 and develops into a sterile male, usually of average intelligence, but with undeveloped testes, sparse body hair, unusually long arms and legs, and enlarged breasts *(gynecomastia).* The aneuploidy often goes undiagnosed until puberty, when failure to develop the typical sex characteristics may result in referral to a genetic counselor or medical geneticist.

The other 22 pairs of chromosomes (the *autosomes*) are also subject to nondisjunction. An individual with three copies of an autosome is said to exhibit **trisomy.** Only three trisomies are survivable: those involving chromosomes 13, 18, and 21. The reason is that these three chromosomes have fewer genes compared to other chromosomes. In all other trisomies, the embryo gets a lethal "gene overdose." Nearly all fetuses with trisomy-13 *(Patau syndrome)* and trisomy-18 *(Edwards syndrome)* die before birth. Infants born with these trisomies are severely deformed, and fewer than 5% survive for 1 year.

Trisomy-21 (Down[14] syndrome) is the most survivable of the three, and therefore the most common among children and adults. Its signs include impaired physical development; short stature; a relatively flat face with a flat nasal bridge; low-set ears; *epicanthal folds* at the medial corners of the eyes; an enlarged, often protruding tongue; stubby fingers; and a short broad hand with only one palmar crease (fig. 20.9). People with Down syndrome tend to have outgoing, affectionate personalities. Cognitive challenges are common and sometimes severe, but are not inevitable. About 75% of fetuses with trisomy-21 die before birth, but it occurs in about 1 out of 700 to 800 live births in the United States. Many persons born with it die by age 10 from such causes as immune deficiency and abnormalities of the heart or kidneys, but for those who survive beyond that age, modern medical care has extended life expectancy to about 60 years. After age 40, however, many of these develop early-onset Alzheimer disease, linked to a gene on chromosome 21.

Aneuploidy is far more common in humans than in any other species, and 90% of cases are of maternal rather than paternal origin. These facts seem to result from the extraordinarily long time it takes for human oocytes to complete meiosis—as long as 50 years (see "Folliculogenesis" in section 19.4c). For various reasons including defects in the mitotic spindle and in chromosomal crossing over, aging eggs become less able to separate their chromosomes into two identical sets. This is evident in the statistics of Down syndrome: The chance of having a child with Down syndrome is about 1 in 3,000 for a woman under 30, 1 in 365 by age 35, and 1 in 9 by age 48.

---

[13]Harry F. Klinefelter Jr. (1912–90), American physician
[14]John Langdon H. Down (1828–96), British physician

Figure 20.9 Down Syndrome. (a) A child with Down syndrome (right) and her sister. (b) The chromosomes of a person with Down syndrome, showing the trisomy of chromosome 21. (c) Characteristics of the hand in Down syndrome. (d) The epicanthal fold over the medial commissure (canthus) of the left eye.
a: © Science Source; b: Biophoto Associates/Science Source

? What was the sex of the person from whom the karyotype in part (b) was obtained? How do you know?

Teratogens

Environmental agents that cause birth defects are called **teratogens.** They fall into three major categories: radiation, infectious diseases, and chemicals. For this reason, diagnostic X-rays should be avoided by pregnant women. Several microorganisms can cross the placenta and cause serious congenital anomalies, stillbirths, or neonatal death. Teratogenic viruses include herpes simplex, rubella, cytomegalovirus, and human immunodeficiency virus. Many infants were born in 2016, especially in Brazil, with *microcephaly* (abnormally small heads) attributed to the mosquito-borne Zika virus. Congenital bacterial infections include gonorrhea and syphilis. *Toxoplasma,* a protozoan contracted from meat, unpasteurized milk, and housecats, is another common cause of fetal deformity. Fetuses are more vulnerable than adults to these infectious agents because of their immature immune systems.

Alcohol causes more birth defects than any other drug. Alcohol abuse during pregnancy can cause **fetal alcohol syndrome (FAS),** characterized by a small head, malformed facial features, cardiac and central nervous system defects, stunted growth, and behavioral symptoms such as hyperactivity, nervousness, and poor attention span. If a woman consumes even one drink a day during pregnancy, there may be noticeable effects on fetal and childhood development, some of which are not noticed until a child begins school. Cigarette smoke also contains potent teratogens that contribute to fetal and infant mortality, ectopic pregnancy, anencephaly (failure of the cerebrum to develop), cleft palate and lip, and cardiac abnormalities.

The effect of a teratogen depends on the genetic susceptibility of the embryo, the dosage of the teratogen, and the time of exposure. Teratogen exposure during the first 2 weeks usually doesn't cause birth defects; it's more likely to abort the embryo. The period of 3 to 8 weeks is when exposure to teratogens is most likely to result in a child being born with birth defects. Different organs have different critical periods for vulnerability to teratogenesis. For example, limb abnormalities are most likely to result from teratogen exposure at 24 to 36 days, and brain abnormalities from exposure at 3 to 16 weeks.

Before You Go On

Answer these questions from memory to test your understanding of the previous section. Reread the material if you can't answer the questions.

12. How does inflation of the lungs at birth affect the route of blood flow through the heart?
13. How is an infant protected from infection up to the time when it produces enough antibodies of its own?
14. Why is the regulation of body temperature so challenging for an infant, and even more so for a premature infant?
15. Why is respiratory distress syndrome common in premature infants?
16. Why does immaturity of the liver of a premature infant lead to such a broad spectrum of problems?
17. Genetic birth defects can result from either mutation or nondisjunction. Explain the difference and give examples of each.
18. What is a teratogen? What are the three classes of teratogens? Give examples of each.

Clinical Application 20.2

THE THALIDOMIDE TRAGEDY

Perhaps the most notorious teratogenic drug is thalidomide, a sedative first marketed in West Germany in 1957. Thalidomide was taken by many women in early pregnancy, even before they knew they were pregnant, as a sleeping aid or to relieve morning sickness. By the time it was removed from the world market in 1961, it had affected an estimated 10,000 to 20,000 babies worldwide, many of them born with unformed arms or legs (fig. 20.10) and often with defects of the ears, heart, and intestines. The U.S. Food and Drug Administration never approved thalidomide for market, but many American women obtained it by participation in clinical drug trials or from foreign sources. Thalidomide has recently been reintroduced and used under tightly controlled conditions for more limited purposes such as treating leprosy. In some Third World countries, however, people take it in a misguided attempt to treat AIDS and other diseases, resulting in an upswing in severe birth defects. A general lesson to be learned from the thalidomide tragedy and other cases is that pregnant women should avoid all sedatives, barbiturates, and opiates. Even the acne medicine isotretinoin (Accutane) has caused severe birth defects.

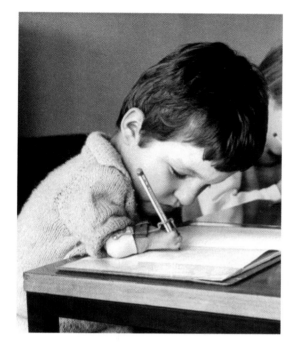

Figure 20.10 Schoolboy Showing the Effect of Thalidomide on Upper Limb Development.
© Time & Life Pictures/Getty Images

20.4 Aging, Senescence, and Death

Expected Learning Outcomes

When you have completed this section, you should be able to

a. distinguish between aging and senescence;
b. discuss the relationship of exercise to senescence;
c. summarize some current theories of senescence;
d. explain why evolution is unlikely to eliminate genes for the diseases of old age; and
e. distinguish between life expectancy and life span.

Like Ponce de León searching for the legendary fountain of youth in Florida, people yearn for a way to preserve their youthful appearance and function. Our real concern, however, is not aging but **senescence,** the degeneration that occurs in an organ system after the age of peak functional efficiency. It includes a gradual loss of reserve capacities, reduced ability to repair damage and compensate for stress, and increased susceptibility to disease.

Senescence is an issue of paramount importance for health-care providers. One in seven Americans is 65 or older. As the average age of the population rises, health-care professionals find themselves increasingly occupied by the prevention and treatment of the diseases of older age. The leading causes of death change markedly with age. Accidents, homicide, and suicide figure prominently in the deaths of people 18 to 34 years old, whereas the major causes of death after age 55 are clearly related to senescence of the organ systems: heart disease, cancer, stroke, diabetes, and lung disease. The causes of senescence, however, remain as much a scientific mystery today as cancer was 50 years ago and heredity was 100 years ago. The scientific and clinical study of aging is called **gerontology,**[15] and the clinical diagnosis and management of the medical conditions of old age is called **geriatrics.**[16]

This subject raises many points relevant to personal health and fitness choices that can lessen the rate and effects of senescence. In addition, the study of senescence calls renewed attention to the multiple interactions among organ systems. As we have already seen in earlier chapters, the senescence of one organ system typically contributes to the senescence of others.

20.4a Exercise and Rates of Senescence

The organ systems don't all degenerate at the same rate or to the same extent. For example, from ages 30 to 80, the speed of nerve conduction declines only 10% to 15%, but the number of functional glomeruli in the kidneys declines about 60%. Some physiological functions show only moderate changes at rest but more pronounced differences when tested under exercise conditions. The organ systems also vary widely in the age at which senescence becomes noticeable. There are forerunners of atherosclerosis, for example, even in infants, and visual and auditory sensitivity begin to decline soon after puberty. By contrast, the female reproductive system shows no conspicuous senescence until menopause, and then declines relatively abruptly. Aside from these unusual examples, most physiological measures of performance peak between the late teens and age 30 and then decline at a rate influenced by the level of use of the organs.

The rate of senescence depends on one's habits of use or abuse of an organ system. Obesity and insufficient exercise hasten the process of senescence. Conversely, good nutrition and exercise are the best ways to slow its progress. There is clear evidence that exercise improves the quality of life in old age. It maintains endurance, strength, and joint mobility; it also reduces the incidence and severity of hypertension, osteoporosis, musculoskeletal pain, obesity, and diabetes mellitus. It improves cognitive function and reduces the risk of Alzheimer disease. The greatest benefit derives from beginning a program of regular exercise early in life and making a lasting habit of it. If you stop exercising regularly, the body rapidly becomes deconditioned; appreciable reconditioning can be achieved, however, even when an exercise program is begun late in life. A person in his or her 90s can increase muscle strength two- or threefold in 6 months with as little as 40 minutes of isometric exercise a week. The improvement results from a combination of muscle hypertrophy and neural efficiency.

Resistance training may be the most effective way of reducing accidental injuries such as bone fractures, whereas endurance training reduces body fat and increases cardiac output, maximum oxygen uptake, and insulin sensitivity. Ideal endurance training requires three to five periods of aerobic exercise per week, each 20 to 60 minutes long and vigorous enough to reach 60% to 90% of your maximum heart rate. The maximum is best determined by a stress test but averages about 220 beats per minute minus your age in years.

An exercise program should ideally be preceded by a complete physical examination. Warm-up and cool-down periods are important in avoiding soft tissue injuries. Because of their lower capacity for thermoregulation, older people must

---

[15]*geronto* = old age; *logy* = study of
[16]*ger* = old age; *iatr* = treatment of

be careful not to overdo exercise, especially in hot weather. At the outset of a new exercise program, it is best to "start low and go slow."

20.4b Theories of Senescence

Why do our organs wear out? Why must we die? There still is no general theory on this. The question actually comes down to two issues: (1) What are the mechanisms that cause the organs to deteriorate with age? (2) Why hasn't natural selection eliminated these and produced bodies capable of longer life?

Mechanisms of Senescence

Numerous theories have been proposed and discarded to explain why organ function degenerates with age. Some authorities maintain that senescence is governed by inevitable or even preprogrammed changes in cellular function. Others attribute senescence to environmental factors that progressively damage our cells over the course of a lifetime.

There is good evidence of a hereditary component to life expectancy. Unusually long and short lives tend to run in families. Monozygotic twins are more likely than dizygotic twins to die at a similar age. Some genetic diseases such as *progeria*[17] (fig. 20.11) are associated with rapid, early senescence and premature death. Such syndromes differ from normal senescence, but nevertheless demonstrate that many changes associated with old age can be brought on by a genetic anomaly.

Knowing that senescence is partially hereditary doesn't answer the question about why tissues degenerate. Quite likely, no one theory explains all forms of senescence, but we can briefly examine some of them.

Replicative Senescence Normal organ function depends on a rate of cell renewal that keeps pace with cell death. There is a limit, however, to how many times cells can divide. Cultured human fibroblast cells divide 80 to 90 times if taken from a fetus, but only 20 to 30 times if taken from older people. After their maximum number of divisions, the cells degenerate and die. This decline in mitotic potential with age is called **replicative senescence.**

The reason for this may lie in the **telomere,**[18] a "cap" on the end of each chromosome analogous to the plastic tip of a shoelace. In humans, it consists of a noncoding nucleotide sequence CCCTAA repeated 1,000 times or more. The enzymes that replicate DNA can't reach the very ends of the molecule. If there were functional genes at the end, they would not get duplicated. The telomere may therefore provide a bit of "disposable" DNA at the end. Every time DNA is replicated, 50 to 100 bases are lost from the telomeres. Once the telomere is used up, the terminal genes may fail to replicate, making old chromosomes more vulnerable to damage and causing old cell lines to be increasingly dysfunctional. The "immortality" of cancer cells results from an enzyme, lacking from healthy cells, that repairs telomere damage and enables the cells to escape the limit on number of cell divisions.

DNA Damage Theory Telomere damage and replicative senescence are clearly not the entire answer to why organs degenerate. Skeletal muscle fibers and brain neurons exhibit pronounced senescence, yet these cells are nonmitotic. Their senescence obviously is not a result of repeated mitosis and cumulative telomere damage. Another leading theory of senescence is unrepaired damage to the coding regions of DNA. Our DNA suffers 10,000 to 100,000 damaging events per day, especially oxidative stress from free radicals generated by the cell itself. Most of this is fixed

Figure 20.11 Progeria. This is a genetic disorder in which senescence appears to be greatly accelerated. The individuals here, from left to right, are 15, 12, and 26 years old. Few people with progeria live as long as the woman on the right.
© Bettmann/Getty Images

---

[17]*pro* = before; *ger* = old age
[18]*telo* = end; *mer* = piece, part, segment

by DNA repair enzymes, but this isn't 100% efficient. Some damage persists and accumulates as cells age, especially in nondividing cells like neurons, cardiomyocytes, and skeletal muscle fibers. Such cumulative damage has been shown to cause age-related declines in liver and kidney function, cardiac and muscular strength, and brain functions related to neuron survival, synaptic plasticity, learning, and memory.

Protein Abnormalities Collagen and many other proteins exhibit increasingly abnormal structure in older tissues and cells. The changes are not in amino acid sequence—therefore not attributable to DNA mutations—but lie in the way the proteins are folded, cross-linked with each other through disulfide bridges, or bound to components such as carbohydrates. This is another reason that cells accumulate more dysfunctional proteins as they age. Collagen cross-linking makes proteins less soluble and more stiff, and is thought to be a factor in several of the most noticeable changes of the aging body such as stiffening of the joints, lenses, and arteries.

Autoimmune Theory Some of the altered macromolecules described previously may be recognized as foreign antigens, stimulating lymphocytes to mount an immune response against the body's own tissues. Autoimmune diseases such as rheumatoid arthritis do, in fact, become more common in old age.

Evolution and Senescence

If certain genes contribute to senescence, why doesn't natural selection eliminate them, much as it tends to eliminate other harmful genes from the population? Biologists once postulated that senescence and death were for the good of the species—a way for older, worn-out individuals to make way for younger, healthier ones. We can see the importance of death by imagining that science had put an end to senescence and people of all ages died at the same rate that 18-year-olds do now: 1 per 1,000 per year. If so, the median age of the population would be 163, and 13% of us would live to be 2,000 years old. The implications for world population and competition for resources would be staggering. Thus, it is easy to understand why scientists once interpreted death as a self-sacrificing phenomenon for the good of the species.

But this hypothesis has several weaknesses. One is the fact that natural selection works exclusively through the effects of genes on the reproductive rates of individuals. A species evolves only because some members reproduce more than others. A gene that doesn't affect reproductive rate can be neither eliminated nor favored by natural selection. Genes for many disorders of old age have little or no effect until a person is past reproductive age. Our prehistoric and even fairly recent ancestors usually died of accidents, predation, starvation, weather, and infectious diseases at an early age. Few people were affected by colon cancer, atherosclerosis, or Alzheimer disease, diseases associated with both age and modern lifestyle. Natural selection is "blind" to such death-dealing genes, which escape the selection process and remain with us today.

20.4c Life Expectancy and Death

Life expectancy, the average length of life in a given population, has increased impressively in industrialized countries. People born in the United States at the beginning of the twentieth century had a life expectancy of only 45 to 50 years; nearly half of them died of infectious disease. The average boy born today can expect to live 76 years and the average girl 81 years. This is due mostly to victories over infant and child mortality, not to advances at the other end of the life span. **Life span,** the maximum age attainable by humans, has not increased significantly for many centuries and there seems to be little prospect that it ever will. Ninety-nine percent of us will die before age 100, and there is little chance that this outlook will change anytime soon. We don't presently foresee any "cure for old age" or significant extension of the human life span. The real issue is to maintain the best possible quality of life, and when the time comes to die, to be entitled and able to do so with dignity and comfort.

Before You Go On

Answer these questions from memory to test your understanding of the previous section. Reread the material if you can't answer the questions.

19. Define and distinguish between *aging* and *senescence*.
20. Explain why both endurance and resistance exercises are important in old age. Support your argument with specific examples.
21. Summarize five mechanisms that may be responsible for senescence.
22. Explain why natural selection tends to remove many harmful genes from a population, but is unlikely to affect genes for the diseases of old age.
23. Distinguish between life expectancy and life span, and explain why one of these has changed significantly over the past century while the other one has not.

CAREER SPOTLIGHT

Genetic Counselor

Genetic counselors provide information, support, and advocacy for families with members who have genetic disorders or birth defects and couples concerned about their probability of transmitting genetic disorders. They trace family histories of disease, identify risk factors, interpret information for their clients, advise them on testing and prevention options and coping strategies, and refer affected individuals to community support services. Among the numerous disorders studied by genetic counselors are cystic fibrosis, sickle-cell disease, breast cancer, Down syndrome, cleft palate, and hemophilia. Genetic counselors also advise women who are pregnant or planning to become so, and who are concerned about their age or exposure to risk factors such as certain drugs, infectious agents, or radiation. Genetic counselors come from backgrounds as diverse as biology, nursing, and social work, and work in hospitals, clinics for pediatric care and adult genetics, diagnostic laboratories and biotechnology companies, research institutions, private practice, and government agencies.

To become a certified genetic counselor, one must first earn a bachelor's degree in a relevant biological or social science, then gain a master's degree and pass a board examination. See appendix B for links to further information.

© MCT/Getty Images

Study Guide

Assess Your Learning Outcomes

To test your knowledge, discuss the following topics with a study partner or in writing, ideally from memory.

20.1 Fertilization and Preembryonic Development

1. The distinction between a preembryo and embryo, and the age of transition from one to the next
2. Where the sperm and egg meet, why the sperm must travel most of the distance, how many sperm make it, and why so many of them do not
3. Why a freshly ejaculated sperm cannot fertilize an egg, and what must happen to make it capable of doing so
4. The time frame around ovulation and ejaculation when fertilization is possible
5. How a sperm enters the egg and how the egg usually prevents more than one sperm from doing so
6. Events that occur between the moment of sperm penetration and the existence of a true zygote
7. Distinguishing aspects of the three clinical trimesters of pregnancy
8. The three major processes that occur in the preembryonic stage of development
9. Distinctions between a zygote, blastomere, morula, and blastocyst; the two cell masses of the blastocyst; and the destiny of each of those cell masses
10. The two functions of the trophoblast
11. The two ways in which twins are produced and the distinction between monozygotic and dizygotic twins
12. The mechanism of implantation of a conceptus in the uterine wall
13. The defining processes of embryogenesis—how the three primary germ layers are created and how the embryoblast becomes an embryo

20.2 The Embryonic and Fetal Stages

1. The term for the overall process that transforms an embryo into a fetus; how we define when a fetus exists; and at what gestational age this is attained
2. The three modes of prenatal nutrition; what distinguishes the trophoblastic from the placental phases of nutrition; and the timetable on which these develop and overlap
3. How the placenta is produced, and its five functions in pregnancy
4. The anatomy of the placenta and umbilical cord, including the origin of the umbilical arteries and destination of the umbilical vein
5. The anatomy and functions of the amnion, yolk sac, allantois, and chorion, and the functions of amniotic fluid
6. How the flat embryonic disc becomes a cylindrical body; how its primitive gut, thoracic cavity, peritoneal cavity, neural tube, and somites come into being; and what mature tissues arise from the somites
7. How the route of blood flow in the fetus differs from its route after birth, including the names and anatomy of the three circulatory shunts
8. The meanings of the following features of the fetus—*digital rays, meconium, lanugo,* and *vernix caseosa*—and the meanings of *quickening* and the *vertex position* of the fetus

20.3 The Neonate

1. The time frame of the neonatal period
2. How the neonate's respiratory and circulatory systems change and influence each other at birth
3. What gives the neonate immunity to pathogens until it can form its own antibodies
4. Why a neonate is at risk of hypothermia and fluid loss; how its body is adapted to minimize heat loss; and the implications of these risks for neonatal care
5. The criterion for a premature infant and the reasons why premature infants have such problems with respiration, thermoregulation, feeding, and liver functions
6. The subject of teratology and why it is of such public-health importance
7. The importance of mutation as a cause of birth defects, and some examples of genetic birth defects
8. The meaning of *nondisjunction* and *aneuploidy;* the role of aneuploidy in producing birth defects; and methods for the prenatal diagnosis of aneuploidy
9. Examples of birth defects related to the nondisjunction of sex chromosomes and autosomes
10. The causation, characteristics, and prognosis for persons with Down syndrome
11. The three basic categories of teratogens; examples of each; and why the effects of a teratogen can vary according to the gestational age at exposure

20.4 Aging, Senescence, and Death

1. The distinction between aging and senescence; why senescence is so relevant to the health-care professional's work; and why it has such personal relevance

2. Examples of variation in the rate of senescence from one organ system to another
3. The extent to which regular exercise can have an impact on senescence and life expectancy
4. Evidence for a hereditary influence on longevity
5. The possible role of telomere damage in replicative senescence, and the limitations of the telomere theory
6. Other theories of senescence related to molecular cross-linking, other changes in protein structure, free radical damage, and autoimmunity
7. The theory of why evolution by natural selection has not eliminated genes for the diseases of old age, even though it does tend to eliminate other harmful genes from a population
8. The difference between life expectancy and life span, and why one of these has changed so much and the other has changed so little over the past century or more

Testing Your Recall

1. A 32-celled stage that has developed from a fertilized egg is best classified as
   a. a morula.
   b. a blastocyst.
   c. a zygote.
   d. an embryo.
   e. an embryoblast.

2. Which of these tissues does *not* exist prior to embryonic folding?
   a. the trophoblast
   b. the mesoderm
   c. the endoderm
   d. the epidermis
   e. the ectoderm

3. A newly ejaculated sperm cannot fertilize an egg because it has not yet undergone
   a. polyspermy.
   b. capacitation.
   c. the mobilization reaction.
   d. quickening.
   e. cleavage.

4. The fertilization membrane around a zygote is a product of
   a. the acrosome.
   b. the zona pellucida.
   c. the cortical reaction.
   d. gastrulation.
   e. embryogenesis.

5. All of these occur in the preembryonic stage *except*
   a. cleavage.
   b. implantation.
   c. trophoblastic nutrition.
   d. organogenesis.
   e. embryogenesis.

6. Early in pregnancy, the corpus luteum remains active under the influence of _____ from the conceptus.
   a. estrogen
   b. progesterone
   c. meconium
   d. lanugo
   e. human chorionic gonadotropin

7. The lanugo is
   a. fetal hair.
   b. a sperm enzyme used to penetrate the egg.
   c. contents of the fetal colon at full term.
   d. a form of neonatal respiratory distress.
   e. a circulatory shunt that bypasses the liver.

8. Fetal alcohol syndrome results from
   a. a mutagen.
   b. a teratogen.
   c. a carcinogen.
   d. nondisjunction.
   e. aneuploidy.

9. One theory of replicative senescence is that it results from depletion of the _____ at the ends of the chromosomes.
   a. telomeres
   b. sarcomeres
   c. centromeres
   d. somites
   e. acrosomes

10. Another theory of senescence is that we suffer more and more tissue damage from _____ as our antioxidant levels decline with age.
    a. autoimmune disease
    b. cross-linking
    c. polyspermy
    d. nondisjunction
    e. free radicals

11. Endoderm and mesoderm are formed by the migration of cells down into the _____ of the embryonic disc.

12. The process of migration described in question 11, resulting in three primary germ layers, is called _____.

13. Muscle, bone, and blood arise from the primary germ layer called _____.

14. The oldest verifiable age to which any human being has lived is called the human _____.

15. Prior to implantation, a blastocyst is nourished by a secretion from the _____ glands.

16. In trophoblastic nutrition, the conceptus digests and consumes _____ cells of the endometrium.

17. Treelike processes called _____ contain fetal blood and project into the placental sinus, where they are surrounded by maternal blood.

18. The "breaking of the waters" as a woman goes into labor is a release of the _____ fluid in which the fetus has been suspended for most of its gestation.

19. In the fetus, a shunt between the pulmonary trunk and the aorta called the _____ diverts blood from the nonfunctional lungs.

20. Any chemical, infectious organism, or radiation capable of causing a birth defect is called a/an _____.

Answers in Appendix A

What's Wrong with These Statements?

Briefly explain why each of the following statements is false, or reword it to make it true.

1. Most birth defects result from exposure to teratogens in the first two weeks of gestation.

2. Cleavage divides a morula into more and more blastomeres, so the morula grows several times as large as the zygote.

3. Twins result from disjunction, the separation of cells of the morula into two masses that each produce an embryo.

4. By the time implantation is complete, the blastocyst is attached to the mother by a placenta and umbilical cord.

5. Trophoblastic nutrition of the embryo must end before placental nutrition can begin.

6. An ovulated secondary oocyte must complete meiosis II before it can be fertilized.

7. The ductus venosus allows most fetal blood to bypass the lungs.

8. Gerontologists are optimistic that the human life span will increase significantly in the twenty-first century.

9. Freshly ejaculated sperm are more capable of fertilizing an egg than are sperm 5 or 6 hours old.

10. The developing individual is not considered a fetus until it enters the third trimester of pregnancy.

Answers in Appendix A

Testing Your Comprehension

1. Some health-supplement vendors market the enzyme superoxide dismutase (SOD) as an oral antioxidant to retard senescence. Explain why it would be a waste of your money to buy it.

2. In some children, the ductus arteriosus fails to close after birth—a condition that eventually requires surgery. Predict how this condition would affect (a) pulmonary blood pressure, (b) systemic diastolic pressure, and (c) the right ventricle of the heart.

3. Only one sperm is needed to fertilize an egg, yet a man who ejaculates fewer than 10 million sperm is usually infertile. Explain this apparent contradiction. Supposing 10 million sperm were ejaculated, predict how many would come within close range of the egg. How likely is it that any one of these sperm would fertilize it?

Design Element Credits: (Camping): monkographic/Shutterstock; (Abseiling): Image Source; (Stethoscope): MedicalRF.com; (Clouds): Pakhnyushchy/Shutterstock.com; (Doctors): Purestock/SuperStock; (Mountain climber): McGraw-Hill Education.

Appendix A
ANSWER KEYS

This appendix provides answers to questions in the figure legends and in the end-of-chapter "Testing Your Recall" and "What's Wrong with These Statements?" sections.

Chapter 1

Figure Legend Questions

1.1 A PET scan requires an injection of radioisotopes, which is an invasion of the body surface. No invasive procedure is needed for an MRI.
1.2 Radiography uses X-rays, a high-energy radiation that can cause mutations and birth defects. Sonography uses only sound waves, not known to cause any harm to a fetus.
1.5 Yes, childbirth could be regarded as a homeostatic response to pregnancy that culminates in ending the pregnancy and returning the body to its more usual state.

Testing Your Recall

1. a
2. b
3. b
4. e
5. c
6. a
7. c
8. b
9. e
10. c
11. palpation
12. computed tomography
13. homeostasis
14. organ
15. wrist, ankle
16. medial
17. parietal, visceral
18. negative feedback loop
19. anatomical position
20. proximal

What's Wrong with These Statements?

1. A median section of the head would pass vertically between the eyes, but not through both eyes at once.
2. The diaphragm is inferior to the lungs.
3. A PET scan requires an injection of radioactive glucose and is therefore invasive.
4. The pleural and pericardial cavities are lined by serous membranes.
5. Abnormal skin color or dryness is one piece of information that could be obtained by inspection.
6. Sonography is simpler than CT and is safer for a fetus because it does not involve ionizing radiation.
7. Histopathology is a subdiscipline of histology.
8. Negative feedback is usually beneficial because it corrects deviations from a physiological set point.
9. There are more organelles than cells in the human body.
10. The pericardial sac is external to the heart, separated from it by the space of the pericardial cavity.

Chapter 2

Figure Legend Questions

2.6 Relatively little energy is required to make an N_2 molecule vaporize because it has no attraction to other molecules in the liquid state. Thus, N_2 boils at a low temperature.

Testing Your Recall

1. a
2. c
3. a
4. c
5. a
6. e
7. b
8. c
9. b
10. d
11. cation
12. free radicals
13. catalyst, enzymes
14. anabolism
15. dehydration synthesis
16. -ose, -ase
17. phospholipids
18. ATP
19. catabolism
20. substrate

What's Wrong with These Statements?

1. The monomers of a polysaccharide are monosaccharides.
2. Water molecules are polar.
3. ATP is a very short-lived molecule, not a long-term energy-storage molecule.
4. Enzymes are catalysts that enable chemical reactions to go faster but are not consumed by the reaction.
5. Proteins are polymers formed of amino acids linked by peptide bonds.
6. A saturated fat is defined as a fat to which no more hydrogen can be added.
7. Anabolism is an energy-storing oxidation process that combines small, low-energy molecules into large, high-energy molecules.
8. Humans cannot digest cellulose and therefore get no calories from it.
9. Two isomers do not have identical chemical properties. Ethanol and ether, for example, are isomers with very different properties.
10. A solution of pH 7 has 10 times the hydrogen ion concentration of a solution with pH 8.

Chapter 3

Figure Legend Questions

3.2 The phospholipid heads are hydrophilic and thus attracted to the water inside and outside the cell.
3.4 The photo shows an intestinal absorptive cell, which has prominent microvilli for nutrient absorption and a glycocalyx for protection from acid and enzymes of the digestive tract.
3.7 Gap junctions
3.8 Simple diffusion
3.13 The nucleus, endoplasmic reticulum, and Golgi complex participate directly in protein synthesis; mitochondria participate indirectly because protein synthesis requires the ATP that they provide.
3.19 Chromatids are also composed of protein.

Testing Your Recall

1. e
2. d
3. b
4. b
5. e
6. a
7. d
8. a
9. d
10. b
11. micrometers
12. receptor
13. gates
14. glycocalyx
15. anaphase
16. phospholipids
17. nucleus, mitochondria
18. smooth ER, peroxisomes
19. phagocytosis
20. pseudopods

What's Wrong with These Statements?

1. A cell specialized for absorption would be expected to have a high density of microvilli on its surface.
2. DNA replication occurs in the S phase, not during mitosis.
3. A cell releases secretory products by exocytosis, not phagocytosis or pinocytosis.
4. In the plasma membrane, it is the fatty acid tails of the phospholipid molecules that cluster together in the middle of the membrane, not the phosphate heads, which are pointed toward the ICF and ECF.
5. Cells of the digestive glands release enzymes by the process of exocytosis into the digestive tract when needed to digest food and do not store them in lysosomes.
6. Movement of substances down a concentration gradient requires no ATP.
7. Osmosis is a passive process, not active transport.
8. White blood cells can move about in the tissues by means of pseudopods, not cilia.
9. Desmosomes are just spots of intercellular attachment; they have no channels for transferring material from cell to cell.

10. Ribosomes synthesize proteins, whereas the Golgi complex modifies and packages them.

Chapter 4

Figure Legend Questions

4.12 Exfoliation of squamous cells also occurs in the epidermis, oral cavity, esophagus, anal canal, and a few other places.

4.29 Holocrine glands require a high rate of stem-cell mitosis to replace the cells that continually disintegrate to become the secretion.

4.30 Mucous membranes are also found throughout the digestive, reproductive, and urinary tracts and other regions of the respiratory tract in addition to the trachea cited. Serous membranes are also found in the pericardium, pleurae, parietal peritoneum, mesenteries, and serosae of many other abdominal organs in addition to the small intestine cited.

Testing Your Recall

1. a
2. b
3. d
4. e
5. c
6. e
7. b
8. a
9. b
10. c
11. necrosis
12. simple squamous
13. lacunae
14. periosteum
15. collagen
16. glial cells
17. basement membrane
18. matrix
19. holocrine
20. simple

What's Wrong with These Statements?

1. The esophageal epithelium is nonkeratinized.
2. Mucus-secreting goblet cells are only found in simple epithelia.
3. Not all skeletal muscle attaches to bones. For example, muscles of facial expression attach to skin.
4. Ground substance and fibers make up the matrix of a connective tissue.
5. Adipose tissue is an exception to this rule.
6. Adipocytes also occur in areolar connective tissue.
7. A single exocrine gland can be both serous and mucous.
8. Weight lifting creates new muscle tissue by the process of hypertrophy.
9. All living cells are electrically excitable, but nerve and muscle cells have this developed to the highest degree.
10. The tissue degeneration that occurs in gangrene is an example of necrosis, not apoptosis.

Chapter 5

Figure Legend Questions

5.3 Keratinocytes

5.4 Basal cell carcinoma is the most common but melanoma is the most dangerous.

5.5 Shampoo only contacts the dead shafts of the hairs, which therefore cannot be nourished. The only nourishment a hair can receive is from the dermal papilla at the base of the hair root, from blood-borne nutrients.

Testing Your Recall

1. d
2. b
3. d
4. a
5. e
6. b
7. c
8. a
9. a
10. e
11. insensible perspiration
12. arrector muscle
13. dermato-, cutane-
14. cyanosis
15. dermal papillae
16. earwax
17. sebaceous glands
18. keratin, collagen
19. vitamin D
20. keratinocytes

What's Wrong with These Statements?

1. Exfoliation is the flaking off of dead cells from epithelial surfaces, such as occurs in the stratum corneum.
2. The term *integumentary system* refers to the skin, hair, nails, and cutaneous glands.
3. The dermis is mostly collagen; keratin occurs in the epidermis.
4. Vitamin D is synthesized by collaboration of epidermal keratinocytes, the liver, and the kidneys.
5. Cells of the stratum basale multiply rapidly to produce the thick stratum corneum of the skin surface.
6. Hair growth results from the mitosis of cells in the hair matrix. Hair cells die as they are pushed upward away from the blood supply in the dermal papilla.
7. The hypodermis is not part of the skin.
8. People of all ethnicities have about equal densities of melanocytes.
9. Pallor is a temporary paleness, not a genetic lack of pigment.
10. Eccrine sweat glands help cool the body; apocrine sweat glands are activated at puberty and produce pheromones.

Chapter 6

Figure Legend Questions

6.1 Bone marrow

6.2 Osteoblasts engage actively in synthesizing the collagen of the bone matrix and require a lot of rough endoplasmic reticulum to carry this out. Osteogenic cells do not synthesize proteins for export from the cell.

6.3 Spongy bone has more surface area and is therefore more strongly affected by osteoclasts.

6.11 The appendicular skeleton includes the pectoral and pelvic girdles, which are within the axial region of the body.

6.14 The temporal bone articulates with the occipital, parietal, frontal, sphenoid, and zygomatic bones and the mandible.

6.22 Rupture of this ligament could allow the head and atlas to slip posteriorly and thus cause the dens of the axis to rip through the cervical spinal cord.

6.23 Transverse foramina and a forked spinous process

6.35 Interphalangeal joints are not subject to shock or pressure like the jaw and knee joints, and therefore require no pressure-absorbing menisci.

Testing Your Recall

1. e
2. a
3. a
4. d
5. b
6. b
7. d
8. e
9. a
10. c
11. temporal
12. canaliculi
13. ligament
14. synovial fluid
15. osteopenia
16. osteoblasts
17. kinesiology
18. menisci
19. sutures
20. anulus fibrosus

What's Wrong with These Statements?

1. There are eight carpal bones and seven tarsal bones.
2. The growth zone of a child's bone is the metaphysis (or epiphyseal plate).
3. The zygomatic bones do not articulate with the parietal bones.
4. It is not possible to palpate the muscles in the subscapular fossa but in a living person one may palpate those of the supraspinous fossa.
5. This action involves hyperextension of the shoulder.
6. The cranial bones that enclose the brain develop from fetal membranous tissue, not from cartilage.
7. The bursae contain synovial fluid but do not secrete it; the synovial membrane does.
8. The arm contains only one bone (the humerus), but the leg contains two (tibia and fibula).
9. Parathyroid hormone causes osteoclasts to break down bone mineral rather than stimulating bone deposition.
10. Climbing a flight of stairs requires strong, alternating flexion and extension of the hip joint.

Chapter 7

Figure Legend Questions

7.1 Skeletal muscle cells are long and threadlike; cardiac and smooth myocytes are not.

7.4 A bands are a region of thick and thin filament overlap, so there is a greater density of protein here than in the I bands, which have only thin filaments.

7.9 I bands would be absent because there is no region of the sarcomere occupied by thin filaments only.

7.13 For the peristaltic contractions of swallowing
7.15 The triceps brachii would be the prime mover, whereas the biceps brachii and brachialis muscles would exert a braking action.
7.17 The pterygoid muscles approach the mandible obliquely and pull it from side to side, whereas the temporalis muscles approach it from directly above and can only pull the mandible upward.
7.20 The hip and the knee

Testing Your Recall

1. a
2. b
3. e
4. c
5. b
6. c
7. a
8. d
9. c
10. d
11. myosin
12. acetylcholine
13. muscle tone (tonus)
14. fascicle
15. calcium
16. involuntary
17. sphincter
18. quadriceps femoris
19. latissimus dorsi
20. gastrocnemius

What's Wrong with These Statements?

1. A motor neuron branches out to anywhere from 3 to 1,000 muscle fibers or so.
2. Calcium binds to troponin, not to myosin.
3. Slow-twitch fibers resist fatigue; fast-twitch fibers fatigue relatively quickly.
4. Myofilaments slide over each other; they do not shorten.
5. Compared to other muscle types, cardiac muscle has the largest and most numerous mitochondria, which makes cardiac muscle more resistant to fatigue.
6. The zygomaticus major and minor muscles attach to the zygomatic bone and the orbicularis oris at the corner of the mouth.
7. The serratus anterior is the prime mover in forward thrusts and pushing, whereas the trapezius retracts the scapulae.
8. In isometric contraction, a muscle does not shorten or lengthen, even though it develops tension.
9. The iliopsoas is a hip flexor.
10. The brachialis muscles are the prime mover of elbow flexion, and provide more power than the biceps brachii muscles do.

Chapter 8

Figure Legend Questions

8.6 Schwann cells
8.11 Proximal to the action potential, the axon is in its refractory period and cannot be stimulated immediately to fire again.

Testing Your Recall

1. a
2. c
3. a
4. e
5. d
6. b
7. e
8. b
9. c
10. e
11. dura mater
12. central
13. muscle spindles
14. brachial
15. node of Ranvier
16. bipolar
17. ependymal
18. action potential
19. neurotransmitters
20. gray matter

What's Wrong with These Statements?

1. Motor signals travel the spinal cord by way of its descending, not ascending, tracts.
2. The arachnoid mater lies between the dura mater and pia mater.
3. The outermost layer of a nerve is the epineurium.
4. There are no spinal nerve plexuses in the thoracic region.
5. Some somatic reflex arcs pass through the brainstem.
6. Nerve signals slow down slightly at the nodes of Ranvier. The signal travels faster in myelinated axons because the signal travels more quickly along the internodes.
7. A neuron at rest has a negative internal charge relative to the charge outside its plasma membrane.
8. A nerve fiber is just the axon of a neuron, not a synonym for *neuron,* the entire cell.
9. The myelin sheath in the brain is produced by oligodendrocytes.
10. A bipolar neuron has one axon and one dendrite.

Chapter 9

Figure Legend Questions

9.3 The cranial dura mater is pressed against the surrounding bone, and even attached to it at places, with no intervening epidural space or fat; it has two layers, separated in some places by dural sinuses; and it folds inward to form partitions between parts of the brain.
9.8 The amygdala
9.9 One would lose somatosensory sensations such as touch and pain from the lower limb on the opposite side of the body.
9.17 Such an injury could prevent the heart from accelerating in response to fear because it would disrupt the conduction of signals from the brain (fear centers) down the spinal cord to the origins of the cardiac nerves. It would not affect gastric secretion, because that is mediated through the vagus nerves, which originate above the point of injury.

Testing Your Recall

1. c
2. c
3. a
4. e
5. a
6. b
7. d
8. d
9. e
10. e
11. corpus callosum
12. ventricles, cerebrospinal
13. arbor vitae
14. cholinergic
15. choroid plexus
16. precentral
17. proprioceptors
18. association
19. occipital
20. Broca area

What's Wrong with These Statements?

1. The longitudinal fissure separates the cerebral (not cerebellar) hemispheres.
2. Cerebrospinal fluid is produced not only by the choroid plexuses but also by filtration through the brain surface and by the ependymal cells that line the ventricles.
3. The limbic system is associated with emotions, memory, and learning, whereas the medulla oblongata regulates respiratory and cardiac rhythms.
4. Each half of the brain has its own thalamus, but the hypothalamus and epithalamus are single, median structures.
5. The sympathetic nervous system inhibits digestion.
6. Hearing is a function of the temporal lobe; the occipital lobe is visual.
7. The white matter is organized into tracts.
8. The parasympathetic division has long cholinergic preganglionic fibers.
9. Cranial nerve X (the vagus) innervates numerous thoracic and abdominal organs.
10. The optic nerve does not move the eye; the oculomotor, abducens, and trochlear nerves do.

Chapter 10

Figure Legend Questions

10.1 Small receptive fields; the lips are very sensitive to fine two-point discrimination.
10.6 Vestibulocochlear, CN VIII
10.12 Inner hair cells. The outer hair cells tune the cochlea to enable the brain to better discriminate one pitch of sound from another.
10.14 The eyes would be more prone to watering because the tears could not drain away into the nasal cavity.
10.17 It is the subject's right eye. The optic disc is medial to the fovea (the optic nerves converge medially from the two eyes), so the subject's nose would be toward page right, making this the subject's right eye.
10.20 The inner segment, because this is where the mitochondria are concentrated.
10.22 By tracing fibers from the upper optic radiation in this figure to the retinas, it can be seen that one would lose vision from the lateral retina of the right eye and medial retina of the left eye. These two retinal areas cover the left peripheral vision, so the effect would be blindness in the left half of the visual field.

Testing Your Recall

1. b
2. e
3. a
4. c
5. b
6. d

7. b
8. a
9. d
10. d
11. lamellar corpuscle
12. fovea centralis
13. optic chiasm
14. lacrimal
15. saccule
16. accommodation
17. olfactory bulbs
18. hair cells
19. cone
20. proprioceptors

What's Wrong with These Statements?

1. Pain has its own specialized receptors (nociceptors) and is not the result of overstimulation of touch receptors.
2. The olfactory hairs in the nose are modified cilia, whereas the hairs on cochlear hair cells are modified microvilli.
3. Differences in the color of light depend on which cones are stimulated and the pitch of a sound depends on which part of the basilar membrane vibrates; differences in the firing frequencies of the sensory nerve fibers determine the sense of stimulus intensity such as brightness or loudness, not color or pitch.
4. A hair receptor is a nerve fiber associated with a hair follicle, not to be confused with hair cells of the inner ear.
5. Most refraction occurs when light rays pass from air into the cornea.
6. The right and left occipital lobes receive visual input from both eyes because some of the fibers of the optic nerve cross to the opposite side of the brain at the optic chiasma.
7. Deep bass tones stimulate different cochlear hair cells than high treble tones.
8. The auditory tube lets air into the middle ear, not the inner ear.
9. After passing across the eye, tears drain into the nasal cavity via the nasolacrimal duct.
10. Lingual papillae are visible bumps on the tongue. Most types contain taste buds, but they are not the same as the taste buds.

Chapter 11

Figure Legend Questions

11.1 Liver, kidneys, stomach, heart, skin, small intestine, bones, adipose tissue, placenta
11.4 Mechanism (b), because testosterone is a steroid
11.6 The posterior lobe
11.8 The C cells secrete calcitonin.
11.10 The exocrine cells secrete digestive enzymes.

Testing Your Recall

1. b
2. d
3. a
4. c
5. c
6. b
7. e
8. b
9. e
10. d
11. cortisol
12. hypothalamus, posterior pituitary gland
13. leptin
14. ghrelin
15. anterior pituitary gland
16. interstitial endocrine cells
17. ovaries, testes
18. pineal
19. thymus
20. oxytocin

What's Wrong with These Statements?

1. Increased percentage of body fat is associated with greater risk of insulin insensitivity.
2. Follicle-stimulating hormone and luteinizing hormone target both the ovaries and the testes.
3. Blockage of this system would cut off the hypothalamic stimulus for the pituitary to secrete FSH and LH, and without those gonadotropins, the ovaries and testes could not function.
4. The pineal gland and thymus shrink (involute) with age.
5. The adrenal cortex secretes steroids.
6. Insulin is released immediately after you have eaten a meal.
7. The second stage of the stress response is dominated by the hormone cortisol.
8. The infundibulum is not a duct and no hormones are secreted through ducts.
9. Thyroxine raises metabolic rate and heat production.
10. A steroid hormone can freely enter a target cell, but a peptide hormone cannot.

Chapter 12

Figure Legend Questions

12.1 Nuclei
12.3 The nucleus of the stem cell line was here; the cell caves in at the center when the nucleus is lost.
12.4 To the iron atom at the center of the heme group
12.11 Although superabundant, these cells are immature and unable to perform their defensive roles.

Testing Your Recall

1. e
2. a
3. b
4. d
5. e
6. c
7. b
8. c
9. d
10. a
11. hematopoiesis
12. hematocrit
13. macrophages
14. hemostasis
15. serum
16. hemophilia
17. leukopenia
18. albumin
19. polycythemia
20. erythropoiesis

What's Wrong with These Statements?

1. Blood plasma typically constitutes about 55% of the volume.
2. An elevated albumin concentration raises the osmolarity of blood, causing osmotic retention or absorption of water into the bloodstream and raising the blood pressure.
3. A low blood oxygen level is the result of anemia, not the cause.
4. Neutrophils are the most important bacteria-fighting WBCs.
5. Erythrocytes and platelets lack nuclei.
6. The most abundant WBCs are neutrophils.
7. Coagulation and agglutination are entirely different processes. Coagulation is the formation of sticky fibrin threads and entrapment of RBCs, platelets, and WBCs in the mesh. Agglutination is a process in which RBCs or foreign cells become stuck together by antibody molecules.
8. Lymphocytes arise in the red bone marrow from the same hemopoietic stem cells as other formed elements, although they also proliferate in other sites such as the thymus and lymph nodes after leaving the bone marrow.
9. RBCs live longer than most granulocytes in spite of their lack of a nucleus.
10. Leukemia is the malignant overproduction of WBCs.

Chapter 13

Figure Legend Questions

13.1 Both; they receive pulmonary blood from the pulmonary arteries and systemic blood from the bronchial arteries.
13.2 It mostly lies to the left of the median plane.
13.8 The right atrium begins contracting a little sooner than the left because the pacemaker is in the right atrium, so cardiomyocytes of the right atrium are first to receive its signal.
13.10 Phase 4 is the point at which the aortic valve opens and the left ventricle begins ejecting blood into the aorta.
13.11 Being closer to the heart, the arteries are subjected to greater blood pressure surges and must be more elastic to withstand this.
13.14 Vasodilation lowers blood pressure.
13.21 Aorta, left common carotid artery, left external carotid artery

Testing Your Recall

1. c
2. d
3. b
4. b
5. d
6. c
7. b
8. a
9. e
10. e
11. systole, diastole
12. sphygmomanometer
13. celiac trunk
14. vagus
15. cerebral arterial circle
16. hypertension
17. baroreceptors
18. myocardial infarction
19. T wave
20. left coronary artery

What's Wrong with These Statements?

1. The systemic circuit includes the coronary circulation.
2. It is the brachiocephalic trunk.

3. Aldosterone causes salt retention and tends to sustain or raise blood pressure rather than lowering it.
4. The pulmonary veins carry blood from the lungs to the heart.
5. Valves are found in medium veins, not in arteries.
6. A doubling of vessel radius increases blood flow 16-fold.
7. It passes through two capillary beds—one in the digestive tract and one in the liver.
8. An ECG is a composite recording from all cardiomyocytes of the heart, and is not an individual action potential.
9. The two ways to increase cardiac output are to raise heart rate and increase stroke volume.
10. The cords attach to the AV valves; there are no tendinous cords attached to the semilunar (pulmonary and aortic) valves.

Chapter 14

Figure Legend Questions

14.2 Lymphatic capillaries have larger gaps between their endothelial cells than blood capillaries do.
14.4 Axillary lymph nodes are the first place where cells metastasizing from a breast tumor would lodge. The absence of malignant cells here may indicate that the cancer was diagnosed and treated before it had metastasized.
14.5 Without lymphatic valves, there could be no steady one-way flow of lymph toward the subclavian veins.

Testing Your Recall

1. b
2. c
3. b
4. c
5. d
6. c
7. b
8. e
9. a
10. d
11. pathogen
12. lacteals
13. regulatory T (T_R) cells
14. interferons
15. plasma
16. antigen-presenting cells
17. palatine tonsils
18. thymus
19. neutrophils
20. histamine

What's Wrong with These Statements?

1. T_C cells are involved only in adaptive immunity.
2. Perforins are employed both in innate immunity by NK cells and in adaptive immunity by T_C cells.
3. Lymphatic capillaries are more permeable than most blood capillaries because of large gaps between their endothelial cells.
4. Helper T (T_H) cells are involved in both humoral and cellular immunity.
5. A person who has previously encountered an antigen will mount a rapid secondary response.
6. Both B and T cells reside in lymph nodes.
7. HIV infects primarily helper T (T_H) cells.
8. Most plasma cells develop in the lymph nodes.
9. Complement proteins are produced by the liver.
10. The thymus and spleen do not have lymphatic nodules.

Chapter 15

Figure Legend Questions

15.2 The line would be drawn across the upper end of the trachea.
15.5 The right main bronchus is wider and more vertical than the left, so it is easier for aspirated objects to fall into the right one.
15.12 Without a nose clip, much of the air would flow through the nose rather than the spirometer mouthpiece and therefore would not be recorded.

Testing Your Recall

1. b
2. b
3. e
4. b
5. e
6. d
7. d
8. a
9. d
10. c
11. spirometry
12. ventral respiratory group
13. thyroid
14. alveolar macrophages
15. parietal pleura
16. carbonic acid
17. chemoreceptors
18. mucociliary escalator
19. hilum
20. asthma

What's Wrong with These Statements?

1. Tidal volume averages about 500 mL.
2. This passage describes obstructive lung diseases.
3. Most CO_2 transport is in the form of carbonic acid and bicarbonate ions.
4. The palatine tonsils are located at the fauces where the oral cavity opens into the oropharynx.
5. Each lung has about 150 million alveoli.
6. Surfactant is produced by the great (type II) alveolar cells.
7. P_{O_2} normally has very little effect on breathing rate and depth.
8. Acidosis increases respiratory rate and depth as a mechanism for blowing off excess CO_2 and raising the pH.
9. Freshly inspired alveolar air has a P_{O_2} of about 104 mm Hg.
10. Air intake depends on intrapulmonary pressure falling below atmospheric pressure.

Chapter 16

Figure Legend Questions

16.5 The afferent arteriole is larger than the efferent arteriole. Because it has a large inlet and small outlet, the glomerulus has an unusually high blood pressure (BP) compared to other capillaries. This high BP drives the glomerular filtration process.
16.13 This sphincter serves to prevent reflex of semen into the bladder during ejaculation, and is thus limited to males.
16.14 The urine volume would decrease.

Testing Your Recall

1. a
2. d
3. b
4. b
5. d
6. e
7. e
8. d
9. c
10. a
11. CO_2
12. glomerular filtration rate
13. ureters
14. macula densa
15. podocytes
16. nephron
17. water, salt
18. internal urethral
19. sodium, potassium
20. fibrous capsule

What's Wrong with These Statements?

1. The DCT is the most significant target of aldosterone.
2. The skin also loses water by cutaneous transpiration.
3. Many nephrons drain into each collecting duct, so nephrons outnumber the collecting ducts.
4. Sodium is reabsorbed in the PCT, nephron loop, and DCT.
5. Natriuretic peptides promote sodium excretion and lower blood volume.
6. Dilation of the afferent arteriole increases blood flow into the glomerulus and thus increases glomerular filtration rate.
7. Angiotensin II is a potent vasoconstrictor and this effect raises overall blood pressure.
8. Urine can be as dilute as 50 mOsm/L.
9. The events described, causing urination, are effects of the parasympathetic nervous system.
10. The normal GFR is about 105 mL/min. in women and 125 mL/min. in men.

Chapter 17

Figure Legend Questions

17.5 The two premolars and the third molar of each side of the jaw are not yet present at age 3.
17.7 The root of the tongue blocks the oral cavity and the soft palate rises to block the nasal cavity so food cannot be forced into either one from the pharynx.
17.8 The muscularis externa has three layers in the stomach, only two in the esophagus. The esophagus lacks the oblique layer. The addition of a

layer in the stomach is related to the intense churning contractions that occur in this organ.

17.17 Sugars and amino acids are absorbed into the blood, which goes from the small intestine directly to the liver via the hepatic portal system. Dietary fat is absorbed by the intestinal lacteals and travels the route: lacteal → lymphatic trunk → thoracic duct → left subclavian vein → left brachiocephalic vein → superior vena cava → right side of heart → lungs → left side of heart → aorta → celiac trunk → common hepatic artery → liver.

17.18 The internal anal sphincter is controlled by the autonomic nervous system because it is composed of smooth muscle. The external anal sphincter is composed of skeletal muscle and therefore controlled by the somatic nervous system.

Testing Your Recall

1. b
2. a
3. a
4. b
5. d
6. a
7. e
8. c
9. b
10. a
11. dentin
12. duodenum, jejunum, ileum
13. falciform ligament
14. chief
15. colon
16. enteric
17. amylase, saliva
18. parotid, submandibular, sublingual
19. palate
20. peristalsis

What's Wrong with These Statements?

1. Chylomicrons are absorbed into lacteals; micelles remain in the small intestine.
2. Taeniae coli are features of the large intestine.
3. The pyloric valve regulates the passage of chyme from stomach to duodenum; the ileocecal valve regulates the passage from small intestine to large intestine (ileum to cecum).
4. Vitamin D promotes absorption of calcium, not iron.
5. The small intestine absorbs glucose, fructose, and galactose.
6. Proteases digest proteins and smaller peptides.
7. Swallowing begins with the oral phase.
8. Most of the water is absorbed by the colon.
9. Intrinsic factor enables the absorption of vitamin B_{12}.
10. The jejunum is a segment of the small intestine.

Chapter 18

Figure Legend Questions

18.1 HDLs remove excess cholesterol from the tissues, whereas LDLs deliver it to the tissues to be deposited. A high HDL level therefore indicates a high rate of clearance of excess cholesterol from the body.

18.3 Niacin and riboflavin, because they are the dietary precursors of the coenzymes NAD^+ and FAD, the coenzymes that import electrons into the mitochondrial pathways of aerobic respiration.

Testing Your Recall

1. a
2. c
3. b
4. e
5. b
6. d
7. e
8. a
9. d
10. d
11. incomplete
12. high-density lipoproteins, low-density lipoproteins
13. gluconeogenesis
14. urea
15. liver
16. insulin
17. core temperature
18. hypothalamus
19. lactate
20. kilocalorie, kcal

What's Wrong with These Statements?

1. Leptin inhibits the appetite.
2. Water is classified as a nutrient because it is absorbed, becomes part of the body's tissues, and performs necessary metabolic roles.
3. Fat contains 9 kcal/g and sugar contains 4 kcal/g.
4. Most of the body's cholesterol is made by the body itself.
5. Most liver functions are nondigestive; only the secretion of bile acids and lecithin aids digestion.
6. The mitochondria cannot produce ATP without oxygen.
7. The mitochondrial membrane reactions produce more energy than the combined production of glycolysis and the mitochondrial matrix reactions.
8. Gluconeogenesis occurs in the fasting state.
9. Ketone bodies come from fatty acid oxidation and are thus more likely to occur in excess in someone on a high-fat, low-carbohydrate diet.
10. At a comfortable air temperature, the body loses most of its heat by radiation.

Chapter 19

Figure Legend Questions

19.11 Crossing-over in prophase I exchanges portions of the paternal and maternal chromosomes of each pair and thus creates genetically novel chromosomes to be passed to the offspring.

19.14 One egg per primary oocyte; four sperm per primary spermatocyte. It is important to produce a large egg with adequate nutrients to support early development and cytoplasm to be divided among multiple blastomeres, so the oocyte should not be divided into small cells.

Testing Your Recall

1. b
2. d
3. a
4. c
5. c
6. a
7. d
8. e
9. a
10. d
11. zygote
12. first polar body
13. antrum
14. luteal (postovulatory)
15. seminal vesicle
16. oxytocin
17. seminiferous tubules
18. colostrum
19. luteinizing hormone
20. glans

What's Wrong with These Statements?

1. Interstitial endocrine cells secrete testosterone; nurse cells secrete inhibin.
2. Sperm carry the DNA in the nucleus, located in the head.
3. The seminal vesicles produce about 60% of the semen.
4. The clitoris has no corpus spongiosum.
5. Androgens are also responsible for development of the pubic and axillary hair and apocrine glands in girls and for the libido of both sexes.
6. The vagina has no mucous glands.
7. Lactiferous ducts exist even in the nonlactating breast.
8. Men remain fertile after climacteric.
9. Spermiogenesis is the postmeiotic transformation of spermatids to mature sperm; it does not include meiosis II.
10. Climacteric is a midlife change in reproductive function, not to be confused with climax (orgasm).

Chapter 20

Figure Legend Questions

20.2 An unfertilized oocyte would already be dead by the time it arrived in the uterus; it does not live long enough to survive the 3-day migration.

20.4 Answers may vary but could include bones, muscles, and blood vessels (any mesodermal derivative).

20.9 The chromosomes shown came from a female, as we can see from the presence of two X chromosomes.

Testing Your Recall

1. a
2. d
3. b
4. c
5. d
6. e
7. a
8. b
9. a
10. e
11. primitive groove
12. gastrulation
13. mesoderm
14. life span
15. uterine
16. decidual
17. chorionic villi
18. amniotic
19. ductus arteriosus
20. teratogen

What's Wrong with These Statements?

1. Teratogen exposure at that early stage is more likely to cause embryonic death (abortion) than a defect that would persist until birth.

2. A morula is about the same size as the zygote, but divided into smaller and more numerous cells.
3. Disjunction refers to chromosomal separation in meiosis, not the separation of morula cells into two embryos.
4. Implantation occurs on day 3; the placenta doesn't begin to form until day 11, and the umbilical cord still later.
5. Trophoblastic and placental nutrition overlap until about week 12.
6. Fertilization triggers the completion of meiosis II; if unfertilized, an oocyte never completes meiosis.
7. The ductus venosus bypasses the fetal liver.
8. Despite significant improvements in life expectancy, there is little expectation of significantly increasing life span.
9. The sperm's ability to fertilize an egg increases in the hours after ejaculation because of capacitation.
10. The transition from embryo to fetus does not correspond to the trimesters of pregnancy. It occurs around the end of week 8, whereas the third trimester doesn't begin until week 25.

Appendix B
HEALTH SCIENCE CAREERS

Further information is available from the following organizations for each of the health-science careers featured in the Career Spotlight essays. Relevant information may be found by entering the organizational names and a key word such as *careers* or *jobs* in any search engine, or by going to the organization's home page and looking for a career or job link. Education and licensing requirements vary from state to state, so you may also wish to search by career title and your own state for specific information on what may be required of you to enter one of these careers.

This chapter-by-chapter list is followed by some general sources of health-career information and a list of many more career ideas than the ones addressed in the chapter Career Spotlight essays.

Chapter 1—Radiologic Technologist
American Society of Radiologic Technologists

Chapter 2—Medical Technologist
American Medical Technologists
American Society for Clinical Laboratory Science

Chapter 3—Cytotechnologist
American Society for Cytotechnology

Chapter 4—Histotechnician
National Society for Histotechnology

Chapter 5—Dermatology Nurse
Dermatology Nurses' Association

Chapter 6—Orthopedic Nurse
National Association of Orthopaedic Nurses

Chapter 7—Massage Therapist
American Massage Therapy Association
National Certification Board for Therapeutic Massage and Bodywork

Chapter 8—Occupational Therapist
American Occupational Therapy Association

Chapter 9—Electroneurodiagnostic Technologist
ASET–The Neurodiagnostic Society

Chapter 10—Optician
Opticians Association of America

Chapter 11—Diabetes Educator
American Association of Diabetes Educators
National Certification Board for Diabetes Educators

Chapter 12—Phlebotomist
National Phlebotomy Association
American Medical Technologists

Chapter 13—Electrocardiographic Technician
American Society of Radiologic Technologists
Radiological Society of North America

Chapter 14—Public Health Nurse
Association of Public Health Nurses
American Public Health Association

Chapter 15—Respiratory Therapist
American Association for Respiratory Care

Chapter 16—Dialysis Technician
Explore Medical Careers
National Kidney Foundation

Chapter 17—Dental Hygienist
American Dental Hygienists' Association

Chapter 18—Dietitian
American Dietetic Association

Chapter 19—Midwife
American College of Nurse Midwives

Chapter 20—Genetic Counselor
National Society of Genetic Counselors
The American Board of Genetic Counseling

Additional Careers
Numerous other health-science and health-care professions are available, going far beyond the 20 that we have described in connection with individual chapters. Here we list some sources of general information that cover many more of these professions, and a list of professions you may wish to explore. The requirements for entering these professions range from programs that extend only 1 to 2 years beyond high school, to associate degrees, bachelor's degrees, and graduate degrees. We have not listed doctoral-level careers (physician, dentist, etc.) or the many specialties within registered nursing, which generally require a B.S.N. degree followed by experience and specialized training. An Internet search of any of these career names plus a search term such as *careers, jobs,* or *programs* should yield useful information on training and entry requirements. In some cases, we combine two similar careers on one line with designations such as "technologist/technician" and "assistant/aide." Your career research will show the differences between the two.

General Sources
These organizations can be found online by typing the name of each into a search engine. Their websites provide information

about educational and training requirements, pay ranges, and what different jobs entail.

ExploreHealthCareers.org

Healthcare Occupations–Bureau of Labor Statistics

Careers

Acupuncturist
Allied dental educator
Anesthesiologist assistant
Athletic trainer
Biomedical equipment technician
Blood bank technology specialist
Cardiopulmonary rehabilitation specialist
Cardiovascular technologist/technician
Clinical assistant
Clinical laboratory technologist/technician
Community health worker
Cytogenetic technologist
Dental assistant
Dental informatics specialist
Dental laboratory technician
Diagnostic medical sonographer
Diagnostic molecular scientist
Dietetic technician
Dosimetrist
Electrocardiograph technician
Emergency medical dispatcher
Emergency medical technician/paramedic
Environmental health specialist
Health-care administrator
Health-care documentation specialist
Health-care interpreter
Health counselor
Health educator
Health information technician
Health unit coordinator
Home care assistant/aide
Hospital biller
Kinesiotherapist
Licensed practical nurse
Licensed vocational nurse
Lymphedema therapist
Long-term care specialist
Mammographer
Medical animation specialist
Medical assistant
Medical biller
Medical coder
Medical dosimetrist
Medical illustrator
Medical insurance specialist
Medical laboratory technician
Medical librarian
Medical office administrator
Medical photographer
Medical transcriptionist
Music therapist
Nuclear medicine technologist
Nurses aide/assistant
Nursing informatics specialist
Nutritionist
Occupational health and safety specialist
Occupational therapy assistant/aide
Ophthalmic medical technician
Optical laboratory technician
Optometric technician
Orientation and mobility specialist
Orthopedic technician
Orthoptist
Orthotist/prosthetist
Pathologists' assistant
Perfusionist
Pharmacy technician
Physical therapist assistant
Physician assistant
Psychiatric technician/aide
Radiation therapist
Radiographer
Recreational therapist
Rehabilitation counselor
Sleep technologist
Sonographer
Speech–language therapist
Surgical technologist
Vascular technologist
X-ray technologist

Appendix C

SYMBOLS, WEIGHTS, AND MEASURES

UNITS OF LENGTH

m	meter
cm	centimeter (10^{-2} m)
mm	millimeter (10^{-3} m)
μm	micrometer (10^{-6} m)
nm	nanometer (10^{-9} m)

UNITS OF MASS

g	gram
kg	kilogram (10^3 g)
mg	milligram (10^{-3} g)

UNITS OF VOLUME

L	liter
dL	deciliter (10^{-1} L) (= 100 mL)
mL	milliliter (10^{-3} L)
μL	microliter (10^{-6} L) (= 1 mm^3)

GREEK LETTERS

α	alpha (as in α helix)
β	beta (as in β hemoglobin)
γ	gamma (as in γ globulin)
Δ	delta (uppercase)
δ	delta (lowercase) (as in a small positive charge, δ+)
μ	mu (as in micrometer, μm)

CONVERSION FACTORS

1 cm = 0.394 in	1 in = 2.54 cm
1 mL = 0.034 fl oz	1 fl oz = 29.6 mL
1 L = 1.057 qt	1 qt = 0.946 L
1 g = 0.0035 oz	1 oz = 28.38 g
1 kg = 2.2 lb	1 lb = 0.45 kg
°C = (5/9)(°F−32)	°F = (9/5)(°C) + 32

UNITS OF CONCENTRATION

Chemical concentrations—the amounts of solute in a given volume of solution—are expressed in different ways for different scientific or clinical purposes. Some of these are explained here, particularly those used in this book.

Weight per Volume

A simple way to express concentration is the weight of solute in a given volume of solution. For example, intravenous (I.V.) saline typically contains 8.5 grams of NaCl per liter of solution (8.5 g/L). For many biological purposes, however, we deal with smaller quantities such as milligrams per deciliter (mg/dL). For example, a typical serum cholesterol concentration may be 200 mg/dL, also expressed as 200 mg/100 mL or 200 milligram-percent (mg-%).

Percentages

Percentage concentration is also simple to compute, but it is necessary to specify whether the percentage refers to the weight or to the volume of solute in a given volume of solution. For example, if we begin with 5 g of dextrose (an isomer of glucose) and add enough water to make 100 mL of solution, the resulting concentration will be 5% weight per volume (w/v). A commonly used intravenous fluid is D5W, which stands for 5% w/v dextrose in distilled water.

If the solute is a liquid, such as ethanol, percentages refer to volume of solute per volume of solution. Thus, 70 mL of ethanol diluted with water to 100 mL of solution produces 70% volume per volume (70% v/v) ethanol.

Percentage concentrations are easy to prepare, but that unit of measurement is inadequate for many purposes.

Molarity

The physiological effect of a chemical depends on how many molecules of it are present in a given volume, not the weight of the chemical. Five percent glucose, for example, contains almost twice as many sugar molecules as the same volume of 5% sucrose. Each solution contains 50 g of sugar per liter, but glucose has a molecular weight (MW) of 180 and sucrose has a MW of 342. Since each molecule of glucose is lighter, 50 g of glucose contains more molecules than 50 g of sucrose.

To produce solutions with a known number of molecules per volume, we must factor in the molecular weight. If we know the MW and weigh out that many grams of the substance, we have a quantity known as its **gram molecular weight, or 1 mole.** One mole of glucose is 180 g and 1 mole of sucrose is 342 g. Each quantity contains the same number of molecules of the respective sugar—a number known as Avogadro's number, 6.023×10^{23} molecules per mole.

Molarity (M) is the number of moles of solute per liter of solution. A one-molar (1.0 M) solution of glucose contains 180 g/L, and a 1.0 M solution of sucrose contains 342 g/L. Both have the same number of solute molecules in a given volume. Body fluids and laboratory solutions usually are less concentrated than 1 M, so biologists and clinicians more often work with **millimolar (mM, 10^{-3} M)** and **micromolar (μM, 10^{-6} M)** concentrations.

Osmolarity and Osmolality

The osmotic concentration of body fluids has a great effect on cellular function. If it is greater than the osmotic concentration within cells, cells lose water and shrivel; if it is less, cells absorb water and swell. Either of these can cause potentially fatal cellular breakdown or dysfunction. Osmotic concentration differences also influence such things as intestinal nutrient absorption, the sense of thirst, urinary water loss, total body water volume, and blood pressure. Thus, it is important to quantify osmotic concentrations in physiology and in clinical practice (as when giving I.V. fluid therapy).

One **osmole** is 1 mole of dissolved particles. If a solute does not ionize in water, then 1 mole of the solute yields 1 osmole (osm) of dissolved particles. A solution of 1 M glucose, for example, is also 1 osm/L. If a solute does ionize, it yields two or more dissolved particles in solution. A 1 M solution of NaCl, for example, contains 1 mole/L of sodium ions and 1 mole/L of chloride ions. Both ions equally affect osmosis and must be separately counted in a measure of osmotic concentration. Thus, 1 M NaCl = 2 osm/L. Calcium chloride ($CaCl_2$) would yield three ions if it dissociated completely (one Ca^{2+} and two Cl^-), so 1 M $CaCl_2$ = 3 osm/L.

Osmolality is the number of osmoles of solute per kilogram of water, and **osmolarity** is the number of osmoles per liter of solution. Most clinical calculations are based on osmolarity, since it is easier to measure the volume of a solution than the weight of water it contains. The difference between osmolality and osmolarity can be important in experimental work, but at the concentrations of human body fluids, there is less than 1% difference between the two, and the two are essentially interchangeable for clinical purposes.

All body fluids and many clinical solutions are mixtures of many chemicals. The osmolarity of such a solution is the total osmotic concentration of all of its dissolved particles.

A concentration of 1 osm/L is substantially higher than we find in most body fluids, so physiological concentrations are usually expressed in terms of **milliosmoles per liter (mOsm/L)** (1 mOsm/L = 10^{-3} osm/L). Blood plasma, tissue fluid, and intracellular fluid measure about 300 mOsm/L.

Milliequivalents per Liter

Electrolyte concentrations are typically expressed in **milliequivalents per liter (mEq/L)**. Electrolytes are important for their chemical, physical (osmotic), and electrical effects on the body. Their electrical effects, which determine such things as nerve, heart, and muscle actions, depend not only on their concentration but also on their electrical charge. A calcium ion (Ca^{2+}) has twice the electrical effect of a sodium ion (Na^+), for example, because it carries twice the charge. In measuring electrolyte concentrations, one must take the charges into account.

One **equivalent (Eq)** of an electrolyte is the amount that would electrically neutralize 1 mole of hydrogen ions (H^+) or hydroxide ions (OH^-). For example, 1 mole (58.4 g) of NaCl yields 1 mole, or 1 Eq, of Na^+ in solution. Thus, an NaCl solution of 58.4 g/L contains 1 equivalent of Na^+ per liter (1 Eq/L). One mole (98 g) of sulfuric acid (H_2SO_4) yields 2 moles of positive charges (H^+). Thus, 98 g of sulfuric acid per liter would be a solution of 2 Eq/L.

The electrolytes in human body fluids have concentrations less than 1 Eq/L, so we more often express their concentrations in milliequivalents per liter (mEq/L). If you know the millimolar concentration of an electrolyte, you can easily convert this to mEq/L by multiplying it by the charge on the ion:

$$1 \text{ mM } Na^+ = 1 \text{ mEq/L}$$

$$1 \text{ mM } Ca^{2+} = 2 \text{ mEq/L}$$

$$1 \text{ mM } Fe^{3+} = 3 \text{ mEq/L}$$

Acidity and Alkalinity (pH)

Acidity is expressed in terms of **pH,** a measure derived from the molarity of H^+. Molarity is represented by square brackets, so the molarity of H^+ is symbolized $[H^+]$. pH is the negative logarithm of hydrogen ion molarity; that is, pH = $-\log [H^+]$.

In pure water, 1 in 10 million molecules of H_2O ionizes into hydrogen and hydroxide ions: $H_2O \rightleftharpoons H^+ + OH^-$. Pure water has a neutral pH because it contains equal amounts of H^+ and OH^-. Because 1 in 10 million molecules ionize, the molarity of H^+ and the pH of water are as follows:

$$[H^+] = 0.0000001 \text{ molar} = 10^{-7} \text{ M}$$

$$\log [H^+] = -7$$

$$pH = -\log [H^+] = 7$$

The pH scale ranges from 0 to 14 and is logarithmic, so each integer up or down the scale represents a 10-fold difference in $[H^+]$. This is exemplified by the following three strongly acidic pH values:

If $[H^+]$ = 0.1 M   pH = $-\log 10^{-1}$ = 1.0

If $[H^+]$ = 0.01 M   pH = $-\log 10^{-2}$ = 2.0

If $[H^+]$ = 0.001 M   pH = $-\log 10^{-3}$ = 3.0

The less concentrated the H^+, the higher the pH. pH values below 7.0 are considered acidic. Above 7.0, they are basic or alkaline. The following three values lie at the basic end of the pH scale:

If $[H^+]$ = 0.000000000001 M   pH = $-\log 10^{-12}$ = 12.0

If $[H^+]$ = 0.0000000000001 M   pH = $-\log 10^{-13}$ = 13.0

If $[H^+]$ = 0.00000000000001 M   pH = $-\log 10^{-14}$ = 14.0

Appendix D
BIOMEDICAL WORD ROOTS, PREFIXES, AND SUFFIXES

a- no, not, without (atom, agranulocyte)
ab- away (abducens, abduction)
acetabulo- small cup (acetabulum)
acro- tip, extremity, peak (acromion, acromegaly)
ad- to, toward, near (adsorption, adrenal)
adeno- gland (lymphadenitis, adenohypophysis)
aero- air, oxygen (aerobic)
af- toward (afferent)
ag- together (agglutination)
-al pertaining to (parietal, pharyngeal)
ala- wing (ala nasi)
albi- white (albicans, linea alba, albino)
algi- pain (analgesic, myalgia)
aliment- nourishment (alimentary)
allo- other, different (allele)
amphi- both, either (amphipathic, amphiarthrosis)
an- without (anaerobic, anemic)
ana- 1. up, buildup (anabolic, anaphylaxis). 2. apart (anaphase, anatomy). 3. back (anastomosis)
andro- male (androgen, andropause)
angi- vessel (angiogram, hemangioma)
ante- before, in front (antebrachium)
antero- forward (anterior, anterograde)
anti- against (antidiuretic, antibody, antagonist)
apo- from, off, away, above (apocrine, aponeurosis)
arbor- tree (arborization, arbor vitae)
artic- 1. joint (articulation). 2. speech (articulate)
-ary pertaining to (axillary, coronary)
-ase enzyme (polymerase, amylase)
ast-, astro- star (aster, astrocyte)
-ata plural of **-a** (carcinomata, chiasmata)
-ate possessing, like (hamate, cruciate)
athero- fat (atheroma, atherosclerosis)
atrio- entryway (atrium, atrioventricular)
auri- ear (auricle, binaural)
auto- self (autorhythmic, autoimmune)
axi- axis, straight line (axial, axoneme, axon)
baro- pressure (baroreceptor, hyperbaric)
bene- good, well (benign, beneficial)
bi- two (bipedal, biceps)
bili- bile (biliary, bilirubin)
bio- life, living (biology, biopsy, microbial)
blasto- precursor, producer (fibroblast, blastomere)
brachi- arm (brachium, brachial)
brady- slow (bradycardia, bradypnea)
bucco- cheek (buccal, buccinator)
burso- purse (bursa, bursitis)
calc- calcium, stone (calcaneus, hypocalcemia)
callo- thick (callus, callosum)
calori- heat (calorie, calorigenic)
calvari- bald, skull (calvaria)
calyx cup, chalice (glycocalyx, renal calyx)
capito- head (capitis, capitulum)
capni- smoke, CO_2 (hypocapnia)
carcino- cancer (carcinogen, carcinoma)
cardi- heart (cardiology, pericardium)
carot- 1. carrot (carotene). 2. stupor (carotid)
carpo- wrist (carpus, metacarpal)
case- cheese (caseosa, casein)
cata- down, break down (catabolism)
cauda- tail (cauda equina, caudate lobe)
-cel little (pedicel)
celi- belly, abdomen (celiac)
centri- middle (centromere, centriole)
cephalo- head (cephalic, encephalitis)
cervi- neck, narrow part (cervix, cervical)
chiasm- cross, X (optic chiasm)
choano- funnel (choana)
chole- bile (cholecystokinin)
chondro- 1. grain (mitochondria). 2. cartilage, gristle (chondrocyte)
chromo- color (chromatin, cytochrome)
chrono- time (chronotropic, chronic)
cili- eyelash (cilium, supraciliary)
circ- about, around (circumduction)
cis- cut (incision, incisor)
clast- break down, destroy (osteoclast)
clavi- hammer, key (clavicle, supraclavicular)
-cle little (tubercle, ossicle)
cleido- clavicle (sternocleidomastoid)
cnemo- leg (gastrocnemius)
co- together (coenzyme, cotransport)
collo- 1. hill (colliculus). 2. glue (colloid, collagen)
contra- opposite (contralateral)
corni- horn (cornified, corniculate)
corono- crown (coronary, corona)
corpo- body (corpus luteum)
corti- bark, rind (cortex, cortical)
costa- rib (intercostal, subcostal)
coxa- hip (os coxae, coxal)
crani- helmet (cranium, epicranius)
cribri- sieve, strainer (cribriform plate)
crino- separate, secrete (holocrine, endocrine)
crista- crest (crista ampullaris, crista galli)
crito- to separate (hematocrit)
cruci- cross (cruciate ligament)

-cul small (canaliculus, trabecula)
cune- wedge (cuneiform, cuneatus)
cutane-, cuti- skin (subcutaneous, cuticle)
cysto- bladder (cystitis, polycystic)
cyto- cell (cytology, monocyte)
de- down (defecate, dehydration)
demi- half (demifacet, demilune)
den-, denti- tooth (dentition, dens)
dendro- tree, branch (dendrite, oligodendrocyte)
dermo- skin (dermatology, hypodermic)
desmo- band, bond, ligament (desmosome, syndesmosis)
dia- across, through, separate (diaphragm, dialysis)
dis- 1. apart (dissect, dissociate). 2. opposite, absence (disinfect, disability)
diure- pass through, urinate (diuretic)
dorsi- back (dorsal, latissimus dorsi)
duc- to carry (duct, adduction, abducens)
dys- bad, abnormal, painful (dyspnea, dystrophy)
e- out (ejaculate, eversion)
-eal pertaining to (hypophyseal, pineal)
ec-, ecto- out, external (ectopic, ectoderm, splenectomy)
ef- out of (efferent, effusion)
-el, -elle small (fontanel, organelle)
electro- electricity (electrocardiogram, electrolyte)
em- in, within (embolism, embedded)
emesi-, emeti- vomiting (emetic, hyperemesis)
-emia blood condition (anemia, hypoxemia)
en- in, into (enzyme, parenchyma)
encephalo- brain (encephalitis, electroencephalogram)
enchymo- poured in (mesenchyme, parenchyma)
endo- within, into, internal (endocrine, endocytosis)
entero- gut, intestine (mesentery, myenteric)
epi- upon, above (epidermis, epiphysis, epididymis)
ergo- work, energy, action (allergy, adrenergic)
eryth-, erythro- red (erythema, erythrocyte)
esthesio- sensation, feeling (anesthesia, somesthetic)
eu- good, true, normal, easy (eupnea, aneuploidy)
exo- out (exocytosis, exocrine)
facili- easy (facilitated)
fasci- band, bundle (fascia, fascicle)
fenestr- window (fenestrated)
fer- to carry (efferent, uriniferous)
ferri- iron (ferritin, transferrin)
fibro- fiber (fibroblast, fibrosis)
fili- thread (myofilament)
flagello- whip (flagellum)
foli- leaf (folic acid, folia)
-form shape (cuneiform, fusiform)
fove- pit, depression (fovea)
funiculo- little rope, cord (funiculus)
fusi- 1. spindle (fusiform). 2. pour out (perfusion)
gamo- marriage, union (gamete, monogamy)
gastro- belly, stomach (digastric, gastrointestinal)
-gen, -genic, -genesis producing, giving rise to (pathogen, carcinogenic, glycogenesis)
germi- 1. sprout, bud (germ cell). 2. microbe (germicide)

gero- old age (geriatrics, gerontology)
gesto- 1. to bear, carry (ingest). 2. pregnancy (gestation, progesterone)
glia- glue (neuroglia, glioma)
globu- ball, sphere (globulin, hemoglobin)
glom- ball (glomerulus)
glosso- tongue (hypoglossal, glossopharyngeal)
glyco- sugar (glycogen, hypoglycemia)
gono- 1. angle, corner (trigone). 2. seed, sex cell, generation (gonad, oogonium, gonorrhea)
gradi- walk, step (retrograde, gradient)
-gram recording of (sonogram, electrocardiogram)
-graph recording instrument (sonograph, electrocardiograph)
-graphy recording process (sonography, radiography)
gravi- severe, heavy (gravid, myasthenia gravis)
gyro- turn, twist (gyrus)
hallu- great toe (hallux, hallucis)
hemi- half (hemidesmosome, hemisphere)
-hemia blood condition (polycythemia)
hemo- blood (hemoglobin, hematology)
hetero- different, other, various (heterozygous, heterograft)
histo- tissue, web (histology, histone)
holo- whole, entire (holistic, holocrine)
homeo- constant, unchanging, uniform (homeostasis, homeothermic)
homo- same, alike (homologous, homozygous)
hyalo- clear, glassy (hyaline, hyaluronic acid)
hydro- water (dehydration, hydrolysis, hydrophobic)
hyper- above, above normal, excessive (hyperkalemia, hypertonic)
hypo- below, below normal, deficient (hypodermis, hyponatremia)
-ia condition (anemia, hypocalcemia)
-ic pertaining to (isotonic, antigenic)
-icle, -icul small (ossicle, canaliculus, reticular)
ilia- flank, loin (ilium, iliac)
-illa, -illus little (bacillus)
-in protein (trypsin, fibrin, globulin)
infra- below (infraspinous, infrared)
insulo- island (insula, insulin)
inter- between (intercellular, intervertebral)
intra- within (intracellular, intraocular)
iono- ion (ionotropic, cationic)
ischi- to hold back (ischium, ischemia)
-ism process, state, condition (metabolism, rheumatism)
iso- same, equal (isometric, isotonic, isomer)
-issimus most, greatest (latissimus, longissimus)
-ite little (dendrite, somite)
-itis inflammation (dermatitis, gingivitis)
jug- to join (conjugated, jugular)
juxta- next to (juxtamedullary, juxtaglomerular)
kali- potassium (hypokalemia)
karyo- seed, nucleus (megakaryocyte, karyotype)
kerato- horn (keratin, keratinocyte)
kine- motion, action (kinetic, kinase, cytokinesis)
labi- lip (labia majora, levator labii)

lacera- torn, cut (laceration)
lacrimo- tear, cry (lacrimal gland, nasolacrimal)
lacto- milk (lactose, lactation, prolactin)
lamina- layer (lamina propria)
latero- side (bilateral, ipsilateral)
lati- broad (fascia lata, latissimus dorsi)
-lemma husk (sarcolemma, neurilemma)
lenti- lens (lentiform)
-let small (platelet)
leuko- white (leukocyte, leukemia)
levato- to raise (levator labii, elevation)
ligo- to bind (ligand, ligament)
line- line (linea alba, linea aspera)
litho- stone (otolith, lithotripsy)
-logy study of (histology, hematology)
lucid- light, clear (stratum lucidum, zona pellucida)
lun- moon, crescent (lunate, lunule, semilunar)
lute- yellow (macula lutea, corpus luteum)
lyso-, lyto- split apart, break down (lysosome, hydrolysis, electrolyte, hemolytic)
macro- large (macromolecule, macrophage)
macula- spot (macula lutea, macula densa)
mali- bad (malignant, malformed)
malle- hammer (malleus, malleolus)
mammo- breast (mammary, mammillary)
mano- hand (manus, manipulate)
manubri- handle (manubrium)
masto- breast (mastoid, mastectomy)
medi- middle (medial, mediastinum)
medullo- marrow, pith (medulla)
mega- large (megakaryocyte, hepatomegaly)
melano- black (melanin, melanocyte)
meno- month (menstruation, menopause)
mento- chin (mental, mentalis)
mero- part, segment (isomer, centromere)
meso- in the middle (mesoderm, mesentery)
meta- beyond, next in a series (metaphase, metacarpal)
metabolo- change (metabolism, metabolite)
-meter measuring device (spirometer, sphygmomanometer)
metri- 1. length, measure (isometric, emmetropic). 2. uterus (endometrium)
micro- small (microscopic, microglia)
mito- thread, filament (mitochondria, mitosis)
mono- one (monocyte, monogamy)
morpho- form, shape, structure (morphology, amorphous)
muta- change (mutagen, mutation)
myelo- 1. spinal cord (poliomyelitis, myelin). 2. bone marrow (myeloid, myelocytic)
myo-, mysi- muscle (myosin, epimysium)
natri- sodium (hyponatremia, natriuretic)
neo- new (neonatal, gluconeogenesis)
nephro- kidney (nephron)
neuro- nerve (aponeurosis, neurosoma)
nucleo- nucleus, kernel (nucleolus, nucleic acid)
ob- 1. life (aerobic, microbe). 2. against, toward, before (obstetrics, obturator, obstruction)

oculo- eye (oculi, oculomotor)
odonto- tooth (odontoblast, periodontal)
-oid like, resembling (colloid, sigmoid)
-ole small (arteriole, bronchiole)
oligo- few, a little, scanty (oligopeptide, oliguria)
-oma tumor, mass (carcinoma, hematoma)
omo- shoulder (omohyoid, acromion)
onycho- nail, claw (hyponychium)
oo- egg (oogenesis, oocyte)
op- vision (optics, myopia, photopic)
-opsy viewing, to see (biopsy, rhodopsin)
or- mouth (oral, orbicularis oris)
orbi- circle (orbicularis, orbit)
organo- tool, instrument (organ, organelle)
ortho- straight (orthodontics, orthopedics)
-ose 1. full of (adipose). 2. sugar (sucrose, glucose)
-osis 1. process (osmosis, exocytosis). 2. condition, disease (cyanosis, thrombosis). 3. increase (leukocytosis)
osmo- push (osmosis)
osse-, oste- bone (osseous, osteoporosis)
oto- ear (otolith, otitis, parotid)
-ous 1. full of (nitrogenous, edematous). 2. pertaining to (mucous, nervous). 3. like, characterized by (squamous, filamentous)
ovo- egg (ovum, ovary, ovulation)
oxy- 1. oxygen (oxyhemoglobin, hypoxia). 2. sharp, quick (oxytocin)
palli- pale (pallor, globus pallidus)
palpebro- eyelid (palpebrae)
pan- all (pancreas, panhysterectomy)
papillo- nipple (papilla, papillary)
par- birth (postpartum, parturition)
para- next to (parathyroid, parotid)
parieto- wall (parietal)
patho- 1. disease (pathology, pathogen). 2. feeling (sympathetic)
pecto- 1. chest (pectoralis). 2. comblike (pectineus)
pedi- 1. foot (bipedal, pedicle). 2. child (pediatrics)
pelvi- basin (pelvis, pelvic)
-penia deficiency (leukopenia, osteopenia)
penna- feather (unipennate, bipennate)
peri- around (periosteum, peritoneum, periodontal)
phago- eat (phagocytosis, macrophage)
philo- loving, attracted (hydrophilic)
phobo- fearing, repelled (hydrophobic)
phragm- partition (diaphragm)
phreno- diaphragm (phrenic nerve)
physio- nature, natural cause (physiology, physician)
-physis growth (diaphysis, hypophysis)
pilo- hair (piloerector)
pino- drink, imbibe (pinocytosis)
planto- sole of foot (plantaris, plantar wart)
plasi- growth (hyperplasia)
plasm- shaped, molded (cytoplasm, endoplasmic)
plasti- form (thromboplastin)
platy- flat (platysma)
pnea- breath, breathing (eupnea, dyspnea)

pneumo- air, breath, lung (pneumonia, pneumothorax)
podo- foot (pseudopod, podocyte)
poies- forming (hemopoiesis, erythropoietin)
poly- many, much, excessive (polypeptide, polyuria)
primi- first (primary, primipara)
pro- 1. before, in front, first (prokaryote, prophase, prostate). 2. promote, favor (progesterone, prolactin)
pseudo- false (pseudopod)
ptero-, pterygo- wing (pterygoid)
-ptosis dropping, falling, sagging (apoptosis, nephroptosis)
puncto- point (puncta)
pyro- fire (pyrogen, antipyretic)
quadri- four (quadriceps, quadratus)
quater- fourth (quaternary)
radiat- radiating (corona radiata)
rami- branch (ramus)
recto- straight (rectus abdominis, rectum)
reno- kidney (renal, renin)
reti- network (reticular, rete testis)
retinac- retainer, bracelet (retinaculum)
retro- behind, backward (retroperitoneal, retrovirus)
rhombo- rhombus (rhomboideus)
rubo-, rubro- red (bilirubin)
rugo- fold, wrinkle (ruga, corrugator)
sacculo- little sac (saccule)
sarco- flesh, muscle (sarcoplasm, sarcomere)
scala- staircase (scala tympani, scalene)
sclero- hard, tough (sclera, sclerosis)
scopo- see (microscope, endoscopy)
secto- cut (section, dissection)
semi- half (semilunar, semimembranosus)
sepsi- infection (asepsis, septicemia)
-sis process (diapedesis, amniocentesis)
sole- sandal, sole of foot (sole, soleus)
soma-, somato- body (somatic, somatotropin)
spheno- wedge (sphenoid)
spiro- breathing (inspiration, spirometry)
spleno- 1. bandage (splenius capitis). 2. spleen (splenic artery)
squamo- scale, flat (squamous, desquamation)
stasi-, stati- put, remain, stay the same (hemostasis, homeostatic)
steno- narrow (stenosis)
ster-, stereo- solid, three-dimensional (steroid, stereoscopic)
sterno- breast, chest (sternum, sternocleidomastoid)
stria- stripe (striated, corpus striatum)
sub- below (subcutaneous, subclavicular)
sulc- furrow, groove (sulcus)
supra- above (supraspinous, supraclavicular)

sura- calf of leg (triceps surae)
sym- together (sympathetic, symphysis)
syn- together (synostosis, synovial)
tachy- fast (tachycardia, tachypnea)
tarsi- ankle (tarsus, metatarsal)
tecto- roof, cover (tectorial membrane)
telo- last, end (telophase, telencephalon)
tempo- time (temporal)
terti- third (tertiary)
theli- nipple, female, tender (epithelium, thelarche)
thermo- heat (thermogenesis, hypothermia)
thrombo- blood clot (thrombosis, thrombin)
thyro- shield (thyroid, thyrotropin)
-tion process (circulation, pronation)
toci- birth (oxytocin)
tomo- 1. cut (tomography, atom, anatomy). 2. segment (dermatome, myotome, sclerotome)
tono- force, tension (isotonic, tonus, myotonia)
topo- place, position (isotope, ectopic)
trabo- plate (trabecula)
trans- across (transpiration, transdermal)
trapezi- 1. table, grinding surface (trapezium). 2. trapezoid (trapezius)
tri- three (triceps, triglyceride)
tricho- hair (trichosiderin, hypertrichosis)
trocho- wheel, pulley (trochlea)
troph- 1. food, nourishment (trophic, trophoblast). 2. growth (dystrophy, hypertrophy)
tropo- to turn, change (metabotropic, gonadotropin)
tunica coat (tunica intima, tunica vaginalis)
tympano- drum, eardrum (tympanic, tensor tympani)
-ul small (tubule, capitulum, glomerulus)
-uncle, -unculus small (homunculus, caruncle)
uni- one (unipennate, unipolar)
uri- urine (glycosuria, urinalysis, diuretic)
utriculo- little bag (utriculus)
vagino- sheath (tunica vaginalis)
vago- wander (vagus)
vaso- vessel (vascular, vas deferens)
ventro- belly, lower part (ventral, ventricle)
vermi- worm (vermis)
vertebro- spine (vertebrae, intervertebral)
vesico- bladder, blister (vesical, vesicular)
villo- hair, hairy (microvillus)
vitre- glass (in vitro, vitreous humor)
zygo- union, join, mate (zygomatic, zygote, azygos)

Appendix E
PERIODIC TABLE OF THE ELEMENTS

Nineteenth-century chemists discovered that when they arranged the known elements by atomic weight, certain properties reappeared periodically. In 1869, Russian chemist Dmitri Mendeleev published the first modern periodic table of the elements, leaving gaps for those that had not yet been discovered. He accurately predicted properties of the missing elements, which helped other chemists discover and isolate them.

Each row in the table is a *period,* and each column is a *group (family).* Each period has one electron shell more than the period above it, and as we progress from left to right within a period, each element has one more proton and electron than the one before. The dark steplike line from boron (5) to astatine (85) separates the metals to the left of it (except hydrogen) from the nonmetals to the right. Each period begins with a soft, light, highly reactive *alkali metal,* with one valence electron, in family IA. Progressing from left to right, the metallic properties of the elements become less and less pronounced. Elements in family VIIA are highly reactive gases called *halogens,* with seven valence electrons. Elements in family VIIIA, called *noble (inert) gases,* have a full valence shell of eight electrons, which makes them chemically unreactive.

Ninety-one of the elements occur naturally on earth. Physicists have created elements up to atomic number 118 in the laboratory, but the International Union of Pure and Applied Chemistry has established formal names only through element 112 to date.

The 24 elements with normal roles in human physiology are color-coded according to their relative abundance in the body (see chapter 2). Others, however, may be present as contaminants with very destructive effects (such as arsenic, lead, and radiation poisoning).

Glossary

A

abdominal cavity The body cavity between the diaphragm and pelvic brim. (fig. 1.12)

abduction (ab-DUC-shun) Movement of a body part away from the median plane, as in raising an arm away from the side of the body. (fig. 6.37)

absorption 1. Process in which a chemical passes through a membrane or tissue surface and becomes incorporated into a body fluid or tissue. 2. Any process in which one substance passes into another and becomes a part of it.

acetylcholine (ACh) (ASS-eh-till-CO-leen) A neurotransmitter released by somatic motor fibers, parasympathetic fibers, and some other neurons, composed of choline and an acetyl group.

acetylcholinesterase (AChE) (ASS-eh-till-CO-lin-ESS-ter-ase) An enzyme that breaks down acetylcholine, thus halting signal transmission at a synapse.

acid A proton (H^+) donor; a chemical that releases protons into solution.

acidosis An acid–base imbalance in which the blood pH is lower than 7.35.

acinus (ASS-ih-nus) A sac of secretory cells at the inner end of a gland duct. (fig. 4.28)

actin A filamentous intracellular protein that provides cytoskeletal support and interacts with other proteins, especially myosin, to cause cellular movement; important in muscle contraction and membrane actions such as phagocytosis, ameboid movement, and cytokinesis.

action The movement produced by the contraction of a particular muscle.

action potential A rapid voltage change in which a plasma membrane briefly reverses electrical polarity; has a self-propagating effect that produces a traveling wave of excitation in nerve and muscle cells.

active site The region of a protein that binds to another molecule, such as the substrate-binding site of an enzyme or the hormone-binding site of a receptor.

active transport Transport of particles through a selectively permeable membrane, up their concentration gradient, with the aid of a carrier that consumes ATP.

adaptive immunity Defenses against specific pathogens or other disease agents, resulting from prior exposure, immune memory, and rapid response upon reexposure; divided into T lymphocyte–based cellular immunity and antibody-based humoral immunity.

adduction (ah-DUC-shun) Movement of a body part toward the median plane, such as bringing the feet together from a spread-legged position. (fig. 6.37)

adenosine triphosphate (ATP) (ah-DEN-oh-seen tri-FOSS-fate) A molecule composed of adenine, ribose, and three phosphate groups that functions as a universal energy-transfer molecule; yields adenosine diphosphate (ADP) and an inorganic phosphate group (Pi) upon hydrolysis. (fig. 2.19)

adipocyte (AD-ih-po-site) A fat cell.

adipose tissue A connective tissue composed predominantly of adipocytes; fat.

aerobic respiration Oxidation of organic compounds in a reaction series that requires oxygen and produces ATP.

afferent (AFF-uh-rent) Carrying toward, as in *afferent neurons,* which carry signals toward the central nervous system, and *afferent arterioles,* which carry blood toward a tissue.

agglutination (ah-GLUE-tih-NAY-shun) Clumping of cells or molecules by antibodies. (fig. 14.22)

agranulocyte Either of the leukocyte types (lymphocytes and monocytes) that lack specific granules and thus have a relatively clear cytoplasm. (fig. 12.9)

albumin (al-BYU-min) A class of small proteins constituting about 60% of the protein fraction of the blood plasma; plays roles in blood viscosity, osmosis, and solute transport.

aldosterone (AL-doe-steh-RONE, al-DOSS-teh-rone) A steroid hormone secreted by the adrenal cortex that acts on the kidneys to promote sodium retention and potassium excretion; indirectly promotes water retention and thereby opposes a drop in blood pressure.

alkalosis An acid–base imbalance in which the blood pH is higher than 7.45.

alveolus (AL-vee-OH-lus) 1. A microscopic air sac of the lung. 2. A tooth socket.

amino acids Small organic molecules with an amino group and a carboxyl group; the monomers of which proteins are composed.

anabolism (ah-NAB-oh-lizm) Any metabolic reactions that consume energy and construct more complex molecules with higher free energy from less complex molecules with lower free energy; for example, the synthesis of proteins from amino acids. *Compare* catabolism.

anaerobic fermentation (AN-err-OH-bic) A reduction reaction independent of oxygen that converts pyruvate to lactate and enables glycolysis to continue under anaerobic conditions.

anatomical position A reference posture that allows for standardized anatomical terminology. A subject in anatomical position is standing with the feet flat on the floor, arms down to the sides, and the palms and eyes directed forward. (fig. 1.7)

androgen (AN-dro-jen) Testosterone or a related steroid hormone. Stimulates body changes at puberty in both sexes, adult libido in both sexes, development of male anatomy in the fetus and adolescent, and spermatogenesis.

angiotensin II (AN-jee-oh-TEN-sin) A hormone produced from angiotensinogen (a plasma protein) by the kidneys and lungs; raises blood pressure by stimulating vasoconstriction and stimulating the adrenal cortex to secrete aldosterone.

anion (AN-eye-on) An ion with more electrons than protons and consequently a net negative charge.

antagonistic effect Effect in which two hormones, or both divisions of the autonomic nervous system, produce opposite effects on the same target organ or cell.

antebrachium (AN-teh-BRAY-kee-um) The region from elbow to wrist; the forearm.

anterior Pertaining to the front (facial–abdominal aspect) of the body; ventral.

anterior root The branch of a spinal nerve that emerges from the anterior side of the spinal cord and carries efferent (motor) nerve fibers; often called the *ventral root.* (fig. 8.17)

antibody A protein of the gamma globulin class that reacts with an antigen and aids in protecting the body from its harmful effect; found in the blood plasma, in other body fluids, and on the surfaces of certain leukocytes and their derivatives.

antidiuretic hormone (ADH) (AN-tee-DYE-you-RET-ic) A hormone released by the posterior lobe of the pituitary gland in response to low blood pressure; promotes water retention by the kidneys. Also known as *vasopressin.*

antigen (AN-tih-jen) Any large molecule capable of binding to an antibody or immune cell and triggering an immune response; usually a protein, polysaccharide, glycoprotein, or glycolipid.

antigen-presenting cell (APC) A cell that phagocytizes an antigen and displays fragments of it on its surface for recognition by other cells of the immune system; chiefly macrophages and B lymphocytes.

apical surface The uppermost surface of an epithelial cell, usually exposed to the lumen of an organ. (fig. 3.1)

apocrine 1. A mode of secretion in which the cell product is produced in the cytosol, migrates to the cell surface without an enclosing membrane (vesicle), and buds from the surface coated with a thin layer of cytoplasm and plasma membrane. (fig. 4.29b) 2. A term that persists for certain glands, such as axillary sweat glands, that have unusually large lumens and relatively thick secretions, formerly thought to employ the apocrine mode of secretion. (fig. 5.7a)

apoptosis (AP-oh-TOE-sis) Programmed cell death; the normal death of cells that have completed their function.

appendicular (AP-en-DIC-you-lur) Pertaining to the limbs and their supporting skeletal girdles. (fig. 6.10)

areolar tissue (AIR-ee-OH-lur) A fibrous connective tissue with loosely organized, widely spaced fibers and cells and an abundance of fluid-filled space; found under nearly every epithelium, among other places. (fig. 4.14)

arrector pili A bundle of smooth muscle cells associated with a hair follicle, responsible for erection of the hair; also called a *pilomotor muscle*. (fig. 5.5)

arteriole (ar-TEER-ee-ole) A small artery that empties into a metarteriole or capillary.

arteriosclerosis (ar-TEER-ee-oh-sclare-OH-sis) Stiffening of the arteries correlated with age or disease processes, caused primarily by cumulative free radical damage and tissue deterioration. *Compare* atherosclerosis.

articular cartilage A thin layer of hyaline cartilage covering the articular surface of a bone at a synovial joint, serving to reduce friction and ease joint movement. (fig. 6.34)

articulation A skeletal joint; any point at which two bones meet; may or may not be movable.

atherosclerosis (ATH-ur-oh-skleh-ROE-sis) A degenerative disease of the blood vessels characterized by the presence of atheromas, often leading to calcification of the vessel wall and obstruction of coronary, cerebral, or other vital arteries. *Compare* arteriosclerosis.

atrophy (AT-ro-fee) Shrinkage of a tissue due to age, disuse, or disease.

autoantibody An antibody that fails to distinguish the body's own molecules from foreign molecules and thus attacks host tissues, causing autoimmune diseases.

autoimmune disease Any disease in which antibodies fail to distinguish between foreign and self-antigens and attack the body's own tissues; for example, rheumatoid arthritis and type 1 diabetes mellitus.

autonomic nervous system (ANS) (AW-toe-NOM-ic) A motor division of the nervous system that innervates glands, smooth muscle, and cardiac muscle; consists of sympathetic and parasympathetic divisions; and functions largely without voluntary control. *Compare* somatic nervous system.

axial (AC-see-ul) Pertaining to the head, neck, and trunk; the part of the body excluding the appendicular portion. (fig. 6.10)

axillary (ACK-sih-LERR-ee) Pertaining to the armpit.

axon A fibrous extension of a neuron that transmits action potentials; also called a *nerve fiber*. There is only one axon to a neuron, and it is usually much longer and less branched than the dendrites. (fig. 8.2)

axon terminal The swollen tip at the distal end of an axon; the site of synaptic vesicles and neurotransmitter release. (figs. 8.2, 8.12)

B

B lymphocyte A lymphocyte that functions as an antigen-presenting cell and, in humoral immunity, differentiates into an antibody-producing plasma cell; also called a *B cell*.

baroreceptor (BAR-oh-re-SEP-tur) Pressure sensor located in the heart, aortic arch, or carotid sinus that triggers autonomic reflexes in response to fluctuations in blood pressure.

base 1. A chemical that binds protons from solution; a proton acceptor. 2. Any of the purines or pyrimidines of a nucleic acid (adenine, thymine, guanine, cytosine, or uracil) serving in part to code for protein structure. 3. The broadest part of a tapered organ such as the uterus or heart, or the inferior aspect of an organ such as the brain.

basophil (BASE-oh-fill) A granulocyte with coarse cytoplasmic granules that produces heparin, histamine, and other chemicals involved in inflammation.

body 1. Part of a cell, such as a neuron, containing the nucleus and most other organelles. 2. The largest or principal part of an organ such as the stomach or uterus; also called the *corpus*.

brachial (BRAY-kee-ul) Pertaining to the arm proper, the region from shoulder to elbow.

brainstem The stalklike lower portion of the brain, composed of all of the brain except the cerebrum and cerebellum. (Many authorities also exclude the diencephalon and regard only the medulla oblongata, pons, and midbrain as the brainstem.) (fig. 9.5)

bronchiole (BRONK-ee-ole) A pulmonary air passage that is usually 1 mm or less in diameter and lacks cartilage but has relatively abundant smooth muscle, elastic tissue, and a simple cuboidal, usually ciliated, epithelium.

bronchus (BRONK-us) A relatively large pulmonary air passage with supportive cartilage in the wall; any passage beginning with the main bronchus at the fork in the trachea and ending with segmental bronchi, from which air continues into the bronchioles.

brush border A fringe of microvilli on the apical surface of an epithelial cell, serving to enhance surface area and promote absorption. (fig. 3.4)

C

calorie The amount of thermal energy that will raise the temperature of 1 g of water by 1°C. Also called a *small calorie*.

Calorie *See* kilocalorie.

capillary (CAP-ih-LERR-ee) The narrowest type of vessel in the cardiovascular and lymphatic systems; engages in fluid exchanges with surrounding tissues.

capillary exchange The process of fluid transfer between the bloodstream and tissue fluid.

capsule The fibrous covering of a structure such as the spleen or a synovial joint.

carbohydrate A hydrophilic organic compound composed of carbon and a 2:1 ratio of hydrogen to oxygen; includes sugars, starch, glycogen, and cellulose.

carcinogen (car-SIN-oh-jen) An agent capable of causing cancer, including certain chemicals, viruses, and ionizing radiation.

cardiovascular system An organ system consisting of the heart and blood vessels, serving for the transport of blood. This term does not include the blood.

carpal Pertaining to the wrist (carpus).

carrier A protein in a cellular membrane that performs carrier-mediated transport.

carrier-mediated transport Any process of transporting materials through a cellular membrane that involves reversible binding to a transport protein; includes active transport and facilitated diffusion.

catabolism (ca-TAB-oh-lizm) Any metabolic reactions that release energy and break relatively complex molecules with high free energy into less complex molecules with lower free energy; for example, digestion and glycolysis. *Compare* anabolism.

cation (CAT-eye-on) An ion with more protons than electrons and consequently a net positive charge.

caudal (CAW-dul) 1. Pertaining to a tail or narrow tail-like part of an organ. 2. Relatively distant from the forehead, especially in reference to structures of the brain and spinal cord; for example, the medulla oblongata is caudal to the pons. *Compare* rostral.

central nervous system (CNS) The brain and spinal cord.

centriole (SEN-tree-ole) An organelle composed of a short cylinder of nine triplets of microtubules, usually paired with another centriole perpendicular to it; origin of the mitotic spindle; identical to the basal body of a cilium or flagellum. (fig. 3.13)

cephalic (seh-FAL-ic) Pertaining to the head.

cerebellum (SER-eh-BEL-um) A large portion of the brain posterior to the brainstem and inferior to the cerebrum, responsible for equilibrium, motor coordination, certain perceptual abilities, and memory of learned motor skills. (fig. 9.6)

cerebrospinal fluid (CSF) (SERR-eh-bro-SPY-nul, seh-REE-bro-SPY-nul) A liquid that fills the ventricles of the brain, the central canal of the spinal cord, and the space between the CNS and dura mater.

cerebrum (seh-REE-brum, SER-eh-brum) The largest and most superior part of the brain, divided into two convoluted cerebral hemispheres separated by a deep longitudinal fissure.

cervical (SUR-vih-cul) Pertaining to the neck or any cervix.

cervix (SUR-vix) 1. The neck. 2. A narrow or necklike part of an organ such as the uterus and gallbladder. (fig. 19.6)

channel protein A protein in the plasma membrane that has a pore through it for the passage of materials between the cytoplasm and extracellular fluid. (fig. 3.3)

chemoreceptor An organ or cell specialized to detect chemicals, as in the carotid bodies and taste buds.

cholecystokinin (CCK) (CO-leh-SIS-toe-KY-nin) A polypeptide employed as a hormone and neurotransmitter, secreted by some brain neurons and cells of the small intestine. In the digestive system, stimulates contraction of the gallbladder, release of bile, and secretion of pancreatic enzymes.

cholesterol (co-LESS-tur-ol) A steroid that functions as part of the plasma membrane and as a precursor for all other steroids in the body.

chondrocyte (CON-dro-site) A cartilage cell. (fig. 4.19)

chronic 1. Long-lasting. 2. Pertaining to a disease that progresses slowly and has a long duration. *Compare* acute.

chronic bronchitis A chronic obstructive pulmonary disease characterized by damaged and immobilized respiratory cilia, excessive mucus secretion, infection of the lower respiratory tract, and bronchial inflammation; caused especially by cigarette smoking. *See also* chronic obstructive pulmonary disease.

chronic obstructive pulmonary disease (COPD) Certain lung diseases (chronic bronchitis and emphysema) that result in long-term obstruction of airflow and substantially reduced pulmonary ventilation; one of the leading causes of death in old age.

cilium (SIL-ee-um) A hairlike process, with an axoneme, projecting from the apical surface of an epithelial cell; often motile and serving to propel matter across the surface of an epithelium, but sometimes nonmotile and serving a sensory role. (fig. 3.5)

circulatory shock A state of cardiac output inadequate to meet the metabolic needs of the body.

circulatory system An organ system consisting of the heart, blood vessels, and blood; the cardiovascular system plus the blood.

cisterna (sis-TUR-nuh) A fluid-filled space or sac, such as the cisterna chyli of the lymphatic system and a cisterna of the endoplasmic reticulum or Golgi complex. Also called a *cistern.* (figs. 3.13, 7.2)

clone A population of cells that are mitotically descended from the same parent cell and are identical to each other genetically or in other respects.

coagulation (co-AG-you-LAY-shun) The clotting of blood, lymph, tissue fluid, or semen.

collagen (COLL-uh-jen) The most abundant protein in the body, forming the fibers of many connective tissues in places such as the dermis, tendons, and bones.

columnar A cellular shape that is significantly taller than wide. (fig. 4.2)

commissure (COM-ih-shur) A bundle of nerve fibers that crosses from one side of the brain or spinal cord to the other.

complement 1. To complete or enhance the structure or function of something else, as in the coordinated action of two hormones. 2. A system of plasma proteins involved in defense against pathogens.

concentration gradient A difference in chemical concentration from one point to another, as on two sides of a plasma membrane.

conception The fertilization of an egg, producing a zygote.

conceptus All products of conception, ranging from a fertilized egg to the full-term fetus with its embryonic membranes, placenta, and umbilical cord. *Compare* embryo, fetus, preembryo.

condyle (CON-dile) An articular surface on a bone, usually in the form of a knob (as on the mandible), serving to smooth the motion of a joint. (fig. 6.9)

conformation The three-dimensional structure of a protein that results from interaction among its amino acid side groups, its interactions with water, and the formation of disulfide bonds.

congenital Present at birth; for example, an anatomical defect, a syphilis infection, or a hereditary disease.

connective tissue A tissue usually composed of more extracellular than cellular volume and usually with a substantial amount of extracellular fiber; forms supportive frameworks and capsules for organs, binds structures together, holds them in place, stores energy (as in adipose tissue), or transports materials (as in blood).

cooperative effect Effect in which two hormones, or both divisions of the autonomic nervous system, work together to produce a single overall result.

coronal plane *See* frontal plane.

coronary circulation A system of blood vessels that serves the wall of the heart. (fig. 13.5)

corpus Body or mass; the main part of an organ, as opposed to such regions as a head, tail, or cervix.

cortex (plural, *cortices*) The outer layer of some organs such as the adrenal gland, cerebrum, lymph node, and ovary; usually covers or encloses tissue called the *medulla.*

costal (COSS-tul) Pertaining to the ribs.

cranial (CRAY-nee-ul) Pertaining to the cranium of the skull.

cranial nerve Any of 12 pairs of nerves connected to the base of the brain and passing through foramina of the cranium.

cranium The complex of 8 bones that enclose the brain; together with 14 facial bones, these form the skull.

crista An anatomical crest or ridge, such as the crista galli of the ethmoid bone or the crista of a mitochondrion.

cross section A cut perpendicular to the long axis of the body or an organ.

cuboidal (cue-BOY-dul) A cellular shape that is roughly like a cube or in which the height and width are about equal; typically looks squarish in tissue sections. (fig. 4.2)

current A moving stream of charged particles such as ions or electrons.

cutaneous (cue-TAY-nee-us) Pertaining to the skin.

cyanosis (SY-uh-NO-sis) A bluish color of the skin and mucous membranes due to ischemia or hypoxemia.

cyclic adenosine monophosphate (cAMP) A cyclic molecule produced from ATP by the removal of two phosphate groups; serves as a second messenger in many hormone and neurotransmitter actions.

cytoplasm The contents of a cell between its plasma membrane and its nuclear envelope, consisting of cytosol, organelles, inclusions, and the cytoskeleton.

cytoskeleton A system of protein microfilaments, intermediate filaments, and microtubules in a cell, serving in physical support, cellular movement, and the routing of molecules and organelles to their destinations within the cell. (fig. 3.12)

cytosol A clear, featureless, gelatinous colloid in which the organelles and other internal structures of a cell are embedded.

cytotoxic T cell A T lymphocyte that directly attacks and destroys infected body cells, cancerous cells, and the cells of transplanted tissues.

D

daughter cells Cells that arise from a parent cell by mitosis or meiosis.

decomposition reaction A chemical reaction in which a larger molecule is broken down into smaller ones. *Compare* synthesis reaction.

decussation (DEE-cuh-SAY-shun) The crossing of nerve fibers from the right side of the central nervous system to the left or vice versa, especially in the spinal cord, medulla oblongata, and optic chiasm.

deep Relatively far from the body surface; opposite of *superficial*. For example, most bones are deep in the skeletal muscles.

denaturation A change in the three-dimensional conformation of a protein that destroys its enzymatic or other functional properties, usually caused by extremes of temperature or pH.

dendrite Extension of a neuron that receives information from other cells or from environmental stimuli and conducts signals to the soma. Dendrites are usually shorter, more branched, and more numerous than the axon and are incapable of producing action potentials. (fig. 8.2)

dendritic cell An antigen-presenting cell of the epidermis and mucous membranes. (fig. 5.2)

dense connective tissue A connective tissue with a high density of fiber, relatively little ground substance, and scanty cells; seen in tendons and the dermis, for example.

depolarization A shift in the electrical potential across a plasma membrane toward a value less negative than the resting membrane potential, associated with excitation of a nerve or muscle cell.

dermis The deeper of the two layers of the skin, underlying the epidermis and composed of fibrous connective tissue.

desmosome (DEZ-mo-some) A patchlike intercellular junction that mechanically links two cells together. (fig. 3.7)

diabetes mellitus (DM) (mel-EYE-tus) A form of diabetes that results from hyposecretion of insulin or from a deficient target cell response to it; signs include hyperglycemia and glycosuria.

diaphysis (dy-AF-ih-sis) The shaft of a long bone. (fig. 6.4)

diarthrosis (DY-ar-THRO-sis) *See* synovial joint.

diastole (dy-ASS-toe-lee) A period in which a heart chamber relaxes and fills with blood; especially ventricular relaxation.

differentiation Development of a relatively unspecialized cell or tissue into one with a more specific structure and function.

diffusion Spontaneous net movement of particles from a place of high concentration to a place of low concentration (down a concentration gradient).

diploid ($2n$) In humans, having 46 chromosomes in 23 homologous pairs; in any organism or cell, having paired chromosomes of maternal and paternal origin.

disaccharide (dy-SAC-uh-ride) A carbohydrate composed of two simple sugars (monosaccharides) covalently bonded together; for example, lactose, sucrose, and maltose. (fig. 2.9b)

distal Relatively distant from a point of origin or attachment; for example, the wrist is distal to the elbow. *Compare* proximal.

disulfide bond A covalent bond that links two cysteine residues through their sulfur atoms (—S—S—), serving to link one peptide chain to another or to hold a single chain in its three-dimensional conformation.

diuretic (DY-you-RET-ic) A chemical that increases urine output.

dorsal Toward the back (spinal) side of the body; in humans, usually synonymous with *posterior*.

dorsal root *See* posterior root.

dorsiflexion (DOR-sih-FLEC-shun) A movement of the ankle that reduces the joint angle and raises the toes. (fig. 6.42)

duodenum (dew-ODD-eh-num, DEW-oh-DEE-num) The first portion of the small intestine extending for about 25 cm from the pyloric valve of the stomach to a sharp bend called the duodenojejunal flexure; receives chyme from the stomach and secretions from the liver and pancreas. (fig. 17.14)

E

eccrine gland An exocrine gland that releases its product by exocytosis.

edema (eh-DEE-muh) Abnormal accumulation of tissue fluid resulting in swelling of the tissue.

effector A molecule, cell, or organ that carries out a response to a stimulus.

efferent (EFF-ur-ent) Carrying away or out, such as a blood vessel that carries blood away from a tissue or a nerve fiber that conducts signals away from the central nervous system.

elastic fiber A connective tissue fiber, composed of the protein elastin, that stretches under tension and returns to its original length when released; responsible for the resilience of organs such as the skin and lungs.

elasticity The tendency of a stretched structure to return to its original dimensions when tension is released.

electrolyte A salt that ionizes in water and produces a solution that conducts electricity; loosely speaking, any ion that results from the dissociation of such salts, such as sodium, potassium, calcium, chloride, and bicarbonate ions.

electron micrograph A photograph made with a scanning or transmission electron microscope rather than with a light microscope.

elevation A joint movement that raises a body part, as in hunching the shoulders or closing the mouth.

embolism (EM-bo-lizm) The obstruction of a blood vessel by an embolus.

embolus (EM-bo-lus) Any abnormal traveling object in the bloodstream, such as agglutinated bacteria or blood cells, a blood clot, or an air bubble.

embryo A developing individual from the sixteenth day of gestation when the three primary germ layers have formed, through the end of the eighth week when all of the organ systems are present. *Compare* conceptus, fetus, preembryo.

endocardium All tissue deep to the muscle of the heart wall, lining the heart chambers and valve surfaces; consists of a simple squamous epithelium and thin layer of areolar tissue, and is coated with prostacyclin, which repels blood platelets and prevents clotting within the heart.

endochondral ossification (EN-doe-CON-drul OSS-ih-fih-CAY-shun) A process in which a hyaline cartilage precursor (model) is replaced by calcified tissue to form a bone; employed for the production of most bones of the body except for flat bones of the skull and part of the clavicle.

endocrine gland (EN-doe-crin) A ductless gland that secretes hormones into the bloodstream; for example, the thyroid and adrenal glands. *Compare* exocrine gland.

endocytosis (EN-doe-sy-TOE-sis) Any process in which a cell forms vesicles from its plasma membrane and takes in large particles, molecules, or droplets of fluid; for example, phagocytosis and pinocytosis.

endometrium (EN-doe-MEE-tree-um) The mucosa of the uterus; the site of implantation and source of menstrual discharge.

endoplasmic reticulum (ER) (EN-doe-PLAZ-mic reh-TIC-you-lum) An extensive system of interconnected cytoplasmic tubules or channels; classified as rough ER or smooth ER depending on the presence or absence of ribosomes on its membrane. (fig. 3.13)

endothelium (EN-doe-THEEL-ee-um) A simple squamous epithelium that lines the lumens of the blood vessels, heart, and lymphatic vessels.

enteric (en-TERR-ic) Pertaining to the small intestine, as in enteric hormones.

eosinophil (EE-oh-SIN-oh-fill) A granulocyte with a large, often bilobed nucleus and coarse cytoplasmic granules that stain with eosin; phagocytizes antigen–antibody complexes, allergens, and inflammatory chemicals and secretes enzymes that combat parasitic infections.

epicardium All tissue of the heart wall superficial to its muscle, the myocardium. Consists of a simple squamous epithelium and thin layer of areolar tissue, collectively equivalent to the visceral pericardium; in some places also has an adipose layer, considered by some to be *subepicardial fat* and not part of the epicardium proper.

epidermis A stratified squamous epithelium that constitutes the superficial layer of the skin, overlying the dermis. (fig. 5.1)

epinephrine (EP-ih-NEFF-rin) A catecholamine that functions as a neurotransmitter in the sympathetic nervous system and as a hormone secreted by the adrenal medulla; also called *adrenaline*.

epiphysis (eh-PIF-ih-sis) 1. The head of a long bone. (fig. 6.4) 2. The pineal gland (epiphysis cerebri).

epithelium A type of tissue consisting of one or more layers of closely adhering cells with little intercellular material and no blood vessels; forms the coverings and linings of many organs and the secretory tissue and ducts of the glands.

erectile tissue A tissue that functions by swelling with blood, as in the penis and clitoris.

erythrocyte (eh-RITH-ro-site) A red blood cell.

erythropoiesis (eh-RITH-ro-poy-EE-sis) The production of erythrocytes.

erythropoietin (eh-RITH-ro-POY-eh-tin) A hormone that is secreted by the kidneys and liver in response to hypoxemia and stimulates erythropoiesis.

estrogen (ESS-tro-jen) Collective name of three similar steroid sex hormones (estriol, estrone, estradiol) known especially for producing female secondary sex characteristics and regulating various aspects of the menstrual cycle and pregnancy.

excitability The ability of a cell to respond to stimuli, especially the ability of nerve and muscle cells to produce membrane voltage changes in response to stimuli; irritability.

excitation–contraction coupling Events that link the synaptic stimulation of a muscle cell to the onset of contraction.

excretion 1. A waste product produced by the body, including feces, urine, and sweat. 2. The process of eliminating metabolic waste products from a cell or from the body. *Compare* secretion.

exocrine gland (EC-so-crin) A gland that secretes its products into another organ or onto the body surface, usually by way of a duct; for example, salivary and gastric glands. *Compare* endocrine gland.

exocytosis (EC-so-sy-TOE-sis) A process in which a vesicle in the cytoplasm of a cell fuses with the plasma membrane and releases its contents from the cell; used in the elimination of cellular wastes and in the release of gland products and neurotransmitters.

expiration 1. Exhaling. 2. Dying.

extension Movement of a joint that increases the angle between articulating bones (straightens the joint). (fig. 6.36) *Compare* flexion.

extracellular fluid (ECF) Any body fluid that is not contained in the cells; for example, blood, lymph, and tissue fluid.

extrinsic (ec-STRIN-sic) 1. Originating externally, such as extrinsic blood-clotting factors; exogenous. 2. Not fully contained within an organ but acting on it, such as the extrinsic muscles of the hand and eye. *Compare* intrinsic.

F

facilitated diffusion The process of transporting a chemical through a cellular membrane, down its concentration gradient, with the aid of a carrier that does not consume ATP; enables substances to diffuse through the membrane that would do so poorly, or not at all, without a carrier.

fascia (FASH-ee-uh) A layer of connective tissue between the muscles or separating the muscles from the skin. (fig. 7.14)

fascicle (FASS-ih-cul) A bundle of muscle or nerve fibers ensheathed in connective tissue; multiple fascicles bound together constitute a muscle or nerve as a whole. (figs. 7.14, 8.17)

fat 1. A triglyceride molecule. 2. Adipose tissue.

fatty acid An organic molecule composed of a chain of an even number of carbon atoms with a carboxyl group (—COOH) at one end and a methyl group (—CH$_3$) at the other; one of the structural subunits of triglycerides and phospholipids.

fenestrated (FEN-eh-stray-ted) Perforated with holes or slits, as in fenestrated blood capillaries and the elastic sheets of large arteries.

fetus In human development, an individual from the beginning of the ninth week when all of the organ systems are present, through the time of birth. *Compare* conceptus, embryo, preembryo.

fibrin (FY-brin) A sticky fibrous protein formed from fibrinogen in blood, tissue fluid, and lymph; forms the matrix of a blood clot.

fibroblast A connective tissue cell that produces collagen fibers and ground substance; the only type of cell in tendons and ligaments.

fibrosis Replacement of damaged tissue with fibrous scar tissue rather than by the original tissue type; scarring. *Compare* regeneration.

fibrous connective tissue Any connective tissue with a preponderance of fiber, such as areolar, reticular, dense regular, and dense irregular connective tissues.

filtrate A fluid formed by filtration, as at the renal glomerulus and other capillaries.

filtration A process in which hydrostatic pressure forces a fluid through a selectively permeable membrane (especially a capillary wall).

fire To produce an action potential, as in nerve and muscle cells.

fix To hold a structure in place, for example, by fixator muscles that prevent unwanted joint movements.

flexion A joint movement that, in most cases, decreases the angle between two bones. (fig. 6.36) *Compare* extension.

fluid balance A state in which the average daily gain and loss of water are equal and water is properly distributed among the body's fluid compartments.

fluid compartment Any of the major categories of fluid in the body, separated by selectively permeable membranes and differing from each other in chemical composition. Primary examples are the intracellular fluid, tissue fluid, blood, and lymph.

follicle (FOLL-ih-cul) A small space, such as a hair follicle, thyroid follicle, or ovarian follicle.

follicle-stimulating hormone (FSH) A hormone secreted by the anterior pituitary gland that stimulates development of the ovarian follicles and egg cells in females and promotes sperm production in males.

foramen (fo-RAY-men) A hole through a bone or other organ, in many cases providing passage for blood vessels and nerves.

fossa (FOSS-uh) A depression in an organ or tissue, such as the fossa ovalis of the heart or a cranial fossa of the skull.

fovea (FOE-vee-uh) A small pit, such as the fovea capitis of the femur or fovea centralis of the retina.

free radical A particle derived from an atom or molecule, having an unpaired electron that makes it highly reactive and destructive to cells; produced by intrinsic processes such as aerobic respiration and by extrinsic agents such as chemicals and ionizing radiation.

frontal plane An anatomical plane that passes through the body or an organ from right to left and superior to inferior, such as a vertical plane that separates the anterior portion of the chest from the back; also called a *coronal plane*. (fig. 1.7)

fundus The base, the broadest part, or the part farthest from the opening of certain organs such as the stomach and uterus.

fusiform (FEW-zih-form) Spindle-shaped; elongated, thick in the middle, and tapered at both ends, such as the shape of a smooth muscle cell or a muscle spindle.

G

gamete (GAM-eet) An egg or sperm cell.

ganglion (GANG-glee-un) A cluster of nerve cell bodies in the peripheral nervous system, often resembling a knot in a string.

gangrene Tissue necrosis usually resulting from ischemia and often involving infection.

gap junction A junction between two cells consisting of a pore surrounded by a ring of proteins in the plasma membrane of each cell; allows solutes to diffuse from the cytoplasm of one cell to the next, thereby serving for cell-to-cell electrical and chemical communication in tissues such as cardiac and smooth muscle. (fig. 3.7)

gastric Pertaining to the stomach.

gate A protein channel in a cellular membrane that can open or close in response to chemical, electrical, or mechanical stimuli, thus controlling when substances are allowed to pass through the membrane.

gene An information-containing segment of DNA that codes for the production of a molecule of RNA, which in most cases goes on to play a role in the synthesis of one or more proteins.

germ cell An egg or sperm cell (gamete) or any precursor cell destined to become a gamete.

gestation (jess-TAY-shun) Pregnancy.

globulin (GLOB-you-lin) A globular protein such as an enzyme, antibody, or albumin; especially a family of proteins in the blood plasma that includes albumin, antibodies, fibrinogen, and prothrombin.

glucagon (GLUE-ca-gon) A hormone secreted by alpha cells of the pancreatic islets in response to hypoglycemia; promotes glycogenolysis and other effects that raise blood glucose concentration.

glucocorticoid (GLUE-co-COR-tih-coyd) Any hormone of the adrenal cortex that affects carbohydrate, fat, and protein metabolism; chiefly cortisol and corticosterone.

gluconeogenesis (GLUE-co-NEE-oh-JEN-eh-sis) The synthesis of glucose from noncarbohydrates such as fats and amino acids.

glucose A monosaccharide ($C_6H_{12}O_6$) also known as blood sugar. Glycogen, starch, cellulose, and maltose are made entirely of glucose, and glucose constitutes half of a sucrose or lactose molecule. The isomer involved in human physiology is also called *dextrose*.

glycocalyx (GLY-co-CAY-licks) A layer of carbohydrate molecules covalently bonded to the phospholipids and proteins of a plasma membrane; forms a surface coat on all human cells.

glycogen (GLY-co-jen) A glucose polymer synthesized by liver, muscle, uterine, and vaginal cells that serves as an energy-storage polysaccharide.

glycogenesis (GLY-co-JEN-eh-sis) The synthesis of glycogen.

glycogenolysis (GLY-co-jeh-NOLL-ih-sis) The hydrolysis of glycogen, releasing glucose.

glycolipid (GLY-co-LIP-id) A phospholipid molecule with a carbohydrate covalently bonded to it, found in the plasma membranes of cells.

glycolysis (gly-COLL-ih-sis) A series of anaerobic oxidation reactions that breaks a glucose molecule into two molecules of pyruvate and produces a small amount of ATP.

glycoprotein (GLY-co-PRO-teen) A protein molecule with a smaller carbohydrate covalently bonded to it; found in mucus and the glycocalyx of cells, for example.

goblet cell A mucus-secreting gland cell, shaped somewhat like a wineglass, found in the epithelia of many mucous membranes. (fig. 4.6)

Golgi complex (GOAL-jee) An organelle composed of several parallel cisternae, somewhat like a stack of saucers, that modifies and packages newly synthesized proteins and synthesizes carbohydrates. (fig. 3.13)

Golgi vesicle A membrane-bounded vesicle pinched from the Golgi complex, containing its chemical product; may be retained in the cell as a lysosome or become a secretory vesicle that releases the product by exocytosis.

gonad An organ that produces gametes; the ovary or testis.

gonadotropin (go-NAD-oh-TRO-pin) A pituitary hormone that stimulates the gonads; specifically FSH and LH.

gradient A difference or change in any variable, such as pressure or chemical concentration, from one point in space to another; provides a basis for molecular movements such as gas exchange, osmosis, and facilitated diffusion, and for bulk movements such as blood flow and air flow.

granulocyte Any of the leukocyte types (neutrophils, eosinophils, basophils) that possess specific granules and thus have a relatively coarse-grained cytoplasm. (fig. 12.9)

gray matter A zone or layer of tissue in the central nervous system where the neuron cell bodies, dendrites, and synapses are found; forms the cerebral cortex and basal nuclei; cerebellar cortex and deep nuclei; nuclei of the brainstem; and core of the spinal cord. (figs. 9.3, 9.4)

gross anatomy Bodily structure that can be observed without magnification.

growth factor A chemical messenger that stimulates mitosis and differentiation of target cells that have receptors for it; important in such processes as fetal development, tissue maintenance and repair, and hemopoiesis; sometimes a contributing factor in cancer.

growth hormone (GH) A hormone of the anterior pituitary gland with multiple effects on many tissues, generally promoting tissue growth; also called *somatotropin*.

gustation (gus-TAY-shun) The sense of taste.

gyrus (JY-rus) A wrinkle or fold in the cortex of the cerebrum or cerebellum.

H

hair cell Sensory cell of the cochlea, semicircular ducts, utricle, and saccule, with a fringe of surface microvilli that respond to the relative motion of a gelatinous membrane at their tips; responsible for the senses of hearing, body position, and motion.

hair follicle An epithelial pit that contains a hair and extends into the dermis or hypodermis.

haploid (n) In humans, having 23 unpaired chromosomes instead of the usual 46 chromosomes in homologous pairs; in any organism or cell, having half the normal diploid (2*n*) number of chromosomes for that species. *Compare* diploid.

helper T cell A type of lymphocyte that performs a central coordinating role in humoral and cellular immunity; target of the human immunodeficiency virus (HIV).

hematocrit (he-MAT-oh-crit) The percentage of blood volume that is composed of erythrocytes; also called *packed cell volume*.

hematoma (HE-muh-TOE-muh) A mass of clotted blood in the tissues; forms a bruise when visible through the skin.

heme (pronounced "heem") The nonprotein, iron-containing component of a hemoglobin or myoglobin molecule; oxygen binds to its iron atom. (fig. 12.4)

hemoglobin (HE-mo-GLO-bin) The red gas-transport pigment of an erythrocyte.

hemopoiesis (HE-mo-poy-EE-sis) Production of any of the formed elements of blood.

hemopoietic stem cell A cell of the red bone marrow that can give rise, through a series of intermediate cells, to leukocytes, erythrocytes, platelets, and various kinds of macrophages.

heparin (HEP-uh-rin) A polysaccharide secreted by basophils and mast cells that inhibits blood clotting.

hepatic (heh-PAT-ic) Pertaining to the liver.

hepatic portal system A network of blood vessels that connect capillaries of the intestines to capillaries (sinusoids) of the liver, thus delivering newly absorbed nutrients directly to the liver.

high-density lipoprotein (HDL) A lipoprotein of the blood plasma that is about 50% lipid and 50% protein; functions to transport phospholipids and cholesterol from other organs to the liver for disposal. A high proportion of HDL to low-density lipoprotein (LDL) is desirable for cardiovascular health.

hilum (HY-lum) A point on the surface of an organ where blood vessels, lymphatic vessels, or nerves enter and leave, usually marked by a depression and slit; the midpoint of the concave surface of any organ that is roughly bean-shaped, such as the lymph nodes, kidneys, and lungs. (fig. 15.6b)

histamine An amino acid derivative secreted by basophils, mast cells, and some neurons; functions as a neurotransmitter or local chemical signal to stimulate effects such as gastric secretion, bronchoconstriction, and vasodilation.

histology 1. The microscopic structure of tissues and organs. 2. The study of such structure.

homeostasis (HO-me-oh-STAY-sis) The tendency of a living body to maintain relatively stable internal conditions in spite of changes in its external environment.

hormone A chemical messenger that is secreted by an endocrine gland or isolated gland cell, travels in the bloodstream, and triggers a physiological response in distant cells with receptors for it.

host cell Any cell belonging to the human body, as opposed to foreign cells introduced to it by such causes as infections and tissue transplants.

human chorionic gonadotropin (HCG) (CORE-ee-ON-ic go-NAD-oh-TROPE-in) A hormone of pregnancy secreted by the chorion that stimulates continued growth of the corpus luteum and secretion of its hormones. HCG in urine is the basis for pregnancy testing.

human immunodeficiency virus (HIV) A virus that infects human helper T cells and other cells, suppresses immunity, and causes AIDS.

hyaline cartilage (HY-uh-lin) A form of cartilage with a relatively clear matrix and fine collagen fibers but no conspicuous elastic fibers or coarse collagen bundles as in other types of cartilage; the most widespread type of cartilage in the human body.

hydrogen bond A weak attraction between a slightly positive hydrogen atom on one molecule and a slightly negative oxygen or nitrogen atom on another molecule, or between such atoms on different parts of the same molecule; responsible for the cohesion of water and the coiling of protein and DNA molecules, for example.

hydrolysis (hy-DROL-ih-sis) A chemical reaction that breaks a covalent bond in a molecule by adding an —OH group to one side of the bond and —H to the other side, thus consuming a water molecule.

hydrophilic (HY-dro-FILL-ic) Pertaining to molecules that attract water or dissolve in it because of their polar nature.

hydrophobic (HY-dro-FOE-bic) Pertaining to molecules that do not attract water or dissolve in it because of their nonpolar nature; such molecules tend to dissolve in lipids and other nonpolar solvents.

hydrostatic pressure The physical force generated by a liquid such as blood or tissue fluid, as opposed to osmotic and atmospheric pressures.

hypercalcemia (HY-pur-cal-SEE-me-uh) An excess of calcium ions in the blood.

hyperextension A joint movement that increases the angle between two bones beyond 180°. (fig. 6.36)

hyperglycemia (HY-pur-gly-SEE-me-uh) An excess of glucose in the blood.

hyperkalemia (HY-pur-ka-LEE-me-uh) An excess of potassium ions in the blood.

hypernatremia (HY-pur-na-TREE-me-uh) An excess of sodium ions in the blood.

hyperplasia (HY-pur-PLAY-zhuh) The growth of a tissue through cellular multiplication, not cellular enlargement. Compare hypertrophy.

hypertension Excessively high blood pressure; criteria vary but it is often considered to be a condition in which resting systolic pressure exceeds 140 mm Hg or diastolic pressure exceeds 90 mm Hg.

hypertonic Having a higher osmotic pressure than human cells or some other reference solution and tending to cause osmotic shrinkage of cells.

hypertrophy (hy-PUR-truh-fee) The growth of a tissue through cellular enlargement, not cellular multiplication; for example, the growth of muscle under the influence of exercise. Compare hyperplasia.

hypocalcemia (HY-po-cal-SEE-me-uh) A deficiency of calcium ions in the blood.

hypodermis (HY-po-DUR-miss) A layer of connective tissue deep to the skin; also called *superficial fascia, subcutaneous tissue,* or when it is predominantly adipose, *subcutaneous fat.*

hypoglycemia (HY-po-gly-SEE-me-uh) A deficiency of glucose in the blood.

hypokalemia (HY-po-ka-LEE-me-uh) A deficiency of potassium ions in the blood.

hyponatremia (HY-po-na-TREE-me-uh) A deficiency of sodium ions in the blood.

hypothalamic thermostat (HY-po-thuh-LAM-ic) A nucleus in the hypothalamus that monitors body temperature and sends afferent signals to hypothalamic heat-promoting or heat-losing centers to maintain thermal homeostasis.

hypothalamus (HY-po-THAL-uh-mus) The inferior portion of the diencephalon of the brain, forming the walls and floor of the third ventricle and giving rise to the posterior pituitary gland; controls many fundamental physiological functions such as appetite, thirst, and body temperature, and exerts many of its effects through the endocrine and autonomic nervous systems. (fig. 9.2)

hypothesis An informed conjecture that is capable of being tested and potentially falsified by experimentation or data collection.

hypotonic Having a lower osmotic pressure than human cells or some other reference solution and tending to cause osmotic swelling and lysis of cells.

hypoxemia (HY-pock-SEE-me-uh) A deficiency of oxygen in the blood.

hypoxia (hy-POCK-see-uh) A deficiency of oxygen in any tissue.

I

immunity The ability to ward off a specific infection or disease, usually as a result of prior exposure and the body's production of antibodies or lymphocytes against a pathogen.

immunoglobulin (IM-you-no-GLOB-you-lin) An antibody; a protein of the globulin class that is produced by plasma cells; found in blood plasma, other body fluids, and lymphocyte plasma membranes; and functions in humoral immunity.

implantation The attachment of a conceptus to the endometrium of the uterus.

inclusion Any visible object in the cytoplasm of a cell other than an organelle or cytoskeletal element; usually a foreign body or a stored cell product, such as a virus, dust particle, lipid droplet, glycogen granule, or pigment.

infarction (in-FARK-shun) 1. The sudden death of tissue from a lack of blood perfusion. 2. An area of necrotic tissue produced by this process; also called an *infarct.*

inferior Lower than another structure or point of reference from the perspective of anatomical position; for example, the stomach is inferior to the diaphragm.

inflammation A complex of tissue responses to trauma or infection serving to ward off a pathogen and promote tissue repair; recognized by the cardinal signs of redness, heat, swelling, and pain.

inguinal (IN-gwih-nul) Pertaining to the groin.

innate immunity Nonspecific defenses against infection or disease that are present or functional from birth, work equally against multiple disease agents, and do not require prior exposure; examples include the skin, mucous membranes, inflammation, and fever. *See also* adaptive immunity.

innervation (IN-ur-VAY-shun) The nerve supply to an organ.

inspiration 1. Inhaling. 2. The stimulus that resulted in this book.

integral protein A protein that extends through a plasma membrane and contacts both the extracellular and intracellular fluid. (fig. 3.2)

intercalated disc (in-TUR-kuh-LAY-ted) A complex of mechanical and electrical junctions that joins two cardiac muscle cells end to end, microscopically visible as a dark line that helps to histologically distinguish this muscle type; functions as a mechanical and electrical link between cells. (fig. 13.6)

intercellular Between cells.

intercostal (IN-tur-COSS-tul) Between the ribs, as in the intercostal muscles, arteries, veins, and nerves.

interdigitate (IN-tur-DIDJ-ih-tate) To fit together like the fingers of two folded hands; for example, at the dermal–epidermal boundary, intercalated discs of the heart, and foot processes of the podocytes in the kidney. (fig. 16.8)

interleukin (IN-tur-LOO-kin) A hormonelike chemical messenger from one leukocyte to another, serving as a means of communication and coordination during immune responses.

interneuron (IN-tur-NOOR-on) A neuron that is contained entirely in the central nervous system and, in the path of signal conduction, lies anywhere between an afferent pathway and an efferent pathway.

interosseous membrane (IN-tur-OSS-ee-us) A fibrous membrane that connects the radius to the ulna and the tibia to the fibula along most of the shaft of each bone. (fig. 6.28)

interphase That part of the cell cycle between one mitotic phase and the next, from the end of cytokinesis to the beginning of the next mitosis.

interstitial (IN-tur-STISH-ul) 1. Pertaining to the extracellular spaces in a tissue. 2. Located between other structures, as in the interstitial endocrine cells of the testis.

intervertebral disc A cartilaginous pad between the bodies of adjacent vertebrae.

intracellular Within a cell.

intracellular fluid (ICF) The fluid contained in the cells; one of the major fluid compartments.

intramembranous ossification (IN-tra-MEM-bruh-nus OSS-ih-fih-CAY-shun) A process in which a sheet of mesenchyme becomes calcified to form a bone, employed primarily for the production of flat bones such as those of the cranium.

intravenous (I.V.) 1. Present or occurring within a vein, such as an intravenous blood clot. 2. Introduced directly into a vein, such as an intravenous injection or I.V. drip.

intrinsic (in-TRIN-sic) 1. Arising from within, such as intrinsic blood-clotting factors; endogenous. 2. Fully contained within an organ, such as the intrinsic muscles of the hand and eye. *Compare* extrinsic.

involuntary Not under conscious control, including tissues such as smooth and cardiac muscle and events such as reflexes.

involution (IN-vo-LOO-shun) Shrinkage of a tissue or organ by autolysis, such as involution of the thymus after childhood and of the uterus after pregnancy.

ion A chemical particle with unequal numbers of electrons or protons and consequently a net negative or positive charge; it may have a single atomic nucleus as in a sodium ion or a few atoms as in a bicarbonate ion, or it may be a large molecule such as a protein.

ionic bond The force that binds a cation to an anion.

ionizing radiation High-energy electromagnetic rays that eject electrons from atoms or molecules and convert them to ions, frequently causing cellular damage; for example, X-rays and gamma rays.

ischemia (iss-KEE-me-uh) Insufficient blood flow to a tissue, typically resulting in metabolite accumulation and sometimes tissue death.

isometric contraction A form of contraction in which a muscle develops increasing tension but does not shorten because of an overriding external resistance; a normal phenomenon at the beginning of any muscle contraction working against an external load.

isotonic Having the same osmotic pressure as human cells or some other reference solution.

J

jaundice (JAWN-diss) A yellowish color of the skin, corneas, mucous membranes, and body fluids due to an excessive concentration of bilirubin; usually indicative of a liver disease, obstructed bile secretion, or hemolytic disease.

K

ketone (KEE-tone) Any organic compound with a carbonyl (C=O) group covalently bonded to two other carbons; ketones are produced by fat oxidation and are a cause of diabetic acidosis.

ketonuria (KEE-toe-NEW-ree-uh) The abnormal presence of ketones in the urine; a sign of diabetes mellitus but also occurring in other conditions that entail rapid fat oxidation.

kilocalorie The amount of heat energy needed to raise the temperature of 1 kg of water by 1°C; 1,000 calories. Also called a *Calorie* (with a capital *C*). *See also* calorie.

L

labium (LAY-bee-um) A lip, such as those of the mouth and the labia majora and minora of the vulva.

lactate A small organic acid produced as an end product of the anaerobic fermentation of pyruvate; also called *lactic acid*.

lacuna (la-CUE-nuh) A small cavity or depression in a tissue such as bone, cartilage, and the erectile tissues.

lamella (la-MELL-uh) A little plate or layer, such as the lamellae of bone. (fig. 6.3)

lamina (LAM-ih-nuh) A thin layer, such as the lamina of a vertebra or the lamina propria of a mucous membrane. (fig. 6.20)

lamina propria (PRO-pree-uh) A thin layer of areolar tissue immediately deep to the epithelium of a mucous membrane. (fig. 17.2)

larynx (LAIR-inks) A cartilaginous chamber in the neck containing the vocal cords; colloquially called the *voice box*.

lateral Away from the midline of an organ or median plane of the body; toward the side. *Compare* medial.

law A verbal or mathematical description of a predictable natural phenomenon or of the relationship between variables; for example, the law of complementary base pairing.

leukocyte (LOO-co-site) A white blood cell.

leukopoiesis (LOO-co-poy-EE-sis) The production of white blood cells.

libido (lih-BEE-do) Sex drive; the psychological motivation to seek and engage in sexual activity.

ligament A cord or band of tough collagenous tissue binding one organ to another, especially one bone to another, and serving to hold organs in place; for example, the cruciate ligaments of the knee and falciform ligament of the liver.

linea (LIN-ee-uh) An anatomical line, such as the linea aspera of the femur.

lingual (LING-gwul) Pertaining to the tongue, as in lingual papillae.

lipase (LY-pace) An enzyme that hydrolyzes a triglyceride into fatty acids and glycerol.

lipid A hydrophobic organic compound composed mainly of carbon and a high ratio of hydrogen to oxygen; includes fatty acids, fats, phospholipids, steroids, and prostaglandins.

lipoprotein (LIP-oh-PRO-teen) A protein-coated lipid droplet in the blood plasma or lymph, serving as a means of lipid transport; for example, chylomicrons and high- and low-density lipoproteins.

load 1. To pick up a gas for transport in the bloodstream. 2. The resistance acted upon by a muscle.

lobule (LOB-yool) A small subdivision of an organ or of a lobe of an organ, especially of a gland.

long bone A bone such as the femur or humerus that is markedly longer than wide and that generally serves as a lever.

longitudinal Oriented along the longest dimension of the body or of an organ.

loose connective tissue A type of fibrous connective tissue characterized by a seemingly random array of fibers of different types (collagenous, elastic, and reticular), with abundant ground substance between them, and widely scattered cells.

low-density lipoprotein (LDL) A bloodborne droplet of about 20% protein and 80% lipid (mainly cholesterol) that transports cholesterol from the liver to other tissues for their use.

lumbar Pertaining to the lower back and sides, between the thoracic cage and pelvis.

lumen (LOO-men) The internal space of a hollow organ such as a blood vessel or the esophagus, or a space surrounded by cells as in a gland acinus.

lymph The fluid contained in lymphatic vessels and lymph nodes, produced by the absorption of tissue fluid.

lymph node A small organ found along the course of a lymphatic vessel that filters the lymph and contains lymphocytes and macrophages, which respond to antigens in the lymph. (fig. 14.9)

lymphatic system (lim-FAT-ic) An organ system consisting of lymphatic vessels, lymph nodes, tonsils, spleen, red bone marrow, and thymus; functions include tissue fluid recovery and immunity.

lymphocyte (LIM-fo-site) A relatively small agranulocyte with numerous types and roles in nonspecific defense, humoral immunity, and cellular immunity.

lysosome (LY-so-some) A membrane-bounded organelle containing a mixture of enzymes with a variety of intracellular and extracellular roles in digesting foreign matter, pathogens, and expired organelles.

lysozyme (LY-so-zime) An enzyme found in tears, milk, saliva, mucus, and other body fluids that destroys bacteria by digesting their cell walls.

M

macromolecule Any molecule of large size and high molecular weight, such as a protein, nucleic acid, polysaccharide, or triglyceride.

macrophage (MAC-ro-faje) Any cell of the body, other than a leukocyte, that is specialized for phagocytosis; usually derived from a blood monocyte and often functioning as an antigen-presenting cell.

macula (MAC-you-luh) A patch or spot, such as the *macula lutea* of the retina.

malignant (muh-LIG-nent) Pertaining to a cell or tumor that is cancerous; capable of metastasis.

mast cell A connective tissue cell, similar to a basophil, that secretes histamine, heparin, and other chemicals involved in inflammation; often concentrated along the course of blood capillaries.

matrix 1. The extracellular material of a tissue. 2. The fluid within a mitochondrion containing enzymes of the citric acid cycle. 3. The substance or framework within which other structures are embedded, such as the fibrous matrix of a blood clot. 4. A mass of epidermal cells from which a hair root or nail root develops.

medial Toward the midline of an organ or median plane of the body. *Compare* lateral.

median plane The sagittal plane that divides the body or an organ into equal right and left halves; also called *midsagittal plane*. (fig. 1.7)

mediastinum (ME-dee-ah-STY-num) The thick median partition of the thoracic cavity that separates one pleural cavity from the other and contains the heart, great blood vessels, esophagus, trachea, and thymus. (fig. 1.12)

medulla (meh-DULE-uh, meh-DULL-uh) Tissue deep to the cortex of certain two-layered organs such as the adrenal glands, lymph nodes, hairs, and kidneys.

medulla oblongata (meh-DULL-uh OB-long-GAH-ta) The most caudal part of the brainstem, immediately superior to the foramen magnum of the skull, connecting the spinal cord to the rest of the brain. (figs. 9.2, 9.5)

meiosis (my-OH-sis) A form of cell division in which a diploid cell divides twice and produces four haploid daughter cells; occurs only in gametogenesis.

melanocyte A cell of the stratum basale of the epidermis that synthesizes melanin and transfers it to the keratinocytes.

memory T cells A class of lymphocytes descended from cytotoxic T cells, serving to remember a specific antigen and enable the body to respond more quickly to it upon reexposure.

meninges (meh-NIN-jeez) (singular, *meninx*) Three fibrous membranes between the central nervous system and surrounding bone: the dura mater, arachnoid mater, and pia mater. (fig. 9.3)

menstruation (men-STRAY-shun, MEN-stru-AY-shun) Periodic vaginal discharge of blood and tissues from the functional layer of the endometrium of the nonpregnant uterus.

mesentery (MESS-en-tare-ee) A serous membrane that binds the intestines together and suspends them from the abdominal wall; the visceral continuation of the peritoneum. (figs. 1.14, 17.18)

mesoderm (MES-oh-durm) The middle layer of the three primary germ layers of an embryo; gives rise to muscle and connective tissue.

metabolism (meh-TAB-oh-lizm) The sum of all chemical reactions in the body.

metabolite (meh-TAB-oh-lite) Any chemical produced by metabolism, especially a breakdown product or waste.

microbiome The collection of bacteria and other microorganisms that inhabit the skin, mucosal surfaces, and gut.

microtubule An intracellular cylinder of protein, forming centrioles, the axonemes of cilia and flagella, and part of the cytoskeleton.

microvillus An outgrowth of the plasma membrane that increases the surface area of a cell and functions in absorption and some sensory processes; distinguished from cilia and flagella by its smaller size and lack of an axoneme.

milliequivalent One-thousandth of an equivalent, which is the amount of an electrolyte that would neutralize 1 mole of H^+ or OH^-. Electrolyte concentrations are commonly expressed in milliequivalents per liter (mEq/L). *See* appendix C.

mitochondrion (MY-toe-CON-dree-un) An organelle specialized to synthesize ATP, enclosed in a double unit membrane with infoldings of the inner membrane called cristae.

mitosis (my-TOE-sis) A form of cell division in which a cell divides once and produces two genetically identical daughter cells; sometimes used to refer only to the division of the genetic material or nucleus and not to include cytokinesis, the subsequent division of the cytoplasm.

molarity A measure of chemical concentration expressed as moles of solute per liter of solution. *See* appendix C.

mole The mass of a chemical equal to its molecular weight in grams, containing 6.023×10^{23} molecules. *See* appendix C.

monocyte An agranulocyte specialized to migrate into the tissues and transform into a macrophage.

monomer (MON-oh-mur) 1. One of the identical or similar subunits of a larger molecule in the dimer to polymer range; for example, the glucose monomers of starch, the amino acids of a protein, or the nucleotides of DNA. 2. One subunit of an antibody molecule, composed of four polypeptides.

monosaccharide (MON-oh-SAC-uh-ride) A simple sugar, or sugar monomer; chiefly glucose, fructose, and galactose.

motor end plate *See* neuromuscular junction.

motor neuron A neuron that transmits signals from the central nervous system to any effector (muscle or gland cell); its axon is an efferent nerve fiber, carrying signals away from the CNS.

motor protein Any protein that produces movements of a cell or its components owing to its ability to undergo quick repetitive changes in conformation and to bind reversibly to other molecules; for example, myosin.

motor unit One motor neuron and all the skeletal muscle fibers innervated by it.

mucosa (mew-CO-suh) A membrane that lines an anatomical tract that is open to the exterior (the respiratory, digestive, urinary, and reproductive tracts) and that secretes mucus. Also called a *mucous membrane*. (fig. 4.30a)

muscle fiber One skeletal muscle cell.

muscle tone A state of continual, partial contraction of resting skeletal or smooth muscle.

muscular system An organ system composed of the skeletal muscles, specialized mainly for maintaining postural support and producing movements of the bones.

muscularis externa The external muscular wall of certain viscera such as the esophagus and small intestine. (fig. 17.2)

mutagen (MEW-tuh-jen) Any agent that causes a mutation, including viruses, chemicals, and ionizing radiation.

mutation Any change in the structure of a chromosome or a DNA molecule, often resulting in a change of organismal structure or function.

myelin (MY-eh-lin) A lipid sheath around a nerve fiber, formed from closely spaced spiral layers of the plasma membrane of a Schwann cell or oligodendrocyte. (fig. 8.6)

myocardium (MY-oh-CAR-dee-um) The middle, muscular layer of the heart.

myocyte A muscle cell, especially a cell of cardiac or smooth muscle.

myoepithelial cell An epithelial cell that has become specialized to contract like a muscle cell; important in dilation of the pupil and ejection of secretions from gland acini.

myofilament A protein microfilament responsible for the contraction of a muscle cell, composed mainly of myosin or actin. (fig. 7.3)

myoglobin (MY-oh-GLO-bin) A red oxygen-storage pigment of muscle; supplements hemoglobin in providing oxygen for aerobic muscle metabolism.

myosin A motor protein that constitutes the thick myofilaments of muscle and has globular, mobile heads that bind to actin molecules.

N

necrosis (neh-CRO-sis) Pathological tissue death due to such causes as infection, trauma, or hypoxia. *Compare* apoptosis.

negative feedback A self-corrective mechanism that underlies most homeostasis, in which a bodily change is detected and responses are activated that reverse the change and restore stability and preserve normal body function.

negative feedback inhibition A mechanism for limiting the secretion of a pituitary hormone. The pituitary hormone stimulates another endocrine gland to secrete its own hormone, and that hormone inhibits further release of the pituitary hormone.

neonate (NEE-oh-nate) An infant up to 6 weeks old.

neoplasia (NEE-oh-PLAY-zhuh) Abnormal growth of new tissue, such as a tumor, with no useful function.

nephron One of approximately 1 million blood-filtering, urine-producing units in each kidney; consists of a glomerulus, glomerular capsule, proximal convoluted tubule, nephron loop, and distal convoluted tubule. (fig. 16.6)

nerve A cordlike organ of the peripheral nervous system composed of multiple nerve fibers ensheathed in connective tissue.

nerve fiber The axon of a single neuron.

nerve impulse A wave of self-propagating action potentials traveling along a nerve fiber.

nervous tissue A tissue composed of neurons and neuroglia.

neural pool A group of interconnected neurons of the central nervous system that performs a single collective function; for example, the vasomotor center of the brainstem and speech centers of the cerebral cortex.

neural tube A dorsal hollow tube in the embryo that develops into the central nervous system. (fig. 20.6)

neuroglia (noo-ROG-lee-uh) All cells of nervous tissue except neurons; cells that perform various supportive and protective roles for the neurons.

neuromuscular junction (NMJ) A synapse between a nerve fiber and a muscle fiber; also called a *motor end plate*. (fig. 7.6)

neuron (NOOR-on) A nerve cell; an electrically excitable cell specialized for producing and transmitting action potentials and secreting chemicals that stimulate adjacent cells.

neurotransmitter A chemical released at the distal end of an axon that stimulates an adjacent cell; for example, acetylcholine, norepinephrine, or serotonin.

neutrophil (NEW-tro-fill) A granulocyte, usually with a multilobed nucleus, that serves especially to destroy bacteria by means of phagocytosis, intracellular digestion, and bactericidal secretions.

nitrogenous base (ny-TRODJ-eh-nus) An organic molecule with a single or double carbon–nitrogen ring that forms one of the building blocks of ATP, other nucleotides, and nucleic acids; the basis of the genetic code. (fig. 2.17)

nitrogenous waste Any nitrogen-containing substance produced as a metabolic waste and excreted in the urine; chiefly ammonia, urea, uric acid, and creatinine.

nociceptor (NO-sih-SEP-tur) A nerve ending specialized to detect tissue damage and produce a sensation of pain; pain receptor.

norepinephrine (nor-EP-ih-NEF-rin) An organic molecule that functions as a neurotransmitter secreted by neurons, especially in the sympathetic nervous system, and as a hormone secreted by the adrenal gland.

nuclear envelope (NEW-clee-ur) A pair of membranes enclosing the nucleus of a cell, with prominent pores allowing traffic of molecules between the nucleus and cytoplasm. (fig. 3.13)

nucleic acid (new-CLAY-ic) An acidic polymer of nucleotides found or produced in the nucleus, functioning in heredity and protein synthesis; of two types, DNA and RNA.

nucleotide (NEW-clee-oh-tide) An organic molecule composed of a nitrogenous base, a monosaccharide, and a phosphate group; examples include ATP and the monomers of nucleic acids.

nucleus (NEW-clee-us) 1. A cell organelle containing DNA and surrounded by a double membrane. 2. A mass of neurons (gray matter) surrounded by white matter of the brain, including the basal nuclei and brainstem nuclei. 3. The positively charged core of an atom, consisting of protons and neutrons. 4. A central structure, such as the nucleus pulposus of an intervertebral disc.

O

olfaction (ole-FAC-shun) The sense of smell.

oocyte (OH-oh-site) In the development of an egg cell, any haploid stage between meiosis I and fertilization.

oogenesis (OH-oh-JEN-eh-sis) The production of a fertilizable egg cell through a series of mitotic and meiotic cell divisions; female gametogenesis.

opposition A movement of the thumb in which it approaches or touches any fingertip of the same hand.

orbit The eye socket of the skull.

organ Any anatomical structure that is composed of at least two tissue types, has recognizable structural boundaries, and has a discrete function different from the structures around it. Many organs are microscopic and many organs contain smaller organs, such as the skin containing numerous microscopic sense organs.

organelle Any structure within a cell that carries out one of its metabolic roles, such as mitochondria, centrioles, endoplasmic reticulum, and the nucleus; an intracellular structure other than the cytoskeleton and inclusions.

osmolarity (OZ-mo-LERR-ih-tee) The molar concentration of dissolved particles in a solution.

osmosis (oz-MO-sis) The net flow of water through a selectively permeable membrane, resulting from either a chemical concentration difference or a mechanical force across the membrane.

osseous (OSS-ee-us) Pertaining to bone.

osteoblast Bone-forming cell that arises from an osteogenic cell, deposits bone matrix, and eventually becomes an osteocyte.

osteoclast Macrophage of the bone surface that dissolves the matrix and returns minerals to the extracellular fluid.

osteocyte A mature bone cell formed when an osteoblast becomes surrounded by its own matrix and entrapped in a lacuna.

osteon A structural unit of compact bone consisting of a central canal surrounded by concentric cylindrical layers of matrix. (fig. 6.3)

osteoporosis (OSS-tee-oh-pore-OH-sis) A degenerative bone disease characterized by a loss of bone mass, increasing susceptibility to spontaneous fractures, and sometimes deformity of the vertebral column; causes include aging, estrogen hyposecretion, and insufficient resistance exercise.

ovulation The release of a mature oocyte by the bursting of an ovarian follicle.

ovum Any stage of the female gamete from the conclusion of meiosis I until fertilization; a primary or secondary oocyte; an egg.

oxidation A chemical reaction in which one or more electrons are removed from a molecule, lowering its energy content; opposite of reduction and always linked to a reduction reaction.

P

pancreatic islet (PAN-cree-AT-ic EYE-let) A small cluster of endocrine cells in the pancreas that secretes insulin, glucagon, somatostatin, and other intercellular messengers; also called *islet of Langerhans*. (fig. 11.10)

papilla (pa-PILL-uh) A conical or nipplelike structure, such as a lingual papilla of the tongue or the papilla of a hair bulb.

papillary (PAP-ih-lerr-ee) 1. Pertaining to or shaped like a nipple, such as the papillary muscles of the heart. 2. Having papillae, such as the papillary layer of the dermis.

parasympathetic nervous system (PERR-uh-SIM-pa-THET-ic) A division of the autonomic nervous system that issues efferent fibers through the cranial and sacral nerves and exerts cholinergic effects on its target organs.

parathyroid hormone (PTH) A hormone secreted by the parathyroid glands that raises blood calcium concentration by stimulating bone resorption by osteoclasts, promoting intestinal absorption of calcium, and inhibiting urinary excretion of calcium.

parietal (pa-RY-eh-tul) 1. Pertaining to a wall, as in the parietal cells of the gastric glands and parietal bone of the skull. 2. The outer or more superficial layer of a two-layered membrane such as the pleura, pericardium, or glomerular capsule. *Compare* visceral. (fig. 1.13)

pathogen Any disease-causing microorganism.

pedicle (PED-ih-cul) A small footlike process, as in the vertebrae and the renal podocytes; also called a *pedicel*.

pelvic cavity The space enclosed by the true (lesser) pelvis, containing the urinary bladder, rectum, and internal reproductive organs.

pelvic girdle A ring of three bones—the sacrum and two hip bones—that enclose the pelvis and link the lower limbs to the vertebral column. (fig. 6.30)

pelvis 1. The basinlike enclosure of the inferior abdominopelvic cavity, composed of the pelvic girdle and its associated body-wall ligaments and muscles. 2. A basinlike structure such as the urine-collecting funnel near the hilum of the kidney. (fig. 16.2)

peptide Any chain of two or more amino acids. *See also* polypeptide, protein.

peptide bond A group of four covalently bonded atoms (a —C=O group bonded to an —NH group) that links two amino acids in a protein or other peptide. (fig. 2.14)

perfusion The amount of blood supplied to a given mass of tissue in a given period of time.

perichondrium (PERR-ih-CON-dree-um) A layer of fibrous connective tissue covering the surface of hyaline or elastic cartilage.

perineum (PERR-ih-NEE-um) The region between the thighs bordered by the coccyx, pubic symphysis, and ischial tuberosities; contains the orifices of the urinary, reproductive, and digestive systems. (fig. 19.9)

periosteum (PERR-ee-OSS-tee-um) A layer of fibrous connective tissue covering the surface of a bone. (fig. 6.3)

peripheral (peh-RIF-eh-rul) Away from the center of the body or of an organ, as in peripheral vision and peripheral blood vessels.

peripheral nervous system (PNS) A subdivision of the nervous system composed of all nerves and ganglia; all of the nervous system except the central nervous system.

peristalsis (PERR-ih-STAL-sis) A wave of constriction traveling along a tubular organ such as the esophagus or ureter, serving to propel its contents.

peritoneum (PERR-ih-toe-NEE-um) A serous membrane that lines the peritoneal cavity of the abdomen and covers the mesenteries and viscera.

phagocytosis (FAG-oh-sy-TOE-sis) A form of endocytosis in which a cell surrounds a foreign particle with pseudopods and engulfs it.

pharynx (FAIR-inks) A muscular passage in the throat at which the respiratory and digestive tracts cross.

phospholipid An amphipathic molecule composed of two fatty acids and a phosphate-containing group bonded to the three carbons of a glycerol molecule; composes most of the molecules of the plasma membrane and other cellular membranes.

pinocytosis (PIN-oh-sy-TOE-sis) A form of endocytosis in which the plasma membrane sinks inward and imbibes droplets of extracellular fluid.

plantar (PLAN-tur) Pertaining to the sole of the foot.

plaque A small scale or plate of matter, such as dental plaque, the fatty plaques of atherosclerosis, and the amyloid plaques of Alzheimer disease.

plasma The noncellular portion of the blood.

plasma membrane The membrane that encloses a cell and controls the traffic of molecules into and out of the cell. (fig. 3.2)

platelet A formed element of the blood known especially for its role in stopping bleeding, but with additional roles in dissolving blood clots, stimulating inflammation, promoting tissue growth and maintenance, and destroying bacteria.

pleura (PLOOR-uh) A double-walled serous membrane that encloses each lung.

plexus A network of blood vessels, lymphatic vessels, or nerves, such as a choroid plexus of the brain or brachial plexus of nerves.

polymer A molecule that consists of a long chain of identical or similar subunits, such as protein, DNA, or starch.

polypeptide Any chain of more than 10 or 15 amino acids. *See also* protein.

polysaccharide (POL-ee-SAC-uh-ride) A polymer of simple sugars; for example, glycogen, starch, and cellulose.

polyuria (POL-ee-YOU-ree-uh) Excessive output of urine.

popliteal (pop-LIT-ee-ul) Pertaining to the posterior aspect of the knee.

positron emission tomography (PET) A method of producing a computerized image of the physiological state of a tissue using injected radioisotopes that emit positrons.

posterior Near or pertaining to the back or spinal side of the body; dorsal.

posterior root The branch of a spinal nerve that enters the posterior side of the spinal cord and carries afferent (sensory) nerve fibers; often called *dorsal root*. (fig. 8.17)

postganglionic (POST-gang-glee-ON-ic) Pertaining to a neuron that conducts signals from a ganglion to a more distal target organ.

postsynaptic Pertaining to a neuron or other cell that receives signals from the presynaptic neuron at a synapse. (fig. 8.12)

potential A difference in electrical charge from one point to another, especially on opposite sides of a plasma membrane; usually measured in millivolts.

preembryo A developing individual from the time of fertilization to the time, at 16 days, when the three primary germ layers have formed. *Compare* conceptus, embryo, fetus.

preganglionic (PRE-gang-glee-ON-ic) Pertaining to a neuron that conducts signals from the central nervous system to a ganglion.

presynaptic (PRE-sih-NAP-tic) Pertaining to a neuron that conducts signals to a synapse. (fig. 8.12)

primary germ layers The ectoderm, mesoderm, and endoderm; the three tissue layers of an early embryo from which all later tissues and organs arise.

prime mover The muscle that produces the most force in a given joint action; also called the *agonist*.

programmed cell death *See* apoptosis.

pronation (pro-NAY-shun) A rotational movement of the forearm that turns the palm downward or posteriorly. (fig. 6.41)

proprioception (PRO-pree-oh-SEP-shun) The nonvisual perception, usually subconscious, of the position and movements of the body, resulting from input from proprioceptors and the vestibular apparatus of the inner ear.

proprioceptor (PRO-pree-oh-SEP-tur) A sensory receptor of the muscles, tendons, and joint capsules that detects muscle contractions and joint movements.

prostaglandin (PROSS-ta-GLAN-din) A modified fatty acid with a five-sided carbon ring in the middle of a hydrocarbon chain, playing a variety of roles in inflammation, neurotransmission, vasomotion, reproduction, and metabolism.

prostate (PROSS-tate) A male reproductive gland that encircles the urethra immediately inferior to the bladder and contributes to the semen. (fig. 19.1)

protein A large polypeptide; while criteria for a protein are somewhat subjective and variable, polypeptides over 50 amino acids long are generally classified as proteins.

proximal Relatively near a point of origin or attachment; for example, the shoulder is proximal to the elbow. *Compare* distal.

pseudopod (SOO-doe-pod) A temporary cytoplasmic extension of a cell used for locomotion (ameboid movement) and phagocytosis. (fig. 3.6)

pseudostratified columnar A type of epithelium with tall columnar cells reaching the free surface and shorter basal cells that do not reach the surface, but with all cells resting on the basement membrane; creates a false appearance of stratification. (fig. 4.7)

pulmonary circuit A route of blood flow that supplies blood to the pulmonary alveoli for gas exchange and then returns it to the heart; all blood vessels between the right ventricle and the left atrium of the heart.

pyrogen (PY-ro-jen) A fever-producing agent.

R

ramus (RAY-mus) An anatomical branch, as in a nerve or of the pubis.

receptor 1. A cell or organ specialized to detect a stimulus, such as a taste cell or the eye. 2. A protein molecule that binds and responds to a chemical such as a hormone, neurotransmitter, or odor molecule.

receptor-mediated endocytosis A process in which certain molecules in the extracellular fluid bind to receptors in the plasma membrane, these receptors gather together, the membrane sinks inward at that point, and the molecules become incorporated into vesicles in the cytoplasm.

reduction A chemical reaction in which one or more electrons are added to a molecule, raising its energy content; opposite of oxidation and always linked to an oxidation reaction.

reflex A stereotyped, automatic, involuntary response to a stimulus; includes somatic reflexes, in which the effectors are skeletal muscles, and visceral (autonomic) reflexes, in which the effectors are usually visceral muscle, cardiac muscle, or glands.

reflex arc A simple neural pathway that mediates a reflex; involves a receptor, an afferent nerve fiber, usually one or more interneurons, an efferent nerve fiber, and an effector.

refractory period 1. A period of time after a nerve or muscle cell has responded to a stimulus in which it cannot be reexcited by a threshold stimulus. 2. A period of time after male orgasm when it is not possible to reattain erection or ejaculation.

regeneration Replacement of damaged tissue with new tissue of the original type. *Compare* fibrosis.

regulatory T cells A class of T lymphocytes that moderate the activity of cytotoxic T cells to reduce the immune response after a pathogen is defeated and to prevent autoimmune diseases.

renin (REE-nin) An enzyme secreted by the kidneys in response to hypotension; converts the plasma protein angiotensinogen to angiotensin I, leading indirectly to a rise in blood pressure.

repolarization Reattainment of the resting membrane potential after a nerve or muscle cell has depolarized.

residue Any one of the amino acids in a protein or other peptide.

resistance 1. A nonspecific ability to ward off infection or disease regardless of whether the body has been previously exposed to it. *Compare* immunity. 2. A force that opposes the flow of a fluid such as air or blood. 3. A force, or load, that opposes the action of a muscle or lever.

resting membrane potential (RMP) A stable voltage across the plasma membrane of an unstimulated nerve or muscle cell.

reticular cell (reh-TIC-you-lur) A delicate, branching phagocytic cell found in the reticular connective tissue of the lymphatic organs.

reticular fiber A fine, branching collagen fiber coated with glycoprotein, forming part of the framework of lymphatic organs and some other tissues and organs.

reticular tissue A connective tissue composed of reticular cells and reticular fibers, found in bone marrow, lymphatic organs, and in lesser amounts elsewhere.

ribosome A granule found free in the cytoplasm or attached to the rough endoplasmic reticulum and nuclear envelope, composed of ribosomal RNA and enzymes; specialized to read the nucleotide sequence of messenger RNA and assemble a corresponding sequence of amino acids to make a protein.

risk factor Any environmental factor or characteristic of an individual that increases one's chance of developing a particular disease; includes such intrinsic factors as age, sex, and race and such extrinsic factors as diet, smoking, and occupation.

rostral Relatively close to the forehead, especially in reference to structures of the brain and spinal cord; for example, the frontal lobe is rostral to the parietal lobe. *Compare* caudal.

ruga (ROO-ga) **1.** An internal fold or wrinkle in the mucosa of a hollow organ such as the stomach and urinary bladder; typically present when the organ is empty and relaxed but not when the organ is full and stretched. **2.** Tissue ridges in such locations as the hard palate and vagina. (fig. 17.8)

S

saccule (SAC-yule) A saclike receptor in the inner ear with a vertical patch of hair cells, the macula sacculi; senses the orientation of the head and responds to vertical acceleration, as when riding in an elevator or standing up. (fig. 10.9)

sagittal plane (SADJ-ih-tul) Any plane that extends from anterior to posterior and cephalic to caudal and that divides the body into right and left portions. *Compare* median plane.

sarcomere (SAR-co-meer) In skeletal and cardiac muscle, the portion of a myofibril from one Z disc to the next, constituting one contractile unit. (fig. 7.4)

sarcoplasmic reticulum (SR) The smooth endoplasmic reticulum of a muscle cell, serving as a calcium reservoir. (fig. 7.2)

scanning electron microscope (SEM) A microscope that uses an electron beam in place of light to form high-resolution, three-dimensional images of the surfaces of objects; capable of much higher magnifications than a light microscope.

sclerosis (scleh-RO-sis) Hardening or stiffening of a tissue, as in multiple sclerosis of the central nervous system or atherosclerosis of the blood vessels.

sebum (SEE-bum) An oily secretion of the sebaceous glands that keeps the skin and hair pliable.

second messenger A chemical that is produced within a cell (such as cAMP) or that enters a cell (such as calcium ions) in response to the binding of a messenger to a membrane receptor, and that triggers a metabolic reaction in the cell.

secondary active transport A mechanism in which solutes are moved through a plasma membrane by a carrier that does not itself use ATP but depends on a concentration gradient established by an active transport pump elsewhere in the cell.

secretion **1.** A chemical released by a cell to serve a physiological function, such as a hormone or digestive enzyme. **2.** The process of releasing such a chemical, often by exocytosis. *Compare* excretion.

selectively permeable membrane A membrane that allows some substances to pass through while excluding others; for example, the plasma membrane and dialysis membranes.

semicircular ducts Three ring-shaped, fluid-filled tubes of the inner ear that detect angular acceleration of the head; each is enclosed in a bony passage called the semicircular canal. (fig. 10.10)

semilunar valve A valve that consists of crescent-shaped cusps, including the aortic and pulmonary valves of the heart. (fig. 13.4)

senescence (seh-NESS-ense) Degenerative changes that occur with age.

sensation Conscious perception of a stimulus; pain, taste, and color, for example, are not stimuli but sensations resulting from stimuli.

sensory nerve fiber An axon that conducts information from a receptor to the central nervous system; an afferent nerve fiber.

sensory neuron A neuron that transmits signals from a sense organ or other receptor to the brain or spinal cord; its axon is an afferent nerve fiber, carrying signals toward the CNS.

serosa (seer-OH-sa) A membrane such as the peritoneum, pleura, or pericardium that lines a body cavity or covers the external surfaces of the viscera; composed of a simple squamous epithelium and a thin layer of areolar connective tissue. Also called a *serous membrane*. (fig. 4.30b)

serous fluid (SEER-us) A watery, low-protein fluid similar to blood serum, formed as a filtrate of the blood or tissue fluid or as a secretion of serous gland cells; moistens the serous membranes.

sex chromosomes The X and Y chromosomes, which determine a person's sex.

shock In the circulatory sense, a state of cardiac output that is insufficient to meet the body's physiological needs, with consequences ranging from fainting to death.

sign An objective manifestation of illness that any observer can see, such as cyanosis or edema. *Compare* symptom.

simple epithelium An epithelium in which all cells rest directly on the basement membrane; includes simple squamous, cuboidal, and columnar types, and pseudostratified columnar. (fig. 4.3)

sinus **1.** An air-filled space in the cranium. **2.** A modified, relatively dilated vein that lacks smooth muscle and is incapable of vasomotion, such as the dural sinuses of the cerebral circulation and coronary sinus of the heart. **3.** A small fluid-filled space in an organ such as the spleen and lymph nodes. **4.** Pertaining to the sinoatrial node of the heart, as in *sinus rhythm*.

somatic **1.** Pertaining to the body as a whole. **2.** Pertaining to the skin, bones, and skeletal muscles as opposed to the viscera. **3.** Pertaining to cells other than germ cells.

somatic nervous system A division of the nervous system that includes afferent fibers mainly from the skin, muscles, and skeleton and efferent fibers to the skeletal muscles. *Compare* autonomic nervous system.

somatosensory **1.** Pertaining to widely distributed *general senses* in the skin, muscles, tendons, joint capsules, and viscera, as opposed to the *special senses* found in the head only. **2.** Pertaining to the cerebral cortex of the postcentral gyrus, which receives input from such receptors. Also called *somesthetic*.

somite One segment in a linear series of mesodermal masses that form on each side of the embryonic neural tube and give rise to trunk muscles, vertebrae, and dermis. (fig. 20.7a, b)

spermatogenesis (SPUR-ma-toe-JEN-eh-sis) The production of sperm cells through a series of mitotic and meiotic cell divisions; male gametogenesis.

spermatozoon (SPUR-ma-toe-ZOE-on) A sperm cell.

sphincter (SFINK-tur) A ring of muscle that opens or closes an opening or passageway; found, for example, in the eyelids, inferior to the urinary bladder, and at the beginning of a blood capillary.

spinal nerve Any of the 31 pairs of nerves that arise from the spinal cord and pass through the intervertebral foramina.

spindle **1.** An elongated structure that is thick in the middle and tapered at the ends (fusiform). **2.** A football-shaped complex of microtubules that guide the movement of chromosomes in mitosis and meiosis. (fig. 3.18) **3.** A stretch receptor in the skeletal muscles.

spine **1.** The vertebral column. **2.** A pointed process or sharp ridge on a bone, such as the styloid process of the cranium and spine of the scapula.

squamous (SKWAY-mus) Having a flat, scaly shape; pertains especially to a class of epithelial cells. (figs. 4.2–4.4)

stem cell Any undifferentiated cell that can divide and differentiate into more functionally specific cell types such as blood cells and germ cells.

stenosis (steh-NO-sis) The narrowing of a passageway such as a heart valve or uterine tube; a permanent, pathological constriction as opposed to normal physiological constriction of a passageway.

steroid (STERR-oyd, STEER-oyd) A lipid molecule that consists of four interconnected carbon rings; cholesterol and several of its derivatives.

stimulus A chemical or physical agent in a cell's surroundings that is capable of creating a physiological response in the cell; especially agents detected by sensory cells, such as chemicals, light, and pressure.

stratified epithelium A type of epithelium in which some cells rest on top of others instead of on the basement membrane; includes stratified squamous, cuboidal, and columnar types, and transitional epithelium. (fig. 4.3)

stress 1. A mechanical force applied to any part of the body; important in stimulating bone growth, for example. 2. A condition in which any environmental influence disturbs the homeostatic equilibrium of the body and stimulates a physiological response, especially involving the increased secretion of certain adrenal hormones.

stroke volume The volume of blood ejected by one ventricle of the heart in one contraction.

subcutaneous (SUB-cue-TAY-nee-us) Beneath the skin.

substrate 1. A chemical that is acted upon and changed by an enzyme. 2. A chemical used as a source of energy, such as glucose and fatty acids.

sulcus (SUL-cuss) A groove in the surface of an organ, as in the cerebrum or heart.

summation A phenomenon in which multiple muscle twitches occur so closely together that a muscle fiber cannot fully relax between twitches but develops more tension than a single twitch produces. (fig. 7.10)

superficial Relatively close to the surface; opposite of deep. For example, the ribs are superficial to the lungs.

superior Higher than another structure or point of reference from the perspective of anatomical position; for example, the lungs are superior to the diaphragm.

supination (SOO-pih-NAY-shun) A rotational movement of the forearm that turns the palm so that it faces upward or forward. (fig. 6.41)

surfactant (sur-FAC-tent) A chemical that reduces the surface tension of water and enables it to penetrate other substances more effectively. Examples include pulmonary surfactant and bile acids.

sympathetic nervous system A division of the autonomic nervous system that issues efferent fibers through the thoracic and lumbar nerves and usually exerts adrenergic effects on its target organs; includes a chain of ganglia adjacent to the vertebral column, and the adrenal medulla.

symphysis (SIM-fih-sis) A joint in which two bones are held together by fibrocartilage; for example, between bodies of the vertebrae and between the right and left pubic bones.

symptom A subjective manifestation of illness that only the ill person can sense, such as dizziness or nausea. *Compare* sign.

synapse (SIN-aps) 1. A junction at the end of an axon where it stimulates another cell. 2. A gap junction between two cardiac or smooth muscle cells at which one cell electrically stimulates the other; called an *electrical synapse*.

synaptic cleft A narrow space between the axon terminal and the adjacent cell, across which a neurotransmitter diffuses. (fig. 8.12)

synaptic vesicle A spheroid organelle in an axon terminal containing neurotransmitter. (fig. 8.12)

syndrome A suite of related signs and symptoms stemming from a specific pathological cause.

synergist (SIN-ur-jist) A muscle that works with the prime mover (agonist) to contribute to the same overall action at a joint.

synergistic An effect in which two agents working together (such as two hormones) exert an effect that is greater than the sum of their separate effects. For example, neither follicle-stimulating hormone nor testosterone alone stimulates significant sperm production, but the two of them together stimulate production of vast numbers of sperm.

synovial fluid (sih-NO-vee-ul) A lubricating fluid similar to egg white in consistency, found in the synovial joint cavities and bursae.

synovial joint A point where two bones are separated by a narrow, encapsulated space filled with lubricating synovial fluid; most such joints are relatively mobile. Also called a *diarthrosis*.

systemic (sis-TEM-ic) Widespread or pertaining to the body as a whole, as in the systemic circulation.

systemic circuit All blood vessels that convey blood from the left ventricle to all organs of the body and back to the right atrium of the heart; all of the cardiovascular system except the heart and pulmonary circuit.

systole (SIS-toe-lee) The contraction of any heart chamber; ventricular contraction unless otherwise specified.

systolic pressure (sis-TOLL-ic) The peak arterial blood pressure measured during ventricular systole.

T

T cell A type of lymphocyte involved in nonspecific defense, humoral immunity, and cellular immunity; occurs in several forms including helper, cytotoxic, and regulatory T cells and natural killer cells.

T lymphocyte A lymphocyte that matures in the thymus and functions in innate immunity, cellular immunity, and humoral immunity; includes cytotoxic, helper, and memory T cells.

T tubule Transverse tubule; a tubular extension of the plasma membrane of a muscle cell that conducts action potentials into the sarcoplasm and excites the sarcoplasmic reticulum. (fig. 7.2)

target cell A cell acted upon by a nerve fiber, hormone, or other chemical messenger.

tarsal Pertaining to the ankle (tarsus).

tendon A collagenous band or cord associated with a muscle, usually attaching it to a bone and transferring muscular tension to it.

tetanus 1. A state of sustained muscle contraction produced by summation as a normal part of contraction; also called *tetany*. 2. Spastic muscle paralysis produced by the toxin of the bacterium *Clostridium tetani*.

thalamus (THAL-uh-muss) The largest part of the diencephalon, located immediately inferior to the corpus callosum and bulging into each lateral ventricle; a point of synaptic relay of nearly all signals passing from lower levels of the CNS to the cerebrum. (figs. 9.2, 9.5)

theory An explanatory statement, or set of statements, that concisely summarizes the state of knowledge on a phenomenon and provides direction for further study; for example, the fluid-mosaic theory of the plasma membrane and the sliding filament theory of muscle contraction.

thermogenesis The production of heat, for example, by shivering or by the action of thyroid hormones.

thermoreceptor A neuron specialized to respond to heat or cold, found in the skin and mucous membranes, for example.

thermoregulation Homeostatic regulation of the body temperature within a narrow range by adjustments of heat-promoting and heat-losing mechanisms.

thorax A region of the trunk between the neck and the diaphragm; the chest.

threshold The minimum voltage to which the plasma membrane of a nerve or muscle cell must be depolarized before it produces an action potential.

thrombosis (throm-BO-sis) The formation or presence of a thrombus.

thrombus A clot that forms in a blood vessel or heart chamber; may break free and travel in the bloodstream as a thromboembolus. *Compare* embolus.

thyroid hormone Either of two similar hormones, thyroxine and triiodothyronine, synthesized from iodine and tyrosine.

thyroid-stimulating hormone (TSH) A hormone of the anterior pituitary gland that stimulates the thyroid gland; also called *thyrotropin*.

thyroxine (thy-ROC-seen) The thyroid hormone secreted in greatest quantity, with four iodine atoms; also called *tetraiodothyronine*.

tight junction A region in which adjacent cells are bound together by fusion of the outer phospholipid layer of their plasma membranes; forms a zone that encircles each cell near its apical pole and reduces or prevents flow of material between cells. (fig. 3.7)

tissue An aggregation of cells and extracellular materials, usually forming part of an organ and performing some discrete function for it; the four primary classes are epithelial, connective, muscular, and nervous tissue.

trabecula (tra-BEC-you-la) A thin plate or layer of tissue, such as the calcified trabeculae of spongy bone or the fibrous trabeculae that subdivide a gland. (fig. 6.3)

trachea (TRAY-kee-uh) A cartilage-supported tube from the inferior end of the larynx to the origin of the main bronchi; conveys air to and from the lungs; colloquially called the *windpipe.*

translation The process in which a ribosome reads an mRNA molecule and synthesizes the protein specified by the genetic code.

transmission electron microscope (TEM) A microscope that uses an electron beam in place of light to form high-resolution, two-dimensional images of ultrathin slices of cells or tissues; capable of extremely high magnification.

triglyceride (try-GLISS-ur-ide) A lipid composed of three fatty acids joined to a glycerol; also called a *neutral fat.* (fig. 2.10)

trunk 1. That part of the body excluding the head, neck, and appendages. 2. A major blood vessel, lymphatic vessel, or nerve that goes for only a short distance and gives rise to smaller branches; for example, the pulmonary trunk and spinal nerve trunks.

tunic (TOO-nic) A layer that encircles or encloses an organ, such as the tunics of a blood vessel or eyeball.

tympanic membrane The eardrum.

U

ultraviolet radiation Invisible, ionizing, electromagnetic radiation with shorter wavelength and higher energy than violet light; causes skin cancer and photoaging of the skin but is required in moderate amounts for the synthesis of vitamin D.

unmyelinated (un-MY-eh-lih-nay-ted) Lacking a myelin sheath.

urea (you-REE-uh) A nitrogenous waste produced from two ammonia molecules and carbon dioxide; the most abundant nitrogenous waste in the blood and urine.

uterine tube A duct that extends from the ovary to the uterus and conveys an egg or conceptus to the uterus; also called *fallopian tube* or *oviduct.*

utricle (YOU-trih-cul) A saclike receptor in the inner ear with a nearly horizontal patch of hair cells, the macula utriculi; senses the orientation of the head and responds to horizontal acceleration, as when riding in a car at varying speeds. (fig. 10.9)

V

varicose vein A vein that has become permanently distended and convoluted due to a loss of competence of the venous valves; especially common in the lower limb, esophagus, and anal canal (where they are called *hemorrhoids*).

vas (vass) (plural, *vasa*) A vessel or duct.

vascular Pertaining to blood vessels.

vasoconstriction (VAY-zo-con-STRIC-shun) The narrowing of a blood vessel due to muscular constriction of its tunica media.

vasodilation (VAY-zo-dy-LAY-shun) The widening of a blood vessel due to relaxation of the muscle of its tunica media and the outward pressure of the blood exerted against the wall.

vasomotion (VAY-zo-MO-shun) Collective term for vasoconstriction and vasodilation.

vasomotor center A nucleus in the medulla oblongata that transmits efferent signals to the blood vessels and regulates vasomotion.

ventral Pertaining to the front of the body, the regions of the chest and abdomen; anterior.

ventral root *See* anterior root.

ventricle (VEN-trih-cul) A fluid-filled chamber of the brain or heart.

venule (VEN-yool) The smallest type of vein, receiving drainage from capillaries.

vertebra (VUR-teh-bra) One of the bones of the vertebral column.

vertebral column (VUR-teh-brul) A posterior series of usually 33 vertebrae; encloses the spinal cord, supports the skull and thoracic cage, and provides attachment for the limbs and postural muscles. Also called the *spine* or *spinal column.*

vesicle (VESS-ih-cul) A fluid-filled tissue sac or an organelle such as a synaptic or secretory vesicle.

vesicular transport The movement of particles or fluid droplets through the plasma membrane by the process of endocytosis or exocytosis.

viscera (VISS-er-uh) (singular, *viscus*) The organs contained in the body cavities, such as the brain, heart, lungs, stomach, intestines, and kidneys.

visceral (VISS-er-ul) 1. Pertaining to the viscera. 2. The inner or deeper layer of a two-layered membrane such as the pleura, pericardium, or glomerular capsule. *Compare* parietal. (fig. 1.13)

visceral muscle A form of smooth muscle found in the walls of blood vessels and the digestive, respiratory, urinary, and reproductive tracts.

viscosity The resistance of a fluid to flow; the thickness or stickiness of a fluid.

voluntary muscle Muscle that is usually under conscious control; skeletal muscle.

vulva The female external genitalia; the mons, labia majora, and all superficial structures between the labia majora.

W

water balance *See* fluid balance.

white matter White myelinated nervous tissue deep to the cortex of the cerebrum and cerebellum and superficial to the gray matter of the spinal cord. (figs. 9.3, 9.4)

X

X chromosome The larger of the two sex chromosomes; males have one X chromosome, and females have two, in each somatic cell.

X-ray 1. A high-energy, penetrating electromagnetic ray with wavelengths in the range of 0.1 to 10 nm; used in diagnosis and therapy. 2. A photograph made with X-rays; radiograph.

Y

Y chromosome Smaller of the two sex chromosomes, found only in males and having little genetic function except development of the testis.

yolk sac An embryonic membrane that encloses the yolk in vertebrates that lay eggs; serves in humans as the origin of the first blood and germ cells.

Z

zygomatic arch An arch of bone anterior to the ear, formed by the zygomatic processes of the temporal, frontal, and zygomatic bones; origin of the masseter muscle.

zygote A single-celled, fertilized egg.

Index

Note: Page numbers in *italics* indicate illustrations and their captions; page numbers followed by *t* indicate material in tables.

A

A bands, *200*, 202, *202*
ABCD rule, of skin cancer, 132
Abdominal aorta, *445*, 446, 452, *453*
Abdominal cavity, 18, *18*, 18*t*
Abdominal region, 16, *18*
  arteries of, 452, *453*
  muscles, 227–228, 229*t*
  quadrants of, 16, *16*
  regions of, 16, *17*
  veins of, 452–455, *454*
Abdominopelvic cavity, *18*, 18–20, 18*t*
Abduction, 184, *185*
ABO blood group, 394, *394*
Abscess, 476
Absorption, 557, 580–584
  carbohydrates, 580, *581*
  fats, *581*, 582–583
  in large intestine, 586
  lipid, 465
  proteins, 580–582, *581*
  transdermal, 129
Absorptive cells (enterocytes), 577, *578*
Absorptive (fed) state, 611
Accessory organs, 557, *557*
  of skin, *127*, 134–138
Acclimatization, 515
Acetylcholine (ACh), 203, *204*, 206, 262, 312, 568–569, 575
  in cardiac physiology, 425, 429
Acetylcholinesterase (AChE), 203, *204*
Acetyl groups, 609
ACh receptors, 203, *204*
Acid–base balance, 496, 548–550
Acidic solutions, 39
Acidity and alkalinity (pH), A–11
Acidosis, 429, 516, 548
Acid(s), 39, 39–40. *See also specific acids*
Acinus (acini), of exocrine gland, 116, *116*
Acromial end, of clavicle, 171, *171*
Acromial region, *17*
Acromion, *171*, 172
Acrosome, 647
Actin, 75, *201*, 202
Action potential, 206
  in cardiac muscle, 425
  generating, 259, *259*
Active immunity, 487–488
Active site, 49
Active transport, *70*, 72
Acute gastrointestinal illness, 571
Acute leukemia, 400
Adam's apple, 446

Adaptation, bipedalism, 164, 189, 191
Adaptive immunity, 477–482
  memory, 478
  specificity, 478
Adduction, 184, *185*
Adenine, 50
Adenocarcinoma, 60
Adenosine, caffeine and, 312, *313*
Adenosine diphosphate (ADP), 52
Adenosine triphosphate (ATP), 52, *52*
  as energy currency, 52
  in erythrocytes, 392
  in membrane transport, 72, 73
  in mitochondria, 80
  in muscle metabolism, 211–213
  production during exercise, 212, *212*
  production of, 80
Adhesion, 38, 474
  cell-adhesion molecules, 65, *65*
  water, 38
Adipocytes, 24, 105, 108, *108*, 108*t*
Adipose tissue, 44, 105, *108*, 108–109, 109*t*
  aging and, 380
  breast, *108*
  brown fat, 108, 109
  endocrine function of, 376
  hypertrophy of, 119
  subcutaneous, 130
  white fat, 108
Adolescence, 634
Adrenal cortex, 374–375
  hormones of, 374–375
  layers of, *374*, 374–375
Adrenal glands, 307, *358*, 373–375
  embryonic development of, 374
  gross anatomy of, 374, *374*
  histology of, 374, *374*
Adrenaline. *See* Epinephrine
Adrenal medulla, 307, 374, *374*
  embryonic development of, 374
Adrenocorticotropic hormone (ACTH), 366, 367*t*, *368*, 375
  aging and, 380
  in stress response, 379
Adult stem cells, 115
Adventitia, 558, *558*
Aerobic fermentation, 603
Aerobic respiration, 211, 212
  matrix reactions, 604–605
  membrane reactions, 605–606
Afferent arterioles, 526, *527*
Age, as physiological variation, 13
Agglutination, 394, 404
Aggregated lymphoid nodules, 576, *577*
Aging
  of circulatory system, 457
  of digestive system, 588
  and endocrine system, 380
  of immune systems, 490

  and integumetary system, 139
  of lymphatic system, 490
  of muscular system, 243
  of nervous system, 313
  of reproductive system, 663
  of respiratory system, 517
  of sense organs, 351
  of skeletal system, 192
  of urinary system, 550
Agranulocytes, 396, *397*, 398, *399*
AIDS, 489
Air, composition of, 512–513
Alarm reaction, 379
Albinism, 133
Albumin, 388–389, 436
Aldosterone, 375, 533, 538
  in blood pressure control, 438
  in renal regulation, 375
  in stress response, 379
Alimentary canal. *See* Digestive tract
Alkaline solutions, 39
Alkalosis, 516, 548
Allergies, 479
Alpha (α) cells, of pancreatic islets, *372*, 373
Alpha (α) chains, of hemoglobin, 391
Alpha (α) helix, 48, *48*
Alveolar gas exchange, *513*, 513–514
Alveolar sacs, *442*
Alveolus, 504, *505*, *561*
Alzheimer disease (AD), 198, 264, *264*
Amino acid(s), *47*, 47–48, 612
  monoamine synthesis from, 360, *360*
Amino acid sequence, 48, 81–83
Amino group, 47
Ammonia, 39
Amoeba, 68
Amphetamine, 428
Amphiarthrosis. *See* Cartilaginous joints
Amphipathic substances, 45
Amygdala, 294
Amylase, 49
Anabolic steroids, 51
Anabolism, 55, 55*t*, 603
Anaerobic fermentation, 211–212, 603
  in erythrocytes, 392
Anaerobic respiration, 604
Anal canal, 585, *585*
Anaphase, in mitosis, 87, 86, 88
Anaphylactic shock, 479
Anastomoses, 434, *434*
  arterial, 434
  arteriovenous, *432*, 434, *434*
  cerebral, 447
  venous, 434
Anatomical planes, 15, *15*, *16*
Anatomical position, 15, *15*
  planes, 15, *15*, *16*
Anatomical terminology, 21, 24–26, 25*t*
  analyzing, 24–25

I-1

Anatomical terminology (*continued*)
   directional, 25–26, 25*t*
   precision in, importance in, 21, 24
   singular and plural forms in, 25, 25*t*
Anatomy, 2–7
   comparative, 3
   definition of, 3
   gross, 3
   hierarchy of complexity in, *14,* 14–15
   radiologic, 4
   regional, 3
   surface, 3
   systemic, 3
   variations in, 6, *7,* 9
Androgen-insensitivity syndrome (AIS), 378
Androgens, 375
Andropause, 634
Anemia, 399–400
   hemolytic, 400
   hemorrhagic, 400
   iron-deficiency, 399
   pernicious, 399
   sickle-cell, 392
Angina pectoris, 421
Angiogenesis, 437
Angiography, *4, 410*
Angioplasty, 421
Angiotensin-converting enzyme (ACE), 438
Angiotensin II, 376, 438, 533
Angiotensinogen, 376
Angle (bone), mandible, 163
Anions, 33
Ankle, 17, *17*
   bones of, 178–180, *179*
Ankylosis, 155, *155*
Anomic aphasia, 300
Antagonist, *221,* 222
Antagonistic effects, of hormones, 361
Antagonistic pair, 222
Antebrachial region (antebrachium), 17, *17. See also* Forearm
Anterior, 15, 25–26, 25*t*
Anterior cruciate ligament (ACL), 191
Anterior fontanel, *164, 164*
Anterior interventricular sulcus, *415,* 417
Anterior pituitary gland
   anatomy of, *364, 365*
   embryonic development of, *364, 364*
   hormones of, *365,* 366
   hypothalamic control of, *367, 368*
Antibodies, 387, 480–481, 481*t*
   blood type, 394, *394*
Anticodon, 83
Antidiuretic hormone (ADH), 364, *365,* 367, 533, 539–540
   in blood pressure control, 438
Antigen-presenting cells (APCs), 481–482, *482*
Antigens, 478
   blood type, 394, *394*
Antihemophiliac factor, 404
Antioxidants, 34
Antrum, 650
Anulus fibrosus, *166,* 168–169
Aorta, 412, *412, 413, 415, 416, 443,* 445–446
   abdominal, *445,* 446, 452, *453*
   anatomical variations in, *7*
   aortic arch, *415, 443,* 445, *445, 449*
   ascending, *415,* 445, *445*
   descending, *445,* 445–446, *449,* 452, *453*
   thoracic, *445,* 446, 447
Aortic hiatus, *445, 453*
Aortic valve, *416,* 417
Apex, of heart, 412, *413, 415*
Aphasia, 299, 300
Apical surface, of epithelial tissue, 97
Apocrine glands, 117, *117*
Apocrine sweat glands, 117, *127, 135,* 136, *137,* 137–138
Aponeurosis, 220
Apoptosis (programmed cell death), 80, 120–121
Appendages. *See* Lower limb; Upper limb; *specific anatomy*
Appendages (accessory organs), of skin, *127,* 134–138
Appendicular region, *16,* 16–17
   blood vessels of, 450–451
   muscles of, *230–231,* 230–235, 240*t*–242*t*
Appendicular skeleton, 157, *157, 171,* 171–180
Appendix, 9, 584, *585*
Appetite regulation, 614–615
Apulla, 334, *335*
Aqueous humor, *341,* 341–342, 341–342, *342*
Arachnoid granulations, 287
Arachnoid mater, 270, 284, *285*
Arches
   aortic, *415, 443,* 445, *445, 449*
   fallen, 180
   zygomatic, *160,* 161, 163
Areola, 641
Areolar glands, 641
Areolar tissue, 105, *106,* 106*t*
Arginine, *47*
Aristotle, 8, 411
Arm, 17, *17*
Armpit, *17*
Arrhythmias, 428, 457
Arterial anastomosis, 434
Arteries, 430–432
   abdominopelvic region, 452, *453*
   arcuate, 526, *527*
   axillary, *443, 446,* 447, *449, 449,* 450, 451
   basilar, *446,* 447
   blood flow in, 434
   blood pressure in, 435
   brachial, 442, *443, 449,* 450, 451
   branches of, 445
   bronchial, 441, *442, 449,* 505
   carotid
      common, *443,* 446, *446, 449*
      common, left, 445, *445,* 446
      common, right, 445, *445,* 446
      external, *443,* 446, *446*
      internal, 161, *443, 446,* 447
   celiac trunk, 452, *453*
   cerebral, 447
      anterior, *446,* 447
      arterial circle, *446,* 447
      middle, *446,* 447
      posterior, *446,* 447
   conducting, *431, 431*
   coronary, *419,* 445, *445*
      angiogram of, *410*
      anterior interventricular branch of, 418, *419*
      circumflex branch of, 418, *419*
      left, 418, *419*
      left anterior descending branch of, 418
      left marginal branch of, 418, *419*
      posterior interventricular branch of, 419, *419*
      right, 418–419, *419*
      right marginal branch of, 418–419, *419*
   cortical radiate, 526, *527*
   disorders of
      arteriosclerosis, 420, 457
      atheirosclerosis, 378, 420, 457
      coronary artery disease, 420–421
   distributing, *431,* 431–432
   dorsal pedal, *443,* 455
   elastic, *431, 431*
   femoral, 442, *443,* 455
   femoral, deep, *443*
   fibular, *443*
   gastric, left, 452, *453*
   gonadal, *443,* 452, *453*
   great, 412
   head and neck, *446,* 446–447
   hepatic, common, 452, *453*
   iliac
      common, *443,* 452, *453,* 455
      external, *443,* 452, *453,* 455
      internal, *443,* 452, *453,* 455
   internal carotid, 287
   large, *431, 431*
   lower limb, 455
   major systemic, *443*
   medium, *431,* 431–432
   mesenteric
      inferior, *443,* 452, *453*
      superior, *443,* 452, *453*
   muscular, *431,* 431–432
   nomenclature for, 445
   ovarian, 452, *453*
   popliteal, *443,* 455
   pulmonary, 412, *412,* 441, *442*
      left, *415, 416, 442*
      right, *415, 442*
   radial, *443, 449,* 451
   renal, *443,* 452, *453,* 526, *527*
   resistance (small), 432
   splenic, 452, *453*
   subclavian, *443,* 445, *446,* 447, *449, 449,* 450, 451
      left, 445, *445, 449,* 451
      right, 445, *445, 449,* 451
   systemic circuit, 412, *412,* 442–456, *443*
   testicular, 452, *453*
   thorax, 447, *449, 449*
   thyroid, superior, *370*
   tibial
      anterior, *443,* 455
      posterior, *443,* 455
   tunics of, 430
   ulnar, 442, *443, 449,* 451
   upper limb, *449,* 450–451
   vertebral, 167, 287, *443, 446, 446, 449*
Arterioles, *431,* 432, *432,* 433
Arteriosclerosis, 420, 457
Arteriovenous anastomosis, *432,* 434, *434*
Arthritis, 154–155, *155*
   gouty, 155
   osteoarthritis, 154–155
   rheumatoid, 155, *155*
Arthrology, 144
Arthroscopy, 191

Articular cartilage, *148,* 149, 151
Ascending aorta, *415,* 445, *445*
Ascending colon, 584, *585*
Ascorbic acid (vitamin C), 34, 153
Aspirin, 40, 570
Asthma, 504
Astigmatism, 350
Astrocytes, 254, *254*
Atheromas, 420
Atherosclerosis, 378, 420, 457
Atherosclerotic plaques, 420, *420,* 457
Athletes
    bone density and mass in, 152
    genetic predisposition, 213
    Marfan syndrome in, 108
    steroid use by, 51
Atlantoaxial joint, *167*
Atlas (C1 vertebra), *165,* 167
Atmospheric (barometric) pressure, 506, *507*
alpha α-tocopherol (vitamin E), 34
Atomic structure, 32–33
Atom(s), 15, 32–33
ATP synthase, 606
Atria, of heart, *415, 416,* 416–417
    blood flow through, *416,* 418
    contraction and relaxation of, 426–428, *427*
Atrial diastole, 427, *427*
Atrial flutter, 428
Atrial systole, *427*
Atrioventricular (AV) bundle, *424,* 424–425
Atrioventricular node, 423, *424,* 424–425
Atrioventricular (AV) valves, *416,* 417, 427
Atrophy, 120
Auditory canal, 330, *331*
Auditory ossicles, 331, *331*
Auditory projection pathway, 338
Auditory (eustachian) tube, 499
Auditory (pharyngotympanic/eustachian) tube, 331, *331*
Aura, 325
Auricle (pinna), 330, *331*
Auscultation, 4, 428
Autonomic nervous system (ANS), 250, 304–311
    dual innervation, 309–310, 311*t*
    higher-level regulation of, 310
    neural pathways, 305, *306*
    neurotransmitters, 309, 309*t*
    parasympathetic division, 250, 306–307, 307–309, 311*t*
    receptors, 309, 309*t*
    subdivisions, 306–309
    sympathetic division, 250, 306, 307, 311*t*
    visceral reflexes, 305, *305*
Axiallary tail, 640
Axial region, 16, *16*
Axial skeleton, 155–170, *157*
Axillary region, *17*
Axis (C2 vertebra), *165,* 167
Axon (nerve fiber), 113, *113,* 252
Axon terminal, 203, *204*
Azygos system, 450, *450*

B

Bacteria, intestinal, 586
Ball-and-socket joints, 182, *183*
Balloon angioplasty, 421
Baroreceptors, 11, 305, 437, *437*
Baroreflexes, 305, 437
Bartholin gland, 640
Basal cell carcinoma, 132, *132*
Basal lamina, 203
Basal layer, 638
Basal metabolic rate (BMR), 613
Basal nuclei, 293–294
Basal surface, of epithelial tissue, 97
Base
    anatomical, of heart, 412, *413*
    of bone
        metacarpal, 174, *175*
        patella, *177,* 178
    chemical, 39–40
    nitrogenous, *49,* 50
Basement membrane, *62, 76,* 97
    blood vessel, *431*
Base pairs, 50
Basic (alkaline) solutions, 39
Basilar membrane, 337, *338*
Basophils, *388, 397,* 397–398, 473
    production of, 398, *399*
B cells. *See* B lymphocytes (B cells)
Bed sore, 120
Behavioral thermoregulation, 619
Benign prostatic hyperplasia (BPH), 632
Benign tumor, 89
Beta (β) adrenergic agonists, 504
Beta (β) cells, of pancreatic islets, *372,* 373
Beta (β) chains, of hemoglobin, 391
Bicuspid valve. *See* Mitral valve
Bile, *574,* 574–575
Bile acids (bile salts), 46, 574–575
Bile duct, *573,* 574
Bilirubin, 393, 574
Biochemistry, 41
Biomedical word roots, A-12–A-15
Bipedalism, 164, 189, 191
Bipolar cells, 345
Bipolar neurons, 252, *252*
Birth. *See* Childbirth
Bitter taste, 326
Bladder. *See* Gallbladder; Urinary bladder
Blagden, Charles, 12
Bleaching, 346
Blink reflex, 136
Blister, 130
Blood, 111–112, 385–409
    centrifuging, *388, 388*
    circulation of (*See* Circulation)
    coagulation (*See* Coagulation)
    components of, 387–388, *388,* 389*t*
    composition of, 111–112, *112*
    deoxygenated, 412, 418
    disorders of, 399–400
        anemia, 399–400
        coagulation, 404
        hemophilia, 404
        leukemia, 400
        sickle-cell disease, 392, *392*
        thrombosis, 404, 457
    effects on other systems, 406
    formed elements of, 111, *112, 385,* 387–388, *388*
        disorders of, 399–400
        as percentage of whole blood, *385,* 387, 388
        production of, 389–390
        separation from plasma, 387, *388*
    functions of, 111, 387
    general properties of, 387–388, 387*t*
    historical and cultural view of, 386, 411
    osmolarity of, 389
    oxygenated, 412, 418
    pH of, 39–40, 387*t,* 429
    production of, 389–390
    regulatory function of, 387
    study of (hematology), 386
    thermoregulation by, 387
    transport in, 111, 387
        carbon dioxide, 387, 390
        oxygen, 387, 390
    viscosity of, 388–389, 436
    volume of, 387*t,* 388, 436
Blood–brain barrier (BBB), 287
Blood flow, 435–438
    blood pressure and, 435–436
    blood viscosity and, 389, 436
    hormonal control of, 437–438
    local control of, 437
    neural control of, 437
    regulation of, 437–438
    resistance to, 436
    thermoregulation in, 387
    variations in routes, 434, *434*
    vessel length and, 436
    vessel radius and, 436
Bloodletting, 386
Blood pressure (BP), 435–438
    definition of, 435
    diastolic, 435
    elevated (hypertension), 72, 436
    low, 436
    measurement of, 435
    normal, 435
    regulation of, 437–438
        baroreflexes/baroreceptors in, 11, *11,* 437
        hormonal control, 437–438
        local control, 437
        negative feedback in, 11, *11*
        neural control, 437
        renal, 438
    resistance and, 436
    systolic, 435
Blood smear, *112,* 396
Blood sugar, 596
Blood supply, brain, 287
Blood transfusions
    blood type in, 394–395
    exchange, in hemolytic disease of newborn, 395
Blood types, 66, *394,* 394–395
Blood vessels. *See also* Arteries; Capillaries; Vein(s)
    constriction and dilation of, 436
    endothelium of, 416, 430–431, *431*
    evolution of, 386
    general anatomy of, 430–434
    histology of, 430–431, *431*
    historical and cultural view of, 411
    length of, 436
    nomenclature for, 445
    physiology of, 435–441
    radius of, 436
    regulation of, 436

Blood vessels (*continued*)
    tunics of, 430–431, *431*
    wall of, 430–431
B lymphocytes (B cells), 480, *480*
Body cavities, *18*, 18–20, 18*t*. *See also specific cavities*
Body heat and thermoregulation
    body temperature, 617
    disturbances of thermoregulation, 620
    hair and retention of, 134–136
    heat production and loss, 617–618
    thermoregulation, 618–619
Body of nail, *136*
Body openings and passages, control by muscles, 199
Body regions, 16–17, *16–17*. *See also specific regions*
Body temperature
    aging and, 139
    negative feedback and, 11
    regulation of, 11, 127, 387
    set point of, 11
Bohr effect, 515
Bond(s). *See* Chemical bond(s)
Bone cells, 109, 111, *145*, 145–146
Bone markings, 156, *156*, 156*t*
Bone marrow, 144, *147*, 149
    cancer of, 400
    distribution of, 149, *149*
    hemopoiesis in, 389
    long bone, 111
    red (*See* Red bone marrow)
    sinusoids of, *402*
    yellow, *148*, 149, *149*
Bone mass, 152
Bone(s), 109, 111
    acetabulum, *176*, 177
    adaptations to bipedalism, 164
    appendicular skeleton, 157, *157*, *171*, 171–180
    auditory ossicles, 163
    axial skeleton, 155–170, *157*
    calcification of, 151
    capitate, 174, *175*
    carpal, *157*, 172, 174, *175*
    clavicle, *157*, 171, *171*
    coccyx, *157*, 165, 168, *169*
    composition of, 109, 111
    cranial, 150, 158
    development of, 144, 150–153
    ethmoid, *159*, 161–162
    facial, 158, 162
    femur, *148*, 149, *157*, *177*, 177–178
        epicondyles, *177*, 177–178
        head of, *156*, 177, *177*
        muscles of, 236–237, *236–237*
        neck of, 177, *177*
        patellar surface, *177*, 178
        trochanters of, *156*, 177, *177*
    fibula, *157*, 178, *178*
        interosseous membrane, 178, *178*
        lateral malleolus, 178, *178*
    flat, 149, *149*
    frontal, *157*, *159*, *160*, *162*, *164*
    functions of, 109, 111, 144–145
    gross anatomy of, 148–149, *148–149*
    growth of, 152
    hamate, 174, *175*
    hip (os coxae), *157*, 175, *176*, 176–177
    humerus, *157*, 172, 172–173
        epicondyles of, *156*

        head of, *156*
        markings of, *156*
    hyoid, 163
    ilium, *176*, 176–177
        anterior superior spine of, 16
    irregular, 149
    ischium, *176*, *176*, 177
    lower limb, *157*
    lunate, 174, *175*
    mandible, *157*, *159*, 163
        angle of, 163
        body of, 163
        infant, *164*
        ramus of, *159*, 163
    markings of, 156, *156*, 156*t*
    maxillae, *157*, *159*, *160*, *162*, *162*
        infant, *164*
    metabolism of, 152–153
    metacarpal, 151, *151*, *157*, 172, 174, *175*
    metatarsal, *157*
    nasal, *159*, 163, 164
    number of, 157
    occipital, *157*, 158–161, *159*, *160*, *162*, *164*
    orbit, 158
    ossification of, 150–152, *151*
    palatine, *159*, *160*
    parietal, *157*, 158, *159*, *160*, *162*, *164*
    patella, *157*, *177*, 178
        apex, *177*, 178
        articular facets, *177*, 178
        base, *177*, 178
    pectoral girdle, *157*, *170*, 171–175
    pelvic girdle, 175–180, *176*
    phalanges, 172
        of foot, *157*, *179*, 179–180
        of hand, *157*, 174–175, *175*
    pisiform, 174, *175*
    pubic, *176*, *176*, 177
    radius, *157*, *173*, 173–174
    remodeling of, 152
    ribs, *157*, 169–170, *170*
        costal cartilage, 169, *170*
        false ribs, 170, *170*
        floating ribs, 170, *170*
        true ribs, 170, *170*
    sacrum, *157*, 165, 175, *176*
    scaphoid, 174, *175*
    scapula, *157*, *171*, 171–172
        markings of, *156*
        spine of, *156*
    short, 149
    skull, 156–164, 157, *157*, 158–164
    sphenoid, *160*, 161, *162*
        infant, *164*
        wings of, 161
    sternum, *157*, 169, *170*
        body of, 169, *170*
        manubrium, 169, *170*
        xiphoid process, 169, *170*
    study of (osteology), 144
    tarsal, *157*, 178, *179*
    temporal, *160*, 161, *162*
        infant, 164
        styloid process of, *159*
    tibia, *157*, 178, *178*
        anterior border, 178, *178*
        crural region, 178, *178*

        leg, 178, *178*
        medial malleolus, 178, *178*
    trapezium, 174, *175*
    trapezoid, 174, *175*
    triquetrum, 174, *175*
    ulna, *157*, *173*, 174
    vomer, *159*, *160*, 163
    zygomatic, *159*, *160*, *162*, 163, *164*
Bone tissue, 145–148
    cells of, 109, *111*, *145*, 145–146
    compact, 109, 111, *111*, 145, 147, *147*, *149*, *151*
    endocrine function of, 376
    histology of, 109, 111, *111*, *145*, 145–148, *147*
    homeostasis in, 153–153
    matrix of, 146, 152
    spongy, 111, 145, 146–147, *147*, 149
Bony collar, 151, *151*
Bony palate. *See* Hard palate
Brachial artery, 442, *443*
Brachialis, 221, *221*
Brachial region (brachium), 17, *17*, 172
Brachiocephalic trunk, *443*, 445, *445*, 446, 449
Brain
    blood–brain barrier, 287
    blood supply, 287
    brainstem, 288–291, *289*
    cerebellum, 291, *291*
    cerebrospinal fluid, 285–287, *286*
    cerebrum, 292–294
    circulation in, *446*, 447, *448*
    divisions, 288–294
    gray and white matter, 285
    landmarks, 281–284
    medial aspect, *283*
    meninges, 158, 284, *285*
    multiregional functions, 295–301
        cerebral lateralization, 299–300, *300*
        cognition, 296–297
        electroencephalogram, 300, *301*
        language, 299, *299*
        sensation, 296, *297*
        voluntary movement, 297–299, *298*
    overview, 281–287
    surface anatomy, *282*
    ventricles, 285–287, *286*
Brain death, 9
Brainstem, 284, 288–291, *289. See also* Medulla oblongata
Branches, of arteries, 445
Braxton Hicks contractions, 659
Breast milk, 662
Breast(s), 640
    blood supply to, 447
    enlarged male, 51
Breathing. *See* Pulmonary ventilation
Brittle bone disease, 51
Bronchial arteries, 441, *442*
Bronchial tree, 501–504
Bronchioles, 503, *503*
Bronchoconstriction, 503, *503*
Bronchodilation, 503, *503*
Brown fat, 108, 109
Brush border, *66*
    small intestine, 578, *578*
Buccal (oral) cavity, 158, 560, *560*
Buffer(s), 40
    chemical, 548

renal, 548–549
respiratory, 548
Buffy coat, 388, *388*
Bulbospongiosus muscle, 656
Bulbourethral glands, 632
Bulb(s), 634
    hair, *127,* 134, *135*
Bulimia, 568
Bundle branches, 424, *424*
Bunion, 182
Burns, 132
Bursae, synovial joints, 182
Bursitis, 182

C

Cadaver, 3
Caffeine, adenosine and, 312, *313*
Calcaneal region, *17*
Calcaneal (Achilles) tendon, 239, *239*
Calcification of bone, 151
Calcitonin, 153, 371
Calcitriol (vitamin D), 133, 152–153, 376, 524
Calcium
    body content of, 31, 32*t*
    in cardiac physiology, 429
    deficiency of, 152, 153
    excess of, 152
    homeostasis of, 152–153
    parathyroid hormone and, 153, 371
    storage in bones, 145
Calcium channel blockers, 72
Calcium chloride, 34*t*
Callus, 154
Calorie (cal), 38
Calvaria, *161*
Canaliculi
    bone, *111,* 146, *147*
    platelet, open system of, 402, *402*
Canal(s), 156*t*
    alimentary (*See* Digestive tract)
    auditory, *160,* 161
    carotid, *160,* 161
    central, of bone, 109, 111, *147,* 148
    hypoglossal, *159*
    optic, 161, *162*
    sacral, 168, *169*
    vertebral, 18, *18,* 18*t*
Cancer, 89
    adenocarcinoma, *60*
    blood, 399–400
    bone marrow, 400
    leukemia, 400, *400*
    lung, 511, *511*
    metastasis of, 89
    metastatic, 471
    skin, 132, 139
Canines, 561, *561*
Cannon, Walter, 10, 12
Capillaries, blood, 430, *432,* 432–433
    alveolar, *442*
    blood flow in, 433
    continuous, 433
    fenestrated, 433
    walls of, 431, *431*
Capillary beds, *432,* 433

Capillary exchange, 70, 433, 438–440, *439*
Capitulum, *172,* 173
Capsular space, 527, *529*
Capsule, gland, 116, *116*
Carbohydrate metabolism
    aerobic fermentation, 604–606
    anaerobic fermentation, 604
    ATP production, 606
    glucose catabolism, 603–604
    glycogen metabolism, 606–607
    glycolysis, 604
Carbohydrates, 41–43, 611, 612
    digestion and absorption, 580, *581*
    functions of, 42, 43*t*
    hydrophilic nature of, 41
    naming of, 41
    structure of, *42*
Carbon
    atomic structure of, *32*
    body percentage of, 31, 32*t*
    isotopes of, 33
Carbon-12, 33
Carbon (organic) compounds, 40–53
Carbon dioxide
    molecule, 35
    transport in blood, 387, 390
Carbonic acid, 514
    reaction, 54
Carbonic anhydrase, 49, 514
Carboxyl group, 47
Carcinogens, 89
Cardiac arrhythmia, 428, 457
Cardiac conduction system, 422, 423–425, *424*
Cardiac cycle, 426–428, *427*
Cardiac muscle, 113, *114,* 114*t, 215,* 215–216, 422, *423*
    action potentials in, 425
    cells of, 422, *423*
    contraction of, 425, 426–428, *427*
    electrical activity of, 425
    involuntary, 216
    relaxation of, 426–428, *427*
    striations of, 113, *114,* 422, *423*
Cardiac orifice, 563
Cardiac output, 429
Cardiology, 412
Cardiomyocytes, 422, *423,* 425
Cardiovascular disease (CVD), 457
Cardiovascular system, 386, 411–414. *See also*
    Blood; Blood vessels; Heart
    components of, 386, 411
    disorders of
        angina pectoris, 421
        arrhythmias, 428, 457
        arteriosclerosis, 420, 457
        atherosclerosis, 378, 420, 457
        cardiovascular disease (CVD), 457
        congestive heart failure, 439
        coronary artery disease, 420–421
        heart murmurs, 428
        Marfan syndrome, 108
        myocardial infarction, 13, 120
        valvular prolapse, 417
    effects on other systems, 458
    pulmonary circuit in, 412, *412,* 441, *442*
    systemic circuit in, 412, *412,* 442–456, *443–444*
Career spotlight
    cytotechnologist, 90

    dental hygienists, 588
    dermatology nurse, 139
    diabetes educator, 380
    dialysis technicians, 550
    dietitian, 621
    electrocardiographic technician, 457
    electroneurodiagnostic (END) technologists, 314
    genetic counselors, 697
    histotechnician, 121
    massage therapist, 243
    medical technologis, 56
    midwife, 664
    occupational therapist (OT), 276
    optician, 352
    orthopedic nurse, 192
    phlebotomist, 405
    public health nurses, 490
    radiologic technologist, 26
    respiratory therapist, 517
Carotene, 131
Carotenoids, 34
Carotid sinus, *446*
Carpal region, 17, *17*
Carpal tunnel syndrome, 234
Carrier-mediated transport, 72
Carrier proteins, 72
Cartilage, 109, *110,* 110*t*
    articular, *148,* 149, *151*
    costal, 16, *157*
    elastic, 109, *110,* 110*t*
    endochondral ossification in, 151, *151*
    fibrocartilage, 109, *110,* 110*t*
    functions of, 109
    hyaline, 109, *110,* 110*t*
        articular, 151
        endochondral ossification in, *151,* 151–152
    thyroid, 370, *370*
Cartilaginous joints, 181
Catabolism, 55, 55*t,* 603
Catalysts, enzyme, 49–50
Cataract, 350
Cations, 33
Caudate lobe, liver, 572, *573*
Cavities
    body, *18,* 18–20, 18*t*
    joint (articular), 181, *181*
    medullary, 149, 151, *151*
    skull, 158
C (clear) cells, 371
Cecum, 584, *585*
Celiac trunk, *443,* 452, *453*
Cell-adhesion molecules, 65, *65*
Cell body, 113, *113*
Cell counters, electronic, 401
Cell cycle, *85,* 85–86
Cell death, programmed, 80, 120–121
Cell division
    cytokinesis, 86, 88
    meiosis, 86
    mitosis, 86–89, *87*
Cell-identity markers, 65, *65*
Cell junctions, 68–69, *69*
Cell nests, 109, *110*
Cell(s), *14,* 14–15
    basic components of, 62
    as characteristic of life, 9–10

Cell(s) (*continued*)
  components of, 61–62, *62*
  cytoskeleton of, 61, 75, *76*
  interior of, 75–84
  life cycle of, 85–88
  microscopy of, 61
  organelles of, 14–15, 61, 77–80, *78*, 81*t* (*See also specific organelles*)
  protective, 472–474, *473*
  shapes of, 95–96, *96, 98*
  size of, 62–63
  structure of, 61–63, *62*
  study of (cytology), 14, 61
  surface extensions of, 66–68
  surface of, *62*, 63–74
Cells per cubic millimeter, 391
Cell suicide, 80
Cellular immunity, 483–485, *484*
Cellular (cell-mediated) immunity, 478
Cellulose, 42, 43*t*
Cement, *561*, 562
Center, integrating, 11
Central nervous system (CNS), 249, 250
Central vein (hepatic), 572, *573*
Centrioles, *62*, 75, *78*, 80, 81*t*, 85, *87, 88*
Centromeres, 86–88, *87*
Centrosomes, *62, 76*, 80
Cephalic phase, 568, *569*
Cephalic region, 16, *17*
Cerebellum, 284, 291, *291*
Cerebral angiogram, *4*
Cerebral arterial circle, *446,* 447
Cerebral cortex, 292, *293*
Cerebral hemispheres, 281, *282*
Cerebral infarction, 120
Cerebral peduncles, 290
Cerebrospinal fluid (CSF), 285–287, *286*
Cerebrum, 281, *282*, 292–294, *293*
  frontal lobe, 292
  insula, 293
  occipital lobe, 293
  parietal lobe, 292–293
  temporal lobe, 293
  tissues, 292, *293*
Cerumen, 138, 331, *331*
Ceruminous glands, 136, 138
Cervical canal, 638
Cervical cancer, 639
Cervical curvature, 164–166, *165*
Cervical region, 16, *17*
Cervical vertebrae, 164, *165,* 167, *167*
  atlas (C1), *165,* 167
  axis (C2), *165,* 167
Cervix, 637
Chancre, 657
Channel proteins, *64, 65,* 65
Chemical bond(s), 35–37, 37*t*
  covalent, *35,* 35–36, 37*t*
  hydrogen, 35, 36–37, 37*t*
  ionic, *33,* 34, 35, 37*t*
  peptide, *47,* 47–48
Chemical buffers, 548
Chemical digestion, 557, 580–584
Chemical equation, 54
Chemical reactions
  in body (metabolism), 9–10, 53, 55
  classes of, 54, *54*
  definition of, 54
  energy and, 53–55, 55*t*
  equilibrium in, 54
  reversible, 54
  water's reactivity in, 38
Chemical senses, 326–329
Chemoreceptors, 326
Chemotaxis, 474
Chewing, muscles of, 225–227, *226*
Chief cells, 566
Childbirth, 659–660
  positive feedback in, 12, *12*
Chlorine, body percentage of, 32*t*
Chlorine ions, *33,* 33–34
Choanae, 497, *498*
Cholecystokinin (CCK), 569, 575, 614
Cholesterol, 43, *45,* 45–46
  cardiovascular effects of, 46, 421
  dietary, 46
  in plasma membrane, *64,* 65
  steroid synthesis from, 360, *360*
Cholinesterase inhibitors, 205
Chondrocytes, 109, *110*
  in bone growth, 152
  in endochondral ossification, 151, *151*
Chordae tendineae, 417
Choroid plexus, 287
Chromatids, 86–88, *87, 88*
Chromatin, 77
Chromium, 32*t*
Chromosome(s), 77
  mitosis and, 86–88, *87, 88*
Chronic leukemia, 400
Chronic obstructive pulmonary disease (COPD), 511
Chronic vomiting, 568
Chylomicrons, 583
Chymotrypsin, 575
Cilia, 66, *67*
  defects in, 68
  primary, 66
  uterine tube, 66, *67,* 93
Ciliary body, 341, *341*
Circadian rhythms, 369
Circular folds, small intestine, 577, *577*
Circulation
  abdominopelvic region, 452–455, *453, 454*
  coronary, 418–421, *419,* 442
  head and neck, *446,* 446–447
  physiology of, 435–441
  pulmonary, 412, *412,* 441, *442*
  renal, 452, *453*
  systemic, 412, *412,* 442–456, *443–444*
  thoracic, 447, 449, *449, 450*
  upper limb, *449,* 450–451
  variations in routes, 434, *434*
Circulatory shock, 440
Circulatory system, 21, *23, 665. See also* Blood; Blood vessels; Heart
  aging of, 457
  components of, 386, 411
  effects on other systems, 406, 458
  evolution of, 386–387
  functions of, 386–387
  during pregnancy, 658
Circumduction, *185,* 185–186
Cisternae, 77–79, *78*
Citric acid (Krebs2) cycle, 604

Clathrin-coated vesicle, *84*
Clear (C) cells, 371
Cleavage furrow, *87,* 88
Climacteric, 642
Clitoris, 640
Clonal selection, 483
Closed reduction of fracture, 154
Clotting, *403,* 403–404
Clotting factors, 402, *403,* 403–404
Coagulation, 403–404
  agglutination *vs.,* 404
  disorders of, 404
Cobalt, 32*t*
Coccygeal vertebrae, 164, 168, *169*
Cochlea, *336,* 336–338
Cochlear duct, *336,* 336–337
Codons, *82,* 83
Coenzymes, 600
Cognition
  brain function, 296–297
  defined, 296
Cohesion, 38
Collagen, *48,* 49, 104–105, *105*
  aging and, 139
  in bone tissue, 146, *147,* 149, 150
  in cartilage, 109, *110*
  in ligaments, 144
  in skin, 115, 129–131
Collagen diseases, 51, 146
Collagenopathies, 51
Collecting ducts, *466,* 467, 529, *530,* 539, *539*
Collecting vessels, 467
Colon, 584, *585*
Colony-forming units (CFUs), 393, *393,* 398, *399*
Color blindness, 348
Color vision, 348, *348*
Colostrum, 662
Columnar cells, 96, *96, 98*
Communication
  nonverbal, 127
  respiratory system, 496
Compact bone, 109, 111, *111,* 145, 147, *147,* 149, *151*
Compaction, 557
Comparative anatomy, 3
Comparative physiology, 8
Compartment syndrome, 218, 220
Complement system, 474
Complete blood count (CBC), 401
Complete proteins, 599
Complete tetanus, 210, *211*
Compound(s), 35
  inorganic, 41
  organic, 40–53
Computed tomography (CT), *4,* 5, *5*
Concentration
  solution, 39
  units of, A–10
Concentration gradients, *70,* 70–72
Concentric contraction, skeletal muscles, 211
Concentric lamellae, *147*
Conception, 658
Conchae, nasal, 161–162, *163*
Condensed structural formula, 35
Conducting arteries, 431, *431*
Conducting zone, 503
Conduction, 617–618

Conduction system, of heart, 422, 423–425, *424*
Conductive deafness, 350
Condylar (ellipsoid) joints, 182, *183*
Condyles, *156,* 156t
   femoral, *156*
   femur, *177,* 178
   humeral, *156*
   mandibular, 163
   occipital, *160,* 161
   tibia, 178, *178*
Conformation, 49
Congestive heart failure, 439
Conjunctiva, 340
Connecting tissues, 665
Connective tissue, 94, 94t, 104–112
   areolar, 105, *106,* 106t
   dense, 105, *107,* 107t
      irregular, *107,* 107t
      regular, *107,* 107t
   endomysium, 218
   epimysium, 218, *219*
   fibrous, 104–105, *106–107,* 106t–107t
   functions of, 104
   loose, 105, *106,* 106t
   muscle attachments and, 218–221, *219*
   overview of, 104
   perimysium, 218, *219*
   reticular, 105, *106,* 106t
   supportive, 109, 111
Connective tissue disease, 108
Continuous capillaries, 433
Contraction(s)
   skeletal muscles, 207–209, *208, 209*
      concentric, 211
      eccentric, 211
      isometric, 210–211
      isotonic, 210–211
      muscle twitch and tetanus, 210, *211*
      whole-muscle, 210–211
   of small intestine, 579, *579*
Control center, in negative feedback, 11
Contusion, 220
Convection, 618
Conversion factors, A–10
Copper, 32t
Core temperature, 617
Cornea, 341, *341*
Coronal plane, 15, *15, 16*
Coronary angiogram, *410*
Coronary arteries, 418–419, *419,* 445, *445*
   angiogram of, *410*
   anterior interventricular branch of, 418, *419*
   circumflex branch of, 418, *419*
   left, 418, *419*
   left anterior descending branch of, 418
   left marginal branch of, 418, *419*
   posterior interventricular branch of, 419, *419*
   right, 418–419, *419*
   right marginal branch of, 418–421, *419*
Coronary artery disease (CAD), 420–421
Coronary bypass surgery, 421
Coronary circulation, 418–421, *419,* 442
Coronary sinus, *415,* 419, *419*
Coronary sulcus, *415,* 417
Coronoid fossa, *172,* 173
Corpora cavernosa, 633
Corpus callosum, 282, *282*

Corpus luteum, 366, 650, 652
Corpus spongiosum, 633
Corticosteroids, 374–375
Corticotropin-releasing hormone (CRH), *368*
Cortisol, 46, 375, 379
Costal cartilage, 169, *170*
Costal facets, 167–168, *168*
Cough reflex, 500
Covalent bonds, 35–36, *35–36,* 37t
Cow milk, 662
Coxal region, *17*
Cramps, muscle, 220
Cranial cavity, 18, *18,* 18t, 158
Cranial nerves, 302–304, 302t, *303*
Cranium, 158
Creatine phosphate, 212, *212*
Creatinine, 524
Cremaster, 630
Crest, of bone, *156,* 156t
Cristae, of mitochondria, *78,* 80
Crista galli, *159, 160,* 162
Cross-bridge, 208, *208*
Cross-training, 214
Crown, *561,* 562
Crural region, 17, *17*
Cryptorchidism (undescended testes), 631
CT scans, *4, 5,* 5
Cuboidal cells, 96, *96, 98*
Cupula, 334, *335*
Curare, 205
Curvatures, of vertebral column, 164–166, *165*
Cusps, of heart valves, 417
Cutaneous glands, 136–140, *137*
Cuticle
   of hair, 134
   of nail, *136*
Cyanide, 52
Cyanosis, 131
Cyclic adenosine monophosphate (cAMP), 361, *361*
Cytokinesis, 86, 88
Cytology, 14, 61
Cytology, exfoliate, 101
Cytoplasm, 61
Cytosine, 50
Cytoskeleton, 61, 75, *76*
Cytosol, 62, 75
Cytotechnologist, 90
Cytotoxic T (T_C) cells, 483

D

Dander, 129
Dartos fascia, 630
Daughter cells, 85–86, *87,* 88
Daughter chromosome, *87,* 88
Day (photopic) vision, 347
DCT. *See* Distal convoluted tubule (DCT)
Deafness, 350
Deaminated amino acids, 609
Death, 9
Decomposition reactions, 54, *54,* 55, 55t
Decubitus ulcers, 120
Deep, 25t
Deep artery, 654
Defecation, 557, 587
Defecation reflexes, 587

   intrinsic, 587
   parasympathetic, 587
Defibrillator, 428
Degenerative joint disease, 154–155
Deglutition (swallowing), 563, *563*
   esophageal phase, 563
   oral phase, 563
   peristalsis, 563
   pharyngeal phase, 563
   swallowing center, 563
Dehydration synthesis, 38, 41, *41*
Dehydroepiandrosterone (DHEA), 375
Delayed onset muscle soreness (DOMS), 220
Deltoid tuberosity, *172,* 173
Denaturation, 50
Dendrites, 113, *113,* 252
Dendritic cells, 128, *128*
Dens, 167, *167*
Dense (compact) bone, 109, 111, *111,* 145, 147, *147, 149, 151*
Dense connective tissue, 105, *107,* 107t
   irregular, *107,* 107t
   regular, *107,* 107t
Dental hygienists, 550, 588
Dentin, *561,* 562
Dentition, 561, *561*
Deoxygenated blood, 412, 418
Deoxyribonuclease, 575
Deoxyribose, 50
Depolarization, 259
Deposition of minerals, 152
Depression
   management, 312
   movement, 185
Dermal papillae, *127, 128,* 130, 134, *135*
Dermatology, 132
Dermatology nurse, 139
Dermis, 126, *127,* 129–130
   collagenous fibers of, 105
   dense connective tissue of, *107,* 107t
Descending aorta, *445,* 445–446, *449,* 452, *453*
Descending colon, 584, *585*
Desmosomes, *62,* 68–69, *69, 76,* 215, 216, 422, *423*
Detrusor, 542, *543*
Deuterium, 33
Development, 10
Dextrocardia, 9
Diabetes, 198
Diabetes educator, 380
Diabetes insipidus, 378
Diabetes mellitus, 378–379
   aging and, 380
   type 1, 378
   type 2, 362, 378
Diabetic neuropathy, 378
Dialysis, 541
Diapedesis, 474
Diaphoresis, 137
Diaphragm, 508, *508*
Diaphysis, *148,* 149
Diarrhea, 571
Diarthrosis, 181
Diastole, 426–427
Diastolic pressure, 435
Diencephalon, *289,* 290–291
Dietary fiber, 42
Dietitian, 621, 621b

Differential WBC count, 400, 401
Diffusion
　　capillary, 438
　　facilitated, 70, 72
　　simple, 70, 70
Digestion, 557
　　carbohydrates, 580, 581
　　chemical, 557, 580–584
　　fats, 581, 582–583
　　mechanical, 557
　　proteins, 580–582, 581
　　starch, 49
Digestive system, 20, 21, 23, 555–592, 665
　　aging of, 588
　　anatomy, 557–559, 557–559
　　bile, 574, 574–575
　　blood supply to, 452, 453, 453
　　chemical digestion and absorption, 580–584
　　digestive tract (alimentary canal), 557, 557
　　effects on other organ systems, 589
　　esophagus, 563
　　functions, 557
　　gallbladder, 574, 574–575
　　large intestine, 584–587, 585
　　liver, 572–574, 573
　　mouth, 560–561, 560–562
　　overview, 556–560
　　pancreas, 574, 575
　　pharynx, 562
　　during pregnancy, 658
　　saliva and salivary glands, 562, 562
　　small intestine, 576–579, 577–579
　　stomach, 564–570, 565
　　swallowing, 563
Digestive tract, 557, 557
　　mesenteries, 559, 559
　　mucosa (mucous membrane), 557–558, 558
　　muscularis externa, 558, 558
　　serosa (serous membrane), 558, 558
　　submucosa, 558, 558
　　tissue layers, 557–558, 558
Digital rectal examination (DRE), 632
Digit(s), 17
　　fingers, 17
　　toes, 17
Dilation, 659
　　pregnancy, 659
Dimetapp, 312
Dipeptides, 47
Diploid cell, 644
Directional terms, 25–26, 25t
Disaccharides, 42, 42, 43t
Disodium phosphate, 34t
Dissection, 3–4
Distal, 25t
Distal convoluted tubule (DCT), 529, 530, 537–538
Distributing arteries, 431, 431–432
Diuresis, 544
DNA, 50
　　double helix of, 50, 51
　　errors in, 89
　　functions of, 50
　　genetic code in, 48
　　hydrogen bonds of, 36–37
　　mitochondrial, 80
　　mutations of, 10, 89
　　nucleotides of, 49, 50

replication of, 85, 86
transcription of, 81–83, 82
DNA polymerase, 85, 86
Dominant follicle, 651
Dopamine, 358
Dorsal, 15, 25t
Dorsal artery, 654
Dorsiflexion, 187, 187
Double covalent bonds, 36, 37t
Double helix, of DNA, 50, 51
Drugs
　　nervous system and, 312, 313
　　pH and, 39–40
　　pharmacogenetics of, 89
　　transdermal, 129
Duchenne muscular dystrophy (DMD), 220
Ducts, 116, 116
Ductus deferens, 630
Duodenum, 576, 577
Dural sinuses, 284, 285, 447, 448
Dura mater, 269, 284, 285
Duration, sensory information, 322
Dwarfism, 362
Dynamic equilibrium, 11, 333

E

Ear
　　anatomy of, 330–333, 331
　　auditory projection pathway, 338
　　cochlea and hearing, 336–337, 336–338, 336–338
　　elastic cartilage of, 110, 110t
　　equilibrium, 333–336, 334–335
　　tissue engineering of, 115
Earwax, 138
Eccentric contraction, skeletal muscles, 211
Echocardiography, 6
Edema, 439, 440, 466, 467
Effacement, 659
Effector, 11
Efferent arteriole, 526, 527
Egg (ovum), 626
Ehlers–Danlos syndrome (EDS), 51
Einthoven, Willem, 426
Ejaculation, 655
Ejaculatory duct, 630
Elastic arteries, 431, 431
Elastic cartilage, 109, 110, 110t
Elastic fibers, 105, 110, 130
Elastic filaments, 201, 202
Elastic tissue, 105
Elastin, 105, 139
Elbow, 17, 17
　　synovial joints of, 188–189, 189
Electrocardiogram (ECG, EKG), 425–426, 426, 427, 457
Electrocardiographic technician, 457
Electroencephalogram, 300, 301
Electrolyte balance, 547–548
Electrolyte(s), 34, 34t, 536
Electron acceptor (oxidizing agent), 55
Electron donor (reducing agent), 55
Electroneurodiagnostic (END) technologists, 314
Electron microscopes, 61
Electron(s), 32, 32–33
　　sharing, in covalent bonds, 35, 35–36

transfer, in ionic bonds, 35
valence, 33
Electron shells, 32–33
Elements, 31–32, 32, 32t
Elevation, movement, 185
Elongation of bone, 152
Embolus/embolism, 404, 420
Embryonic stem cells, 115
Emetic center, 568
Emission, 655
Emmetropia, 344, 345
Emulsification droplets, 583
Enamel, 561, 562
Endergonic reactions, 55, 55t
Endocannabinoids, 615
Endocardium, 412, 414–416
Endochondral ossification, 151, 151–152
Endocrine cells, interstitial, 375, 376
Endocrine disorders, 357, 378. See also specific disorders
　　androgen-insensitivity syndrome, 378
　　diabetes insipidus, 378
　　diabetes mellitus (See Diabetes mellitus)
　　diabetic neuropathy, 378
　　hypo- and hypersecretion, 378
　　pheochromocytoma, 378
　　thyroid disorders, 371, 378
Endocrine glands, 116, 116–117, 357. See also specific endocrine glands
　　exocrine glands vs., 357
　　major, 358
Endocrine system, 21, 23, 356–384, 665. See also specific glands
　　aging and, 380
　　communication in, 358, 359
　　components of, 357, 358
　　effects on other systems, 381
　　embryonic development of, 364, 364
　　negative feedback in, 362
　　negative feedback inhibition in, 367–368
　　nervous system vs., 358–359, 359, 359t
　　overview of, 357–358
　　physiology of, 360–363
　　positive feedback in, 362, 367
Endocrinology, 8
Endocytosis, 73, 73–74
Endometrium, 638
Endomysium, 218
Endoplasmic reticulum, 77–79, 78
　　protein processing and secretion in, 83
　　rough, 62, 77–79, 78, 81t
　　smooth, 62, 79, 81t
Endosteum, 148, 149
Endothelial cells, 432
Endothelium, of blood vessels, 416, 430–431, 431
Endpiece, 647
Endurance, 212
Endurance (aerobic) exercise, 214
Energy, 53–56
　　ATP as currency of, 52
　　carbohydrates as source of, 42
　　in chemical reactions, 53–55, 55t
　　definition of, 53
　　kinetic, 54
　　potential, 53–54
　　transfer in body (metabolism), 9–10, 53, 55
Energy balance, 614

Energy levels (electron shells), 32–33
Enteric hormones, 376
Enteric nervous system (ENS), 558–559
Enteroendocrine cells, 566
Enzymes, 49–50, 575
    active site of, 49
    brush border, 578
    digestive (See specific entries)
    intestinal crypts, 578
    lock-and-key model of, 50
    lysosomal, 79–80
    naming of, 49
    pH and, 50
    in plasma membrane, 65, *65*
    reaction products of, 49
    substrate of, 49
    temperature and, 50
Eosinophils, *388*, 397, *397*, 398, *399*, 473
Ependymal cells, 253, *254*
Epicardium, 412, *413*, 414
Epicondyles, *156*, 156t
    femoral, *156*
    humeral, *156*
    lateral, *172*, 173
    medial, *172*, 173
Epidermal cells, *125*
Epidermal water barrier, 127
Epidermis, 101, *127*, *128*, 128–129
Epididymis, 630
Epigastric region, 24
Epiglottis, *499*, 500
Epimysium, 218, *219*
Epinephrine, 360, *360*
    adrenal secretion of, 374
    in blood pressure control, 437–438
    in cardiovascular function, 429
    in stress response, 379
Epiphyseal closure, 152
Epiphyseal lines, *148*, 149
Epiphyseal plate, *151*, 151–152
Epiphysis, *148*, 149
Epithalamus, 291
Epithelial tissue, 94, 94t, 97–103
    apical surface of, 97
    basal surface of, 97
    histology of, 97–103
    keratinized, *101*, *102*, 102t
    of kidney, *105*, 105t
    lateral surfaces of, 97
    nonkeratinized, *101*, *102*, 102t
    pseudostratified columnar, 98, *98*, *100*, 100t
    of renal tubules, *99*, 99t
    simple, 98, *98*, 99–100, 99t–100t
        columnar, 98, *98*, *100*, 100t
        cuboidal, 98, *99*, 99t
        squamous, 98, *99*, 99t
    of small intestine, 100t, *101*
    sole of foot, *102*, 102t
    squamous
        simple, 98, *99*, 99t
        stratified, 101, *102*, 102t
    stratified, 98, *98*, 101, *102–103*, 102t–103t
        columnar, 101
        cuboidal, 101, *103*, 103t
        squamous, 101, *102*, 102t
        transitional, 101, *103*, 103t
    of trachea, 100t, *101*
        transitional, 101, *103*, 103t
    of vagina, *102*, 102t
EPOC. *See* Excess postexercise oxygen consumption (EPOC)
Equilibrium
    in chemical reactions, 54
    dynamic, 11
    in internal state of body, 10–13
Equilibrium, ear and, 333–336, *334–335*
Equivalent (Eq), A–11
Erectile tissues, 633
Erection, 655
Errors, in DNA, 89
Erythema, 131
Erythroblastosis fetalis, 395
Erythrocyte(s), 111, *112, 385*, 388, *388*, 390–395, *432*
    agglutination of, 394, 404
    anaerobic fermentation in, 392
    antigens, in blood types, *394*, 394–395
    count, 387t, 400, 401
    death and disposal of, 393
    deficiency of, 390, 399–400
    disorders of, 399–400
    functions of, 390
    hemolysis of, 393, 394
    life cycle of, 393, *393*
    as percentage of whole blood, 388, *388*
    production of, 389–390, 393, *393*
    quantities of, 387t, 390–391, 401
    sickle-cell disease and, 392, *392*
    structure of, 390–391, *391*
    tonicity and, 71, *71*
    viscosity and, 436
Erythropoiesis, 393
Erythropoietin (EPO), 376, 393, 399, 524
Esophageal phase, swallowing, 563, *563*
Esophagus, 557, *557*, 563
Essential amino acids, 599
Essential fatty acids, 597
Essential nutrients, 595
Estradiol, 360, *360*, 651
Estrogen receptors, 362
Estrogen(s), 43, 46, 375, 642
    in bone physiology, 153, 154
    permissive effect with progesterone, 361
Ethanol, structural isomer of, 35, *35*
Ethmoid sinus, 158, *158*, 161
Ethyl ether, structural isomer of, 35, *35*
Eumelanin, 135
Evaporation, 618
Eversion, *187*, 187–188
Evolution, 10
    as characteristic of life, 10
    of circulatory system, 386–387
Evolutionary medicine, 10
Excess postexercise oxygen consumption (EPOC), 213
Exchange transfusion, for hemolytic disease of newborn, 395
Exchange vessels, capillaries as, 433, 438–440, *439*
Excitability, 10
Excitable tissues, 112–114
Excitation, skeletal muscles, 206–207, *207*
Excretion, 116, 524
Exercise
    ATP production during, 212, *212*
    endurance (aerobic), 214
    resistance, 214

Exergonic reactions, 55, 55t
Exfoliate cytology, 101
Exfoliation, 101, *104*, 129
Exhaustion stage, in stress response, 379
Exocrine glands, 116–117, *116–117*, 357
Exocytosis, *73*, 74
Exophthalmia, 371
Expiration, 506, *507*
Expiratory reserve volume, 509, *509*
Exploratory surgery, 4
Expulsion, 656
    of abdominal contents, 496
    during pregnancy, 659
Extension, 184, *184*, 232
External anal sphincter, *585*, 586
External genitalia, 627
External intercostal muscles, 508, *508*
External urethral orifice, 543
External urethral sphincter, 543
Extracellular compartment, 546
Extracellular face, of plasma membrane, *64*
Extracellular fluid (ECF), 62, 71, 94, 387
Extracellular material (matrix), 94
Eyeball, 341–343, *341–343*
Eyebrows, 339, *340*
Eyelashes, 340
Eyelids (palpebrae), 339–340, *340*

F

Face, muscles, 225, *226*
Facet, of bone, 156t
Facial bones, 158, 162
Facial expression, 127, 136
Facial region, *17*
Facilitated diffusion, *70*, 72
Factor VIII, 404
FAD (flavin adenine dinucleotide), 604
Falciform ligament, liver, 572, *573*
Fallen arches, 180
False ribs, 170, *170*
Fasciae, 218, *219*
    superficial (hypodermis), *127*
Fascicles, 218, *219*
Fast-twitch fibers, 213
Fat cells (adipocytes), 24, 105, 108, *108*
Fatigue, muscle, 212
Fat(s), 43–44, 611, 612
    digestion and absorption, *581*, 582–583
    polyunsaturated, 44
    saturated, 43–44
    trans, 46, *46*
    unsaturated, 43, 44
Fat-soluble vitamins, 600
Fat (adipose) tissue, 44
Fatty acids, 43–44
    polyunsaturated, 44
    saturated, 43–44, *44*
    structure of, *44*
    unsaturated, 43, 44, *44*
Fecal incontinence, 587
Feedback loops, 11–13
Female reproductive system, 20, 21, *23*
    breasts, 640–642
    external genitalia, 640
    menopause in, 154

Female reproductive system (*continued*)
  ovaries and reproductive tract, 636–640
  from puberty to menopause, 642
Females, erythrocytes in, 391
Femoral region, 17, *17*
Femoral triangle, 455
Fenestrated capillaries, 433
Fermentation, 55
  anaerobic, 392
Fetal sonogram, 6
Fetus, hyaline cartilage of, 110, 110*t*
Fever, 12, *476*, 476–477
Fiber, dietary, 42
Fibrillin, 108
Fibrin, 389, 403–404
Fibrinogen, 389, 403–404
Fibroblasts, 105
Fibrocartilage, 109, *110*, 110*t*
Fibrosis, 120
Fibrous cells, 96, *96*
Fibrous connective tissue, 104–105, *106–107*, 106*t*–107*t*
Fibrous joints, 181
Fibrous layer, 341, *341*
Fibrous proteins, *48*, 48–49
Fibrous skeleton, of heart, 416
Filtration, 70, *70*
  capillary, 70, 438, *439*
Filtration pores, 433
Fimbriae, 637
Fingerprints, 130
Finger(s), 17
First-degree burns, 132
First gap phase, in cell cycle, 85, *85*
First heart sound, 427
First messenger, 361
First polar body, 648
Fissures, 156*t*
  inferior orbital, 162, *162*
  longitudinal, 282, *283*
  superior orbital, 161, *162*
Fitness training, 198
Fixator, 222
Flaccid paralysis, 205
Flagella, 67
  defects in, 68
  of sperm, 67
Flat bones, 149, *149*
Flat feet, 180
Flatus, 586
Flavor, 327
Flexion, 184, *184*, 232
Flexor reflex, 275, *275*
Floating ribs, 170, *170*
Fluent aphasia, 300
Fluid balance, 546–547, *546–547*
Fluid compartments, 546
Fluid recovery, 465
Fluid(s). *See also specific fluids*
  extracellular, 62, 71, 94, 387
  intracellular, 62, 71
Fluorine, 32*t*
Fluoxetine, 312
Folia, 284
Follicle(s), 636
  hair, *127*, 134–135, *135*
  thyroid, 370, *370*

Follicle-stimulating hormone (FSH), 366, 367t, *368*, 634, 651
  inhibin and, 375
  synergism with testosterone, 362
Follicular cells, 370, *370*, 375
Follicular phase, 651
Folliculogenesis, 650
Fontanels, 150, 163, *164*
Foot, 17, *17*
  bones of, *157*, 178–180, *179*
  calcaneu, 178, *179*
  cuboid, 179, *179*
  cuneiforms, 179, *179*
  metatarsals, 179, *179*
  muscles of, 238–239, *239*
  navicular, 178, *179*
  sole, epithelium of, *102*, 102t
  talus, 178, *179*
Foramina, *156*, 156*t*
  intervertebral, 166, *166*
  jugular, *159, 160*, 161, 447
  magnum, 158–161, *160*
  mandibular, *160*, 161
  nasal, 162
  obturator, *176*, 177
  optic, *160*
  sacral, 168, *169*
  of skull, 158
  transverse, 167, *167*
  vertebral, 167, *170*
Forearm, 17, *17*
  muscles, 232–233, *234–235*
Formed elements, of blood, 111, *112*, 385, 387–388, *388*. *See also specific elements*
  disorders of, 399–400
  as percentage of whole blood, 388, *388*
  production of, 389–390
  separation from plasma, 388, *388*
Fossae, *156*, 156*t*
  coronoid, *172*, 173
  iliac, 177
  infraspinous, *171*, 172, 232
  mandibular, *160*, 161
  nasal, 162
  olecranon, *172*, 173
  popliteal, 238
  radial, *172*, 173
  subscapular, *171*, 172
  supraspinous, *171*, 172, 232
Fovea centralis, 343
Fractures, 154
Free edge, of nail, 136, *136*
Free nerve endings, 323
Free radicals, 34
Free ribosomes, *62*
Frenulum, 633
Friction rub, 412
Frontal plane, 15, *15, 16*
Frontal sinus, 158, *158, 159*
Frostbite, 620
Fructose, 41, 43*t*
Fuel, 595
Full-thickness burns, 132
Functional layer, 638
Functional morphology, 3
Functional MRI (fMRI), 5
Fundus, 637
Fundus (rear), of the eye, *343*
Fusiform cells, 96, *96*

G

Galactose, 41, 43*t*
Gallbladder, 557, *557*, 574, 574–575
  anatomical variations in, 9
  blood supply to, 452, 453, *453*, 454
Gametes, 626
Gametogenesis, 644
Gamma globulins, 480
Ganglion cells, 345
Gangrene, 120, *120*
Gap junctions, 69, *69*, 216, 422, *423*
Gap phases of cell cycle, 85, *85*
Gas exchange, 496, 512–516
  alveolar, *513*, 513–514
  composition of air, 512–513
  systemic, 515
Gas transport, 514
Gastric glands, 566
Gastric juice, 567
Gastric lipase, 567
Gastric motility, 568
Gastric phase, 568–569, *569*
Gastric pits, 566
Gastric rugae, 565
Gastric secretions, 567, 567*t*
Gastrin, 376, 567, 569
Gastritis, 570, *570*
Gastrocolic reflex, 587
Gastroenterology, 25
Gastrointestinal (GI) tract, 557
Gates (gated channels), 65, *65*
Gene sequencing, 89
Gene-substitution therapy, 89
Genetic code, 48, 81–83, *82*
Genetic engineering, 115
Genetic(s), 85–89
Genomic medicine, 89
Germ cells, 375, 628
Ghrelin, 376, 614
Gingiva, *561*, 562
Glands, 116–117. *See also specific entries*
  cutaneous, 136–140, *137*
  endocrine (*See* Endocrine glands)
  exocrine (*See* Exocrine glands)
  gastric, 566
  salivary, 562, *562*
  secretions of, 116–117, *117*
Glans, 632, 640
Glaucoma, 342, 350
Glenoid cavity, *171*, 172
Glial cells (neuroglia), 113, *113*
Globins, 391–392
Globular proteins, 48–49
Globulins, 389
Glomerular capsule, 527, *529*
Glomerular filtration, *531–532*, 531–535
  basement membrane, 532, *532*
  endothelium of the capillary, 532, *532*
  filtration slits, 532, *532*
  hormonal control, 533, *534*
  membrane, 532, *532*
  neural control, 533
  pressure, 532–533
  rate of, 533
  regulation of, 533–535, *534*

Glomerular filtration rate (GFR), 533
Glomerulus, 527, *529*
Glottis, 500, *500*
Glucagon, 359, 361, 373, *373*
Glucocorticoids, 375
Gluconeogenesis, 373, *373*, 607
Glucose, 41, *42*, 43*t*, 536
    blood levels of, 373, *373*, 378
    regulation of, 359, 361, 373, *373*, 378
Glucose catabolism, 603–604
Glucose concentration, 596
Gluteal region, *17*
Gluteus maximus, 236, *237*
Glycemic control, 199
Glycerol, 43, *44*
Glycocalyx, 66, *66*
Glycogen, 41, 42, *42*, 43*t*, 200, 606
Glycogenesis, 606
Glycogen–lactate system, 212, *212*
Glycogen metabolism, 606–607
Glycogenolysis, 373, *373*, 606
Glycolipids, *64*, 65, 66
Glycolysis, 603, 604
Glycoproteins, *64*, 65, *65*, 67, 360
Glycosuria, 378, 537
Goblet cells, 98, *100*, 577, *578*
Goiter, toxic, 371
Golgi complex, *62*, *78*, 79, 81*t*, 83, *84*
Golgi vesicles, *62*, *78*, 79
Gonadotropin-releasing hormone (GnRH), *368*, 634
Gonadotropin(s), 366
Gonad(s), *358*, 375, 375–376, 626. *See also* Ovaries; Testes
Gouty arthritis, 155
Gram molecular weight, A–10
Granules, 396, 402, *402*
Granulocytes, 396–398, *397*, *399*
Granulosa cells, 650
Graves disease, 371
Gray matter, 256–257, 285
    spinal, 268
Greater omentum, *19*, 559
Greater sciatic notch, 177
Greater vestibular gland, 640
Great vessels, 412
Greek letters, A–10
Gross anatomy, 3
Ground substance, 94, 105
Growth, 10
    bone, 152
    hair, 135
Growth factors, 153, 402, 404
Growth hormone (GH), 153, 362, 366, *368*
    clearance of, 362
Growth hormone-releasing hormone (GHRH), *368*
Guanine, 50
Guard hairs, 136, 497, *498*
Gustation, 326–328, *327*
Gut-brain peptides, 567, 614
Gynecomastia, 51
Gyrus, 281–282, *282*

H

Hair bulb, *127*, 134, *135*
Hair cells, 333, *334*

Hair color, 135
Hair follicle, *127*, 134–135, *135*
Hair matrix, 134, *135*
Hair receptors, *127*, 134–135, *135*, 323
Hair root, 134, *135*
Hair(s), *127*, 134–136
    aging and, 139
    functions of, 135–136
    growth and loss of, 135
    guard, 136
    microscopy of, *125*
    structure of, 134–135, *135*
Hand, 17, *17*
    bones of, *157*
    muscles of, 233–235, *234–235*
Haploid, 644
Haptens, 478
Hard callus, 154
"Hardening of the arteries," 420. *See also* Atherosclerosis
Hard palate, 162
Hashimoto disease, 371
Haustra, 585, *585*
Haustral contractions, 586
Head, 647
    of bone, *148*, 149
        of femur, *156*, 177, *177*
        of humerus, *156*
        metacarpal, 174, *175*
        ulna, *173*, 174
    muscles of, 227, 229*t*
    of pancreas, 372
    of phospholipids, 45, *45*, 64
    of thick filaments, *201*, 202
Head (cephalic region), 16, *17*
    arteries of, *446*, 446–447
    radiologic images of, *4*
    veins of, 447, *448*
Healing, 120, 154
Hearing, *336–337*, 336–338. *See also* Ear
Hearing disorders, 350
Heart
    anatomical variations in, *7*, *9*
    atria of, *415*, *416*, 416–417, 418, 426–428, *427*
    autorhythmic, 216
    autorhythmicity of, 422, 423
    blood flow through, *412*, 412, 418
    cardiac cycle in, 426–428, *427*
    chambers of, 416–417, 418
    conduction system of, 422, 423–425, *424*
    coronary circulation of, 418–421, *419*, 442
    electrical and contractile activity of, 425
    electrocardiogram of, 425–426, *426*
    endocrine function of, 376
    energy for, 422
    evolution of, 386–387
    external anatomy of, *415*
    fibrous skeleton of, 416
    gross anatomy of, 414–421, *415*, *416*
    historical and cultural view of, 411
    internal anatomy of, *416*
    muscles, *215*, 215–216
    nerve supply to, 425
    physiology of, 421–429, 435–441
    position of, 412, *413*
    shape of, 412
    size of, 412

    sonographic examination of, 6
    study of (cardiology), 411
    venous drainage of, 419, 421
    venous return to, *440*, 440–441
    ventricles of, *413*, *415*, *416*, 417, 418, 426–428
Heart attack (myocardial infarction), 13, 120
Heart disorders
    angina pectoris, 421
    arrhythmias, 428, 457
    cardiovascular disease (CVD), 457
    congestive heart failure, 439
    coronary artery disease, 420–421
    heart murmurs, 428
    myocardial infarction, 13, 120
    valvular prolapse, 417
Heart murmurs, 428
Heart rate (HR), 426, 429
Heart sounds, 427–428
Heart valves, *416*, 417
    blood flow through, 418
    cusps (leaflets) of, 417
    operation of, 417
    prolapse of, 417
    repair or replacement of, 428
Heart wall, 414–416, *415*
Heat cramps, 620
Heat exhaustion, 620
Heat generation, by muscles, 199
Heat-losing center, 618
Heat-promoting center, 618
Heatstroke (sunstroke), 620
Heel strike, 187
Helicobacter pylori, 570
Helper T (T_H) cells, 483
Hematocrit, 387*t*, 388, 390–391, 401
Hematology, 386
Hematoma, 131–133
    fracture, 154
Hematopoiesis, 389–390
    erythrocyte, 393, *393*
    leukocyte, 398, *399*
    platelet, *402*, 402–403
Hematopoietic stem cells (HSCs), 389, 393, *393*, 398, *399*
Hematuria, 542, 544
Heme, 391, *392*
Heme group, 48, *48*
Hemicrania, 325
Hemidesmosomes, *62*, 76
Hemodialysis, 541
Hemoglobin, 48, *48*, 131, 387*t*, 391–392
    concentration, 391–392, 401
    sickle-cell (HbS), 392
    structure of, 391–392, *392*
Hemoglobin S (HbS), 392
Hemolysis, 393, 394
Hemolytic anemia, 400
Hemolytic disease of the newborn (HDN), 395
Hemophilia, 404
Hemopoiesis, 145, 389–390
    erythrocyte, 393, *393*
    leukocyte, 398, *399*
    platelet, *402*, 402–403
Hemorrhagic anemia, 400
Hemostasis, *403*, 403–404
Heparin, 397–398
Hepatic lobules, 572, *573*

Hepatic macrophages, 572, *573*
Hepatic portal system, 453, *454*
Hepatic sinusoids, 455, 572, *573*
Hepatitis, clinical application, 610b
Hepatocytes, 572, *573*
Hepatopancreatic ampulla, *573*, 574
Hepcidin, 376
Herniated disc, *166*, 169
Hiatus, aortic, *445*, 453
High-altitude physiology, 515
High-density lipoproteins (HDLs), 598
Hilum, 525, *526*
Hinge joints, 183, *183*
Hip, *17*
    bones of (os coxae), *157*, 175, *176*, 176–177
    joints, 189, *189*
    muscles of, 236–237, *236–237*
Hippocampus, 294
Hippocrates, 89
Histamine, 397, 567, 569
Histological sections, 95
Histology, 3, 93–124
    of blood vessels, 430–431, *431*
    connective tissue, 104–112
    definition of, 94
    epithelial tissue, 97–103
    glands, 116–117
    membrane, 118, *118*
    muscular tissue, 113–114, *114*, 114*t*
    nervous tissue, 113, *113*
    stem cell, 115
    tissue growth, 119–120
Histopathology, 3
Histotechnician, 121
Holocrine glands, 117, *117*, 138
Homeostasis, 10–13. *See also specific substances and processes*
    blood's role in, 387
    definition of, 10
    hormonal control of, 357, 362–363
    negative feedback in, 11
    positive feedback in, 12
Homologous chromosomes, 644
Hooke, Robert, 61
Hormone receptors, *361*, 361–362
Hormone(s), 116–117, 357–363. *See also specific hormones*
    action and effects of, 361, *361*
    aging and, 380
    antagonistic effects of, 361
    in bone physiology, 153
    clearance of, 362
    definition of, 357
    enteric, 376
    as first messenger, 361
    function of, 361
    gastric, 567
    glomerular filtration, 533, *534*
    in homeostasis, 357, 362–363
    hypersecretion of, 378
    hyposecretion of, 378
    inaction of, 378
    inhibiting, 367
    interactions of, 361–362
    negative feedback inhibition of, 367–368
    negative feedback of, 362
    permissive effects of, 361
    positive feedback of, 362
    releasing, 367, *368*
    secretion of, 357–359, *359*
    structure of, 360, *360*
    synergistic effects of, 362
    transport of, 387
Horseshoe kidney, 7
Hot flashes, 642
Human body
    anatomical planes of, 15, *15*, 16
    anatomical position of, 15, *15*
    anatomical variations in, 6, *7*, 9
    anatomy of, 2–7
    characteristics of life, 8–10
    examination of, 3–4
    homeostasis in, 10–13
    MRI full-body image of, *1*
    negative feedback in, 11
    physiological variations in, 13
    plan of, 13–21
    positive feedback in, 12
    regions of, 16–17, *16–17*
    sections of, 15, *16*
    structural hierarchy of, *14*, 14–15
Human immunodeficiency virus (HIV), 489, *489*
Humerus, muscles, *231*, 231–232
Humoral immunity, 478, 485–487, *486*
Hunchback, 166
Hunger contractions, 615
Hunger hormone, 614
Hyaline cartilage, 109, *110*, 110*t*
    articular, *151*
    endochondral ossification, 151, *151*
Hydrochloric acid (HCl), 567
Hydrogen
    body percentage of, 31, 32*t*
    covalent bonds of, 35, *35*
    isotopes of, 33
Hydrogen bonds, 35, 36–37, 37*t*
Hydrogen ions, 39
Hydrolysis, 38, 41, *41*
Hydrophilic heads, of phospholipids, 45, *45*, 64
Hydrophilic substances, 38, 41
Hydrophobic substances, 38, 43
Hydrophobic tails, of phospholipids, 45, *45*, 64
Hydroxide ions, 39
Hyman, Flo, 108
Hymen, 640
Hypercalcemia, 152
Hyperextension, 184, *184*
Hyperglycemia, 378
Hyperkalemia, 429, 548
Hyperkyphosis, *165*, 166
Hyperlordosis, *165*, 166
Hypernatremia, 548
Hyperopia, 350, *351*
Hyperplasia, 119
Hypersecretion, of hormones, 378
Hypersensitivity, 479
Hypertension, 72, 436
Hyperthermia, 620
Hyperthyroidism, 371, 378
Hypertonic solutions, 71, *71*
Hypertrophy, 119
Hypocalcemia, 152
Hypodermis, 126, *127*, 130
Hypogastric region, *17*, 24
Hypoglycemia (glucose deficiency), 595
Hypokalemia, 548
Hyponatremia, 24, 548
Hypophyseal portal system, 364, *365*
Hyposecretion, of hormones, 378
Hypotension, 436
Hypothalamus, 290, *358*, 363–369
    anatomy of, 364, *365*
    embryonic development of, *364*
    functions of, 363
    hormones of, *365*
    in pituitary regulation, 363, 367, *368*
Hypothermia, 620
Hypothyroidism, 371, 378
Hypotonic solutions, 71, *71*
Hypoxemia, 514
Hypoxia, 400

I

I bands, *200, 202,* 202–203
Involution, 652
Ileocecal junction, 577, *577*
Ileocecal valve, 577, *577*
Ileum, 576, *577*
Iliac crest, 177
Iliac fossa, 177
Imaging, 4–6
Immune system, 463, *665*
    aging of, 490
    effects on other organ systems, 491
    physical barriers, 472
    protective cells, 472–474, *473*
    protective proteins, 474
Immunity, 465
    active, 487–488
    adaptive, 477–482
    cellular, 483–485, *484*
    connective tissue and, 104
    humoral, 478, 485–487, *486*
    innate, 472–477
    passive, 487–488
    skin and, 126–127
Immunoglobulins (Ig), 480
Immunology, 463
Incisors, 561, *561*
Inclusions, cellular, 61, 75, 77
Incomplete proteins, 599
Incomplete tetanus, 210, *211*
Incus, 331, *331*
Induced pluripotent stem cells (iPS), 115
Infarction, 120, 404
    cerebral, 120
    myocardial, 13, 120
Inferior, 25*t*
Infertility (sterility), 647
Inflammation, 474
    neutrophil behavior in, 474–475, *475*
Infraspinous fossa, *171*, 172, 232
Infundibulum, 637
Ingestion, 557
Inguinal canal, 630
Inguinal ligament, 228, 455
Inguinal region, *17*
Inhibin, 375, 651
Inhibiting hormones, 367

Injuries, muscle, 220
Innate immunity, 472–477
Inner ear, 332–333, *333*
Inner-ear cavity, 158
Inorganic compounds, 41
Inorganic phosphate, 52
Insensible perspiration, 137
Inspection, 3
Inspiration, 506, *507*
Inspiratory reserve volume, 509, *509*
Insulin, 360, *360*, 361, 373, *373*, 612, 615
    aging and, 380
    in bone physiology, 153
    in diabetes mellitus, 362, 378
    receptor-mediated endocytosis of, 74
    in stress response, 379
Insulin-like growth factor I (IGF-I), 376
Insulin receptors, 361, 362
Insulin resistance, 378
Integral proteins, *64*
Integrating (control) center, 11
Integumentary system, 20, 21, *22*, 125–142, *665*.
    See also Hair; Nail; Skin
    aging of, 139
    components of, 126
    diagnostic value of, 126
    effects on other organ systems, 140
    during pregnancy, 658
    study and treatment of (dermatology), 132, 139
Intensity, sensory information, 322
Intercalated discs, 215, *216*, 422, *423*
Intercellular clefts, *432*, 433
Intercellular space, *62*
Intercostal veins, 450
Interferons, 474
Intermediate filaments, *62*, 75, *76*
Internal anal sphincter, *585*, 586
Internal genitalia, 627
Internal intercostal muscles, *508*, 509
Internal urethral sphincter, 543
Interneurons, 253, *253*
Internodal conduction, cardiac, 424
Internodes, 255
Interosseous membrane, *173*, 174
Interphase, *85*, 86
Interscapular region, *17*
Interstitial endocrine cells, *375*, 376, 627
Interstitial (tissue) fluid, 62
Intertubercular sulcus, *172*, 173
Interventricular sulcus, *415*, 417
Intervertebral discs, 164, *166*, 168–169
    fibrocartilage of, 110, 110t, 164, *166*
    herniated, *166*, 169
Intervertebral foramen, 166, *166*
Intestinal bacteria, 586
Intestinal phase, 569, *569*
Intracellular compartment, 546
Intracellular face, of plasma membrane, *64*
Intracellular fluid (ICF), 62, 71
Intramembranous ossification, 150
Intraperitoneal organs, 20
Intrapulmonary pressure, 506, *507*
Intrinsic defecation reflexes, 587
Intrinsic factor, 399, 567
Invasive imaging techniques, 4
Inversion, *187*, 187–188
In vitro fertilization (IVF), 115

Involuntary muscle, 113
Involution, 659
Iodine, 32t
Ionic bonds, *33*, 34, 35, 37t
Ionization, *33*, 33–34
Ionizing radiation, 33
Ion(s)
    formation of, *33*, 33–34, 34t
    valence of, 33
Ion trapping, 40
Iris, 341, *341*
Iron
    body percentage of, 32t
    oxidation of, 55
Iron-deficiency anemia, 399
Irregular bones, 149
Ischial tuberosity, *176*, 177
Islets, pancreatic, 372, *372*
Isomers, 35, *35*
Isometric contraction, 210–211
Isotonic contraction, 210–211
Isotonic solutions, 71, *71*
Isotopes, 33
Isthmus, of thyroid gland, 370, *370*
Ivar the Boneless, 51

J

Jaundice, 133, 393
Jejunum, 576, *577*
Joint(s), 180–191
    articular cavity, 181, *181*
    atlantoaxial, *167*
    ball-and-socket, 182, *183*
    cartilaginous, 181
    classification, *181*, 181–184, *183*
    condylar (ellipsoid), 182, *183*
    degenerative disease of, 154–155
    elbow, 188–189, *189*
    fibrous, 181
    hinge, 183, *183*
    hip, 189, *189*
    knee, *190*, 190–191
    movement, 184–188
    pivot, 183, *183*
    plane (gliding), 183, *183*
    sacroiliac, 175, *176*
    saddle, 182–183, *183*
    shoulder, 188, *188*
    study of (arthrology), 144
    sutures, 149, 158, *159*, *160*, *161*, 164
    synovial, *181*, 181–184, *183*
    temporomandibular, 163, 184
Jugular foramen, 447
Juxtaglomerular apparatus, 534, *535*

K

Kartagener syndrome, 68
Keratin, 75, 101, 126–129
Keratinized epithelium, *101*, *102*, 102t
Keratinocytes, *128*, 128–131
Ketoacidosis, 378, 609
Ketone bodies, 609
Ketone(s), diabetes mellitus and, 378

Ketonuria, 378
Kidney disease, anemia in, 399
Kidney(s), *524*
    anatomical variations in, 6, *7*, 9
    blood pressure and, 438
    circulation, 526, *527*
    circulation of, 452, *453*
    endocrine function of, 376
    epithelial tissue of, *99*, 99t, *103*, 103t
    functions, 523–524
    gross anatomy, 525–526, *526*
    nephron, 527–529, *528–530*
    renin-angiotensin-aldosterone (RAA) system, 375
Kidney stone, 542
Killing zone, 473
Kilocalorie (kcal), 38, 595
Kinesin, 76
Kinesiology, 144
Kinetic energy, 54
Knee, *17*
    anterior cruciate ligament (ACL), 191
    cruciate, 191
    injuries and arthroscopic surgery, 191
    joints, *190*, 190–191
    menisci, 191
    muscles of, 238
    popliteus muscle, 191
    posterior cruciate ligament (PCL), 191
Kyphosis, 164, *165*

L

Labia majora, 640
Labia minora, 640
Labor contractions, 659
Lacrimal apparatus, 340
Lacrimal (tear) gland, 340
Lacrimal sac, 340
Lactate (lactic acid), 604
Lactation, 659, 660, 662
Lacteal, 577, *578*
Lactiferous ducts, 642
Lactose, 42, 43t
Lactose intolerance, 582
Lacunae
    of bone, 109, 111, *111*, 146, *147*
    of cartilage, 109, *110*
    cavernous spaces, 633
Lamella(e), *110*, 111, *147*, 148
Lamellar corpuscles, *127*
Lamellar (pacinian) corpuscles, 324
Language, brain function in, 299, *299*
Large arteries, 431, *431*
Large intestine, 557, *557*, 559, *559*, 584–587, *585*
    absorption and motility, 586
    anatomy, 584–587, *585*
    blood supply to, 452, *453*
    defecation, 587
    endocrine function of, 376
    microbiome, 586
Large veins, *431*, 434
Laron dwarfism, 362
Laryngeal prominence, 446
Laryngopharynx, 499, 562
Laryngospasm, 152
Larynx, 499–500, *499–500*

Lateral, 25t
Lateralization, of cerebral functions, 299–300, *300*
Lateral longitudinal arch of foot, 180
Lateral surfaces, of epithelial tissue, 97
Law of complementary base pairing, 50, 85
Leaflets, of heart valves, 417
Leeuwenhoek, Antony van, 61
Left atrioventricular valve, 417, 418
Left atrium, *415*, 416, *416*, 418
Left bundle branch, 424
Left lobe, liver, 572, *573*
Left lower quadrant (LLQ), 16, *16*
Left upper quadrant (LUQ), 16, *16*
Left ventricle, *415*, *416*, 417, 418
Leg, 17. *See also* Lower limb
Length, units of, A–10
Lens, 342
Leptin, 376, 615
Lesser omentum, *19*, 559
Lesser vestibular glands, 640
Lethal hit, 484
Leukemia, 400, *400*
Leukocyte(s), 105, 111, *112*, 388, *388*, 396–401.
    See also specific types
  disorders of, 400, *400*
  functions of, 387, 396
  life cycle, 398, *399*
  as percentage of whole blood, *388*
  phagocytosis by, 74
  production of, 389–390, 398, *399*
  pseudopods of, 68, *68*
  quantities (count) of, 387t, 400, 401
  structure of, 396, *396*
  types, 396–398, *397*
Leukocytosis, 400
Leukopenia, 400
Leukopoiesis, 398, *399*
Leydig cells, 376
Libido, 634
Life
  characteristics of, 8–10
  defining, 8
  end of (death), 9
  essential functions of, 8–10
Ligaments, 105, 144
Ligand-gated channel, 259
Limbic system, 294, *294*
Lincoln, Abraham, 108
Linea alba, 228
Lines of bone, *156*
  epiphyseal, *148*, 149
  linea aspera, *177*, *177*
Lingual frenulum, *560*, 561
Lingual lipase, 562
Lingual papillae, 326, *327*, *560*, 561
Linoleic acid, 44, *44*
Lipid absorption, 465
Lipid bilayer, 64, *64*
Lipid(s), 41, 43–46
  blood lipoproteins, 598
  composition of, 43
  dietary fat and cholesterol, 597–598
  fat and energy storage, 597
  functions of, 43t
  metabolism, 608–609
  in plasma membrane, 64, 64–65
Lipogenesis, 608

Lipolysis, 608
Lipoproteins, 598
Liver, 557, *557*, 572–574, *573*
  anatomy, 572–574, *573*
  blood supply to, 452, *453*
  endocrine function of, 376
  functions, 610t
Lobar (secondary) bronchi, 503, *503*
Lobes
  of glands, 116
  of pituitary gland, 364, *365*
Local chemical messengers, 567
Local control (renal autoregulation), 533–535
Local potential, 259
Location, sensory information, *321*, 321–322
Lochia, 659
Lock-and-key model, 50, 361
Long bones, 111, *148*, 148–149
Long-wavelength (L) cones, 348
Loop of Henle (nephron loop), 528–529, *530*, 537
Loose connective tissue, 105, *106*, 106t
Lordosis, 164, *165*
Low-density lipoproteins (LDLs), 598
Lower esophageal sphincter, 563
Lower limb, 17, *17*
  arteries of, 455
  bones of, *157*, 175–180, *176*
  veins of, 455–456
Lower respiratory tract, 497
Lubb-dupp, 428
Lumbar curvature, 166
Lumbar region, 17
Lumbar vertebrae, 164, 168, *168*, *169*
Lumen, 97
  of blood vessels, *431*
Lung cancer, 511, *511*
Lung(s), 501–504, *502–503*
  blood supply, 412, *412*, 505
  heart's anatomic relationship to, 412, *413*
Luteal phase, 652
Luteal (postovulatory) phase, 652
Luteinizing hormone (LH), 366, 367t, *368*, 634, 652
Lymph, 464, *465–466*, 465–467
Lymphatic nodules, 467
Lymphatic organs, 465, 468–471
  lymph nodes, 470–471, *471*
  spleen, 469–470, *470*
  thymus, 468–469, *469*
  tonsils, 468, *468*
Lymphatic system, 21, *22*, 463–471, *464*, 665
  aging of, 490
  capillaries, 465, *465*
  effects on other organ systems, 491
  functions, 465
  thoracic region, *466*, 467
Lymphatic tissue, 465, 467, *467*
Lymphatic trunks, *466*, 467
Lymphatic vessels, 464, *465–466*, 465–467
Lymph nodes, 470–471, *471*
  metastatic cancer and, 471
Lymphoblasts, 398, *399*
Lymphocytes, *112*, 385, 388, *397*, 398, 473. *See also*
    B lymphocytes; T lymphocytes
  production of, 398, *399*
Lymphoid leukemia, 400
Lymphoma, 471

Lysosomal enzymes, 79–80
Lysosomes, *62*, *76*, *78*, 79–80, 81t, *84*, 396, *396*, 402

M

Macromolecules, 41
Macronutrients, 595
Macrophages, 105, 398, 473, *473*
  in erythrocyte disposal, 393
  phagocytosis by, 74
  pseudopods of, 68, *68*
Macula, 333, *334*
Macula lutea, 343
Magnesium, 32t
Magnesium chloride, 34t
Magnetic resonance imaging (MRI), 4, 5, *5*
  full-body scan, *1*
  functional, 5
  lumbar, pelvic, and upper femoral region
      muscles, *197*
  transverse sections in, *16*
Main (primary) bronchi, 502, *503*
Major calyx, 526, *526*
Male climacteric, 634
Male reproductive system, 20, 21, *23*
  anatomy, 627–634
  from puberty to andropause, 634–635
Males, erythrocytes in, 391
Malignant tumor, 89
Malleus, 331, *331*
Maltose, 42, 43t
Mammary glands, 117, *117*, 136, 640
Manganese, 32t
Marfan syndrome, 108
Marrow cavity, *148*, 151, *151*
Mass, units of, A–10
Massage therapist, 243
Mass movements, 586
Mastication, muscles of, 225–227, *226*
Mastoid fontanel, 163, *164*
Matrix
  of bone tissue, 146, 152
  of cartilage, 109, *110*
  extracellular, 94
  of hair, 134, *135*
  of mitochondria, *78*, 80
  of nail, 136
Matrix reactions, 604
Mature (graafian) follicles, 650
Maxillary sinus, 158, *158*
Mechanical digestion, 557
Mechanoreceptors, 330
Medial, 25t
Medial longitudinal arch of foot, 180
Median plane, 15
Mediastinum, 18, *18*, 501, 502
Medical imaging, 4–6
Medical technologist (MT), 56
Medical terminology, 21, 24–26, 25t
Medium arteries, *431*, 431–432
Medium veins, *431*, 433–434
Medium-wavelength (M) cones, 348
Medulla, of hair, 134, *135*
Medulla oblongata, 288, *289*
  in blood pressure control, 437, *437*
  vasomotor center of, 437, *437*

Medullary cavity, 149, 151, *151*
Megakaryocytes, 402, *402,* 403
Meiosis, 86, 644
Melanin, 128, 130–131, *131,* 135
Melanocortin, 615
Melanocytes, *128,* 128–131
Melanoma, 132, *132*
Melatonin, 369–370
Membrane proteins, 64, *64, 65*
Membrane reactions, 604
Membrane(s), 18–20, 118, *118. See also specific entries*
Membrane skeleton (terminal web), *62,* 75, *76*
Membrane transport, 69–74, *70*
    active, *70,* 72
    carrier-mediated, 72
    by filtration, 70, *70*
    by osmosis, 70–71, *71*
    by simple diffusion, 70, *70*
    tonicity in, 71, *71*
    vesicular, 72–74, *73*
Membranous labyrinth, 332, *333*
Memory, 487
Memory B cells, 487
Memory T (T_M) cells, 483
Menarche, 642
Meninges, 18, 284, *285*
    brain, 158
    spinal, 269–270
Meningitis, 332, *332*
Menopause, 154, 642
Menstrual cycle, 375
Menstrual fluid, 653
Menstrual phase (menses), 653
Menstruation, 650
Merocrine glands, 117, *117*
Merocrine sweat glands, *127,* 136–137, *137*
Mesenteries, 19–20, *19–20,* 106, 106t, 559, *559*
Messenger RNA (mRNA), 81–83
    DNA transcription to, 81–83, *82*
    translation of, *82,* 83
Metabolic rate, 613
Metabolic states and metabolic rate
    absorptive state, 610–612
    metabolic rate, 613
    postabsorptive state, 612
Metabolic waste, 524
Metabolism, 8, 9–10, 53, 55, 603. *See also under specific entries*
    bone, 152–153
    inevitable effect of, 523
    and nutritional needs during pregnancy, 658–659
    skeletal muscles, 211–213
Metaphase, in mitosis, 86–88, *87*
Metastasis, 89
Metastatic cancer, 471
Micelles, 583
Microbiome, 463
    large intestine, 586
Microfilaments, *62,* 75, *76,* 76
Microglia, 253, *254*
Micrometer (μm), 62
Micronutrients, 595
Microscopes, 61
Microscopic anatomy, 3
Microtubules, *62,* 66–67, *67,* 75, *76,* 78
Microvilli, *62,* 66, *66,* 76, 578, *578*
Micturition, 544

Micturition center, 544
Micturition reflex, 544
Midbrain, 290
Middle ear, 331–332
Middle-ear cavity, 158
Midpiece, 647
Midsagittal plane, 15
Midwife, 664
Migraine headaches, 325
Milk ejection, 662
Milliosmoles per liter, A–11
Mineralocorticoids, 375
Mineral(s), 32
    deposition of, 152
    digestion and absorption, *581,* 583
    homeostasis, in bone, 152–153
    resorption of, 152
    storage in bones, 145
    and vitamin requirements, sources, 601t
Minor calyx, 525, *526*
Mitochondria, *62, 76, 78,* 80, 81t, 593
    ATP production in, 80
Mitochondrial DNA, 80
Mitochondrial electron-transport chain, 605
Mitochondrial matrix, *78,* 80
Mitosis, 86–89, *87*
Mitotic phase of cell cycle, *85,* 85–86
Mitotic spindle, 75, *87,* 88
Mitral valve, *416,* 417, 418
Mixed glands, 117
Modality, sensory information, 321
Molarity (M), A–10
Molars, *561,* 562
Molecular formula, 35
Molecules, 15, 35–37
Monoamines, 312, 360, *360*
Monoblasts, 398, *399*
Monocytes, *112, 388, 397,* 398, *399,* 473
Monomers, 41
Monosaccharides, 41, *42,* 43t
Mons pubis, 640
Morning sickness, 658
Morphology, functional, 3
Motility, gastric, 568
Motor association area, 297
Motor (efferent) division, 250
Motor neurons, 203
Motor (efferent) neurons, 253, *253*
Motor unit, 203, *203*
Mouth, 557, *557, 560–561,* 560–562
Movement
    muscles, 199
    study of (kinesiology), 144
    voluntary, 297–299, *298*
MRI, *4,* 5, *5*
    full-body scan, *1*
    functional, 5
    transverse sections in, *16*
Mucin, 117
Mucociliary escalator, 500, *500*
Mucosa-associated lymphatic tissue (MALT), 467
Mucous cells, 566
Mucous glands, 117
Mucous membranes, 18, 118, *118*
    digestive tract, 557–558, *558*
    of trachea, *100*
    of vagina, *102,* 102t

Mucus, 98, 117
Multiple sclerosis12 (MS), 257
Multipolar neurons, 252, *252*
Multiregional functions, brain, 295–301
    cerebral lateralization, 299–300, *300*
    cognition, 296–297
    electroencephalogram, 300, *301*
    language, 299, *299*
    sensation, 296, *297*
    voluntary movement, 297–299, *298*
Mumps, 562, *562*
Murmurs, heart, 428
Muscle-connective tissue relationships, 218–221, *219*
Muscle fibers, 114, *114*
    of cardiac muscle, 422
    fast-twitch, 213
    slow-twitch, 213
Muscle(s)
    abdominal, 227–228, 229t
    acromion, *231*
    acting on humerus, *231,* 231–232
    adductor longus, *223*
    adductor magnus, *224*
    adductors, *223*
    adductors (thigh), 236–237, *237*
    antagonist, *221,* 222
    anterior group (leg), 238, *239*
    appendicular, *230–231,* 230–235, 240t–242t
    arrector pili, *127,* 135, *135*
    attachments, 228
    axial, 228t
    of back, 228, 229t
    belly of, 221
    biceps brachii, 213, *223, 230,* 232–233, *234*
    biceps femoris, *224,* 238
    brachialis, *223, 230, 231,* 232, 233, *234*
    brachioradialis, *223, 230,* 232, 233, *234*
    buccinator, 225, *226,* 229t
    calcaneal tendon, *224*
    calcaneal (Achilles) tendon, 213, 239, *239*
    cardiac, *215,* 215–216, 422, *423*
    of chewing, 225–227, *226*
    classification, 221–222
    clavicle, *230, 231*
    compartments, 218, *219*
    conditioning, 214
    control of body openings and passages, 199
    coordinated action, *221,* 221–222
    coracobrachialis, *231*
    cramps, 220
    deltoid, *223, 230,* 232
    diaphragm, 18, *18*
    endurance, 212
    erector spinae, *224,* 228, 229t
    extensor carpi radialis brevis, 235, *235*
    extensor carpi radialis longus, 235, *235*
    extensor carpi ulnaris, *224,* 235, *235*
    extensor digitorum, *224,* 235, *235*
    extensor digitorum longus, 238, *239*
    extensor indicis, 235, *235*
    extensor retinaculum, 233, *235*
    external intercostal, 508, *508*
    external oblique, *223, 224,* 227, 228, 229t
    extrinsic, 340
    of face, 225, *226*
    fascicles, 218, *219*

Muscle(s) (continued)
  fatigue, 212
  of femur, 236–237, *236–237*
  fixator, 222
  flexor carpi radialis, *223*, 233, *234*
  flexor carpi radialis brevis, 222
  flexor carpi ulnaris, 233, *234*
  flexor digitorum profundus, *223*, 234
  flexor digitorum superficialis, 233, *234*, *235*
  flexor retinaculum, 233
  foot, 238–239, *239*
  forearm, 232–233, *234–235*
  frontalis, *223*, *225*, *226*, 229t
  functions of, 199
  gastrocnemius, 213, *224*, 239, *239*
  gluteal, 236, *237*
  gluteus maximus, *224*
  gluteus medius, *224*, 236, *237*
  gluteus minimus, *224*
  glycemic control, 199
  gracilis, *223*, *224*
  gross anatomy, 218–242
  hamstrings, 238
  of head, 227, 229t
  heat generation by, 199
  of hip, 236–237, *236–237*
  iliacus, *176*, 177, 236, *236*
  iliopsoas, 236, *236*
  iliotibial tract, *224*
  infraspinatus, *224*, *230*, *231*, 232
  injuries, 220
  insertion, 221
  intercostal, 227
  internal intercostal, *508*, 509
  internal oblique, *223*, 227, 228, 229t
  involuntary, 113
  of knee, 238
  lateral rotators, *224*
  latissimus dorsi, *224*, *230*, *231*, 231–232
  levator scapulae, *224*, *230*, 230–231
  of lower limb, 236–242, *239*
  masseter, *223*, *226*, 226, 229t
  of mastication, 225–227, *226*
  metabolism, 211–213
  minimus, 236, *237*
  movement, 199
  muscle-connective tissue relationships, 218–221, *219*
  naming, 222, 222t, *223–224*
  of neck, 227, 229t
  occipital protuberance, 227
  orbicularis oculi, *223*, *225*, *226*, 229t
  orbicularis oris, *223*, *225*, *226*, 229t
  origin, 221
  pain, 220
  palmaris longus, 6, 233, *234*
  papillary, 417
  patellar ligament, 238
  of pectoral girdle, *230–231*, 230–235
  pectoralis major, *223*, 227, *230*, *231*, 231
  pectoralis minor, *223*, 230
  of pelvic girdle, *236–237*, 236–242
  piloerector, *127*, 135, *135*
  plantaris, 6
  popliteus, 191
  posterior group (leg), 239, *239*
  prime mover (agonist), 221, *221*
  pronator quadratus, *223*, 232, 233
  pronators, 232, 233
  pronator teres, *231*, 232, 233, *234*
  psoas major, 236, *236*
  pterygoids, *226*, 226–227, 229t
  quadriceps femoris, 238
  quadriceps (patellar) tendon, 238
  recovery period, 213
  rectus abdominis, *223*, 227, 228, 229t
  rectus femoris, *223*, 238
  red, 213
  rhomboids, *224*, 231
  roator cuff, *231*, 232
  rotators (thigh), 237, *237*
  sartorius, *223*, 238
  scapular, 232
  semimembranosus, 190, *224*, 238
  semitendinosus, *224*, 238
  serratus anterior, *223*, *224*, 230
  skeletal (*See* Skeletal muscle(s))
  smooth, *216*, 216–217
  soleus, 213, *224*, 239, *239*
  sphincter, 199
  spine of scapula, *230*
  splenius capitis, *224*
  stability, 199
  stapedius, 331, *331*
  sternocleidomastoid, *223*, *224*, 227, 229t, *230*
  sternum, *230*
  strength, 214
  striations of, 113, *114*, 422, *423*
  subscapularis, *231*, 232
  supinator, *223*, *231*, 232, 233
  supraspinatus, *224*, *230*, *231*, 232
  synergist, *221*, 221–222
  temporalis, *225*, *226*, 229t
  tensor tympani, 331, *331*
  teres major, *224*, *230*, 232
  teres minor, *224*, *230*, *231*, 232
  tibialis anterior, *223*, 238, *239*
  transverse abdominal, *223*, 227, 228, 229t
  trapezius, *223*, *224*, 227, 229t, *230*
  triceps brachii, *224*, *230*, 232, 233, *234*
  of trunk, 227–228, 229t
  of upper limb, *230–231*, 230–235
  vastus intermedius, *223*, 238
  vastus lateralis, *223*, 238
  vastus medialis, *223*, 238
  visceral, 114, *216*, 217
  voluntary, 114
  white, 213
  wrist and hand, 233–235, *234–235*
  zygomaticus major, *223*, *225*, *226*, 229t
  zygomaticus minor, *225*, *226*, 229t
Muscle tone (tonus), 209
Muscle twitch, 210, *211*
Muscular arteries, *431*, 431–432
Muscular dystrophy, 220
Muscularis externa, digestive tract, 558, *558*
Muscular system, 21, *22*, 197–247, 665
  aging of, 243
  anatomy of, 218–242
  anterior view, *223*
  defined, 198
  effects on other organ systems, 244
  functions, 199
  muscular tissue and cells, 198–205
  nerve–muscle relationship, 203, *203, 204*
  overview, 198
  posterior view, *224*
  tissue and cells, 198–205
Muscular tissue, 94, 94t, 198–205
  excitability of, 113
  histology of, 113–114, *114*, 114t
  striations of, 113, *114*, 422, *423*
Mutagens, 89
Mutations, 10, 89
Myelination, 256, *256*
Myelin sheath, 255, *256*
Myeloblasts, 398, *399*
Myeloid leukemia, 400
Myocardial infarction, 13, 120
Myocardium, *413*, 414
Myocytes, 114, *114*
Myofibrils, 200, *200*
Myofilaments, *201*, 201–203, *202*
  elastic, *201*, 202
  thick, *201*, 202
  thin, *201*, 202
Myoglobin, 200
Myometrium, 638
Myopia, 350, *351*
Myosin, *201*, 202, 208, 209

N

NAD+ (nicotinamide adenine dinucleotide), 604
Nail bed, 136
Nail body, *136*
Nail fold, 136, *136*
Nail matrix, 136
Nail plate, 136, *136*
Nail(s), 136, *136*
Nares, 497, *498*
Nasal apertures, posterior, *160*, 161
Nasal cavity, 158, 497, *498*
Nasal conchae, 161–162, *163*, 497, *498*
  inferior, 162, 163
  middle, 162
  superior, 162
Nasal septum, 497, *498*
Nasolacrimal duct, 340
Nasopharynx, 499, 562
Natriuretic peptides, 376, 538
  in blood pressure control, 438
Natural killer (NK) cells, *462*, *473*, 473–474
Near point, 344, *345*
Neck (cervical region), 16, *17*
  arteries of, *446*, 446–447
  muscles of, 227, 229t
  veins of, 447, *448*
Necrosis, 120
Negative feedback, 11
  in blood pressure control, 11, *11*
  in endocrine system, 362
  in thermoregulation, 11
Negative feedback inhibition, 367–368
Neonate
  skull of, 163, *164*
  thymus of, 371, *371*
Neoplasia, 119
Neoplasm, 89
Nephron, 526, 527–529, *528*, 528–530

blood circulation in, 526, *528*
renal corpuscle, 527, *529*
renal tubule, 528–529, *530*
Nephron loop (Loop of Henle), 528–529, *530,* 537
Nerve endings
free, 323
simple, 323–324, *324*
Nerve–muscle relationship, 203, *203, 204*
Nerve(s)
cardiac, 425
cranial, 302–304, 302*t, 303*
defined, 270
spinal, 270–272, *271–272,* 273*t*
Nerve signal, 259–263, *260, 261–263*
Nervous system, 21, *23,* 665
aging of, 313
autonomic (*See* Autonomic nervous system (ANS))
brain (*See* Brain)
central nervous system (CNS), 249, 250
drugs and, 312, *313*
endocrine system *vs.,* 358–359, *359,* 359*t*
forms of nervous tissue, 255–257
myelin sheath, 255, *256*
neuroglia, 253–254, *254,* 255*t*
neurons (nerve cells), 250–253, *251*
overview, 249–250, *250*
peripheral nervous system (PNS), 250
Nervous tissue, 94, 94*t*
excitability of, 113
forms of, 255–257
histology of, 113, *113*
Neural control, of breathing, 510, *510*
Neuroendocrine cells, 358–359
Neuroglia, 253–254, *254,* 255*t*
Neuromuscular junction (NMJ), 203, *204*
Neuromuscular toxins, paralysis and, 205
Neurons (nerve cells), 113, *113,* 250–253, *251*
action potential, generating, 259, *259*
bipolar, 252, *252*
interneurons, 253, *253*
motor (efferent), 253, *253*
multipolar, 252, *252*
nerve signal, 259–263, *260, 261, 263*
physiology of, 258–263
resting membrane potential (RMP), *258,* 258–259
sensory (afferent), 253, *253*
structure, 250–252, *251*
unipolar, 252, 252–253
Neuropathy, diabetic, 378
Neuropeptide Y (NPY), 615
Neurophysiology, 8
Neurosoma, 113, *113,* 250, *251*
Neutral solution, 39
Neutrons, *32,* 32–33
Neutrophil extracellular traps (NETs), 397, 473
Neutrophil(s), *112, 388,* 397, *397,* 473, *473*
production of, 398, *399*
pseudopods of, *68*
Night blindness, 346
Night (scotopic) vision, 347
Nitrogen, 31, 32*t*
Nitrogenous base, *49,* 50
Nitrogenous wastes, 524, 536
Nitroglycerine patches, 129
Nociceptors, 324
Nodes of Ranvier, 255

Nonfluent aphasia, 300
Noninvasive imaging, 4
Nonkeratinized epithelium, *101, 102,* 102*t*
Nonpolar covalent bonds, 36, *36,* 37*t*
Nonspecific granules, 396
Nonsteroidal anti-inflammatory drugs (NSAIDs), 570
Nonverbal communication, 127
Norepinephrine
adrenal secretion of, 374
in blood pressure control, 437–438
cardiovascular effects of, 425, 429
as hormone, 358–359
as neurotransmitter, 358
in stress response, 379
Nose, 497–499, *498*
Nostrils, 497, *498*
Nuchal region, *17*
Nuclear envelope, *62,* 77, *78*
Nuclear medicine, 6
Nuclear pores, 77, *78*
Nuclei
of atoms, 32
of brain, *366,* 367
of cells, *62,* 77, *78*
Nucleic acids, 41, 50
Nucleolus, *62,* 77, *78*
Nucleotides, DNA, *49,* 50
Nucleus, 81*t*
Nucleus pulposus, *166,* 168
Nurse cells, 375, *375*
supporting cells, sustentocytes, 628
Nutrient, 595
Nutrition
calories, 594–595
carbohydrates, 595–596
fiber, 596–597
lipids, 597–598
minerals and vitamins, 600–601
nutrients, 595
proteins, 598–600

O

Obesity, 602
Obstructive disorders, 510
Obturator foramen, *176,* 177
Occupational therapist (OT), 276
Oils, 43–44
Olecranon, *173,* 174
Olecranon bursitis, 189, *189*
Olecranon fossa, *172,* 173
Olfaction, 328–329, *329*
Olfactory cells, 328, *329*
Olfactory epithelium, 499
Olfactory hairs, 328, *329*
Olfactory mucosa, 328, *329*
Oligodendrocytes, 253, *254*
Oliguria, 544
Omentum
greater, *19*
lesser, *19*
Oncogenes, 89
Onhibin, 634
Oogenesis, 648
Oogonia, 648

Open canalicular system, of platelets, 402, *402*
Open reduction of fracture, 154
Opposition, joint movement, 187
Optic chiasm, 348, *349*
Optic disc, 343
Optician, 352
Optic nerve, 343
Oral (buccal) cavity, 158, 560, *560*
Oral phase, swallowing, 563, *563*
Ora serrata, 343
Orbit, 158
Orbital fat, 340
Orbital region, *162*
anatomy of, 339–340, *340*
eyeball, 341–343, *341–343*
visual image formation, 344, *345*
visual nerve signal, generating, 345–348, *346–348*
Organelles, 14–15, 61, 75, 77–80, *78,* 81*t*
Organic chemistry, 41
Organic compounds, 40–53
Organism, 14, *14*
Organ(s), 14, *14*
intraperitoneal, 20
retroperitoneal, *19,* 20
Organ systems, 14, *14,* 20–21, *22–23*
Orgasm (climax), 655
Oropharynx, 499, 562
Orthopedic clinical nurse specialist (OCNS-C), 192
Orthopedic nurse, 192
Osmolality, A–11
Osmolarity, A–11
of blood, 389
Osmole, A–11
Osmoreceptors, 547
Osmosis, 70–71, *71*
capillary, 70, 438, *439*
imbalances in, 71
Ossification, 150–152, *151*
Ossification center
primary, 151, *151*
secondary, 151, *151*
Osteoarthritis (OA), 154–155
Osteoblasts, *145,* 146, 151, 152, 154, 192, 376
Osteocalcin, 376
Osteoclasts, *145,* 146, 151, 153, 154, 192
Osteocytes, 111, *145,* 146, *147*
Osteogenesis imperfecta (OI), 51
Osteogenic cells, *145,* 145–146
Osteoid tissue, 152
Osteology, 144
Osteomalacia, 146
Osteon(s), *111, 147,* 147–148
Osteoporosis, 154
Oteopenia, 192
Otitis media, 332, *332,* 499
Otolithic membrane, 333, *334*
Otoliths, 333, *334*
Outer ear, 330–331
Oval window, 332, *333*
Ovarian cycle, 375, 648, 650–652
Ovaries, *358,* 636
endocrine function of, 375
histology of, 375, *375*
Ovoid cells, 96, *96*
Ovulation, 652
Oxidation, 55, 55*t*
Oxidizing agent, 55

Oxygen
  body content of, 31, 32*t*
  molecule, 35
  transport in blood, 387, 390
Oxygenated blood, 412, 418
Oxygen debt, 213
Oxyhemoglobin, 514
Oxytocin (OT), 12, 360, *360*, 364, *365*, 366–367, 368, 659

P

Pacemaker, 424, *424*, 425
Pacinian (lamellar) corpuscles, 127
Packed cell volume (hematocrit), 387*t*, 388, 390–391
Paganini, Nicoloì, 51
Pain, *324*, 324–325
Palate, 497, *498*, *560*, 561
  hard, 162
Pallor, 133
Palmar region, *17*
Palmitic acid, *44*
Palpation, 4
Pampiniform plexus, 631
Pancreas, 372–373, *557*, *557*, *574*, 575
  blood supply to, 452, 453, *453*, *454*
  endocrine function of, 373
  exocrine function of, 373
  gross anatomy of, 372, *372*
  histology of, 372, *372*
Pancreatic amylase, 575
Pancreatic duct, *372*, *574*, 575
Pancreatic islets, 372, *372*
Pancreatic juice, 575
Pancreatic lipase, 575
Papillae, dermal, *127*, *128*, 130, 134, *135*
Papillary layer, of dermis, 130
Pap smear, 101
Parafollicular cells, *370*, 371
Paralysis
  flaccid, 205
  neuromuscular toxins and, 205
  spinal cord injury and, 269
Paranasal sinuses, 158
Parasagittal planes, 15
Parasympathetic defecation reflexes, 587
Parasympathetic division, of ANS, 425
Parathyroid glands, 9, *358*, 371
Parathyroid hormone (PTH), 153, 371
Paraurethral glands, 640
Parietal cells, 566
Parietal pericardium, 412, *413*
Parietal peritoneum, 19, *19*, 559
Parietal pleura, 501
Parkinson disease (PD), 265
Parotid glands, 562, *562*
Partial pressure, 512
Partial-pressure gradient, 513, 514
Partial-thickness burns, 132
Parturition, 659
Passive immunity, 487–488
Patellar region, *17*
Pathogens, 464
Pathological fracture, 154
Pathophysiology, 8
PCT. *See* Proximal convoluted tubule (PCT)

Pectoral girdle
  bones of, *157*, *170*, 171–175
  muscles of, *230–231*, 230–235
Pectoral region, *17*
Pedal region, *17*
Pelvic cavity, 18, *18*, 18*t*
Pelvic curvature, 164, *165*
Pelvic girdle, bones of, 175–180, *176*
Pelvic inlet, 175
Pelvic outlet, 175
Pelvic region, 16
  arteries of, 452, *453*
  veins of, 452–455, *454*
Pelvis, *157*, 175–180, *176*
  comparative anatomy of, 3
  sexual dimorphism of, 175–176
Pemphigus vulgaris, 69
Penicillin, 40
Penile artery, 654
Penis, 632
Pepsin, 49, 567
Peptic ulcer, 570, *570*
Peptide bonds, *47*, 47–48
Peptide hormones, 360–361, *360–361*
Peptide(s), 47–48
  natriuretic, 376, 438
Peptide YY (PYY), 376, 614
Perception, 320
Percussion, 4
Perfusion, 435, 437
Pericardial cavity, 18, *18*, 18*t*, 412, *413*
Pericardial fluid, 412
Pericardial sac, 412, *413*
Pericarditis, 412
Pericardium, 18, 412, *413*
  parietal, 412, *413*
  visceral, 412, *413*, 414
Perichondrium, 109, *110*, *151*
Pericytes, *432*, 433
Perimetrium, 638
Perimysium, 218, *219*
Perineal region, *17*
Perineum, 630
Periodic table, A–16
Periosteum, 111, *147*, *148*, 149
Peripheral nervous system (PNS), 250
  motor (efferent) division, 250
  sensory (afferent) division, 250
  somatic motor division, 250
Peripheral proteins, in plasma membrane, *64*
Peripheral resistance, 436
Peristalsis, 563, 568, 579, *579*
Peritoneal cavity, *19*, 19–20
Peritoneum, 19–20
  parietal, 19, *19*
  visceral, *19*, 19–20
Peritonitis, 20
Peritubular capillaries, 526, *527*
Permeability, membrane, 69
Permissive effects, of hormones, 361
Pernicious anemia, 399
Peroxisomes, 80, 81*t*
Persistent vegetative state, 9
Personalized medicine, 89
Perspectives on health
  acute gastrointestinal illness, 571
  AIDS, 489

  Alzheimer disease (AD), 264, *264*
  anabolic steroids, 51
  arthritis, 154–155, *155*
  atherosclerosis, 420
  blood disorders, 399–400
  burns, 132
  cancer, 89
  chronic obstructive pulmonary disease (COPD), 511
  collagen diseases, 51
  contraception, 687
  diabetes mellitus, 378
  dialysis, 541
  drugs and nervous system, 312, *313*
  endocrine disorders, 378–379
  fractures, 154
  genomic medicine, 89
  hearing disorders, 350
  hemodialysis, 541
  human immunodeficiency virus (HIV), 489, *489*
  lung cancer, 511, *511*
  muscle injuries, 220
  muscle pain, 220
  muscular dystrophy, 220
  obesity, 602
  osteoporosis, 154
  Parkinson disease (PD), 265
  renal insufficiency, 541
  sexually transmitted diseases (STDs), 657
  situs inversus and anatomical variations, 9
  skin cancer, 132
  smoking and respiratory health, 511, *511*
  stem-cell controversy, 115
  tissue engineering, 115
  visual disorders, 350
PET scans, *4*, *5*, 6, 33
pH, 39, A–11
  of blood, 387*t*, 429
  buffers and, 40
  and drug action, 39–40
  and enzyme action, 50
  of extracellular fluid, 387
  partitioning, 40
  and physiology, 39–40
  scale, 39, *39*
Phagocytes, 473, *473*
Phagocytosis, 73, *73*, 74, 463
Phagosome, *73*
Phalanx (phalanges)
  foot, *157*, *179*, 179–180
  hand, *157*, 174–175, *175*
Phantom pain, 325
Pharmacogenetics, 89
Pharyngeal constrictors, 562
Pharyngeal phase, swallowing, 563, *563*
Pharynx, 499, *557*, *557*, 562
pH balance, 548
Phenobarbital, 40
Pheochromocytoma, 378
Pheomelanin, 135
Pheromones, 138
Phlebotomist, 405
Phosphagen system, 212, *212*
Phosphate, 52
Phospholipids, 43, 45, *45*
  in plasma membrane, *64*, 64–65
Phosphorus, 31, 32*t*, 145

Photoaging, 139
Photopsin, 346
Photopupillary reflexes, 344
Photoreceptors, 339, 345
Physical barriers, 472
Physiology, 8–13
   characteristics of life in, 8–10
   comparative, 8
   definition of, 3, 8
   homeostasis in, 10–13
   negative feedback in, 11
   positive feedback in, 12
   subdisciplines of, 8
   variations in, 13
Pia mater, 270, 284, *285*
Pilocarpine, 312
Pineal gland, 291, *358*, 369–370
Pineal sand, 370
Pinocytosis, 73, *73*, 74
Pinocytotic vesicle, *432*
Pituitary dwarfism, 362
Pituitary gland, *358*, 363–369
   anatomy of, 363–364, *365*
   embryonic development of, 364, *364*
   hormones of, *365*, 366–367
   hypothalamic control of, 363, 367, *368*
   lobes of, 364, *365*
   negative feedback inhibition of, 367–368
   positive feedback control of, 367–368
Pituitary stalk, 364, *364*, *365*
Pivot joints, 183, *183*
Placental stage during pregnancy, 659
Plane (gliding) joints, 183, *183*
Planes, anatomical, 15, *15*, *16*
Planes of section, 95
Plantar flexion, 187, *187*
Plaques, atherosclerotic, 420, *420*, 457
Plasma, 111, 387–390, *388*
   composition of, 388–389, 389*t*
   as percentage of whole blood, *388*
   proteins in, 388–389, 389*t*
   separation form formed elements, *388*, *388*
   viscosity of, 388–389
Plasma (cell) membrane, 14, 61, *62*, *64*, 64–65
   extracellular face of, *64*
   functions of, 64
   intracellular face of, *64*
   lipids of, 64–65
   permeability of, 69
   proteins of, *64*, 65, *65*
   transport across, 69–74, *70*
Plasmin, 404
Platelet count, 387*t*, 401
Platelet-derived growth factor (PDGF), 404
Platelet plugs, 402, 403
Platelet(s), 111, *112*, *385*, 387, *388*
   functions of, 387, 402
   in hemostasis, 401–404
   open canalicular system of, 402, *402*
   as percentage of whole blood, *388*
   production of, 389–390, *402*, *403*
   structure of, 402, *402*
Plate(s)
   epiphyseal, *151*, 151–152
   nail, 136, *136*
Pleura, 18, 501–504
Pleural cavity, 18, *18*, 18*t*, 502, *503*

Pleural fluid, 502, *503*
Plural forms, of medical terms, 25, 25*t*
Pluripotent stem cells, 115
Podocytes, 527, *529*
Polar covalent bonds, 36, *36*, 37*t*
Polycythemia, 399
Polydipsia, 378
Polygonal cells, 96, *96*
Polymerization, 41, *41*
Polymer(s), 41
   nucleic acids as, 50
   protein as, 47
Polyphagia, 378
Polysaccharides, 42, *42*, 43*t*
Polyunsaturated fats, 44
Polyunsaturated fatty acids, 44
Polyuria, 378, 544
Pons, 290
Popliteal fossa, 238
Portal system, 434, *434*
   hepatic, 453, *454*
   hypophyseal, 364, *365*
Position, anatomical, 15, *15*
Positive feedback, 12
   in childbirth, 12, *12*
   in endocrine system, 362, 367
   in harmful processes, 12
Positron emission tomography (PET), 4, 5, 6, 33
Postabsorptive (fasting) state, 611
Posterior, 15, 25*t*, 26
Posterior cruciate ligament (PCL), 191
Posterior fontanel, 163, *164*
Posterior interventricular sulcus, *415*, 417
Posterior nasal apertures, *160*, 161, 497, *498*
Posterior pituitary gland
   anatomy of, 364, *365*
   embryonic development of, 364, *364*
   hormones of, *365*, 366–367
Postganglionic neuron, 305
Postpartum, 659
Postpartum changes, 660
Postsynaptic neuron, 262, *263*
Potassium
   body content of, 31, 32*t*
   in cardiac physiology, 429
   cardiovascular effects of, 429
Potassium chloride, 34*t*
Potential energy, 53–54
Power stroke, 209
Prebone (osteoid) tissue, 152
Precapillary sphincter, *432*, 433
Precentral gyrus, 297
Precision in terminology, 21, 24
Prefixes, 24, A–12–A–15
Preganglionic neuron, 305
Pregnancy, 658–659
Premenstrual phase, 653
Premolars, *561*, 562
Prepuce, 640
   foreskin, 632
Presbyopia, 344, *345*
Presynaptic neuron, 262, *263*
Primary auditory cortex, 338
Primary cilium, 66
Primary follicles, 650
Primary marrow cavity, 151
Primary motor cortex, 297

Primary olfactory cortex, 328, *329*
Primary oocytes, 648
Primary ossification center, 151, *151*
Primary sensory areas, 296
Primary sex organs, 626
Primary spermatocyte, 646
Primary visual cortex, 348, *349*
Prime mover (agonist), 221, *221*
Primordial follicles, 650
Principal piece, 647
Processes of bone, *156*, 156*t*
   condylar, *159*, 163
   coracoid, *171*, 172, *231*
   coronoid, *159*, 163, *173*, 174
   mastoid, *159*, *160*, 161
   scapular, *156*, 232
   spinous, 166, *166*, 167
   styloid, *173*, 174
      of temporal bone, *159*, *160*, 161
   transverse, 166, *166*, 167
   zygomatic, of temporal bone, *159*, 161
Products
   of chemical reactions, 54
   of enzyme reactions, 49
Progesterone, 46, 375, 652
   permissive effect with estrogen, 361
Programmed cell death, 80, 120–121
Prolactin, 366, *368*
Prolactin-inhibiting hormone (PIH), 367
Proliferative phase, 653
Pronation, 186–187, *187*, 232
Prone, definition of, 15
Propanolol, 312
Prophase, 86–88, *87*
Proplatelets, *402*, 403
Proprioceptors, 323
Prostate, 631
Prostate cancer, 632
Prostate diseases, 632
Prostate-specific antigen (PSA), 647
Proteases, 580
Proteasomes, 80
Protective cells, 472–474, *473*
Protective proteins, 474
Protein(s), 41, 46–50, 612
   carrier, 72
   channel, *64*
   composition of, *47*, 47–48
   conformation of, 49
   denaturation of, 49
   digestion and absorption, 580–582, *581*
   enzymes as, 49–50
   fibrous, *48*, 48–49
   functions of, 47*t*
   globular, 48–49
   hydrogen bonds of, 36–37
   integral, *64*
   membrane, *64*, 65, *65*
   metabolism, 609
   origin of word, 46
   peptide bonds of, *47*, 47–48
   peripheral, *64*
   plasma, 388–389, 389*t*
   processing of, 83
   protective, 474
   receptor, 65, *65*
   secretion of, 83, *84*

Protein(s) (*continued*)
    shapes of, *48,* 48–49
    synthesis of, 50, 80–84, *82, 84*
Proton acceptor, base as, 39
Proton donor, acid as, 39
Proton(s), *32,* 32–33
Protraction, 185, *185*
Provitamins, 600
Proximal, 25*t*
Proximal convoluted tubule (PCT), 528, *530,* 536–537, *537*
PSA test, 632
Pseudopods, 68, *68, 73,* 402, *402*
Puberty, 634
    to menopause, 634, 642
Pubic region, *17*
Pubic symphysis, 175, *176,* 177
Pubis, bones of, 176, *176,* 177
Public health nurses, 490
Pulled hamstrings, 220
Pulmonary circuit, 412, *412,* 441, *442*
Pulmonary trunk, 412, *412, 413, 415,* 418, 441
Pulmonary valve, 417, 418
Pulmonary veins, 441, *442*
Pulmonary ventilation, 506–510
    breathing, mechanics of, 506–509, *507, 508*
    measurements, *509,* 509–510, *510*
    neural control of breathing, 510, *510*
Pulp, *561,* 562
Pulp cavity, *561,* 562
Pumps
    skeletal muscle, 440, *440*
    sodium–potassium, 72
    thoracic, 440
Pupil, 341, *341*
Pupillary constriction, 344, *345*
Pupillary constrictor, 344
Pupillary dilator, 344
Purkinje fibers, 424
Pus, 476
P wave, *426,* 426–427, *427*
Pyloric sphincter, 565, *565*
Pylorus, 565, *565*
Pyrogen, 476
Pyruvate, 212, 604
Pyuria, 544

Q

QRS complex, *426,* 426–427, *427*
Quadrants, of abdominopelvic region, *16,* 16–17
Quadrate lobe, liver, 572, *573*

R

Radial fossa, *172,* 173
Radial notch, *173,* 174
Radial tuberosity, 173, *173*
Radiation, 617
    ionizing, 33
Radioactive decay, 33
Radioactivity, 33
Radiograph, 4
Radiography, *4,* 4–5, 33
Radioisotopes, 33

Radiologic anatomy, 4
Radiologic technologists, 4, 26
Radiologists, 4
Radiotherapy, 33
Radius, 172
Ramus (rami), of mandible, *159,* 163
Reactants, 54
Receptive field, 321, *321*
Receptor-mediated endocytosis, *73,* 74
Receptor(s), 11, 320. *See also specific types*
    hormone, 361, *361*
    protein, 65, *65*
Recommended daily allowance (RDA), 595
Recovery period, muscles, 213
Recovery stroke, 209
Rectum, 585, *585*
Rectus sheath, 228
Red blood cell(s). *See* Erythrocyte(s)
Red bone marrow, *148,* 149
    distribution of, *149*
    hemopoiesis in, 145, 389
Red muscles, 213
Reduced erythropoiesis, 399–400
Reducing agent, 55
Reduction, 55, 55*t*
Reduction division, 644
Referred pain, 324, *324*
Reflex arc, 273–274, *274*
Reflex(es)
    absence, in brain death, 9
    blink, 136
    flexor, 275, *275*
    somatic, 273–275, *274*
    spinal cord, 266
    stretch, *274,* 274–275
Refraction, 344
Regeneration, 120
Regenerative cells, 566
Regional anatomy, 3
Region(s), of body, 16–17, *16–17*
Registered dietitian (RD), 621
Regulation of the absorptive state, 612
Relaxation, skeletal muscles, 209, *209*
Releasing hormones, 367, *368*
Remodeling, bone, 152
Renal artery, 526, *527*
Renal autoregulation, 533–535
Renal buffer, 548–549
Renal calculus, 542
Renal circulation, 452, *453,* 526, *527*
Renal columns, 525, *526*
Renal corpuscle, 527, *529*
Renal cortex, 525, *526*
Renal insufficiency, 541
Renal medulla, 525, *526*
Renal papilla, 525, *526*
Renal pelvis, 526, *526*
Renal pyramids, 525, *526*
Renal sinus, 525, *526*
Renal tubules, 99, 99*t,* 528–529, *530*
Renal vein, 526, *527*
Renin, 533
Renin-angiotensin-aldosterone (RAA) system, 375, 533–535, *534*
Repair, tissue, 120
Replication, DNA, 85, *86*
Repolarization, 259

Reposition, joint movement, 187
Reproduction, 10
Reproductive hormones, 375. *See also specific hormones*
Reproductive system, 20, 21, *23,* 626
    female (*See* Female reproductive system)
    male (*See* Male reproductive system)
Residual volume, 509, *509*
Resistance
    to blood flow, 436
    peripheral, 436
Resistance arteries, 433
Resistance exercise, 214
Resistance stage, in stress response, 379
Resorption, 152
Respiratory buffer, 548
Respiratory cycle, 506, *507*
Respiratory epithelium, 499
Respiratory membrane, 504
Respiratory system, 20, 21, *22,* 495–521, *665*
    aging of, 517
    alveoli, 504, *505*
    anatomy, 496–505, *497, 498*
    bronchial tree, 501–504
    cilia of, 66
    effects on other organ systems, 518
    functions, 496
    gas exchange and transport, 512–516
    larynx, 499–500, *499–500*
    lower respiratory tract, 497
    lungs, 501–504, *502–503*
        blood supply, 505
    nose, 497–499, *498*
    pharynx, 499
    pleurae, 501–504
    during pregnancy, 658
    pulmonary ventilation, 506–510
    trachea, 500, *501*
    upper respiratory tract, 497
Respiratory therapist, 517
Respiratory zone, 503
Response, in feedback loop, 11
Resting membrane potential (RMP), *258,* 258–259, 425
Restrictive disorders, *509,* 509–510
Reticular fibers, 105, *106,* 106*t,* 129, 130
Reticular layer, of dermis, 130
Reticular tissue, 105, *106,* 106*t*
Reticulocytes, 393
Retina, 342, 346
Retraction, 185, *185*
Retroperitoneal organs, *19,* 20
Reversible reactions, 54
Rh blood group, 395
Rheumatism, 155
Rheumatoid arthritis (RA), 155, *155*
Rheumatoid factor, 155
Rheumatologist, 155
Rh-negative, 395
Rhodopsin, 346
Rh-positive, 395
Ribonuclease, 575
Ribose, 50
Ribosomal RNA, 83
Ribosomes, *62,* 77, 79, 81*t*
Ribs, 169–170, *170*
    costal cartilage, 169, *170*

false ribs, 170, *170*
floating ribs, 170, *170*
true ribs, 170, *170*
Rickets, 133, 146, 153, *153*
Right atrioventricular valve, *416*, 417, 418
Right atrium, *415*, 416, *416*, 418
Right bundle branch, 424
Right lobe, liver, 572, *573*
Right lower quadrant (RLQ), 16, *16*
Right lymphatic duct, *466*, 467
Right upper quadrant (RUQ), 16, *16*
Right ventricle, *415*, *416*, 417, 418
Rigor mortis, 210
RNA, 50, 79
    DNA transcription to, 81–83, *82*
    messenger (*See* Messenger RNA)
    ribosomal, 83
    transfer, *82*, 83
    translation of, *82*, 83
Rod and cone cells, *347*, 347–348
Root, word, 21
Root canal, *561*, 562
Root(s), 634
    of hair, 134, *135*
    of tooth, *561*, 562
Root sheath, of hair follicle, 134, *135*
Rotation, joint movement, 186, *186*
    lateral (external), 186
    left, 186
    medial (internal), 186
    right, 186
Rotator cuff, 188, 220
Rough endoplasmic reticulum, *62*, 77–79, *78*, 81*t*
Round ligament, liver, 572, *573*
Ruffled border, of osteoclast, *145*, 146
Rust, 55

S

Saccule, 333, *334*
Sacral canal, 168, *169*
Sacral foramina, 168, *169*
Sacral region, *17*
Sacral vertebrae, 168, *169*
Sacroiliac joints, 175, *176*
Sacrum, 168, *169*, 175, *176*
Saddle joints, 182–183, *183*
Sagittal plane, 15, *15*, *16*
Sagittal sinus, superior, 447
Saliva, 562, *562*
Salivary amylase, 562
Salivary glands, 557, *557*, 562, *562*
Salts, 34
Salty taste, 326
Sarcolemma, 200, *200*
Sarcomeres, *202*, 203, 217
    relaxed and contracted, 209, *209*
Sarcoplasmic reticulum (SR), *200*, 200–201
Satellite cells, 254, *254*
Saturated fats, 44
Saturated fatty acids, 43–44, *44*
Scapular region, *17*
Scarification, 120
Scars, 120
Schwann cells, 254, *254*
Sclera, 341, *341*

Scoliosis, *165*, 166
Scrotum, 630
Sebaceous glands, 117, *117*, 127, *135*, 136, *137*, 138
Sebum, 138
2nd lumbar vertebra (L2), *166*
Secondary follicles, 650
Secondary marrow cavity, *151*
Secondary oocytes, 648
Secondary ossification center, 151, *151*
Secondary sex organs, 626
Secondary spermatocytes, 647
Second-degree burns, 132
Second gap phase, in cell cycle, 85, *85*
Second heart sound, 428
Second messengers, 361, *361*
Second polar body, 648
Secretin, 569, 575
Secretion(s), 116–117, *117*
    gastric, 567, 567*t*
    small intestine, 579, *579*
    tubular, 536, 538
Secretory phase, 653
Secretory vesicles, *62*, 76, 79, *116*, 432
Sections
    of body, 15, *16*
    histological, 95
Segmental (tertiary) bronchi, 503, *503*
Segmentation, small intestine, 579, *579*
Segment(s), of, 17
Selective permeable membrane, 69
Selective serotonin reuptake inhibitor (SSRI), 312
Selenium, 34
Sella turcica, *159*, *160*, 161, 363
Semicircular ducts, 334, *335*
Semilunar valves, 417, 427
Seminal vesicles (seminal glands), 631
Seminiferous tubules, 375, *375*, 627
Sensation, 320
    brain function, 296, *297*
    skin, 127
    taste, 326–327, *327*
Sense of smell, 496
Sense organs, 319–355
    aging of, 351
    defined, 320
Senses
    chemical, 326–329
    general, 296, 322, 323–325
    of smell (olfaction), 328–329, *329*
    somatosensory, 296
    special, 296, 322
    of taste (gustation), 326–328, *327*
Sensorineural (nerve) deafness, 350, 351
Sensory adaptation, 322
Sensory (afferent) division, 250
Sensory information
    duration, 322
    intensity, 322
    location, *321*, 321–322
    type (modality), 321
Sensory (afferent) neurons, 253, *253*
Sensory receptors
    free nerve endings as, 323
    simple nerve endings as, 323–324, *324*
Septum (septa)

interatrial, 417
interventricular, *413*, *415*, *416*, 417
nasal, 161–162
Serine protease, 647
Serotonin, 403, *403*, 567
Serous fluid, 118, *118*
Serous glands, 117
Serous membranes (serosa), 18, 18*t*, 118, *118*
    digestive tract, 558, *558*
    meninges, 18, 18*t*
    mesenteries, 19–20, *19–20*
    pericardial, 18
    peritoneal, 18*t*, *19*, 19–20
    pleural, 18*t*
Serum, 388
Set point, 11
Sex, as physiological variation, 13
Sex cells, production and union
    folliculogenesis, 650
    meiosis and gametogenesis, 644
    menstrual cycle, 652–654
    oogenesis, 648
    ovarian cycle, 650–652
    semen, 647–648
    sperm, 647
    spermatogenesis, 646–647
    uniting the gametes, 654–656
Sex hormones, 375. *See also specific hormones*
Sexual arousal, 655
Sexual dimorphism, pelvis, 175–176
Sexually transmitted diseases (STDs), 657
Sexual reproduction, 626–627
Shaban, Nabil, 51
Shaft, 632
    of bone, 148, *149*
    of hair, 134, *135*
Shell temperature, 617
Shinsplints, 220
Shock, circulatory, 440
Short bones, 149
Short-wavelength (S) cones, 348
Shoulder, synovial joints of, 188, *188*
Shunts, 434, *434*
Sickle-cell disease, 392, *392*
Sigmoid colon, 585, *585*
Simple diffusion, 70, *70*
Simple nerve endings, as sensory receptors, 323–324, *324*
Simple sugars, 41
Single covalent bond, 36, 37*t*
Singular forms, of medical terms, 25, 25*t*
Sinuatrial node, 423, 424, *424*
Sinuses, *156*, 156*t*
    carotid, *446*
    coronary, *415*, 419, *419*
    dural, 447, *448*
    ethmoid, 158, *158*, 161
    frontal, 158, *158*, *159*
    maxillary, 158, *158*
    paranasal, 158
    sagittal, superior, 447
    sphenoid, 158, *158*, *159*
    straight, 448
    transverse, 447, *448*
Sinusoids, bone marrow, *402*
Situs inversus, 9
Situs perversus, 9

Skeletal muscle fibers, *199*, 199–203, 213
    fast-twitch, 213
    myofibrils, 200, *200*
    myofilaments, *201,* 201–203, *202*
    slow-twitch, 213
    smooth endoplasmic reticulum of, 200–201
    striations, 199, *199,* 201–203, *202*
    structure, *200,* 200–201
Skeletal muscle pump, 440, *440*
Skeletal muscle(s), 113, *114,* 114t
    ATP production during exercise, 212, *212*
    conditioning, 214
    contraction, 207–209, *208, 209*
    defined, 199
    excitation, 206–207, *207*
    fast-twitch fibers, 213
    fatigue and endurance, 212
    fibers (*See* Skeletal muscle fibers)
    flexor carpi radialis brevis, 222
    hypertrophy of, 119
    isometric and isotonic contraction, 210–211
    metabolism, 211–213
    muscle twitch and tetanus, 210, *211*
    physiology, 206–215
    recovery period, 213
    relaxation, 209, *209*
    slang terms for, 222t
    slow-twitch fibers, 213
    strength, 214
    striated, 199
    striations of, 113, *114,* 199, *199,* 201–203, *202*
    voluntary, 199
    whole-muscle contraction, 210–211
Skeletal system, 21, *22,* 143–196, *665*
    adaptations to bipedalism, 164
    aging of, 192
    bones of, 157, *157*
    components of, 144
    effects on other systems, 193
    functions of, 144–145
Skeleton, *157*
    appendicular, 157, *157*
    axial, 155–170, *157*
Skene glands, 640
Skin
    absorption through, 129
    accessory organs of, *127,* 134–138
    aging and, 139
    cancer, 132
    collagen of, 115, 129–131
    epithelial tissue of, 101
    functions of, 126–127
    glands of, 136–138, *137*
    healing (repair) of, 122
    immune function of, 126–127
    nonverbal communication by, 127
    sensation in, 127
    structure of, 126, *127,* 127–133
    thermoregulation by, 127
    ultraviolet radiation and, 132, 133, 139
Skin cancer, 132, *132,* 139
Skin color, 130–133, *131*
Skull, *157–160,* 158–164
    bone markings of, *156*
    bones associated with, 163
    cavities of, 158
    infant, 163, *164*

Skullcap (calvaria), *161*
Sleep, melatonin and, 369
Sliding filament model, 208, *208*
Slow-twitch fibers, 213
Small arteries, 433
Small intestine, 557, *557,* 559, *559,* 576–579, *577–579*
    anatomy, 576–579, *577–579*
    contractions of, 579, *579*
    endocrine function of, 376
    epithelial tissue of, 100t, *101*
    peristalsis, 579, *579*
    secretion and motility, 579, *579*
    segmentation, 579, *579*
Small veins, 433
Smell, sense of (olfaction), 328–329, *329*
Smoking, respiratory health and, 511, *511*
Smooth endoplasmic reticulum, 62, 79, 81t
    of muscle fiber, 200–201
Smooth muscle, 113, *114,* 114t, *216,* 216–217
Smooth muscle tone, *216,* 217
Sodium
    atomic structure of, *32*
    body percentage of, 32t
    deficiency of, 24
    renal reabsorption of, 375
Sodium bicarbonate, 34t
Sodium chloride (NaCl), *33,* 33–34, 34t
Sodium hydroxide, 39
Sodium ion(s), *33,* 33–34. See also Sodium
Sodium–potassium pump, 72
Soft callus, 154
Soft palate, 497, *498*
Soleus, 213
Solute, 38–39
Solution(s)
    acidic, 39
    basic (alkaline), 39
    concentration of, 39
    neutral, 39
    tonicity of, 71, *71*
Solvent, 38
Solvent drag, 438
Soma (cell body), *113*
Somatic motor division, 250, 297
Somatic pain, 324
Somatic reflexes, 273–275, *274*
Somatostatin, 567
Somatotropin. *See* Growth hormone
Sonogram, 6
Sonography, 6, *6*
Sonography, fetal, 6
Sour taste, 326
Spastic paralysis, 205
Specific granules, 396
Specificity, of hormone, 361
Sperm, 626
    tail of, 67
Spermatic cord, 630
Spermatids, 647
Spermatogonia, 646
Spermatozoon (sperm cell), 647
Sperm count, 647
Spermiogenesis, 647
Sphenoid fontanel, 163, *164*
Sphenoid sinus, 158, *158, 159*
Spheroidal cells, 96, *96*

Sphincter muscles, 199
Sphincters, precapillary, *432,* 433
Sphygmomanometer, 435
Spinal cord, 265–269, *266*
    conduction, 265
    cross-sectional anatomy of, *268,* 268–269
    gray matter, 268
    injury and paralysis, 269
    locomotion, 265
    meninges, 269–270
    neural integration, 265
    reflexes, 266
    white matter, 268
Spinal gating, 325
Spinal meninges, 269–270
Spinal nerves, 270–272, *272*
Spindle fibers, *87,* 88
Spine of bone, *156,* 156t
    subscapular, *171,* 172
Spirometry, 509, *509*
Spleen, 469–470, *470*
    anatomical variations in, 6, 9
    blood supply to, 452, *453*
    erythrocyte death in, 393
    reticular tissue, *106,* 106t
Spongy bone, 111, 145, 146–147, *147,* 149
Squamous cell carcinoma, 132
Squamous cell(s), 95, *96,* 98
Squamous epithelial tissue, simple, 98, *99,* 99t
Stability, muscles, 199
Stains, 95
Stalk, of pituitary gland, 364, *365*
Stapedius, 331, *331*
Stapes, 331, *331*
Starch, 41, 42, 43t
    digestion of, 49
Static equilibrium, 333
Stearic acid, *44*
Stem cell(s)
    adult, 115
    controversy over, 115
    embryonic, 115
    epidermal, 128, *128*
    hemopoietic, 389, 393, *393,* 398, *399*
    induced pluripotent, 115
    pluripotent, 115
    transport of, 387
Stereocilia, 333, *334*
Sternal end, of clavicle, 171, *171*
Sternal region, *17*
Sternum, 169, *170*
Steroid hormones, 360–361, *360–361,* 374–375
Steroids, 43, *45,* 45–46
    anabolic, 51
Stomach, 557, *557*
    anatomy, 564–566, *565–566*
    blood supply to, 452, *453*
    body, 564–565
    cardiac part, 564
    endocrine function of, 376
    fundus, 564
    gastric motility, 568
    gastric secretions, 567, 567t
    protection of, 569–570
    pyloric part, 565
    regulation of gastric function, 568–569, *569*
        cephalic phase, 568, *569*

gastric phase, 568–569, *569*
intestinal phase, 569, *569*
Stop codon, 83
Straight sinus, *448*
Stratum basale, *128*, 129
Stratum corneum, *128*, 129
Stratum granulosum, *128*, 129
Stratum lucidum, *128*, 129
Stratum spinosum, *128*, 129
Stress, 379
Stress fracture, 154
Stressors, 379
Stretch reflex, *274*, 274–275
Striations
cardiac muscle, 113, *114*, 422, *423*
skeletal muscle fibers, 113, *114*, 199, *199*, 201–203, *202*
skeletal muscles, 113, *114*
Stroke volume (SV), 428
Structural formula, 35
Structural hierarchy, *14*, 14–15
Structure–function relationships, 3
Subcutaneous fat, 130
Subcutaneous tissue (hypodermis), 126, *127*, 130
Subendocardial conducting network, 424
Sublingual glands, 562, *562*
Submandibular glands, 562, *562*
Submucosa, digestive tract, 558, *558*
Subscapular fossa, *171*, 172
Substantia nigra, 290
Substrate, 49
Sucrose, 42, *42*, 43*t*
Sudafed PE, 312
Suffixes, 24, A–12–A–15
Sugar(s), 41–42
Sulcus (sulci)
of brain, 282
of heart, 417
coronary, *415*, 417
interventricular, *415*, 417
Sulfur, body percentage of, 32*t*
Summation, 210
Superficial, 25*t*
Superior, 15, 25*t*
Superior articular facets, 167, *167*
Superomedial, definition of, 25, 25*t*
Superoxide anion, 34
Supination, 186–187, *187*, 232
Supine, definition of, 15
Supportive connective tissue, 109, 111
Supraspinous fossa, *171*, 172, 232
Surface anatomy, 3
Surgery, exploratory, 4
Sustentocytes (nurse cells), 375, *375*
Sutures, *149*, 158
coronal, 158, *159*, *161*, *164*
lambdoid, 158, *159*, *160*, *161*, *164*
sagittal, 158, *161*, *164*
squamous, *164*
Swallowing. *See* Deglutition
Swayback, 166
Sweat duct, *128*
Sweat glands, *103*, 103*t*, 117, *127*, 135, 136–138, *137*
Sweat pores, *127*, *128*
Sweet taste, 326
Sympathetic chain, 307
Sympathetic division, of ANS, 425

Synapse, 203, *204*, 262, *263*
Synaptic cleft, 203, *204*, 262, *263*
Synaptic vesicles, 203, *204*
Synarthrosis, 181
Syncope, 440
Synergist, *221*, 221–222
Synergistic effects, of hormones, 362
Synovial fluid, *181*, 181–182
Synovial joints, *181*, 181–184, *183*
articular disc, 182
bursa, 182
of elbow, 188–189, *189*
fibrous capsule, *181*, 182
hip, 189, *189*
joint (articular) capsule, *181*, 182
joint (articular) cavity, *181*, *181*
knee, *190*, 190–191
meniscus, 182
movement, 184–188
abduction, 184, *185*
adduction, 184, *185*
circumduction, *185*, 185–186
depression, 185
dorsiflexion, *187*, 187
elevation, 185
eversion, *187*, 187–188
extension, 184, *184*
flexion, 184, *184*
hyperextension, 184, *184*
inversion, *187*, 187–188
opposition, 187
plantar flexion, 187, *187*
pronation, 186–187, *187*
protraction, 185, *185*
reposition, 187
retraction, 185, *185*
rotation, 186, *186*
supination, 186–187, *187*
of shoulder, 188, *188*
structure, *181*, 181–182
synovial fluid, *181*, 181–182
synovial membrane, *181*, 182
tendons, 182
types of, 182–184, *183*
Synthesis phase of cell cycle, 85, *85*, 86
Synthesis reactions, 54, *54*, 55*t*
Systemic anatomy, 3
Systemic circuit, 412, *412*, 442–456, *443–444*
Systemic gas exchange, 515
Systole, 426–427
Systolic pressure, 435

T

T_3 (triiodothyronine), 370
T_4 (thyroxine), *360*, 370
Tactile cells, 128, *128*
Tactile corpuscles, *127*
Tactile (Meissner) corpuscles, 323
Tactile (Merkel) discs, 323
Taenia coli, 585, *585*
Tail, 647
of pancreas, *372*
of phospholipids, 45, *45*, 64
of sperm, 67
of thick filaments, *201*, 202

Target cells, of hormones, 361, *361*
Tarsal glands, 340
Tarsal region, 17, *17*
Taste
primary sensation, 326–327, *327*
sense of (gustation), 326–328, *327*
Taste buds, 326, *327*
Taste cells, 326, *327*
Teeth, 557, *557*
deciduous (baby), *561*
permanent, *561*
structure, *561*, 561–562
Telophase, in mitosis, 86–88, *87*
Temporomandibular joints, 184
Tendinous cords, 417
Tendinous intersections, 228
Tendon(s), 144, 220. *See also* Muscle(s)
aponeurosis, 220
calcaneal (Achilles) tendon, 239, *239*
collagenous fibers of, 105, *105*
dense connective tissue of, *107*, 107*t*
synovial joints, 182
Tensor tympani, 331, *331*
Terminal cisterns, *200*, 201
Terminal web, *62*, 75, *76*
Terminology, medical, 21, 24–26, 25*t*
Tertiary follicles, 650
Testes, *358*
endocrine function of, 375
histology of, 375, *375*
Testes (testicles), 627
Testosterone, 46, 360, *360*, 375
in bone physiology, 153
synergism with FSH, 362
Tetanus, 205, 210, *211*
complete, 210, *211*
incomplete, 210, *211*
muscle, 152, 425
Tetrahydrocannabinol (THC), 615
Tetraiodothyronine (T4, thyroxine), *360*, 370
Thalamus, 290
Theca, 636, 650
Thermal stability of water, 38
Thermoregulation, 617, 618–619
aging and, 139
by blood, 387
negative feedback in, 11
by skin, 127
"The Strangest Viking" (documentary), 51
Thick filaments, *201*, 202
Thigh, 17, *17*
muscles of, 236–237, *236–237*
Thin filaments, *62*, *76*, 201, *202*
Third-degree burns, 132
Thoracic aorta, *445*, 446, 447
Thoracic cage, 157, 169–17, 169–170, *170*
Thoracic cavity, 18, *18*, 18*t*
Thoracic curvature, 164, *165*
Thoracic duct, *466*, 467
Thoracic pump, 440
Thoracic region (chest), 16, *16*, 17
arteries of, 447, 449, *449*
heart position in, 412, *413*
lymphatics of, *466*, 467
veins of, 449–450, *450*
Thoracic vertebrae, 164, *165*, 167–168, *168*

Threshold, 259
Thromboembolism, 404
Thrombopoiesis, 403, *403*
Thrombosis, 404, 457
Thrombus, 404
Thumb, 179
Thymine, 50
Thymopoietin, 371
Thymosin, 371
Thymus, *358*, 371, 468–469, *469*
    hormones of, 371
    infant *vs.* adult, 371, *372*
    lymphocyte maturation in, 371
Thyroid cartilage, *499*, 500
Thyroid disorders, 371, 378
Thyroid follicles, 370, *370*
Thyroid gland, *356*, *358*, *370*, 370–371
    embryonic development of, *364*
    gross anatomy of, 370, *370*
    histology of, *370*, 370–371
Thyroid hormone (TH), 153, *356*, 362, 367, 370
Thyroid-stimulating hormone (TSH), 366, 367, 367t, 368
Thyrotropin. *See* Thyroid-stimulating hormone
Thyrotropin-releasing hormone (TRH), 367, *368*
Thyroxine, *360*, 370
Tibial tuberosity, 178, *178*
Tidal volume, 509, *509*
Tight junctions, 68, *69*
    in capillaries, *432*
Tissue engineering, 115
Tissue fluid, 62
Tissue(s), 14, *14*
    definition of, 94
    growth of, 119–120
    primary classes of, 94, 94t
    repair of, 120
    sections, 95, *95*
    shrinkage and death of, 120–121
    study of, 94–96
T lymphocytes (T cells), 479–480, *480*
    maturation of, 371
Toe(s), 17, *17*
    bones of, *179*, 179–180
Tongue, *557*, *557*, *560*, 560–561
Tonicity, 71, *71*
Tonsils, 468, *468*
Total metabolic rate, 613
Touch receptor, *127*
Toxic goiter, 371
Toxins, neuromuscular, 205
Trabeculae, 633
    of bone, 146, *147*, *149*
Trace elements, 31–32, 32t
Trachea, 500, *501*
    epithelial tissue of, *100*
    mucosa of, *100*
Transcription, 81–83, *82*
Transdermal absorption, 129
Transducers, 322
Trans fats, 46, *46*, 597
Transfer RNA (tRNA), *82*, 83
Translation, *82*, 83
Transport
    active, *70*, 72
    by blood, 111
    carrier-mediated, 72

circulatory, 387
    by filtration, 70, *70*
    gas, 514
    membrane, 69–74, *70*
    by osmosis, 70–71, *71*
    by simple diffusion, 70, *70*
    tonicity in, 71, *71*
    vesicular, 72–74, *73*
Transporters. *See* Carrier proteins
Transport vesicles, 83, *84*
Transverse arch of foot, 180
Transverse colon, 584, *585*
Transverse costal facet, 168, *168*, 170
Transverse foramen, 167, *167*
Transverse plane, 15, *15*, *16*
Transverse sinuses, 447, *448*
Transverse (T) tubules, 200
Tributaries, of veins, 442
Triceps brachii, 209
Tricuspid valve, *416*, 417, 418
Triglycerides, 43–44, *44*
    trans fats, 46, *46*
Triiodothyronine (T_3), 370
Trimesters, 658
Tripeptides, 47
Tristearin, 43
Tritium, 33
Trochanters, *156*
    of femur, *156*
Trochlea, *172*, 173
Trochlear notch, *173*, 174
Tropomyosin, *201*, 202
Troponin, *201*, 202
True ribs, 170, *170*
Trunk (of body), 16
    arteries of, 447, 449, *449*
    blood vessel, 445
    brachiocephalic, *443*, 445
    celiac, *443*
    defined, 445
    muscles of, 227–228, 229t
    pulmonary, *412*, *412*, *413*, *415*, 418, 441
    veins of, 449–450, *450*
Trypsin, 49, 575
Tryptophan, 360
Tubercle, of bone, *156*, 156t
Tuberosities, *156*, 156t
Tubular fluid, 531
Tubular reabsorption, 536–538, *537*
Tubular secretion, 536, 538
Tubules
    renal, *99*, 99t
    seminiferous, 375, *375*
Tumors, 89
    benign, 89
    malignant, 89
    metastasizing, 89
Tumor-suppressor (TS) genes, 89
Tunica externa, 430–431, *431*
Tunica interna, 430–431, *431*
Tunica media, 430–431, *431*
Tunics, of blood vessels, 430–431, *431*
T wave, 426, *426*, *427*
Tympanic cavity, 331, *331*
Tympanic membrane, 331, *331*
Tympanostomy, 332, *332*
Type 1 diabetes mellitus, 378

Type 2 diabetes mellitus, 362, 378
Tyrosine, *47*, 360

U

Ulcer(s)
    decubitus, 120
    peptic, 570, *570*
Ulna, 172
Ultrasound (sonography), 6, *6*
    fetal, *6*
Ultraviolet radiation, and skin, 132, 133, 139
Umami taste, 326
Umbilical region, *17*
Umbrella cells, 101
Unipolar neurons, *252*, 252–253
Unsaturated fats, 44
Unsaturated fatty acids, 43, 44, *44*
Upper limb, 17, *17*
    arteries of, *449*, 450–451
    bones of, *157*, 171–175
    muscles of, *230–231*, 230–235
    veins of, *450*, 451–452
Upper respiratory tract, 497
Up the gradient, *70*, 72
Uracil, 50
Urea, *30*, 524
Uremia, 544
Ureters, 6, 523, *524*, 542
Urethra, 523, *524*, 543–544, 630
Uric acid, 524
Urinary bladder, 523, *524*, 542, *543*
    tissue engineering of, 115
Urinary system, 20, 21, *22*, 522–554, *665*
    acid–base balance, 548–550
    aging of, 550
    defined, 523
    effects on other organ systems, 551
    electrolyte balance, 547–548
    fluid balance, 546–547, *546–547*
    functions, 522–525
    glomerular filtration, *531–532*, 531–535
    kidneys (*See* Kidney(s))
    during pregnancy, 658
    transitional epithelium of, 101
    tubular reabsorption and secretion, 536–538
    urine storage and elimination, 541–545
    water conservation, 539–540
        collecting duct, 539, *539*
        control of loss, 539–540
Urinary tract infection (UTI), 543
Urine
    appearance, 544
    chemical composition, 545, 545t
    defined, 531
    elimination, 541–545
    odor, 544
    osmolarity, 545
    pH, 545
    specific gravity, 545
    stages of formation, *531*, 531–532
    storage, 541–545
    voiding, 544
    volume and properties, 544–545
Uterine (fallopian) tube, 637
    cilia of, 66, *67*, *93*

Uterus, 637
Utricle, 333, *334*

V

Vaccination, 488
Vagina, 639
    epithelial tissue of, *102,* 102t
    exfoliation of, 101, *104*
    mucosa of, *102,* 102t
Valence, of ions, 33
Valence electrons, 33
Valves
    heart (*See* Heart valves)
    venous, *431,* 434, *440,* 440–441
Valvular prolapse, 417
Varicose veins, 440
Vasa recta, 526, *527*
Vasa vasorum, 430
Vascular layer, 341, *341*
Vascular spasm, 403, *403*
Vas deferens, 630
Vasoconstriction, 436
Vasoconstrictors, 402, 403
Vasodilation, 436
Vasomotor center, 437
Vasopressin. *See* Antidiuretic hormone (ADH)
Vegetative state, persistent, 9
Vein(s), 430, 433–434
    abdominopelvic region, 452–455, *454*
    arcuate, 526, *527*
    axillary, *444,* *448,* *450,* 452
    azygos, 450, *450*
    basilic, *444, 450,* 451
    blood flow in, 433–434
    brachial, *444, 450,* 452
    brachiocephalic, *444, 448,* 450, *450*
    bronchial, 505
    cardiac
        great, *419,* 421
        middle, 421
    cephalic, *444, 450,* 451
    coronary sinus, *415, 419, 419*
    femoral, 442, *444,* 456
    femoral, deep, *444*
    fibular (peroneal), 456
    gonadal, *444,* 453, *454*
    great, 412
    head and neck, 447, *448*
    hepatic, *444, 454, 455*
    hepatic portal, 453, *454, 455*
    iliac
        common, *444,* 452, *454,* 456
        external, *444,* 456
        internal, *444,* 452, *454*
    intercostal, 450, *450*
    interventricular, posterior, *419,* 421
    jugular
        external, *444,* 447, *448, 450*
        internal, 161, *444,* 447, *448, 450*
    large, *431,* 434
    lower limb, 455–456
    lumbar, ascending, *450*
    major systemic, *444*
    median cubital, *444, 450,* 451
    medium, *431,* 433–434
    mesenteric
        inferior, 453, *454*
        superior, *454,* 455
    nomenclature for, 445
    ovarian, 453, *454*
    popliteal, *444,* 456
    pulmonary, 412, *412,* 441, *442*
        left, *415, 416*
        right, *415, 416*
    radial, *444, 450,* 451
    renal, 442, *444,* 453, *454,* 526, *527*
    saphenous
        great (long), *444,* 455–456
        small, 433
    splenic, *454,* 455
    subclavian, *444, 448,* 449, *450,* 452
    testicular, 453, *454*
    thorax, 449–450, *450*
    thyroid
        inferior, *370*
        superior, *370*
    tibial
        anterior, *444,* 456
        posterior, *444,* 456
    tributaries of, 445
    tunics of, 430–431
    ulnar, *444, 450,* 452
    upper limb, *450,* 451–452
    varicose, 440
    vena cava
        inferior, 412, *412, 415, 416,* 418, *444,* 453, *454*
        superior, 412, *412, 413, 415, 416,* 418, 440, *444, 450, 450*
    vertebral, *448,* 450
Venous anastomosis, 434
Venous drainage, 419, 421
Venous pooling, 440
Venous return, *440,* 440–441
Venous sinuses
    coronary, *415, 419, 419*
    dural, *447,* 448
Venous valves, *431,* 434, *440,* 440–441
Ventilation, pulmonary, 506–510
    breathing, mechanics of, 506–509, *507, 508*
    measurements, *509,* 509–510, *510*
    neural control of breathing, 510, *510*
Ventral, 15, 25t, 26
Ventral respiratory group (VRG), 510
Ventricles
    of brain, 285–287, *286*
    of heart, 413, *415, 416,* 417
        blood flow through, *416,* 418
        contraction and relaxation of, 426–428, *427*
Ventricular diastole, 427, *427*
Ventricular ejection, 427–428
Ventricular fibrillation, 428
Ventricular systole, 427, *427*
Venules, *431, 432,* 433
Vertebrae, *157,* 164–169, *165*
    body (centrum) of, 166
    cervical, 164, *165,* 167, *167*
        atlas (C1), *165,* 167
        axis (C2), *165,* 167
        axis of rotation, *167*
    coccygeal, 164, 168, *169*
    fractures of, 154
    general structure of, 166, *166*
    inferior articular processes, 166, *166*
    lumbar, 164, 168, *168, 169*
    regional characteristics, 167–168, *167–169*
    sacral, 168, *169*
    spinous process of, 166
    superior articular processes, 166, *166*
    thoracic, 164, *165,* 167–168, *168*
    transverse process of, 166, 167
Vertebral arteries, 167
Vertebral column, 164–169, *165*
    adaptations to bipedalism, 164
    bones of, *157,* 164–169
    curvatures of, 164–166, *165*
    function of, 164
Vertebral region, *17*
Very-low-density lipoproteins (VLDLs), 598
Vesicles, 72–73
    clathrin-coated, *84*
    Golgi, *62, 78*
    pinocytic, *432*
    secretory, *62, 76,* 79, *116, 432*
    transport, 83, *84*
Vesicular transport, 72–74, *73*
Vestibular bulbs, 640
Vestibule, 333, *333,* 560, *560,* 640
Vibrissae, 497, *498*
    guard hairs, 136
Villi
    microvilli, *62,* 66, *66, 76*
    small intestine, 577, *578*
Viscera, 18, 18t
Visceral motor division, 305
Visceral muscle, 114
Visceral pain, 324
Visceral pericardium, 412, *413,* 414
Visceral peritoneum, *19,* 19–20
Visceral pleura, 501
Viscosity
    blood/plasma, 388–389, 436
    definition of, 388
Vision, 339. *See also* Orbital region
Visual association area, 348, *349*
Visual disorders, 350, *351*
Visual image, formation, 344, *345*
Visual nerve signal, generating, 345–348, *346–348*
Visual pigments, 345
Visual projection pathway, 348–349, *349*
Vital capacity, 509, *509*
"Vital force," 411
Vitamin A, 153, *581,* 583
    deficiency, 600–601
Vitamin B_{12}, 399
Vitamin C, 34, 153
Vitamin D, 127, 133, 153, 192, *581,* 583
Vitamin E, 34
Vitamin K, *581,* 583
Vitamins
    digestion and absorption, *581,* 583
    requirements, sources, 601t
Vitreous body, 342
Vocal cords, 500, *500*
Voiding urine, 544
Voltage-gated channel, 259

Volume, units of, A–10
Voluntary muscles, 113
Vomiting, 568
Vulva, 640

W

Waste, 524
 excretion of, 116
 nitrogenous, 524, 536
Water, 38–39, 536
 adhesion of, 38
 in body fluids, 39
 chemical reactivity of, 38
 cohesion of, 38
 conservation of, 539–540
  collecting duct, 539, *539*
  control of loss, 539–540
 digestion and absorption, *581,* 583
 hydrogen bonds of, 36, *36*
 osmosis of, 70–71, *71*
 properties of, 38
 as solvent, 38
 thermal stability of, 38
Water barrier, epidermal, 127
Water-insoluble fiber, 597
Water molecule, 36, *36,* 38
Water-soluble fiber, 596
Water-soluble vitamins, 600
White blood cells. *See* Leukocyte(s)
White fat, 108
White matter, 255–256, 285
 spinal, 268
White muscles, 213
Whole-muscle contraction, 210–211
Wings, of sphenoid bone, 161
The Wisdom of the Body (Cannon), 10
Work, energy as capacity to do, 53
Wound healing, 120
Wrist, 17, *17*
 muscles of, 233–235, *234–235*

X

X-ray(s), *4,* 4–5

Y

Yellow bone marrow, *148,* 149, *149*

Z

Z disc, *200, 202,* 203, 217
Zona fasciculata, 374, *374*
Zona glomerulosa, 374, *374*
Zona pellucida, 650
Zona reticularis, 374, *374*
Zygote, 626